Particle Acceleration and Detection

The series "Particle Acceleration and Detection" is devoted to monograph texts dealing with all aspects of particle acceleration and detection research and advanced teaching. The scope also includes topics such as beam physics and instrumentation as well as applications. Presentations should strongly emphasize the underlying physical and engineering sciences. Of particular interest are

- contributions which relate fundamental research to new applications beyond the immediate realm of the original field of research
- contributions which connect fundamental research in the aforementioned fields to fundamental research in related physical or engineering sciences
- concise accounts of newly emerging important topics that are embedded in a broader framework in order to provide quick but readable access of very new material to a larger audience

The books forming this collection will be of importance to graduate students and active researchers alike.

Jie Wei • Leo R. Dalesio • Alberto Facco •
Venkatarao Ganni • Yue Hao •
Peter N. Ostroumov • Ting Xu •
Yoshishige Yamazaki

High Power Hadron Accelerators

Springer

Jie Wei
Facility for Rare Isotope Beams Laboratory
Michigan State University
East Lansing, MI, USA

Alberto Facco
Facility for Rare Isotope Beams Laboratory
Michigan State University
East Lansing, MI, USA

INFN-Laboratori Nazionali di Legnaro
Padova, Italy

Yue Hao
Facility for Rare Isotope Beams Laboratory
Michigan State University
East Lansing, MI, USA

Ting Xu
Facility for Rare Isotope Beams Laboratory
Michigan State University
East Lansing, MI, USA

Leo R. Dalesio
Osprey Distributed Control Systems
Ocean City, MD, USA

Venkatarao Ganni
Facility for Rare Isotope Beams Laboratory
Michigan State University
East Lansing, MI, USA

Peter N. Ostroumov
Facility for Rare Isotope Beams Laboratory
Michigan State University
East Lansing, MI, USA

Yoshishige Yamazaki
Facility for Rare Isotope Beams Laboratory
Michigan State University
East Lansing, MI, USA

ISSN 1611-1052 ISSN 2365-0877 (electronic)
Particle Acceleration and Detection
ISBN 978-3-032-17661-5 ISBN 978-3-032-17662-2 (eBook)
https://doi.org/10.1007/978-3-032-17662-2

This Springer imprint is published by the registered company Springer Nature Switzerland AG
The registered company address is: Gewerbestrasse 11, 6330 Cham, Switzerland

Preface

The purpose of this book is to provide a reference to this flourishing field of accelerator science: high-power hadron accelerators. We intend to introduce the physics and technology of particle accelerators in the beam power frontier; these machines serve not only as tools to fundamental high-energy and nuclear physics studies but also as platforms beneficial to our society in general, including neutron sources for scientific research and industrial applications, isotope harvesting, medical therapy, nuclear waste transmutation, and applications.

It took 11 years since 2014 for the authors to complete this book, a time period that almost coincided with the establishment of the Facility for Rare Isotope Beams (FRIB), a project with which we have all been associated. Prior to FRIB, we were either leaders for the design and construction of major accelerator projects or subject matter experts on key accelerator physics and technologies. Some of us participated in the entire cycle of design and construction of accelerator facilities such as the Spallation Neutron Source (SNS) and the Japan Proton Accelerator Research Complex (J-PARC). Most of us are members of the advisory committees of numerous accelerator projects around the world. We intend to share our decades of experience, insights, and reflections practicing in the field of accelerator science.

This book aims at both readers of general interest in the concepts and approaches of accelerators and professionals who may use it as a handbook with accessible citations. We survey the evolution of accelerators on the hadron beam power frontier from 1970 to now and toward future years based on the main accelerator types of cyclotrons, linacs, and synchrotron rings. As these types of accelerators are still being proposed as the main drivers to modern facilities, we highlight in Chap. 1 technology selections, performance-limiting mechanisms, and design philosophy.

Accelerator design and technical considerations are summarized in Chap. 2, again covering cases of cyclotrons, normal-conducting and superconducting RF linacs, and synchrotrons. In each case, we attempt to outline important design aspects and figures of merit, rationale of main parameter choices, and commonly adopted design practices and considerations—the major ingredients for the conceptual design of high-power accelerator facilities.

Chapter 3 presents examples of performance-limiting mechanisms and related beam dynamics. Contents that may not be directly related to present high-power accelerator development are listed in the appendices of this chapter.

Much emphasis is placed on summarizing technologies used in high-power accelerators. In particular, we elaborate on large-scale cryogenics (Chap. 4), superconducting RF acceleration (Chap. 5), magnets (Chap. 6), and controls (Chap. 7). Each of these chapters was written by a pioneer or a seasoned expert in the field. Chapter 8 goes through other technologies essential to high-power performance.

Typical high-power accelerator projects take more than 10 years through the cycle of establishment of conception, design, construction, and commissioning. It goes beyond accelerator physics and technology to deliver a successful accelerator project. Chapter 9 describes the main phases and milestones of such accelerator projects throughout the cycle. As examples, Chap. 10 summarizes the experiences and lessons learned during the establishment of SNS, J-PARC, and FRIB facilities.

Given our busy day-time job, this book has been drafted mostly during weekends and on late evenings. The homework problems are selected from material taught in the US Particle Accelerator School classes. Solutions are provided to illustrate the logic, approaches, and implications.

East Lansing, MI, USA — Jie Wei
Ocean City, MD, USA — Leo R. Dalesio
Padova, Italy — Alberto Facco
East Lansing, MI and Yorktown, VA, USA — Venkatarao Ganni
East Lansing, MI, USA — Yue Hao
East Lansing, MI, USA — Peter N. Ostroumov
East Lansing, MI, USA — Ting Xu
Tsuchiura, Japan — Yoshishige Yamazaki
November 2025

Acknowledgments The authors are deeply grateful to John Thomason for his thorough proofreading of the opening and closing chapters of the manuscript, as well as for his insightful advice, suggestions, and corrections; Rick Baartman for corrections and enlightening suggestions regarding the cyclotron sections of the manuscript; Hiromi Okamoto for corrections on space-charge and crystalline-beam-related presentations; Shinji Machida for contributing to the FFA section; Wolfram Fischer and Oliver Kester for reading and correcting Chap. 1; and Yoichi Sato for reading and amending the accelerator software. We thank Sarah Cousineau for checking the SNS section, Michikazu Kinsho for checking the J-PARC section, Osamu Kamigaito for checking and amending the RIKEN-related section, Robin Ferdinand for checking the SPIRAL2-related section, Juan Knaster for checking the IFMIF-related sections, Ralf Gebel for amending work and references from Germany, Russia, and Europe, Simone S. Gilardoni for checking, amending, and contributing to the CERN-related sections, Alexander Valishev for checking and amending the FNAL-related sections, Sheng Wang for checking the CSNS-related sections, Yuan He for checking the IMP-sponsored projects, Hyoek-Jung Kwon for checking the KOMAC section, Yeon-Sei Chung for checking and amending the RAON-related sections, Håkan Danared for checking and amending the ESS- and CRYRING-related sections, Todd Satogata for checking the CEBAF section, Nusair Hasan for assistance with the cryogenics sections, Felix Marti for verifying the simplified Bethe-Bloch formula, Hsiu-Chuan Hsueh for checking and amending the vacuum related sections, Kent Holland for checking and contributing to the power supply section, Peter Manwiller for checking and amending the survey and alignment section, Masanori Ikegami for checking and correcting the PPS, MPS, and EPS sections, Sang-hoon Kim for checking and correcting the SRF sections, and Shen Zhao for checking and correcting the RF sections.

Materials from the class taught at the US Particle Accelerator School were developed jointly with Yannis Papaphilippou and Alexei Fedotov around 2004.

We thank Alexander W. Chao, Yong-Sub Cho, Tadashi Koseki, Chris Prior, Michael Seidel, Richard Talman, Joe Tuozzolo, Zhijun Wang, Jiawen Xia, Wenlong Zhan, and Frank Zimmermann for their valuable comments and encouragement.

We are grateful to many colleagues, institutes, and publishers for permission to use their published materials; we attempt to include acknowledgments in the corresponding figures and tables. We are indebted to our colleagues, in particular, those who were associated with the SNS, J-PARC, and FRIB projects, for their years of dedication and teamwork.

Finally, we thank Ling Wang for her dedicated and highly efficient editorial efforts in preparing the manuscript, and Xing Rao for assisting with figure drawings.

Competing Interests The authors have no competing interests to declare that are relevant to the content of this manuscript.

Contents

Acronyms

1-D	One dimensional
1GSR	1st generation light source
2-D	Two dimensional
2GSR	2nd generation light source
3-D	Three dimensional
3GSR	3rd generation light source
4GSR	4th generation light source
2-q, 3-q, 5-q	Two-charge-state, three-charge-state, 5-charge-state, respectively
6-D	Six dimensional
A	Ampere
AA	Antiproton Accumulator, CERN
AAA	Advanced Accelerator Applications program, LANL
AC	Alternating current
ACO	Electron-positron collider at Orsay, France
ACS	Annular-ring coupled structure
AD	Antiproton Decelerator, CERN
AdA	Annelli di Accumulazione, electron-positron collider at Frascati, Italy
ADONE	Electron positron collider, Frascati, Italy
ADS	Accelerator-driven subcritical systems
AG	Alternating gradient
AGS	Alternating Gradient Synchrotron, BNL
AHF	Advanced Hydrotest Facility, LANL
AI	Artificial intelligence
AISI	American Iron and Steel Institute
ALARA	As low as reasonably achievable
ALICE	Accelerators and Lasers In Combined Experiments, Daresbury Laboratory, UK
ALS	Advanced Light Source, LBNL
ALSU	Advanced Light Source Upgrade, LBNL

APT	Accelerator Production of Tritium, LANL
ANL	Argonne National Laboratory, USA
AR	Accumulator ring
ARIS	Advanced rare isotope separator
ARR	Accelerator readiness review
ASD	Accelerator systems division
ASE	Accelerator safety envelope
ASME	American Society of Mechanical Engineers
ASTRID	Aarhus STorage RIng in Denmark, Aarhus University, Denmark
A-TAC	Accelerator Systems Advisory Committee, J-PARC
ATCA	Advanced telecommunications computing architecture
ATLAS	Argonne Tandem Linac Accelerator System, ANL
ATW	Accelerator Transmutation of Waste project, LANL
AVF	Azimuthally varying fields
BACnet	Building automation and control network
BARC	Bhabha Atomic Research Centre, India
BASE	Berkeley Accelerator Space Effects facility, LBNL
BBU	Beam break-up
BCM	Beam current monitor
BCS	Bardeen–Cooper–Schrieffer
BDF	Beam Dump Facility, CERN
BDS	Beam delivery system, FRIB
BEAST	Best ever alarm system toolkit
BEPC	Beijing Electron Positron Collider, IHEP
BESAC	Basic Energy Sciences Advisory Committee, USA
BIG	Beam-in-gap
BINP	Budker Institute of Nuclear Physics, Russia
BNCT	Boron neutron capture therapy
BNL	Brookhaven National Laboratory, USA
BO	Bayesian optimization
BOD	Beneficial occupancy date
BOE	Basis of estimate
BPM	Beam position monitor
bps	Bits per second
BPVC	Boiler and pressure vessel code, ASME
BVR	Built-in volume ratio
CAA	Clean Air Act, EPA
CAC	Credited administrative control
CADS	China Accelerator Driven Sub-critical System, CAS
CARIBU	Californium Rare Isotope Breeder Upgrade, ANL
CAS	Chinese Academy of Sciences
CBX	Electron-positron collider at SLAC
CC	Cold compressor
CCDB	Central configuration database

CCF	Coupled Cyclotron Facility, MSU
CCL	Coupled-cavity linac
CD	Critical Decision
CDR	Conceptual design report
CEA	Commissariat à l'Énergie Atomique et aux Energies Alternatives, France
CEA Bypass	Cambridge Electron Colliding Beam Facility, Harvard University, USA
CEBAF	Continuous Electron Beam Accelerator Facility, JLab
CEC	Credited engineered control
CEPC	Circular Electron Position Collider, CAS
CELSIUS	Cooling with ELectrons and Storing of Ions from the Uppsala, Uppsala University, Sweden
CERN	European Organization for Nuclear Research, Switzerland
CESR	Cornell Electron Storage Ring
CHA	Chilled water for activated environment
CHW	Chilled water
CF	Conventional facility
CAFe	China ADS front End, IMP
CAFE2	China Accelerator Facility for Superheavy Elements, IMP
CH	Crossbar H-mode resonator
CiADS	China initiative Accelerator Driven System, IMP
CICC	Cable in conduit conductor
CLIC	Compact Linear Collider
CLWR	Commercial light-water reactors
c.m.	Center of mass
CM	Cryomodule
CMM	Coordinate measurement machine
CNGS	CERN Neutrinos to Gran Sasso project, CERN
CO	Conduct of operations
COP_{INV}	Inverse performance coefficient
COSY	Cooler Synchrotron, Institut für Kernphysik des Forschungszentrums Jülich, Germany
COTS	Commercial off the shelf
	CP Chemical polishing
CP violation	Violation of charge conjugation parity symmetry
CPHS	Compact Pulsed Hadron Source, Tsinghua University, China
C-S	Courant-Snyder
CSNS	China Spallation Neutron Source, IHEP
CSR	Cooler Storage Ring, IMP
CSR	Charge Stripper Ring, RIKEN
CS-Studio	Control System Studio, or CSS
CRYRING	Manne Siegbahn Laboratory, Stockholm University, Sweden
CW	continuous-wave
DAFNE	Electron positron collider, Frascati, Italy

DAQ	Data acquisition
DC	Direct current
DCC	Document Control Center
DCCT	Direct-current current transformer
DCI	Electron positron collider, Orsay, France
DEE	Cyclotron D-shaped chamber
DEMO	A conceptual demonstration power plant for a fusion reactor
DESY	Deutsches Elektronen-Synchrotron, Germany
DI	Deionized
Diamond	Diamond Light Source, Harwell Science and Innovation Campus, United Kingdom
DOE	U.S. Department of Energy, U.S.
DORIS	Electron positron collider, DESY
DRR	Device readiness review
DSR	Double-spoke resonator
D-station	Diagnostics station
DTL	Drift-tube linac
DUNE	Deep Underground Neutrino Experiment, FNAL
EAC	Estimate at complete
EB	Electron beam
EBIS	Electron beam ion source
EBIT	Electron beam ion trap
EBW	Electron beam welding
ECR	Electron-cyclotron-resonance ion source
EI	Expected improvement
ELENA	Extra Low ENergy Antiproton ring, CERN
ELS	Electronic log book service
EM	Electromagnetic
EMI	Electromagnetic interference
EMMA	Electron Model for Many Applications, STFC
e-p	Electron-proton
EP	Electropolishing
EPA	Environment Protection Agency, USA
EPICS	Experimental Physics and Industrial Control System
ES	experimental system
ESR	Experimental Storage Ring, GSI
ESH&Q	Environment, safety, health, and quality
ESS	European Spallation Source, Sweden
ETC	Estimate to complete
EU	European Union
ev	Electron-volt
EVEDA	Engineering Design and Engineering Validation Activities, IFMIF
EVG	Event generator
EVMS	Earned value management system

EVR	Event receiver
F4E	Fusion for Energy, EU
FAIR	Facility for Antiproton and Ion Research, GSI
FCC	Future Circular Collider, CERN
FCC-ee	Future Circular electron-positron Collider, CERN
FCC-eh	Future Circular electron-hadron Collider
FE	Front end
FEA	Finite element analysis
FEL	Free electron laser
FET	Field-effect transistor
FFA	Fixed-field alternating-gradient accelerator
FGPDB	FRIB general purpose digital board
FMEA	Failure mode and effect analysis
FODO	Focusing/defocusing lattice structure with drift sections
FNAL	Fermi National Accelerator Laboratory, USA
FPC	Fundamental power coupler
FPGA	Field programmable gate arrays
FPS	Fast machine protection system
FRIB	Facility for Rare Isotope Beams, MSU
FS	Folding segment
FSEE	FRIB Single Event Effects beam line, MSU
FWHM	Full width at half maximum
FX	Fast-extraction
FZJ	Forschungszentrum Jüelich, Germany
MI	Main Injector, FNAL
G	10^9
G	Gauss
$g_\mu - 2$	Anomalous magnetic dipole moment of muon
GANIL	Large Heavy Ion National Accelerator (Grand Accélérateur National d'Ions Lourds), France
Gbps	Giga bits per second GBps Giga bytes per second
GFR	Good field region
GPS	Global Positioning System
GS	Gaussian process
GSI	Gesellschaft für Schwerionenforschung, Germany
h	Hour
HA	Hazard Analysis
HBS	High Brilliance Neutron Source project, FZJ
HDF5	Hierarchical data format version 5
HEBT	High energy beam transport
HEDP	High Energy Density Physics experiment, GSI
HEPA	High efficiency particulate air
HERA	Hadron–Electron Ring Accelerator (Hadron-Elektron-Ringanlage), DESY

HIAF	Heavy Ion Accelerator Facility, IMP
HIDIF	Heavy Ion Driven Ignition Facility
HIMM	Heavy Ion Medical Machine, IMP
HLA	High level application
HLI	High Charge State Injector, GSI
HNB	ITER heating neutral beam injector
HOM	High order modes
HP	High power
HP	High pressure (Chap. 4)
HPHA	High-power hadron accelerator
HPR	High pressure rinsing
HRS	High Rigidity Spectrometer project, FRIB, MSU
HTS	High-temperature superconducting
HUNS	Hokkaido University Neutron Source, Japan
HV	High vacuum (Sect. 8.7)
HV	High voltage
HVPS	High voltage power supply
HVAC	Heating, ventilation, air conditioning
HVCM	High-voltage converter modulator
HWR	Half-wave resonator
HW	Hardware
HWR	Half-wave resonator
HX	Heat exchanger
Hz	Hertz
IBS	Intrabeam scattering
IBS, Korea	Institute for Basic Science, Korea
IBSt	Intrabeam stripping
IC	Inner conductor
ICD	Interface control document
ICF	Inertial confinement fusion
ID	Identification
IF	In-flight fragmentation separation
IFMIF	International Fusion Materials Irradiation Facility
IFMIF-DONES	IFMIF-DEMO-Oriented Neutron Source
IFMIF-EVEDA	IFMIF engineering validation and engineering design activities
IGBT	Insulated gate bipolar transistor
IH	Interdigital H-Structures
IHEP	Institute of High Energy Physics, CAS
IHEP	Institute of High Energy Physics, Protvino, Russia
ILC	International Linear Collider
IMP	Institute of Modern Physics, CAS
INFN	Istituto Nazionale di Fisica Nucleare, Italy
INS	Institute of Nuclear Studies, RAS
INS	Institute of Nuclear Study, Japan

I/O	Input/output
IOC	Input/output controller
IOT	Inductive output tube
IOTA	Integrable Optics Test Accelerator, FNAL
IPM	Ionization profile monitor
IPS	Instant phase setting application
IRIS	Institute for Rare Isotope Science, IBS, Korea
IS	Ion source
ISAC-II	Isotope Separator and ACcelerator II, TRIUMF
ISIS	Spallation neutron facility at RAL, UK
ISO	International Organization for Standardization
ISOL	Isotope separation on-line
ISOLDE	On-Line Isotope Mass Separator, CERN
ISR	Intersecting Storage Rings, CERN
ITER	International Thermonuclear Experimental Reactor, France
IUCF	Indiana University Cyclotron Facility, Indiana University, USA
J	Joule
JAERI	The Japan Atomic Energy Research Institute, Japan
JET	Joint European Torus, UK
JINR	Joint Institute for Nuclear Research, Russia
JHF	Japan Hadron Facility project, Japan
JLab	Thomas Jefferson National Accelerator Facility, USA
JNC	Japan Nuclear Cycle Development Institute, Japan
J-PARC	Japan Proton Accelerator Research Complex, Japan
JT	Joule-Thompson
k	10^3
K	Kelvin, unit of measurement for temperature
K500, K1200	SC cyclotrons at NSCL, MSU
KAMINI	Kalpakkam Mini reactor, Indira Gandhi Center for Atomic Research in Kalpakkam, India
KEK	High Energy Accelerator Research Organization, Japan
KEK-B	Electron-positron B-factory two-ring collider, KEK
KEK-DA	KEK Digital Accelerator, KEK
KEK PS	KEK Proton Synchrotron, KEK
KOMAC	Korea Multi-purpose Accelerator Complex, Korea
KPP	Key performance parameters
LAL	Laboratoire de l'Accélérateur Linéaire, France
LAMPH	Los Alamos Meson Physics Facility, LANL
LANL	Los Alamos National Laboratory, USA
LANSCE	Los Alamos Neutron Science Center, LANL
LASREF	Los Alamos Spallation Radiation Effects Facility, LANL
LBE	Lead-Bismuth Eutectic
LBNE	Long Baseline Neutrino Experiment, FNAL
LC	Inductor-capacitor resonant circuit

LCLS-II	Linear Coherent Light Source -II, SLAC
LCA	Low conductivity water for activated environment
LCW	Low conductivity water
LEAR	Low Energy Antiproton Ring, CERN
LEBT	Low-energy beam transport
LEDA	The Low Energy Demonstration Accelerator, LANL
LEDP	Low energy differential pumping
LENS	Low Energy Neutron Source, Indiana University
LEP	Large Electron-Positron Collider, CERN
LET	Linear energy transfer
LFD	Lorentz force detuning
LHC	Large Hadron Collider, CERN
LHeC	Large Hadron-Electron Collider, CERN
linac	Linear accelerator
LIPAc	Linear IFMIF Prototype Accelerator, IFMIF/EVEDA, Japan
LIS	Laser on source
LLNL	Lawrence Livermore National Laboratory, USA
LLRF	Low-level radio frequency
LNF	Frascati National Laboratory, INFN
LNL	Legnaro National Laboratory, INFN
LOTO	Lock out, tag out
LS	Linac segment
LTS	Low-temperature superconducting
m	10^{-3}
μ	10^{-6}
M	10^{6}
MA	Magnetic alloy
MaRIE	Matter-Radiation Interactions in Extremes, LANL
MASAR	Machine snapshot, archiving, and retrieve
MAX IV	Synchrotron light source facility in Lund, Sweden
MCE	Maximum credible event
MD	Molecular dynamics
MDT	Mean downtime
MEBT	Medium energy beam transport
MEGAPIE	Megawatt pilot target experiment
MERIT	Mercury intense target experiment
MEVVA	Metal vapor vacuum arc ion source
MGB	Multi-gap buncher
MHB	Multi-harmonic buncher
MICE	Muon ionization cooling experiment
MicroTCA, μTCA	Micro telecommunications computing architecture
MIOSHA	Michigan Occupational Safety and Health Administration
ML	Machine learning
MLF	Materials and Life Science Experimental Facility, J-PARC
MLI	Multi-layer insulation

MMF	Moscow Meson Factory, INS, RAS
MP	Multipacting
MPI	Maximum probability of improvement
MPIK	Max-Planck Institute for Nuclear Physics, Heidelberg, Germany
MPQ	Max-Planck Institute for Quantum Optics, München, Germany
MPS	Machine protection system
MR	Main Ring
MRF	Micro Research Finland Oy
MRI	Magnetic resonance imaging
MSU	Michigan State University, USA
MTBF	Mean time between failures
MTE	Multi-turn extraction
MURA	Mid-western Universities Research Association
MYRRHA	Multi-purpose hYbrid Research Reactor for High-tech Applications, Belgium
n	10^{-9}
NAAQS	National Ambient Air Quality Standards, EPA
NAP-M	Storage ring used for early experiments on electron cooling, BINP
NAS	National Academy of Science, USA
NASA	National Aeronautics and Space Administration, USA
NASA-JSC	NASA Johnson Space Center
Nb	Niobium
NC	Normal-conducting
N-doping	Nitrogen doping
NEC	National Electrical Code
NEPA	National Environmental Policy Act, USA
NFPA	National Fire Protection Association
NIF	National Ignition Facility, LLNL
NRC	National Research Council, USA
NRC	Nuclear Regulatory Commission, USA
NSAC	Nuclear Science Advisory Committee
NSLS II	National Synchrotron Light Source II, BNL
NSP	Neutron Science Project, Japan
NSRL	NASA Space Radiation Laboratory, BNL
NSCL	National Superconducting Cyclotron Laboratory, MSU
NSF	National Science Foundation, USA
NTP	Network time protocol
NTU	Number of transfer units
OC	Outer conductor
ODH	Oxygen deficiency hazard
OFC	Oxygen-free copper
OFHC	Oxygen-free, high-conductivity copper

OIC	Operator in charge
OLog	Operations log book
OPA	Office of Project Assessment, DOE
ORNL	Oak Ridge National Laboratory, USA
OSC	Optical stochastic cooling
OSE	Operational safety envelope
OSHA	Occupational Safety and Health Administration, USA
p	10^{-12}
P&ID	Piping and instrumentation diagram
PALLAS	RF quadrupole storage ring for acceleration and laser cooling, MPQ
PCR	Project change request
PDR	Pre-conceptual design report
PEL	Permissible exposure limit
PEP	Project execution plan
PEP, PEP-II	Positron electron collider, SLAC
PETRA	Electron positron collider, DESY
PFL	Pulse-forming line
PFN	Pulse-forming network
PIC	Particle-in-cell
PIG	Penning ionization gauge
PIP-II	The Proton Improvement Plan II, FNAL
PISL	π-mode stabilizing loop
PLC	Programmable logic controller
PMQ	Permanent magnet quadrupoles
PPE	Personal protective equipment
ppmv	Parts per million by volume
ppp	Particles per pulse
PRIMA	Padova Research on ITER Megavolt Accelerator
PPS	Personnel protection system
PPU	Proton Power Upgrade project, ORNL
Project X	Proposed multi-megawatt proton accelerator facility, FNAL
PRP	Polyimide resin pre-impregnated
PRTRF	Power Reactor Thoria Reprocessing Facility, India
PS	Power supply
PSB	Proton Synchrotron Booster, CERN
PSI	Paul Scherrer Institute, Switzerland
PSR	Proton Storage Ring, LANL
PTP	Precision time protocol
PV	Process Variable
Q disease	A significant degradation of the cavity's quality factor (Q) and accelerating gradient
QRD	Quench recovery Dewar
QWR	Quarter-wave resonator
r	Radian

R2A2	Roles, responsibilities, authorities, and accountabilities
R&D	Research and development
RAL	Rutherford-Appleton Laboratory, UK
RAM	Random access memory
RAON	Rare isotope Accelerator complex for ON-line experiments, IRIS, IBS
RAS	Russian Academy of Sciences
RATS	Receiving, acceptance, test and storage
RCS	rapid-cycling synchrotron
rem	Roentgen equivalent man
ReA	Reaccelerator ReA3 and ReA6 at MSU
REBCO	Rare earth barium copper oxide
RF	Radio frequency
RFQ	Radio-frequency quadrupole
RGA	Residual gas analyzer
RHIC	Relativistic Heavy Ion Collider, BNL
RI	Radioisotope
RIA	Rare Isotope Accelerator project
RIBF	Radioactive Isotope Beam Factory, RIKEN
RIKEN	Institute of Physical and Chemical Research (Rikagaku Kenkyusho), Japan
rms	Root-mean-squared
rpm	Revolution per minute
RPS	Run permit system
RRR	Residual resistivity ratio
RTBT	Ring-to-target beam transport
RTD	Resistance temperature detector
s	Second
SAD	Safety assessment document
SARAF	Soreq Applied Research Accelerator Facility, Israel
SC	Superconducting
SCADA	Supervisory control and data acquisition
SCL	Superconducting linac
SCS	Side-coupled structure
SDDS	Self describing data sets
SDTL	Separated drift tube linac
SEE	Single event effects
SEY	Secondary-emission yield
SGR	Single-gap resonator
SHiP	Search for Hidden Particles experiment, CERN
SI	Système International, based on meter, kilogram, second, ampere (MKSA)
SIL	Safety integrity level
SINQ	Swiss Institute for Nuclear Research, PSI
SLC	Stanford Linear Collider, SLAC

SLS	Swiss Light Source
S-LSR	Laser equipped cooler / storage ring, Kyoto University, Japan
SME	Subject matter expert
SNQ	Spallation Neutron Source project, FZJ
SNS	Spallation Neutron Source, ORNL
SPEAR	Stanford Positron Electron Asymmetric Ring, SLAC
SPIRAL2	Système de Production d'Ions RAdioactifs en Ligne de 2e génération (Second-generation radioactive ion production with linear accelerator), GANIL
SPL	Superconducting Proton Linac, CERN
S-POD	Simulator for Particle Orbit Dynamics, Hiroshima University
SPPC	Super Proton Proton Collider, CAS
SPring-8	Super Photon ring - 8 GeV
SPS	Super Proton Synchrotron, CERN
SRD	System requirements document
SRF	Superconducting radiofrequency
SRS	Synchrotron Radiation Source, Daresbury Laboratory, UK
SRC	Superconducting Ring Cyclotron, RIKEN
SS	Stainless steel
SSC	Superconducting Super Collider
SSCL	Superconducting Super-Collider Laboratory
SS316L	Stainless Steel 316L (a low-carbon version of stainless steel)
SSPA	Solid-state power amplifier
SSR	Single-spoke resonator
STFC	Science and Technology Facilities Council, UK
SuperKEKB	Upgrade project of KEK-B
Sv	Sievert; 100 rem
SW	Software
SX	Slow-extraction
T	10^{12}
T2K	Tokai to Kamioka experiment
TCP/IP	Transmission control protocol/internet protocol
TDR	Technical design report
TE	Transverse electric
TEC	Total estimated cost
TEM	Transverse electric and magnetic
Tevatron	Superconducting collider at FNAL
TM	Transverse magnetic
TPC	Total project cost
TRISTAN	Electron-positron collider at KEK
TRIUMF	Tri-University Meson Facility, Canada
TSR	Heavy ion Test Storage Ring, MPIK
TSR	Triple-spoke resonator
TTF	Transit time factor

TARN	Test Accumulation Ring for Numatron, University of Tokyo, Japan
U1.5	Rapid-cycling booster synchrotron at IHEP, Protvino, Russia
U70	Proton synchrotron at IHEP, Protvino, Russia
UA	Net thermal rating
UCB	Upper confidence bound
UDP	User datagram protocol
UHV	Ultra-high vacuum
UNILAC	Universal Linear Accelerator, GSI
UOF	Unscheduled off
UPS	Uninterruptible power supply
USI	Unreviewed safety issue procedure
V	Volt
VD	Virtual diagnostics
VEP-1	Electron-electron collider at BINP
VENUS	28 GHz ECR ion source, LBNL
VEPP-	Electron-position colliders at BINP
VME	Versamodule Eurobus
W	Watt
WAN	Wide area network
WBS	Work breakdown structure
XFEL	X-ray free electron laser
XFEL-O	X-ray free electron laser oscillator
XFEL-SASE	X-ray free electron laser self-amplified spontaneous emission
XML	Extensible markup language
YBCO	Yttrium barium copper oxide
Ω	Ohm

Symbols

Roman letters:

a	Average beam radius
a_2	Skew quadrupole strength
a_i, b_i, d_i	Intrabeam scattering geometric factors
a_n	$n = 0,\ 1,\ 2, \ldots$; Magnetic-field skew-multipole coefficient
a_w	Waveguide longer-side width
A	Atomic mass number: total number of protons plus neutrons in an isotope
A_{s1}, A_{s2}	Lorentz stripping coefficients
A_P	Projectile atomic mass number
A_T	Absorber (or target) atomic mass number
b	Average radius of vacuum pipe
b_n	$n = 0,\ 1,\ 2, \ldots$; Magnetic-field normal-multipole coefficient
B	Magnetic flux density
$\dot{B}$	Rate of change of magnetic flux density
B_0	Nominal guiding magnetic-field flux density
$B_0\rho$	Rigidity of a particle
B_f	Bunching factor; ratio of average and peak beam density
B_p	Peak magnetic field
B_{sol}	Solenoid magnetic field
B_{RF}	RF magnetic induction field
B_x	Horizontal component of magnetic field
B_y	Vertical component of magnetic field
c	speed of light
C_n	$n = 1,\ 2,\ \ldots$; Multipole tune-shift coefficients
C_l	Envelope-oscillation coefficient
$c_{v,T}$	Specific heat of material
df/dE_a^2	Lorentz de-tuning coefficient
df/dP	Frequency detuning coefficient to pressure
d_m	Molecular density of gas for ion species m

d_w	Thickness of vacuum pipe wall
d_y :	: Vertical deviation from the ellipsoid surface in survey
D	Lattice dispersion function
D_{IBS}	Fokker-Planck diffusion coefficient due to intrabeam scattering
D_{SC}	Fokker-Planck diffusion coefficient due to stochastic cooling
D_x or D_H	Horizontal dispersion function
D_y or D_V	Vertical dispersion function
D^+	Maximum dispersion
dP_r/ds	Power generated in a resistive loop
e	Unit electric charge
E	Electric field strength
E_T	Total energy of particle or photon
E_{eff}	Effective electric field
E_a	Average accelerating electric field strength
E_{cm}	Particle energy in the center-of-mass frame
E_i	Injection energy
E_k	Kinetic energy
E_p	Peak electric field
E_s	Synchronous energy of the beam
E_{su}	Beam space charge field, $u = x, y, z$
E_y	Young's modulus
D_w	Characteristic dimension of cooling channel
ΔE_t	Energy gain per turn in cyclotron
f	Frequency
F	Cyclotron flatter parameter
f_c	Waveguide low-frequency cutoff
f_e	Electron flux ratio: ratio of the number of electrons striking the vacuum pipe in one turn to the number of protons stored in the ring
F_{IBS}	Fokker-Planck dynamic drift coefficient due to intrabeam scattering
F_L	Lorentz force
f_M	Mechanical eigen-frequency
f_{rep}	Repetition rate
f_{RF}	RF frequency
f_{sc}	Space-charge form factor
F_{SC}	Fokker-Planck dynamic drift coefficient due to stochastic cooling
$F_{\parallel}$	Longitudinal stability bunch form factor in ring
$F_{\perp}$	Transverse stability bunch form factor in ring
g	gap height of dipole magnet
g_0	Space-charge geometric factor
g_q	Transverse to longitudinal quench velocity ratio
g_r	Carlson elliptic integral
G	focusing gradient of quadrupole
h_P	Planck constant
h_{RF}	RF harmonic number
h_w	Heat transfer coefficient

H	magnetic field strength
H_C	Critical magnetic field
h_r	Height of a resistive loop
I	beam current
I_0	Average beam current in ring
$\hat{I}$	Peak beam current
I_C	Superconductor critical current
I_{op}	Operating current
j	$\sqrt{(-1)}$
J_L	Longitudinal action
$J_{x,y}$	Transverse action
k	Excited resonance harmonics
K	Cyclotron K value in non-relativistic limit
k_B	Boltzmann constant
k_c	Field exponent index
K_c	Coupling constant
K_{CYC}	Cyclotron K value
$K_{\mathrm{sc}\,x,y,z}$	Linac space-charge de-focus parameter in x, y, and z planes
$K_{\mathrm{sc,3D}}$	Linac 3-D space-charge parameter
K_{sp}	The generalized perveance for transverse space charge effect
$K_\perp$	Transverse-focusing strength
$\Delta K_\perp$	Deviation in transverse-focusing strength
$K_\perp^{(l)}$	Strength of $2l$th pole error
l	resonance order
L_0	Air-cored inductance
L_a	Dimension of accelerator
L_c	FODO cell length
$\mathcal{L}_c$	Intrabeam scattering Coulomb logarithm
L_{int}	Distance between the beam ports in cavity
L_{max}	Maximum length occupied by the EM field of the cavity in the direction of the beam line
l_{q}	Length of quadrupole magnet
m, n	integers
m_0	Atomic mass unit, amu
m_e	Rest mass of electron
m_u	Atomic mass unit
M	Number of lattice superperiods
M_m	Material mass
n_g	Number of accelerating gaps
n_t	Turn number in cyclotron
n_{th}	Epithermal neutron
n_p	Volume density of the proton beam
N_0	Number of protons per bunch
P	pressure

P_{cav}	RF power dissipated in cavity
P_b	Beam power
P_{ext}	Power emitted through input power coupler, or external power
P_f	Forward RF power
P_{inst}	Instantaneous power
P_{loss}	Total energy consumption power
P_r	Reflected total RF power
P_{tr}	Power transmitted through RF pick-up
q	Electric-charge number
Q	Quality factor
Q_0	Unloaded quality factor
Q_b	Total charge in a bunch
Q_L	Loaded quality factor
r_0	Classical radius of proton; $e^2/4\pi\epsilon_0 m_0 c^2$
r_e	Classical radius of electron; $e^2/pi\epsilon_0 m_e c^2$
r_r	Beam radius in radial direction in cyclotron
r_u	Beam radius, $u = x, y, z$
R	Beam orbit radius under magnetic field
R_0	Ring circumference divided by 2π
R_c	Cyclotron orbit radius
R_e	Average earth's radius, $R_e \approx$6,378 km
R_s	Surface resistance of cavity walls
$R_{s,\text{BCS}}$	BCS surface resistance
$R_{s,\text{NC}}$	Surface resistance of normal-conducting material
$R_{s,\text{res}}$	Residual surface resistance
$R_{s,\text{SC}}$	Surface resistance of superconducting material
R_{sh}	Shunt impedance
s	Azimuthal displacement along ring
S :	Longitudinal phase-space area of the bunch
S	entropy (Chap. 4)
s_b	Center-to-center distance between subsequent bunches
S_p	Stopping power
S_{rms}	rms longitudinal bunch area
t	time in laboratory frame
T_B	Beam temperature, laboratory frame
T_c	Characteristic transition-crossing time
T_C	Superconductor critical temperature
T_L, T_H	Low and high temperature, respectively
t_{mps}	MPS response time
T_m	Tension in material
T_{mw}	Duration time of microwave instability near transition
T_N	Beam temperature, normalized
T_{nl}	Chromatic nonlinear time near transition
T_{op}	Operating temperature
U	RF stored energy in resonator

U_m	Material stored energy
V_{RF}	RF peak voltage
	V_{acc}:Maximum accelerating voltage on axis
V_t	Peak RF voltage per turn in cyclotron
v_w	Water velocity
w_r	Width of a resistive loop
W	Kinetic energy
	W_i Kinetic energy at injection
W_s	Kinetic energy of synchronous particle
x	horizontal displacement
x'	dx/ds
x_{sep}	Transverse displacement at the septum
X	normalized horizontal displacement; $X \equiv x/\sqrt{\beta_x}$
X'	$dX/d\mu_x \equiv (\alpha_x x + \beta_x x')/\sqrt{\beta_x}$
y	Vertical displacement
y'	dy/ds
y_{cod}	Vertical closed-orbit displacement during ring extraction
Y	normalized vertical displacement; $Y \equiv y/\sqrt{\beta_y}$
Y'	$dy/d\mu_y \equiv (\alpha_y y + \beta_y y')/\sqrt{\beta_y}5$
Y_{ee}	Electron secondary-emission yield upon electron impact
Y_{ep}	Electron secondary-emission yield upon proton impact
$Y_{\mathrm{ep}}^{\mathrm{max}}$	Maximum electron secondary-emission yield upon proton impact
z	Longitudinal displacement
Z	Atomic number; number of protons in an isotope
$Z_{\parallel}$	Longitudinal coupling impedance
$Z_{\perp}$	Transverse coupling impedance
Z_0	$(\varepsilon_0 c)^{-1}$=377 Ω
$Z_{\parallel}^{\mathrm{rw}}$, $Z_{\perp}^{\mathrm{rw}}$	Coupling impedances due to resistive wall
$Z_{\parallel}^{\mathrm{sc}}$, $Z_{\perp}^{\mathrm{sc}}$	Coupling impedances due to space charge
Z_P	Projectile atomic number
$Z_{P,\mathrm{eff}}$	Effective projectile atomic number
Z_T	Absorber (or target) atomic number

Greek letters and other symbols:

$\lvert\ \rvert$	absolute value
$\langle\ \rangle$	Average value
α	Availability
α_0	Momentum compaction factor, $\alpha_0 \equiv \gamma_T^2$
α_1	First-order nonlinear momentum compaction factor
$\alpha_{\perp}$	Transverse a lattice function
α_l	Material linear expansion coefficient
α_x	Horizontal α lattice function
$\alpha_{x,i}$	Horizontal lattice function of injecting beam
β	Relativistic factor; βc is beam velocity
β_0	Reference or optimal β of the SRF cavity

β'	$d\beta/ds$
$\beta_\perp$	Transverse amplitude function
$\beta_\perp^+$	Maximum transverse amplitude function in FODO cell
$\beta_\perp^-$	Minimum transverse amplitude function in FODO cell
β_c, $\beta_{c,\text{opt}}$	Cavity RF coupling factor and the optimum value
β_e	Electron relativistic factor; $\beta_e c$ is the electron velocity
β_{kick}	Transverse amplitude function at the kicker
β_{sep}	Transverse amplitude function at the septum
β_L	Longitudinal amplitude function
β_x or β_H	Horizontal amplitude function
$\beta_{x,i}$	Horizontal amplitude function of injecting beam
β_y, or β_V	Vertical amplitude function
γ	Relativistic factor; $\gamma \equiv \sqrt{1-\beta^2}$
$\dot{\gamma}$	rate of γ variation in time
γ_T	Relativistic factor corresponding to the transition energy
$\dot{\gamma}_T$	Rate of γ_T variation in time during transition jump
$\gamma_{x,y}$	Horizontal and vertical lattice functions
δ_{RF}	RF loss in SRF
δ_s	Skin depth
$\left(\frac{\Delta E}{E_s}\right)_{\text{FWHM}}$	Full-width, half-maximum relative energy spread
$\Delta p/p$	Relative momentum deviation
ΔQ_L, ΔQ_H	Heat load at low and high temperature, respectively
ΔR	Cyclotron orbit turn-to-turn separation
ΔS	Change in longitudinal phase-space area of the bunch
ΔW_{ext}	Energy gain per turn at cyclotron extraction
$\Delta\mu$	Betatron phase advance
$\Delta\nu$	Transverse tune shift
$\Delta\nu_{\text{sc}}$	Transverse incoherent space-charge tune shift
$\Delta\nu_{x,y}$	Horizontal or vertical tune shift
ΔV_{RF}	Amount of RF volume change
ε	Permittivity
ε_0	Permittivity in vacuum
$\epsilon_{x,y}$	Transverse unnormalized (or geometric) emittance
$\epsilon_{N,\text{rms}}$	Transverse normalized rms beam emittance
ϵ_x	Horizontal unnormalized (or geometric) emittance
$\epsilon_{x,i}$	Horizontal unnormalized (or geometric) emittance of injecting beam
ζ_m	Multibunch electron-cloud multipacting parameter
ζ_s	Single-bunch electron-cloud multipacting parameter
η	Phase-slip factor
$\eta_{x,y}$	Transverse dispersion function
θ_g	Grazing angle
θ_{kick}	Transverse momentum deflection from the kicker
κ	Resonance strength

κ_c	Cyclotron field exponent index
λ	Wavelength
λ_B	Beam line density
λ_c	Waveguide cutoff wavelength
λ_{RF}	RF wavelength
λ_s	Mean decay length due to Lorentz stripping
μ	Permeability
μ_0	Permeability in vacuum
μ_{0u}	Phase advance of transverse or longitudinal oscillations in phase space per focusing period, $u = x, y, z$
μ_c	Betatron phase advance per FODO cell
ν_0	Unperturbed transverse tune
$\nu_{x,y}$	Transverse tunes
$\nu_{y,r}$	Cyclotron tunes
ξ_B	Characteristic distance between charged particles under Coulomb interaction in an accelerator
ξ_c	Cyclotron magnet pole spiral angle
ξ_{gap}	Cyclotron gap-to-envelope coefficient
$\xi_\perp$	Transverse chromaticity of the ring
ρ_m	Material density
ρ_r	Volume resistivity
ρ_T	Absorber (or target) material density
σ_{ion}	Ionization cross section
σ_m	Material tensile stress
$\sigma_{m,y}$	Material yield stress
σ_{SB}	Stefan-Boltzmann constant
σ_{sm}	Interaction cross section with ion species m
σ_p	rms spread in relative momentum deviation $\Delta p/p$
σ_x, σ_y	Horizontal and vertical rms beam radius
σ_ϕ	rms spread in RF phase
ϕ	RF phase
ϕ_c	Bending angle per FODO cell
ϕ_s	Synchronous RF phase
ψ	Cavity optimum detuning angle
Ψ	Density function
φ_x	Horizontal betatron-oscillation phase
φ_y	Vertical betatron-oscillation phase
ω	Angular frequency
ω_c	Cyclotron angular-revolution frequency
ω_e	Electron bounce frequency
ω_{RF}	Angular RF frequency
ω_s	Synchronous angular-revolution frequency
τ_m	H^- lifetime
τ_n	Decay time of mode n
$\mathcal{F}_{\text{IBS}}$	Intrabeam scattering evolution function
$\mathcal{L}_C$	Intrabeam scattering Coulomb logarithm

Chapter 1
Introduction

Abstract This chapter introduces the energy/luminosity, brightness/coherence, and power/intensity frontiers of charged-particle accelerators. Evolution of accelerators on the hadron beam power frontier is surveyed from 1970 to now and toward future years based on the main accelerator types of cyclotrons, linacs, and synchrotron rings as the main drivers to modern facilities. Consequently, technology selections, performance-limiting mechanisms, and design philosophy are highlighted. Specific examples of major applications of power-frontier accelerators are given covering high-energy and nuclear physics, accelerator-based multi-user platforms, accelerator-driven sub-critical systems, accelerators servicing fusion research, hadron therapy and medicine, accelerators for security and defense, and compact hadron accelerators for research and prototyping.

Particle accelerators are machines that use electromagnetic fields to propel charged particles, confined in well-defined beams, to nearly the speed of light. Beams of charged particles are used for scientific research and industrial applications, including high-energy physics, nuclear physics, isotope production, synchrotron-radiation light sources, spallation neutron sources, and industrial and medical applications. Currently, there are more than 30,000 accelerators operating worldwide, ranging from large accelerators primarily located at national laboratories and institutes to smaller accelerators that are more widely distributed [1]. This chapter introduces the frontiers of accelerators and elaborates on high-power hadron accelerator concepts, design philosophies, and applications.

1.1 A Brief History of Particle Accelerators

In 1928, Cockcroft and Walton began to design an 800 kV generator upon the encouragement of Rutherford. Following the predictions by Gurney and Gamov on tunneling, they succeeded in using this machine to split the lithium nucleus with 400 keV protons [2]. Such electrostatic accelerators typically produce an

J. Wei et al., *High Power Hadron Accelerators*, Particle Acceleration and Detection,
https://doi.org/10.1007/978-3-032-17662-2_1

output beam of low current, with energy below 30 MeV limited by the voltage ceiling imposed by electrical discharge. These limitations were overcome by electrodynamic acceleration using either nonresonant magnetic induction or resonators excited by oscillating radio-frequency (RF) fields.

Efforts in electrodynamic acceleration started in 1924 when Ising proposed time-varying fields across drift tubes for "resonant acceleration"; thus, a RF linear accelerator (linac) [3]. In 1928, Wideröe demonstrated Ising's principle with a 1 MHz, 25 kV oscillator to produce 50 keV potassium ions [4]. Inspired by Wideröe and Ising, Lawrence conceived the cyclotron concept in 1929. Subsequently, Livingston demonstrated the cyclotron by accelerating hydrogen ions to 80 keV [5]. Lawrence's cyclotron produced 1.25 MeV protons, and he also split the nucleus just a few weeks after Cockcroft and Walton in 1932. Almost a century later, both linacs and cyclotrons are actively being developed as drivers for high-power hadron facilities.

The other method of electrodynamic acceleration is non-resonant magnetic induction. Wideröe initiated the betatron design along with the conditions for radial stability but failed to build a functional machine. Kerst re-invented the betatron concept and built the first working machine for 2.2 MeV electrons in 1940 and the world's largest betatron of 300 MeV in 1950 [6].

In 1944, McMillan [7] and Veksler [8] discovered the principle of phase stability and invented the synchrotron. In a synchrotron, the guide field increases with particle energy to keep the beam orbit stationary, as in the betatron; however, acceleration is applied with an RF voltage via a gap or cavity. Subsequently, Courant, Livingston, Snyder [9] and Christofilos [10] proposed strong focusing, also known as alternating-gradient (AG) focusing. This principle revolutionized synchrotron design, allowing for smaller magnets to be used and higher energies to be envisioned. It also enabled cyclotrons to reach relativistic energies.

In parallel to the development of circular accelerators, advances in microwave frequency technology during World War II stimulated new developments in linac accelerator structures. In 1946, Alvarez built the first linear proton accelerator of 32 MeV [11]. The Alvarez accelerators have since become popular both as injectors for synchrotrons and as full-energy machines.

Superconducting (SC) technology has played an important role in both circular accelerators, which increase the peak field in electrodynamic magnets, and linear accelerators, which increase the peak accelerating gradient of the resonators [12]. The first particle acceleration by means of an SC cavity was achieved at Stanford University in 1965, using a lead-plated copper resonator and an electron beam. In 1977, the first SC linac, a 27 m long structure consisting of multicell 1.3 GHz pillbox cavities made of niobium welded by an electron beam, was able to achieve 50 MV acceleration in the CW mode at Stanford. The first application in a circular accelerator was the 2.86 GHz muffin-tin structure that accelerated electrons to 4 GeV in the Cornell Electron Synchrotron in 1975. After discovering that multipacting can be eliminated by using elliptical shapes rather than pill-box shapes, single- and multi-cell elliptical cavities became the reference for superconducting electron accelerators, both linear and circular.

For protons and heavy ions, where low-velocity particles are involved in the initial part of accelerators, different shapes and lower frequencies must be pursued. In 1960, a superconducting niobium helical cavity was developed, followed by a lead plated copper split ring at Caltech in 1974. The first SC heavy-ion linac was ATLAS in Argonne in 1978 based on split-ring resonators. Many other SC heavy-ion linacs have been built worldwide in subsequent years, mostly based on quarter-wave resonators (QWR). New shapes and technologies have been developed to extend SC resonators for high-current and high-energy proton and ion beam production. An important milestone has been set by the Spallation Neutron Source (SNS), which started operation in 2006 and, for the first time, accelerated an H^- beam to 1 GeV energy with beam power above 1 MW using an SC linac.

A collider is a type of particle accelerator that brings two opposing particle beams into collision. The drawback in the energy of early fixed-target accelerators is overcome by the invention of colliders, where two counter-moving beams collide nearly head-on [13]. Colliders may be either ring accelerators or linear accelerators. In 1943, Wideröe filed a patent for the collider concept [14]. This idea was further developed by Kerst and researchers at the Midwestern Universities Research Association (MURA) [15] and by O'Neill [16]. In the late 1950s, three teams began working on colliding beams [17]. Since the 1960s, colliders have been used as a research tool in particle physics to reach high center-of-mass(c.m.) energies. Analysis of the by-products of these collisions offers insight into the structure of the subatomic world and the laws of nature governing it.

In recent decades, the concepts of collective acceleration have been developed since the pioneering work of [8] and [18]. These developments have enabled the achievement of substantially higher accelerating gradients compared to traditional methods, in which particles gain energy from an external field. This innovative approach encompasses techniques such as laser-driven and beam-driven acceleration within plasma or other accelerating structures, representing a notable evolution in particle-acceleration technology.

1.2 Frontiers of Particle Accelerators

During the past century, accelerator physics and technology have continuously developed to advance the achievable beam properties, not only in particle energy but also in intensity and brightness. This section introduces three representative frontiers of accelerator development: energy and luminosity, brightness and coherence, and power and intensity.

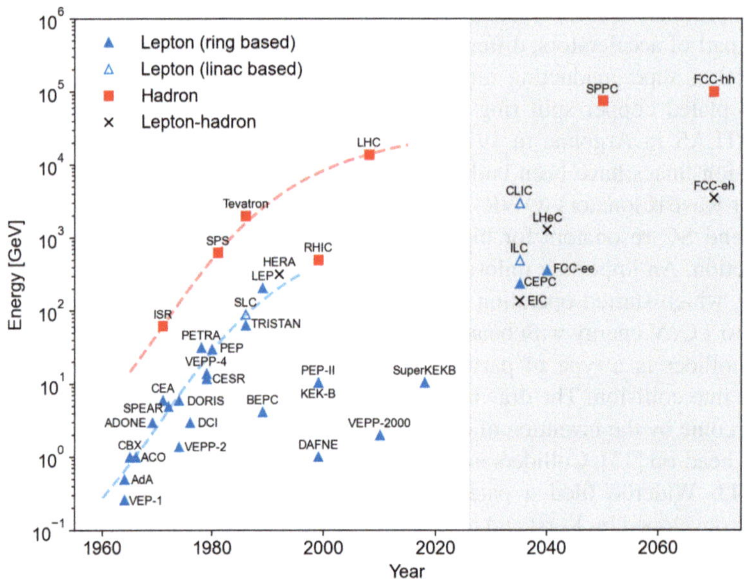

Fig. 1.1 Center-of-mass energy as a function of year that is either attained (before 2025) or planned with lepton-lepton, hadron-hadron, and lepton-hadron colliders. The dashed lines indicate the corresponding energy frontiers

1.2.1 Energy and Luminosity Frontier Accelerators

Energy and luminosity frontier accelerators are machines designed to accelerate hadron or lepton beams for particle collisions at high energies for fundamental physics experiments. In these experiments, the figures of merit are the energy in the center-of-mass frame and the luminosity, which determines collision events. Since 1960, storage-ring colliders and linear colliders have dominated the high-energy physics field.

Figure 1.1 shows the evolution of the center-of-mass energy of particles attained with such colliders. This "Livingston chart" shows the succession of new ideas and new technologies pushing accelerator beam energies over five decades at a rate of more than one and a half orders of magnitude per decade [19–21]. However, the present technology no longer supports the traditional exponential growth of particle energy that occurred over the past several decades.

1.2.1.1 Circular Hadron and Lepton Colliders

Single-ring colliders using electrons and positrons were the first type of collider built in the late 1950s to early 1960s, with the AdA (Annelli di Accumulazione) project at Frascati, Italy, and with the Princeton-Stanford Experiment Collider

(CBX) in the USA. Also, in the early 1960s, the VEP-1 electron-electron collider was independently developed and built at the Soviet Institute of Nuclear Physics. Proton-antiproton colliders were only made possible by the invention of stochastic cooling by van der Meer [22] for the accumulation of antiprotons in CERN's Super Proton Synchrotron (SPS) in 1984. The highest energy single-ring proton-antiproton collider is the Tevatron at FNAL, with a center-of-mass energy of nearly 2 TeV [23].

The first double-ring proton-proton collider was the CERN ISR (Intersecting Storage Rings). The electron-proton collider HERA was built at DESY [24]. The Relativistic Heavy Ion Collider (RHIC) was built at BNL to collide any combination of ions, from protons to gold, for nuclear physics research [25].

Circular proton-proton colliders are the main experimental tools available for exploring particle physics in the energy range of tens of TeV. Presently, the Large Hadron Collider (LHC) [26] at CERN operates at 14 TeV c.m. energy with countercirculating protons in a tunnel of 26.7 km [26]. The proposed Future Circular Collider (FCC) is designed to increase the collision energy to 100 TeV, using bending magnets of 16 T with advanced superconductors in a tunnel of about 100 km circumference [21]. The same tunnel infrastructure could accommodate a high-luminosity circular e^+e^- collider (FCC-ee), operating at 90–350 GeV (Fig. 1.1). In parallel, the Super Proton Proton Collider (SPPC) at 70 TeV c.m. and the "Higgs factory" Circular Electron Positron Collider (CEPC) at 240 GeV c.m. energy are proposed to be placed in a tunnel of about 54 km circumference using similar technologies [27].

1.2.1.2 Linear Lepton Colliders

The collision of fundamental particles, such as electrons and positrons, is preferred over hadrons for precision high-energy experiments. However, the attainable energy of leptons in circular accelerators is much lower in comparison due to synchrotron radiation energy loss, which is inversely proportional to the fourth power of the particle mass. The Large Electron-Positron Collider (LEP) at CERN is the highest-energy lepton ring built, where the penalty to be paid in RF power to compensate for the synchrotron radiation loss is nearly prohibitive. The solution is to change to linear electron colliders; a solution that was foreseen in 1965 by Tigner [28]. The Stanford Linear Collider (SLC) reached a c.m. energy of 100 GeV in a tunnel length of approximately 4 km using normal-conducting (NC) RF accelerating structures [29]. The proposed International Linear Collider (ILC) is baselined to be a 200–500 GeV (c.m.) high luminosity linear electron-positron collider of approximately 31 km in length, based on 1.3 GHz superconducting radio-frequency (SRF) accelerating technology [30]. The proposed Compact Linear Collider (CLIC) adopts a novel two-beam acceleration concept for efficient RF power transfer from a low energy, high-intensity drive beam to the NC accelerating structures for the main beam, accelerating the electrons and positrons to up to 3 TeV (c.m.) in a tunnel of about 50 km in length [31].

1.2.1.3 Muon Collider

The concept of a muon collider was initially proposed by physicists Tikhonin [32] and Budker [33] in the late 1960s. A muon collider retains the advantages of lepton colliders, colliding fundamental particles. With a mass much heavier than that of the electron's, synchrotron radiation is greatly reduced. Such key advantages of cleaner collisions and negligible synchrotron radiation allow for circular accelerators that are much smaller than equivalent electron or proton colliders [34, 35].

The main challenges include the muons' short lifetime, which requires rapid acceleration before they decay, and the muons' high-emittance generation process, which necessitates rapid beam cooling [36, 37]. Encouraging R&D efforts include the muon ionization cooling experiment (MICE) [38, 39] (Chap. 3.1, Appendix II.A) and the mercury intense target experiment using a free-surface mercury jet as the target material [40, 41] (Sect. 8.9). Other key technologies include a high-power proton driver capable of delivering short, intense proton pulses to the production target, and high-field, large acceptance solenoids to capture the widely spread muons produced from pion decay, possibly using high-temperature superconductor technology.

1.2.1.4 Advanced Concepts for High Gradient Acceleration

Based on so-called advanced accelerator concepts such as laser-plasma acceleration and plasma wakefield acceleration, conceptual facility proposals exist. For example, laser-plasma linear colliders aim to achieve 1 TeV c.m. energy at a distance of 100 m for each beam [42]. In 2006, a high-quality electron beam was accelerated with a laser-plasma accelerator to about 1 GeV in a plasma structure 3 cm long [43]. In 2019, an electron beam was accelerated by laser-plasma to 7.8 GeV in a 20 cm long capillary [44].

1.2.2 Brightness and Coherence Frontier Accelerators

Brightness-and-coherence-frontier light sources are application platforms driven by electron beams. Early generation facilities utilized storage rings, where electrons, deflected by bending magnets, emitted synchrotron radiation. Subsequent generations introduced dedicated achromat lattices to achieve low-emittance electron beams, along with insertion devices such as undulators and wigglers to enhance brightness.

The introduction of a multibend achromat lattice [45] and a hybrid 7-bend achromat lattice [46] significantly increased the brightness of the synchrotron radiation light sources, as shown in Fig. 1.2 [47], since the electron emittance is reduced to less than 50 pm and approaches the diffraction limit of X-ray radiation.

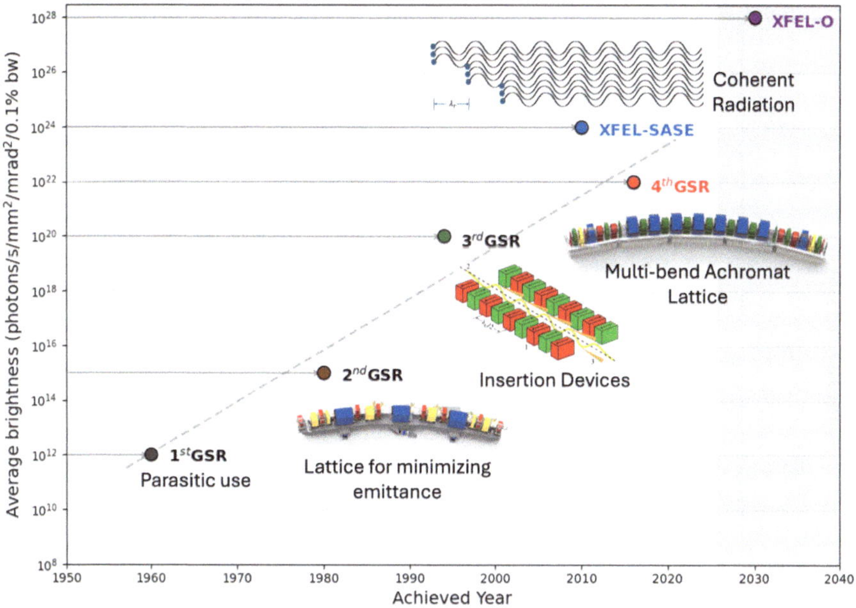

Fig. 1.2 The brightness comparison of the first to the fourth generation synchrotron radiation (4GSR) light sources and X-ray free electron lasers [47]

Simultaneously, linear electron accelerator facilities have been developed worldwide to achieve even higher brightness, laser-like coherence, and ultra-short pulse durations through the principle of free electron lasers (FELs). The Linear Coherent Light Source II (LCLS-II) is an X-ray free-electron laser designed to deliver photons between 200 eV and 5 keV at repetition rates of up to 1 MHz using an SRF linac [48]. The X-ray radiation generated through the FEL process from a fresh, high-energy electron beam in a linac generally exhibits good transverse quality. However, the photon energy, or equivalently the radiation frequency, still shows relatively large fluctuations at the $1 \times 10^{-}4$ level because the FEL amplification process starts from the intrinsic shot noise of the electron beam. Alternative approaches, such as the X-ray FEL oscillator (XFEL-O) [49] and harmonic-generation schemes [50, 51], are attracting increasing attention as pathways toward fully coherent X-ray radiation. In addition, steady-state microbunching schemes [52] in storage rings are also being actively explored to achieve high-average-power radiation.

1.2.3 Power and Intensity Frontier Accelerators

Power and intensity frontier accelerators refer to machines that accelerate moderate energy, high intensity charged beams for neutron sources, as well as for nuclear and

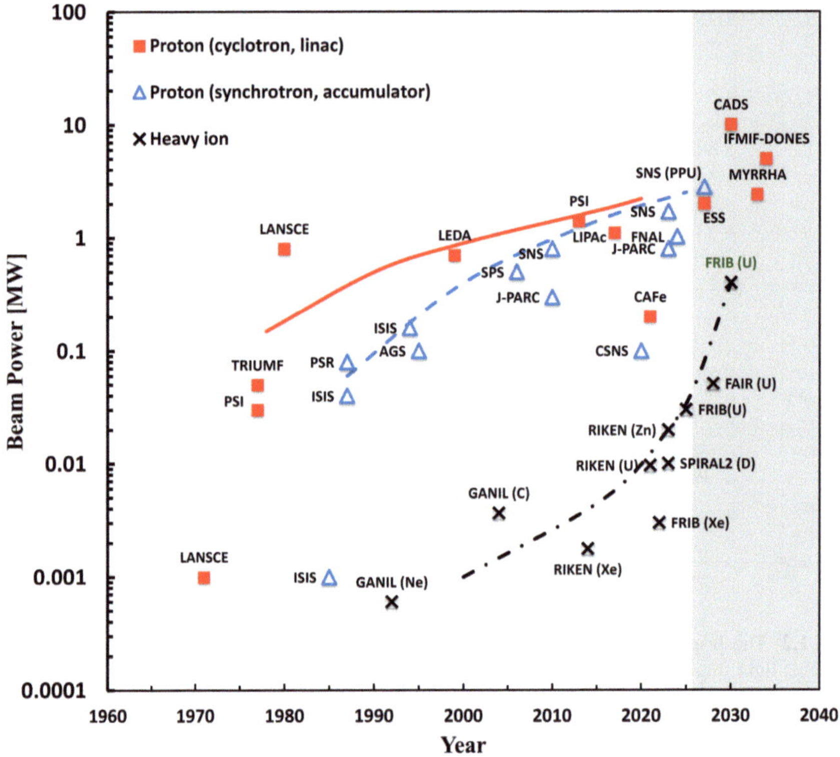

Fig. 1.3 Beam power on target as a function of year that is either attained (before 2025) or planned (shaded area) with cyclotrons and linacs for continuous-wave or long pulsed mode, and synchrotrons and accumulator rings for short pulsed mode, for both proton and heavy ion primary beams

high energy physics research and applications. Accelerated beams bombarding a fixed target generate secondary beams (neutrons, muons, kaons, neutrinos, and rare isotopes). In a desired energy range, the yield of the secondary beam is proportional to the beam power of the primary beam.

Figure 1.3 shows the evolution of the average beam power for proton and heavy-ion beam accelerators. During the past several decades, accelerator-based neutron- and neutrino-generating facilities such as SNS [53], J-PARC [54], FNAL [55], PSI [56] and LANSCE [57] advanced the frontier of proton beam power to the level 1 MW, as shown in Fig. 1.4 with beam-on-target power as the product of the average beam current and the beam kinetic energy. Compared with proton beams, the advancement of power for heavy-ion beams is much more challenging due to the enhancement of stopping power density (Sect. 8.2.1.1). The Facility for Rare Isotope Beams (FRIB) plans to advance the heavy-ion power frontier by more than an order of magnitude to 400 kW [58].

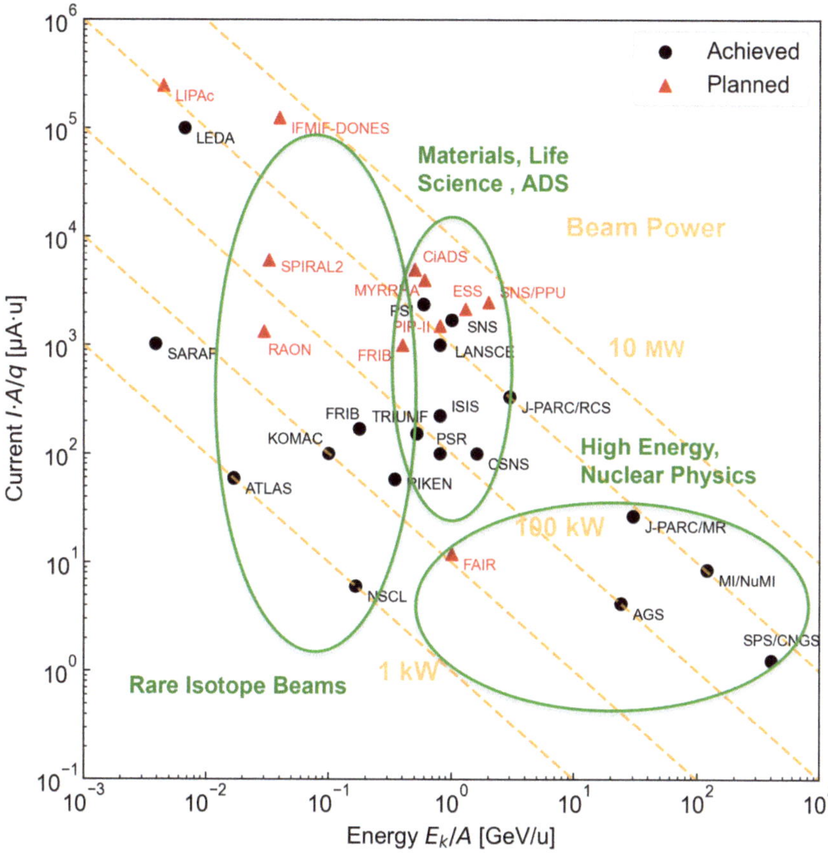

Fig. 1.4 Beam kinetic energy, average beam current, and power of some of the world's high-power hadron accelerators [59]. Beam power in existing high-power proton accelerators is near 1 MW. Beam power in existing high-power heavy ion accelerators is near 10 kW. Circles indicate types of accelerator applications

Figure 1.4 shows accelerators at the "frontier" of beam power, which is the product of the average beam current and the particle kinetic energy. The cutting-edge technologies continuously developed for accelerator systems have sustained continuous growth in beam intensity and power (Fig. 1.3). High-power operations have been made possible by various types of accelerators: linac, cyclotron, synchrotron, and accumulator. During the past decade, SRF-related technology has become indispensable for next-generation machines.

High-power or high-intensity hadron accelerators may be classified into two categories: (1) continuous-wave (CW) or long-pulsed ones, and (2) short-pulsed ones. Accelerators in the first category are generally in two configurations, as illustrated in Fig. 1.5 (a) (linac driven) and (b) (cyclotron driven). Accelerators in

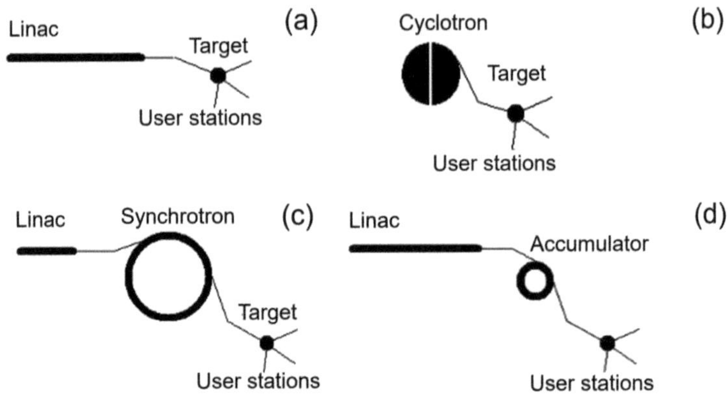

Fig. 1.5 Typical types of high-power accelerator configurations of (**a**) linac driven long-pulse; (**b**) cyclotron-driven continuous-wave; (**c**) linac and synchrotron driven short-pulse, and (**d**) full-energy-linac and accumulator driven short pulse facilities

the second category are generally in two configurations, as illustrated in Fig. 1.5 (c) (linac and synchrotron driven) and (d) (full-energy linac and accumulator driven).

Table 1.1 shows examples of some high-power hadron accelerators in the design, construction, and operation stages. They are intended for high-energy physics (AGS [60], SPS [61], FNAL MI [55], J-PARC/MR [62], MMF [63], and PIP-II [64] for neutrino, kaon, and muon physics), nuclear physics (TRIUMF/ISAC [65, 66], RIKEN [67], GANIL/SPIRAL2 [68, 69], FAIR [70], FRIB, HIAF [71], and RAON [72] for rare isotope physics; FAIR for antiproton physics; LANSCE [73]), basic energy science and applications (LANSCE, PSI [56], SNS [74], J-PARC/RCS [62], ISIS [75], SARAF [76], GANIL/SPIRAL2, CSNS [77], and ESS [78, 79], for neutron sources; KOMAC [80, 81] for proton applications), radioisotope production (SARAF), material neutron irradiation (IFMIF [82], its validation prototype LIPAc [83–85], and DEMO-oriented neutron source IFMIF-DONES [86]), and accelerator-driven subcritical systems (CiADS [87] and MYRRHA [88] for nuclear waste transmutation and power generation). Other proposed, operating, or recently retired facilities in the world include ISOLDE [89], GANIL [90], GSI [91], NSCL [92, 93], LEDA [94], LANSCE/PSR [95], and CPHS [96, 97].

1.2.3.1 Continuous-Wave and Long-Pulsed Accelerators

Continuous-wave (CW) or long-pulse hadron facilities use cyclotrons or linacs where high average beam power is desired, while the beam-time structure is not important. In the 1980s, the LANSCE linac reached a power level of 1 MW for nuclear research and neutron applications. The PSI cyclotron has a CW power record of 1.4 MW as a neutron source. The European Spallation Source (ESS) is being constructed to generate 5 MW beam power at 2 GeV using mainly SRF

Table 1.1 Major parameters of some proton and heavy ion accelerators at design, construction, and operation stage [59]

Project	Status	Primary beam	Second. beam	Accel. type	f_{rep} [Hz]	Beam duty	Target type	Energy [MeV/u]	Power [MW]
AGS	Achieve	p	μ, K	LN/SR	0.5	5e-7;.5 ^{t}	Ni; Pt	24000	0.1
SPS	Achieve	p	ν	LN/SR	0.17	3.5e-6 ^{t}	C	400000	0.5
FNAL MI	Achieve	p	ν	LN/SR	0.94	1.e-5 ^{t}	C	120000	1.02
J-PARC MR	Achieve	p	π, ν	LN/SR	0.74	4e-6	C	30000	0.8
			K		0.19	0.3	Au		0.08
LANSCE	Achieve	p, H^-	π, μ, n	LN	100	0.15	C ^{r}	800	0.8
PSR	Achieve	p	n	LN/AR	20	0.08 ^{i}	W	800	0.1
MMF	Achieve	p	π, μ, n	LN	50	0.01	W	502	0.075
	Design	(H^-) p	π, μ, n	LN/SR	100	0.01	W	600	0.6
RIKEN	Achieve	d to U	RIB	LN/CY	CW	1	Be ^{r}	345–400	0.01–0.02
	Goal	d to U	RIB	LN/CY	CW	1	Be ^{r}	345–400	0.08 (U)
PSI	Achieve	p	n	CY	CW	1	Pb	590	1.4
			μ				C ^{r}		
TRIUMF	Achieve	(H^-) p	RIB	CY	CW	1	ISOL	520	0.05
			π, μ				Be, C		0.06
SNS	Achieve	(H^-) p	n	LN/AR	60	0.06 ^{i}	Hg ^{l}	1000	1.7
	Goal	(H^-) p	n	LN/AR	60	0.06 ^{i}	Hg ^{l}	1300	2.8
J-PARC RCS	Achieve	(H^-) p	n	LN/SR	25	0.02 ^{i}	Hg ^{l}	3000	1.0
			μ				C ^{r}		
ISIS	Achieve	(H^-) p	n, μ	LN/SR	50; 10	0.01 ^{i}	W, C	800	0.19
	Goal	(H^-) p	n, μ	LN/SR	50; 10	0.01 ^{i}	W, C	800	0.24
CSNS	Achieve	(H^-) p	n	LN/SR	25	0.01 ^{i}	W	1600	0.17
	Goal	(H^-) p	n	LN/SR	25	0.01 ^{i}	W	1600	0.5
SARAF	Achieve	p; d	; –	LN	CW	1	SST;Li ^{l}	3.9; 2.8	0.004
	Goal	p, d	n, RIB	LN	CW	1	Li ^{l}; Be	40; 20	0.2
KOMAC	Achieve	p	–	LN	10–30	0.005–0.03	–	100	0.01–0.03
FRIB	Achieve	p to U	RIB	LN	CW	1	C ^{r}	>200	0.03
	Goal	p to U	RIB	LN	CW	1	C ^{r}	>200	0.4
SPIRAL2	Achieve	p,d, $A/q \leq 3$	RIB, n	LN	CW	1	Be ^{r}, Li, C An	33,20, 14	0.17, 0.2, 0.045
	Goal	$A/q \leq 7$							
FAIR	Constru.	p to U	RIB	LN/SR	0.2	≤0.25 ^{i}	M ^{r}	1e3	0.012
		p to U	p̄		0.5		Ni	3e4	0.001
RAON	Constru.	p to U	RIB	LN/CY	CW	1	C	>200	0.4
HIAF	Constru.	p to U	RIB	LN/SR	CW; 5	1 ^{i}	C ^{r}, M ^{r}	834, 9300	0.009; 0.01
LIPAc	Constru.	d	n	LN	CW	1	Li ^{l}	4.5	1.1
PIP-II Bst	Constru.	p	ν, μ	LN/SR	20	0.01 ^{i}	C; W	8000	0.167
PIP-II MI	Constru.	p	ν	LN/SR	0.83	1.e-5 ^{t}	C	120000	1.2
ESS	Constru.	p	n	LN	14	0.04	W ^{r}	2000	5
IFMIF-DONES	Constru.	d	n	LN	CW	1	Li ^{l}	20	5
CiADS	Constru.	p	n	LN	CW	1	Pb-Bi ^{l}	600	3–6
MYRRHA	Design	p	n	LN	CW	1	Pb-Bi ^{l}	600	1.5–2.4

Notations: LN for linac; CY for Cyclotron; SR for Synchrotron; AR for Accumulator; C for graphite; An for antinides; M for metal; RIB for rare isotope beams; Superscripts r for rotating and l for liquid targets, i for linac beam duty, and t for beam duty on target, respectively accelerating cavities (Fig. 1.22). The China Accelerator Driven Subcritical System (CADS) project proposed aims for 15 to 30 MW beam power for the transmutation of nuclear waste and power generation. Acceleration becomes increasingly difficult with heavier particles. FRIB is driven by a heavy ion linac to accelerate all stable ion species to $>$200 MeV/u with the ultimate design beam power of 400 kW (Fig. 1.14).

1.2.3.2 Short-Pulsed Accelerators

Synchrotrons and accumulator rings are typically used to stack the beam produced by linacs to reach high peak intensity and power for pulsed applications. The Spallation Neutron Source (SNS) holds the pulsed power record of 1.7 MW with a proton pulse length of 1 μs at 60 Hz repetition rate bombarding a liquid mercury target. A full-energy linac consisting of NC and SC RF cavities accelerates the H^- beam to 1.3 GeV for injection into an accumulator ring for stacking. The Japan Particle Accelerator Research Complex (J-PARC) uses an NC linac to partially accelerate the H^- beam to 400 MeV and inject it into a rapid-cycling synchrotron to reach a beam power of $\sim$1 MW at 3 GeV to drive a spallation neutron source and a muon source for materials and life science applications. Part of the 3 GeV proton beam is also further accelerated in the main ring to 30 GeV to create heavy hadronic particles, such as pions and kaons. The main ring beam is also used to create neutrino beams for analysis at the Kamioka detector, located approximately 300 km to the west. The FNAL accelerator complex uses an NC linac to accelerate H^- beam to 400 MeV for charge-stripping injection into the Booster, a 15 Hz rapid-cycling synchrotron that accelerates the resulting protons to 8 GeV for delivery to the short-baseline neutrino and muon research facilities. Most of the 8 GeV proton beam is further accelerated in the Main Injector (MI) to 120 GeV at an average beam power of 1 MW for the long-baseline (800 km) experiment NOvA.

1.3 Power Frontier Technology Selections

For power and intensity frontier hadron facilities, the selection of accelerator technology is mainly determined by the desired beam energy, pulse structure, and beam power. High-power facilities that do not require a specific time structure in the output beam often choose continuous-wave (CW) beams. Facilities that need to reach multiple levels of high energy often choose a cluster of synchrotrons along with a linac injector.

1.3.1 Continuous-Wave and Long-Pulsed Beam Facilities

Both cyclotrons and linacs have been used to produce CW or long-pulsed beams for high-power fixed-target user facilities. Cyclotrons have been used to accelerate high-power proton beams to energies of up to 590 MeV and heavy-ion beams to about 350 MeV/u. Linacs have been used to accelerate proton beams to energies greater than 1 GeV and heavy-ion beams greater than 200 MeV/u. For higher energies, synchrotrons are often used, where beams are slow-extracted for CW applications.

Compared with linacs, cyclotrons are typically compact and power-efficient. However, the maximum energy is limited to about 1 GeV as a result of the diminishing turn-to-turn separation for high-energy particles and magnetic stripping for H^- ions. To keep the orbits separate, the energy gain per turn is proportional to $\beta\gamma^2$, where β and γ are the relativistic factors [98] (see Problem 2.2 and the Solution). A larger circumference can increase the orbit separation in ring cyclotrons proposed for the acceleration of protons up to 2 GeV, as discussed in recent publications [99, 100]. The state-of-the-art design may allow an average beam current of up to $\sim$5 mA with tolerable beam loss and hands-on maintenance capability. Practical challenges remain: very high energy gain per turn, large RF resonators, high beam loading, and high voltage gain per cavity. Extraction losses and space-charge effects must also be studied. On the other hand, a linac can, in principle, be extended to any energy, limited only by economic constraints. Unlike a cyclotron whose size scales with $\beta\gamma^2$, the size of a linac is proportional to γ, so linacs become more economical beyond some energy (in the GeV range for protons). A linac is also more economical at any energy if a high CW beam current (above 3–5 mA) is needed. The beam loss can be minimized, provided that care is taken with the acceptance of the machine and the structural design. For heavy ion acceleration, the linac advantage is further extended with robustness and flexibility in charge stripping and multi-charge-state simultaneous acceleration, along with large machine acceptance.

Superconducting technology is particularly suited for CW and long-pulsed accelerators. Low RF loss on the cavity walls not only reduces operational costs but also allows for high-gradient operation for any duty cycle. Due to water cooling difficulties, NC cavities are limited to either a high gradient but low duty cycle or a low gradient ($\sim$1 MV/m) CW operations. On the other hand, SC cavities can sustain substantially higher gradients (typically exceeding 5–10 MV/m at low- and intermediate-β, and 20 MV/m at high-β) for both pulsed and CW operations. For CW operations, SC technology allows for shorter linacs with less installed RF power and lower operating costs compared to linacs using NC technology.

1.3.2 Short-Pulsed Beam Facilities

Pulsed beam facilities generally rely on circular accelerators to stack beams that reach high peak intensities. The injector to these circular machines often accelerates the H^- beam so that the phase-space density of the circulating beam can be continuously increased through the charge-exchange process.

Accumulator rings (AR) are most appropriate if the facility is built for a single purpose with modest final energy. Operating at a fixed energy, the sole purpose of the ring is to stack the beam for high peak intensity. Due to the high cost of a full-energy linac and the H^- stripping issue, the highest energy currently foreseen for an AR with H^- beam injection is ~1.3 GeV. In principle, permanent magnets can be used to save construction and operating costs. In practice, electromagnets are used instead to maintain the flexibility that accommodates varying injection energies. FNAL is a notable exception that has successfully implemented a permanent-magnet accumulator ring. The Recycler is an 8 GeV fixed-energy storage ring that is used to accumulate and double the intensity of the proton beam injected into the Main Injector. As the ring is designed to withstand occasional beam loss, scraped beams, and secondary emission, it is best to avoid SC technology for both magnets and RF cavities, as maintaining a particulate-free environment would be challenging. Large-bore NC electromagnets are usually used along with ferrite-loaded NC RF cavities for longitudinal beam containment. The core of the magnets can be made from solid steel or laminated steel, as the eddy current is not an issue with DC power supplies. The particle beams that circulate in the ring encounter resonances and collective instabilities that limit the intensity and power of the beam. An accumulator design shortens the time that the beam spends in the ring, eliminates variations in the beam's orbit caused by mismatches in synchronization, avoids reductions in the stable longitudinal phase-space area during acceleration, and reduces the power needed for the magnet system.

Synchrotrons are most appropriate if the facility is intended for multi-stage, multi-purpose use and if the final beam energy is high. Synchrotrons are used to both accumulate and accelerate beams. Compared with an AR, the requirements for the injection linac are significantly reduced. As most beam loss occurs at the injection energy, the tolerable beam loss is also much higher. The main challenges are due to the rapid-cycling requirements needed to reach high average beam power at modest energies (e.g., several GeV). Although a programmable power supply or dual-harmonic resonance power supply can optimize the ramping cycle, the associated cost is usually prohibitively high for high-power (needed for large-bore magnets) and rapid-cycling (e.g., 25 Hz or higher) power supplies. Single-harmonic resonance power supplies are thus used for the main lattice magnets in rapid-cycle synchrotrons. For slower cycling (e.g., less than 10 Hz) and higher output energy (e.g., greater than 10 GeV), synchrotrons typically use programmable power supplies for flexible ramping. For slower cycling, higher-energy synchrotrons, slow extraction is sometimes added to extract DC-structured beams.

Almost two decades after the start of operations, SNS and J-PARC demonstrated that AR and RCS can be successfully established to meet high-power operational requirements, respectively. While SNS has kept the uncontrolled beam loss at the 10^{-4} level as designed, J-PARC has maintained the uncontrolled loss at the 10^{-3} level, allowing high-power operations at 1 MW and above.

1.4 Power-Frontier Low-Loss Design Philosophies

The primary concern in designing high-power hadron facilities is that radio-activation caused by uncontrolled beam loss can limit machine availability and maintainability. Uncontrolled beam loss has been the dominant performance-limiting factor for the cyclotron, linac, and rings over the past decades.

Based on past operational experience for a proton accelerator operating around GeV energy, hands-on maintenance ($\sim$1 mSv/h or $\sim$100 mrem/h at 30 cm from the surface, 4 hours after shut-down) requires an average uncontrolled beam loss not exceeding about 1 W of beam power per tunnel meter [101, 102]. LANSCE operated successfully with an average fractional uncontrolled beam loss of 5×10^{-4} over a linac length of 800 m to reach 0.8 MW beam power. Before the rings of SNS and J-PARC were constructed, many operating synchrotrons were limited in beam power due to excessive uncontrolled beam losses. Typical fractional beam losses were 3×10^{-3} in the PSR accumulator ring and more than 10% in rapid-cycle synchrotrons. Figure 1.6 shows the growth of beam power over the years of operation in several major hadron facilities. The final achievable beam power is frequently limited by the level of radio-activation that compromises hands-on maintenance of the accelerator complex.

The SNS accelerator complex was designed to deliver 1.4 MW beam power. Its full-energy linac is designed with an average fractional uncontrolled beam loss of 10^{-6} per meter. Similarly, the 248 m circumference SNS accumulator ring was designed for a fractional uncontrolled beam loss of 10^{-4}. This 1 W/m design guideline on average uncontrolled beam loss has been extended to heavy-ion facilities, such as the FRIB SRF linac. Even though the radio-activation caused by lost heavy ions is of much less concern compared to the same wattage of lost protons, challenges in machine protection against lost heavy ions, whose range is short at low energy, and the increase in cryogenic heat load dictate the need.

1.4.1 Cyclotron Low-Loss Design

Beam-loss minimization is fundamentally important in high-power cyclotron designs. Cyclotron-like accelerators include the classical cyclotron, isochronous cyclotron, separated sector cyclotron, cyclotron accelerating H^- beams, cyclotron using SC magnets, synchrocyclotron and fixed-field alternating-gradient accelerator

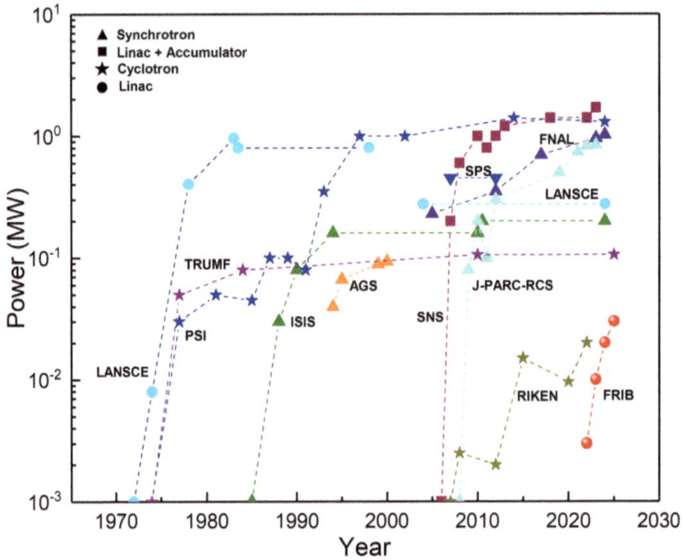

Fig. 1.6 Achieved beam power at some major hadron facilities. Upgrade plans exist for most facilities

(FFA) [103]. The low-loss design philosophy may be illustrated by the experience and upgrade path in the separated sector ring cyclotron at PSI, which achieves a proton beam power record of over 1.4 MW [56, 104, 105]. In the PSI Ring Cyclotron, the proton beam is accelerated to a kinetic energy of 590 MeV with the fractional beam loss controlled to approximately 2×10^{-4}. As the top energy of a cyclotron is intrinsically limited, the pursuit of higher beam power is based on the increase in beam current. Correspondingly, the fractional beam losses in the cyclotrons and transport lines must be proportionally reduced to maintain radio-activation at the same acceptable level.

In general, uncontrolled losses in cyclotrons are attributed to:

(1) complication at extraction caused by tight turn separation and high particle density;
(2) complications at injection caused by phase-space mismatches;
(3) longitudinal and transverse space-charge effects;
(4) RF beam loading due to high beam intensity; and
(5) magnetic stripping at high beam energy for H^- beams (Sect. 3.7).

A low-loss design or upgrade attempts to rectify these problems. The following lists some examples:

1. As cyclotrons are susceptible to beam loss at extraction due to reduced turn separation with radius, the key to a low-loss design is an advanced RF system with high energy gain to decrease the number of turns and increase separation, particularly at the extraction radius. With a higher energy gain per turn, the

total number of turns in the cyclotron is reduced, leading to a larger separation between the last and second-to-last turns in the cyclotron at the location of the extraction element, which reduces extraction beam losses.
2. An external ion source with a possible multistage injector is used to allow flexible choices of ion sources and to mitigate matching issues. For heavy-ion acceleration, multistage injection also facilitates charge stripping for higher acceleration efficiency.
3. Unless a charge-exchange process is used, the beam is extracted in a single turn so that beam loss is minimized during extraction. When the H^- beam is accelerated in a cyclotron, it can be easily extracted by charge stripping through a foil. However, excessive magnetic fields that lead to magnetic stripping at high beam energy must be avoided (Sect. 3.7.2). For example, the TRIUMF cyclotron is capable of operating at 520 MeV output energy, where uncontrolled beam loss would be $\sim$10%; it typically runs at 480 MeV where stripping losses are less than 1%.
4. The higher harmonic RF system is used in combination to flatten the total RF waveform and reduce the energy spread and the peak beam density.
5. Higher accelerating voltage is, furthermore, advantageous for controlling beam loading and longitudinal space-charge effects.

In addition to these measures to minimize beam losses, high-power cyclotron facilities are designed to localize beam loss to specific shielded regions and are equipped with cleaning and collection capabilities. At PSI, special transportable enclosures were designed to allow the exchange and handling of critical components, such as injection/extraction elements in the ring cyclotron, as well as graphite targets and absorbers. On the accelerator site, there was a hot cell workshop available, which allowed us to work on highly activated components using remotely controlled manipulators. The professional handling and disposal of activated components and radioactive waste has been important.

1.4.2 Linac Low-Loss Design

The low-loss design is of primary importance in the establishment of high-power linacs. The low loss linac design philosophy has been formed and extended during the preparation for the high-power proton linac projects at LANL [101], the SNS SRF H^- linac [106], and the FRIB heavy-ion linac. The lessons learned during high-power operations of facilities such as LANSCE, SNS, J-PARC, and FRIB linacs are valuable. In general, uncontrolled losses in linacs are attributed to:

(1) incomplete bunching action in the LEBT upon exiting the ion source for accelerators without a radio frequency quadrupole (RFQ);
(2) mismatch upon abrupt change of linac RF frequency, lattice and acceptance;
(3) imperfect bunch chopping;
(4) beam-induced transient effects for pulsed operations;

(5) space-charge effects including envelope and parametric resonances;
(6) limited physical and momentum acceptance upon emittance growth and halo formation;
(7) various uncontrolled charge-stripping mechanisms for an H^- beam (Sect. 3.7); and
(8) contaminant beam (unwanted beam of the same charge-to-mass ratio) and scattering and energy straggling at the charge stripper for a heavy-ion beam.

The following are examples of low-loss designs to rectify these problems:

1. Use RFQ for simultaneous bunching, acceleration, and transverse focusing, along with transverse emittance filtering and matching.
2. Maintain the same transverse and longitudinal focus strength through the accelerating structure transition for beam-current-independent matching [101].
3. Reserve space for multistage chopping and beam-in-gap cleaning.
4. Plan RF control with feedback and feedforward to mitigate transient effects.
5. Avoid lower-order parametric resonance and the unstable region in the Hofmann chart [107].
6. Consider using superconducting cavities with a larger aperture for beam halo containment.
7. Choose design parameters to mitigate uncontrolled charge stripping (Sect. 3.7).
8. Design a well-shielded charge stripper (Sect. 8.8.2) and charge selector with control of scattering and collection of contaminants (Sect. 8.10.2.2).

The emphasis is also placed on the flexibility, tuneability, reliability, and upgradeability of the machine. The SRF linac allows for operation with easy tuning in the event of a failure of the cavity or klystron. Design reserves and redundancy, such as spare cryomodules and power supplies, ensure high availability. The reserved linac tunnel spaces have allowed SNS to increase the injection energy of the ring from 1 to 1.3 GeV, doubling the design beam-on-target power to 2.8 MW, and capable of supplying a second target station more than 20 years after the initial operation. The reserved linac tunnel spaces will also allow FRIB to double the beam energy on target in the future from >200 to >400 MeV/u for enhanced scientific reach.

1.4.3 Ring Low-Loss Design

Before the establishment of the SNS and J-PARC accelerator facilities, most proton rings experienced beam losses as high as several tens of percent. These losses primarily occurred during beam injection, during the initial capture by the accelerating system, at the start of the acceleration ramp, and at the time of transition-energy crossing when the motion of longitudinal particles is non-adiabatic. Losses also occur through the development of instabilities in the transverse or longitudinal motion of the particles. Before SNS, the lowest fractional beam loss attained was approximately 3×10^{-3}, achieved at a beam intensity of about 5×10^{13} protons per pulse in the Proton Storage Ring (PSR) accumulator at Los Alamos National Laboratory [108]. Uncontrolled beam losses are usually attributed to

(1) a high space-charge tune shift (0.25 or larger) at injection, resulting in resonance crossing;
(2) limited geometric and momentum acceptance;
(3) premature H^- and H^0 stripping and injection-foil scattering;
(4) large errors in the magnetic field and alignment, and dipole-quadrupole tracking errors during acceleration ramping;
(5) instabilities (e.g., head-tail instability, coupled-bunch instability, negative mass and microwave instability, electron-cloud instability) in particular for AR of high beam intensity;
(6) complications upon transition crossing for RCS (Sect. 3.8);
(7) accidental beam loss (such as malfunction of the ion source and linac or misfiring of the extraction kickers); and
(8) beam-halo loss during fast extraction.

A low-loss design or upgrade attempts to rectify these problems. The following lists some examples:

1. The beam is "painted" in the transverse phase space to keep the space-charge tune shift low (below 0.15) and "painted" in the longitudinal phase space to reduce the peak current and damp instabilities (Sect. 2.3.4.3).
2. A large transverse aperture is designed to contain not only the circulating beam but also the beam halo, at the cost of increased magnet size and power supply. Additionally, a large longitudinal acceptance is realized to encompass the halo and tail of the beam momentum by optimizing the chromatic properties of the lattice and using multi-harmonic RF systems with large voltages (Sect. 2.3.3).
3. The orbit of the beam in the injection region is programmed to avoid excessive foil hits, and the magnetic fields at injection are designed to prevent premature stripping of H^- and H^0 (Sect. 2.3.4).
4. A moderate main magnetic field avoids saturation effects, and shimmed pole tips in both dipole and quadrupole magnets help compensate for the effects of fringe fields (Sect. 2.3.9).
5. To avoid instabilities in the motion of the particles, efforts are made to reduce the coupling impedance caused by the beam environment: the vacuum pipe steps are tapered; the connecting bellows, vacuum ports, and instrumentation ports are shielded; and the inner surface of the vacuum chamber is coated with a material having a low secondary emission rate of electrons to limit the production of an electron cloud (Sects. 3.4 and 3.5).
6. Avoid RCS transition crossing with an optical lattice that has a low or negative momentum-compaction factor (Sect. 3.8).
7. The pre-injection transport is designed with beam collimation systems to prevent accidental malfunction of the ion source and the linac, and redundancy is built into the extraction system to prevent beam loss after the accidental misfiring of the extraction kickers (Sect. 2.3.6). Subsequent sections will discuss these design and upgrade issues in detail.

In addition to these measures to minimize beam losses, modern accelerator facilities are designed to localize beam loss to specific shielded regions and are equipped with cleaning and collection capabilities. With a large transverse and momentum aperture, multistage collimation and momentum cleaning can be incorporated.

A clean gap between subsequent beam bunches, key to low-loss beam extraction, is ensured by multistage beam cleaning that includes multi-step chopping at low energies and beam-in-gap sweeping with collimator collection at top energy.

Flexibility and reliability are important aspects that must be addressed early in the design stage. For example:

- Rings are designed with electromagnets to accommodate variations in the beam injection energy.
- Separate dipole and quadrupole magnets are preferred instead of combined function magnets to allow independent orbit steering and focusing adjustments.
- Multifamily power supplies are employed for quadrupole magnets, permitting a wide adjustment of the tunes of transverse motion.
- Programmable orbit bumps are designed to create different particle distributions through painting.
- The beam scraping and collimation systems are designed to adjust for changes in the beam tail and halo.
- Engineering design considers redundancy, radiation resistance, "active maintenance" capability, and "hot" spares [109].

1.5 Performance Limiting Mechanisms and Technological Challenges

The sustained advancement of high-power accelerators during the past several decades has been made possible by persistent efforts to understand performance-limiting mechanisms and technological challenges and, correspondingly, to develop new mitigation methods and technological breakthroughs. This section summarizes the performance-limiting mechanisms and technological challenges for the present cyclotrons, NC linacs, SC linacs, and rings. Detailed elaborations will be provided in subsequent chapters.

1.5.1 Cyclotron Mechanisms and Challenges

Performance limitations in cyclotrons are primarily associated with uncontrolled beam loss, which prevents hands-on maintenance and affects machine availability. Usually, the most effective mitigation method is the implementation of an advanced

RF acceleration system to enhance turn separation, increase extraction efficiency, suppress the effects of longitudinal space charge, and facilitate beam loading control (Sect. 2.1).

The ion source, the design of the injector chain, and phase-space matching are major challenges in controlling beam loss at injection (Sect. 2.1.2). Due to the compactness of typical cyclotron layouts, the design of magnets is complicated. Sophisticated beam diagnostics are often lacking. Complications in extraction caused by tight turn separation and high particle density require careful design optimization and elaborate tuning for beam loss control. Accelerating an H^- beam instead of a proton beam simplifies extraction. However, the maximum magnetic field allowed is limited, keeping Lorentz stripping at a tolerable level (Sect. 3.7). Both the transverse and longitudinal space charge effects need to be analyzed (Sect. 3.4.1). The target system is often a bottleneck for high-power operations. Non-stationary or liquid targets are often used to extend the lifetime of the target (Sect. 8.9). Excess radio-activation near the target demands remote handling, radiation-resistant magnets, accelerator sub-components, and quick-access vacuum and utility connections.

1.5.2 Normal-Conducting Linac Mechanisms and Challenges

Performance limitations of NC linacs are primarily associated with their ion sources (Sect. 8.4). Given an affordable machine aperture, both the beam current and the brightness are important. At a normalized rms emittance of $\epsilon_{N,rms} \sim 0.2\,\pi$mm·mr, the maximum peak proton beam current is $\sim$100 mA with the present technology [94]. The H^- peak beam currents of the SNS linac and J-PARC linac are between 50 mA and 100 mA with a duty factor of 10% and 1.5%, respectively. The heavy-ion beam current is much lower, e.g., $<$25 emA for pulsed and 1 emA for DC uranium beams.

The funnel system was designed, although not tested, to increase the beam current beyond the range of each ion source [101]. Another possible current doubler is to use funneling along with the simultaneous acceleration of protons and H^- ions downstream of the RFQ. Such a scheme is only possible in cases where the linac RF frequency transition is a multiple of an odd number. Both the J-PARC linac and MMF linac designs [110] maintain the possibility of future upgrades to the beam current. However, the LANSCE experience indicated operational challenges of the combined acceleration of H^+ and H^- [101].

Various uncontrolled charge-stripping mechanisms limit the practical length and, thus, the output beam energy of an H^- linac (Sect. 3.7). An 8 GeV H^- linac was once proposed but not built [111].

For heavy-ion acceleration, FRIB preserves the usable beam and effectively enhances the source current by simultaneously accelerating multiple charge states of the beam (Sect. 3.6) [112]. For a uranium beam, two charge states may be accelerated from the ion source, and up to five charge states may be accelerated downstream of the charge stripper simultaneously, dramatically increasing the usable beam current [113].

Charge stripping technology potentially imposes limitations on high-power heavy-ion facilities. Recent breakthroughs using liquid metal technologies [114] and inert gas [115] are promising (Sect. 8.8.2).

1.5.3 Superconducting Linac Mechanisms and Challenges

Performance limitations of SC cavities are related to gradient and RF power losses, above which part of the cavity wall becomes NC, causing a cavity quench. During a quench, a small spot initially turns to the NC state, increasing the local RF power losses by several orders of magnitude. If not sufficiently cooled, it grows in size almost instantaneously, absorbing all the electromagnetic (EM) stored energy of the cavity and dropping its gradient to zero. A quench can be caused by locally exceeding the SC critical temperature or magnetic field (which in SRF also depends on local overheating and surface quality). Local overheating can be caused by the bombardment of primary electrons coming from field emission (extracted from the surface by a high electric field) and secondary electrons from multipacting (generated by the resonant trajectories of electrons), and even by beam losses impinging on the SRF surface. These phenomena impose rather strict limits on the peak surface electric and magnetic fields achievable in operation.

If a magnetic field is present at the cavity surface during the NC-to-SC transition, it can be trapped in the superconductor, creating small spots of NC material that increase RF losses. Earth's magnetic field, if not properly shielded, would intolerably degrade cavity performance. Other sources of magnetic fields can be cryomodule components and SC solenoids if operated during cavity cooldown. The quality and cleanliness of the surface and SC material are of utmost importance to achieve high performance and must be maintained rigorously during linac operation. Contamination from vacuum loss or improper handling can drastically reduce cryomodule performance. Multipacting, or resonant field emission, can be enhanced to critical values by flawed design and improper RF conditioning of cavities and couplers.

Mechanicals instabilities causing cavities to detune and making it difficult or impossible to lock resonators in phase and amplitude can also limit the operating gradient, requiring that a significant part of the available RF power be devoted to controlling a cavity that is detuned most of the time. Another potential limitation arises from higher-order modes (HOMs) excited by the beam in cavities, which cause additional RF losses and beam degradation. SC resonators are particularly sensitive to HOMs because their low RF losses allow them to survive for a long time. In the particular case of high-power hadron accelerators, fortunately, HOMs have been of little concern until now, but they might become an issue in future projects. These aspects will be treated in more detail in Chaps. 2 and 5 .

1.5.4 Ring Mechanisms and Challenges

Performance limitations in high-intensity circular hadron accelerators are primarily associated with uncontrolled beam loss, which prevents hands-on maintenance and machine availability. Usually, the most effective mitigation method is a large aperture in the beam chamber. A large beam aperture, in turn, increases technical challenges as well as facility construction and operating costs.

For hadron accelerators that rely on charge-exchange processes, charge-stripping efficiency is often a bottleneck. For rings with H^- beam injection, particle scattering on the stripping foil leads to uncontrolled beam loss and radio-activation (Sect. 3.1). Prior to charge-exchange injection, undesired charge-stripping mechanisms impose constraints: Lorentz stripping (Sect. 3.7.2) limits the maximum magnetic field allowed in the beamline; intrabeam stripping (Sect. 3.7.4) limits the phase-space density; black body stripping imposes an upper limit on beam energy (Sect. 3.7.3); and ionization stripping demands good vacuum.

The accumulator ring is relatively simpler to engineer because of the elimination of acceleration and ramping. The main beam dynamics challenges are due to high beam intensity, as for a given beam power, the beam energy is relatively low: beam instabilities induced by a broad-band impedance source such as extraction kickers, electron cloud effects including electron-proton instability and electron-induced pressure rises, and other collective effects including space charge (Sect. 3.4.3.2). Space-charge-induced tune spread driven by magnetic nonlinearities often leads to resonance beam loss. The beam collimator must handle the full beam power at the final beam energy (Sect. 2.3.6).

Rapid-cycling synchrotron carries engineering challenges mostly associated with fast ramping: a relatively high gradient wide-band RF system for fast acceleration (Sect. 2.3.5.1); magnets with eddy current mitigation (Sect. 8.6.2); bending and focusing magnets tracking each other closely along the ramp cycle; and a vacuum chamber that avoids eddy current heating (Sect. 2.3.5.2). In addition, the crossing transition energy should be avoided at high beam intensity (Sect. 3.8). As the beam is injected at a relatively low energy, the space-charge tune spread is usually significant. Fortunately, most beam loss occurs during the initial ramping when the beam energy is low. Together with a beam intensity that is typically lower than in accumulator rings, a higher fractional beam loss can be tolerated in rapid-cycle synchrotrons.

Synchrotrons for high-energy acceleration usually have a large machine circumference. Resistive wall impedance often causes complications. Again, transition-energy crossing is better avoided for high-beam intensity operations.

Pulsed beams fast-extracted from accumulator rings or synchrotrons produce a high peak impact on the secondary beam production target that demands active mitigation (Sect. 8.9). Excess radio-activation near the target demands remote handling, radiation-resistant magnets and accelerator sub-components, and quick access vacuum and utility connections (Sect. 8.9.3).

1.6 Major Applications of Power-Frontier Accelerators

This section discusses major applications of high-power accelerators, ranging from discovery science in traditional high-energy and nuclear physics to multi-user platforms of spallation neutron sources, muon sources, and proton sources, as well as accelerators for the environment and energy, medicine, security and defense, and industry. Existing accelerator facilities and proposed accelerator projects are used to illustrate the current approaches.

1.6.1 Accelerators for High-Energy and Nuclear Physics

High-energy particle physics and nuclear physics are two traditional branches of the discovery sciences pursued with accelerators.

1.6.1.1 High-Energy Physics Facilities

High-energy physics (also called particle physics) is a field of physics that studies the nature of the particles that constitute matter and radiation. The pivotal developments of high-energy physics have occurred within three correlating frontiers of research: the energy, power (or intensity), and cosmic frontiers. For two of them, the energy and power frontiers, particle accelerators are essential tools for exploration.

At the energy frontier, mainstream particle physicists use high-energy colliders to discover new particles and probe the fundamental forces at an ever smaller scale. Hadron colliders are usually used in the search for new phenomena, while lepton colliders are employed as precision probes of these phenomena. Achievements of hadron colliders include the discovery of W and Z bosons at the Super Proton Synchrotron (SPS), the top quark at the Tevatron, and the Higgs boson at the Large Hadron Collider (LHC) (Fig. 1.9). Achievements of lepton colliders include the discovery of the J/Ψ meson at the Stanford Positron Electron Asymmetric Ring (SPEAR) and the Large Electron-Positron Collider (LEP), the discovery of many charmonium states and the tau lepton at SPEAR, the studies of Z and W bosons at LEP, and the B meson factory at KEKB and at PEP-II.

At the power frontier, physicists use intense beams that strike fixed targets to study new forces, new properties of matter, and new dimensions. In particular, high-power hadron accelerators are used to study rare processes, precision measurements, and neutrino physics, exploring high-energy physics beyond the Standard Model of particles and forces. The J-PARC Main Ring (MR) was established using high-power protons to produce high-intensity secondary beams for particle physics studies. As shown in Fig. 1.7, the proton beam accelerated in the MR can be slow-extracted to the hadron experimental area, where the beam strikes the target to create heavy hadronic particles such as pions and kaons. The beam can also be

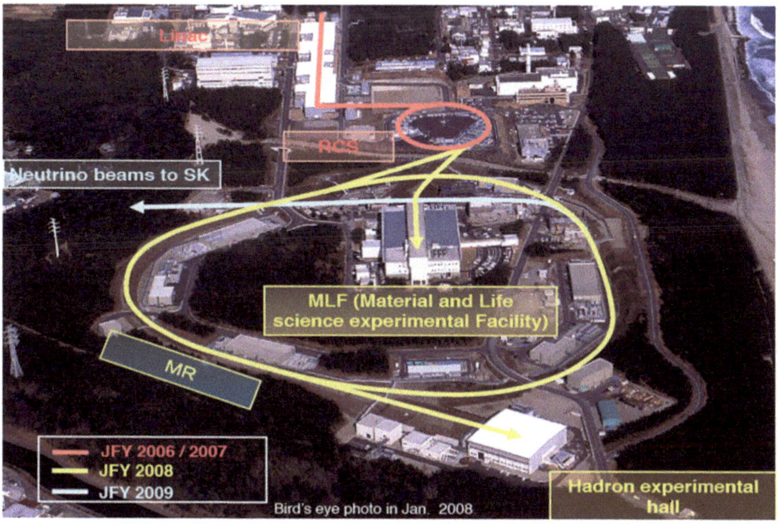

Fig. 1.7 Aerial view of the Japan Particle Accelerator research Complex. (Courtesy J-PARC, Japan)

fast extracted to strike the neutrino production target, creating neutrino beams for analysis at the Kamioka laboratory, located approximately 300 km to the west (that is, Tokai to Kamioka, T2K experiment).

Other intensity-frontier particle physics programs include neutrino physics studies at the CERN Neutrinos to Gran Sasso (CNGS) using beams from the SPS (Fig. 1.9), the Deep Underground Neutrino Experiment (DUNE) and PIP-II at FNAL (Fig. 1.8); rare-process studies using high-intensity negative muon beams produced by protons striking targets at FNAL and J-PARC; and rare-process studies with kaons at AGS at BNL, SPS at CERN, FNAL and J-PARC; as well as the spallation source of ultra-cold neutrons and isotopes for intensity frontier research.

1.6.1.2 Nuclear Physics Facilities

Nuclear physics is the field of physics that studies atomic nuclei, their constituents, and their interactions. Accelerator-based nuclear facilities can also be classified as energy frontiers and power (or intensity) frontiers. Representative nuclear physics facilities at the energy frontier include the Relativistic Heavy Ion Collider (RHIC) at BNL (Fig. 1.10), which studies quark-gluon plasma physics [117], and the Large Hadron Collider, which also collides heavy ions to study nuclear matter under extreme conditions [118].

Facilities at the power (intensity) frontier can be further classified into electron, proton, and heavy-ion facilities. Examples of the electron power-frontier facilities are the Continuous Electron Beam Accelerator Facility (CEBAF) [116, 119], which

Fig. 1.8 Aerial view of the PIP-II project under construction at the FNAL complex. (Courtesy FNAL, USA)

uses the electromagnetic interaction to probe the structure of the nucleon, the production and decay of light mesons, and aspects of the interactions of nucleons in the atomic nucleus, and SPring-8, which uses laser Compton backscattering to provide a photon beam for nuclear and hadron physics studies. As shown in Fig. 1.11, with an increase in output energy from 6 to 12 GeV, the CEBAF linear accelerator complex provides continuous electron beams reaching a beam power of 1 MW at 12 GeV.

At the power frontier for proton accelerators were meson factories at the Los Alamos Meson Physics Facility (LAMPF), TRIUMF (Tri-University Meson Facility) [120], the Paul Scherrer Institute (PSI, formerly the Swiss Institute for Nuclear Physics, SIN), and the Moscow Meson Factory (MMF) at the Institute for Nuclear Research (INR) in Troitsk [63]. Figure 1.12 shows the TRIUMF facility, which is based on proton beams of 520 MeV produced by a cyclotron. The MMF was designed to accelerate protons and H^- with a total average current of 1 mA [121], followed by a proton storage ring based on isochronous optics to compress the 100 μs linac pulse to 330 ns and stretch it to deliver continuous beams to the muon experiments [122]. Its current applications include neutron research, isotope production, and proton beam therapy.

Heavy-ion accelerators at the power frontier are primarily used for the study of unstable isotopes. Stable heavy ions up to uranium are accelerated to strike a fixed target to produce projectile fragments, which are separated by mass and charge to yield radioisotope (RI) beams for experiments. The main operating facilities include the Online Isotope Mass Separator (ISOLDE) at CERN, SIS18 at the GSI Helmholtz Center for Heavy Ion Research, the Radioactive Isotope Beam Factory (RIBF) at RIKEN (Fig. 1.13), SPIRAL1 in GANIL, the Cooler Storage Ring (CSR) at IMP

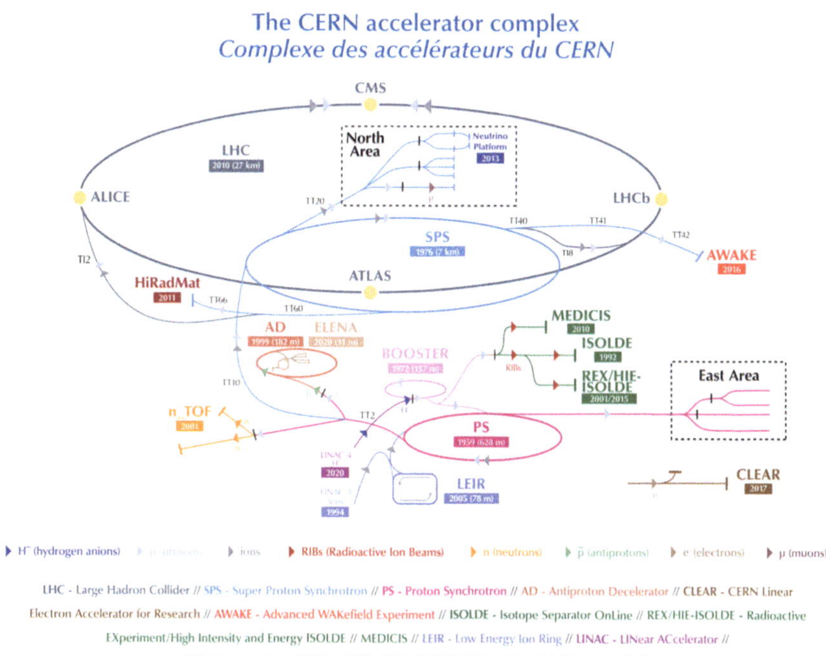

Fig. 1.9 Aerial view and the schematics of the CERN accelerator complex. (Courtesy CERN, Switzerland)

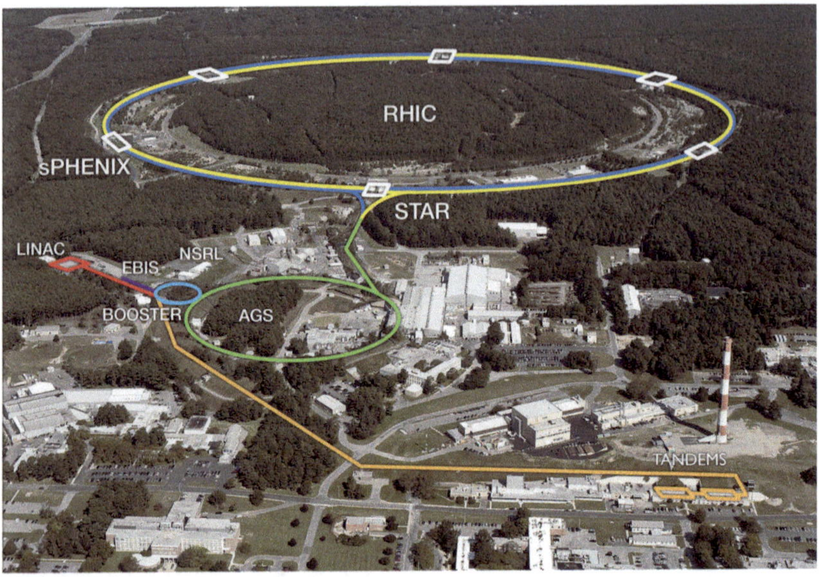

Fig. 1.10 Aerial view of Relativistic Heavy Ion Collider accelerator complex. (Courtesy BNL, USA)

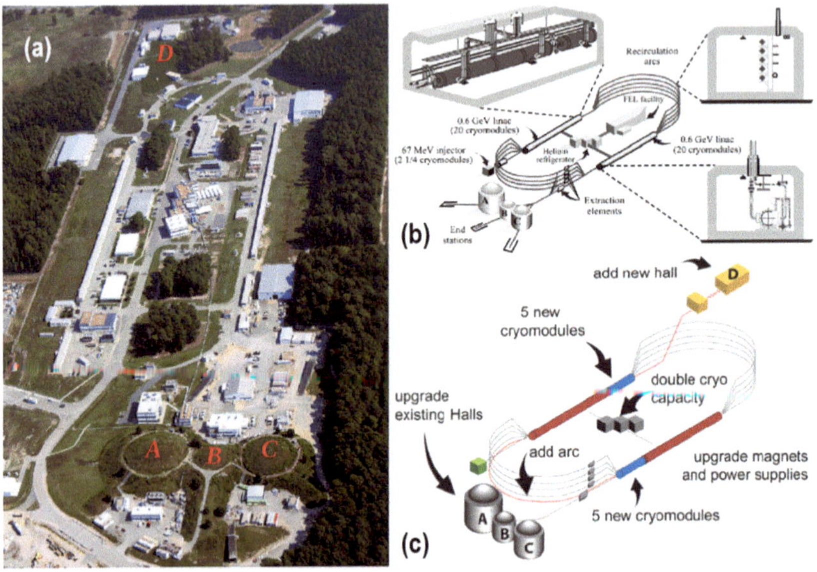

Fig. 1.11 (**a**) Aerial view of the folded linear accelerator and the experimental halls (marked in red) after the 12 GeV energy upgrade at the Continuous Electron Beam Accelerator Facility; (**b**) schematics of the 6 GeV CEBAF before the upgrade; (**c**) schematics of the 12 GeV upgrade project technical scope [116]. (Courtesy JLab, USA)

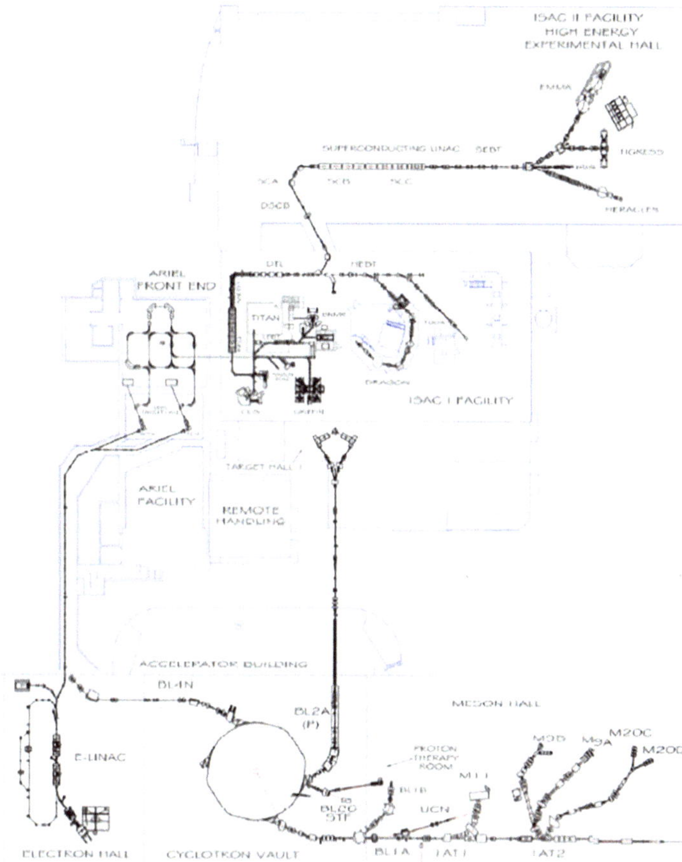

Fig. 1.12 TRIUMF schematic layout [65]. (Courtesy TRIUMF, Canada)

[123], the Argonne Tandem Linac Accelerator System (ATLAS) at ANL, and the Facility for Rare Isotope Beams (FRIB) at MSU (Fig. 1.14).

Compared with power-frontier electron and proton accelerators, which have an average beam-on-target power at approximately 1 MW level, heavy-ion accelerators typically deliver much lower beam power. RIBF at RIKEN and FRIB at MSU presently hold power records of about 20 kW for light ions and 10 kW for the heaviest ion, uranium (Fig. 1.6). Recent projects under construction or programs actively proposed include SPIRAL2 at GANIL (Fig. 1.15), the RIBF upgrade at RIKEN, the Facility for Antiproton and Ion Research (FAIR) at GSI, the Heavy Ion Accelerator Facility (HIAF) at IMP (Fig. 1.24), and the Rare Isotope Science Project (RAON) at IBS (Fig. 1.16).

Figure 1.14 shows the schematic layout of the FRIB accelerator complex. The SRF driver linac accelerates all types of stable ions to energies $\sim$200 MeV/u,

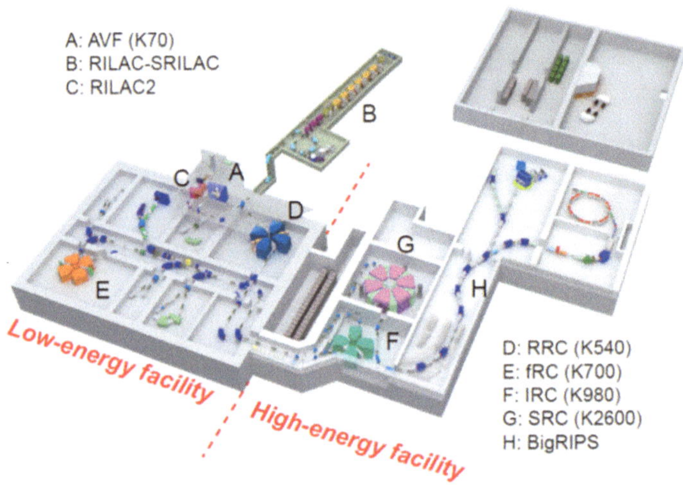

Fig. 1.13 Schematic of the RIKEN RI Beam Factory (RIBF) [124]. By combining three injectors (AVF, RILAC, RILAC2) with four ring cyclotrons (RRC, fRC, IRC, SRC), RIKEN can accelerate all ions up to uranium to 345 MeV/u. (Courtesy O. Kamigaito, RIKEN, Japan)

striking the production target for rare-isotope user programs, including the study of nuclear structure, reactions, astrophysics, and isotope applications.

1.6.2 Accelerator-Based Multi-user Platforms

Since the first storage ring dedicated to synchrotron radiation at SRS Daresbury, light sources based on electron accelerators have served as multi-user platforms, offering photons from infrared to hard X-rays. Scientific and industrial users span multiple disciplines of research fields, benefiting from the ever-growing brightness and coherence.

User platforms based on intensity frontier hadron accelerators have similarly served scientific and industrial users that extend across research fields such as physics, chemistry, biology, and medicine, and have been used in applications beneficial to society. Figure 1.17 shows the ISIS spallation neutron and muon source and the Diamond Light Source, co-located at the Rutherford Appleton Laboratory (RAL). The neutron- and X-ray-based facilities complement each other in the study of the structure and dynamics of a variety of systems in both solid and liquid phases.

1.6.2.1 Neutron Sources

Neutrons are powerful probes because of their important features: charge neutrality, sensitivity to isotopes, non-zero magnetic moment, and carrying spin. Neutron

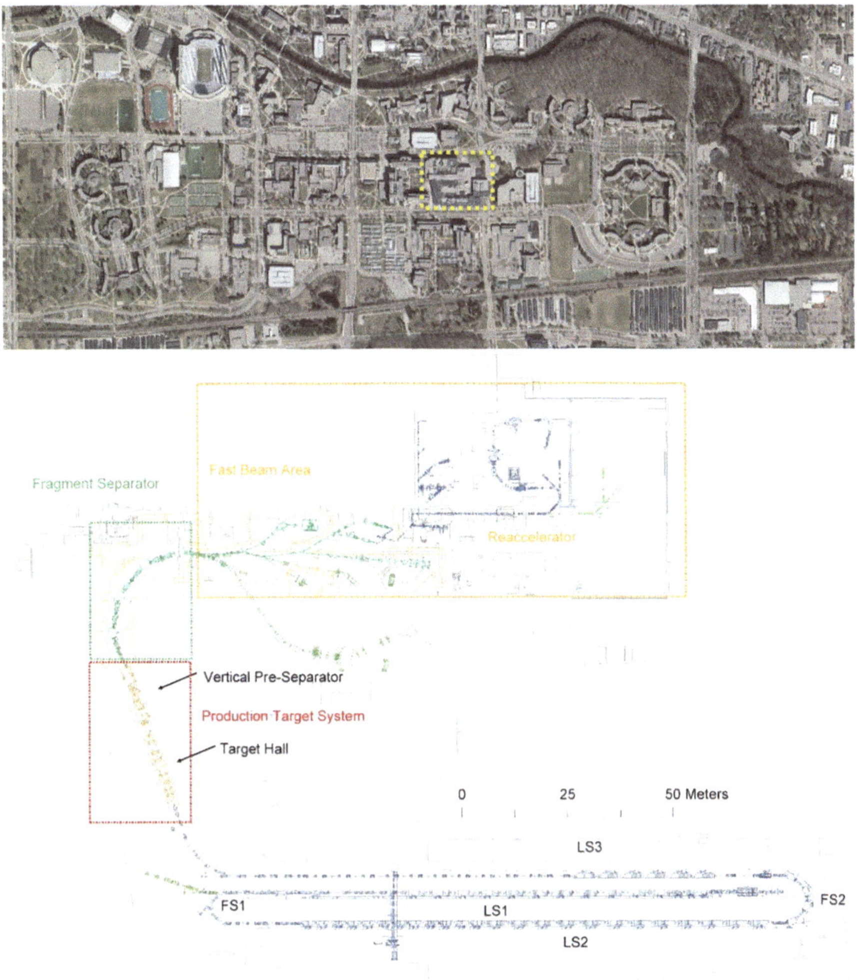

Fig. 1.14 Aerial view of the Facility for Rare Isotope Beams (yellow dashed box) in the middle of the campus of the Michigan State University and the schematic layout of FRIB. (Courtesy MSU, USA). The underground driver linac consists of three straight segments, Linac Segment 1 through Linac Segment 3 (LS1, LS2, LS3), and two folding segments (FS1, FS2), with charge stripping in FS1. The beam delivery system (BDS) delivers beam to the Target Hall at the linac elevation. The vertical pre-separator takes isotope beams to the ground level for transport through the fragment separator to the experimental areas

beams are used in physics, engineering, medicine, nuclear weapons, petroleum exploration, biology, chemistry, and nuclear power. Accelerator-based neutron sources consist of medium-scale facilities driven by electrons and light ions, and large-scale facilities driven by protons.

Large-scale neutron sources include nuclear fission reactors and accelerator-driven spallation sources. Since the late 1940s, the large fluxes of thermal neutrons

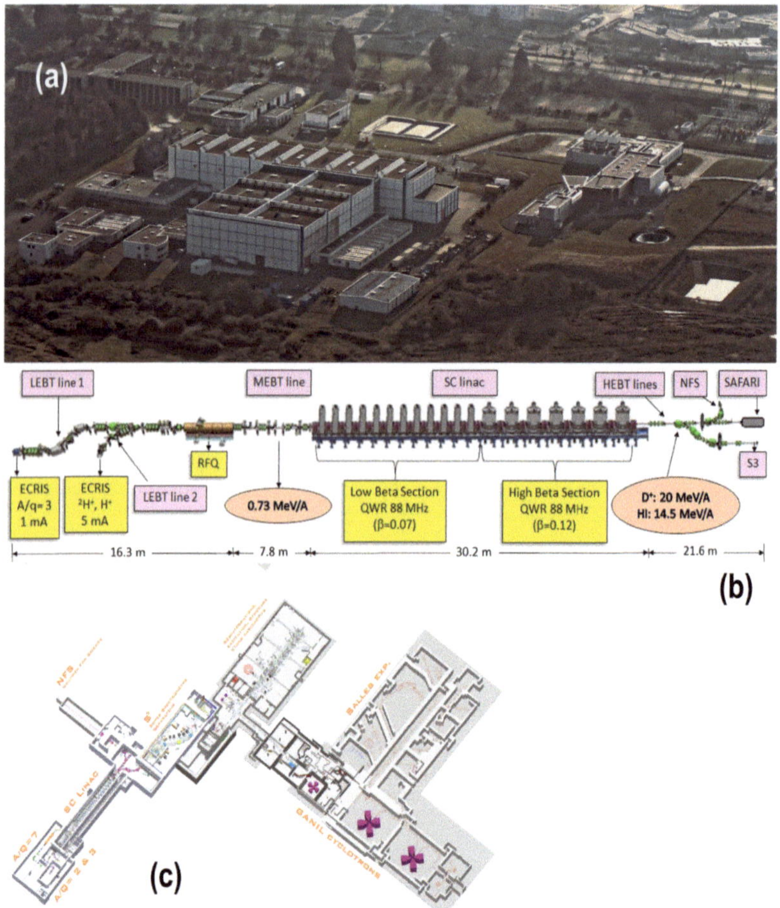

Fig. 1.15 (**a**) Aerial view of the The GANIL accelerator complex consists of 5 cyclotrons in operation since 1983 for studying properties of matter at the atomic scale, the SPIRAL2 superconducting LINAC in operation since 2020, containing the NFS experiment rooms for experiments in fundamental nuclear physics and applied physics with intense neutron beam and Super Separator Spectrometer (S3) for studying heavy and super heavy ion beams. DESIR is a bridge between the two installations, combining spectroscopy, mass spectrometry, and ion trap measurement techniques, to address questions in nuclear physics, weak interaction physics, and astrophysics. (**b**) SPIRAL2 SC linac layout [69]. (**c**) GANIL schematics. (Courtesy R. Ferdinand, GANIL)

produced within fission reactors have been exploited for neutron-scattering experiments around the world. Large-scale nuclear fusion experiments exist but have not yet become neutron sources: the National Ignition Facility (NIF) in the United States, JET in the UK, and the ITER experiment currently under construction in France.

Fig. 1.16 Aerial view of the RAON heavy-ion accelerator complex [72]. (Courtesy RAON/IRIS, IBS, Korea)

Fig. 1.17 Aerial view of the Rutherford Appleton Laboratory with the ISIS neutron and muon source shown in the lower half of the picture and the Diamond Light Source shown in the upper half. (Courtesy RAL, STFC, UK)

A spallation source is a high-flux source in which protons are accelerated to GeV-level energies and strike a target to produce neutrons through the highly effective spallation process:

$$p + (\text{target}) \rightarrow (\text{fragments}) + k \cdot n. \tag{1.1}$$

where the number of neutrons produced per proton is in the range of 20∼30. Neutrons produced from the target are moderated to a useful energy range and transported through neutron guides to spectrometer stations. Each target often supports more than 20 such stations that form a large-scale platform for scientific and industrial users [125].

Among short-pulsed spallation sources, Fig. 1.18 shows the layout of ISIS at RAL. The linac injects the 70 MeV H^- beam into the rapid-cycling synchrotron, which then strips and accelerates the proton beam to 800 MeV, offering 190 kW of beam power split between two production targets, with about 30 neutron research instruments. More than 1200 experiments are performed by 3000 users each year. A second target station has been in operation since 2007, providing low energy neutrons to study soft condensed matter, biological systems, advanced composites, and nano-materials. The Spallation Neutron Source (SNS) has been operating with an average beam power of 1.7 MW (Fig. 1.19 and Sect. 10.1). The SNS target station also offers ∼20 instruments that serve approximately 1400 users per year. SNS upgrade programs are being pursued, raising the beam power to 2.8 MW (Sect. 10.1.5) and offering a second target station. The J-PARC rapid-cycling synchrotron provides a beam to the Material and Life Science Experimental Facility (Fig. 1.7), enabling a large user group, among which 25% are industrial users (Sect. 10.2). All of these pulsed neutron sources use rings for beam compression. At CSNS, the beam is accelerated first by the linac and then compressed and further accelerated by the rapid-cycling synchrotron to an energy of 1.6 GeV. Phase 1 of the project goal has been achieved with a proton beam-on-target power of 160 kW. Phase 2 is planned to increase the beam power to 500 kW (Fig. 1.21).

Spallation neutron sources driven by continuous wave (CW) or long-pulse beams from linacs and cyclotrons have reached a power level of 1 MW since around 1980 (Fig. 1.6). At the Los Alamos Neutron Science Center (LANSCE), protons are accelerated through an 800 MeV linac to supply long-pulse neutrons (Fig. 1.20). Subsequently, the Proton Storage Ring (PSR) was built to compress proton pulses from 750 μs to a quarter of a microsecond for short-pulse neutrons [95]. Users conduct research at five facilities at LANSCE: Isotope Production Facility, Lujan Neutron Scattering Center, Proton Radiography Facility, Ultra-cold neutrons, and Weapons Neutron Research Facility. Paul Scherrer Institute (PSI) operates a spallation neutron source user facility driven by its 590 MeV proton cyclotron, which operates at a world-record beam power above 1.4 MW, complemented by the Swiss Light Source (SLS) with a 2.4 GeV electron storage ring since 2011. The European Spallation Source (ESS) is based on a 2 GeV SC proton linac that will deliver 5 MW average beam power with 2.86 ms long pulses at 14 Hz. Unlike SNS, ESS is classified as a long-pulse neutron source where no accumulator ring is used. Instead, the linac beam directly hits a rotating solid tungsten target wheel, and shorter neutron pulses are obtained using neutron choppers (Fig. 1.22) [79]. The linac was commissioned to 870 MeV in 2025, and spallation neutrons are expected in 2026.

Many countries and institutions have been pursuing accelerator-driven neutron sources. For example, Germany's engagement can be traced back to the Spallation

Fig. 1.18 Layout of the ISIS spallation neutron and muon source at Rutherford Appleton Laboratory showing the 70 MeV linac, the 800 MeV rapid-cycling synchrotron, and two neutron target stations, and many user stations. (Courtesy ISIS, STFC, UK)

Neutron Source (SNQ) project at Jülich in the late 1970s and early 1980s [126, 127]. Although the SNQ was never built, its design studies and technical groundwork positioned Jülich as a leading laboratory in this field [128, 129]. Around 2000, Jülich again played a central role when comprehensive scientific and technical volumes for the European Spallation Source (ESS) [130–132] were compiled as

Fig. 1.19 Aerial view of the Spallation Neutron Source. (Courtesy ORNL, USA)

Fig. 1.20 Aerial view of the Los Alamos Neutron Science Center and additional facilities. (Courtesy LANL, USA): (1) Lujan center; (2) weapons neutron research; (3) proton radiography; (4) isotope production facility; (5) ultracold neutrons; (6) materials test station; and (7) linear accelerator

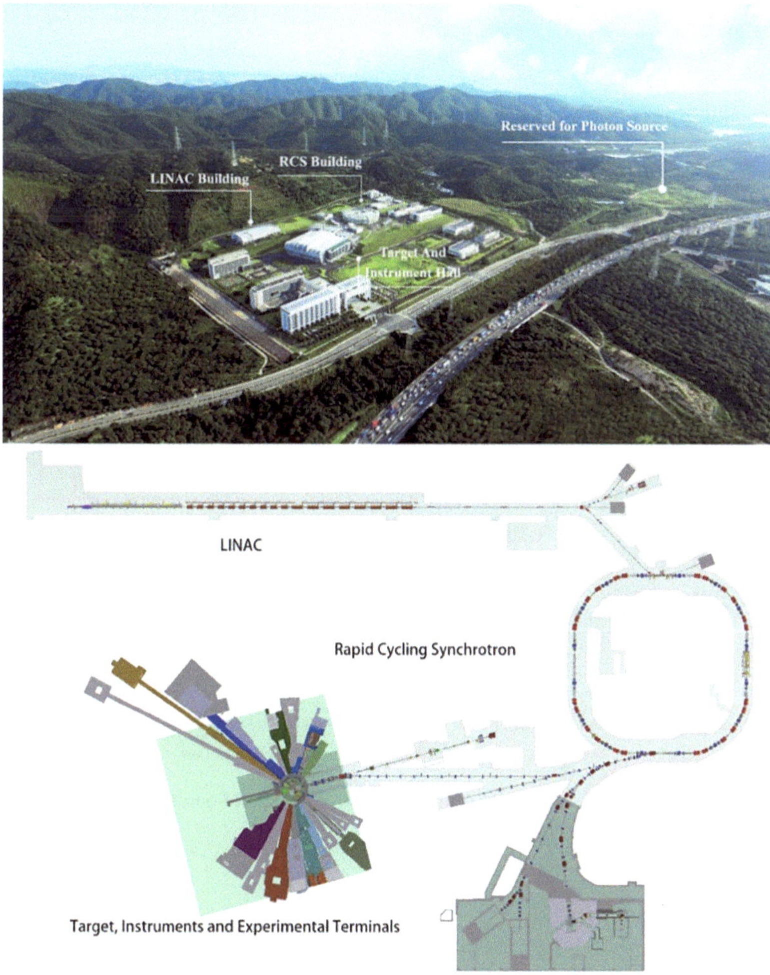

Fig. 1.21 Aerial view and schematics of the China Spallation Neutron Source. (Courtesy CSNS, China)

part of the German hosting proposal [133–135], followed by substantial in-kind contributions to the Lund-based facility after 2013 [136]. Building on this legacy, FZJ has advanced the High Brilliance Neutron Source (HBS) concept for a compact, high-current accelerator-driven facility [137–139].

The neutron time-of-flight research facility at CERN, n_TOF, has been a platform for neutron cross-section measurements since 2001, with three distinct experimental areas, each tailored to address a specific range of neutron energies [140]. The PS proton beam has been directed onto a Pb target at 20 GeV to produce neutrons (Fig. 1.9).

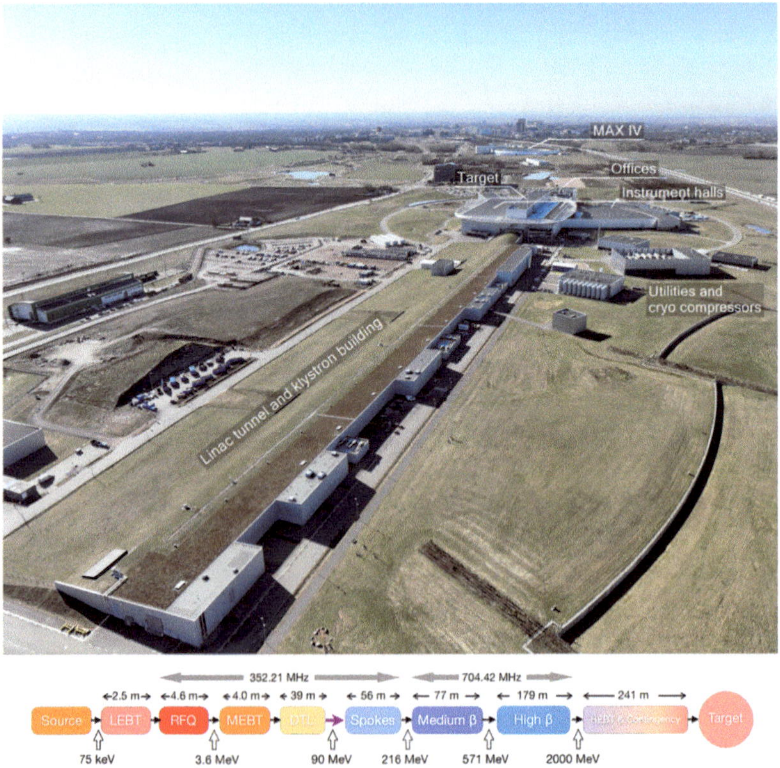

Fig. 1.22 Aerial view and schematics of the European Spallation Source with the MAX IV light source in the background. (Courtesy ESS, Sweden)

1.6.2.2 Muon Sources

Although cosmic muons have been used in scientific research, accelerator-based facilities offer high-intensity muon sources through hadron-hadron collisions to produce pion beams, which then quickly decay into muon beams over short distances. At several high-power hadron facilities, muon sources are established parasitically to the neutron source by passing the high energy (>500 MeV) proton beam through a target containing nuclei of intermediate mass (carbon, beryllium): ISIS, J-PARC, PSI, and TRIUMF. Depending on the energy of the protons, processes produce either single pions, e.g.

$$\mathrm{p} + \mathrm{p} \rightarrow \mathrm{p} + \mathrm{n} + \pi^{+}; \quad \mathrm{p} + \mathrm{n} \rightarrow \mathrm{p} + \mathrm{p} + \pi^{-}, \tag{1.2}$$

Or, at higher energies, two pions can be produced, e.g.

$$\mathrm{p} + \mathrm{p} \rightarrow \mathrm{p} + \mathrm{p} + \pi^{+} + \pi^{-}; \quad \mathrm{p} + \mathrm{n} \rightarrow \mathrm{p} + \mathrm{n} + \pi^{+} + \pi^{-}. \tag{1.3}$$

The threshold for the proton energy E_k to produce single pions is $E_k > 280$ MeV, and production peaks between 500 and 1000 MeV. The threshold for double pion production is $E_k > 600$ MeV and production increases to $E_k \approx 1.5$ GeV, beyond which it saturates. Pions decay into muons and neutrinos via the two-body reactions [141]

$$\pi^+ \to \mu^+ + \nu_\mu; \quad \pi^- \to \mu^- + \bar{\nu}_\mu. \tag{1.4}$$

Such muon sources possess the same time structure as the neutron source.

ISIS and J-PARC offer pulsed muon sources with a time structure determined by the post-synchrotron proton pulse width and pion lifetime. PSI and TRIUMF offer continuous-wave muon sources based on protons extracted from the cyclotrons. The Swiss Muon Source at PSI is the world's most intense continuous beam muon source today, where one of the main applications of these intense muon beams is muon spin spectroscopy.

The muon $g - 2$ experiment, a collaborative project involving CERN, BNL, and FNAL, aims to precisely measure the muon's magnetic moment to test the Standard Model of particle physics. These muons have been produced by first colliding high-energy protons with a target, which creates pions. These pions then decay into muons and neutrinos. Strong magnetic fields have been used to steer and select the muons, focusing them into a beam that is injected into the storage ring where the experiment's measurements are taken.

1.6.2.3 Proton Sources

Although neutron and muon sources are secondary beam user platforms, primary beams, such as protons, can also serve as user platforms that offer multiple beamlines for scientific and industrial users. Figure 1.23 shows the layout of the dedicated proton user facility of the Korea Multipurpose Accelerator Complex (KOMAC). The 100 MeV linear accelerator feeds the proton beams to two groups of beam lines at 20 and 100 MeV, respectively. These beamlines are currently being used for materials, energy, environment, and nanoscience applications.

1.6.3 Accelerator-Driven Sub-Critical Systems

Accelerators have the potential to address the energy and environmental challenges facing twenty-first century society. Specifically, high-power hadron accelerators are capable of advancing nuclear energy technology in two aspects: the management of spent nuclear fuel [142–144] and providing a sustainable energy source [145, 146]. As high-power accelerators efficiently produce neutrons by spallation, this process can be linked to nuclear reactor technology in an accelerator-driven subcritical system (ADS) to transmute long-lived radioisotopes in spent nuclear fuel into shorter-lived fission products (Sect. 1.6.3.1).

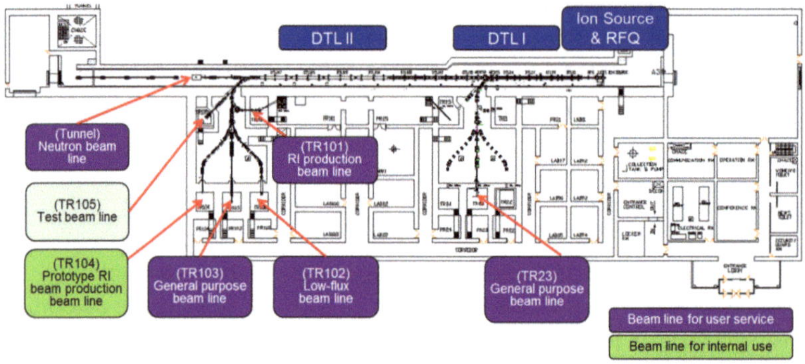

Fig. 1.23 Schematics of the Korean proton facility KOMAC [81]. (Courtesy KOMAC, Korea)

1.6.3.1 Accelerator Driven Subcritical Systems for Nuclear Waste Transmutation

A key challenge facing the nuclear fuel cycle is reducing the radio-toxicity and lifetime of spent nuclear fuel. Both reactors with a fast neutron spectrum and accelerator-driven subcritical systems offer transmutation capabilities, acting as waste incinerators that destroy longer-lived fission products such as Technetium-99 ($^{99}_{43}$Tc) and Iodine-129 ($^{129}_{53}$I) (213,000 years and 16 million years half-lives, respectively). These isotopes can acquire a neutron to become short-lived $^{100}_{43}$Tc and $^{130}_{53}$I, respectively.

Effective material partitioning must accompany the transmutation process, as reliable separation is needed to ensure that stable isotopes are not transmuted into radioactive ones. Multiple recycling of the fuel would be necessary to achieve a significant reduction in radio-toxicity.

An accelerator-driven subcritical system for transmutation consists of a high-power proton accelerator driver, a spallation target for neutron production, and a subcritical reactor containing partitioned waste to be burned. A continuous-wave accelerator with an output energy of several GeV is preferred for high efficiency, high flux neutron production.

The MYRRHA (Multi-purpose hYbrid Research Reactor for High-tech Applications) project is based on a 600 MeV, 2.5 mA CW proton linear accelerator and a spallation target embedded in a Lead-Bismuth Eutectic (LBE) cooled reactor capable of operating in subcritical and critical modes.

CiADS (Fig. 1.24), MYRRHA, as well as the ATW project have chosen linac-based technology to achieve ultimate high power. Worldwide research and development have been conducted not only on linac-based programs but also on cyclotron- and FFA-based ones.

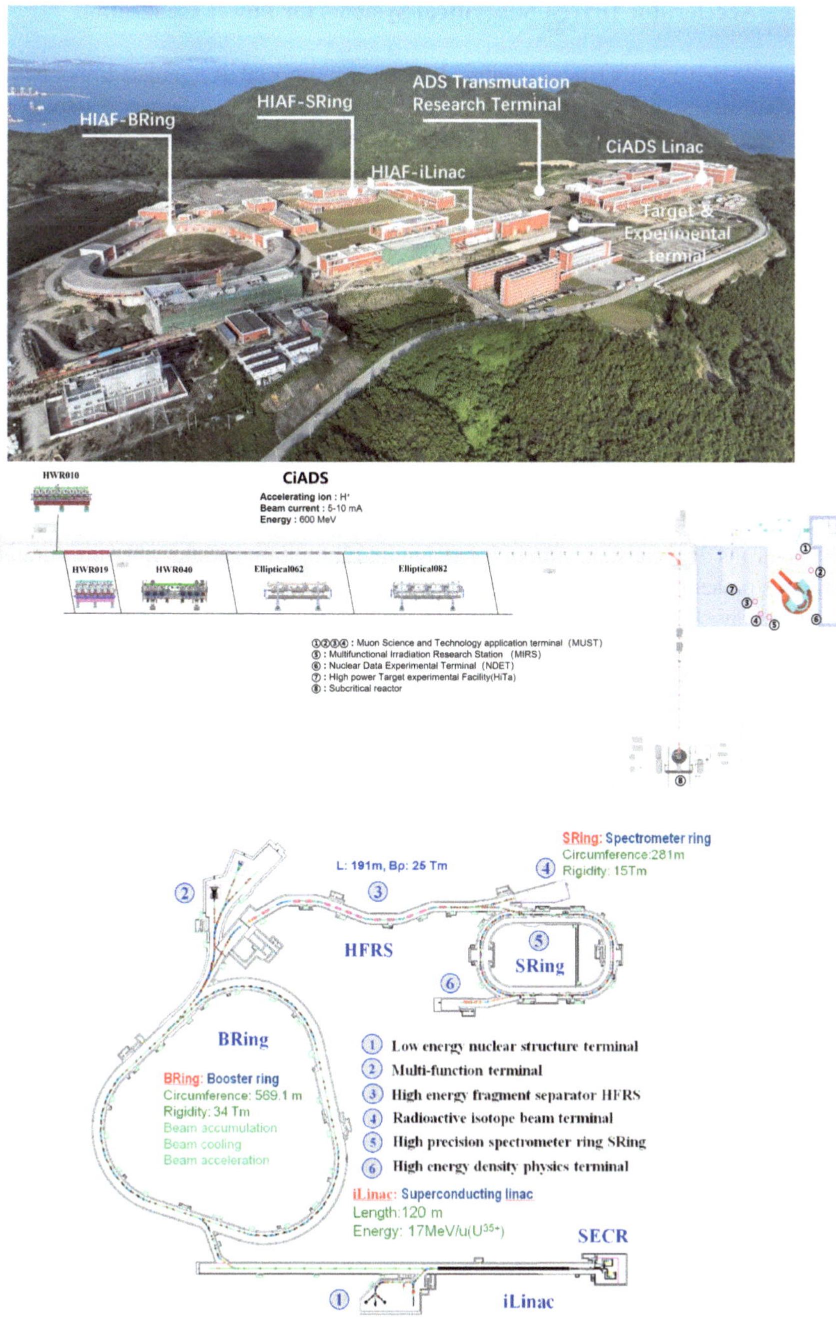

Fig. 1.24 Aerial view and schematics of the China initiative Accelerator Driven System (CiADS) and Heavy Ion Accelerator Facility (HIAF). (Courtesy IMP, China)

1.6.3.2 Accelerator Driven Subcritical Systems for Power Generation

Accelerator-driven subcritical systems have been proposed as a means to bootstrap a source of fissile material from fertile feed. In particular, the systems are associated with using thorium (instead of uranium in conventional light water reactors) as the fuel because it is much more abundant in nature.

The high-power accelerator supplies continuous wave protons down a beam pipe to the spallation target inside the core. The spallation target of e.g. molten lead or lead-bismuth provides neutrons to drive the reactor loaded with fertile fuel that cannot support a self-sustaining chain reaction. The thorium core is located at the bottom of the reactor, with circulating molten lead or lead-bismuth as the primary coolant. The neutrons enter the fuel and transmute thorium to protactinium, which soon decays to uranium-233 ($^{233}_{92}$U) which is fissile. The production of fissile material occurs in the following reaction:

$$^{232}_{90}\text{Th} + ^{1}_{0}\text{n} \rightarrow ^{233}_{90}\text{Th} + \gamma \rightarrow ^{233}_{91}\text{Pa} + ^{0}_{-1}\beta \rightarrow ^{233}_{92}\text{U} + ^{0}_{-1}\beta$$

The neutrons also cause fission in the uranium, plutonium, and possibly transuranic elements present, releasing energy.

There have been programs and proposals to develop prototypes of such reactors. The Power Reactor Thoria Reprocessing Facility (PRTRF) has been commissioned at the Bhabha Atomic Research Center (BARC) to breed and extract $^{233}_{92}$U from thorium assemblies that are irradiated in another reactor. The obtained $^{233}_{92}$U is used as fuel for a 30 kWth research reactor, KAMINI.

In comparison with conventional nuclear power plants, power generation based on accelerator-driven subcritical systems has the advantages of abundant fuel resources, minimal possibility of runaway reactions, significantly reduced production of long-lived actinides, and efficient burning of minor actinides. However, the requirement for accelerator availability is unprecedented. The system design is complex; in particular, the beam pipe may protrude from the reactor containment barrier.

1.6.4 Accelerators Servicing Fusion Research

In providing sustainable energy, accelerators play a role in both subcritical fission (Sect. 1.6.3.2) and controlled fusion systems (Sect. 1.6.4).

1.6.4.1 Accelerators for Fusion Material Testing

The International Thermonuclear Experimental Reactor (ITER) project aims to become the first research reactor capable of demonstrating energy production with a positive balance of nuclear fusion. The scope of the International Fusion Material

Irradiation Facility (IFMIF) is to test the materials that will be used for the future commercial fusion reactor under operating conditions with typical fusion reactor-like neutron fluxes.

Since 2007, activities have been pursued through the IFMIF Engineering Design and Engineering Validation Activities (IFMIF-EVEDA) project that aim to produce an integrated IFMIF engineering design and validate the main technological challenges by completing the engineering validation activities of its three main facilities: the test facility, the lithium target facility, and the accelerator facility (LIPAc) [147, 148].

The LIPAc accelerator, located in Rokkasho, Japan, is a prototype accelerator for the IFMIF and is being constructed and commissioned as a joint project between Europe and Japan [85]. The LIPAc linac was designed to accelerate a 125 mA continuous wave deuteron beam to an energy of 4.5 MeV/u. The linac consisted of a 9.8 m long RFQ that accelerates the deuteron beam to 2.5 MeV/u, followed by cryomodules containing 175 MHz SC half-wave resonators (HWR) and SC solenoids. Recent activities included the successful re-installation and testing of upgraded RFQ couplers after resolving a vacuum leak issue, and the resumption of RF conditioning in 2023. The next steps involve preparing for high duty cycle and CW operation with the RFQ and the cryomodule containing 8 SC HWR and solenoids.

The DEMO-oriented [149] neutron source IFMIF-DONES is an international facility under construction in Spain that will provide a high intensity flux of neutrons for research on materials for future fusion reactors and other complementary applications [86] (Fig. 1.25). The generation of those neutrons is driven by stripping reactions between energetic deuteron particles and a liquid lithium jet. The requirements for the accelerator are to deliver a 20 MeV/u deuteron beam with a CW current of 125 mA.

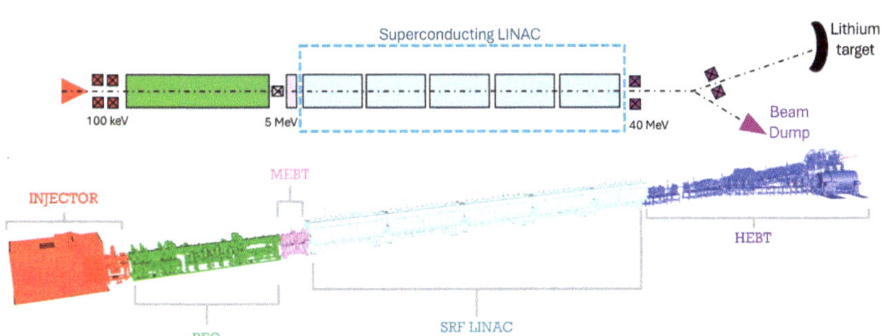

Fig. 1.25 Schematic layout and mock-up of the IFMIF-DONES accelerator beamline [86]

1.6.4.2 Accelerators as Injectors for Plasma Devices

Neutral beams are widely used in experimental magnetic fusion devices, heating the plasma in the reactor by injecting high-energy neutral particles. Charged particles are accelerated to a high energy outside the reactor and then stripped of their charge, becoming neutral atoms. These neutral particles can pass through the magnetic cage that confines the plasma and transfer their energy to it through collisions, raising its temperature needed for nuclear fusion.

During the past several decades, several efforts have been pursued worldwide. The ITER heating neutral beam injectors (HNB) are designed to operate under conditions and constraints similar to those that will be encountered in a fusion reactor [150]. The injectors are designed to accelerate 40 A of CW D^- beam to 1 MeV energy for pulse lengths of up to 3600 s. The Padova Research on ITER Megavolt Accelerator (PRIMA), as the ITER neutral beam test facility hosted in Padova, Italy, includes two experiments: one for the full-scale prototype of the ITER heating neutral beam injector and the other for the full-size RF negative-ion source [151]. The BINP has developed and produced more than 50 neutral beam injectors for plasma heating and diagnostics over the last 30 years. The current activities are focused on the development of ion sources and injectors with a power of up to 10 MW and continuous operation [152, 153].

1.6.4.3 Accelerators for Inertial Fusion Energy Production

Inertial confinement fusion (ICF) research attempts to initiate nuclear fusion reactions by heating and compressing a fuel target, typically in the form of a pellet that contains a mixture of deuterium and tritium. Energy is delivered to the target using high-power beams of laser light, electrons, or ions. The largest operational ICF experiment is the National Ignition Facility (NIF) that uses laser beams for attempted ignition.

Heavy-ion-driven inertial confinement fusion research dates back to the 1970s [154]. The performance characteristics desired to support an inertial-fusion energy facility include a high average beam power (>0.1 GW) and a high instantaneous beam power (> TW). The proposed European Heavy Ion Driven Ignition Facility (HIDIF) project is based on a high-power heavy-ion driver consisting of a full-energy linear accelerator, multiple storage rings, transfer lines with multiple induction linac bunchers, and a final focus target station. Three ion species with different masses but the same charge and momentum are configured in a "telescoping arrangement". The full-energy linac, with a length exceeding 3 km, is assumed to start with 16 ion sources for each of the three ion species, funneled in four stages with trees of RFQs and drift tube linacs (DTL) into a high-current main DTL linac with a peak current of 400 mA. Multiple bunch synchronization delay lines are planned, along with separate induction linac bunchers for each ion species. A project like HIDIF is complex and challenging, not only in the accelerator system design but also in the coupling with the hydrodynamically unstable target.

1.6.5 Hadron Therapy and Medicine

Since their invention, accelerators have long been workhorses in medical applications. Tens of millions of patients receive accelerator-based diagnoses and therapy each year in hospitals and clinics around the world. In general, accelerators play three important roles in this field: producing radioisotopes for nuclear medicine and diagnostics, producing particle beams for therapy treatment, and providing accelerator-based light and neutron sources that unravel the complex structures and dynamics of biological systems and develop new pharmaceuticals.

1.6.5.1 Radioisotopes and Radiopharmaceuticals

Radioisotopes emitting X-rays, γ rays, and positrons have served as medical diagnostic probes, while radioisotopes emitting β and α particles have been used for short-range cancer therapy [155]. Examples include: fluorodeoxyglucose (^{18}FDG) as a carrier of the positron emitter ^{18}F for positron-emission tomography; thallous chloride (^{201}Tl) and molybdenum-99 as a technetium-99m generator (^{99}Mo / ^{99m}Tc) for nuclear cardiography; as well as α emitters ^{211}At and ^{213}Bi and β emitters ^{90}Sr, ^{103}Pd, and ^{192}Ir for brachytherapy applications.

Both reactors and accelerators have been used for the production of radiopharmaceuticals [156]. The high beam power of accelerators, such as the cyclotrons at PSI and TRIUMF, has enabled the production of large inventories of abundant radionuclides. These proton cyclotron facilities use isotope separation on-line (ISOL) methods, driving high-power proton beams onto thick targets to produce radioisotopes through mass separation.

At CERN, the MEDICIS facility at ISOLDE produces radionuclides for research in nuclear medicine, specifically for cancer therapy and diagnosis. The facilities use ISOLDE-type targets, irradiated parasitically in front of one of the ISOLDE dumps [157]. The targets are then mounted on a dedicated front end, from which the extracted radionuclides are selected by laser ionization and mass separation in a magnetic spectrometer [158]. A non-exhaustive list of radionuclides for the MEDICIS programme includes: ^{44}Sc/^{47}Sc, ^{128}Ba/^{128}Cs, ^{129}Cs, ^{149,152,155}Tb, ^{165}Tm/^{165}Er, ^{167}Tm, ^{169}Er, ^{223}Ra, ^{224}Ra/^{212}Pb, and ^{225}Ra/^{225}Ac.

1.6.5.2 Hadron Therapy

Radiation therapy with external beams typically uses photon (X-ray), neutron, proton, or carbon ion beams produced by accelerators. Most of these radiation treatments are performed with commercial RF linacs that produce electrons in the energy range from 5 to 30 MeV, striking a heavy metal target to produce photons through the bremsstrahlung process. About 5,000 of these units are operating throughout the world.

Fast neutron therapy uses neutrons that are typically between 50 and 70 MeV produced by reactors, cyclotrons, and linear accelerators. Compared with photons, fast neutrons have the advantage of high "linear energy transfer (LET)" radiation.

BNCT [159] is based on nuclear capture and decay reactions that occur when non-radioactive boron-10, is irradiated with epithermal neutrons (n_{th}) to produce excited boron-11 ($^{11}B^*$). This undergoes radioactive decay to produce high-energy alpha particles (^4He) and high-energy lithium-7 (^7Li) nuclei,

$$^{10}B + n_{th} \rightarrow [^{11}B^*] \rightarrow ^4He + ^7Li + 2.31\,\text{MeV}. \tag{1.5}$$

Although early treatment used neutrons from research reactors [160], accelerators are now becoming the primary source of epithermal neutrons for clinical BNCT using reactions like $^7Li(p,n)^7Be$ or $^9Be(p,n)^9B$ [161].

Charged hadron radiation therapy typically utilizes protons and carbon ions accelerated by cyclotrons and synchrotrons. The chief advantage of charged hadron therapy over other types of external beam radiation therapy is that the dose is deposited over a narrow range around the "Bragg peak" and there is a minimal exit dose [162].

Many of the nearly 100 operating and under-construction proton facilities use cyclotrons to accelerate protons to fixed extraction energies of up to 250 MeV. Degraders are used to reduce the proton energy in accordance with the desired treatment depth. Cyclotrons are usually compact and easy to operate. However, degraders used for energy reduction cause undesired radio-activation.

Newly proposed rapid-cycling medical synchrotrons use single-turn fast extraction to provide hadron pulses of variable energy, position, and intensity, offering more precise three-dimensional spot scanning treatment.

Almost 10 operating and under-construction carbon ion radiation treatment facilities use synchrotrons to accelerate carbon ions to variable energies of up to 450 MeV/u. Synchrotrons have the advantage of being easily adjustable in extraction energy, which matches the desired treatment depth. The beam is usually slow-extracted from the synchrotron and transversely collimated to provide irradiation. Figure 1.26 shows the schematics of the first-generation Heavy Ion Medical Machine (HIMM) facility in Wuwei, China, established in 2019 [163]. The second-generation HIMM facilities are under development with a much more compact footprint, as the synchrotron circumference has been reduced to 36.6 m for accelerating $^{12}C^{6+}$ ions up to 400 MeV/u.

1.6.6 Accelerators for Security and Defense

The accelerator has long played an important role in national defense and security. Topics range from the historical separation of isotopes using industrial-scale accelerator mass spectrometry to the modern defense application of free-electron laser technology. In this section, a few examples related to hadron-beam applications are selected.

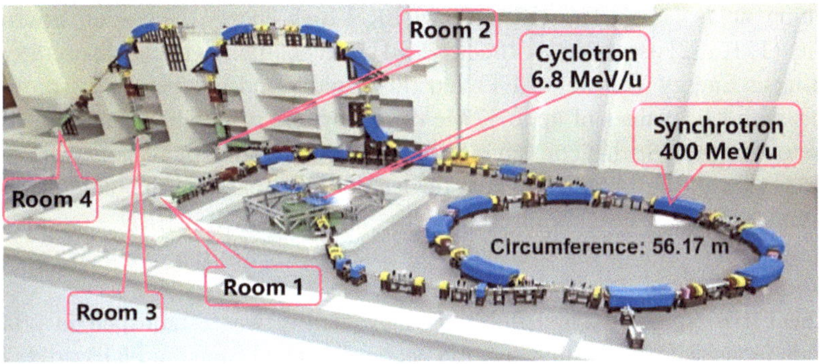

Fig. 1.26 Schematics of the first-generation Heavy Ion Medical Machine (HIMM) facility in Wuwei, China. (Courtesy IMP, China) $^{12}C^{6+}$ ions are accelerated to a range of 120–400 MeV/u for cancer therapy treatment

1.6.6.1 Neutron Imaging and Radiography

Although electron linac-based X-ray scanners provided by industry dominate the cargo inspection market, neutron scanners offer complementary advantages: they are sensitive to light isotopes, are highly penetrating, and possess magnetic moments. Neutrons are attenuated differently by different materials, thus providing information on both the composition and structure of the sample [164]. Commercial neutron imaging or radiography is usually based on a compact driver linac that accelerates a proton or deuteron beam striking a light target, such as beryllium, to produce neutrons that can be moderated for imaging. The linac energy is often modest (for example, below 10 MeV) to avoid excessive radio-activation of the equipment.

1.6.6.2 Irradiation and Single-Event Effects

Single Event Effects (SEE) induced by heavy ions, protons, and neutrons have become an increasing limitation on the reliability of electronic components, circuits, and systems and have stimulated efforts to improve our understanding and develop mitigation techniques [165]. Dedicated and parasitic accelerator-based SEE facilities have been developed, including the Berkeley Accelerator Space Effects (BASE) facility using the LBNL 88-inch cyclotron [166], the Radiation Effects Facility at Texas A&M University's Cyclotron Institute [167], the NASA Space Radiation Laboratory (NSRL) at BNL using the AGS Booster synchrotron [168], the Single Event Upset (SEU) Test Facility at BNL using the Tandem Van de Graaff accelerator, and the Crocker Nuclear Laboratory at the University of California, Davis using UC-Davis' 76-inch cyclotron, [169].

The CHARM facility at CERN utilizes the slow-extracted proton beam from the CERN-PS at 24 GeV sent to a dedicated target to create the appropriate radiation fields [170]. The HEARTS@CERN high-energy, heavy-ion irradiation facility [171, 172] uses a slow-extracted lead beam configured to enable users to tune the Linear Energy Transfer (LET) and flux rapidly and independently, providing beams with LET values of up to 37 MeV·cm^2/mg and penetration ranges in silicon of up to ~50 mm. Such a combination allows for the testing of advanced electronic technologies and full boards without device de-lid preparation, including complex 3-D and packaged systems directly in the air. HEARTS@CERN supports both scientific and proprietary access and is optimized for space-relevant heavy-ion SEE studies.

The FRIB Single Event Effects (FSEE) Facility was established in 2022 to provide industrial users with heavy ion beams produced from the FRIB driver linac for accurate characterization of beam parameters and LET, easy-to-use controls, and prompt reporting of dosimetry results [173]. Beams ranging from ^{16}O to ^{209}Bi can typically be prescribed with a beam energy ranging from 15 to 40 MeV/u and LET at the Bragg peak ranging from 7 to 100 MeV·cm^2/mg. In 2025, the dedicated K500 SEE user facility was established using MSU's K500 cyclotron.

1.6.6.3 Proton Flash Radiography

Flash radiography is a technique for recording images of objects moving at extremely high speeds to measure their dynamic properties and internal density profiles. Beam-based flash radiography typically uses photon (X-ray) or proton beams.

Proton flash radiography provides imaging in rapid succession (e.g., 200 ns between frames) on thick samples (e.g., 100 g/cm^2). Since protons interact through both Coulomb and nuclear forces, they can measure material shape, density, and composition. The proton energy (e.g. 1–50 GeV) can vary so that the proton attenuation length matches the sample thickness.

The previously proposed Advanced Hydrotest Facility (AHF) project [174] planned for the driver accelerator beams to be split multiple times to provide imaging in various directions, allowing 3-D tomographic "movies". Matter-Radiation Interactions in Extremes (MaRIE) [175] plans to simultaneously accommodate proton, electron, and hard photon (X-ray) radiography at the same site, exploring dynamic materials performance and process-aware manufacturing. Leveraging the existing LANSCE proton facility, the new addition includes a high peak, coherent free electron laser producing hard photons of about 50 keV.

1.6.6.4 Accelerator Production of Tritium

For strategic nuclear deterrence, periodic replenishment of tritium inventory is needed, e.g., through commercial light-water reactors (CLWR). The high-power proton accelerator, along with spallation technology, provides an alternative. Tritium can be produced through neutron capture by stable isotopes such as ^{3}He or ^{6}Li with reactions such as

$$^{6}_{3}\text{Li} + \text{n} \rightarrow {}^{4}_{2}\text{He} + {}^{3}_{1}\text{T} \tag{1.6}$$

The previously proposed Accelerator Production of Tritium (APT) reference design consists of a proton linear accelerator, beam transport, target/blanket, and tritium-extraction systems with a production capacity of about 1 kg/year. Neutrons are generated by the proton spallation of heavy-metal nuclei, such as tungsten [176, 177].

1.6.7 Compact Hadron Accelerators for Research and Prototyping

Complementing the large-scale national facilities discussed in the previous sections, compact accelerator facilities have been built that allow for education, research, and technology development, serving as prototypes for technological demonstration or as user facilities for research and industrial applications. This section discusses examples of such compact or intermediate-sized facilities. Smaller commercialized facilities are discussed in Sect. 1.6.8.1.

1.6.7.1 Compact Neutron Sources

Compact neutron sources have been built, driven by electrons and hydrogen isotopes [178]. An example of an electron-driven facility is the Hokkaido University Neutron Source (HUNS) where electrons are accelerated to ∼45 MeV in a linac and strike a lead target to produce photoneutrons through the bremsstrahlung process [179]. Table 1.2 shows some common photon-to-neutron conversion processes.

Examples of compact facilities driven by charged hydrogen isotopes are the Low Energy Neutron Source (LENS) at Indiana University [180] and the Compact

Table 1.2 Common photon-to-neutron conversion targets of electron accelerators operating in the 4–20 MeV range: weight percentages and energy thresholds of (γ, n) and (γ, 2n) reactions per isotope

		Energy threshold (MeV)	
Conversion target	Isotope	(γ, n)	(γ, 2n)
Heavy water	^{2}H	2.22	–
Beryllium	^{9}Be	1.67	20.56
Tantalum	^{181}Ta	7.58	14.22
Tungsten	^{182}W	8.07	14.75
	^{186}W	7.19	12.95
Lead	^{206}Pb	8.09	14.82
	^{208}Pb	7.37	14.11

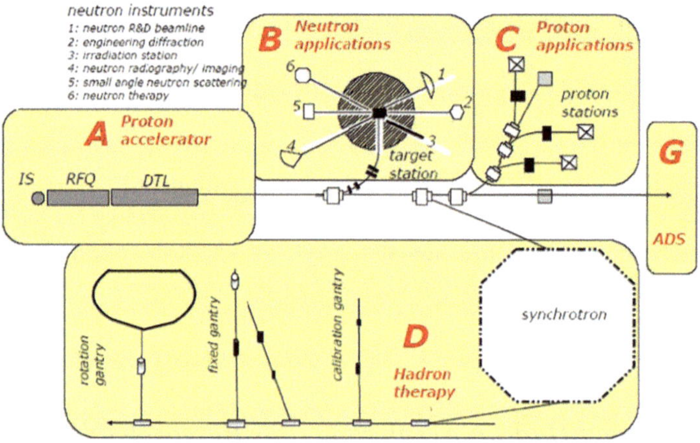

Fig. 1.27 Compact Pulsed Hadron Source and possible technological extensions [97]

Pulsed Hadron Source at Tsinghua University [97]. Pulsed protons are accelerated to 13 MeV to strike beryllium or lithium targets, producing pulsed neutrons,

$$\mathrm{p} + {}_3^7\mathrm{Li} \rightarrow {}_4^7\mathrm{Be} + \mathrm{n}, \quad \mathrm{p} + {}_4^9\mathrm{Be} \rightarrow {}_5^9\mathrm{B} + \mathrm{n}. \tag{1.7}$$

Located on the central campus of the university, CPHS not only serves as a test bed for neutron scattering and imaging research but, more importantly, as an educational center to produce innovators and the workforce required by the hadron community for accelerator-related research and applications (Fig. 1.27) [97].

1.6.7.2 Hadron Facilities for Research and Beam-Dynamics Studies

Many hadron storage rings were built for studies of beam dynamics topics such as beam cooling [181] and crystalline beams [182], including ASTRID [183], CELSIUS [184], CRYRING [185, 186], IUCF Cooler [187], S-LSR [188], and TSR [189] at universities and many other facilities at national laboratories and institutes, including CERN, COSY, FNAL, GSI, IMP, JINR, and KEK. IOTA is a newly constructed storage ring for non-linear beam dynamics studies [190].

1.6.7.3 Prototype Accelerators and Prototyping Experiments

A prototype accelerator is built to demonstrate a solution or technology before full-scale production. It also allows for a more accurate cost estimate of the full production. For example, the primary objectives of LEDA [94] are to verify the design codes, gain fabrication knowledge, understand the operation of the beam,

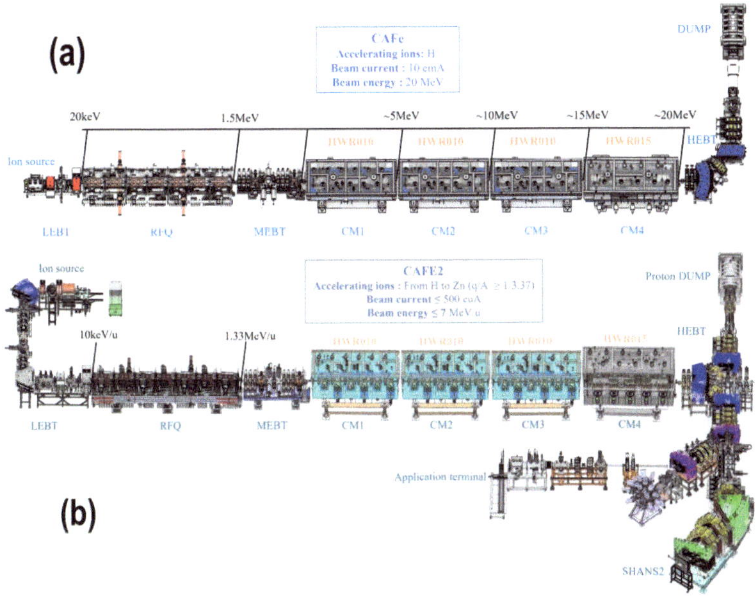

Fig. 1.28 (**a**) Schematics of the China ADS front End (CAFe) test facility consisting of an ECR ion source, an RFQ, an MEBT, three $\beta_0 = 0.10$ HWR cryomodules, a $\beta_0 = 0.15$ HWR cryomodule, an HEBT, and a beam dump, accelerating 10 mA of CW proton beam to 20 MeV. (**b**) Schematics of the China Accelerator Facility for Superheavy Elements (CAFE2) converted from CAFe, re-using the four cryomodules while replacing the ECR ion source, the LEBT, the RFQ, and the user stations, accelerating up to 500 eμA of CW beams of heavy ions from H to Zn with $q/A > 1/3.37$ to up to 7 MeV/u. (Courtesy IMP, China)

measure the characteristics of the output beam, learn how to minimize the frequency of beam trips, and improve the prediction of costs and operational availability for the entire APT accelerator (Sect. 1.6.6.4). The 8 m long 350 MHz LEDA CW RFQ [191, 192] linac delivered a 110 mA proton beam at 6.7 MeV with approximately 94% transmission. Similarly, the LIPAc linac serves as a prototype for the IFMIF project (Sect. 1.6.4.1), and the CAFe (Fig. 1.28) and CiADS (Fig. 1.24) [193] serve as prototypes for a full-scale ADS facility (Sect. 1.6.3.1).

Prototype experiments serve as proof-of-principle for innovative concepts and approaches. EMMA was built as a non-scaling FFA [194, 195] (see Sect. 2.1.1.5), as shown in Fig. 1.29. The renovation of the KEK Digital Accelerator (KEK-DA) [196] is a proof-of-principle that uses an induction synchrotron as a driver for applications such as heavy-ion fusion [197]. The ring-shaped Paul trap PALLAS was used to generate "moving" crystals that led to a 3-D crystalline beam [198]. Compact ion-trap systems (S-POD) were constructed to reproduce the dynamic behavior of space-charge-dominated beams in a local tabletop environment [199, 200] and were also used as a unique ion source that can generate ultralow-emittance nanobeams [201].

a

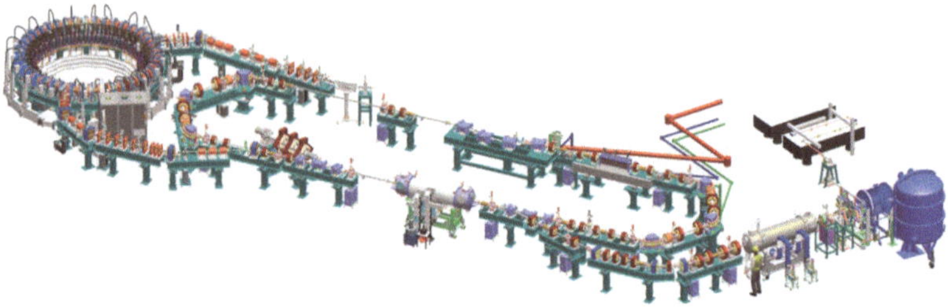

b

Fig. 1.29 (Top) Schematics of the Accelerators and Lasers In Combined Experiments (ALICE) and the Electron Model for Many Applications (EMMA) at Daresbury Laboratory, STFC. (Bottom) EMMA, which was built as the first non-scaling FFA, demonstrated fast acceleration from 12 to 18MeV/c of electrons within a few turns [191]. (Courtesy STFC, UK)

1.6.8 Accelerators and Industry

The development of modern accelerators is closely interwoven with the technological capacities of industry. Innovative accelerator concepts and technologies transferred from national laboratories, universities, and government-funded institutes to industry facilitate the mass production capacity needed for both the construction of large accelerators and the commercialization of accelerator technology. Thus, accelerator science relies on industry to extend its societal impact. The growth in technological capability in industry facilitates the establishment of large-scale accelerators.

1.6.8.1 Industrial Accelerators

More than 99% of the existing ~30,000 accelerators currently at work in the world are for industrial (about 20,000) and medical (about 10,000) applications [202]. The remaining (<1%) of accelerators for science and research are often large in size, have a high impact, and incur high costs. Generally, about 50% of the technical construction expenditure is incurred through procurements with industrial providers. Industry is an integral part of accelerator development.

Industrial accelerators are widely used in material processing and treatment, material analysis, manufacturing, and environmental protection. Electron-based accelerators are used for electron and photon beam irradiation, electron-beam fabrication (e.g., cutting and welding), photon-based non-destructive detection (e.g., cargo inspection), and environmental control (e.g., water purification and treatment, sewage sludge disinfection, and flue gas pollution control). Ion-based accelerators are used for ion implantation (e.g., semiconductor doping, mutations of plants and microorganisms), ion beam analysis (e.g., accelerator mass spectrometry), and neutron-based nondestructive detection (e.g., cargo inspection and mining).

Accelerators for ion implantation are generally classified as medium current (ion beam currents between 10 μA and ~2 mA), high current (ion beam currents up to ~30 mA), high energy (ion energies above 200 keV and up to 10 MeV), and very high dose (efficient implantation of doses greater than 10^{16} ions/cm^2). The accelerators typically consist of an ion source that is sometimes coupled to biased electrodes for the extraction of ions into the beamline, which is then followed by mass selection. The deposition depth corresponds to the "Bragg peak" of the ion beam, which can be easily adjusted by the voltage of the electrodes.

Ion beam analysis typically uses electrostatic accelerators, such as the Tandem Van de Graaff, which accelerates ions to energies of several MV to probe the composition and obtain elemental depth profiles in the near-surface layers of solids. Highly sensitive methods that allow the detection of elements in the sub-monolayer range include Rutherford back-scattering, elastic back-scattering, elastic recoil detection, particle-induced X-ray emission, particle-induced γ-ray emission, nuclear reaction analysis, and channeling.

Accelerator mass spectrometry differs from the usual ion beam analysis in that a small sample of the material to be analyzed is ionized to form the beam. The typically used Tandem Van de Graaff, together with the specially designed beam line for mass and charge differentiation, makes it possible to detect long-lived, naturally occurring radioisotopes such as ^{14}C with sensitivity as high as 10^{-15}.

The pre-accelerated ions are usually separated by a first-stage mass spectrometer before entering the electrostatic tandem accelerator. At the connecting point between the two stages, the ions change their charge from negative to positive by passing through a thin stripper. Molecules such as $^{13}CH^-$ will break apart when stripped, leading to exceptional abundance sensitivity. Upon stripping, the positively charged ions are further accelerated before entering a second stage of the mass spectrometer, which removes the background.

Compared with the large and intermediate/compact sized neutron sources discussed in Sects. 1.6.2.1 and 1.6.7.1, the neutron generators discussed in this section are low-power sources commercially available from the industry. Neutron generators have applications in medicine, well-logging, non-destructive testing, and materials analysis. Thousands of such small, relatively inexpensive systems have been built over the past five decades. While medium and large neutron sources use Be (p, n), Li (d, n), and spallation reactions, these neutron generators typically use fusion T(d, n)^{4}He and D(d, n)^{3}He reactions, with D-T and D-D producing 14.1 and 2.45 MeV neutrons, respectively. Neutron generators contain compact linear accelerators that accelerate either deuterium or tritium onto a metal hydride target that also contains deuterium, tritium, or a mixture of these isotopes.

1.6.8.2 Technology Transfer to Industry

Historically, breakthroughs in accelerator concepts and technologies have most often come from the arena of basic scientific research at universities and laboratories. The construction of modern, large accelerators requires the participation of industrial partners capable of delivering quality components with high fabrication efficiency and capacity. Technology transfer is a key process to establish technically credible vendors that meet the needs.

The two rings of the Relativistic Heavy Ion Collider (RHIC) contain about 1740 SC magnets, including 288 arc dipoles that have an aperture of 8 cm and a length of 9.7 m. These dipoles, designed by physicists and engineers at Brookhaven National Laboratory, provide a modest field of 3.45 T in a cost-effective design. Key features include the use of NbTi cable, a single-layer coil, and cold iron as the yoke and collar. The magnets operate at a nominal temperature of 4.6 K. The RHIC project selected an established industrial partner located near Long Island, New York. Brookhaven staff built the first prototype arc dipole that validated the design. The vendor's staff joined together to build the second Brookhaven prototype as part of the technology transfer. Subsequently, the industrial vendor built the magnets according to the mechanical specifications. All magnets were tested at room temperature, and 20% were tested at superconducting temperature at BNL. A close collaboration between Brookhaven and industrial partners allowed for the delivery of highly reliable, high-field-quality magnets [203, 204].

Build-to-print is often an economical method of procuring customized accelerator components, as the industrial provider typically only needs to fabricate according to the mechanical drawings. A successful build-to-print contract usually requires that (1) the customer completes the engineering design of the accelerator component; (2) the industrial provider possesses the knowledge and skill; and (3) there are frequent communications between the customer and the provider to ensure timely resolution of fabrication issues and to avoid misunderstandings.

With the RHIC magnets discussed above, the industrial provider mass-produced the magnets in accordance with the mechanical drawings provided by RHIC. The

Fig. 1.30 SRF cavities built by an industrial provider installed in the SNS cryomodules. (Courtesy JLab and ORNL, USA)

contract cost was kept to a minimum, as the provider did not need to guaranty magnetic-field quality.

Figure 1.30 shows the SRF cavities built by an industrial provider for the Spallation Neutron Source project. Design and prototyping were carried out at JLab before the build-to-print contract was signed for mass production [205].

Another type of build-to-print can be more economical; that is, detailed mechanical drawings are completed in collaboration between the provider and the customer. The detailed drawings provided by the customer could sometimes impose overly stringent constraints on the production process, the techniques to be used, and the management of mass production. In addition, providers may have better mechanical engineers, designers, and detailers than their customers. The customer specifies the required tolerances for dimensional accuracy in the assembly drawings to meet the necessary performance requirements of the accelerator components. The provider then makes detailed drawings to meet the required accuracy, utilizing familiar techniques and methods that are most economical for the provider. It should be noted that these differ among providers. Therefore, these should not be strongly specified to optimize economic manufacturing. The provider should only guaranty mechanical accuracy, while the performance of the components is the responsibility of the customers, as long as mechanical accuracy is met. Most of the J-PARC accelerator components were purchased in this way, except for the commercial off-the-shelf (COTS) items. Figure 1.31 shows the RF accelerating cavities manufactured thus far [206].

Fig. 1.31 J-PARC NC accelerating cavities fabricated by industry. (Courtesy J-PARC, Japan)

Build-to-specification can alleviate the design burden on the customer, as the provider takes responsibility for designing and building according to the specifications provided by the customer. A successful build-to-specification contract generally requires that (1) the provider has experience designing and fabricating the accelerator device to meet the performance specifications; (2) the customer is knowledgeable about providing rational specifications, requirements, and interfaces, so that the provider's risk is contained; and (3) there is frequent communication between the customer and the provider throughout the design, fabrication, and commissioning.

The FRIB cryogenic system 4 K "cold box," as shown in Fig. 1.32, was purchased with a build-to-specification contract from an industrial provider who had experience supplying similar cryogenic systems for the SNS and the JLab 12 GeV upgrade projects. The FRIB staff worked closely with the provider to produce detailed requirements and interfaces based on prior experience and lessons learned at SNS and JLab.

"Turnkey system" refers to small-scale accelerators and accelerator devices offered to the accelerator customer according to the requirements and interface conditions. Such systems range from accelerators for hadron therapy facilities and boosters to synchrotron radiation light sources and chemistry processing clean rooms. Some providers of these turnkey systems are accelerator builders who later became industrial entrepreneurs.

Fig. 1.32 Cryogenic liquidation "cold box" built by an industrial provider for the FRIB 4 K cryogenic refrigeration system. (Courtesy MSU, USA)

Experienced accelerator builders design their facilities by taking full advantage of available commercial off-the-shelf products. These include a standard-sized helium liquidator/refrigerator, power supplies to NC magnets, RF amplifiers, vacuum flanges, pumps and valves, programmable logic controllers, some beam diagnostic monitors, and cooling-water skids.

References

1. International Atomic Energy Agency, Background: Accelerators. https://www.iaea.org/topics/nuclear-science/nuclear-research/accelerators/background
2. J.D. Cockcroft, E.T.S. Walton, Experiments with high velocity positive ions. R. Soc. Lond. **A129**, 477–489 (1930). https://doi.org/10.1098/rspa.1930.0169
3. G. Ising, Prinzip einer Methode zur Herstellung von Kanalstrahlen hoher Voltzahl. Ark. Mat. Astron. Fys. **18**(30), 1–4 (1924). https://cds.cern.ch/record/433984
4. R. Wideröe, Über ein neues Prinzip zur Herstellung hoher Spannungen. Archiv für Elektrotechnik **21**(4), 387–406 (1928). https://doi.org/10.1007/bf01656341
5. E.O. Lawrence, M.S. Livingston, The production of high speed light ions without the use of high voltages. Phys. Rev. **40**, 19–35 (1932). https://doi.org/10.1103/PhysRev.40.19
6. D.W. Kerst, Acceleration of electrons by magnetic induction. Phys. Rev. **58**, 841–841 (1940). https://doi.org/10.1103/PhysRev.58.841
7. E.M. McMillan, The synchrotron—a proposed high energy particle accelerator. Phys. Rev. **68**, 143–144 (1945). https://doi.org/10.1103/PhysRev.68.143
8. V.I. Veksler, New method for the acceleration of relativistic particles, *Doklady Akademii Nauk USSR*, vol. 43 (1944), pp. 346–348

9. E.D. Courant, M.S. Livingston, H.S. Snyder, The strong-focusing synchroton—a new high energy accelerator. Phys. Rev. **88**, 1190–1196 (1952). https://doi.org/10.1103/PhysRev.88.1190
10. N. Christofilos, Focussing system for ions and electrons (1956). https://ppubs.uspto.gov/dirsearch-public/print/downloadPdf/2736799. US Patent 2,736,799
11. L.W. Alvarez, The design of a proton linear accelerator. Phys. Rev. **70**, 799 (1946)
12. H. Padamsee, J. Knobloch, T. Hays, *RF Superconductivity for Accelerators*. Wiley Series in Beam Physics and Accelerator Technology (Wiley, 1998). https://books.google.com/books?id=2leaQgAACAAJ
13. D. Hartill, Colliders, in *Handbook of Accelerator Physics and Engineering*, ed. by A.W. Chao, M. Tigner, 2nd edn. (World Scientific, Singapore, 2013), pp. 12–14. https://doi.org/10.1142/9789814415859_0001
14. R. Widerøe, Anordnung zur herbeiführung von kernreaktionen (1953). Issued by Deutsches Patentamt, Patentschrift Nr. 876279
15. D.W. Kerst, F.T. Cole, H.R. Crane, L.W. Jones, L.J. Laslett, T. Ohkawa, A.M. Sessler, K.R. Symon, K.M. Terwilliger, N.V. Nilsen, Attainment of very high energy by means of intersecting beams of particles. Phys. Rev. **102**, 590–591 (1956). https://doi.org/10.1103/PhysRev.102.590
16. G.K. O'Neill, Storage-ring synchrotron: device for high-energy physics research. Phys. Rev. **102**, 1418–1419 (1956). https://doi.org/10.1103/PhysRev.102.1418
17. V. Shiltsev, F. Zimmermann, Modern and future colliders. Rev. Modern Phys. **93**(1), 015006 (2021)
18. T. Tajima, J.M. Dawson, Laser electron accelerator. Phys. Rev. Lett. **43**, 267–270 (1979). https://doi.org/10.1103/PhysRevLett.43.267
19. M.S. Livingston, *Particle Accelerators: A Brief History* (Harvard University Press, Cambridge, MA, 1969)
20. V.D. Shiltsev, High-energy particle colliders: past 20 years, next 20 years, and beyond. Phys. Usp. **55**(10), 965 (2012). https://doi.org/10.3367/UFNe.0182.201210d.1033
21. F. Zimmermann, M. Benedikt, D. Schulte, J. Wenninger, Challenges for highest energy circular colliders, in *Proceedings of IPAC'14* (JACoW Publishing, Geneva). https://doi.org/10.18429/JACoW-IPAC2014-MOXAA01. https://jacow.org/IPAC2014/papers/MOXAA01.pdf
22. S. van der Meer, Stochastic damping of betatron oscillations in the ISR. Technical report, CERN, Geneva (1972). https://cds.cern.ch/record/312939
23. R.R. Wilson, The tevatron. Phys. Today **30**(10), 23–30 (1977)
24. P. Schmüser, F. Willeke, 5 the electron-proton collider hera: the largest accelerators and colliders of their time, in *Accelerators and Colliders* (Springer, Berlin, 2013), pp. 428–443
25. H. Hahn, E. Forsyth, H. Foelsche, M. Harrison, J. Kewisch, G. Parzen, S. Peggs, E. Raka, A. Ruggiero, A. Stevens, S. Tepikian, P. Thieberger, D. Trbojevic, J. Wei, E. Willen, S. Ozaki, S. Lee, The RHIC design overview. Nucl. Instrum. Methods Phys. Res., Sect. A **499**(2), 245–263 (2003). https://doi.org/10.1016/S0168-9002(02)01938-1. https://www.sciencedirect.com/science/article/pii/S0168900202019381. The Relativistic Heavy Ion Collider Project: RHIC and its Detectors
26. L.R. Evans, LHC status, in *Proceedings of APAC'07* (JACoW Publishing, Geneva), pp. 46–50. https://jacow.org/a07/papers/TUXMA01.pdf
27. Y. Wang, Future circular colliders (2015). Presented at IPAC2015 in Richmond. Unpublished
28. M. Tigner, A possible apparatus for electron clashing-beam experiments. Nuovo Cim. **37**, 1228–1231 (1965). https://doi.org/10.1007/BF02773204
29. J.T. Seeman, The Stanford linear collider. in *Proceedings of LINAC'90, Linear Accelerator Conference*, (JACoW Publishing, Geneva), pp. 3–7. https://jacow.org/l90/papers/MO101.pdf
30. B. Barish, J.E. Brau, The international linear collider. Int. J. Mod. Phys. A **28**(27), 1330039 (2013). https://doi.org/10.1142/s0217751x13300391
31. P. Lebrun, L. Linssen, A. Lucaci-Timoce, D. Schulte, F. Simon, S. Stapnes, N. Toge, H. Weerts, J. Wells, The CLIC programme: towards a staged e+e- linear collider exploring the terascale: CLIC conceptual design report (2012)

32. F.F. Tikhonin, On the effects with muon colliding beams. Technical Report P2-4120, Joint Institute for Nuclear Research (JINR), Dubna (1968). https://doi.org/10.48550/arXiv.0805.3961. Preprint; English translation available via arXiv: arXiv:0805.3961 [hep-ph]
33. G.I. Budker, Accelerators and colliding beams, in *Proceedings of the 7th International Conference on High-Energy Accelerators of Charged Particles (Proc. HEACC'69)*, vol. 1 (1970), pp. 33–39. https://inspirehep.net/literature/931295
34. D. Schulte, The muon collider, in *Proceedings of IPAC'22, International Particle Accelerator Conference*, no. 13 (JACoW Publishing, Geneva, 2022), pp. 821–826. https://doi.org/10.18429/JACoW-IPAC2022-TUIZSP2. https://jacow.org/IPAC2022/papers/TUIZSP2.pdf
35. N. Pastrone, 1.6.15 muon collider, in *Handbook of Accelerator Physics and Engineering*, ed. by A.W. Chao et al. (World Scientific, Singapore, 2022), pp. 50–53
36. A. Skrinskii, V. Parkhomchuk, Methods of cooling beams of charged particles. Sov. J. Part. Nucl. **12**(3), 223–247 (1981)
37. D.V. Neuffer, Principles and applications of muon cooling. Part. Accel. **14**(FN-378), 75 (1983)
38. M. Bogomilov, Y. Karadzhov, D. Kolev, I. Russinov, R. Tsenov, G. Vankova-Kirilova, L. Wang, F. Xu, S. Zheng, R. Bertoni et al.: The MICE muon beam on ISIS and the beam-line instrumentation of the muon ionization cooling experiment. J. Instrum. **7**(05), P05009 (2012)
39. MICE collaboration, Demonstration of cooling by the muon ionization cooling experiment. Nature **578**(7793), 53–59 (2020)
40. H.G. Kirk et al., The MERIT high-power target experiment at the CERN PS, in *Proceedings of EPAC'08, European Particle Accelerator Conference*, no. 11 (JACoW Publishing, Geneva, 2008), pp. 2886–2888. https://jacow.org/e08/papers/WEPP169.pdf
41. I. Efthymiopoulos et al., The MERIT (nTOF-11) high tntensity liquid mercury target experiment at the CERN PS, in *Proceedings of EPAC'08, European Particle Accelerator Conference*, no. 11 (JACoW Publishing, Geneva, 2008), pp. 262–264. https://jacow.org/e08/papers/MOPC087.pdf
42. C.B. Schroeder, E. Esarey, C.G.R. Geddes, C. Benedetti, W.P. Leemans, Physics considerations for laser-plasma linear colliders. Phys. Rev. ST Accel. Beams **13**, 101301 (2010). https://doi.org/10.1103/PhysRevSTAB.13.101301
43. W.P. Leemans, B. Nagler, A.J. Gonsalves, C. Tóth, K. Nakamura, C.G. Geddes, E. Esarey, C. Schroeder, S. Hooker, GeV electron beams from a centimetre-scale accelerator. Nat. Phys. **2**(10), 696–699 (2006)
44. A. Gonsalves, K. Nakamura, J. Daniels, C. Benedetti, C. Pieronek, T. De Raadt, S. Steinke, J. Bin, S. Bulanov, J. Van Tilborg et al., Petawatt laser guiding and electron beam acceleration to 8 gev in a laser-heated capillary discharge waveguide. Phys. Rev. Lett. **122**(8), 084801 (2019)
45. D. Einfeld, M. Plesko, J. Schaper, First multi-bend achromat lattice consideration. J. Synchrotron Radiat. **21**(5), 856–861 (2014). https://doi.org/10.1107/S160057751401193X
46. M. Borland, V. Sajaev, Y. Sun, Hybrid seven-bend-achromat lattice for the Advanced Photon Source upgrade, in *Proceedings of IPAC'15*, Richmond, 3–8 May 2015, International Particle Accelerator Conference, no. 6 (JACoW, Geneva, 2015), pp. 1776–1779. https://doi.org/10.18429/JACoW-IPAC2015-TUPJE063. http://jacow.org/ipac2015/papers/tupje063.pdf
47. S. Shin, New era of synchrotron radiation: fourth-generation storage ring. AAPPS Bull. **31**(1), 21 (2021). https://doi.org/10.1007/s43673-021-00021-4
48. J.N. Galayda, The new LCLS-II project: status and challenges, in *Proceedings of LINAC'14, Linear Accelerator Conference*, no. 27 (JACoW Publishing, Geneva, 2014), pp. 404–408. https://jacow.org/LINAC2014/papers/TUIOA04.pdf
49. P. Rauer, I. Bahns, B. Friedrich, S. Casalbuoni, M. Di Felice, M. Dommach, I. Freijo Martin, W. Freund, J. Grünert, M. Guetg, I. Karpics, S. Karabekyan, A. Koch, N. Kujala, D. La Civita, J. Liu, T. Maltezopoulos, M. Makita, F. Mayet, L. Müuller, B. Rio, L. Samoylova, S. Schmidtchen, M. Scholz, A. Silenzi, V. Strauch, D. Thoden, T. Wohlenberg, M. Vannoni, F. Yang, W. Decking, J. Rossbach, H. Sinn, Lasing of a cavity-based x-ray source. Nature **650**(8100), 93–96 (2026). https://doi.org/10.1038/s41586-025-10025-x

50. L.H. Yu, L. DiMauro, A. Doyuran, W.S. Graves, E.D. Johnson, R. Heese, S. Krinsky, H. Loos, J.B. Murphy, G. Rakowsky, J. Rose, T. Shaftan, B. Sheehy, J. Skaritka, X.J. Wang, Z. Wu, First ultraviolet high-gain harmonic-generation free-electron laser. Phys. Rev. Lett. **91**, 074801 (2003). https://link.aps.org/doi/10.1103/PhysRevLett.91.074801
51. G. Stupakov, Using the beam-echo effect for generation of short-wavelength radiation. Phys. Rev. Lett. **102**, 074801 (2009). https://link.aps.org/doi/10.1103/PhysRevLett.102.074801
52. D.F. Ratner, A.W. Chao, Steady-state microbunching in a storage ring for generating coherent radiation. Phys. Rev. Lett. **105**, 154801 (2010). https://link.aps.org/doi/10.1103/PhysRevLett.105.154801
53. S. Henderson et al., The SpallationNeutron Source accelerator system design. Nucl. Instrum. Methods Phys. Res., Sect. A **763**, 610–673 (2014). https://doi.org/10.1016/j.nima.2014.03.067. https://www.sciencedirect.com/science/article/pii/S0168900214003817
54. Y. Yamazaki, J-PARC status, in *Proceedings of PAC'09, Particle Accelerator Conference*, no. 23 (JACoW Publishing, Geneva), pp. 18–22. https://jacow.org/PAC2009/papers/MO2BCI03.pdf
55. M.A. Martens et al., Upgrades to the Fermilab NuMI beamline, in *Proceedings of PAC'07, Particle Accelerator Conference*, no. 22 (JACoW Publishing, Geneva, 2007). https://jacow.org/p07/papers/TUPAS028.pdf
56. M. Seidel et al., Production of a 1.3 MW proton beam at PSI, in *Proceedings of IPAC'10* (JACoW Publishing, Geneva), pp. 1309–1313. http://accelconf.web.cern.ch/IPAC10/papers/TUYRA03.pdf
57. D.E. Nagle, Status and performance of LAMPF, in *Proceedings of LINAC'72, Linear Accelerator Conference*, pp. 4–11 (JACoW Publishing, Geneva) https://jacow.org/l72/papers/A02.pdf
58. J. Wei et al., The FRIB project: accelerator challenges and progress, in *Proceedings of HIAT'12* (JACoW Publishing, Geneva), pp. 8–19. https://jacow.org/HIAT2012/papers/MOB01.pdf
59. J. Wei, The very high intensity future, in *Proceedings of IPAC'14* (JACoW Publishing, Geneva), pp. 17–22. https://doi.org/10.18429/JACoW-IPAC2014-MOYBA01. https://jacow.org/IPAC2014/papers/MOYBA01.pdf
60. T. Roser, Production of high brightness proton bunches, in *Proceedings of PAC'01* (JACoW Publishing, Geneva, 2001), pp. 710–714. https://jacow.org/p01/papers/FOAC001.pdf
61. E.K. Elsener, The CERN Neutrino beam to Gran Sasso (Conceptual Technical Design), CERN 98-02. Technical report,s INFN/AE-98/05 (1998)
62. K. Hasegawa, M. Kinsho, H. Oguri, Status of J-PARC accelerators, in *Proceedings of 9th Annual Meeting of Particle Accelerator Society of Japan*, Osaka (2012), pp. 339–343
63. L.V. Kravchuk, INR proton linac operation and applications. Nucl. Instrum. Methods Phys. Res. Sect. A: Accel. Spectrom. Detect. Assoc. Equip. **562**(2), 932–934 (2006)
64. S. Holmes, P. Derwent, V. Lebedev, S. Mishra, D. Mitchell, V.P. Yakovlev, PIP-II status and strategy. Technical report, Fermi National Accelerator Lab (FNAL), Batavia (2015)
65. P.W. Schmor, New achievements at TRIUMF and future plans, in *Proceedings of Cyclotrons'07* (JACoW Publishing, Geneva), pp. 93–98
66. R.E. Laxdal, Completion and operation of ISAC-I and extension to ISAC-II, in *Proceedings of PAC'01* (JACoW Publishing, Geneva), pp. 556–560. https://jacow.org/p01/papers/ROPA001.pdf
67. O. Kamigaito et al., Progress towards high-intensity heavy-ion beams at RIKEN RIBF, in *Proceedings of IPAC'13* (JACoW Publishing, Geneva), pp. 333–335. https://jacow.org/IPAC2013/papers/MOPFI025.pdf
68. R. Ferdinand, P. Bertrand, X. Hulin, M. Jacquemet, E. Petit, Progress of construction and installation of the SPIRAL2 accelerator, in *Proceedings of IPAC'13* (JACoW Publishing, Geneva), pp. 3755–3757. https://jacow.org/IPAC2013/papers/THPWO002.pdf
69. A.K. Orduz, M. Di Giacomo, R. Ferdinand, B. Jacquot, O. Kamalou, J.M. Lagniel, G. Normand, A. Savalle, D. Uriot, Commissioning of a high power linac at GANIL: beam

power ramp-up. Phys. Rev. Accel. Beams **25**, 060101 (2022). https://doi.org/10.1103/PhysRevAccelBeams.25.060101
70. O. Kester et al., Status of the FAIR facility, in *Proceedings of IPAC'13* (JACoW Publishing, Geneva), pp. 1085–1089. https://jacow.org/IPAC2013/papers/TUXB101.pdf
71. J. Yang, J. Xia, G. Xiao, H. Xu, H. Zhao, X. Zhou, X. Ma, Y. He, L. Ma, D. Gao et al., High intensity heavy ion accelerator facility (HIAF) in China. Nucl. Instrum. Methods Phys. Res., Sect. B **317**, 263–265 (2013)
72. Y. Chung, J. Kang, M. Kwon, E. Kwon, J. Kwon, G. Kim, D. Kim, M. Kim, S. Kim, Y. Kim et al., Status of the RAON heavy-ion accelerator. J. Korean Phys. Soc. **87**, 455–465 (2025)
73. R.W. Garnett, E.J. Pitcher, D. Rees, L. Rybarcyk, T. Tajima, High-Power options for LANSCE, in *Proceedings of PAC'11, Particle Accelerator Conference*, no. 24 (JACoW Publishing, Geneva, 2011), pp. 2107–2109. https://jacow.org/PAC2011/papers/THOCN4.pdf
74. J. Galambos, Operations experience of SNS at 1.4 MW and upgrade plans for doubling the beam power, in *Proceedings of IPAC'19* (JACoW Publishing, Geneva), pp. 4380–4384. https://doi.org/10.18429/JACoW-IPAC2019-FRXPLM1. https://jacow.org/ipac2019/papers/FRXPLM1.pdf
75. J. Thomason et al., A 180 MeV injection upgrade design for the ISIS synchrotron, in *Proceedings of IPAC'13* (JACoW Publishing, Geneva), pp. 2678–2680. https://jacow.org/IPAC2013/papers/WEPEA073.pdf
76. D. Berkovits et al., Operational experience and future goals of the SARAF proton/deuteron linac, in *Proceedings of LINAC'12* (2012), pp. 100–104
77. S. Fu, S. Wang, Operation status and upgrade of CSNS, in *Proceedings of IPAC'19* (JACoW Publishing, Geneva), pp. 23–27. https://doi.org/10.18429/JACoW-IPAC2019-MOZPLM1. http://accelconf.web.cern.ch/ipac2019/papers/MOZPLM1.pdf
78. M. Lindroos, M. Eshraqi, D.P. McGinnis, The European spallation source, in *Proceedings of NAPAC'13* (JACoW Publishing, Geneva, 2013), pp. 1448–1452. https://jacow.org/PAC2013/papers/FRYAA2.pdf
79. R. Garoby, A. Vergara, H. Danared, I. Alonso, E. Bargallo, B. Cheymol, C. Darve, M. Eshraqi, H. Hassanzadegan, A. Jansson et al., The European spallation source design. Phys. Scrip. **93**(1), 014001 (2017)
80. Y.S. Cho et al., TheKOMAC accelerator facility, in *Proceedings of IPAC'13* (JACoW Publishing, Geneva), pp. 2052–2054. https://jacow.org/IPAC2013/papers/WEOBB101.pdf
81. H.J. Kwon: Operation status of KOMAC linac, in *26th International Conference on Accelerator and Beam Utilization (ICABU 2024)*, Pohang (2024). https://www.indico.kr/event/59/contributions/942/attachments/699/1588/2024%20ICABU%20Kwon_KOMAC%20FFF.pdf01.pdf
82. J. Knaster, Y. Okumura, Accelerators for fusion materials testing. Rev. Accel. Sci. Technol. **8**, 115–142 (2015)
83. J. Knaster, Y. Okumura, P. Cara, A. Kasughai, M. Sugimoto, Challenges of the high current prototype accelerator of IFMIF/EVEDA, in *Proceeedings of IPAC'16, International Particle Accelerator Conference*, no. 7 (JACoW Publishing, Geneva, 2016), pp. 52–57. https://doi.org/10.18429/JACoW-IPAC2016-MOZB02. https://jacow.org/ipac2016/papers/MOZB02.pdf
84. K. Kondo, T. Akagi, F. Arranz, N. Bazin, L. Bellan, B. Bolzon, B. Branas, P. Cara, Y. Carin, J. Castellanos et al., Validation of the linear IFMIF prototype accelerator (LIPAc) in rokkasho. Fusion Eng. Design **153**, 111503 (2020)
85. H. Dzitko et al., Status and future developments of the Linear IFMIF Prototype Accelerator (LIPAc). Fusion Eng. Des. **168**, 112621 (2021). https://doi.org/10.1016/j.fusengdes.2021.112621
86. I. Podadera, P. Cara, B. Bolzon, C. Caballero, N. Chauvin, M. Comunian, D.D. Cosic, M. Eshraqi, M. Juni Ferreira, V. Hauer et al., Overview and current status of the IFMIFDONES Accelerator Systems. Nucl. Fusion. **65**, 122011 (2025)
87. Y. He, Z. Wang, Z. Qin, H. Xu, L. Yang, Z. Wang, J. Zhang, X. Zhang, L. Gu, W. Zhan, Development of accelerator driven advanced nuclear energy (ADANES) and nuclear fuel recycle, in *Proceedings of IPAC* (2019), pp. 4389–4393

88. H. Aït Abderrahim, P. Baeten, D. De Bruyn, R. Fernandez, MYRRHA—a multi-purpose fast spectrum research reactor. Energy Convers. Manag. **63**, 4–10 (2012)
89. H. Ravn, Status and future development of ion sources for on-line mass separators. Nucl. Instrum. Methods Phys. Res., Sect. B **70**(1–4), 107–117 (1992)
90. S. Gales, The GANIL facility. Scholarpedia **5**(7), 9670 (2010)
91. H. Geissel, P. Armbruster, K. Behr, A. Brünle, K. Burkard, M. Chen, H. Folger, B. Franczak, H. Keller, O. Klepper et al., The GSI projectile fragment separator (FRS): a versatile magnetic system for relativistic heavy ions. Nucl. Instrum. Methods Phys. Res., Sect. B **70**(1–4), 286–297 (1992)
92. A. Gade, C.K. Gelbke, The NSCL laboratory and the FRIB facility. Scholarpedia **5**(1), 9651 (2010)
93. F. Marti, P. Miller, D. Poe, M. Steiner, J. Stetson, X. Wu, Commissioning of the coupled cyclotron system at nscl, in *AIP Conference. Proceedings*, vol. 600 (American Institute of Physics, 2001), pp. 64–68
94. H.V. Smith, J.D. Schneider, R.L. Sheffield, Low-Energy Demonstration Accelerator (LEDA) test results and plans, in *Proceedings of PAC'01* (JACoW Publishing, Geneva), pp. 3296–3298. https://jacow.org/p01/papers/RPPH031.pdf
95. G. Lawrence, R. Hardekopf, A. Jason, P. Clout, G. Sawyer, Los alamos high-current proton siorage ring; a status report. IEEE Trans. Nucl. Sci. **32**(5), 2662–2665 (1985)
96. X. Guan et al., The compact pulsed hadron source status, in *Proceedings of HIAT 2012* (2012), pp. 112–114
97. J. Wei et al., Compact pulsed hadron source—a university-based accelerator platform for multidisciplinary neutron and proton applications, in *Proceedings of PAC'09, Particle Accelerator Conference*, no. 23 (JACoW Publishing, Geneva), pp. 1360–1362. https://jacow.org/PAC2009/papers/TU6PFP035.pdf
98. R.A. Baartman, Space charge limit in separated turn cyclotrons, in *Proceedings of Cyclotrons'13* (JACoW Publishing, Geneva), pp. 305–309. https://jacow.org/CYCLOTRONS2013/papers/WE2PB01.pdf
99. T. Planche, Constant-tune cyclotrons. J. Instrum. **18**(03), P03019 (2023)
100. T. Zhang, W. Fu, C. Wang, S. Pei, Z. Yin, H. Zhou, Z. Yin, S. Zhang, J. Liu, X. Zhu et al., Innovative research and development activities on high energy and high current isochronous proton accelerator. J. Instrum. **18**(03), P03022 (2023)
101. T.P. Wangler et al., Basis for low beam loss in the high-current APT linac, in *Proceedings of LINAC'98, Linear Accelerator Conference* (JACoW Publishing, Geneva), pp. 657–659. https://jacow.org/l98/papers/TU4107.pdf
102. N.V. Mohkov, W. Chou (eds.), Beam halo and scraping, in *Proceedings, 7th ICFA Mini-Workshop on High Intensity High Brightness Hadron Beams*, Lake Como, 13–15 Sept 1999 (2000)
103. M. Craddock, K. Symon, Cyclotrons and fixed-field alternating-gradient accelerators. Rev. Accel. Sci. Technol. **1**(01), 65–97 (2008)
104. W. Joho, High intensity problems in cyclotrons, in *Proceedings of 5th International Conference on Cyclotrons and Their Applications*, Caen (1981)
105. T. Stammbach, S. Adam, H. Fitze, W. Joho, M. Märki, M. Olivo, L. Rezzonico, P. Sigg, U. Schryber, The feasibility of high power cyclotrons. Nucl. Instrum. Methods Phys. Res., Sect. B **113**(1), 1–7 (1996). https://doi.org/10.1016/0168-583X(95)01377-6. https://www.sciencedirect.com/science/article/pii/0168583X95013776. Accelerators in Applied Research and Technology
106. J. Wei, The Spallation Neutron Source project—physical challenges, in *Proceedings of EPAC'02, European Particle Accelerator Conference*, no. 8 (JACoW Publishing, Geneva, 2007), pp. 1064–1066. https://jacow.org/e02/papers/THPLE015.pdf
107. I. Hofmann, Stability of anisotropic beams with space charge. Phys. Rev. E **57**(4), 4713 (1998)
108. R.J. Macek et al., Electron proton two-stream instability at the PSR, in *Proceedings of PAC'01* (JACoW Publishing, Geneva), pp. 688–692. https://jacow.org/p01/papers/FOAB007.pdf

109. J. Wei, Synchrotrons and accumulators for high-intensity proton beams. Rev. Mod. Phys. **75**, 1383–1432 (2003). https://doi.org/10.1103/RevModPhys.75.1383
110. S. Esin, L. Kravchuk, A. Kvasha, P. Ostroumov, V. Serov, MMF linac upgrade possibilities for the pulsed neutron source, in *Proceedings of PAC'95* (1996), p. 1173
111. S.D. Holmes et al., Project X functional requirements specification, in *Proceedings of IPAC'12* (JACoW Publishing, Geneva), pp. 3945–3947. https://jacow.org/IPAC2012/papers/THPPP090.pdf
112. P. Ostroumov, R. Pardo, G. Zinkann, K. Shepard, J. Nolen, Simultaneous acceleration of multiply charged ions through a superconducting linac. Phys. Rev. Lett. **86**, 2798–801 (2001). https://doi.org/10.1103/PhysRevLett.86.2798
113. P.N. Ostroumov, K. Fukushima, T. Maruta, A.S. Plastun, J. Wei, T. Zhang, Q. Zhao, First simultaneous acceleration of multiple charge states of heavy ion beams in a large-scale superconducting linear accelerator. Phys. Rev. Lett. **126**, 114801 (2021). https://doi.org/10.1103/PhysRevLett.126.114801
114. T. Kanemura, M. LaVere, R. Madendorp, F. Marti, T. Maruta, Y. Momozaki, P.N. Ostroumov, A.S. Plastun, J. Wei, Q. Zhao, Experimental demonstration of the thin-film liquidmetal jet as a charge stripper. Phys. Rev. Lett. **128**, 212301 (2022). https://doi.org/10.1103/PhysRevLett.128.212301
115. H. Imao, H. Okuno, H. Kuboki, S. Yokouchi, N. Fukunishi, O. Kamigaito, H. Hasebe, T. Watanabe, Y. Watanabe, M. Kase et al., Charge stripping of U 238 ion beam by helium gas stripper. Phys. Rev. Spec. Top. Accel Beams **15**(12), 123501 (2012)
116. P. Adderley, S. Ahmed, T. Allison, R. Bachimanchi, K. Baggett, M. BastaniNejad, B. Bevins, M. Bevins, M. Bickley, R. Bodenstein et al., The continuous electron beam accelerator facility at 12 gev. Phys. Rev. Accel. Beams **27**(8), 084802 (2024)
117. T. Ludlam, L. McLerran, What have we learned from the relativistic heavy ion collider? Phys. Today **56**(10), 48–54 (2003)
118. E. Waagaard, R. Bruce, R. Alemany Fernández, H. Bartosik, J. Jowett, N. Triantafyllou, Charting the luminosity capabilities of the cern large hadron collider with various nuclear species. Nucl. Instrum. Methods Phys. Res. Sect. A: Accel. Spectrom. Detect. Assoc. Equip. **1083**, 171118 (2026). https://doi.org/10.1016/j.nima.2025.171118. https://www.sciencedirect.com/science/article/pii/S0168900225009209
119. H.A. Grunder, J. Bisognano, W. Diamond, B. Hartline, C. Leemann, J. Mougey, R. Sundelin, R. York, The continuous electron beam accelerator facility. Nucl. Phys. A **478**, 831–846 (1988)
120. E.W. Blackmore, A.S. Dowling, R. Ruegg, M.C. Stenning, Operating experience with meson production targets at TRIUMF, in *Proceedings of PAC'05, Particle Accelerator Conference*, no. 21 (JACoW Publishing, Geneva), pp. 1919–1921. https://jacow.org/p05/papers/RPPE026.pdf
121. B. Murin, A. Fedotov, V. Kulman, G. Batskikh, B. Polyakov, High-current ion linear accelerator for medium-energy physics (meson factory), in *Proceedings of Proton Linac Conference*, vol. 387 (1972), p. 1972
122. M. Grachev, V. Matveev, Y.Y. Stavisky, V. Lobashev, Y.P. Severgin, V. Glukhikh, V. Titov, Experimental area of moscow meson factory. Part. Accel. **27**, 51–58 (1990)
123. J.W. Xia, W.L. Zhan, B.W. Wei, Y.J. Yuan, P. Yuan, W.M. Qiao, D.Q. Gao, X.T. Yang, K.D. Man, G.Q. Xiao et al., HIRFL-CSR complex. Chin. Phys. C **33**(9), 804 (2009)
124. Y. Yano, The RIKEN RI beam factory project: a status report. Nucl. Instrum. Methods Phys. Res., Sect. B **261**(1–2), 1009–1013 (2007)
125. G. Bauer, Physics and technology of spallation neutron sources. Nucl. Instrum. Methods Phys. Res., Sect. **A463**(3), 505–543 (2001). https://doi.org/10.1016/S0168-9002(01)00167-X. https://www.sciencedirect.com/science/article/pii/S016890020100167X. Accelerator driven systems
126. G. Bauer, H. Sebening, J. Vetter, H. Willax, Realisierungsstudie zur spallationsneutronenquelle. Jül-Spez-113, KfK **3175**, 156 (1981)
127. The proposed snq spallation neutron source. CERN Courier **20**, 299 (1980)

128. G. Bauer et al., Targets for neutron beam spallation sources. Technical report, Kernforschungsanlage Juelich GmbH (Germany) (1980)
129. U.S. Department of Energy, New spallation neutron sources: their parameters and prospects, OSTI Report. Technical report (1981)
130. ESS Council, The European spallation source study, volume III: The ESS technical study. Technical Report ESS 96 53 M (1996)
131. ESS Council, The ESS project: volume II—new science and technology for the 21st century. Technical report, Jülich (2002). https://rencurel.essworkshop.org/documents/reports/VolII.pdf. ISBN 3 89336 302 5
132. ESS Council, The ESS project: volume III—technical report. Technical report, Jülich (2002). https://rencurel2006.essworkshop.org/documents/reports/VolIIIeng.pdf. ISBN 3 89336 303 3
133. W. Bräutigam, O. Felden, R. Maier, S. Martin, A. Schnase, G. Schug, R. Stassen, E. Zaplatine, Design considerations for the linac system of the ESS. Nucl. Instrum. Methods Phys. Res. Sect. B: Beam Interact. Mater. Atoms **161**, 1148–1153 (2000)
134. B. Heuel-Fabianek, K. Kühn, J. Vanderborght, M. Herbst, H. Schaal, Modeling the migration of activated radionuclides in groundwater at accelerator sites-an approach to meet safety and regulatory limits-proc, in *ICANS XVI*, Neuss (2003)
135. R. Moormann, R. Hanslik, J. Mertens, D. Findlay, R. Bongartz, S. Reiche-Begemann, H. Schaal, P. Fabi, K. Verfondern, P. Berkvens et al., Safety and licensing of the European Spallation Source (ESS). Technical report, Projekt Europäische Spallationsquelle (international) (2004)
136. S. Peggs (ed.), European spallation source technical design report (TDR). ESS-doc 274. Lund (2013). https://ess.eu/sites/default/files/downloads/2017/09/TDR_online_ver_all.pdf. ISBN 978 91 980173 2 8
137. T. Gutberlet, U. Rücker, E. Mauerhofer, P. Zakalek, T. Cronert, J. Voigt, J. Baggemann, J. Li, P. Doege, S. Böhm et al., Sustainable neutrons for today and tomorrow—the Jülich high brilliance neutron source project. Neutron News **31**(2–4), 37–43 (2020)
138. P. Zakalek et al., High-Brilliance Neutron Source Project, in *Proceedings of HIAT'18* (JACoW Publishing, Geneva), pp. 117–121. https://doi.org/10.18429/JACoW-HIAT2018-WEZAA01. https://jacow.org/hiat2018/papers/WEZAA01.pdf
139. J. Baggemann, M. Droba, O. Felden, T. Gutberlet, H. Kleines, K. Kümpel, S. Lamprecht, E. Mauerhofer, O. Meusel, I. Pechenitzkiy, N. Petry, S. Reimann, U. Rücker, M. Schwarz, P. Zakalek, C. Zhang, Technical Design Report HBS Volume 1—Accelerator, in *Schriften des Forschungszentrums Jülich Reihe Allgemeines*, vol. 9-1. Forschungszentrum Jülich GmbH Zentralbibliothek, Verlag, Jülich (2023). https://doi.org/10.34734/FZJ-2023-03722. https://juser.fz-juelich.de/record/1016730
140. G. Tagliente, O. Aberle, V. Alcayne, S. Amaducci, J. Andrzejewski, V. Babiano-Suarez, M. Bacak, J.B. Correa, A. Bernardes, E. Berthoumieux et al., The n TOF facility at cern, in *EPJ Web of Conferences*, vol. 297 (EDP Sciences, 2024), p. 01013
141. S.J. Blundell, R. De Renzi, T. Lancaster, F.L. Pratt, 200Making muons, in *Muon Spectroscopy: An Introduction* (Oxford University Press, 2021). https://doi.org/10.1093/oso/9780198858959.003.0013
142. P. Bonnaure, P. Mandrillon, H. Rief, H. Takahashi, *Actinide Transmutation by Spallation in the Light of Recent Cyclotron Development*, NEACRP-A-910 (1986). https://www.oecd-nea.org/upload/docs/application/pdf/2020-07/neacrp-a-1988-0910.pdf
143. T. Takizuka, T. Sasa, K. Tsujimoto, H. Takano, Studies on accelerator driven transmutation systems, in *Fifth International Information Exchange Meeting* (Citeseer, 1998), p. 383
144. C. Rubbia, J. Aleixandre, S. Andriamonje et al., A European roadmap for developing accelerator driven systems (ADS) for nuclear waste incineration. ENEA Report (2001)
145. H. Takahashi, The role of accelerators in the nuclear fuel cycle. Technical report, Brookhaven National Lab., Upton (1990)
146. C. Rubbia, C. Roche, J.A. Rubio, F. Carminati, Y. Kadi, P. Mandrillon, J.P.C. Revol, S. Buono, R. Klapisch, N. Fiétier et al., Conceptual design of a fast neutron operated high power energy amplifier. Technical report (1995)

147. J. Knaster, F. Arbeiter, P. Cara, P. Favuzza, T. Furukawa, F. Groeschel, R. Heidinger, A. Ibarra, H. Matsumoto, A. Mosnier et al., IFMIF: overview of the validation activities. Nucl. Fusion **53**(11), 116001 (2013)
148. J. Knaster, A. Ibarra, J. Abal, A. Abou-Sena, F. Arbeiter, F. Arranz, J. Arroyo, E. Bargallo, P. Beauvais, D. Bernardi et al., The accomplishment of the engineering design activities of ifmif/eveda: the European–Japanese project towards a Li (d, xn) fusion relevant neutron source. Nucl. Fusion **55**(8), 086003 (2015)
149. G. Federici et al., Overview of the DEMO staged design approach in Europe. Nucl. Fusion **59**(6), 066013 (2019). https://doi.org/10.1088/1741-4326/ab1178
150. R. Hemsworth, H. Decamps, J. Graceffa, B. Schunke, M. Tanaka, M. Dremel, A. Tanga, H., De Esch, F. Geli, J. Milnes et al., Status of the iter heating neutral beam system. Nucl. Fusion **49**(4), 045006 (2009)
151. V. Toigo, R. Piovan, S. Dal Bello, E. Gaio, A. Luchetta, R. Pasqualotto, P. Zaccaria, M. Bigi, G. Chitarin, D. Marcuzzi et al., The prima test facility: spider and mitica test-beds for iter neutral beam injectors. New J. Phys. **19**(8), 085004 (2017)
152. I. Shikhovtsev, A. Ivanov, V. Davydenko, Y. Belchenko, G. Abdrashitov, V. Belov, T. Akhmetov, V. Amirov, A. Brul, P. Deichuli et al., Overview of neutral beam injectors for plasma heating and diagnostics developed at budker inp. J. Plasma Phys. **90**(1), 155900301 (2024)
153. V. Kuyanov et al., Development of high-power NBIs at BINP and their applications for fusion devices. Plasma Phys. Control. Fusion **47**(5A), A231–A246 (2005). https://doi.org/10.1088/0741-3335/47/5A/018
154. I. Hofmann, Review of accelerator driven heavy ion nuclear fusion. Matter Radiat. Extremes **3**(1), 1–11 (2018). https://doi.org/10.1016/j.mre.2017.12.001
155. J.J.P. de Lima, Radioisotopes in medicine. Eur. J. Phys. **19**(6), 485 (1998)
156. M.R. Kilbourn, P.J. Scott, *Handbook of Radiopharmaceuticals: Methodology and Applications* (Wiley, New Jersey, 2021)
157. CERN-MEDICIS—Medical Isotopes Collected from ISOLDE. https://home.cern/science/experiments/medicis
158. C. Bernerd et al., CERN-MEDICIS: a unique facility for the production of radionuclides for medical research, in *Proceedings of IPAC'25, IPAC'25—16th International Particle Accelerator Conference*, no. 16 (JACoW Publishing, Geneva, 2025), pp. 1071–1074. https://doi.org/10.18429/JACoW-IPAC2025-TUPB047. https://indico.jacow.org/event/81/contributions/8482
159. W.H. Sweet, The uses of nuclear disintegration in the diagnosis and treatment of brain tumor. N. Engl. J. Med. **245**(23), 875–878 (1951)
160. A.D. Chanana, J. Capala, M. Chadha, J.A. Coderre, A.Z. Diaz, E.H. Elowitz, J. Iwai, D.D. Joel, H.B. Liu, R. Ma et al., Boron neutron capture therapy for glioblastoma multiforme: interim results from the phase I/II dose-escalation studies. Neurosurgery **44**(6), 1182–1192 (1999)
161. T.E. Blue, J.C. Yanch, Accelerator-based epithermal neutron sources for boron neutron capture therapy of brain tumors. J. Neuro-Oncol. **62**, 19–31 (2003)
162. R.R. Wilson, Radiological use of fast protons. Radiology **47**(5), 487–491 (1946). https://doi.org/10.1148/47.5.487. PMID: 20274616
163. Q. Li, X. Liu, Z. Dai, P. He, Y. Ma, G. Shen, X. Jin, F. Ye, X. Zheng, T. Zhao et al., Progress in heavy ion cancer therapy at imp and future development. Malignancy Spectrum **1**(2), 91–98 (2024)
164. J. Brenizer, A review of significant advances in neutron imaging from conception to the present. Phys. Proc. **43**, 10–20 (2013)
165. R. Gaillard, *Single Event Effects: Mechanisms and Classification* (Springer, Boston, 2011), pp. 27–54. https://doi.org/10.1007/978-1-4419-6993-4_2
166. M.A. McMahan, The berkeley accelerator space effects (base) facility–a new mission for the 88-inch cyclotron at lbnl. Nucl. Instrum. Methods Phys. Res. Sect. B: Beam Interact. Mater. Atoms **241**(1–4), 409–413 (2005)

167. G. Christian, C.M. Folden III, J. Hardy, Y.W. Lui, D. May, D. Melconian, J. Natowitz, R. Rapp, G. Rogachev, R. Tribble et al., The cyclotron institute at Texas A&M University. Nucl. Phys. News **27**(2), 5–13 (2017)
168. D. Lowenstein, A. Rusek, Technical developments at the nasa space radiation laboratory. Radiat. Environ. Biophys. **46**(2), 91–94 (2007)
169. C. Castaneda, Crocker nuclear laboratory (CNL) radiation effects measurement and test facility, in *2001 IEEE Radiation Effects Data Workshop. NSREC 2001. Workshop Record. Held in conjunction with IEEE Nuclear and Space Radiation Effects Conference* (Cat. No. 01TH8588) (IEEE, 2001), pp. 77–81
170. J. Mekki, M. Brugger, R.G. Alia, A. Thornton, N.D.S. Mota, S. Danzeca, Charm: a mixed field facility at cern for radiation tests in ground, atmospheric, space and accelerator representative environments. IEEE Trans. Nucl. Sci. **63**(4), 2106–2114 (2016)
171. HEARTS@CERN, https://hearts-project.eu/facilities/hearts@cern/
172. D. Söderström et al., Readiness of the HEARTS@CERN facility for space electronics highenergy heavy-ion testing. Presented at IPAC'25, Taipei, Jun 2025, paper TUPB006, Unpublished
173. S. Lidia, S. Cogan, K. Fukushima, T. Gee, S.H. Kim, G. Machicoane, T. Maruta, D. McNanney, P. Ostroumov, A. Plastun, X. Rao, S. Rodriguez, A. Yeck, Q. Zhao, A heavy-ion single-event effects test facility at Michigan State University, in *2022 IEEE Radiation Effects Data Workshop (REDW) (in conjunction with 2022 NSREC)* (2022), pp. 1–7. https://doi.org/10.1109/REDW56037.2022.9921718
174. A.J. Toepfer, A review of accelerator concepts for the advanced hydrotest facility (1998)
175. R.L. Sheffield, C.W. Barnes, J.P. Tapia, Matter-radiation interactions in extremes (MaRIE): project overview. Technical report, Los Alamos National Laboratory (LANL), Los Alamos (2017)
176. M. Wilson, T. Bhatia, F. Guy, G. Neuschaefer, T. Wangler, L. Young, Accelerator for the Production of Tritium (APT), in *Proceedings of the 1989 IEEE Particle Accelerator Conference. Accelerator Science and Technology* (IEEE, 1989), pp. 761–763
177. P.W. Lisowski, Accelerator production of tritium: a review and look ahead (2014)
178. A. Sari, Characterization of photoneutron fluxes emitted by electron accelerators in the 4–20 MeV range using Monte Carlo codes: a critical review. Appl. Radiat. Isot. **191**, 110506 (2023). https://doi.org/10.1016/j.apradiso.2022.110506. https://www.sciencedirect.com/science/article/pii/S0969804322003918
179. M. Furusaka, H. Sato, T. Kamiyama, M. Ohnuma, Y. Kiyanagi, Activity of Hokkaido University Neutron Source, HUNS. Phys. Proc. **60**, 167–174 (2014). https://doi.org/10.1016/j.phpro.2014.11.024. https://www.sciencedirect.com/science/article/pii/S1875389214005719. 3rd International Meeting of the Union for Compact Accelerator-driven Neutron Sources, UCANS III, 31 July–3 Aug 2012, Bilbao & the 4th International Meeting of the Union for Compact Accelerator-driven Neutron Sources, UCANS IV, 23–27 Sept 2013, Sapporo
180. D.V. Baxter, J. Cameron, M. Leuschner, H. Meyer, H. Nann, W. Snow, LENS—a pulsed neutron source for education and research. Nucl. Instrum. Methods Phys. Res., Sect. A **542**(1), 28–31 (2005). https://doi.org/10.1016/j.nima.2005.01.007. https://www.sciencedirect.com/science/article/pii/S0168900205001166. Proceedings of the Fifth International Topical Meeting on Neutron Radiography (Proc. ITMNR-5)
181. D. Möhl, A.M. Sessler, Beam cooling: principles and achievements. Nucl. Instrum. Methods Phys. Res., Sect. A **532**(1), 1–10 (2004). https://doi.org/10.1016/j.nima.2004.06.102. https://www.sciencedirect.com/science/article/pii/S0168900204011477. International Workshop on Beam Cooling and Related Topics
182. J. Wei, H. Okamoto, A.M. Sessler, Necessary conditions for attaining a crystalline beam. Phys. Rev. Lett. **80**, 2606–2609 (1998). https://doi.org/10.1103/PhysRevLett.80.2606
183. J. Hangst, K. Berg-Sørensen, P. Jessen, M. Kristensen, K. Mølmer, J. Nielsen, P. Poulsen, J. Schiffer, P. Shi, Laser cooling of stored ions in ASTRID: a storage ring for ions and electrons. Nucl. Instrum. Methods Phys. Res., Sect. B **68**(1), 17–22 (1992). https://doi.org/10.1016/0168-583X(92)96042-W. https://www.sciencedirect.com/science/article/pii/0168583X9296042W

184. L. Nilsson, The Svedberg Laboratory, Uppsala, Sweden. Nucl. Phys. News **1**(1), 10–12 (1990)
185. C.J. Herrlander, A. Bárány, K.G. Rensfelt, J. Starker, The CRYRING project. Phys. Scr. **1988**(T22), 282 (1988). https://doi.org/10.1088/0031-8949/1988/T22/042
186. K. Abrahamsson, G. Andler, L. Bagge, E. Beebe, P. Carlé, H. Danared, S. Egnell, K. Ehrnstén, M. Engström, C. Herrlander et al., Cryring—a synchrotron, cooler and storage ring. Nucl. Instrum. Methods Phys. Res. Sect. B: Beam Interact. Mater. Atoms **79**(1–4), 269–272 (1993)
187. R. Pollock, T. Ellison, D. Friesel, P. Schwandt, W. Jones, B. Brown, O. Dermois, J. Self, B. Starks, M. Ball et al.: The IUCF Cooler Project (1986)
188. A. Noda, Ion beam cooling at S-LSR project. Nucl. Instrum. Methods Phys. Res., Sect. A **532**(1), 150–156 (2004). https://doi.org/10.1016/j.nima.2004.06.042. https://www.sciencedirect.com/science/article/pii/S016890020401201X. International Workshop on Beam Cooling and Related Topics
189. S. Schröder, R. Klein, N. Boos, M. Gerhard, R. Grieser, G. Huber, A. Karafillidis, M. Krieg, N. Schmidt, T. Kühl, R. Neumann, V. Balykin, M. Grieser, D. Habs, E. Jaeschke, D. Krämer, M. Kristensen, M. Music, W. Petrich, D. Schwalm, P. Sigray, M. Steck, B. Wanner, A. Wolf, First laser cooling of relativistic ions in a storage ring. Phys. Rev. Lett. **64**, 2901–2904 (1990). https://doi.org/10.1103/PhysRevLett.64.2901
190. S. Antipov, D. Broemmelsiek, D. Bruhwiler, D. Edstrom, E. Harms, V. Lebedev, J. Leibfritz, S. Nagaitsev, C. Park, H. Piekarz, P. Piot, E. Prebys, A. Romanov, J. Ruan, T. Sen, G. Stancari, C. Thangaraj, R. Thurman-Keup, A. Valishev, V. Shiltsev, IOTA (Integrable Optics Test Accelerator): facility and experimental beam physics program. J. Instrum. **12**(03), T03002 (2017). https://doi.org/10.1088/1748-0221/12/03/T03002
191. L. Young, L. Rybarcyk, J. Schneider, M. Schulze, H. Smith, High power operations of LEDA. arXiv preprint physics/0008158 (2000)
192. L. Young, 25 years of technical advances in RFQ accelerators, in *Proceedings of PAC'03*, vol. 1 (IEEE, 2003), pp. 60–64
193. Y. He et al., Accelerator driven system—a solution to multiple problems of society, in *Proceedings of IPAC'23, International Particle Accelerator Conference*, no. 14 (JACoW Publishing, Geneva, 2023), pp. 5205–5209. https://doi.org/10.18429/JACoW-IPAC2023-FRYG2
194. R. Barlow et al., EMMA–Theworld's first non-scaling FFAG. Nucl. Instrum. Methods Phys. Res., Sect. A **624**(1), 1–19 (2010). https://doi.org/10.1016/j.nima.2010.08.109. https://www.sciencedirect.com/science/article/pii/S0168900210018978
195. S. Machida, R. Barlow, J. Berg, N. Bliss, R. Buckley, J. Clarke, M. Craddock, R. D'Arcy, R. Edgecock, J. Garland et al., Acceleration in the linear non-scaling fixed-field alternatinggradient accelerator EMMA. Nat. Phys. **8**(3), 243–247 (2012). https://doi.org/10.1038/nphys2179
196. T. Iwashita et al., KEK digital accelerator. Phys. Rev. Spec. Top. Accel. Beams **14**(7), 071301 (2011)
197. K. Takayama, T. Yoshimoto, M. Barata, L.K. Wah, L. Xingguang, T. Iwashita, S. Harada, T. Adachi, T. Arai, D. Arakawa, H. Asao, E. Kadokura, T. Kawakubo, H. Nakanishi, Y. Okada, K. Okamura, K. Okazaki, A. Takagi, S. Takano, M. Wake, Induction acceleration of heavy ions in the kek digital accelerator: demonstration of a fast-cycling induction synchrotron. Phys. Rev. ST Accel. Beams **17**, 010101 (2014). https://doi.org/10.1103/PhysRevSTAB.17.010101
198. T. Schätz, D. Habs, C. Podlech, J. Wei, U. Schramm, Towards crystalline ion beams—the PALLAS ring trap. AIP Conf. Proc. **457**(1), 269–273 (1999). https://doi.org/10.1063/1.57465
199. H. Okamoto, Y. Wada, R. Takai, Radio-frequency quadrupole trap as a tool for experimental beam physics. Nucl. Instrum. Methods Phys. Res., Sect. A **485**(3), 244–254 (2002)
200. H. Okamoto, M. Endo, K. Fukushima, H. Higaki, K. Ito, K. Moriya, S. Yamaguchi, S. Lund, Experimental simulation of beam propagation over long path lengths using radiofrequency and magnetic traps. Nucl. Instrum. Methods Phys. Res. Sect. A: Accel. Spectrom. Detect. Assoc. Equip. **733**, 119–128 (2014)
201. K. Izawa, K. Ito, H. Higaki, H. Okamoto, Controlled extraction of ultracold ions from a linear paul trap for nanobeam production. J. Phys. Soc. Japan **79**(12), 124502 (2010)

202. Wikimedia, Particle accelerator. https://en.wikipedia.org/wiki/Particle_accelerator#cite_note-3
203. P. Wanderer, J. Cottingham, G. Ganetis, M. Garber, A. Ghosh, C. Goodzeit, A. Greene, R. Gupta, J. Herrera, S. Kahn et al., Dipole magnet development for the RHIC accelerator, in *New Techniques for Future Accelerators III: High-Intensity Storage Rings–Status and Prospects for Superconducting Magnets* (Springer, 1990), pp. 175–187
204. J. Wei, M. Harrison, The RHIC project–design, status, challenges, and perspectives, in *Proceedings of 16th RCNP Osaka International Symposium on Multi-GeV High-Performance Accelerators and Related Technology* (1997), pp. 198–206
205. C.H. Rode, The SNS superconducting Linac system, in *Proceedings of PAC'01* (JACoW Publishing, Geneva), pp. 619–623. https://jacow.org/p01/papers/ROPB008.pdf
206. H. Ao et al., Status of mass production of the ACS cavity for the J-PARC linac energy upgrade, in *Proceedings of IPAC'10* (JACoW Publishing, Geneva), pp. 618–620. http://accelconf.web.cern.ch/IPAC10/papers/MOPEC066.pdf

Chapter 2
Accelerator Design and Technical Aspects

Abstract This chapter provides a high-level overview of three major classes of high-power accelerators: cyclotrons and other fixed-field machines, superconducting and normal-conducting linear accelerators, and circular accelerators used for beam accumulation and rapid acceleration. We summarize the underlying design principles and the principal technical considerations that govern injection, acceleration, extraction, and reliable high-intensity operation.

In this chapter, we describe the design principles and main technical aspects of three main types of high-power accelerators: cyclotrons and fixed-field accelerators (Sect. 2.1), linear accelerators (Sect. 2.2), and accumulators and rapid-cycle synchrotrons (Sect. 2.3). The purpose is to provide a coherent "big picture" framework and practical starting points for early design choices, rather than an exhaustive treatment of each subsystem. Detailed descriptions of the most critical topics are developed in subsequent chapters.

2.1 Cyclotrons and Fixed-Field Accelerators

The first acceleration of protons up to 1.2 MeV in the cyclotron was demonstrated by Lawrence and Livingston in 1931 [1]. Since then, the cyclotrons have undergone extensive technical modifications. Many new circular accelerators belonging to the cyclotron family have been invented, developed, built, and are now in operation [2].

A simplified picture of the classical cyclotron is shown in Fig. 2.1. A bunch of ions is injected into the center of the magnet and moves along a circular trajectory under the magnetic field force. The orbit radius depends on the particle momentum along the trajectory and increases as the particle gains energy in the accelerating gaps between the so-called DEEs. The accelerating RF electric field is applied between the DEEs. Therefore, if the RF frequency is equal to the revolution frequency of the particle, the particle energy increases synchronously after each gap is passed. The particles follow a spiral trajectory until they reach the final energy, at

J. Wei et al., *High Power Hadron Accelerators*, Particle Acceleration and Detection,
https://doi.org/10.1007/978-3-032-17662-2_2

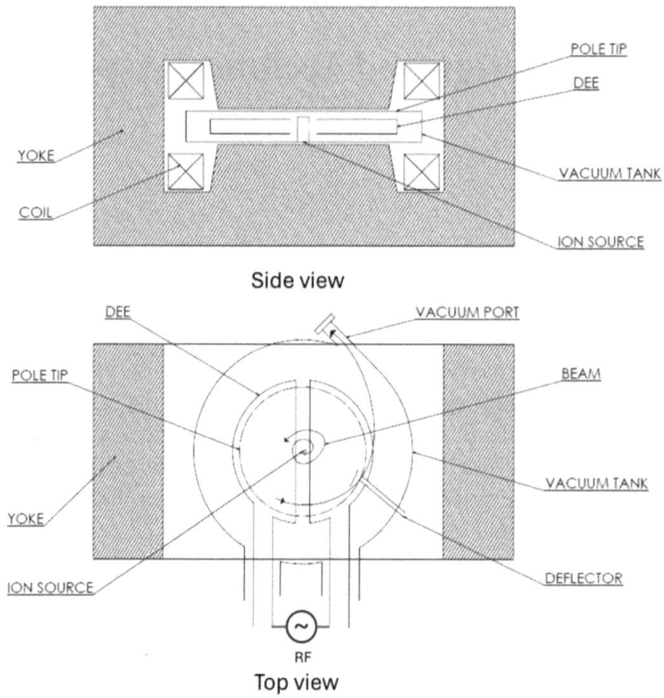

Fig. 2.1 A simplified illustration of the classical cyclotron (weak focusing and without sector structure)

which point the beam is extracted from the magnetic field and sent to the target or to the next stages of acceleration.

Cyclotron-family accelerators have been built in many ways, ranging from small, dedicated accelerators for isotope production or applications in other fields to large facilities for basic research in nuclear and particle physics. It should be noted that the cyclotron held the world record for the highest-power proton beam of 1.4 MW at energies <1 GeV until 2018, when the linac at ORNL began producing the 1.4 MW proton beam routinely [3]. Although the cyclotron is a very efficient accelerator for transforming wall plug power into beam power, several factors can limit the beam power. At the same time, upgrading the existing linac to a higher power is straightforward; for example, ORNL is upgrading the facility to produce a 2.8 MW beam.

The revolution or cyclotron frequency ω_c of a particle with charge qe and rest mass Am_0 in a cyclotron is given by

$$\omega_c = \frac{qeB_z}{Am_0\gamma} , \tag{2.1}$$

where B_z is the vertical component of the magnetic field, A is the mass number, and q is the charge number. In cyclotrons, it is common to use a cylindrical coordinate system r, z, θ to describe the motion of the particles. The frequency of the accelerating voltage should resonate with the revolution frequency,

$$\omega_{\mathrm{RF}} = h\omega_c , \tag{2.2}$$

where the integer harmonic number h defines the number of bunches that can be accelerated in one turn. The orbit radius, R, is defined by the momentum of the particles and the magnetic field:

$$R = \frac{Am_0 c\beta\gamma}{qeB_z} . \tag{2.3}$$

As seen in Eq. 2.1, if the magnetic field B_z is constant as a function of radius, the revolution frequency decreases as the beam energy increases. In the classical cyclotron where $B_z(r)$ and $\omega_{\mathrm{RF}}(t)$ are both constant, the acceleration ceases when the resonance condition (Eq. 2.2) is violated due to the relativistic increase in mass. In order for the cyclotron to remain isochronous beyond about 12 MeV for protons, the magnetic field must increase with radius in such a way as to be proportional to gamma. But this would vertically defocus the circulating beam. This limitation was overcome in the 1950s by incorporating the azimuthal variation of the magnetic field and further splitting it into sectors, spiraling those sectors, thus effectively utilizing strong focusing.

The maximum energy in the cyclotron is defined by the available magnetic field B_z, the outer orbit radius R_c and the charge-to-mass ratio q/A. The maximum kinetic energy $E_{k,\max}$ per nucleon is related to the so-called K value [4],

$$\frac{E_{k,\max}}{A} = K_{\mathrm{CYC}} \left(\frac{q}{A}\right)^2 , \tag{2.4}$$

with

$$K_{\mathrm{CYC}} = \left[\frac{\langle B_z R_c\rangle}{3.1\,[\mathrm{T\,m}]}\right]^2 \frac{931\,[\mathrm{MeV}]}{\gamma + 1} , \tag{2.5}$$

where γ is the relativistic factor of the particle in the exit orbit, and $\langle\ \rangle$ indicates the average value. The non-relativistic K value (K_{CYC} in Eq. 2.5 with $\gamma = 1$) is commonly used to characterize the kinetic energy reach of a cyclotron.

If a cyclotron is designed for the acceleration of heavy ions to certain $E_{k,\max}/A$, the light ions with a greater q/A can be accelerated to substantially higher energies. The same ion beam, with a higher charge state, can also be accelerated to higher energies. This feature is commonly used in heavy-ion accelerator facilities, such as the chain of coupled cyclotrons at MSU [5], and RIKEN [6], where charge stripping is utilized between the cyclotrons.

For a classical cyclotron, the stability of betatron motion is provided by a small negative power law field gradient as a function of radius r, with a reference radius r_0 and field exponent index κ_c.

$$B_z(r) = B_{z0}\left(\frac{r}{r_0}\right)^{-\kappa_c}, \tag{2.6}$$

where B_{z0} is the magnetic field at radius $r = r_0$. The vertical and radial betatron motions are stable if $1 > \kappa_c > 0$ and their betatron tunes are:

$$\nu_z^2 = \kappa_c, \quad \nu_r^2 = 1 - \kappa_c. \tag{2.7}$$

The recent publication of Baartman [4] provides more details on the beam dynamics in fixed-field accelerators and cyclotrons.

2.1.1 *Main Types of Cyclotrons*

Accelerators in the cyclotron family can be defined as circular accelerators where the magnetic field is time-constant (fixed field), in contrast to synchrotrons.

2.1.1.1 Synchrocyclotrons

The modification of classical cyclotrons to higher energies was made possible by introducing synchrocyclotrons. In synchrocyclotrons, the magnetic field decreases as a function of radius, thereby providing vertical stability for particle motion. Acceleration to higher energy occurs by changing the RF frequency to satisfy the resonance condition (Eq. 2.2). Such an acceleration is possible in a pulsed mode, whereas the classical cyclotron is a CW machine. The first synchrocyclotron was built at LBNL in 1946 to exceed the classical cyclotron limit [7] and to accelerate deuterons and α particles to 95 MeV/u. The highest energy synchrocyclotron was built in Saint Petersburg, Russia, to produce 1 GeV protons [8]. Repetition rates for frequency modulation ranged from 50 Hz to 4 kHz with a beam pulse length of 10 to 100 μs. The duty factor and beam intensity were about 1% of that of a CW cyclotron. The longitudinal stability of the beam was provided by conventional phase focusing; therefore, the DEE voltage was low, and the acceleration time was longer than that of classical cyclotrons.

2.1.1.2 Azimuthally-Varying Fields (AVF) or Sector-Focused Cyclotrons

The energy limitation in classical cyclotrons can be overcome by introducing isochronous cyclotrons in which the magnetic field increases as a function of the

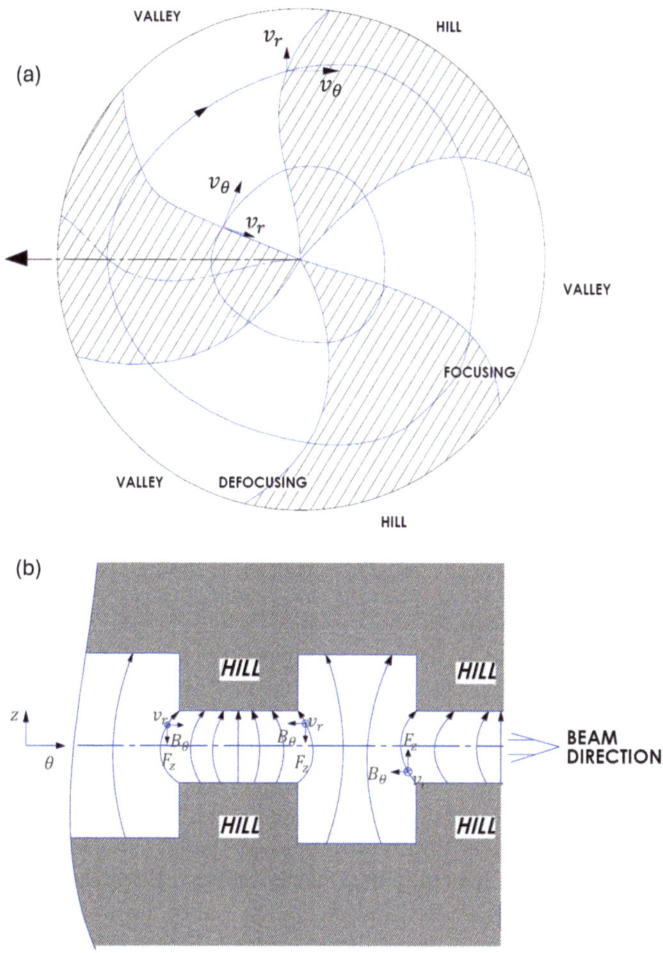

Fig. 2.2 (**a**) Equilibrium orbits in a AVF cyclotron. The inner orbit (i) experiences Thomas focusing from radial sectors; orbit (ii) also experiences strong focusing from spiral sectors. (**b**) Vertical forces seen by orbit (i) at sector boundaries [12]

radius. However, it was quickly realized that this magnetic field cannot provide stability for the vertical motion of particles. Azimuthally-varying fields (AVF) were introduced to ensure focusing, even in the case of radially increasing fields. Most modern cyclotrons, including the highest power PSI cyclotron, are built with azimuthally varying magnetic fields. The AVF concept was proposed by Thomas in 1938 [9], but the first AVF cyclotron was not built and operated until 1958 [10]. In 1954, Kerst [11] invented spirally ridged sector magnets to achieve additional strong focusing.

The concept of AVF with spirally ridged sector magnets can be explained using Fig. 2.2a, which shows the typical pole configuration and vertical focusing forces.

The magnetic field variation along the azimuth θ is introduced by installing spiral-shaped iron sectors on both poles, creating hills and valleys along the beam path, as shown in Fig. 2.2b. The particles have a smaller bending radius in the hills and a larger radius in the valleys. The variation of the magnetic field is characterized by the flutter parameter F defined as

$$F^2 = \frac{\bar{B_z^2} - \bar{B_z}^2}{\bar{B_z}^2} , \tag{2.8}$$

where $\bar{B}_z$ is the average vertical magnetic field in the median plane over the azimuthal range of 360°, and $\bar{B_z^2}$ is the average of the square of this field. The betatron tunes in the AVF can be expressed with the flutter parameter and the field index κ_c

$$\nu_z^2 = \kappa_c + F^2 \left(1 + 2\tan^2 \xi_c\right) ; \tag{2.9}$$

$$\nu_r^2 = 1 - \kappa_c + F^2 \left(1 + \tan^2 \xi_c\right) , \tag{2.10}$$

where ξ_c is the spiral angle of the cyclotron magnet pole. The first terms in parentheses of Eqs. 2.9 and 2.10 are due to Thomas focusing [9]. The $\tan^2 \xi_c$ terms are due to strong focusing, as they arise from the spiral angle. Strong focusing is achieved by having the edges of the sectors alternating between focusing and defocusing [11].

As shown in Fig. 2.2, the shape of the particle orbit generates a radial velocity component v_r at the edges of the sector, which provides focusing in the vertical plane. This focusing term is sufficient to provide stable vertical betatron motion despite a defocusing term related to the radially increasing average magnetic field. Additional focusing for higher energies can be achieved by having the sectors follow a spiral rather than a radial locus, as shown in Fig. 2.2a [13]. The concept of AVF is utilized in many large cyclotrons for the production of high-power and relatively high-energy beams.

2.1.1.3 Separated-Sector Cyclotrons

A crucial step to increase the power of cyclotron beams was a proposal to modify AVFs to separated sector magnets, which means that there is no magnetic field in the valleys [14]. In addition, separated sector cyclotrons provide stronger focusing in both horizontal and vertical planes due to a larger flutter parameter, F, as follows from Eqs. 2.9 and 2.10. The space between the magnets provides the possibility of installing powerful RF resonators to obtain a very high accelerating voltage. The latter strongly suppresses the longitudinal space charge forces, facilitates a large orbit separation, and allows beam extraction to be realized with very low beam

Fig. 2.3 Top view of the PSI K592 separated-sector cyclotron. (Courtesy PSI, Switzerland)

losses. The injection energy to the separated-sector cyclotrons should be higher than that which the DC injectors can provide. Therefore, the separated-sector cyclotrons require an injector-accelerator that can be a smaller cyclotron. The highest beam power of 1.4 MW is routinely produced in the PSI separated-sector cyclotron (see Fig. 2.3). The relative beam extraction losses in this cyclotron are below 10^{-4}.

2.1.1.4 Superconducting Cyclotrons

The cyclotrons, especially those for the acceleration of heavy-ion beams, require large, heavy-weight magnets. The use of SC magnets was a natural way to reduce the weight of the magnet and build compact heavy-ion cyclotrons [15]. The first SC cyclotron, K500, was built at NSCL at MSU and began beam production in 1979 [16]. The weight of the magnet was reduced by a factor of 17 compared to an NC cyclotron at a similar energy. Before the end of operations in 2022, the NSCL housed the coupled cyclotron facility, consisting of two SC cyclotrons, K500 and K1200. This facility can produce up to 25 MeV/u uranium beams and 140 MeV/u light ion beams [17] due to the stripping injection into the second cyclotron.

A more recent successful cyclotron project is the construction of the Superconducting Ring Cyclotron (SRC) at RIKEN, Japan, which delivered the first beam in 2006. Table 2.1 lists selected parameters for large cyclotrons. The photo of the largest SC cyclotron in the world is shown in Fig. 2.4.

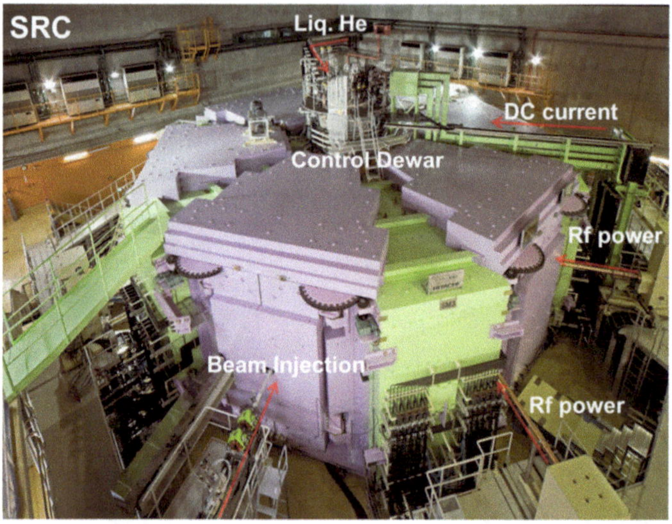

Fig. 2.4 RIKEN's K2600 Superconducting Ring Cyclotron. (Courtesy RIKEN, Japan). The maximum magnetic field is 3.8 T. The RF frequency is 18–38 MHz. The weight is about 8,300 tons. The diameter is about 19 m. The height is about 8 m

Table 2.1 Selected parameters of large sector cyclotrons. The TRIUMF cyclotron uses a single magnet with sector poles. The maximum beam power of the RIKEN cyclotron depends from ion species or operation modes. The maximum demonstrated power of krypton beam is shown in table

Cyclotron	K [MeV]	Type	h	R_{inj} [m]	R_{ext} [m]	Extraction method	Power [kW]	Reference
TRIUMF	520	NC	5	0.25	3.8–7.9	H^- stripping	150	[18]
MSU	1200	SC	1	0.33	1.01	Electrostatic septum	~1	[5]
Dubna	625	NC	2, 4, 6	0.25	2.0	Heavy ion stripping	~1	[19]
PSI	592	NC	6	2.1	4.5	Electrostatic septum	1400	[20]
RIKEN	2600	SC	6	3.6	5.4	Electrostatic septum	13	[6]

2.1.1.5 Fixed-Field Alternating-Gradient Accelerators (FFA)

An FFA is an accelerator that utilizes alternating-gradient focusing, similar to a synchrotron, with fixed field magnets like a cyclotron. An FFA was invented in the 1950s and extensively studied at the Midwestern Universities Research Association (MURA) [2, 21–24]. The first FFA, an electron model named the Michigan Mark I FFA constructed at MURA, accelerated electrons up to 400 keV. However, there was not much progress after MURA was dissolved in the 1960s until the late 1990s, when it was considered a fast accelerator of the muon beam and a possible high-intensity hadron accelerator [25–27]. Figure 2.5 shows a cascade of three FFAs constructed at Kyoto University for proton acceleration [26, 28].

Fig. 2.5 A cascade of three proton FFAs. It was built as a proton driver up to 150 MeV for Accelerator Driven Sub-critical Reactor at the Institute for Integrated Radiation and Nuclear Science at Kyoto University [26, 28]. (Courtesy Y. Ishi, Kyoto University)

Instead of keeping the vertical and radial betatron motions stable simultaneously by imposing κ_c in Eq. 2.6 between 0 and 1, FFA uses a magnet with a large geometrical field index k,

$$k = \frac{r}{B_z}\frac{dB_z}{dr} ,$$

where k can be either positive or negative, but $|k| \gg 1$ is a constant. Alternating-gradient focusing is introduced by the azimuthally varying function $F(\theta)$,

$$B_z(r, \theta) = B_{z0}\left(\frac{r}{r_0}\right)^k F(\theta) . \tag{2.11}$$

In a magnet with a positive k, when $F(\theta)$ is positive, the beam is bent inward and focused in the horizontal direction; when $F(\theta)$ is negative, the beam is bent outward and defocused. When k is negative, the focusing and defocusing are exchanged.

Focusing and defocusing can be enhanced by a spirally ridged sector magnet [11], as in an AVF cyclotron. This is called a spiral sector FFA; and the one without a spiral ridge is called a radial sector FFA. In both FFA, Eq. 2.11 is satisfied; however, a spiral sector FFA measures the azimuthal coordinate not along the angle of the cylindrical coordinate system, but rather a new angle defined as $\vartheta = \theta - \tan\xi_c \ \ln(r/r_0)$, where ξ_c is a spiral angle. A spiral sector FFA whose defocusing relies on the spiral ridge can eliminate a magnet that bends the beam outward. That helps reduce the circumference factor, defined as the ratio of the average radius to the bending radius, thus making an accelerator smaller.

A constant geometrical field index k throughout the accelerator lattice makes the equilibrium orbits, corresponding to each momentum, photographic enlargement, and effective focusing strength independent of the beam momentum. As a result, the transverse tunes are constant during acceleration; this is ideal for avoiding a nearby resonance crossing. This type of FFA is called a scaling (radial or spiral sector) FFA. On the other hand, the momentum compaction factor is determined by k as $\alpha_c = 1/(k+1)$ and the time of flight depends on the beam momentum. An FFA no longer satisfies the isochronous condition. The RF frequency for acceleration must be modulated like a synchrocyclotron, and the duty factor is lower than that of a CW cyclotron. To recover the beam intensity reduced by the duty factor, a design study to increase the bunch intensity in an FFA is underway [29].

As suggested in the original paper [21, 22], having a magnetic field profile as described by Eq. 2.11 is not only a means of keeping the transverse tunes constant during acceleration. A magnetic field profile that provides the constant tunes can be found through numerically optimized iterative processes of magnet design, as well as orbit and optics calculations [30]. For some operations where the beam does not circulate for many turns, constant tunes are not required because the acceleration process is supposed to complete before any deterioration of the beam occurs due to a resonance crossing. Without the requirements of constant tunes, magnet design can be simpler: a DC quadrupole magnet serves as an extreme case, as EMMA [31] (Fig. 1.29). An FFA that does not follow a field profile of Eq. 2.11 is called a non-scaling FFA.

An FFA, as a high power hadron accelerator, has the potential to add unique characteristics to the beam. Beam stacking, for example, can increase bunch intensity, which is usually limited by space charge effects at injection. By keeping the accelerated beams at top energy and merging the beams from several acceleration cycles into one, high power and high peak intensity are obtained with a low repetition rate. That is possible because space charge effects are inversely proportional to $\beta\gamma^2$ when the normalized emittance remains constant; thus, they are much less at the top energy [32]. From the high power beam dynamics point of view, it is always easier to increase the repetition rate and reduce the bunch intensity than to obtain the same beam power by doing the reverse. On the other hand, the users of the high power beam do not necessarily accept high repetition. An accelerator-driven spallation neutron source is a typical example where the time of flight measurement imposes a neutron pulse rate of less than 50 Hz. An FFA is a possible solution to increase the beam power with high repetition and to deliver the beam to users with low repetition using the beam stacking technique.

There are several challenges to making an FFA as a high power hadron accelerator as well [33]. The injection process is similar to that in a synchrotron. Multi-turn charge exchange injection with phase space painting is planned to create a charge distribution in phase space that reduces space charge effects. As in a synchrotron, a bump orbit destroys the high symmetry of the optics and reduces it to one fold. Because of the inherent nonlinearities in the main lattice magnet, the excitation of nonlinear systematic resonances and the reduction of the dynamic aperture become issues. Transverse tune control during acceleration is achieved through the geometric profile of the magnetic field. Detailed modeling of a magnet in 3-D space using a finite element method has progressed significantly. An FFA that only requires a DC magnet has the advantage of precise modeling. However, it still requires significant effort to design a magnet that provides the ring tune excursion within 0.01. Considering that the space charge tune spread would be several times 0.1, it is critical to operate the machine with that level of tune accuracy. An equilibrium orbit moves with acceleration. That makes it possible to set a collimator aperture following the beam size with adiabatic damping. In principle, continuous collimation works in the vertical direction and during injection and extraction in the horizontal direction. Radiation and thermal analysis of the full energy collimation are underway. In a scaling FFA, transverse tunes are adjusted by the field index k and the function of $F(\theta)$ of Eq. 2.11. Since the orbit excursion is also a function of k,

$$r_{\text{ex}} - r_{\text{in}} = \left[\left(\frac{p_{\text{ex}}}{p_{\text{in}}}\right)^{1/(k+1)} - 1\right] r_{\text{in}} \, , \tag{2.12}$$

where r_{in} and r_{ex} are the orbit radii at injection and extraction, and p_{in} and p_{ex} are the beam momenta at injection and extraction, respectively, changing the optics requires adjusting the position of either the injection or extraction system. Considering the complexity of the system involving the injection painting and bump orbit, it would be better to fix the injection position and provide a small movable mechanism for the extraction system. If the users allow, it is also possible to maintain the same position of the injection and extraction system by slightly adjusting the extraction momentum so that the orbit excursion remains unchanged.

2.1.2 Injection

There are many different methods for beam injection into cyclotrons. The choice of the beam injection method depends on the type and purpose of the cyclotron. Internal ion sources were utilized in the early stages of cyclotron development. Although internal sources are cost-efficient and still widely used in commercial cyclotrons, they have many limitations. These sources can only provide low or moderate beam intensities due to the limited space in the central region of the cyclotron and non-optimal beam injection into the 6-D phase space acceptance. In heavy ion machines, the internal sources cannot provide a high charge state of ion beams. In addition, maintaining internal sources requires venting and opening the cyclotron.

Two basic types of external injection are utilized in cyclotrons: radial and axial. In axial injection, the beam is transported along the central axis of the cyclotron and bent into the median plane by an electrostatic or magnetostatic inflector. The design of such an inflector is a complicated task because of the 3-D trajectory of the beam in the combined static EM fields of the inflector and the cyclotron. An inflector based on an electrostatic mirror was simple to build and was utilized in the first cyclotrons. This type of inflector cannot be used for high current beams due to the presence of a grid. The gridless spiral inflectors originally proposed by Belmont and Pabot [34] soon replaced the electrostatic mirror as the inflector of choice. The spiral inflector consists of two electrodes that create an electric field shaped such that the field is perpendicular to the orbit. Injection of high-intensity beams into the central region of cyclotrons is limited by space charge due to the lack of strong focusing. The details of the beam dynamics in the first turns are discussed in [35], along with the references provided in that publication.

In the radial injection, the beam is injected into the midplane of the cyclotron through a hole in the yoke. The concept of radial injection through the yoke is utilized in the coupled cyclotron facility at MSU for injection into the second cyclotron. Radial injection is supplemented with ion stripping to achieve higher charge states, which helps simplify the design of the injection hole in the yoke. Radial injection is greatly simplified in separated-sector cyclotrons. An electrostatic deflector is utilized to capture the beam in the cyclotron orbit. In large separated cyclotrons, such as the PSI cyclotron, the 72 MeV high-intensity proton beam is transported in a conventional transport system based on quadrupole focusing and bending magnets upstream of the electrostatic inflector (see Fig. 2.6). Due to the large separation of neighboring orbits in the PSI cyclotron, radial injection is possible without notable beam losses.

It should be noted that, regardless of the injection method, the injected beam is highly dispersive. This makes bunching problematic, and often a large amount of dumped beam power must be absorbed or collected at injection. Dumping a beam on the first turn leads to practical intensity limitations. Such dumping can erode the RF electrodes and require frequent maintenance, especially in compact cyclotrons where space is limited [36]. This issue prevents the use of small ($\sim$1 mA, low injection energy), compact cyclotrons as injectors for large projects that require CW beams.

Compared to synchrotrons, injection in a cyclotron is usually easier because the magnetic field is fixed and the injected particles accelerate away from the injection orbit. The first-turn orbit separation is typically many times larger than the beam diameter.

2.1.3 Acceleration

In an isochronous cyclotron, the revolution frequency is the same for all particles at all radii, and there is no phase-focusing effect. A small deviation in the magnetic field along the beam trajectory can cause an excessive change in the particle phase, thus violating the resonance. The energy gain per turn ΔW is given by

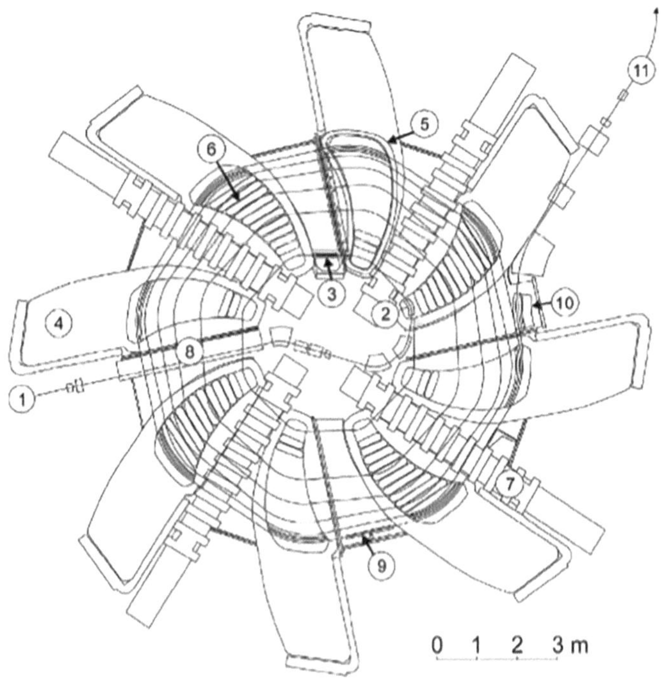

Fig. 2.6 The PSI separated-sector cyclotron. This machine accelerates protons from 72 to 590 MeV. (1) Beam line from 72 MeV injector cyclotron, (2) magnetic inflector channel, (3) electrostatic inflector, (4) one of eight sector magnets, (5) main coil, (6) set of correcting (trim and harmonic) coils wound on a magnet pole, (7) one of four 50 MHz accelerating cavities, (8) 150 MHz "flat-top" cavity, (9) beam diagnostic probe (radial drive), (9) electrostatic deflector, (10) septum magnet, and (11) extraction beam line at 590 MeV. (Courtesy PSI, Switzerland)

$$\Delta W = qeV_0 \cos\phi \,, \tag{2.13}$$

where V_0 is the peak voltage per turn, and ϕ is the RF phase. In order to maintain the resonance condition, Eq. 2.2, during the acceleration process, the particle phase should remain within $\pm 90°$:

$$\sin\phi - \sin\phi_0 = \frac{2\pi h_{\mathrm{RF}}}{qeV_0} \int_{W_i}^{W} \frac{\omega_{c0} - \omega(W)}{\omega(W)} dW. \tag{2.14}$$

The deviation of the revolution frequency from the initial frequency of $\omega_{c0} = qeB_z(0)/(Am_0)$ can occur due to an error in the magnetic field. This results in a phase change that is proportional to the harmonic number and inversely proportional to the maximum energy gain per turn.

Having very high accelerating voltages is especially beneficial in high-power cyclotrons. The effects of space charge in the longitudinal phase space at a given energy are inversely proportional to the third power of the turn numbers, n_t.

Fig. 2.7 The accelerating cavity at the PSI cyclotron. (Courtesy PSI, Switzerland)

Consequently, the maximum achievable beam current in cyclotrons is $\sim 1/(n_t^3)$, as discussed in Sect. 3.4.1.

In the PSI cyclotron, the beam power increased from 0.15 to 1.4 MW over the years by installing four 50.6 MHz resonators capable of producing a gap voltage of up to 1.2 MV. The photo of the accelerating cavity at the PSI cyclotron is shown in Fig. 2.7. The number of turns in the PSI cyclotron was reduced by a factor of 2 over 20 years, resulting in the ability to extract the 2.4 mA beam with tolerable relative beam losses of $\sim 10^{-4}$. The PSI facility is the most efficient accelerator in the world: the overall conversion efficiency of wall plug power to beam power is 18%.

2.1.4 Extraction

The extraction of beams is the most complicated process in the design and operation of cyclotrons. The latter is especially challenging in high beam power and/or very high magnetic-field cyclotrons. The main issue is the proximity of the neighboring orbits at the extraction radius. The turn separation in the isochronous cyclotron at the extraction radius R_{ext} is derived from the cyclotron dynamics as:

$$\Delta R_{\min} = \frac{\Delta W_{\text{ext}}}{qecB_{z,\text{ext}}\beta_{\max}\gamma_{\max}^2}\,. \tag{2.15}$$

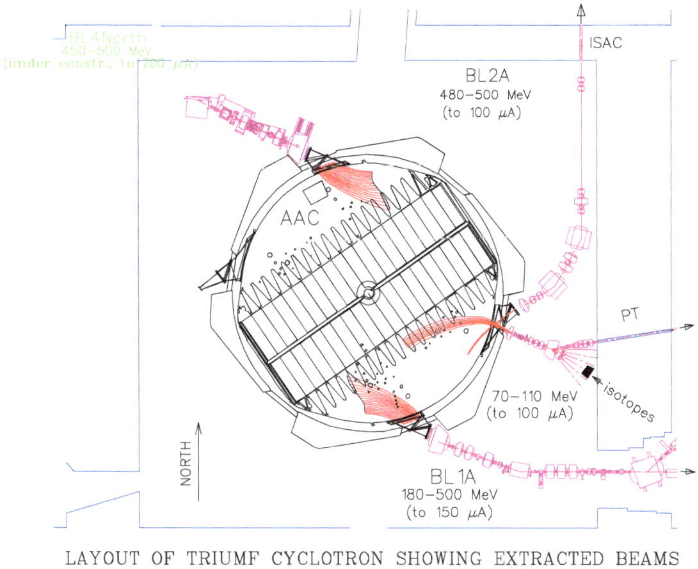

Fig. 2.8 Extraction of proton beams in TRIUMF cyclotron by H^- stripping. (Courtesy TRIUMF, Canada)

where ΔW_{ext} is the energy gain per turn at extraction. Large magnets for limited-energy beams (<1 GeV/u) and a high energy gain per turn facilitate large orbit separation at the extraction radius.

In high-intensity proton and heavy-ion cyclotrons, extraction is provided by a combination of electrostatic and magnetic septa. An additional mechanism, orbit oscillation, is introduced in the PSI cyclotron to increase the orbit separation in the extraction septum by an additional factor of 3. The orbit oscillation is induced by a slight controlled offset of the beam orbit during injection with respect to the reference orbit in the cyclotron. The main factor contributing to the gradual increase in the high-power beam from the PSI cyclotron over the years was the increase in energy gain per turn.

An efficient method to extract high-intensity beams is the stripping of ions with a thin foil. This stripping can be applied to H^- and H_2^+ to extract protons. H^- stripping is effectively utilized in the TRIUMF cyclotron to extract proton beams at different energies simultaneously [37], as shown in Fig. 2.8. Similar stripping extraction can be applied to heavy ions, forcing the ions to move with a significantly smaller radius in the magnetic field and eventually providing a large separation from the circulating beam.

2.2 Linear Accelerators

A linear accelerator, usually shortened as "linac", uses the principle of alternating electric field acceleration, which was first developed by Ising [38] in 1924. The first acceleration was demonstrated by Wideröe [39] in 1928. However, RF acceleration was only possible based on advancements in RF technology during World War II, and Alvarez first accelerated protons using his drift tube linac (DTL) in 1947 [40].

In a linac for hadron acceleration, it is necessary to use different types of accelerating structures and RF frequencies as the beam is accelerated, due to the significant change in the velocity of the particles ($\beta = v/c$). Consider an accelerating structure with a longitudinal length L and an RF frequency ω_{RF}. The phase variation $\Delta\phi$ of a particle traveling through the structure is approximated by

$$\Delta\phi \sim \frac{\omega_{\mathrm{RF}} L}{\beta c} = \frac{2\pi L}{\beta \lambda_{\mathrm{RF}}} \tag{2.16}$$

where λ_{RF} is the wavelength of the RF wave. Therefore, to maintain the accelerating phase of the particles, the product of the accelerator structure length and the RF frequency must be increased as the velocity β increases until the particles become relativistic. In practice, this requirement is met by gradually increasing the length of the accelerating structure as the beam velocity increases. At higher energy, it is possible to switch to a higher harmonic of the RF frequency, allowing for more compact accelerating structures. A linac uses RF resonators as basic accelerating cells, which will be detailed in Sect. 2.2.1.

The particles must be organized into bunches that are separated by the period of the RF wave $2\pi/\omega_{\mathrm{RF}}$ to ensure synchronized acceleration. The bunching is achieved by the longitudinal focusing effect of the RF wave. The dynamics can be briefly described by the following model. Consider a bunch beam entering an RF resonator with the designed phase of the synchronous particle ϕ_s and the kinetic energy W_s. The phase is chosen to be zero at the peak of the field, which is the convention in hadron linacs. The corresponding relativistic factors are $\gamma_s = 1 + W_s/(m_0 c^2)$ and $\beta_s = (1 - 1/\gamma_s^2)^{1/2}$. For one particle with phase and energy deviations $\Delta\phi = \phi - \phi_s$ and W, the longitudinal equations of motion can be written as

$$\begin{cases} \dfrac{dW}{ds} = qeE_{\mathrm{eff}} \left[\cos\left(\phi_s + \Delta\phi\right) - \cos\phi_s\right] \\[2ex] \dfrac{d\phi}{ds} = -\dfrac{\omega_{\mathrm{RF}}}{A m_0 c^3 \beta_s^3 \gamma_s^3} W \end{cases}, \tag{2.17}$$

where qeE_{eff} is the maximum energy gain that a particle can possibly obtain, which includes the time transit factor (see details in Sect. 2.2.1).

Assuming a synchronous particle's energy changes adiabatically, the dynamics of a small amplitude particle can be modeled as:

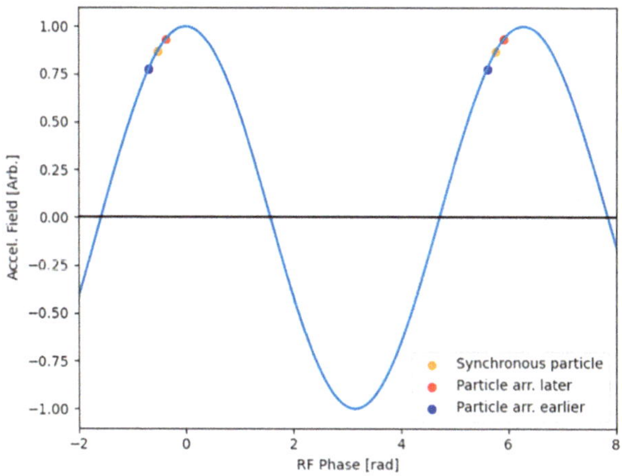

Fig. 2.9 Illustration of choice of synchronous phase and principle of longitudinal focusing

$$\frac{d^2W}{ds^2} = \frac{qeE_{\mathrm{eff}}\omega_{\mathrm{RF}}\sin\phi_s}{Am_0c^3\beta_s^3\gamma_s^3}W\ , \tag{2.18}$$

which indicates that the synchronous phase $\sin\phi_s$ must be negative to ensure stable motion when traversing repeating RF resonator structures. Consequently, the synchrotron phase ϕ_s must lie on the rising side of the sinusoidal wave, as shown in Fig. 2.9. In this configuration, the particle arriving early in the accelerating structure experiences a lower accelerating field, while later particles receive a larger acceleration. This provides a focusing effect in the longitudinal direction, which, in repeating structures, leads to longitudinal oscillations at a frequency ω_l:

$$\omega_l = \omega_{\mathrm{RF}}\frac{qeE_{\mathrm{eff}}\sin(-\phi_s)\lambda_{\mathrm{RF}}}{2\pi Am_0c^2\beta_s\gamma_s^3} \tag{2.19}$$

Therefore, the longitudinal oscillation is relatively fast at low energy, but as the beam becomes more relativistic, this oscillation effectively "freezes".

Meanwhile, Maxwell's equations require that any longitudinal variation in the accelerating field generates transverse kicks. At the entrance of the structure, the beam experiences a focusing force, while at the exit, it encounters a defocusing force. Due to the negative phase of the synchronous particle, it results in an overall defocusing effect throughout the entire structure. For an axially symmetric structure, the focal length can be expressed as

$$f^{-1} = \frac{\pi qeE_{\mathrm{eff}}\sin(-\phi_s)L}{Am_0c^2\beta_s^3\gamma_s^3\lambda_{\mathrm{RF}}}\ . \tag{2.20}$$

The RF transverse focusing strength also decreases as the energy of the particle increases. In addition, the space charge effect also contributes to the defocusing effect of the beam, which scales with the beam energy as $1/\beta_s^2\gamma_s^2$. Therefore, external focusing elements, such as quadrupoles or solenoids, are required to maintain the beam transverse profile in the linac [41].

Historically, the Cockcroft-Walton [42] electrostatic accelerator was built for low-energy proton acceleration; the DTL [39, 43] followed for medium energy, and then the coupled-cavity linacs (CCLs) were used for further acceleration. In the modern design of linacs, electrostatic accelerators have been replaced by RFQ linacs, whereas CCLs have often been replaced by SRF linacs.

In the late 1970s, SC cavities were built as energy boosters for low current, DC heavy-ion beams produced by electrostatic accelerators [44]. Superconductivity allowed the post-acceleration of the beams in CW mode with an efficiency that the NC RF structures could never achieve. In the following years, this technology, initially concentrated around $\beta \sim 0.1$ cavities, was successfully extended with new types of resonators to accommodate any beam velocity above $\beta \sim 0.025$. With the parallel development of NC CW RFQ covering the lower-β range from a few keV/u up to MeV/u energies, electrostatic acceleration could be limited to the extraction of the ion source beam, eventually leading to the current typical configuration of SRF linear accelerators for CW ion beams, which essentially consists of the ion source, RFQ, and SRF linac.

R&D in SRF cavities for high-power proton linacs began around the same period as SC ion boosters [45]. However, its first application occurred a couple of decades later. The pulsed SNS linac [46] consists of $\beta_0 < 1$ elliptical resonators whose design was derived from already successful $\beta_0 = 1$ cavities for electron beams. Preceded by an NC DTL and CCL linac section that accelerates the beam to 186 MeV, the SC section of the SNS linac, which accelerates the beam to ~1 GeV and beyond, is made of elliptical cavities of only two different types, as shown in Fig. 2.10 (Sect. 10.1). This scheme, in particular with the NC section of DRL and CCL replaced by another SC section consisting of non-elliptical, intermediate-β cavities, became a reference for high-power hadron-linac projects.

FRIB, a high-power CW hadron linac, is the first full-scale linac that consists entirely of SC resonators for acceleration above the RFQ energy (Fig. 2.11, Table 8.5, and Sect. 10.3). It is capable of accelerating all types of stable ions, from hydrogen to uranium, above 200 MeV/u energy and delivering 400 kW beam power to the production target [47]. Many other SRF high-power linac projects are underway for different applications, further pushing this technology [48–53].

An SRF linac for hadron beams usually consists of cryomodules interleaved with warm sections. Cryomodules house all components that must be cooled and maintained at cryogenic temperatures by means of liquid helium, including accelerating cavities and, in some cases, SC solenoids, beam steering elements, and diagnostic elements. Typical operating temperature ranges are 4.2–4.5 K and 1.8–2.1 K. In an SRF linac, cavities are typically independently phased and powered. The number of cells in a cavity is kept small to achieve a large velocity acceptance.

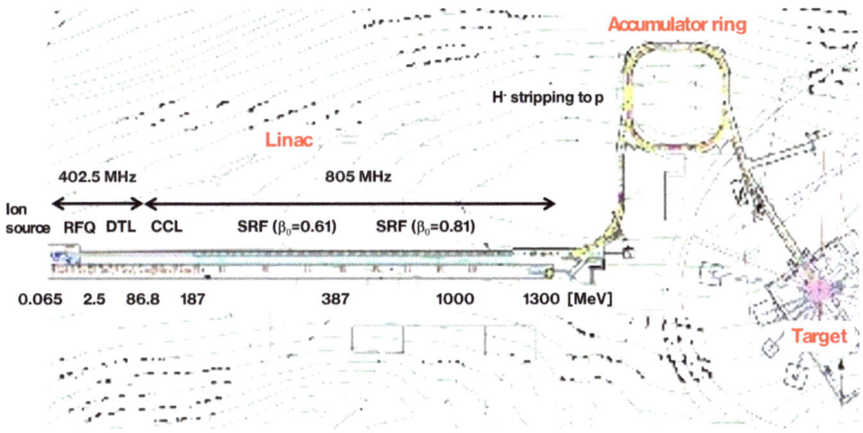

Fig. 2.10 Schematic layout of the SNS accelerator systems with the linac parameters indicated. The energy increase from 1 to 1.3 GeV was the scope of the SNS Proton Power Upgrade Project (Sect. 10.1)

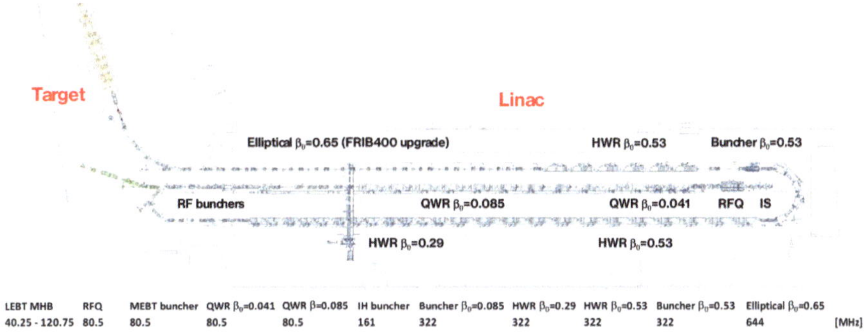

Fig. 2.11 Schematic layout of the FRIB driver linac with the linac parameters indicated. A section of the linac vacant of SRF resonators is reserved for the energy upgrade with future elliptical $\beta_0 = 0.65$ cavities (Sect. 10.3)

Depending on the linac design, the warm sections can host beam focusing and bending elements, steering elements, collimators, vacuum elements, and beam diagnostics. They physically interrupt the cold string of an SC linac and become sources of thermal load, affecting the cryogenic efficiency of the linac. Additionally, warm sections often create unwanted drift spaces in the lattice, potentially causing beam mismatches. Therefore, they are usually kept as short as possible while respecting the lattice periodicity.

On the other hand, cryomodules are constructed to be as long as possible. In linacs where NC quadrupoles in warm sections provide beam focusing, the cryomodule length is determined primarily by the lattice design of the transverse and longitudinal focusing. In contrast, in linacs where SC solenoids inside cryomodules are used for beam focusing, the maximum cryomodule length is constrained by practical considerations, such as assembly and transportation limitations, weight, and the difficulty of maintaining precise alignment in long systems.

2.2.1 Properties of RF Resonators in Linac

The electromagnetic wave form of Maxwell's equations in a vacuum,

$$\left(c^2\nabla^2 - \frac{\partial^2}{\partial t^2}\right)\mathbf{E} = 0\,, \tag{2.21}$$

has a transverse electric and magnetic (TEM) wave solution in free space; i.e., the electric and magnetic fields are perpendicular to each other, and both are perpendicular to the propagation direction of the beam. However, with a boundary condition of an almost perfect conductor defined by the enclosed cavity volume (hollow wake guide), no TEM solution ($E_z = B_z = 0$) can be found without vanishing the wave. Only transverse electric (TE, $E_z = 0$) or transverse magnetic (TM, $B_z = 0$) modes are available. For acceleration, TM modes are necessary in the cavities. These modes are denoted by $\mathrm{TM}_{\mathrm{mnp}}$, where the indices m, n, and p denote the axial, radial, and longitudinal dependencies of the acceleration field, respectively. However, TEM modes of the RF wave may be supported in a coaxial line structure, which consists of an inner conductor wire with a radius a surrounded by a cylindrical outer conductor with a radius b.

The main challenge of using a linac in a high-power hadron accelerator is that the hadron beam never becomes fully relativistic before reaching tens of GeV of energy per nucleon. It is a necessary practice to change the accelerating structures and the accelerating frequencies as the beams are accelerated (the velocity or $\beta = v/c$ increases), since the optimum structure and frequency depend on β. The longitudinal and transverse focusing of the beam also change as β increases, determining the optimal choice of the RF structures.

The RF resonators, or cavities, for high-power hadron linacs can be classified by their optimum beam velocity β_0 into low-β (roughly $\beta_0 < 0.2$), medium-β ($0.2 \leq \beta_0 \leq 0.8$), and high-$\beta$ ($\beta_0 > 0.8$). The energy gain of a charged particle can be expressed as

$$\Delta W = qeE_a L_{\mathrm{eff}} T(\beta)\cos\phi, \tag{2.22}$$

where q is the particle charge number, E_a is the acceleration field, L_{eff} is the effective acceleration length, and ϕ is the particle phase. The transit time factor (TTF) $T(\beta)$, a dimensionless quantity, defines the maximum energy gain of a charged particle traveling through an RF resonator:

$$T(\beta) = \frac{\int_{-\frac{L}{2}}^{\frac{L}{2}} E(z) \cos\left(\frac{\omega z}{\beta c}\right) dz}{\int_{-\frac{L}{2}}^{\frac{L}{2}} E(z) dz}. \tag{2.23}$$

For a rectangular longitudinal field distribution

$$T(\beta) = \frac{\sin\left(\frac{\pi L_{\text{eff}}}{\beta\lambda}\right)}{\left(\frac{\pi L_{\text{eff}}}{\beta\lambda}\right)}, \tag{2.24}$$

where $L_{\text{eff}} \sim \sqrt{(L^2 + a^2)}$ takes into account the penetration of the field in the beam tube due to the beam pipe aperture a [54]. If V_{acc} is the maximum accelerating voltage on the axis, to give the beam a good fraction ($>2/\pi$) of the single-gap accelerating field, the RF wavelength $\lambda = c/f$ should satisfy the relation

$$\beta\lambda > 2L_{\text{eff}} \tag{2.25}$$

In the case of a constant beam velocity inside the cavity, the TTF of a multi-gap resonator is a function of the beam velocity βc and the number of gaps N,

$$T(\beta) = T_g(\beta) \cdot \begin{cases} \frac{1}{N}\left[1 + \sum_{m=1}^{(N-1)/2} (-1)^m 2\cos\left(\frac{m\pi\beta_0}{\beta}\right)\right], & \text{if } N = \text{odd}, \\ \frac{2}{N}\left[\sum_{m=0}^{(N-2)/2} (-1)^m 2\sin\left(\frac{(m+1/2)\pi\beta_0}{\beta}\right)\right], & \text{if } N = \text{even}, \end{cases} \tag{2.26}$$

where $T_g(\beta)$ is the single-gap transit-time factor, which has the same form as in Eq. 2.24, with L_{eff} denoting the effective length of the gap. The parameter β_0 represents the reference (synchronous) velocity that ensures phase synchronization between successive accelerating gaps.

Equation 2.26 remains a good approximation in practical cases when the relative change in velocity within the cavity is small, $\delta\beta/\beta \ll 1$. To guaranty wide acceptance of the velocity in a cavity, it is necessary to keep the number of accelerating gaps low, as shown in Fig. 2.12.

The lower the beam energy, the faster the relative change in velocity from one cavity to the next; therefore, the optimal number of gaps is usually minimal in the lower β_0 cavities and increases with acceleration, reaching up to 5 ~6 gaps in the higher β_0 cavities. If the cavities are short, the number of cavities required to reach the target energy increases, along with the number of cavity interfaces, RF power couplers, tuners, etc., leading to an overall cost increase.

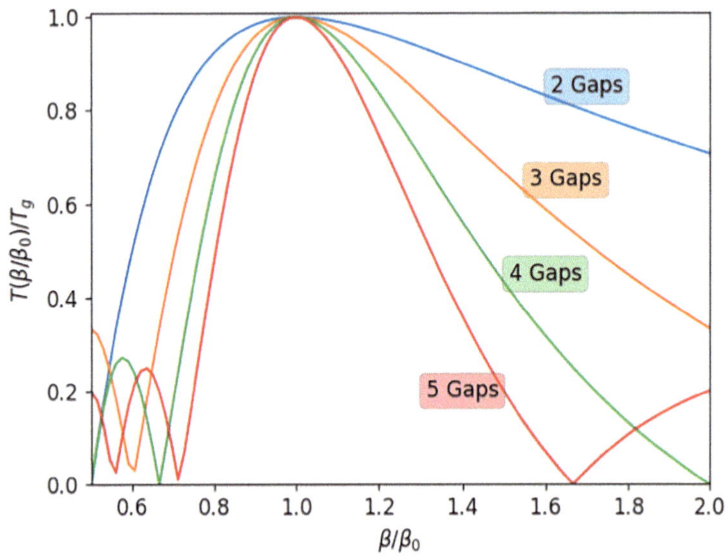

Fig. 2.12 Multi-cell Transit-Time Factor normalized by T_g as a function of the normalized beam velocity β/β_0, for different numbers of accelerating gaps

Upon approaching $\beta \sim 1$, a much larger number of gaps could be utilized, with the drawback of a higher number of dangerous higher-order modes that can be excited by the beam, which could cause an increase in emittance and beam loss. It should be noted that the present applications of high-power hadron linacs are typically limited to beam energies below 2 GeV/u (thus below $\beta_0 \sim 0.95$).

The RF resonator can be characterized by several well-known figures of merit [41]. The quality factor Q measures the ratio of the stored energy U and the energy consumed per RF period,

$$Q = \frac{\omega_{\mathrm{RF}} U}{P_{\mathrm{loss}}} , \tag{2.27}$$

where P_{loss} is the power of energy consumption. Corresponding to the power dissipated by the cavity walls, $P_{\mathrm{loss}} = P_{\mathrm{cav}}$, the cavity (or unloaded) quality factor Q_0 is given by

$$Q_0 = \frac{\omega_{\mathrm{RF}} U}{P_{\mathrm{cav}}} . \tag{2.28}$$

If the power emitted through the input RF power coupler (also called "external" power) P_{ext}, and transmitted through the RF pick-up P_{tr}, is also included $P_{\mathrm{loss}} = P_{\mathrm{cav}} + P_{\mathrm{ext}} + P_{\mathrm{tr}}$, the corresponding loaded quality factor Q_L is given by

$$Q_L = \frac{\omega_{\mathrm{RF}} U}{P_{\mathrm{cav}} + P_{\mathrm{ext}} + P_{\mathrm{tr}}} . \tag{2.29}$$

The shunt impedance R_{sh} characterizes the relationship between the accelerating voltage and the power loss.

$$R_{\text{sh}} = \frac{V_{\text{acc}}^2}{P_{\text{loss}}} = \frac{E_a^2 L_{\text{eff}}^2}{P_{\text{loss}}} \tag{2.30}$$

The shunt impedance, or its "per unit length" version in linacs,

$$R'_{\text{sh}} = \frac{E_a^2}{P_{\text{loss}}/L_{\text{eff}}} \tag{2.31}$$

is usually optimized for higher acceleration voltage when the beam current is not substantial.

The quantity R_{sh}/Q is introduced to make it depend only on the EM stored energy coefficient U/E_a^2, which is solely determined by the cavity geometry and the EM mode considered,

$$\frac{R_{\text{sh}}}{Q} = \frac{L_{\text{eff}}^2}{\omega_{\text{RF}}} \frac{E_a^2}{U} . \tag{2.32}$$

The geometric factor G, another factor that depends only on the geometry of the cavity, relates the cavity quality factor to the resonator's surface resistance R_s,

$$G = Q_0 \cdot R_s . \tag{2.33}$$

These figures of merit are related to directly measurable RF parameters. Geometric quantities U/E_a^2 and G can be calculated quite precisely by means of EM simulation codes.

2.2.2 *Type of RF Structures*

This section discusses the main types of RF structures commonly used in NC and SC linacs for hadron acceleration.

2.2.2.1 TM Mode Cavities

Transverse-Magnetic (TM) mode cavities are axially symmetric, equivalent to cylindrical corrugated waveguides with the beam axis coinciding with the EM wave propagation axis and the magnetic field B_θ perpendicular to it [55]. In multi-cell resonators, the accelerating mode is the π-mode, in which the field distributions in neighboring cells have a phase difference of 180° (Fig. 2.13). Elliptical and reentrant (single or multi-cell) resonators belong to this family and have a typical

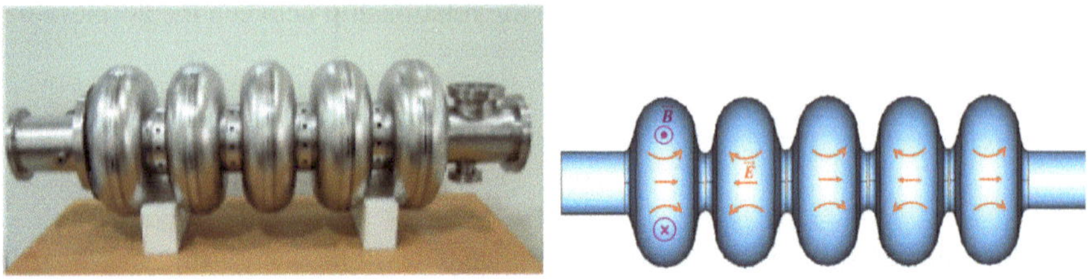

Fig. 2.13 A 5-cell elliptical cavity and sketch of its EM field in the TM01 mode [56]

transverse diameter of approximately $d \sim 0.8\lambda$. The typical aspect ratio per cell (defined as the ratio between the longitudinal and transverse size of the cell, see Sect. 2.2.3.3) is rather low ($\sim 0.6\beta_0$), leading to very large sizes at low frequency. Elliptical cavities are the simplest, most cost-effective, and most efficient resonators at high β_0 and, unlike pillbox cavities, have the virtue of not suffering from high-field multipacting if properly designed [57]. However, below $\beta \sim 0.5$, the shapes of the TM mode cavities become rather flat, with a reduced R_{sh}/Q, making them prone to multipacting and mechanical instabilities, which render their use impractical. Although single cell TM reentrant cavity prototypes have been developed in the past for low-β applications, the only TM cavities currently used in SC high-power hadron linacs are elliptical ones with $0.5 < \beta_0 < 1$.

DTL is another TM mode structure, which is the oldest linac structure for accelerating low-energy hadron beams. It consists of a series of drift tubes suspended within a resonant cavity, typically operating in TM010 mode, as shown in Fig. 2.14. The DTL operates in zero mode, with the electric field oscillating synchronously with the beam, accelerating particles as they traverse the gaps between consecutive drift tubes. When the accelerating field reverses its sign, the charged particles are screened inside the drift tube. Therefore, the length of the drift tube should be

$$L_n = \beta_n \lambda_{RF} \tag{2.34}$$

where L_n is the length of the nth drift tube, β_n denotes the velocity of the reference particle in the tube, and λ_{RF} is the wavelength of the DTL.

2.2.2.2 TEM Mode Cavities

TEM cavities are equivalent to coaxial waveguides using the TEM propagation mode for beam acceleration [55]. Unlike TM cavities, the beam axis is perpendicular to the EM wave propagation axis, creating an accelerating gap on each side of the center conductor. This family includes quarter-wave resonators (QWR) [54] and half-wave resonators (HWR)—either coaxial or spoke type [45]—which represent the large majority of low- and medium-β cavities for heavy ions [59].

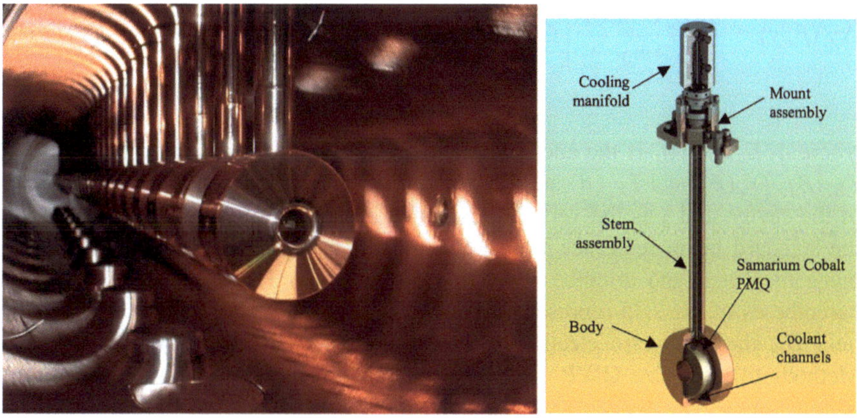

Fig. 2.14 Left: Drift tubes inside DTL tank 3 of SNS; right: Cut away view of a typical Drift tube [58]

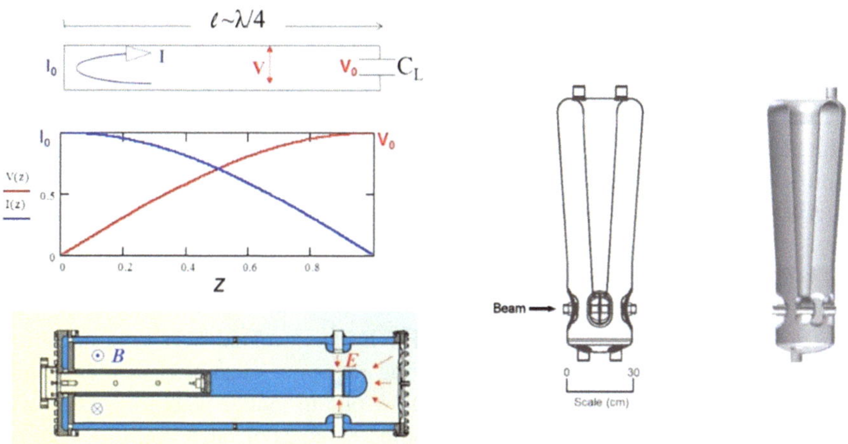

Fig. 2.15 QWR examples. Left, from top down: equivalent circuit; voltage and current distributions along the resonator; schematic of a QWR cavity. Right: bi-conical QWR [60]

In QWRs, the transmission line is about $\lambda/4$ long, with one end shorted. The maximum voltage is created at the open end, where the beam is accelerated in π-mode by the two gaps between the inner and outer conductors, as shown in Fig. 2.15. With two accelerating gaps, it can accelerate a beam in the range of approximately $0.7\beta_0 < \beta < 2\beta_0$ with $T(\beta) > 70\%$. Its large velocity acceptance and its relatively large aspect ratio per gap (of approximately $2\beta_0$) make QWR the most suitable structure for very low-β applications.

Unfortunately, QWRs also cause beam deflection that becomes significant in cavities with optimum velocity $\beta_0 > 0.1$, preventing their use above $\beta \sim 0.2$. The deflection $\Delta y'$ in radians can be expressed as

$$\Delta y'(\beta, y) \sim \frac{qeE_a L \sin\phi_s}{\beta\gamma A m_0 c^2}\left[G_{B_x} T_{B_x}(\beta) + \frac{G_{E_y}}{\beta} T_{E_y}(\beta) + \frac{2\pi f(1-\beta^2)}{2c\beta^2} T_{E_z}(\beta) y\right] \tag{2.35}$$

where $qeE_a L \sin\phi_s$ is the energy gain, G_{B_x} and G_{E_y} are geometric factors, [61], $T_{B_x}(\beta)$, $T_{E_y}(\beta)$, and $T_{E_z}(\beta)$ are the transit time factors depending on the geometry of the cavity, and y is the vertical position of the beam crossing the cavity. (With the standard beam reference frame, the y axis is aligned with the QWR's main mechanical axis and oriented toward its shorted end.) The three terms inside the parentheses of Eq. 2.35 represent the deflection caused by the transverse magnetic field B_x, the transverse electric fields E_y, and the RF defocusing, respectively. Methods to minimize QWR steering (or deflection) are based on modifications of these three terms, either by displacing the cavity along y or by modifying its geometry [62]. These corrections can be very effective in cavities with low β_0, but they become increasingly difficult to apply at $\beta_0 > 0.1$, resulting in the degradation of the cavity characteristics. For $\beta_0 > 0.2$, symmetric and steering-free cavities are needed for beam dynamics reasons, especially for beams with high space charge.

In HWRs, the transmission line is about $\lambda/2$ long and is shorted at both ends. The maximum voltage is at the cavity center, as shown in Fig. 2.16. It is easy to see that an HWR, equivalent to two QWR joined together at their open ends, requires a higher RF power than a QWR of the same frequency and diameter to reach the same accelerating voltage. Thus, in terms of efficiency, QWRs are always preferable to HWRs. However, because of their symmetric design, the HWRs are steering-free. The aspect ratio of the HWR per gap is approximately β_0, half of the QWR but higher than that of the elliptical cavities, making it well suited for medium-β applications above $\beta_0 \sim 0.1$.

The simple coaxial structure of early HWRs has evolved over time into more complex geometries, improving some of the resonator characteristics. Examples include tapered and conical shapes that allow one to reduce peak magnetic fields and increase shunt impedance at the cost of increased cavity size [64].

Half-wave resonators, where the axis of the outer conductor cylinder is parallel to the beam axis and perpendicular to the inner conductor cylinder, are called spoke resonators, as shown in Fig. 2.17 [67]. The lateral size of a single-spoke resonator (SSR) is significantly larger than that of coaxial HWRs (simply called HWRs from now on) for cavities up to $\beta_0 \sim 0.2$, becoming substantially similar for higher β_0 up to ~0.5. Above that value, HWRs start to become less efficient than elliptical ones. Spoke resonators can be stacked together in the same cavity to create multi-gap structures. To maximize EM coupling between cells and EM mode separation, adjacent spokes are oriented 90° to each other. Single, double, and triple spoke resonators (SSR, DSR, and TSR) have been developed for a wide range of β_0.

It can be seen that a multi-spoke cavity, which is a combination of TEM-type elements, is essentially equivalent to a crossbar H-mode (CH) resonator, as shown in Fig. 2.18. In early SC CH cavities, the spokes did not originate directly from the outer conductor but from the girders mounted on it to improve the separation of

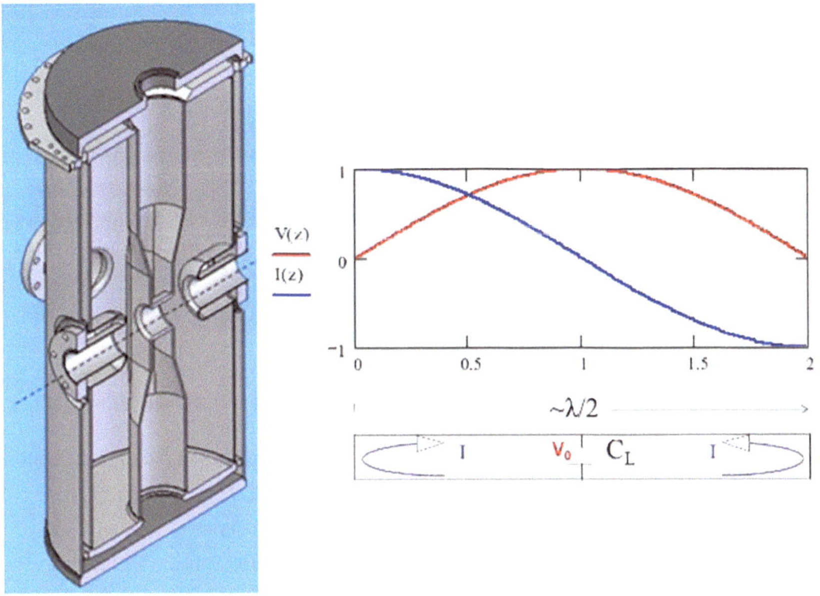

Fig. 2.16 An HWR example. From left: cutaway of a coaxial HWR cavity; equivalent circuit; voltage and current distributions along the resonator [63]

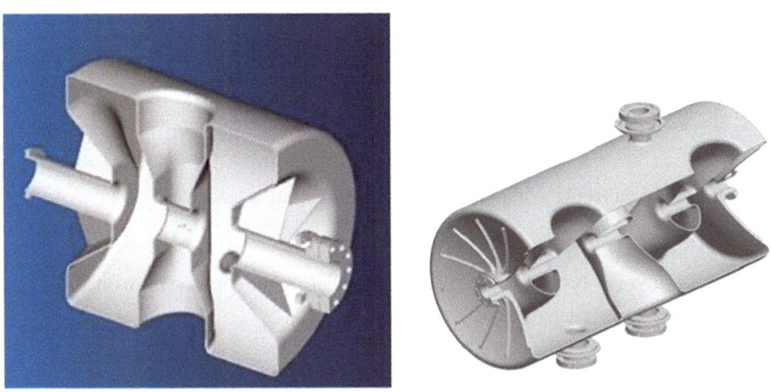

Fig. 2.17 Spoke resonator examples of (left) single [65] and triple spoke [66]

RF modes and the impedance of the shunt. Today, the distinction between spokes and CH can be established only by the number of spokes ($\gg$3 for CH cavities) and by the β_0 (<0.2 for CH). SC CH resonators have been developed for low-β CW applications in different versions, with the number of spokes varying from 7 to 19. They have been proposed in high-power hadron linac projects [69].

A combination of aspect ratio, operational frequency, number of accelerating gaps, and aperture characterizes cavities regarding their possible applications.

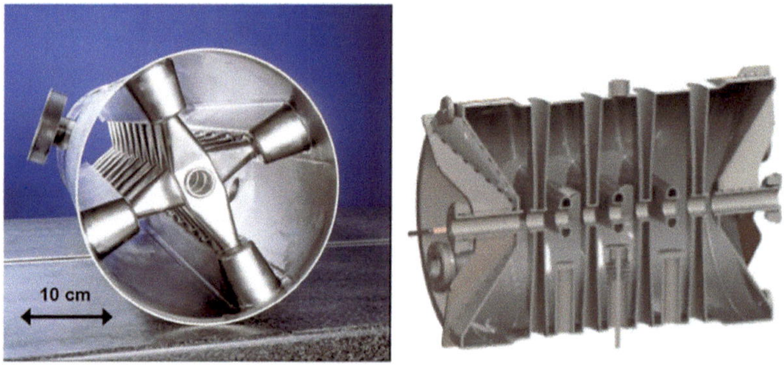

Fig. 2.18 Views of superconducting CH cavities (left) with and (right) without girders [68]

Table 2.2 Examples of SC cavities used in existing or planned high-power hadron accelerators

Optimum β	0.04–0.12	0.05–0.18	0.09–0.16	0.14–0.53	0.45–1
Frequency [MHz]	70–100	217–352	160–176	322–352	650–1300
Cavity type	QWR	CH	HWR	HWR, Spoke	Elliptical
No. of gaps/cells	2	7–19	2	2,3	4–6
Operation mode	C3W	CW	CW	CW, pulsed	CW, pulsed

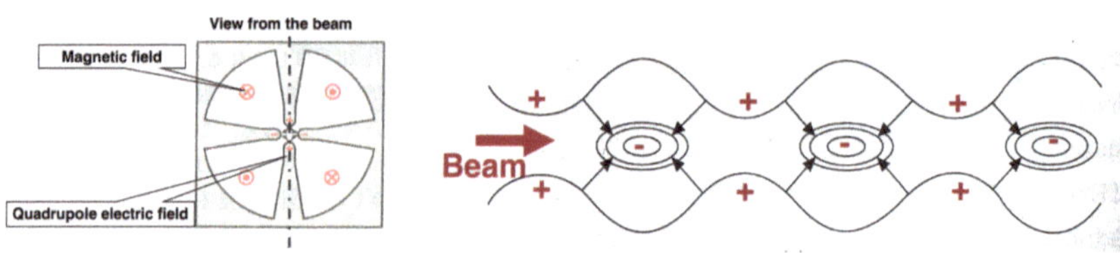

Fig. 2.19 Sketch of an RFQ cavity. Left: beam's view; Right: side view

Table 2.2 lists several types of cavities used in SC high-power hadron linacs, along with the typical range of frequency and velocity β_0.

2.2.2.3 Hybrid Modes Cavities, RFQ

The TM010 mode is commonly used for beam acceleration since there is a longitudinal electric field in this mode along the central axis. On the other hand, the pure transverse electric mode (TE) has only a transverse electric field. No longitudinal electric field exists in the TE modes to accelerate the charged particles. However, it is desirable to use the electric component of the RF field to accelerate, focus, and bunch a continuous low-energy ion beam to maintain beam quality.

The concept of radio frequency quadrupole (RFQ), invented by Teplyakov and Kapchinskii [70], meets this requirement. The TE211 mode of the RFQ, as illustrated in Fig. 2.19 (left), focuses the beam transversely. Longitudinal acceleration is achieved by appropriately modulating the shapes of the electrodes, as shown

Table 2.3 Parameters of the RFQ used in several hadron accelerators

Parameters	SNS [71]	FRIB [72]	J-PARC [73]	HLI [74]	ESS [51]
Beam species	H^-	Heavy ion	H^-	Heavy ion	p
Frequency [MHz]	402.5	80.5	324	108	352.21
Length [m]	3.723	5.04	3.623	3.0	4.6
Injection energy [keV/u]	65	12	50	2.5	75
Output energy [MeV/u]	2.5	0.5	3	0.3	3.62
Peak field [kilpatrick]	1.85	1.6	1.72	~1.6	1.8
RF repetition rate [Hz]	60	CW	50	100	14
Structure type	4 vane	4 vane	4 vane	4 rod	4 vane

in Fig. 2.19 (right). Therefore, it is not a pure TE mode and has sometimes been referred to as a "hybrid" mode. The quadrupole electric field is created by the four vanes near the beam axis, whereas the longitudinal field is generated by the longitudinal modulation of the vane shape. The four vanes can be replaced by four rods, which leads to a different design of the RFQ. However, for high-power applications, 4-rod RFQs face major challenges in thermal management and mechanical stability. Although 4-rod RFQs are often more compact and easier to build, the 4-vane design is generally preferred for high-power linacs. Table 2.3 lists the design parameters of several RFQs used in hadron linacs.

2.2.3 *Linac Design Considerations*

This section presents design considerations shared by both NC and SC linacs. Special SC linac considerations will be covered in Sect. 2.2.4.

2.2.3.1 NC RF for Low and Medium Velocity Beam Acceleration

Ions are extracted by an electrostatic field from the plasma generated and confined in an ion source vacuum chamber. This extraction voltage provides the initial acceleration. It was common to use a Cockcroft-Walton accelerator [42] with a voltage of around 750 kV to accelerate the beam from the source, followed by a DTL, until the 1980s, when facilities like CERN and BNL started to use RFQs as replacements, continuing into the 1990s when new facilities like SNS and J-PARC linacs were built using RFQs instead. As shown in Fig. 2.20, the Cockcroft-Walton accelerator is a gigantic system that uses the voltage multiplier principle to reach high DC voltages. The DC acceleration requires the ion source to be placed on a high-voltage platform. Subsequently, the DC beam must be bunched for later RF acceleration.

Fig. 2.20 KEK Cockcroft Walton accelerator. (Courtesy KEK, Japan)

As discussed in Sect. 2.2.2.3, the RFQ provides bunching, acceleration, and transverse focusing simultaneously. The TE211 mode in the cylindrical cavity generates a quadrupole electric field. The four-vane structure concentrates the electric field near the beam axis, and its longitudinal modulating shape creates a longitudinal accelerating field. Upon entering the RFQ, the hadron beam is transversely matched by the decreasing transverse focusing field and the decreasing RFQ aperture. The beam is then adiabatically bunched by slowly changing the synchronous phase $|\phi_s|$ from 90° to nearly 30°. The modulation is gradually increased towards the end of the bunching stage. The acceleration starts with the maximum modulation and $|\phi_s| \sim 30°$. In an RFQ, approximately two-thirds of its length is used for bunching, while the remaining one-third is used for acceleration. With advances in RFQ beam simulation, such as with PARMTEQ [75] (Table 2.3), the design of 95% transmission for a 50 mA proton beam is possible. The transmitted beam has a good distribution in both the transverse and longitudinal directions for further acceleration.

Following the RFQ, the beam enters the MEBT section, which matches the beam optics and rapid transverse phase advance of the RFQ, minimizing the emittance growth due to mismatch and nonlinear space charge forces. MEBT also incorporates an RF chopper (e.g., the SNS MEBT section in Fig. 2.21 [76]) or a buncher (e.g., the FRIB MEBT [78]) to match the beam to the later accelerating structure in the longitudinal direction.

The drift tube linac (DTL) is usually the first accelerating structure downstream of the MEBT. For the nth drift tube of length $L_n \sim \beta_n \lambda_{\mathrm{RF}}$, the maximum acceleration gradient is approximately

$$\frac{dE}{dL} \sim \frac{qeV_{\max}}{L_n} \sim \frac{qeV_{\max}}{\beta_n \lambda_{\mathrm{RF}}} . \tag{2.36}$$

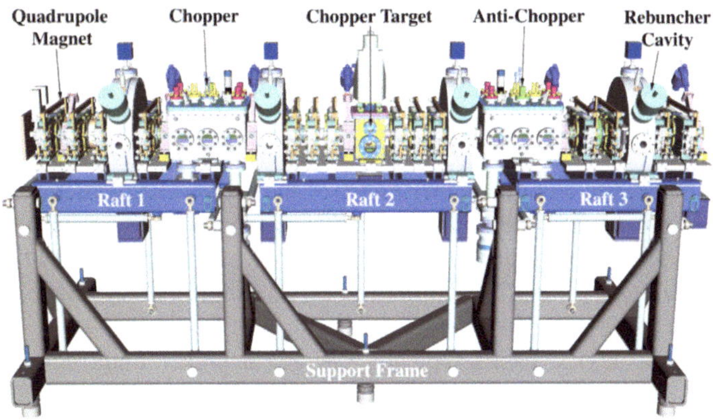

Fig. 2.21 Schematic layout of the SNS MEBT section [76, 77]. (Courtesy LBNL and ORNL, USA)

In the transverse direction, external focusing is required to transport the beam without loss. During RF acceleration, the emittances of the beam are adiabatically damped while the beam is defocused by the RF wave and the intensity-dependent space-charge force. The defocusing force scales with $1/\gamma_s^3$; therefore, it becomes less significant as the beam is accelerated. The external force in the DTL is commonly provided by quadrupoles. The transverse motion of the charged particle is described by Hill's equation [79]:

$$\frac{d^2x}{ds^2} + K(s)x = 0\,, \tag{2.37}$$

where the focusing term $K(s)$ is given by

$$K(s) = \frac{qeG(s)}{Am_0\gamma\beta c} - \frac{\pi q e E_{\mathrm{eff}} \sin(-\phi_s)}{Am_0c^2\beta_s^3\gamma_s^3\lambda_{\mathrm{RF}}} - K_{\mathrm{SC}}(s)\,. \tag{2.38}$$

Here, $G(s)$ is the quadrupole gradient, and K_{SC} is the slope of the linearized space charge force. In the zero-current limit, the strengths of the external focusing elements are designed to maintain the desired beam size. The focusing elements should decrease in strength and be less frequent as the beam is accelerated to keep the normalized emittance nearly constant by minimizing the mismatch of transverse phase space.

In DTLs, each drift tube has the option to house quadrupoles or other electric or magnetic components that focus or steer the beam during acceleration. The SNS DTL adopted a focusing layout "FFODDO" [80], indicating an empty tube (denoted by "O") followed by two cells with focusing ("F") and defocusing ("D") quadrupoles of permanent magnets, creating a transverse focusing period of $6\beta\lambda$.

As seen in Eq. 2.36, DTL is suitable for accelerating low-energy particles, and a short wavelength of the structure is preferred. For higher energy beams, the DTL loses its acceleration efficiency compared to coupled cavity linacs (CCL). The energy for the transition between DTL and CCL is generally chosen to be 50–150 MeV, with a frequency increase of a factor of 2 to 5. This choice is based on the shunt impedance and the quality factor of the accelerating structure, as well as the smooth transition of the transverse and longitudinal focusing of the beam. As an example, the SNS NC linac section consists of six 402.5 MHz DTL tanks that accelerate the beam to 86.8 MeV, followed by four CCL modules with a frequency twice as high at 805 MHz [80].

Compared to DTLs, CCLs typically operate at higher frequencies—typically 800 MHz to 3 GHz. The high frequency is optimum for a higher design β with a cell length $\beta\lambda/2$. CCLs incorporate a large number of cells with a coupling constant of a few percent, which is cost effective by minimizing the number of low-level RF (LLRF) modules, fundamental power couplers, and tuning systems. The coupling between cells is achieved either through electrical coupling, beam apertures, or magnetic coupling via designed coupling slots at the web.

In a $N+1$ coupled cavity system, there are $N+1$ modes with $N+1$ frequencies. The angular frequency of the mth mode ω_m is given by

$$\frac{\omega_m}{\omega_0} - \frac{\omega_0}{\omega_m} = -K_c \cos\frac{m\pi}{N} , \tag{2.39}$$

where K_c is the coupling constant and $m = 0, 1, 2, \ldots, N$. When the coupling is weak, the frequency becomes:

$$\omega_m \approx \omega_0 \left(1 - \frac{K_c}{2}\cos\frac{m\pi}{N}\right) . \tag{2.40}$$

Here, K_c becomes the difference between the maximum and minimum angular frequencies, that is, the "bandwidth" of the coupled cavity cells. The first, middle, and last modes in Eq. 2.40 are the "0", "$\pi/2$", and "π", respectively. The field in the ith mode can be expressed as:

$$E_i \sim \cos\frac{im\pi}{N} e^{i\omega_m t} . \tag{2.41}$$

If the cell length d is chosen as

$$d = \frac{\beta\lambda_{\mathrm{RF}}}{2}\frac{m}{N} , \tag{2.42}$$

then, the RF phase velocity coincides with the beam velocity βc.

As shown in Fig. 2.22, the SNS CCL design adopted a $\pi/2$ mode for its good field stability, low power-flow droop, and negligible power-flow phase shift. The coupling is magnetic; however, there are slots on the walls of the accelerating cells.

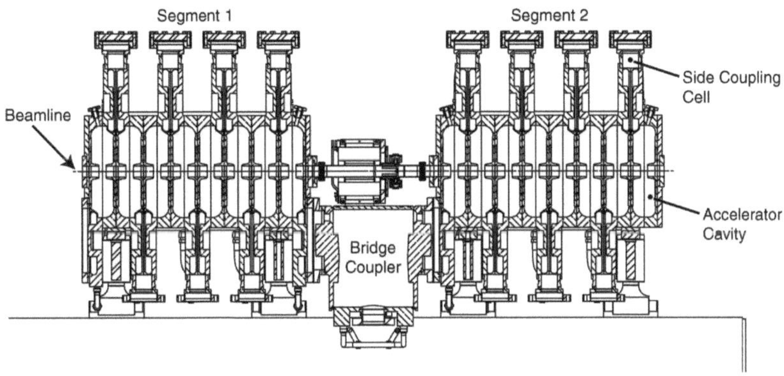

Fig. 2.22 Cross-section view of the first two segments and first bridge coupler of the SNS CCL [81]. (Courtesy N. Bultman)

The coupling cells are placed off-axis to the beam trajectory on both sides of the accelerating cells. Therefore, the beam will experience the same field distribution as in π mode, with an increasing cell length $\beta\lambda/2$.

2.2.3.2 Choice of NCRF or SRF

In NC resonators made of copper, the resistance of the RF surface is classically expressed by the following equation

$$R_{s,\mathrm{NC}} = \sqrt{\frac{\mu_0 \omega}{2\sigma}} \sim 2.6 \times 10^{-7}\sqrt{f}\,, \tag{2.43}$$

where μ_0 is the magnetic permeability of vacuum, σ is the conductivity of copper ($1/\sigma \sim 1.7 \times 10^{-8}\Omega{\cdot}\mathrm{m}$ for pure Cu at room temperature). The power losses are generated by the electrons of the surface currents associated with the EM field, which transfer kinetic energy to the metal lattice through collisions. The EM field penetrates the metal wall with an amplitude that decays exponentially with depth, with a characteristic length, called the skin depth, of

$$\delta_{s,\mathrm{NC}} = \sqrt{\frac{2}{\sigma \mu_0 \omega}}\,. \tag{2.44}$$

Equation 2.44 does not hold for SC resonators. Superconductivity is a quantum mechanical effect that requires a different treatment (for a deeper understanding of RF superconductivity for particle accelerators, see, e.g., [57] and [82]).

In superconductors, the RF current is carried by SC electrons coupled in Cooper pairs, which move without losses in the metal below the critical temperature T_C and below a critical magnetic field H_C, which is given by

$$H_C(T) \sim H_C(0) \times \left[1 - \left(\frac{T}{T_C}\right)^2\right] . \tag{2.45}$$

For niobium material commonly used to build SC resonators, $T_C \sim 9.2\,\text{K}$ and $H_C(0) \sim 0.2\,\text{T}$. The skin depth cannot be defined in the same way as in the NC case (Eq. 2.44); the penetration of the field into the material becomes nearly independent of frequency and extremely small (of the order of $\sim$36 nm for niobium at 4.2 K, to be compared with the NC copper skin depth at room temperature of $\sim 2\,\mu\text{m}$ at 1 GHz). However, even in the SC state, some NC electrons are still present and are accelerated by the oscillating EM field, generating losses similar to those in the NC case, but in an amount proportional to their concentration, which decreases with temperature. The surface resistance can be calculated by means of the BCS theory of superconductivity. A practical, although approximate, expression for $R_{s,\text{BCS}}$ of pure niobium, valid for $T < T_C/2$, can be expressed as [57]

$$R_{s,\text{BCS}} = 2 \times 10^{-4} \frac{1}{T} \left(\frac{f}{1.5}\right)^2 \exp\left(-\frac{17.67}{T}\right) , \tag{2.46}$$

where $R_{s,\text{BCS}}$ is in the unit of Ω, f in the unit of GHz, T in the unit of Kelvin (K), and $T_C = 9.2\,\text{K}$ is the critical temperature of Nb in the ideal case. In real superconductors, however, there are also residual losses generated by the presence of non-SC parts of the material exposed to the RF field (for example, inclusions of foreign material, dust, or other contaminants on the Nb surface, trapped magnetic flux causing NC spots, etc.). The surface resistance R_S of the SC resonator can thus be separated into two components: BCS and residual,

$$R_{s,\text{SC}} = R_{s,\text{BCS}} + R_{s,\text{res}} . \tag{2.47}$$

In a simplified model, useful especially at the design stage, the residual component is assumed to be constant and dependent on the cavity surface quality, while the BCS component represents the ideal theoretical value. Measurements show that in real accelerators, $R_{s,\text{res}}$ amounts to a few nΩ in well-prepared resonators, and typical design specifications for hadron-linac operation are on the order of $R_{s,\text{res}} \sim 10\,\text{n}\Omega$.

The resistances of NC and SC materials depend differently on the RF frequency f, as

$$R_{s,\text{NC}} \propto f^{1/2} ; \quad \text{and} \quad R_{s,\text{SC}} \propto f^2 . \tag{2.48}$$

In both cases, the resonator size scales with $\lambda \propto 1/f$, and their RF surface area scales with $\lambda^2 \propto 1/f^2$. Since the RF power losses at a given EM surface field distribution are proportional to the RF surface area exposed to it, the dissipation of RF power in resonators scales with frequency as $f^{-1/2}$ for NC linacs and as f for SC ones. So, unlike NC linacs, SC linacs favor low frequencies from the perspective of RF losses. Therefore, SC linacs are well suited for high-power hadron accelerators: a lower frequency results in a larger longitudinal acceptance and a larger cavity size and beam aperture, which helps to reduce beam losses from various mechanisms [83].

In addition to frequency scaling, SC resonators provide additional advantages over their NC counterparts. First, the low-power dissipation of SC resonators allows for high gradient operation, which can be sustained for long pulses up to continuous-wave (CW), providing the possibility of obtaining a very high average beam current. This is not possible for NC cavities operated at high RF power, only for short pulses with a low duty cycle, due to the physical limits of water-cooling. Second, the efficiency of NC accelerating structures depends mainly on their geometric shunt impedance R_{sh}/Q, which is made high by long multi-gap resonators with a fixed beam velocity profile. In contrast, the efficiency of SC cavities is mainly determined by their low surface resistance R_s, rather than by their fine geometrical optimization. Resonators can be made with only a few accelerating gaps and, thus, with a large acceptance of velocity.

SC hadron linacs that cover all velocity ranges, from the exit of RFQ to relativistic energies, can be built with only 4 $\sim$5 different types of resonators. This results in a highly modular design, with all related advantages, including a reduction in construction and maintenance costs. A side advantage of a linac made of independently phased, short cavities with large velocity acceptance is the possibility of operating the accelerator in the presence of faulted resonators by switching them off and re-adjusting the RF fields and phases of the remaining cavities to the new configuration. SC cavities, in many cases, can be operated above their design gradient (at the cost of an increased cryogenic load), allowing some compensation for under-performing cavities with over-performing ones. Another obvious advantage of such SRF linacs is the overall plug power efficiency after taking into account cryogenic losses and cryogenic system efficiency. Although its relative importance decreases with increasing beam load, a benefit is always present in SC accelerators, especially large-scale ones [84].

2.2.3.3 Transition Velocity and RF Frequency Choice

With the RFQ as the ideal choice for the first accelerating device of a hadron linac, the selection of the LEBT and MEBT devices is determined by the energy of the hadron source and the voltage or length of the RFQ. A high LEBT energy, the injection energy to an RFQ, is generally preferable because most emittance growth can arise from the space-charge effect, which is more detrimental at low beam energy in a linac. However, the maximum possible LEBT energy is practically constrained by the discharge voltage of the ion source and by the RFQ length, which enables gentle bunching to avoid emittance growth. Typically, a 50 $\sim$100 kV accelerating voltage can be achieved in the ion source; while the length of the RFQ is limited to about 10 meters. The LEBT energy of 50 $\sim$100 keV thus chosen is not high enough to mitigate the growth of the emittance. Therefore, it is a common practice to shorten the LEBT as much as possible; that is, the ion source is located as close to the RFQ as possible. Unfortunately, this introduces undesirable gasses from the ion source to the RFQ (undesirable gasses decrease the discharge limit of the RFQ), and strong differential pumping is required in the limited space for the LEBT. The neutralization of space charge by means of residual gas is another remedial measure. However, that can make the emittance growth caused by the space charge sensitive to partial vacuum pressures. The 50 mA H^- peak current of J-PARC and SNS could be handled by this method thus far, but a higher current requires additional measures.

The choice of the MEBT energy depends on the details required for the linac specifications, including the acceptable beam emittance, the engineering considerations of the RFQ and components in the MEBT, and the efficiency of the downstream linac.

The beam energy for the transition from NC to SC is the result of an overall design optimization and can vary, for different machines, somewhere between $\beta \sim 0.025$ and 0.55. In low duty cycle, pulsed linacs with only one type of beam, NC DTL technology is still competitive at low β, which is supported by the following reasons [84]:

- NC DTL/CCL technology is, nowadays, well established in linacs. The fixed velocity profile allows for the construction of large, high shunt impedance tanks with high RF efficiency and a very regular lattice. The frequent focusing in DTL/CCL machines allows for acceleration with minimal emittance growth of very high current beams (~10–100 mA) in a very high space-charge regime. However, strong focusing requirements relax with increased beam energy and can be fitted rather well with typical SC linac lattices at higher β.
- Operation in short RF pulses allows for reaching high gradients in NC structures without overheating the resonators. Concentrating the beam power P_b in short, high current pulses allows for maximizing P_b/P_{cav}, hence increasing acceleration efficiency in pulsed mode.
- NC low-β cavities are mechanically stable and suitable for pulsed operation. SC low-β cavities are more prone to Lorentz force detuning, and due to their higher Q, they usually operate with a much narrower RF bandwidth, making pulsed operation much more demanding in terms of RF control capabilities.
- The optimal RF frequency in hadron linacs increases with beam energy. NC cavities generally operate at higher RF frequencies than SC cavities at the same β_0. Therefore, the combination of an NC low-β with an SC high-β section allows for minimizing the number of frequency jumps along the linac, thereby reducing the size and cost of the matching sections.

In CW linacs, on the contrary, SRF is winning at low-β regardless of the beam power. The SC section can start just downstream of the RFQ and MEBT, resembling the scheme originally applied for low-current, CW heavy-ion linacs.

The frequency range of the resonator versus the velocity of the beam is determined for reasons related to beam dynamics. If V_{acc} is the maximum accelerating voltage on the axis, to provide the beam with a good fraction ($>2/\pi$) of the accelerating field in a gap of effective length L_{eff}, the RF wavelength $\lambda = c/f$

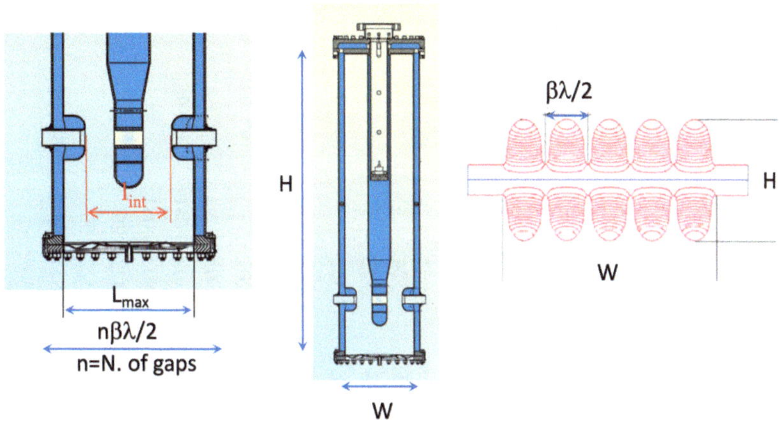

Fig. 2.23 Left: accelerating length definitions in low-β cavities. Center and right: transverse (H) and longitudinal (W) dimensions used in the aspect ratio definition for QWRs and elliptical cavities [63]

should satisfy the relation $\beta\lambda > 2L_{\text{eff}}$ (see Eq. 2.25). Therefore, a low β requires a large λ, especially if a large aperture diameter a is used. As mentioned above, there are several advantages to working at as low a frequency as possible, but the drawback is the size of the cavity, since the volume per MV of acceleration scales with $1/f^2$. Especially for a low-β resonator, geometries with a higher aspect ratio per gap are preferable in terms of cost. As shown in Fig. 2.23, the aspect ratio per cell is defined as W/nH , where W is the size occupied by an accelerating cell along the beam line, H is the transverse size of the resonator, and n is the number of cells. A higher value corresponds to a smaller cavity size [59]. Favorite SRF geometries are QWRs ($W/nH \sim 2\beta$ and $f \sim 70 - 100$ MHz) up to $\beta < 0.2$, HWRs ($W/nH \sim \beta$ and $f \sim 160 - 400$ MHz) up to $\beta < 0.5$, and elliptical cavities ($W/nH \sim 0.6\beta$ and $f \sim 650 - 1300$ MHz) for $\beta > 0.5$.

Table 2.4 lists the transition velocity and RF frequency choices of selected linac projects and facilities at various phases. Also listed are the beam duty cycle and the choice of NCRF versus SRF as a function of beam velocity β.

2.2.3.4 RF Power Coupling and RF Bandwidth

In a typical accelerator cavity system, RF power is produced by a generator, transmitted through a 50 Ω transmission line, and injected into a resonator through a forward power coupler (FPC) (Fig. 2.24). An additional small RF probe called a pick-up (PU) is used to monitor the field in the cavity. The forward RF power P_f reaching the FPC is partly transmitted to the cavity and partly reflected from it. The fraction of power entering the cavity is partly dissipated on its walls (P_{cav}), partly transmitted out through the pick-up port (P_{tr}), and partly emitted back

Table 2.4 Typical parameters of hadron linac examples at various phases (OP: operations, CM: commissioning, PL: planned, CT: construction, RD: R&D). Notation "–" in the "NC-SC transition" column indicates NCRF choice

Facility	Status	Beam type	Beam duty cycle [%]	Pulse[a] current [mA]	Ave.[b] power [MW]	Output energy [MeV/u]	NC-SC transit. [MeV/u]	Cavity type	f_0 [MHz] $/\beta_0$[c]	Reference
SNS	OP	H^-	6	26	1.4	87	186	DTL	402.5	[85, 86]
						187		CCL	805	[81]
						387		E6	805/0.61	
						1000		E6	805/0.81	
(upgrade)	CM	H^-	6	38	2.8[d]	1300		E6	805/0.81	[87, 88]
J-PARC	OP	H^-	1.25	50	0.25	50	–	DTL	324	[89, 90]
						191		SDTL	324	
						400		ACS	972	
Linac4, CERN	OP	H^-	0.1–5	40	0.005	50	–	DTL	352	[91, 92]
						104		CCDTL	352	
						160		PIMS	352	
LANSCE[e]	OP	H^-, p	1.7–10	10	0.8	100	–	DTL	201	[93, 94]
						800		CCL	805	
FRIB	OP	HI	100	≤0.7	0.02		0.5	QWR	80.5/0.041	[95, 96]
								QWR	80.5/0.081	
								HWR	322/0.29	
						>200		HWR	322/0.53	
(upgrade)	PL	HI	100	≤0.7	0.4[d]	>400		E5	644/0.65	[97, 98]

UNILAC	OP	HI	25		36	1.4	–	IH	36	[99, 100]
						11.4		DTL	108	
						13.0		SGR	108	
ESS	CT	p	4	62.5	5[d]	216	90	DSR	352/0.5	[51, 101]
						571		E6	704/0.67	
						2000		E5	704/0.86	
PIP-II	CT	p	1.1–100	2	1.6[d]		2.1	HWR	162.5/0.11	[102]
								SSR	325/0.22	
								SSR	325/0.47	
								E5	650/0.61	
						800		E5	650/0.92	
LIPAc	CT	d	100	125	1.1[d]	9	2.5	HWR	175/0.094	[103]
(IFMIF)	RD	d	100	125	5[d]	40		HWR	175/0.18	[104]
CiADS	CT	p	100	10	2.5[d]		3.2	HWR	162.5/0.10	[105]
								HWR	162.5/0.19	
								HWR	325/0.40	
								E5	650/0.62	
						600		E5	650/0.82	
MYRRHA	RD	p	100	4	2.4[d]		17	SSR	352/0.375	[106, 107]
								E5	704/0.51	
						600		E5	704/0.71	

[a] Average beam pulse current
[b] Average beam power output from the linac
[c] For SRF cavity
[d] Goal beam power value not yet achieved
[e] The CCL frequency is an even integer multiplier of the DTL frequency to allow for both H^- and H^+ (p) acceleration

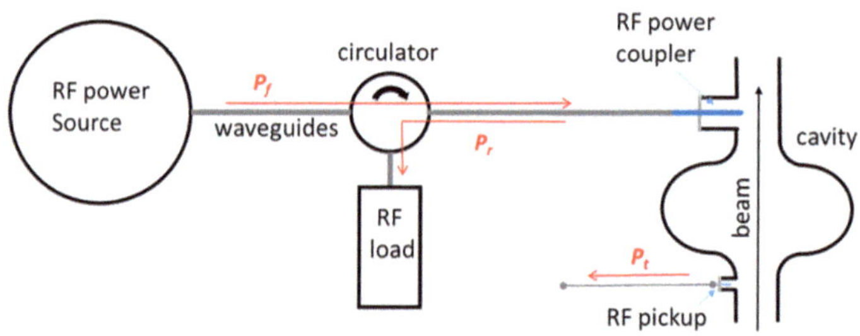

Fig. 2.24 Scheme of a typical cavity-RF system. Further discussion on the linac RF system is given in Sect. 8.5.2

through the FPC (P_{ext}). This emitted signal adds to the reflected one, generating the backward signal with power P_r (usually called reflected power, although somewhat improperly) which travels back along the main transmission line and is finally diverted by an RF circulator to a matched load, where it is dissipated at room temperature with little cryogenic load.

The field in the cavity, that is, the accelerating voltage in a gap, as a function of the frequency f of the driving signal at a constant forward power P_f, follows the typical behavior of the resonance curve,

$$V(f) = \frac{V_0}{\sqrt{1 + 4\left(\frac{f - f_0}{\Delta f_0}\right)^2}}, \tag{2.49}$$

where the full bandwidth $\Delta f_0 = f_0/Q_L$ is completely defined by the loaded quality factor. When $f = f_0 \pm \Delta f_0/2$, the field becomes $V_0/\sqrt{2}$, and the power dissipated in the cavity is half of the power at resonance. The phase shift ϕ between the cavity field and the driving RF signal (ignoring a constant component that depends on the system) is given by

$$\tan\phi = -\frac{2(f - f_0)}{\Delta f_0}. \tag{2.50}$$

Equations 2.49 and 2.50 indicate that any changes in the cavity tune f_0, for example, caused by mechanical deformations of the cavity, will result in changes to the accelerating field, both in amplitude and phase. To restore the correct field and phase for beam acceleration, the RF control system needs extra power. For a given detuning, the narrower the RF bandwidth, the larger the RF power bill will be. The amount of power absorbed, reflected, and emitted by the cavity depends on the coupling factor β_c, defined as

$$\beta_c = \frac{P_{ext}}{P_c} = \frac{Q_0}{Q_{ext}} = \frac{Q_0}{Q_L} - 1, \tag{2.51}$$

where $P_c = P_{\text{cav}} + P_{\text{tr}}$ (the contribution of the beam has not yet been considered), and $Q_{\text{ext}} = \omega U / P_{\text{ext}}$. The cavity is critically coupled if

$$\beta_c = 1, \text{ and } Q_L \sim \frac{Q_0}{2} . \tag{2.52}$$

At resonance and under steady-state conditions, when Eq. 2.52 is met, all forward power coming from the RF transmission line is dissipated inside the cavity, as in a matched resistance. Reflected and emitted signals are still present, but they have the same amplitude and opposite phase, so they perfectly cancel each other out, and no backward power is generated.

If $\beta_c < 1$ (cavity under-coupled), the power emitted by the cavity through the FPC is lower than in the previous case, such that $Q_0 > Q_L > Q_0/2$; the reflected signal is then larger than the emitted one: cancelation is not complete, and part of the RF power is lost backward in the external load. In contrast, if $\beta_c > 1$ (cavity over-coupled), the emitted signal is larger than the reflected one and $Q_L < Q_0/2$; again, backward power is created and lost in the load. When it is out of resonance, $f \neq f_0$, the reflected and emitted signals are always out of phase and cannot be canceled, generating backward power.

The critical coupling conditions are particularly favorable for measuring the cavity figures of merit since $P_r \sim 0$ and $P_{\text{ext}} \sim P_c \sim P_f/2$. By measuring P_f, P_r, and P_{tr} in this configuration, as well as the cavity voltage decay time τ (observing the RF pickup signal after switching off the input power), and denoting $P = P_f - P_r$, one obtains $Q_L = \omega\tau/2$, $U = P\tau/2$, $E_a = [(U/E_a^2)/U]^{1/2}$, and $Q_0 = \omega U/(P/2 - P_{\text{tr}})$. The full 3 dB RF bandwidth of the cavity $\Delta f_0 \sim f_0/Q_L$ and other related parameters can be evaluated [57].

An issue for SRF cavities, due to their extremely high Q_0, is that critical coupling conditions, in the absence of beam loading, i.e., an amount of power transmitted by a cavity to the beam that is significant for cavity operation, can lead to an RF bandwidth that is too narrow to allow for reliable cavity control (see Sect. 5.3.5) in the presence of the detuning forces. In this case, over-coupling is used to decrease Q_L and increase Δf_0 to acceptable values (possibly exceeding the range of fast fluctuations in the cavity f_0). The price to pay is a larger amount of RF power needed to reach the field. Therefore, in the low beam loading regime, just for RF control reasons, the RF system power requirements can significantly exceed the power needed for beam acceleration.

In the presence of beam loading, the forward power to the cavity must include the amount P_b to be delivered to the beam: the coupling strength (measured in the absence of the beam) must be increased accordingly, achieving its optimum value when $\beta_{c,\text{opt}} = 1 + P_b/P_c > 1$.

On the high side of the HPHA beam current range ($\sim$100 mA), the required coupling can push $Q_L \sim \omega U/P_b$ to rather low values, close to those of the NC cavities ($\sim 10^4 - 10^5$). The resulting large bandwidth makes cavity mechanical instabilities negligible, facilitating phase and amplitude control during operation. On the lower side ($\sim$1 mA) cavities may still need some limited over-coupling and some extra RF power available.

The accelerated beam current not only absorbs power from the cavity but also excites the cavity accelerating mode and some higher order modes. The beam-generated signal has an amplitude proportional to the beam current and a phase equal to the beam synchronous phase ϕ_s. This additional RF field adds to the one already present in the cavity and changes its amplitude and phase. The RF control system restores the reference values by adding an opposite amount of signal; however, this prevents complete cancelation of the reflected power and affects efficiency.

However, with a proper frequency detuning of the cavity, it is possible to shift the phase of the cavity signal in the opposite direction to that of the beam signal, again minimizing the reflected power [41]. The optimum detuning angle ψ, at the optimum coupling $\beta_{c,\mathrm{opt}}$, becomes

$$\tan\psi = -2(f - f_0))/(\Delta f_0) = -\frac{P_b}{P_c(1+\beta_{c,\mathrm{opt}})}\tan\phi_s = -\frac{P_b}{2P_c + P_b}\tan\phi_s \tag{2.53}$$

In the case of SC cavities, the condition $P_b \gg P_c$ is easily reached, and the optimum detuning angle always reaches its maximum value $\psi \sim -\phi_s$.

2.2.4 *Special SC Linac Considerations*

SC linacs present typical operational issues that require a specific approach. Some of the most important ones are listed in this subsection.

2.2.4.1 Resonators Mechanical Stability

In order to minimize thermal resistance between the RF surface and the liquid helium, and to save material costs, SC cavities are mainly made of thin (~2−4 mm) niobium sheets. This makes them intrinsically less stable than NC cavities from a mechanical point of view. Working at lower RF frequencies, they also have larger sizes and, consequently, low-frequency mechanical modes that can be easily excited by environmental noise. As pointed out in Sect. 2.2.3.4, their narrow RF bandwidth makes them very sensitive to any force that causes mechanical deformations, which can pull their resonant frequency error out of the useful range $|f - f_0| < \Delta f_0/2$ and make RF control rather challenging. Cavities are designed to reduce these negative effects as much as possible by:

1. minimizing their helium pressure sensitivity coefficient df/dP within a few Hz/mbar;
2. minimizing the Lorentz force detuning (LFD) coefficient df/dE_a^2 within a few Hz/(MV/m)^2; and
3. keeping the lowest mechanical mode frequency as high as possible and well above the most common environmental noise at 50/60 Hz.

In low beam-loaded cavities, the RF bandwidth can be artificially increased by means of over-coupling. Slow helium pressure fluctuations (roughly below 1 mbar/s·Hz or less) can be compensated for by means of a precise mechanical tuning system capable of bringing the frequency error back to approximately zero. Fast detuning, for example, in the case of resonant mechanical modes of the cavity excited by environmental noise, has been addressed in the past by means of electronic fast tuners [108], at some cost to cavity performance. In new resonators, the frequency sensitivity to mechanical vibrations is minimized by careful electromagnetic and mechanical design and, in QWRs, by the use of mechanical dampers [109]. Cryomodules must be designed not to transfer environmental noise to cavities, and the ground noise in accelerator tunnels/vaults must be kept under control by eliminating, shielding, and filtering possible sources (e.g., compressors of the cryogenic system).

LFD is proportional to $-E_a^2$. When going from a low field to an operating gradient, LFD can displace the cavity tune downward by several units of bandwidth, exceeding the capabilities of the RF control system. In CW operation, the cavity gradient can be raised slowly enough to allow the mechanical tuner to track the reference frequency. In pulsed operation, the detuning can be so extensive and rapid that additional fast tuners based on piezoelectric actuators [110] are required. Piezo tuners for pulsed SC linacs are driven, in most cases, in a feedforward mode, i.e., they follow a predefined tuning path that cancels the expected LFD. Feedforward can be of the adaptive type, in which the control system measures the feedforward effect and corrects the tuning path to minimize the residual phase error.

An important issue that can arise, especially when the resonator is phase- and amplitude-locked at high gradients, is the coupling between LFD and RF amplitude control residual error. The Lorentz force deforms the cavity far from its rest position and performs mechanical work that accumulates potential energy. If the field suddenly decreases, for example, due to detuning from mechanical noise, the Lorentz force also decreases, and the cavity starts to move back toward its rest position, finally oscillating at its own mechanical eigen-frequency f_M. This causes periodic detuning and field oscillations at the same frequency f_M. The resulting oscillation of the Lorentz force, if the RF amplitude control system gain is too low and introduces a delay, can feed energy to the mechanical mode, further amplifying it. This self-excited mode, which can unlock the cavity during operation, can be mitigated by increasing either the amplitude control gain or the RF bandwidth through over-coupling. Tuning the cavity slightly lower than the reference frequency can also help reduce the negative effects of LFD oscillations. Any decrease in E_a, and thus in the Lorentz force, will move the cavity frequency up and closer to the reference frequency, increasing the field back and stopping the forced oscillation.

2.2.4.2 Magnetic Field Trapping in SC Cavities

SC cavities are sensitive to magnetic fields, and the SC state cannot exist in the presence of an RF magnetic field exceeding a critical value $H_{c,\mathrm{RF}}$, which depends on the material, temperature, superconductor type, and surface quality [82]. Cavities must be operated below $H_{c,\mathrm{RF}}$ to avoid quenching. Below their critical temperature T_c, ideal superconductors are perfect diamagnets and completely expel DC magnetic fields not exceeding H_c (for type I materials) or H_{c1} (for type II materials like niobium) (the Meissner effect). However, in real superconductors with material impurities and lattice defects, a magnetic field can be trapped through the RF surface and sustained by circular SC currents. This magnetic flux splits into quantum units, each having an NC core that dissipates RF power. The larger the trapped flux, the greater the RF power losses. An estimate of the contribution of an external magnetic field H_{ext} to the surface resistance for the typical high RRR niobium used in SC resonators is given by

$$R_{\mathrm{mag}} \sim 0.3 H_{\mathrm{ext}} \sqrt{f}\ , \tag{2.54}$$

where R_{mag} is in units of nΩ, H_{ext} is in units of mOe, and f is in units of GHz. Magnetic flux trapping typically occurs when an external magnetic field (e.g., an earth magnetic field of the order of ~0.5 Oe, or a field of magnetic components inside the cryomodule) is present through the resonator during the transition from the NC to the SC state. To avoid this, accurate shielding of the earth magnetic field is required down to typically ~1 μT inside the cryomodule, to keep possibly $R_{\mathrm{mag}} \ll R_{\mathrm{BCS}}$. This is obtained by enclosing the resonators in boxes made of material with high magnetic permeability μ [111]. Flux trapping can be strongly reduced by maintaining a high-temperature gradient in the cavity during cooldown through the normal-conducting (NC) to superconducting (SC) transition temperature. This creates a sharply defined border between NC and SC material, which propagates quickly along the resonator and makes it difficult to create circular currents, thus "wiping" all magnetic flux [112] away from the cavity. In cryomodules containing SC solenoids, although provided with magnetic shielding and bucking coils to reduce fringe fields, some internal magnetization of the magnetic shielding material can occur after solenoid powering. Accurate degaussing cycles with the solenoid field can usually mitigate this problem.

2.2.4.3 *Q* Degradation from RF Surface Contamination

The performance of SC cavities depends much more on the RF surface quality and cleanliness than that of NC cavities. During operation, contamination of the resonators can be caused by a bad vacuum, sputtering of material due to beam losses, or, in the worst case, accidental venting in a dirty atmosphere. Particulates, water vapor, and hydrocarbons in cavities can severely degrade cryomodule performance [82]. In situ techniques have been developed to achieve full or partial recovery [113]; however, they are not always applicable. SC linacs require additional care to prevent accidental cavity contamination. In long-term operation, even in the presence of high vacuum, cryo-pumping on the cavity surface can accumulate contaminants and degrade performance [114]. Recovery can usually be obtained by periodic RF reconditioning, in situ plasma processing, warm up, and low temperature baking at $80 - 120\,^\circ\mathrm{C}$ [82].

2.2.4.4 "Q Disease"

Niobium RF performance can be degraded by the presence of excess hydrogen in its bulk, which precipitates during cooldown, creating lossy molecules on the RF surface. Usually, this accumulation of H in the Nb material occurs during chemical polishing or electro-polishing and can be strongly limited by performing these operations while keeping the Nb material below $20\,^\circ\mathrm{C}$. Q disease, when present in an operating cavity, can be cured by warming the cavities and providing a very fast cooldown. However, it can be completely eliminated by preliminary thermal treatments of the resonator in high vacuum at $600 - 800\,^\circ\mathrm{C}$ [57] or above. This operation, now a standard procedure, removes most of the hydrogen from the bulk in the final stages of cavity preparation before their installation in the cryomodules.

2.2.4.5 Multipacting

Multipacting [57] (MP) is a resonant electron emission phenomenon from the RF surface. A free electron in the cavity vacuum is accelerated by the RF field and can hit the surface, causing secondary electron emission.

Secondary electrons can be accelerated; if the trajectories bring electrons back to the same emitting points with the correct RF field and phase, the process can be repeated indefinitely, with an increasing number of electrons limited by the RF power delivered to the cavity. Increasing the RF power under MP conditions does not increase the field in the cavity; it only increases the number of traveling electrons. The MP effect can be simulated using a macroparticle approach in an RF field, as illustrated in the left panel of Fig. 2.25. The conditions for MP to establish are:

1. stable electron trajectories ending on cavity walls that are related to cavity geometry, field distribution, frequency, and field amplitude;
2. secondary-emission yield (SEY, as shown in the right panel of Fig. 2.25) coefficient $Y_{ee} > 1$, related to material properties, surface quality, and the energy of the impinging electrons; and
3. the existence of an initial free electron to be captured by the field initiates the process; this is related to the vacuum level and residual gas ionization.

In practical cases, MP trajectories can end in one or two points.

MP is present and is creating problems in NC cavities as well. However, in SC cavities, it is particularly dangerous and difficult to condition because even a relatively small current of resonant electrons can heat the cavity surface and cause the NC transition and cavity quench. MP conditioning improves the surface quality by reducing SEY; however, in cavities with unfavorable geometry or poor

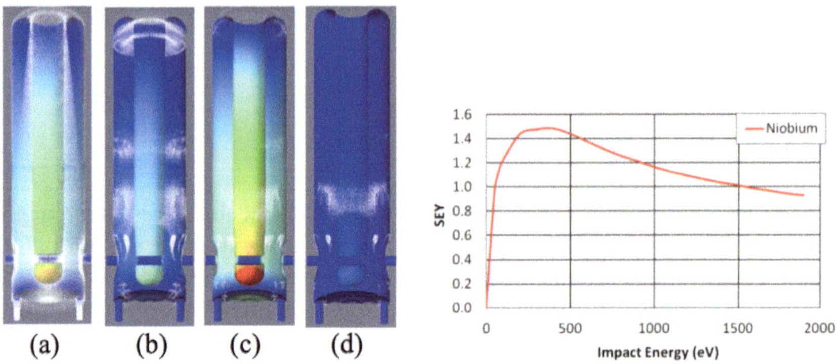

Fig. 2.25 Left: Multipacting simulation of FRIB $\beta = 0.085$ QWR cavity. Figures (**a–d**) represent the evolution of particles survived at increasing RF cycles. Particles survived a large number of RF cycles are considered potential MP particles; Right:Impact energy dependence of SEY for Niobium [115]

cleanliness, it may require a rather long time and must be repeated after every cycle of warm-up and cool-down. MP can also reappear in previously conditioned cavities during operation as a consequence of RF surface contamination by gas or sputtered material. This can cause an interruption of linac operation until the cavity is conditioned again. MP in FPC is troublesome because conditioning must be done at a low enough power to avoid cavity quenching. Dangerous multipacting levels can be eliminated by design in elliptical cavities using 3-D multi-physics simulation codes, but not always in low- and medium-β TEM cavities [59]. Although MP can be predicted, there is not yet a clear way to determine whether it is acceptable. However, geometrical optimization of cavities can eliminate conditions for MP modes or at least move them far from the operational gradient, possibly to a low gradient to ease conditioning at low RF power. MP modes that tend to collect together at "accumulation points" electrons following non-resonant electron trajectories which originate in a large volume, are particularly difficult to overcome and must be avoided by design. Narrowing the gradient interval in which the mode exists and moving the impact energy of electrons to the sides of the SEY curve help reduce the conditioning time.

2.2.4.6 Higher-Order Modes

High-order modes (HOMs) excited by the beam in SC resonators [57] are a critical issue in electron linacs and have also been a concern in high-current hadron linac projects because they can seriously affect and even prevent linac operation [116]. A charged particle traveling through a cavity can excite a cavity mode n with frequency $\omega_n/2\pi$ if it couples to it, i.e., if the shunt impedance of that particular mode, calculated along the trajectory of the particle, is $R_{\mathrm{sh},n} > 0$. The resulting induced voltage is of type $V = V_0 \exp(i\omega_n t)\exp(-t/\tau_n)$, where $\tau_n = 2Q_{L,n}/\omega_n$ is the decay time of the mode itself, and $Q_{L,n}$ is its loaded quality factor. $Q_{L,n}$ takes into account all RF losses that dissipate the energy of the mode, including the power extracted by the RF couplers. Summing together the contributions of all the beam particles that are exciting the cavity at different times and slightly different trajectories gives the steady-state field of the mode in the presence of the beam.

HOMs are dangerous in two aspects. First, when the beam excites an HOM, it dissipates power on the cavity walls, thereby increasing the cryogenic load of the linac. Second, the HOM field generated by the beam interacts with the beam itself (within the single bunch or in following bunches, if it persists longer than the bunch spacing) and can cause emittance growth and beam losses, up to beam breakup (BBU). A rigorous treatment of HOM, as well as formulas for calculating the parameters and effects of HOM, can be found in several publications, e.g., [57] and [41].

It is clear that a mode can be resonantly excited by the beam if the beam frequency ω is a sub-multiple of ω_n (i.e., ω_n/ω is close to an integer) and if the bunch length σ_t of the beam is a short fraction of one RF cycle, roughly $\sigma_t < 1/\omega_n$, resulting in a significant value of its transit time factor. Thus, in designing a cavity, it is necessary to move the frequencies of the HOMs far from the beam frequency or its multiples, and frequencies higher than $1/\sigma_t$ can be neglected. Construction tolerances, mechanical modes, LFD, tuner displacement, and all other possible mechanical deformations must be taken into account. The HOM effects are negligible when they are substantially smaller than the effects of all other specified parameters in the absence of HOMs, such as amplitude and phase stability, alignment tolerances in the linac components, RF losses in the cavity from the accelerating mode, etc. Reduction of τ_n and thus of the steady state HOM field, $Q_{L,n}$, if it cannot be fully accomplished by cavity design, can be achieved by means of additional HOM couplers, which extract from the cavity only the HOM field and not the accelerating one.

In general, hadron beams are less sensitive to HOMs than electron beams because of their much higher rigidity. In the SNS, the first high-power SC linac operating with hadrons, HOM dampers have been initially implemented in the 6-cell SC cavities as a precaution. However, they were not only unnecessary but also detrimental to cavity performance; they were finally removed [117]. Since then, calculations have been performed on several high-power hadron linac projects, showing that dangerous beam-excited HOMs could be eliminated or rendered negligible by design in all cavities [117–119]. This is particularly true in CW operation, where the very low harmonic component of the beam frequency makes HOM excitation in a carefully designed cavity rather unlikely, even when manufacturing errors are taken into account. In QWR, HWR, and spoke cavities, HOM frequencies are usually very well separated from the fundamental one, and HOM effects can be made negligible, at least with hadron beams (although not at all with electron beams [120]). This positive conclusion cannot always be guaranteed in linacs operating with a complex time structure of the accelerated beam, where HOM couplers are still foreseen [121].

2.2.4.7 Beam Losses

Beam losses (see Sect. 3.1) can cause radio-activation and damage to accelerator components, which can become rather severe and even catastrophic, resulting in interruption of operation and downtime for repair [83]. In high-power hadron linacs, these events can reach critical levels in a very short time and must be carefully mitigated by design and counteracted immediately if they occur. SC linacs, in particular, may suffer critical damage to cavities and cryomodules in the case of uncontrolled beam losses in thin-walled cavities immersed in liquid helium. An efficient machine protection system (Sect. 8.2), including sensitive dedicated diagnostics and the ability to perform fast beam inhibition within a few tens of microseconds, is mandatory.

For proton or H^- beams, a widely used "rule of thumb" threshold that should not be exceeded for uncontrolled beam power losses along a linac is 1 W/m up to $\sim$1 GeV proton beam energy, decreasing with increasing beam energy. Up to this value, the resulting activation generally allows for "hands-on" maintenance for a relatively short time (a few hours) after the beam is shut down. In the design stage, the beam dynamics simulation codes for high-power hadron linacs must be sensitive enough to detect, with reasonable statistical confidence, losses of $10^{-6} - 10^{-8}$ of the beam power.

Beam losses can have several origins, as discussed in Sect. 3.1, which are generally not specific to SRF systems. For example, many mechanisms affect all kinds of ion beams. Linac lattice defects, errors in alignment and field settings, mismatches of beam parameters at the injection of linac sections, space-charge forces, and structure resonances can cause unexpected emittance growth and the generation of a beam halo and tails that escape the stable acceleration bucket and are subsequently lost along the linac [41]. In pulsed machines, field transients at the beginning and end of the pulse can generate longitudinal tails, which are usually mitigated by the use of fast choppers. In general, structural resonances (e.g., the well-known envelope instability [122]) can be avoided by a careful beam dynamics design. Other beam losses can occasionally occur after trips in the accelerator components, such as RF cavities, magnets, and ion sources.

A special case where SRF linacs have clear advantages over NC ones is H^- beams, the ions of which can be neutralized and lost along the linac through several specific mechanisms, as elaborated in Sect. 3.7.

H^- ions can lose their extra electrons by interacting with the residual gas (Sect. 3.7.1), especially at low energy. Double stripping (with about 4% of the probability of single stripping) creates H^+ ions, which can be captured, accelerated, and lost at higher energy. In both cases, the losses are proportional to the pressure of the gas. This makes SC linacs superior to NC linacs due to their 1 to 2 orders of magnitude better residual vacuum created by cryogenic pumping of cavity walls.

Intrabeam stripping (IBSt) was found to be an important and unexpected source of beam losses in SNS due to the neutralization of H^- ions [123]. IBSt losses are inversely proportional to the rms bunch sizes σ_x, σ_y, and σ_s, and can thus be reduced by increasing the beam size (Sect. 3.7.4). This can be taken into account at the level of linac design: the considerably larger beam aperture of SC cavities, compared to NC ones, represents an additional advantage.

Lorentz stripping of H^- ions in a magnetic field [124] is discussed in Sect. 3.7.2. As shown in Eq. 3.72, the stripping losses increase very rapidly with beam energy, and even more so with the field strength of the magnets in the linac, putting strict limits on the maximum values of B and β that can coexist at every point of the beam line. High-power H^- beams are used in linacs for injection into synchrotrons, where the beam is generally stripped at energies of up to 1.3 GeV. This is still manageable by means of standard accelerator magnets.

Blackbody radiation inside the beam lines can also cause H^- neutralization, as discussed in Sect. 3.7.3. The beam loss rate at up to 1.3 GeV energy is very low compared to that caused by other mechanisms. However, this mechanism became significant with the previously proposed Project X [125] where an H^- beam was proposed to be accelerated to 8 GeV energy. It might also become an issue in the future, with higher energy accelerator projects using H^- beams [83]. Again, since the blackbody spectrum depends on the temperature, the low temperature inside cryomodules could represent a further advantage of SRF.

2.3 Accumulators and Rapid-Cycling Synchrotrons

Synchrotrons are circular accelerators in which charged particles are constrained to move in fixed orbits by magnets whose fields vary over time, such that the radius of curvature remains constant as the particles gain energy. Accumulators are simplified synchrotrons that do not involve particle acceleration or magnet ramping. Their main function is to accumulate particles to achieve high peak intensity and power upon extraction.

In synchrotrons, charged particles are successively accelerated by alternating electric fields produced by resonating radio frequency (RF) cavities [126]. The motion of the particles is longitudinally stabilized by the RF field [127, 128] and transversely stabilized by alternating gradient quadrupole magnetic fields [129, 130]. Hadron beams are typically formed from ion sources and pre-accelerated by linear accelerators (Sect. 2.2) at low energies to accommodate large variations in velocity and frequency, as well as to overcome space-charge effects.

In a synchrotron, the synchronous condition for a reference particle with charge qe, mass Am_0, and momentum p_s is given by

$$qeB_s\rho_s = p_s = Am_0c\beta_s\gamma_s \;, \tag{2.55}$$

where $B_s\rho_s$ is the magnetic rigidity, and β_s and γ_s are the relativistic factors of the reference particle. Magnetic rigidity can also be expressed in commonly used units as

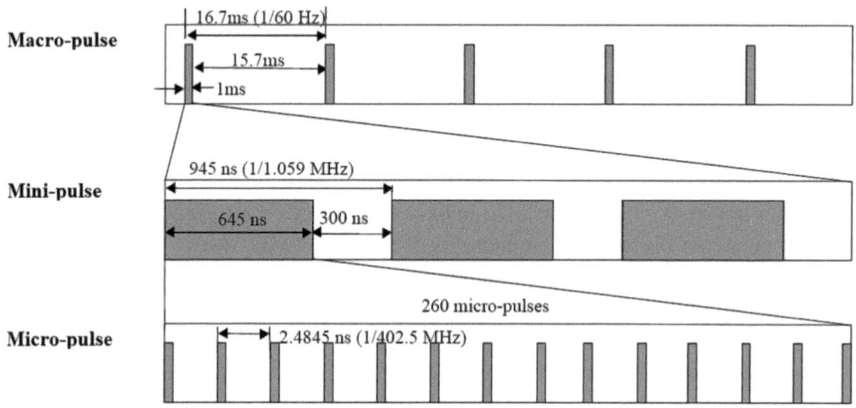

Fig. 2.26 Time structure of beam pulses in the SNS accelerators. The structure of the micro-pulse in the linac is shown at the bottom. Chopping is conducted in the LEBT to create the mini-pulse structure. The final macro-pulse structure shown at the top is seen by the neutron production target

$$B_s \rho_s \text{ [T m]} = \frac{3.3357}{q} p_s \text{ [GeV/c]} = 3.106 \frac{A}{q} \beta_s \gamma_s \, . \tag{2.56}$$

The RF system that accelerates or confines the beam must be synchronized with the revolution of the particles in the ring,

$$\omega_{\text{RF}} = h_{\text{RF}} \omega_s \, , \quad \omega_s = \frac{\beta_s c}{R_0} \, , \tag{2.57}$$

where the integer h_{RF} is the harmonic number of RF, $2\pi R_0$ is the circumference of the ring, and ω_s is the frequency of the angular revolution of the synchronous particle. The increment of the synchronous energy $E_s = A m_0 c^2 \gamma_s$ from the nth to the $(n+1)$th turn of revolution is given by

$$E_{s,n+1} - E_{s,n} = q e \hat{V} \sin \phi_s \, , \tag{2.58}$$

where $\hat{V}$ is the peak RF voltage per turn, and ϕ_s is the synchronous phase.

Figure 2.26 shows the time structure of the SNS hydrogen beam from the injection linac to the accumulator for pulsed operation at a repetition rate of 60 Hz. The structure of the micro-pulse in the linac is shown at the bottom. Chopping is conducted in the LEBT (and MEBT by design) to create the mini-pulse structure to facilitate ring extraction. The final macro-pulse structure shown at the top is observed by the neutron production target.

Figure 2.27 shows the design drawings of the SNS accumulator ring layout. Upon completion of acceleration from the linac, the beam is transferred through the high-energy beam transport (HEBT) for injection into the ring. Upon completion

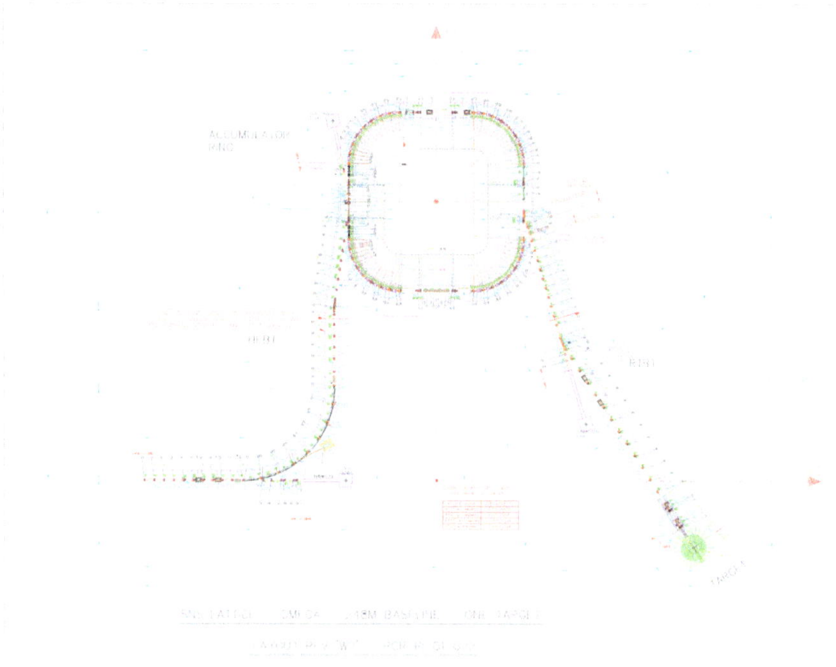

Fig. 2.27 Design drawings of the SNS HEBT, accumulator ring, and RTBT layout. (Courtesy J. Tuozzolo, BNL). The ring circumference is 248 m [131]

of accumulation in the ring, the beam, now stacked in a macro-pulse structure, is transferred through the ring-to-target transport (RTBT) for neutron production.

Table 2.5 lists the typical parameters of some proposed, operating, and retired rings for hadron acceleration and storage. The three types of synchrotron rings shown are (1) accumulators (AR) that only accumulate but do not accelerate beams; (2) rapid-cycling synchrotrons (RCS) that accumulate and accelerate beams; and (3) storage rings that accumulate, accelerate, and store beams for collision experiments.

2.3.1 *Functional Layout*

High-power synchrotrons typically contain arc sections to bend the circulating beam and straight sections to house beam-handling functions that include injection, capture and acceleration, extraction, and collimation. These functions are often distributed in different straight sections to facilitate beam loss control and hands-on maintenance. Figure 2.28 (left) illustrates the layout of the SNS ring [131]. Along the 248-m circumference of the machine, the guiding dipole and focusing quadrupole magnets are arranged to form a lattice with four-fold symmetry. The

Table 2.5 Typical parameters of some operating (unless noted otherwise) hadron ring examples (*AR* accumulator ring, *RCS* rapid-cycling synchrotron, *SR* storage ring)

Facility	Super-period	Beam type	Intensity [10^{13}/pulse]	Rep. rate [Hz]	Flux[a] [10^{20}/y]	Energy [GeV/u]	Power[b] [MW]	Type	Reference
SNS	4	p	14	60	840	1.0	1.4	AR	[131]
SNS PPU[c]	4	p	22	60	1,292	1.3	2.8	AR	[87]
J-PARC RCS	3	p	8	25	200	3	1	RCS	[132]
J-PARC MR	3	p	33	0.74	24	30	0.75	RCS	[133]
ISIS	10	p	2.5	50	125	0.8	0.16	RCS	[134]
CSNS	4	p	2.5	25	62.5	1.6	0.16	RCS	[135]
AGS Booster	6	p	2.3	7.5	17	1.5	0.042	RCS	[136, 137]
AGS	12	p	7	0.5	3.5	24	0.13	RCS	[138]
AGS[d]	12	Au	0.0004			9.8			[139]
RHIC	3	Au	0.0056	$[10\,\mathrm{h}]^{-1}$	$1.6 \cdot 10^{-7}$	100	$4.9 \cdot 10^{-6}$	SR	[140, 141]
PSR	10	p	3	20	60	0.8	0.077	AR	[142]
FNAL Booster	24	p	0.45	15	6.8	8	0.086	RCS	[143]
FNAL MI	2	p	6.5	0.94	6.1	120	1	RCS	[144]
CERN PSB[e]	16	p	1.3			1.4		RCS	[145]
CERN PS[e]	10	p	3.2			25		RCS	[146]
CERN SPS[e]	6	p	4.8	0.17	0.82	400	0.52	RCS	[147, 148]
LHC	1	p	32.3	$[10\,\mathrm{h}]^{-1}$	0.0009	7,000	0.01	SR	[149]
U1.5	12	p	0.09	1	0.09	1.32	0.00019	RCS	[150]
U70	12	p	1.7	0.11	0.19	70	0.021	RCS	[151, 152]
SIS18[f]	12	U	0.0004	2.3	0.0009	0.4	0.00014	RCS	[153, 154]
KEK PSB[g]	8	p	0.24	20	4.8	0.5	0.0038	RCS	[155]
KEK PS[g]	4	p	0.8	0.5	0.4	12	0.0077	RCS	[156]
IPNS[g]	6	p	0.35	30	10.5	0.5	0.0084	RCS	[157, 158]
Tevatron[g]	2	p, $\bar{p}$	1.08	$[10\,\mathrm{h}]^{-1}$	$3 \cdot 10^{-5}$	1,000	$4.8 \cdot 10^{-5}$	SR	[159, 160]
ESS[h,i]	3	p	47	50	2,350	1.334	5	AR	[161]
SSC[h]	6	p	13	$[24\,\mathrm{h}]^{-1}$	$1.5 \cdot 10^{-4}$	20,000	0.0048	SR	[162]

[a] 1 year is taken to be 10^7 s
[b] Average beam power output from the ring
[c] Goal beam power value not yet achieved
[d] In storage ring injection mode at BNL
[e] For CNGS operation at CERN
[f] For HEDP experiment at GSI
[g] Ceased operation
[h] Designed but not constructed
[i] Double ring design; beam power on target combining two rings

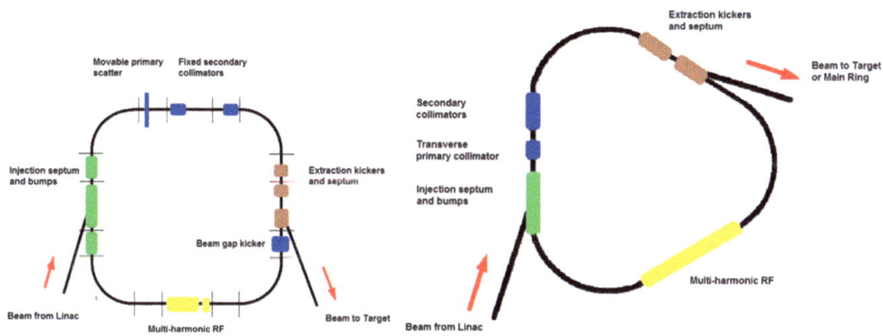

Fig. 2.28 Schematic layout of the (left) SNS accumulator ring and (right) J-PARC rapid-cycling synchrotron

four straight sections are designated for injection, collimation, the RF system for longitudinal containment, and extraction. Figure 2.28 (right) shows the layout of the J-PARC RCS, which has a circumference of 348 m [164]. The lattice superperiodicity is 3. The collimation region, which shares a straight section, follows the injection region. Extraction and RF acceleration occupy their respective straight sections.

2.3.2 *Lattice*

The backbone of a synchrotron is its optical lattice, the periodic magnetic structure that the circulating beam encounters. Lattices of high periodicity are preferred to reduce the impact of lower-order resonances on the transverse motion of particles excited by magnetic imperfections. On the other hand, for a ring of moderate circumference, high periodicity implies a lack of long, uninterrupted straight sections to accommodate dedicated functions. As a compromise, modern high-power synchrotrons often have a lattice periodicity of 2 to 4 (Table 2.5).

The circumference of the machine is determined largely by the space required to accommodate injection, collimation, RF, extraction, diagnostics, and possible upgrades. A large circumference also entails fewer injection turns, reduced foil scattering for H^- injection, and a lower particle density for better beam stability. The details of the lattice design are directly related to the choice of injection, acceleration, cleaning, and extraction schemes, to be discussed in Sects. 2.3.4, 2.3.5, 2.3.6, and 2.3.7, respectively.

The traditional choice is the focusing-drift-defocusing-drift (FODO) structure and its variations. FODO structures require modest quadrupole gradients, and the alternating amplitudes of the transverse beam easily accommodate magnetic correction systems that act selectively in the two transverse directions. An FODO lattice also is relatively insensitive to errors in quadrupole tuning. Figure 2.29 shows

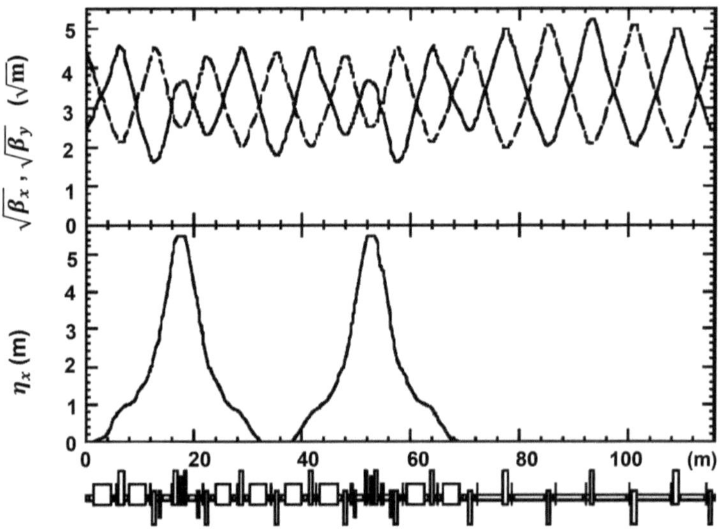

Fig. 2.29 Lattice of a superperiod of the J-PARC RCS. The machine lattice super-periodicity is 3 [163]. (Courtesy S. Machida)

a superperiod of J-PARC RCS consisting of FODO structures [164]. The split quadrupole creates a high-dispersion drift section for scraping the momentum halo and for chromatic adjustments. The lattice also yields a negative momentum slip factor so that the beam does not cross the ring's transition energy (the imaginary γ_T lattice), as elaborated in Sect. 3.8.

The SNS ring adopts a hybrid structure, consisting of FODO arcs and doublet straight sections. It combines the robustness and ease of correction of the FODO structure with the flexibility of the long drift of the doublet structure (12.5 m). The arcs and straights are optically matched to ensure maximum transverse acceptance. Each dipole is centered between two quadrupoles to maximize the vertical acceptance of the dipoles. Figure 2.30 shows one superperiod of the SNS ring consisting of the hybrid structure [131].

A lattice consisting of doublets/triplets has the advantage of encompassing long, uninterrupted straight sections for flexible injection and optimal collimation. Synchrotrons of this structure also have vacuum chambers with fewer segments and joints [165]. Figure 2.31 shows one superperiod of a proposed rapid-cycling synchrotron for the upgrade of the ISIS facility, consisting of doublet and triplet structures [166].

2.3.2.1 FODO Arc Cell

A common lattice base unit is a separated-function FODO structure, as shown in Fig. 2.32. The bending angle ϕ_c per FODO cell is usually determined by the

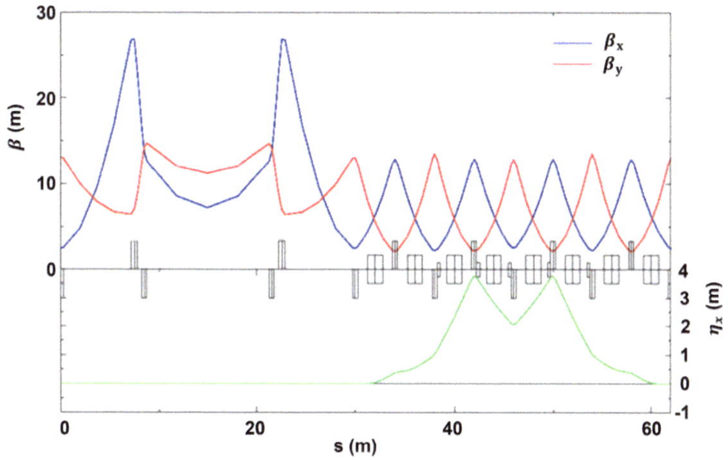

Fig. 2.30 Lattice of a superperiod of the SNS accumulator ring with working point ($\nu_x = 6.40$, $\nu_y = 6.30$). The machine lattice superperiodicity is 4 [131]

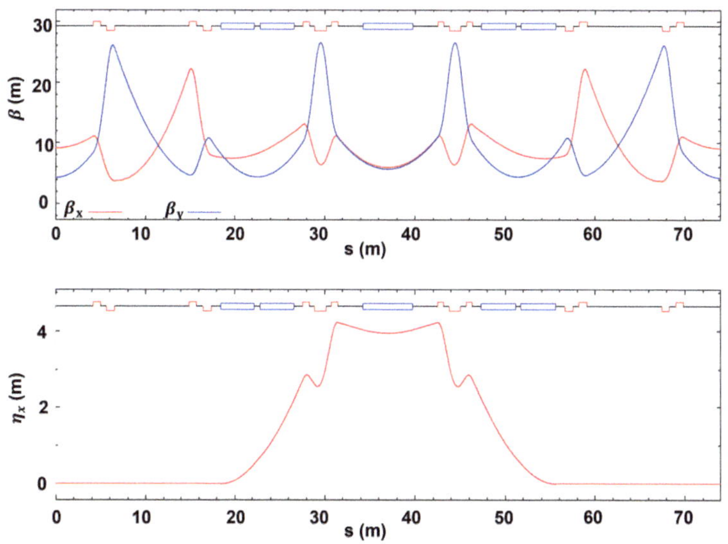

Fig. 2.31 Lattice of a superperiod of a 3.2 GeV rapid-cycling synchrotron for the upgrade of the ISIS neutron facility. The machine lattice superperiodicity is 5 [166]. (Courtesy G. Rees and C. Prior)

geometry of the ring and is often a compromise between the minimum number of arc cells and the maximum tolerable dispersion. As a design guideline, a low $\beta_\perp^+$ is generally preferred to maximize betatron acceptance, a low dispersion to maximize momentum acceptance, and a low $\beta_\perp^+/\beta_\perp^-$ ratio to reduce the possibility

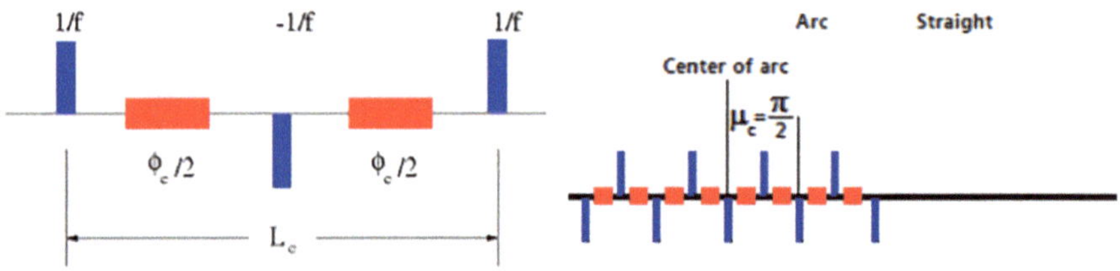

Fig. 2.32 (Left) The FODO lattice cell. (Right) A dispersion-suppressed arc consisting of 4 FODO cells

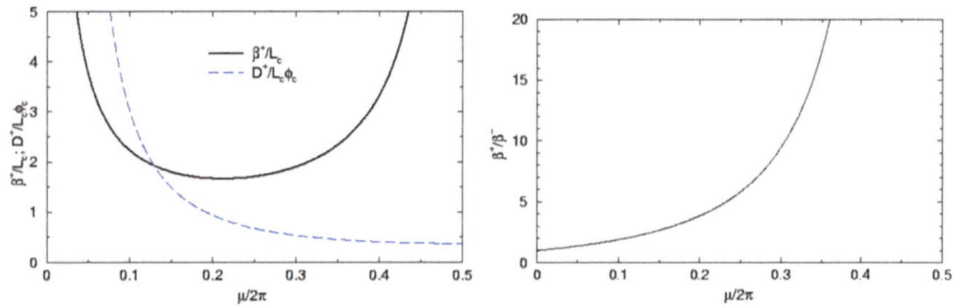

Fig. 2.33 Dependence of (left) maximum amplitude function and dispersion, and (right) maximum-to-minimum amplitude function ratio on the FODO cell phase advance [167]

of generating a beam halo. In the thin-lens approximation, these quantities are expressed as (Fig. 2.33)

$$\frac{\beta_\perp^+}{L_c} = \frac{1+\sin(\mu_c/2)}{\sin\mu_c}, \quad \frac{\beta_\perp^+}{\beta_\perp^-} = \frac{1+\sin(\mu_c/2)}{1-\sin(\mu_c/2)}, \quad \frac{D_\perp^+}{L_c} = \frac{1+\frac{1}{2}\sin(\mu_c/2)}{4\sin^2(\mu_c/2)}\phi_c. \tag{2.59}$$

2.3.2.2 Dispersion Suppressor

Dispersion-free regions are preferred for housing beam injection, extraction, RF acceleration, and transverse collimation, in which the transverse and longitudinal motions of particles are decoupled. Dispersion suppression is used to eliminate dispersion in the straight section without perturbing the lattice functions in the arc section.

A simple method to achieve dispersion-free straight sections is to design the total horizontal phase advance across each arc to be an integer multiple of 2π. Figure 2.32 (right) shows the lattice layout with a horizontal phase advance of 2π across the arc. The advantage of this scheme is that the arc is compact. The disadvantage is that the peak dispersion in the arc is higher than the matched value.

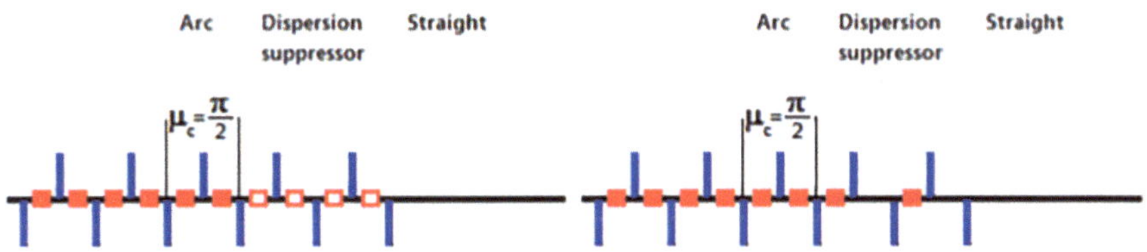

Fig. 2.34 Further examples of dispersion suppression by (left) using dipoles at half of the nominal field (hollow rectangles) when the phase advance per FODO cell is $\pi/2$, and (right) using "missing-dipole" scenario

Dispersion can also be suppressed using a dedicated insertion—the so-called half-field scheme and its variations. By halving the bending strength, a forced dispersion oscillation is launched around half of the matched value. For an FODO arc structure with horizontal phase advance μ_c per cell, the dispersion can be suppressed by π/μ_c FODO cells. If the phase advance per FODO cell in the arc is $\mu_c = \pi/2$, then two FODO cells are needed, as shown in Fig. 2.34 (left).

A half-field kick can be made using half-field dipole magnets for bending or by simply omitting the dipole magnet in one of the two FODO half cells, as shown in Fig. 2.34 (right). This arrangement maximizes usable dispersion-free drift space, avoids the need for different types of dipole magnets in the arc [168], and achieves a low matched peak dispersion in the arc.

2.3.2.3 Working Point

The horizontal and vertical tunes of the synchrotron, i.e. the working point (ν_x, ν_y), is chosen largely to avoid major linear and nonlinear resonances (Sect. 3.3.2). The selection of the fractional part of the tunes also affects the transverse instability threshold (Sect. 3.4.3). When horizontal and vertical motion are to be decoupled, a split-tune working point is usually adopted, with the tunes differing by at least half a unit. In the design stage, it is most important to maintain a wide tuning range (one unit or more) in both directions for operational flexibility (Fig. 2.35).

Transverse tunes are often dynamically varied during the beam cycle in synchrotrons to optimize the performance of the ring. In the ISIS synchrotron [169], the horizontal and vertical tunes are varied during the 10 ms cycle by adjusting the trim quadrupole magnets to (1) compensate for the natural chromaticity and the varying magnetic field at injection; (2) ramp up tunes to minimize the effects of space-charge depressions during beam capture; (3) reduce tunes during the time from 2 to 4 ms after injection to avoid transverse resistive wall instability; and (4) lower tunes to avoid coupling resonances at extraction. Such dynamic tuning is usually not used in accumulators because of the relatively short beam cycle.

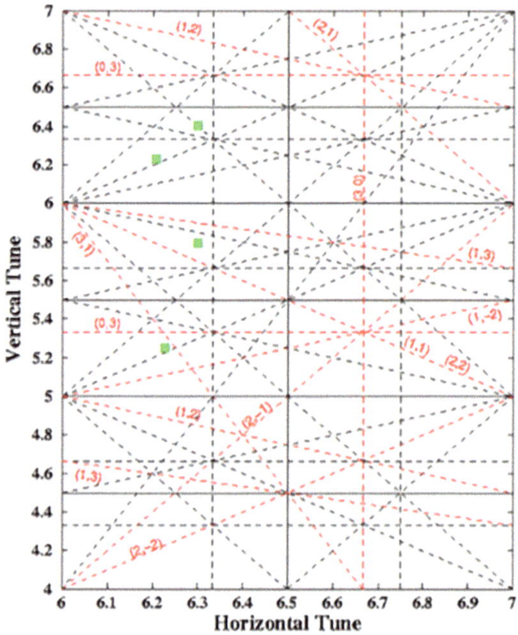

Fig. 2.35 Range of adjustment of the transverse tunes of the SNS ring over two units in the vertical direction and near a unit in the horizontal direction. The green dots indicate four preferred working points [131]

2.3.3 *Acceptance*

2.3.3.1 Transverse Acceptance

Acceptance is defined as the maximum phase-space area within which particles oscillate stably. A large acceptance is critical for high-power accelerators to achieve low-loss operational goals. The transverse acceptance is proportional to the transverse aperture of the accelerator. The longitudinal acceptance is proportional to the momentum aperture of the accelerator. Therefore, large transverse (horizontal and vertical) apertures and momentum acceptance are the most important aspects of low-loss accelerator designs.

The transverse aperture is usually designed to accommodate (1) betatron oscillations; (2) the off-momentum closed orbit due to dispersion (Fig. 2.36, left); (3) design orbit bumps (injection, extraction, diagnostics) (Fig. 2.36, right); (4) closed orbit deviation due to dipole field errors, beam coupling impedance, and coherent image charge effects; (5) an increase in the beam's amplitude function (β beating) due to quadrupole gradient errors; (6) degradation in dynamic acceptance due to nonlinear field errors; and (7) possible diffusion due to intrabeam scattering. For high-power synchrotrons, significantly more additional aperture must be reserved to facilitate beam collimation. For example, with the SNS ring design, the full (unnormalized) beam emittance at the end of injection painting is $\sim 240\,\pi$ mm·mr. The acceptance of the vacuum pipe is $\sim 480\,\pi$ mm·mr for particles with momentum deviation $\Delta p/p = \pm 1\%$ at all working tunes. This acceptance allows particles

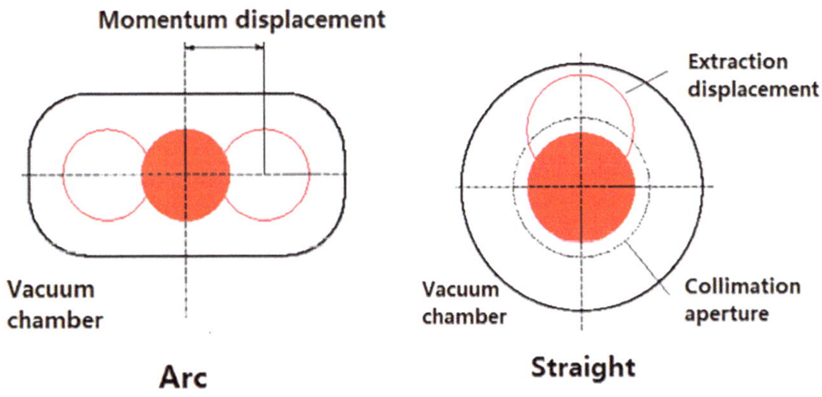

Fig. 2.36 Typical vacuum-chamber cross-sections of synchrotrons showing aperture consideration accommodating (left) off-momentum closed orbit in the dispersive arc region and (right) collimation aperture and extraction displacement in the dispersion-free straight regions

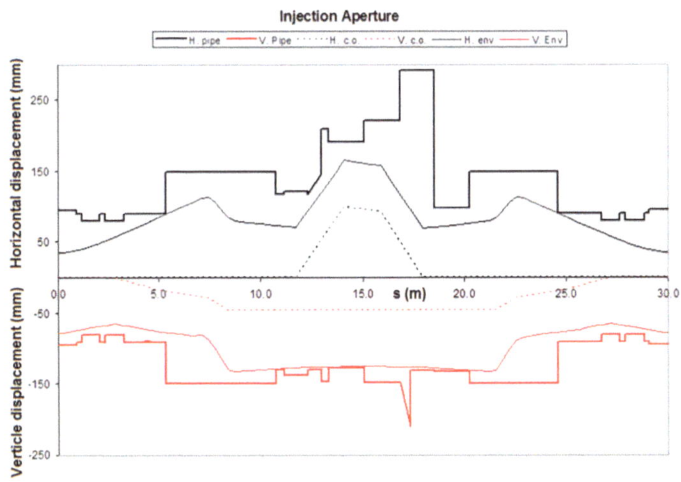

Fig. 2.37 The horizontal and vertical machine aperture relative to the transverse beam envelope in the injection region of the SNS ring. (Courtesy N. Catalan-Lasheras)

scattered by the collimators to return to them without becoming lost elsewhere in the ring. In a dispersive region, the vacuum chamber also accommodates a momentum acceptance of $\Delta p/p = \pm 2\%$ for the 240 πmm·mr beam emittance. The acceptances of the adjustable scraper and fixed collimators are set ~300 πmm·mr in dispersion-free regions. The difference in acceptance between the collimators and the rest of the ring's vacuum pipe ensures that collimation is highly efficient. Figure 2.37 shows the transverse vacuum pipe aperture in the injection region of the SNS ring relative to the transverse beam envelopes superimposed on the closed orbit displacement in the region.

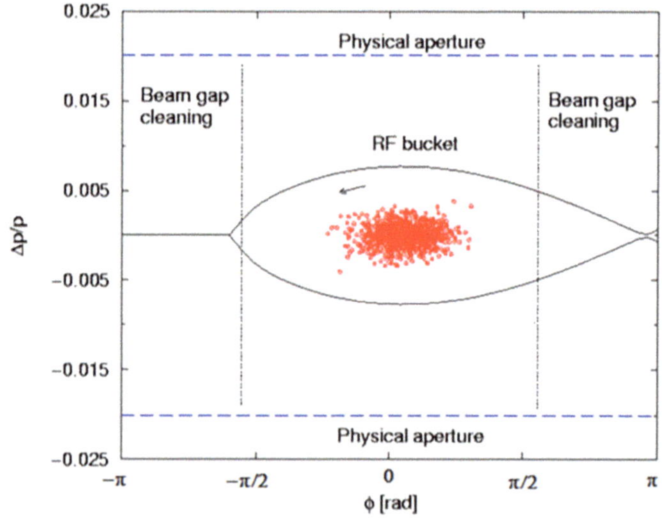

Fig. 2.38 Typical longitudinal phase space showing momentum acceptance determined by the horizontal aperture at dispersive region enclosing the RF bucket of stable synchrotron oscillation, and the space gap allowing beam gap cleaning [131]

2.3.3.2 Longitudinal Acceptance

The longitudinal acceptance is usually designed to accommodate (1) synchrotron oscillation; (2) RF manipulation; and (3) possible diffusion due to noise and intrabeam scattering. For high-power synchrotrons, extra clearance is reserved for multi-turn beam-halo collimation and to prevent accidental beam loss due to the malfunction of injection and the misfiring of the extraction kickers [131]. Figure 2.38 illustrates the longitudinal phase space of the beam contained within an accelerating RF bucket. The vacuum chamber has extra clearance to facilitate longitudinal collimation and beam-in-gap cleaning (Sect. 2.3.6). It also leaves room for future upgrades to the RF system for faster ramping and for damping instabilities (Sect. 3.4.3).

2.3.4 Injection and Phase-Space Painting

Injection refers to the process of transferring the beam from a linac to a ring for accumulation and possible acceleration, or from one ring to another for further acceleration. Linac-to-ring or cyclotron-to-ring injection is usually accomplished by either multi-turn injection using a septum and an orbit bump, or multi-turn charge exchange injection with a stripping foil. Ring-to-ring injection is usually performed by single-turn injection with septum and kicker magnets. High-power rings are prone to radio-activation caused by controlled and uncontrolled beam loss occurring during the injection time. For accumulator rings with charge-exchange injection, injecting particles that strike the stripping medium causes radio-activation in the injection region.

For synchrotrons, beam loss often occurs in the early stages of the acceleration ramp. During acceleration, the transverse amplitudes of oscillation decrease according to the adiabatic damping of the phase space. Thus, the beam size is usually largest at the injection energy. At low energies, space-charge forces produce large shifts in transverse tunes because there is little cancellation between electric and magnetic forces at non-relativistic energies. The combination of these two effects makes injection the most challenging period for preserving beam emittance and minimizing beam loss during the synchrotron acceleration cycle.

A low-loss injection further requires a high-quality injecting beam with little centroid jitter and controlled emittance, adequate uniformity in the magnetic field along the beam path in the septum and subsequent magnets, an adequate fast rise time and flat waveform in the kicker or bump magnets and their power supplies, and an RF system capable of containing the transient beam-induced field ("beam loading"; [170]).

2.3.4.1 Single-Turn Injection

Single-turn injection is typically performed using a septum magnet and a kicker magnet that is $\Delta\mu_x$ apart in the betatron phase, as shown in Fig. 2.39. The matched injection requires the kick strength θ_{kick} to be

$$\theta_{\text{kick}} = \frac{x_{\text{sep}}}{\sqrt{\beta_{\text{sep}}\beta_{\text{kick}}}\sin\Delta\mu_x}, \quad x'_{\text{sep}} = -\frac{\alpha_{\text{sep}} + \cot\Delta\mu_x}{\beta_{\text{sep}}}x_{\text{sep}}. \tag{2.60}$$

Transversely, the injecting beam's profile must be matched to the accelerator lattice to preserve the emittance. To minimize the kicker strength, the phase advance between the septum and the kicker is preferably close to $\pi/2$, and the locations of the large amplitude function β_{sep} and β_{kick} are preferred.

Longitudinally, the frequency of the RF system must be synchronized, and the aspect ratio between the momentum and phase spread needs to be matched to the contours of stable synchrotron oscillations. Such single-turn injection is often used to transfer a bunched beam from one ring to another.

2.3.4.2 Multi-turn Direct Injection

Multi-turn injection is typically accomplished using a septum magnet and several programmable orbit-bump magnets, as shown in Fig. 2.40. The ribbon-like beam of the injecting linac or cyclotron is "wrapped around" the circumference of the ring over many turns. The maximum number of injection turns, typically several tens,

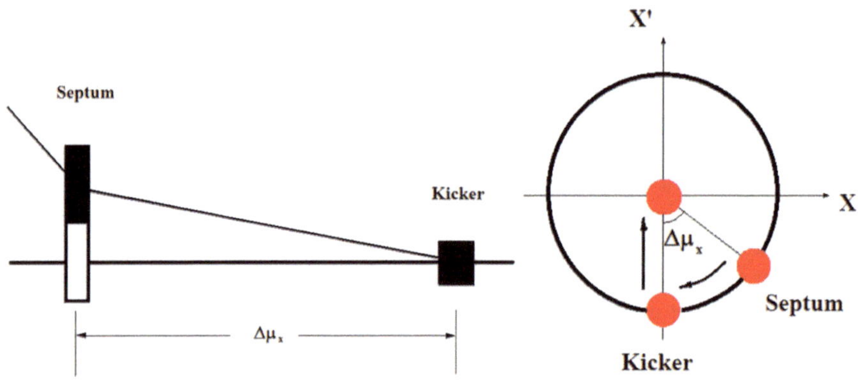

Fig. 2.39 (Left) Single-turn injection using a septum magnet and a kicker magnet. (Right) Illustration of the single-turn injection process with the normalized phase space

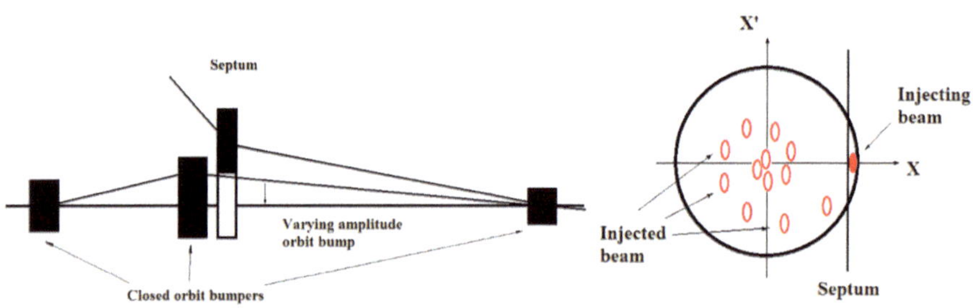

Fig. 2.40 (Left) Multi-turn injection using a septum magnet and four programmable bump magnets and (right) illustration of the injection process with the normalized phase space showing phase-space "painting"

depends on the collapse rate of the orbit bump and the ratio of the ring acceptance to the injecting beam emittance. The efficiency of injection can be optimized by a specific mismatch satisfying the following relations when the curvature of the injecting beam matches that of the circulating beam:

$$\frac{\beta_i}{\beta_x} \approx \left(\frac{\epsilon_i}{\epsilon_x}\right)^{\frac{1}{3}}, \quad \frac{\alpha_i}{\beta_i} = \frac{\alpha_x}{\beta_x} = -\frac{x_i'}{x_i}, \tag{2.61}$$

where the horizontal C-S parameters and emittance for the injected beam are $(\beta_{x,i}, \alpha_{x,i}, \epsilon_{x,i})$ and for the ring $(\beta_x, \alpha_x, \epsilon_x)$, and the input beam's center relative to the instantaneous injection orbit bump is (x, x') [171]. The efficiency of multi-turn injection can be increased by a transverse coupling using skew quadrupole magnets (as in the CERN PSB and the AGS Booster).

Longitudinally, the multi-turn injected beam can be captured adiabatically by raising the voltage of the radio frequency cavities, as shown in Fig. 2.41 [173]. Beam loss often occurs during the injection period, during capture, and also during initial

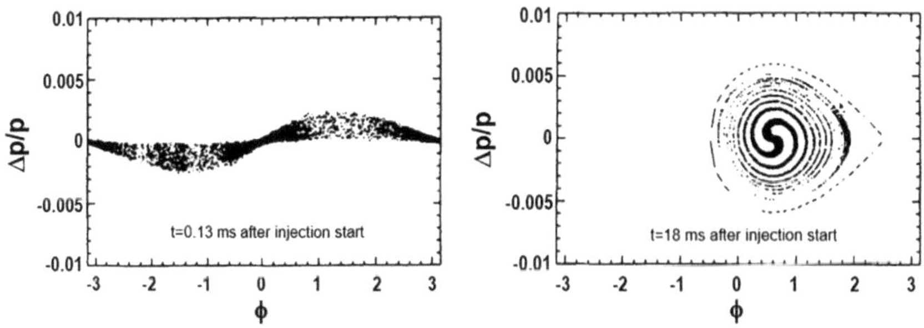

Fig. 2.41 Longitudinal phase-space diagram of RF capture of multi-turn injected $^{197}Au^{+33}$ beam in the AGS Booster. The simulation is performed with codes TIBETAN [172]

acceleration (or "ramping") when the phase-space area of stable longitudinal motion tends to decrease with increasing synchronous phase. The voltage amplitude of the ring RF system must be programmed according to the rising synchronous phase to ensure a monotonically increasing RF bucket area.

2.3.4.3 Multi-turn Charge-Exchange Injection and Phase-Space Painting

High-power rings prefer charge exchange injection because it allows a large number of revolutions (e.g. $\sim$1,000 turns) to be injected and controlled to achieve the desired beam distribution. The constraints imposed by Liouville's theorem on conventional multi-turn injection do not apply since the injecting ions are stripped within the acceptance of the ring. The programmed orbit bumps can be arranged in both horizontal and vertical directions to facilitate "phase-space painting" [131, 174, 175], as shown in Fig. 2.42. The key issues in charge-exchange injection are associated with stripping (Sect. 3.7.2) and collecting the electrons from the injecting beam.

ISIS injects in a high-dispersion region [165, 171]. The injection dipole magnet, part of the periodic lattice structure, separates the injecting and circulating beams without using a septum. Thus, the arrangement of the injection magnets is simplified. The high dispersion at injection also facilitates the collection of the momentum halo (Sect. 2.3.6).

In contrast, SNS injects in a zero-dispersion straight region (Fig. 2.42), thereby allowing independent control of the painting of the phase space both transversely and longitudinally [131]. The injection process is more tolerant of deviations in linac energy. A long, straight section is preferred to contain the entire injection chicane so that the effects of lattice tuning can be minimized. Intentionally mismatched injection can noticeably reduce the foil traverses [176].

Phase space painting during multi-turn injection refers to the process of populating the phase space with the injected beam to achieve a desired distribution. Transverse painting alleviates the space charge limit and controls the uniformity

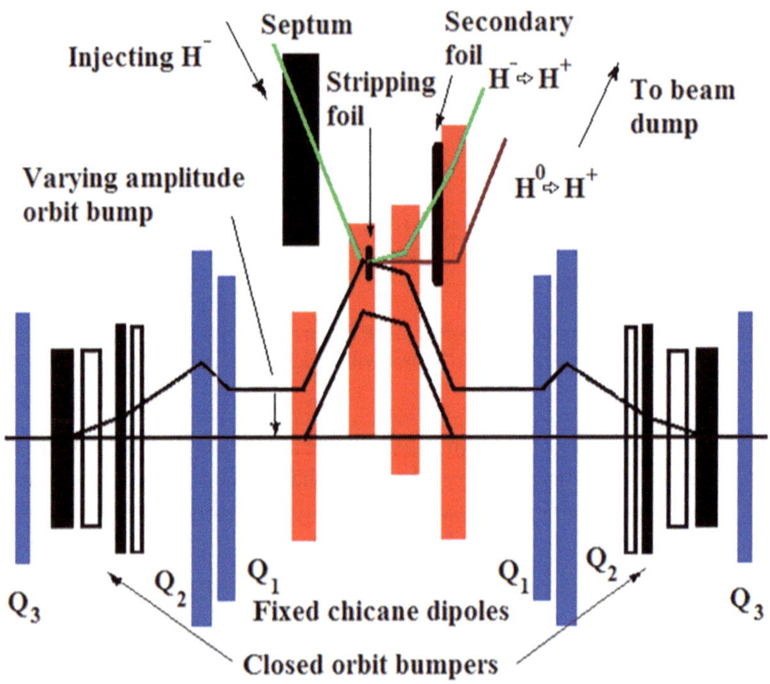

Fig. 2.42 Charge-exchange injection layout in the horizontal direction of H- beam in a dispersion-free region using a fixed chicane (red bars) along with dynamic orbit bump [131]. The blue bars indicate lattice quadrupole magnets. The solid and hollow black bars along the ring beam line indicate horizontal and vertical bumpers, respectively

and shape of the beam profile. Longitudinal painting provides the momentum spread required for beam stability.

Transverse painting can be performed in several ways: by varying the closed orbit bump in the ring; varying the angle of the injecting beam; or varying the energy of the injecting beam upon injection in a dispersive location. As shown in Fig. 2.43, transverse correlated painting using parallel horizontal and vertical orbit bumps generates a rectangular transverse profile. Its advantage is that the beam halo is constantly painted over by the freshly injected beam. SNS design adopts such correlated painting to realize equal transverse emittances, operating near the coupling resonance to achieve a fully coupled, circular distribution in the physical dimension.

The anticorrelated transverse painting shown in Fig. 2.44 utilizes both horizontal and vertical orbit bumps, one with increasing and the other with decreasing closed orbit bump amplitudes, respectively. Ideally, this painting scheme produces a distribution with an elliptical transverse profile and a uniform density distribution. This configuration can also be obtained by painting in one direction and steering in the other (Fig. 2.45), as adopted by the J-PARC RCS.

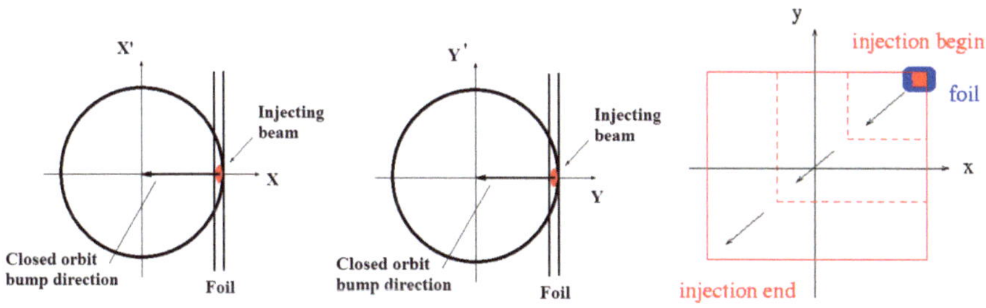

Fig. 2.43 Correlated horizontal and vertical phase-space painting of the charge-exchange injection with collapsing closed orbit in both directions illustrated with (left) the horizontal phase space; (center) the vertical phase space; and (right) the transverse physical space

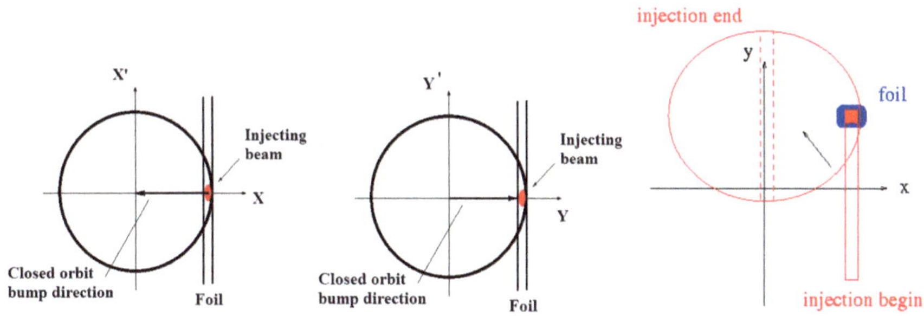

Fig. 2.44 Horizontal and vertical phase-space painting of the charge-exchange injection with anti-correlated closed orbit variations illustrated with (left) the horizontal phase space; (center) the vertical phase space; and (right) the transverse physical space

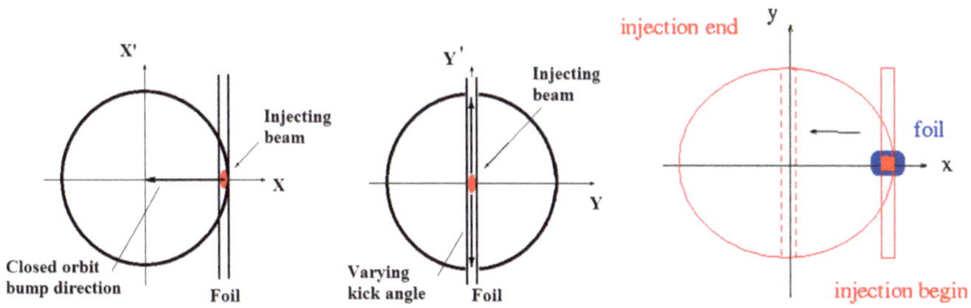

Fig. 2.45 Horizontal and vertical phase-space painting of the charge-exchange injection with collapsing horizontal closed orbit and oscillating-amplitude vertical kick illustrated with (left) the horizontal phase space; (center) the vertical phase space; and (right) the transverse physical space

Longitudinal painting can be performed in several ways: using an "energy spreader" RF cavity to vary the energy of the injecting beam; injecting while changing the synchronous energy of the ring (ramping); or manipulating the longitudinal phase-space distribution of the injected beam in the ring with broad-band devices such as a barrier cavity [177–179] or diffusion cavity [180]. With dispersion-free injection, longitudinal painting can be achieved using an "energy

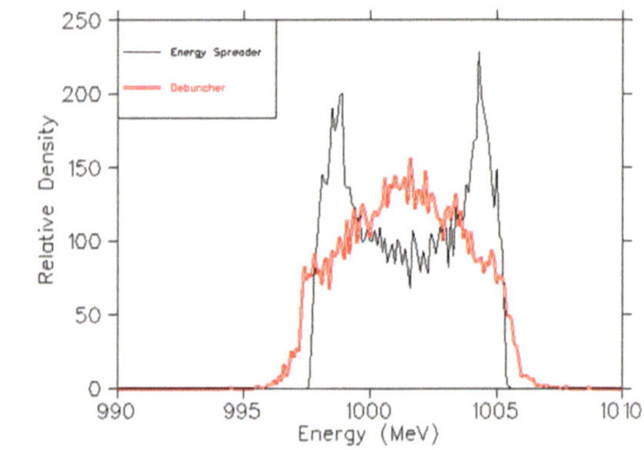

Fig. 2.46 Longitudinal painting provides the momentum spread required for beam stability without introducing an excessive momentum halo [167]. (Courtesy M. Blaskiewicz and D. Raparia, BNL)

spreader" RF cavity that is phase modulated with respect to the linac frequency. To facilitate such a painting scheme, the jitter and spread of the momentum of the injection beam should be controlled, possibly by an "energy corrector" RF cavity synchronized to the linac frequency. The original SNS design uses this arrangement to provide the momentum spread required for beam stability without introducing an excessive momentum halo (Fig. 2.46) [181].

To prepare for charge-exchange injection, the coasting beam is usually pre-chopped at low energy to create a clean gap between successive beam bunches. RF operating at multiple frequencies can be combined to shape the RF bucket and flatten the peak beam density in several ways. For a dual-harmonic system, the frequency of the secondary RF system is chosen to be an integral multiple of the fundamental RF system. Typically, the ratio of the operating voltage amplitude is inversely proportional to the ratio of the RF frequency. With this choice, the frequency of synchrotron oscillation extends from zero in the center of the bucket to a maximum value in its core of approximately 1.2 times the maximum synchrotron frequency of a single harmonic system [182, 183]. Thus, the spread of synchrotron frequency in the bunch is usually greatly enhanced.

2.3.5 Acceleration and RF Gymnastics

Acceleration in rings is performed by RF cavities whose frequency ω_{RF} is synchronized to the varying revolution frequency of the accelerated ions, as shown in Eq. 2.57. The magnetic field is varied proportionally to the momentum of the ions, as shown in Eq. 2.55. Compared with a linear accelerator predominantly composed of accelerating cavities, for which the acceleration gradient is crucial in minimizing the facility footprint and construction cost, a ring lattice structure consists mainly of bending and focusing magnets that confine the beam. The acceleration gradient in a ring becomes demanding when rapid cycling is desired and when the circumference of the ring is limited.

During the acceleration cycle, the longitudinal phase space area of single-particle motion ideally remains constant. To avoid beam loss due to a possible reduction in the phase-space area of stable longitudinal motion, the voltage amplitude and the synchronous phase of the ring RF system must be continuously programmed to ensure a monotonically increasing RF bucket area, similar to the precautions taken during the initial ramping near injection discussed in Sect. 2.3.4.

2.3.5.1 Requirements from Rapid-Cycling

Magnet power supply and RF systems represent the main engineering challenges associated with the rapid cycle. Rapid-cycling synchrotrons require a large amount of electrical power (typically tens of megawatts) for magnet and RF systems. As a general rule, the higher the repetition rate, the larger the peak power drawn from the power grid, and the more likely it is that local energy storage is needed. Thus, most existing rapid-cycling synchrotrons use resonant power systems.

With a resonant power system, the strength of the magnetic field varies sinusoidally, along with the rigidity of the beam. The consequences are the lack of a "flat bottom" for a static injection condition and the demand for high voltage amplitude by the RF system to sustain the largest ramp slope of the sinusoidal waveform. The quality of the magnetic field usually suffers from current-induced effects and saturation. A variation is a dual-harmonic ramp with a slower slope during the up ramp (acceleration) and a faster slope during the down ramp [167].

To satisfy the increasing need to accurately control the injection "flat bottom" and extraction "flat top," and to have a flexible magnet current program, high-frequency AC bridge rectifier systems are being developed, such as those based on IGBT (insulated gate bipolar transistor) rectifiers. A programmable ramp using this technology provides maximum flexibility in manipulating the profile of the rise and fall, thereby reducing the peak ramp rate and current-induced magnetic imperfections.

RF "beam loading" refers to the transient response of the RF system to variations in the beam's current, frequency, and energy [184–186]. Compensation for beam loading aims to maintain a stable accelerating voltage in the presence of varying beam conditions. The effect of beam loading is simplified by using the AR option of a fixed-energy accumulator ring. New challenges are associated with compensating for high-intensity beam loading in the magnetic alloy cavities, where the broadband resonance spans several RF harmonic frequencies (Sect. 8.5.3).

2.3.5.2 Limitation from Acceleration-Induced Eddy Current

Varying magnetic fields in RCS produce eddy currents that cause heating in the magnets and in the beam vacuum chambers beneath the magnets. The heating becomes prohibitively severe with RCS at high repetition rates and large apertures for high-power beams. Sections 8.7.2 and 2.3.9 describe various mitigation methods. Typically, the peak magnetic field at the pole tip is limited to about 1.2 T, and the peak ramp rate is limited to about 50 T/s.

2.3.5.3 RF Gymnastics

RF gymnastics, in general, refers to the manipulation of parameters such as the phase and voltage of one or more RF systems, and of one or more RF harmonics h_{RF} to change the characteristics of the beam in the longitudinal phase space [187]. Commonly used "gymnastics" practices include adiabatic capture at injection (Fig. 2.41); bunch compression and rotation in preparing the beam for transfer between accelerators; controlled blow-up to avoid instabilities; slip-stacking to enhance beam intensity; debunching, rebunching, and splitting; and broadband RF manipulation with barrier/isolated buckets.

2.3.6 Collimation and Cleaning

Beam collimation and cleaning play an indispensable role in high-power rings. The purpose of beam scraping, collimation, cleaning, and collection is to localize beam loss to controlled, shielded areas under both normal operation and fault conditions, thereby reducing uncontrolled radioactivity so that the accelerator facility is maintainable. This section describes collimation processes specific to rings and transport lines. Section 8.10 describes general beam-intercepting systems and hardware.

2.3.6.1 Charge-Exchange Collimation

An efficient method for the collimation of non-fully stripped ions is to use a charge-exchange process to remove unwanted ions. The unwanted beam is usually cleaned with a stripping foil, and then a magnetic field separates the stripped beam from the original beam. Such a collimation can be performed in both the transverse and longitudinal directions.

Figure 2.47 shows the charge-exchange collimation of the H^- beam in the transverse phase space. Collimating foils at two locations about $\pi/2$ apart in phase advance can clean the beam halo in a single pass. Downstream of the collimating foils, shielded collectors are placed to catch derailed, stripped beams. An additional number of collimation locations at optimized phase advances further reduce the escaping radius and improve efficiency. Prior to charge-exchange injection into the ring, the beam can be prepared during transport by cleaning the halo of the beam

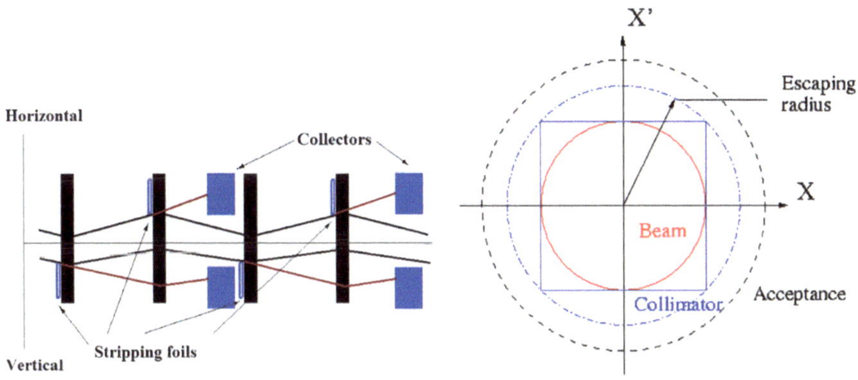

Fig. 2.47 (Left) Schematics of charge-exchange collimation of H^- beam. (Right) Illustration of general transverse collimation in a beam channel using two collimators located $\pi/2$ apart in phase advance

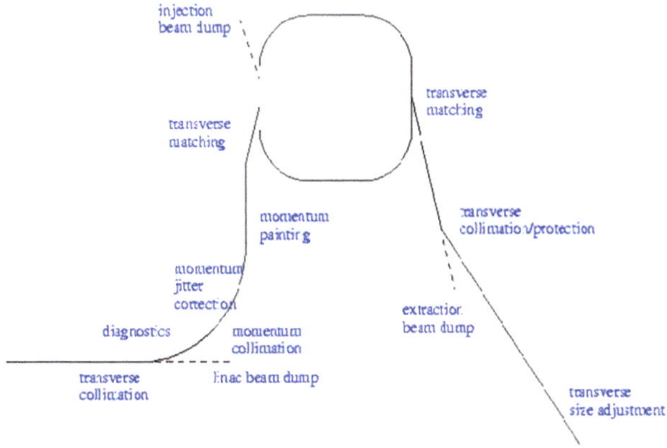

Fig. 2.48 Schematics of the SNS ring and the connecting transport lines from the linac and to the neutron target

from the linac, mitigating the effects of source and linac malfunctions, and thus reducing injection activation (Fig. 2.48).

Charge-exchange collimation can be performed in the longitudinal direction by using a collimating foil at a location of high dispersion along the beam trajectory, followed by a collector down-stream. Figure 2.48 shows the schematic layout of the SNS accumulator ring and the connecting transport lines from the linac to the target. An achromatic bend creates dispersion for cleaning the momentum halo. The charge-exchange momentum (longitudinal) collimation is located near the middle of the 90° bend, where the dispersion is at its maximum.

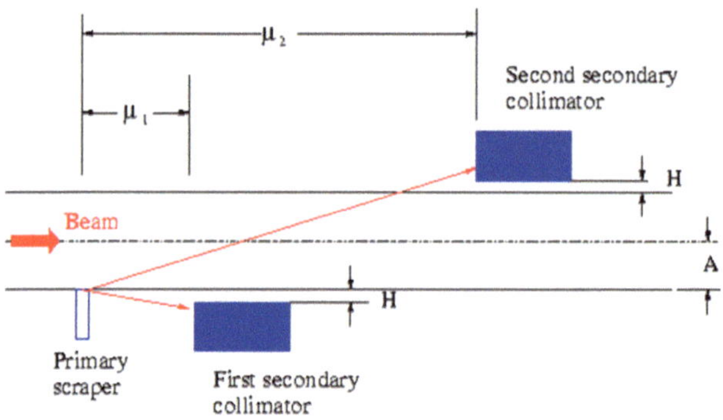

Fig. 2.49 Schematics of two-stage beam collimation for fully stripped ions

2.3.6.2 Two-Stage Transverse Collimation

Two-stage collimation is often used with fully stripped ions, such as protons (Fig. 2.49). The primary scraper, often adjustable, scatters the particles at the edge of the beam and enhances their impact distance in the secondary collectors, thus achieving a high cleaning efficiency [188]. The primary scraper is often placed at a location of high amplitude function, and the secondary collectors can be placed at locations of selected phase advances μ_1 and μ_2 for optimum collimation efficiency. The escape radius can be minimized if the conditions

$$\mu_1 = \cos^{-1}\left(\frac{A}{A+H}\right), \quad \text{and} \quad \mu_2 = \pi - \mu_1 \tag{2.62}$$

are satisfied, where A and $A + H$ are the scraping radius and collection radius, respectively, as illustrated in Fig. 2.50.

Efficient collimation requires an adequate acceptance ratio between the scraper, collimators, and the rest of the vacuum pipe. With the SNS ring, the proton beam injected into the ring is painted to horizontal and vertical emittances in the range from 120 to 160 πmm·mr. The primary scraper can be adjusted around an admittance of 240 πmm·mr. The fixed collectors are placed at admittances of ~300 πmm·mr. The admittance of the vacuum chamber of the machine is ~480 πmm·mr. Transverse collimation can be performed efficiently within a dedicated straight section of the ring.

J-PARC RCS uses a two-stage system for transverse collimation, plus an additional scraper for longitudinal collimation, all located near the injection area, as shown in Fig. 2.51 [189]. The two-stage collimation system consists of a primary scraper and five secondary collectors, all located in the dispersion-free region. The thin primary scraper scatters halo particles, and the secondary collectors absorb those particles. The beam loss can be localized in a restricted region where the phase advance from the primary collimator is less than 180°. The longitudinal

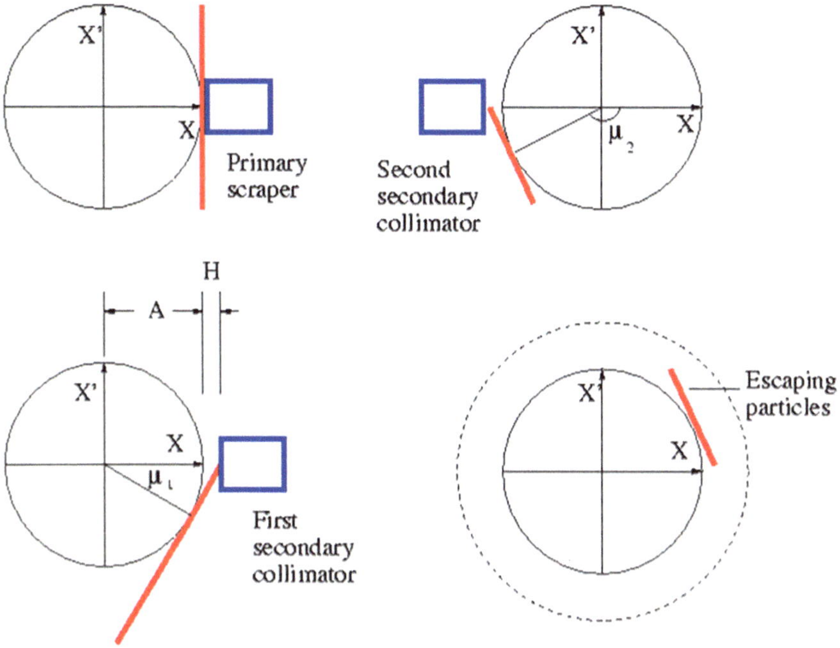

Fig. 2.50 Schematics in normalized transverse phase space of a two-stage beam collimation with a primary scraper and two secondary collectors placed at chosen phase advances [167]

collimation system is composed of a longitudinal primary scraper located near the maximum dispersion location and two-stage transverse collimators. Both primary scrapers have apertures restricting emittance to 324 πmm·mr and 1% momentum deviation. All secondary collector apertures are set at 400 πmm·mr. All other vacuum chambers are designed for a beam with >486 πmm·mr emittance and 1% momentum deviation [190].

2.3.7 *Extraction*

Extraction refers to the process of transferring the beam from the ring to a transport line upon completion of acceleration and/or accumulation, leading to the production target, user stations, or another ring for further acceleration. High-power applications usually employ fast, single-turn extraction with high-repetition acceleration or accumulation. For low-power operations that require more uniform "spill" and can also tolerate a higher fraction of beam loss, multi-turn slow extraction is sometimes used.

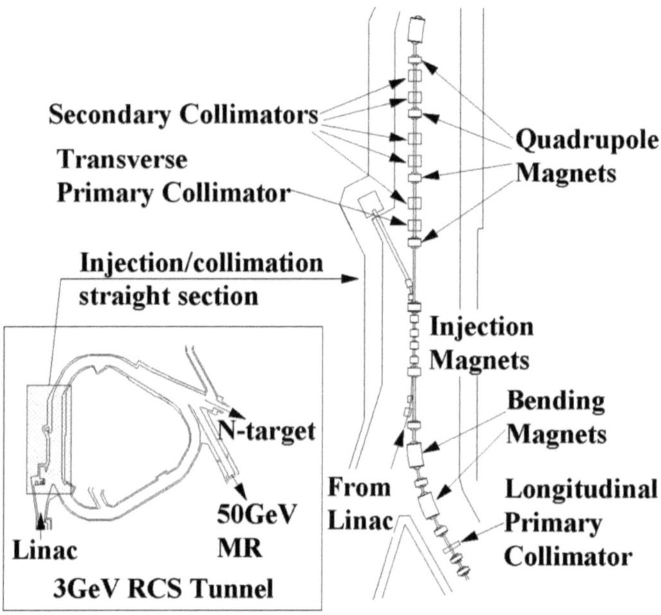

Fig. 2.51 Schematics of the collimation system of the J-PARC RCS [189]. (Courtesy K. Yamamoto, J-PARC)

2.3.7.1 Fast Extraction

The single-turn fast extraction is essentially the reverse process of the single-turn injection described in Sect. 2.3.4.1. The beam is extracted during a single turn by being kicked into the septum magnet extraction channel, as shown in Fig. 2.52. The average deflection in transverse momentum is given by

$$\langle\theta_{\text{kick}}\rangle \approx \frac{x_{\text{sep}}}{\langle\sqrt{\beta_{\text{kick}}}\sin\Delta\mu_x\rangle\sqrt{\beta_{\text{sep}}}}, \tag{2.63}$$

where x_{sep} is the design displacement at the extraction septum, β_{sep} and β_{kick} are the amplitude functions at the septum and kicker, respectively. Similarly to single-turn injection, the optimum average phase advance μ_x between the kicker and the septum is $\pi/2$.

Typical beam loss associated with single-turn extraction is attributed to (1) accidental misfiring of the kicker power supply; (2) residual beam in the beam gap between adjacent bunches; (3) lack of transverse aperture in the extraction channel; (4) lack of momentum aperture in the dispersive sections of the extraction channel; and (5) the slow rise and non-flatness of the kicker pulse waveform. A low-loss design for fast extraction must address these issues.

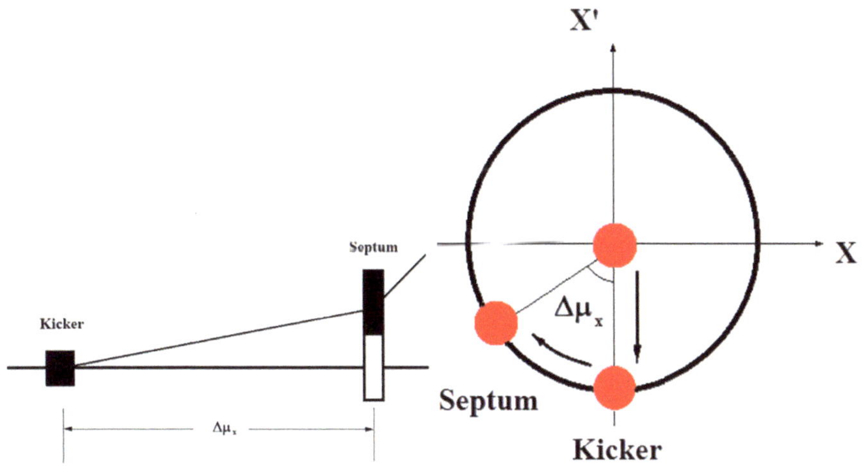

Fig. 2.52 (Left) Single-turn extraction using a kicker magnet and a septum magnet. (Right) Illustration of the single-turn extraction process with the normalized phase space

Figure 2.53 shows the lattice layout and the closed orbit displacement of the single-turn extraction of the SNS accumulator ring. The 700 ns beam pulse is vertically kicked by 14 kicker modules, each contributing about 1.5 mrad of deflection, and is then extracted horizontally with a Lambertson-type septum magnet [191]. The number of multiple kicker modules is chosen so that the beam loss is negligible, even when one module fails. The pulse-forming network (PFN) is installed outside the ring tunnel for easy maintenance. The phase advance between the extraction kicker and the target is also chosen so that the position of the beam on the target does not change if errors develop with the kickers. To avoid complications due to rapid variations in the magnetic field, kicker magnets are made with a single-turn coil winding and are placed inside the vacuum chamber. Good field quality is achieved with a high-permeability ferrite material. The typically high beam coupling impedance associated with the high-permeability material can be reduced, and the undesirable resonance structure can be avoided by selecting the termination in the pulse-forming network (PFN) circuit (Sect. 3.4.3.1). A saturable inductor can isolate the influence of the PFN, shorten the rise time, and improve the flatness of the kicker-pulse waveform [192]. Efforts are made to reduce the transverse coupling impedance, which may lead not only to instabilities but also to intensity-dependent deflection that produces a closed orbit deviation with peak amplitudes varying longitudinally from the center to the edge of the beam pulse.

2.3.7.2 Multi-turn Extraction

The Multi-turn extraction (MTE) extracts a beam of particles over several turns instead of just one by splitting the beam into multiple “beamlets” in the transverse

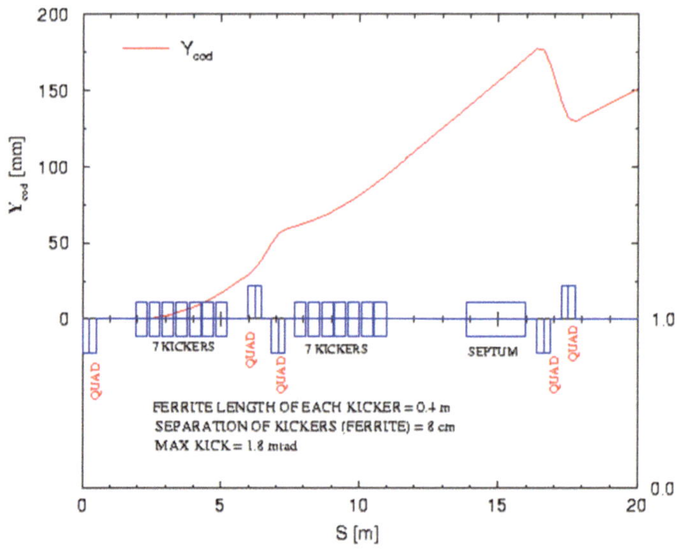

Fig. 2.53 SNS accumulator ring single-turn extraction layout and closed-orbit displacement [191]. (Courtesy N. Tsoupas, BNL)

phase space, thanks to the crossing of a specific transverse resonance excited by non-linear elements that create stable islands in the transverse phase space, in which the beamlets are trapped. Then, the beamlets are extracted sequentially over multiple consecutive turns. This method is particularly useful when transferring a beam between accelerators with different circumferences to minimize fill time. This technique is in use at CERN's PS, injecting five beamlets into the SPS, with the beamlets created in the horizontal plane by crossing a fourth order (0.25) resonance. It was developed to replace the older continuous transfer method because of increased transfer efficiency [193–195], decreasing losses from about 10% to a few percent. In the specific case of the CERN PS, the beam has to be debunched before extraction To further reduce beam losses created during the rise time of the extraction kickers, a barrier bucket has been used to create a gap in the longitudinal distribution. This brought the losses below the % level [196, 197].

2.3.7.3 Slow Extraction

Slow extraction typically uses a half-integer or third-integer resonance of transverse motion to achieve a slow spill of the beam through the unstable region. Third-integer resonance is often preferred for slower and more controllable spill. Higher-order resonances are not used because the angular separation of the separatrices is too small. As shown in Fig. 2.54, the slow bumper moves the beam close to the septum. The horizontal tune is adjusted near the third integer horizontal resonance. The

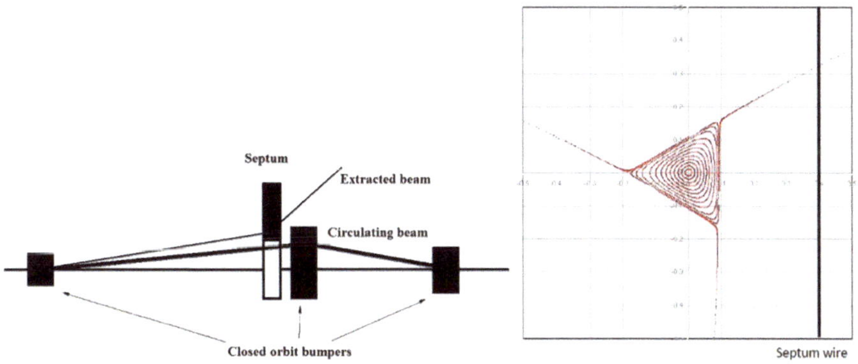

Fig. 2.54 (Left) Slow extraction using the third-integer resonance with a slow bumper, a septum magnet, and a set of sextupole magnets (not shown). (Right) Illustration of the slow extraction process in the horizontal phase space

sextupole magnets are excited to define the stable region in the phase space. The beam that extends along the separatrix across the septum is extracted.

Beam loss is inevitable with resonance-based slow extraction. A low-loss design focuses on optimizing the orbit bump, minimizing beam loss on the electrostatic and magnetic septum, , and optimizing the low-ripple sextupole power supply to maximize extraction efficiency.

Slow extraction using a bent crystal has been successfully demonstrated and utilized at several facilities, including CERN's SPS [198, 199], IHEP's U70 [200, 201], FNAL's Tevatron [202], and LNF's DAFNE [203].

2.3.8 *Transport Lines*

Transport lines not only transfer the beam but also prepare it for the next stage of action. Figure 2.48 shows the transport line connecting the SNS superconducting linac to the accumulator ring and linking the ring to the spallation neutron target.

The transport line that connects the linac to the ring serves the purposes of transverse optical matching, beam diagnostics, fault protection, beam collimation, and injection preparation. Beam collimation in the transport line before ring injection is especially important for accumulators, where both beam energy and beam intensity are high compared to the RCS. The layout of the transport line can be separated into straight sections of zero dispersion and achromatic arcs of high peak dispersion. The straight section is used for transverse beam diagnostics, transverse collimation, momentum spread and jitter correction, and momentum painting, whereas the arc is used for momentum tail cleaning, momentum collimation, and beam energy characterization.

The transport line that connects the ring to the target similarly carries out cleaning and protection functions. Transverse matching is usually performed on lattice amplitude functions, dispersion functions, and their derivatives. The locations of the collimators are strategically chosen with a betatron phase advance of $(n + 1/2)\pi$ (n is an integer) from critical elements such as extraction kickers to capture accidental beam losses due to possible malfunctions. The transverse phase advances between the kickers and the target can be designed to be near an integer multiple of 2π, so that the kicker error does not displace the beam at the target. The achromatic bend serves to reduce the direct "back shine" irradiation of the target. The final set of radiation-resistant quadrupole magnets downstream of the bend adjusts the beam size to the desired value at the target. For both beam dumps and beam targets, adequate beam diagnostics (such as harps, fast beam loss monitors) and fast machine protection systems are necessary.

2.3.9 *Error Correction and Compensation*

As the motion of particles in a ring is mainly governed by steering and focusing magnets, errors and imperfections in the design of magnetic fields often lead to resonance effects that compromise the transverse motion of the circulating particles. Error correction and compensation are integral parts of circular accelerator design.

In comparisons, errors in the amplitude and phase of accelerating electrical fields are much less critical during the acceleration or accumulation cycle due to the small synchrotron ($\nu_s \ll 1$) tune and the lack of resonance conditions. Unlike high-energy colliders, the impact of "RF noise" is usually small during the relatively short period of high-power ring acceleration and accumulation.

2.3.9.1 Imperfections in the Magnetic Field

Magnets are susceptible to geometric magnetic errors. The leading components of error are those allowed by the magnet's symmetry. For a magnet of $2m$ poles ($m = 1$ for a dipole, $m = 2$ for a quadrupole magnet, and so on), the allowed multipoles are

$$2m(2k + 1), \quad k = 0, 1, 2, \ldots \tag{2.64}$$

For a dipole magnet, the allowed multipoles include sextupole, decapole, etc. For a quadrupole magnet, the allowed harmonics are at the 12th pole, the 20th pole, etc. This systematic imperfection can be corrected locally either by shaping the magnet pole profile and shimming or integrally by shaping the magnet ends [204, 205].

Mechanical errors in magnet fabrication and material variations produce random variation in the magnet transfer function (defined as the ratio of the magnetic field to the driving current) and random multipole errors. Mechanical tolerances are specified to satisfy the required magnetic field quality without over-demanding costly fabrication accuracy. The effects of magnetic non-uniformity in material composition are often minimized by randomizing the laminations for laminated-iron magnets and ensuring the use of iron from the same pour for solid-iron magnets. Magnets of high-power rings are usually constructed with a large gap between the pole tips to accommodate large-aperture vacuum chambers. To avoid saturation, the magnets are carefully designed with high uniformity throughout the aperture and with a peak magnetic field typically limited to below approximately 1.2 T. For rapid-cycling magnets, variations in the level of saturation contribute to the tracking errors between different types of magncts.

Ramping magnets are susceptible to further time-dependent effects, hysteresis effects, and eddy current effects. The electric current flows mainly in the "skin" of the conductor, between the outer surface and a level called the skin depth. Technical challenges are posed by the increase in current density and resistive loss in the coil, as the skin depth decreases with the increasing repetition rate, and due to eddy-current losses in the core material, which call for thinner laminations and lower loss iron [206, 207]. The skin depth decreases with increasing repetition rate, according to Eq. 2.44.

Eddy currents induced in the vacuum chamber under a changing magnetic field distort and delay the field, generating multipole errors. For a wide chamber of thickness d_w inside a dipole magnet of gap g, the sextupole field is given by [208]

$$\frac{\partial^2 B_y}{\partial x^2} = \frac{\mu_0 d_w}{\rho_r g}\dot{B}. \tag{2.65}$$

This quantity is linearly proportional to the ramp rate.

2.3.9.2 Magnetic Fringe Field

Magnetic fringe field refers to the longitudinal and transverse field components near the mechanical ends of the magnet that differ from the design field in the magnet body. The effect can be important when large-aperture, short, large-aperture non-solenoidal magnets are used in high-power ring accelerators with moderate circumferences.

The effects of the transverse fringe field component are usually strong functions of the geometry of the magnet's ends. For such iron-dominated magnets operating at room temperature, the leading components of error are again those allowed by the magnet's symmetry (Eq. 2.64). Three-dimensional modeling is usually used for error evaluation and design optimization.

The effects of the longitudinal fringe-field component of the non-solenoidal magnet do not depend on the details of the magnet. The relative impulse from the longitudinal field component on a particle's transverse momentum is approximately equal to the ratio between the beam's emittance and the magnet length [209, 210],

$$\left\langle \frac{|\Delta x'_{fr}|}{|\Delta x'_0|} \right\rangle \sim \frac{\epsilon_\perp}{L}, \tag{2.66}$$

where $\epsilon_{\perp}$ is the rms beam transverse emittance, $\Delta x'_{fr}$ is the change in transverse momentum due to the fringe field, and $\Delta x'_0$ is the change in transverse momentum due to the main magnetic field. This effect is usually small but can become negligible when the beam emittance is intentionally increased to alleviate the space charge force and when the magnet is typically short. For example, the maximum emittance of the beam in the SNS ring is about 240 πmm·mr. The fringe field at each end of a 0.5 m long quadrupole magnet produces an octupole-like transverse kick of about 5×10^{-4} relative to the main quadrupole deflection. Unlike the impulse of the transverse magnetic field component, the impulse of the longitudinal component is independent of the geometry of the magnet's ends. Equation 2.66 implies that, under most circumstances, the impact of non-solenoidal magnets can be evaluated by assuming a 2-D representation of the field in the transverse dimension, expressed in the usual series expansion of magnetic multipoles.

2.3.9.3 Magnetic Interference

Magnetic interference refers to the change in the magnetic field produced by an individual magnet in the presence of other nearby magnetic elements. The effect can be important for high-power rings where short, large-aperture magnets are used, so that magnetic end fields overlap significantly with the iron of nearby elements [211]. The interference usually manifests itself as a change in the magnet transfer function and an enhancement of multipole harmonics. In the presence of magnet iron, the effects may not be obtained by a direct superposition of contributions from individual magnet elements.

Magnetic interference can be alleviated by keeping the minimum distance between the irons of adjacent magnets at least equal to the diameter of the opening of the magnet poles (Fig. 2.55). Measurements and modeling are needed to evaluate

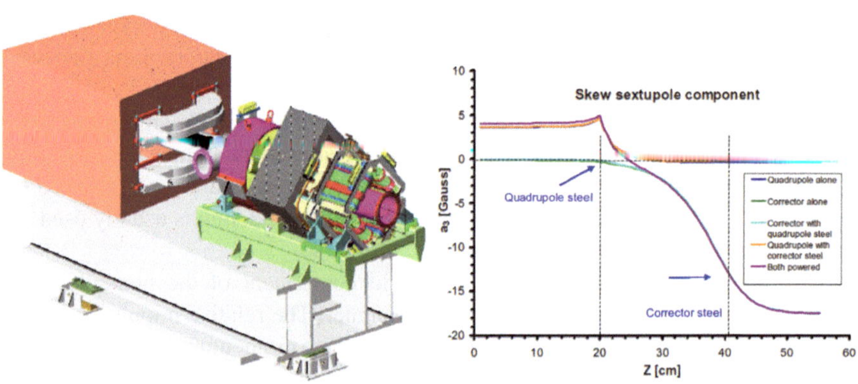

Fig. 2.55 (Left) Half-cell schematic layout of the SNS ring with the main dipole, sextupole, quadrupole and dipole/skew-quadrupole/skew-sextupole corrector. (Right) Skew sextupole field component of the assembly caused by magnet interference [211]. (Courtesy Y. Papaphilippou)

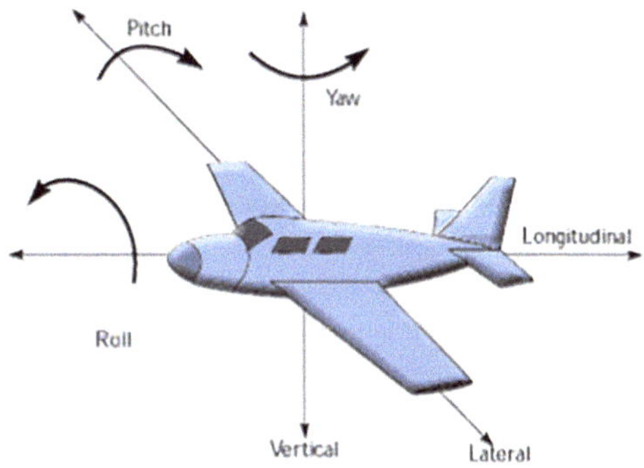

Fig. 2.56 Misalignment of a rigid body by displacement in the horizontal, vertical, and longitudinal directions and rotation about the longitudinal axis (roll), vertical axis (yaw), and lateral axis (pitch)

the effects under operational conditions. For magnets powered by a common power supply, trim magnet coils and trim power supplies are sometimes needed to compensate for the effects of the variation in the magnet's transfer functions.

2.3.9.4 Magnet Misalignments

Misalignment of the magnet as a rigid body can be characterized by six variables: displacement in the horizontal, vertical, and longitudinal directions, and rotation about the longitudinal axis (roll), the vertical axis (yaw), and the lateral axis (pitch), as shown in Fig. 2.56.

Displacement along the horizontal and vertical axes generally produces effects of magnetic multipole "feed-down". For example, the displacement of a dipole magnet merely shifts the median plane of the magnet; displacement of a quadrupole magnet produces a dipole field component; displacement of a sextupole magnet produces a quadrupole field component; etc. Typically, the alignment tolerance for horizontal and vertical displacement is a fraction of a millimeter for magnets in high-power rings. In comparison, ring magnets are far more tolerant of displacement in the longitudinal direction.

Rotation about the longitudinal axis (roll) produces a skew component of the main magnet multipole. For example, a dipole magnet with 1 mr roll produces a skew dipole of 10^{-3} strength; a quadrupole magnet with 1 mr roll produces a skew quadrupole of 10^{-3} strength. In comparison, the effects of rotations about the vertical axis (yaw) and lateral axis (pitch) are usually insignificant for non-solenoidal magnets because the variation of the betatron amplitude is typically small across the magnets in high-power rings (e.g., in contrast to magnets near the interaction region of colliders).

2.3.9.5 Correction Systems

Magnet imperfections can cause deviations from the closed orbit of the beam, variations in its envelope, transverse coupling, chromaticity variations, spread of tune shifts, and resonance excitation. Typically, the relative strength of the

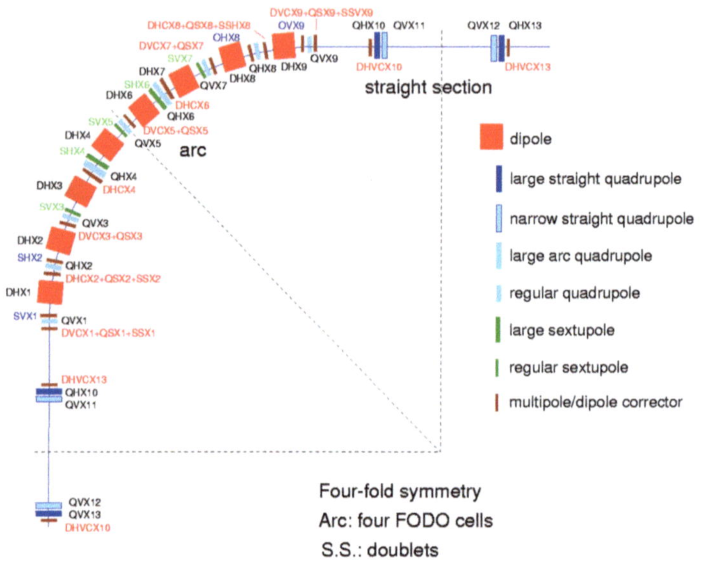

Fig. 2.57 Schematic magnet layout of dipoles quadrupole, sextupoles, and correctors in one superperiod of the SNS ring [131]

multipole field errors needs to be controlled at a 10^{-4} level for lattice dipoles and quadrupoles, at a 10^{-3} level for transfer-line magnets, low-field chicane magnets, and chromaticity sextupoles, and at a 10^{-2} level for correction magnets (Fig. 2.57).

Closed orbit deviations are corrected by minimizing beam displacements at the locations of beam position monitors (BPM) using dipole correctors, preferably located at the places of peak beam amplitude ($\beta_\perp$) function. The correction can be made either globally, using orbit response matrix analysis, or locally, using closed orbit bumps. Ideally, several BPM's and correctors are needed in each betatron wavelength. To prevent radiation damage, electronic elements must be located in areas of low radioactivity outside the accelerator tunnel. Closed-orbit correction during rapid acceleration often requires dynamic correction. Rings that use combined-function gradient dipole magnets usually need separate correction elements.

Trim quadrupoles can correct deviations in the beam envelope. Correction is based on measuring and evaluating the lattice amplitude function $\beta_\perp$ and the betatron phase advance using some of the following methods: (1) measuring the dependence of the tuning on variations in the quadrupole strength to determine $\beta_\perp$, especially at locations of large amplitude (focusing quadrupole). However, the method is susceptible to magnetic hysteresis and coupling; (2) measuring the betatron envelope after perturbing the beam transversely to determine the relative value of $\beta_\perp$; and (3) assessing the betatron phase at the BPM locations after resonantly perturbing the beam at the betatron frequency, determining $\beta_\perp$, especially at small-amplitude locations.

Coupling in transverse motion, originating from skew quadrupole error components, space-charge effects, and solenoidal fields, is usually enhanced by near-equal horizontal and vertical working points in tune space. The typical coupling strength is proportional to the quantity

$$\sum_i a_{2,i}\sqrt{\beta_{x,i}\beta_{y,i}}\sin(\varphi_{x,i}-\varphi_{y,i}), \tag{2.67}$$

where the subscript i indicates the source of error, $\varphi_{x,i}$ and $\varphi_{y,i}$ are the horizontal and vertical betatron phases, and a_2 is the strength of the skew quadrupole error [212–214]. The effect of systematic skew quadrupole errors can usually be alleviated by splitting the horizontal and vertical tunes by at least half a unit, so that variations in phase difference result in randomization. Global decoupling is achieved by at least two families of skew quadrupoles to reduce the minimum tune splitting. Local decoupling is performed by analyzing turn-by-turn data from dual-plane BPM and correcting it with multiple skew quadrupole correctors.

Control of chromaticity is necessary for rings operating above the transition energy to avoid head-tail instability, and it is also desirable for rings operating below (Sect. 3.4.3). For rings operating solely below the transition energy, chromatic sextupoles, arranged according to lattice symmetry and powered in multiple families, offer control of tune spread, damping of instability, and matching of off-momentum optics. The SNS ring uses four-family chromatic sextupole magnets located in high-dispersion regions, complemented by resonance-correction sextupole windings placed in zero-dispersion regions [215].

Resonance correction has been used successfully on several machines, including the CERN PS, the Brookhaven AGS, and the AGS Booster [216]. The principle is summarized in Sect. 3.3.2.5. At the AGS Booster, resonance correction up to normal and skew sextupoles was essential during high-intensity operations [217].

Problems

2.1 (Relativistic Factors of Beam) Let E be the total energy, E_k the kinetic energy and p be the momentum. Assume that the spread ΔE_k in the kinetic energy is much smaller than the kinetic energy. Prove that the relative energy spread and momentum spread satisfy the relations

$$\frac{\Delta E}{E} \approx \beta^2 \frac{\Delta p}{p}, \quad \frac{\Delta p}{p} \approx \frac{\gamma}{1+\gamma}\frac{\Delta E_k}{E_k} \tag{2.68}$$

where β and γ are the relativistic factors. For a proton beam of 1 GeV kinetic energy with a $\pm 1\%$ spread in $\Delta p/p$, how accurate are these relations?

2.2 (Isochronous Cyclotron Scaling) To accelerate a charged particle in an isochronous cyclotron, prove that given a maximum attainable magnetic field at extraction, the needed cyclotron radius is proportional to $\beta\gamma$, where β and γ are the relativistic factors of the particle at extraction. To keep a minimum gap between turns of the orbit trajectory at extraction, prove that the energy gain per turn is proportional to $\beta\gamma^2$.

2.3 (Dispersion Suppression Lattice of a Ring)

1. A ring consists of 4 bending arcs, each a horizontal achromat consisting of 4 identical FODO cells of $\pi/2$ phase advance. The dispersion is suppressed. Evaluate the value and location of peak dispersion of the ring in terms of cell length, and compare it with the minimum achievable peak dispersion of a matched FODO cell.
2. Repeat this exercise by flipping the polarity or the quadrupoles, i.e. DOFO instead of FODO.
3. Replace the dispersion suppression method by the missing-dipole (half-field) scheme. Evaluate the minimum achievable peak dispersion in terms of cell length.
4. Compare the advantage and disadvantage of the above three schemes.

2.4 (Closed Orbit Bump in a Ring)

1. Assume that the betatron phase advance from s_1 to s_2 is $\Delta\mu$. Prove that the transfer matrix for (x, x') can be written as:

$$M(s_2|s_1) = \begin{bmatrix} \sqrt{\frac{\beta_2}{\beta_1}}(\cos\Delta\mu + \alpha_1 \sin\Delta\mu) & \sqrt{\beta_1\beta_2}\sin\Delta\mu \\ \frac{(1+\alpha_1\alpha_2)\sin\Delta\mu + (\alpha_2 - \alpha_1)\cos\Delta\mu}{\sqrt{\beta_1\beta_2}} & \sqrt{\frac{\beta_1}{\beta_2}}(\cos\Delta\mu - \alpha_2\sin\Delta\mu) \end{bmatrix}. \tag{2.69}$$

2. Using Eq. 2.69, prove that the condition for realizing a closed orbit bump from s_1 to s_3 using three dipole magnets is

$$\frac{\theta_1\sqrt{\beta_1}}{\sin\Delta\mu_{32}} = \frac{\theta_2\sqrt{\beta_2}}{\sin\Delta\mu_{13}} = \frac{\theta_3\sqrt{\beta_3}}{\sin\Delta\mu_{21}}. \tag{2.70}$$

Here, θ_i is the dipole kick at location s_i, $\Delta\mu_{ij}$ is the phase advance between s_i and s_j, with $i, j = 1, 2, 3$.

2.5 (Transverse Painting for Ring Injection) Consider the beam density distribution in the normalized phase space during multi-turn injection into a ring. Define $\rho^2 \equiv X^2 + X'^2$ and the quantity $\lambda(\rho)\rho d\rho$ is the number of particles populated within the phase-space circle of radius ρ and width $d\rho$.

1. What is the closed orbit function with time to realize a uniform distribution in phase space with constant density $\lambda(\rho)$ within a radius R during a total injection time of $t_{\max}$?
2. Prove that the closed orbit function to realize a Gaussian density distribution

$$\lambda(\rho) = \frac{2N_0}{\sigma^2}\exp\left(-\frac{\rho^2}{2\sigma^2}\right) \tag{2.71}$$

is approximately

$$X_C(t) = \sqrt{2}\sigma\sqrt{-\ln\left(1 - \frac{t}{t_{\max}}\right)} \tag{2.72}$$

where the injection is performed for a time $0 < t < t_{\max}$.

2.6 (Ring Multi-turn Injection) Upon multi-turn injection of a beam with emittance ϵ_i and Courant-Snyder (C-S) parameters β_i and α_i, injecting with the center of the input beam (x_i, x_i') relative to the instantaneous injection orbit bump. The ring beam emittance is ϵ_x, and ring C-S parameters are β_x and α_x at injection.

1. Prove that the injecting beam ellipse becomes upright and the injection position is optimized when

$$\frac{\alpha_i}{\beta_i} = \frac{\alpha_x}{\beta_x} = -\frac{x_i'}{x_i} . \tag{2.73}$$

2. Assume that $\epsilon_i \ll \epsilon_x$, show that when the condition

$$\frac{\beta_i}{\beta_x} \approx \left(\frac{\epsilon_i}{\epsilon_x}\right)^{\frac{1}{3}} \tag{2.74}$$

is satisfied, the injection efficiency becomes optimal (i.e., minimum phase-space dilution after injection).

2.7 (Two-Stage Collimation in Rings) Define the normalized transverse phase space in terms of the normalized variables X, X' and phase μ:

$$X \equiv \frac{x}{\sqrt{\beta_x}}, \quad X' \equiv \frac{dX}{d\mu} = \frac{\alpha_x x + \beta_x x'}{\sqrt{\beta_x}}, \quad \mu(s) = \int^s \frac{ds'}{\beta(s')}. \tag{2.75}$$

where α_x and β_x are the C-S parameters. Consider a two-stage betatron collimation system consisting of a scraper and two collectors located at betatron phases μ_1 and μ_2 away from the scraper, respectively.

1. Prove that when the following conditions:

$$\mu_1 = \cos^{-1}\left(\frac{A}{A+H}\right), \quad \mu_2 = \pi - \mu_1 \tag{2.76}$$

are satisfied, a minimum number of secondary particles escape the collimation process. Here, the scraper radius A and the collector radius $A + H$ are both defined in terms of the normalized variables.

2. Express the above phase-advance conditions in terms of the physical apertures.

2.8 (Fringe Field Theorem) Transverse particle motion in a circular particle accelerator is governed almost entirely by non-solenoidal magnets, for which the body magnetic field can be expressed as a series expansion of the normal (b_n) and skew (a_n) multipoles.

$$B_y + jB_x = \sum_0^\infty (b_n + ja_n)(x + jy)^n \tag{2.77}$$

where x, y, and z denote horizontal, vertical, and longitudinal (along the magnet) coordinates. Since the magnet length L is necessarily finite, the deflections are actually proportional to "field integrals" such as $\bar{B}L \equiv \int B(x, y; z)dz$, where the integration range starts well before the magnet and ends well after it. For $\bar{a}_n$, $\bar{n}_n$, $\bar{B}_x$, and $\bar{B}_y$ defined this way, Eq. 2.77 is valid, and the "standard" approximation is to neglect any deflections not described by this expansion, despite the fact that Maxwell's equations demand the presence of longitudinal field components at the magnet ends. The purpose of this exercise is to provide a semi-quantitative estimate of the importance of $|\Delta p_\parallel|$, the *transverse* deflection produced by the *longitudinal* component of the fringe field at one end of the magnet relative to $|\Delta p_0|$, and the total deflection produced by passage through the whole magnet. To emphasize the generality and simplicity of the result, it is presented in the form of a theorem.

Theorem 2.1 *For any non-solenoidal magnet,*

$$\left\langle \frac{|\Delta p_\parallel|}{|\Delta p_0|} \right\rangle \approx \frac{\epsilon_\perp}{L}\sqrt{\frac{1 + 0.75\beta_\perp'^2}{8}} \sim \frac{\epsilon_\perp}{L} \tag{2.78}$$

where $\epsilon_\perp$ is the rms beam transverse emittance, $\beta'_\perp = d\beta_\perp/ds$ and $\langle\ \rangle$ denote the average over the betatron phase, and the final approximation must be withdrawn if $\beta'_\perp$ is anomalously large.

References

1. E.O. Lawrence, M.S. Livingston, The production of high speed light ions without the use of high voltages. Phys. Rev. **40**, 19–35 (1932). https://doi.org/10.1103/PhysRev.40.19
2. M. Craddock, K. Symon, Cyclotrons and fixed-field alternating-gradient accelerators. Rev. Accel. Sci. Technol. **1**(01), 65–97 (2008)
3. J. Galambos, Operations experience of SNS at 1.4 MW and upgrade plans for doubling the beam power, in *Proceedings of IPAC'19* (JACoW Publishing, Geneva), pp. 4380–4384 (1989). https://doi.org/10.18429/JACoW-IPAC2019-FRXPLM1. https://jacow.org/ipac2019/papers/FRXPLM1.pdf
4. R. Baartman, Cyclotrons: why/how are their dynamics different? J. Instrum. **18**(03), T03005 (2023). https://doi.org/10.1088/1748-0221/18/03/T03005
5. J.A. Nolen et al., Commissioning experience with the NSCL K1200 superconducting cyclotron, in *Proceedings of Cyclotrons'89* (JACoW Publishing, Geneva), pp. 5–8 (1989). https://jacow.org/c89/papers/A-02.pdf

6. A. Goto et al., Commissioning of Riken RI beam factory, in *Proceedings of Cyclotrons'07* (JACoW Publishing, Geneva), pp. 3–8
7. J.R. Richardson, K.R. Mackenzie, E.J. Lofaren, B.T. Wright, Frequency modulated cyclotron. Phys. Rev. **69**, 669–670 (1946). https://doi.org/10.1103/PhysRev.69.669
8. N. Abrosimov et al., Synchrocyclotron of FTI AN USSR for acceleration of 1 GeV protons, in *Proceedings of 7th International Conference on High Energy Accelerators*, pp. 317–323 (1969)
9. L.H. Thomas, The paths of ions in the cyclotron I. Orbits in the magnetic field. Phys. Rev. **54**, 580–588 (1938). https://doi.org/10.1103/PhysRev.54.580
10. F.A. Heyn, K.K. Tat, operation of a radial sector fixed-frequency proton cyclotron. Rev. Sci. Instrum. **29**(7), 662–662 (1958). https://doi.org/10.1063/1.1716293
11. D.W. Kerst, L.W. Jones, K.R. Symon, K.M. Terwilliger, A fixed field alternating gradient accelerator with spirally ridged poles. Technical Report MURA-042, Mid-Western Universities Research Association (1954). https://inspirehep.net/files/4fe634e2eacb4b0cb5f926179e7c1db2
12. G.H. Mackenzie, P.W. Schmor, H.R. Schneider, Cyclotrons, in *Encyclopedia of Applied High Energy and Particle Physics*, ed. by R. Stock (Wiley-VCH Verlag GmbH & Co. KGaA, Weinheim, 2009)
13. R. Stock (ed.), *Encyclopedia of Applied High Energy and Particle Physics*, Wiley, Weinheim, (2009)
14. H.A. Willax, Proposal for a 500 MeV isochronous cyclotron with ring magnet, in *Proceedings of Cyclotrons'63* (JACoW Publishing, Geneva). https://jacow.org/c63/papers/CYC63H04.pdf
15. H. Blosser, D. Johnson, Focusing properties of superconducting cyclotron magnets. Nucl. Instrum. Methods **121**(2), 301–306 (1974). https://doi.org/10.1016/0029-554X(74)90080-9. https://www.sciencedirect.com/science/article/pii/0029554X74900809
16. H.G. Blosser, Invited paper: The Michigan State University superconducting cyclotron program, in *Proceedings of Cyclotrons'78* (JACoW Publishing, Geneva), pp. 2040–2047. https://jacow.org/c78/papers/C-02.pdf
17. P. Miller et al., Performance of the K1200 cyclotron at MSU, in *Proceedings of Cyclotrons'92* (JACoW Publishing, Geneva), pp. 86–89. https://jacow.org/c92/papers/I-17.pdf
18. J.A. Bagger et al., ARIEL at TRIUMF: science and technology, in *Proceedings of IPAC'18* (JACoW Publishing, Geneva), pp. 6–11. https://doi.org/10.18429/JACoW-IPAC2018-MOXGB2. http://accelconf.web.cern.ch/ipac2018/papers/MOXGB2.pdf
19. G.G. Gulbekyan et al., New development at the JINR heavy ion cyclotron facility, in *Proceedings of Cyclotrons'92* (JACoW Publishing, Geneva), pp. 11–21. https://jacow.org/c92/papers/I-02.pdf
20. M. Seidel, *CERN Accelerator School—Introductory Course, Constanta* (2018)
21. K. Symon, D. Kerst, L. Jones, L. Laslett, K. Terwilliger, Fixed-field alternating-gradient particle accelerators. Phys. Rev. **103**(6), 1837 (1956)
22. D.W. Kerst, K.R. Symon, L.J. Laslett, L.W. Jones, K.M. Terwilliger, Fixed field alternating gradient particle accelerators, in *CERN Symposium on High Energy Accelerators and Pion Physics* (CERN, 1956), pp. 32–35 (1969). https://doi.org/10.5170/CERN-1956-025.32. https://cds.cern.ch/record/107802
23. T. Ohkawa, Bull. Phys. Soc. Jpn. (1953)
24. A.A. Kolomensky, V.A. Petukhov, M. Rabinovich, Lebedev Phys. Inst. Report RF-54. Technical Report, RF-54, Lebedev Physical Institute, Moscow (1953)
25. F. Mills, in *Proceedings of the 4th International Conference on Physics Potential and Development of mu+ mu- Colliders (MUMU 97)* (1998), pp. 696–698
26. Y. Mori, FFAGs: front-end for neutrino factories and medical accelerators, in *Challenges and Goals for Accelerators in the XXI Century* (2016), pp. 767–779. https://doi.org/10.1142/9789814436403_0040. https://www.worldscientific.com/doi/abs/10.1142/9789814436403_0040
27. M. Aiba et al., Development of a FFAG proton synchrotron, in *Proceedings of EPAC'00, European Particle Accelerator Conference*, no. 7 (JACoW Publishing, Geneva, 2000), pp. 581–583. https://jacow.org/e00/papers/MOP1B21.pdf

28. C.H. Pyeon, T. Misawa, J.Y. Lim, H. Unesaki, Y. Ishi, Y. Kuriyama, T. Uesugi, Y. Mori, M. Inoue, K. Nakajima et al., First injection of spallation neutrons generated by high-energy protons into the kyoto university critical assembly. J. Nucl. Sci. Technol. **46**(12), 1091–1093 (2009)
29. S. Machida, FFA design study for a high intensity proton driver, in *Proceedings of IPAC'23, International Particle Accelerator Conference*, no. 14 (JACoW Publishing, Geneva, 2023), pp. 1437–1439. https://doi.org/10.18429/JACoW-IPAC2023-TUPA044
30. S. Brooks, Non-scaling fixed-field proton accelerator with constant tunes, in *Proceedings of IPAC'23, International Particle Accelerator Conference*, no. 14 (JACoW Publishing, Geneva, 2023), pp. 2257–2260. https://doi.org/10.18429/JACoW-IPAC2023-TUPM031
31. S. Machida, R. Barlow, J. Berg, N. Bliss, R. Buckley, J. Clarke, M. Craddock, R. D'Arcy, R. Edgecock, J. Garland et al., Acceleration in the linear non-scaling fixed-field alternating-gradient accelerator EMMA. Nat. Phys. **8**(3), 243–247 (2012). https://doi.org/10.1038/nphys2179
32. T. Uesugi, Y. Ishi, Y. Kuriyama, Y. Mori, C. Jolly, D. Kelliher, J.B. Lagrange, A. Letchford, S. Machida, D.P. de Boer et al., Beam stacking experiment at a fixed field alternating gradient accelerator. Phys. Rev. Accel. Beams **28**(1), 012803 (2025)
33. S. Machida, S. Gallimore, B. Kirk, D. Kelliher, J. Lagrange, C. Rogers, J. Pasternak, I. Gardner, R. Mathieson, T. Kuo et al., Fixed field alternating gradient accelerator (FFA) report. STFC Preprints STFC-P-2025-001 (2025)
34. J.L. Belmont, J.L. Pabot, Study of axial injection for the Grenoble cyclotron. IEEE Trans. Nucl. Sci. **13**(4), 191–193 (1966). https://doi.org/10.1109/TNS.1966.4324204
35. J.G.M. Kleeven, Some examples of recent progress of beam-dynamics studies for cyclotrons, in *Proceedings of Cyclotrons'16* (JACoW Publishing, Geneva), pp. 244–250. https://doi.org/10.18429/JACoW-Cyclotrons2016-WEA01. https://jacow.org/cyclotrons2016/papers/WEA01.pdf
36. T. Planche, R.A. Baartman, H.W. Koay, Y.N. Rao, L. Zhang, Intensity limit in compact h-and h+ 2 cyclotrons. J. Instrum. **18**(03), P03024 (2023)
37. I.V. Bylinskii, R.A. Baartman, K. Jayamanna, T. Planche, Y.N. Rao, Recent improvements in beam delivery with the TRIUMF's 500 MeV cyclotron, in *Proceedings of Cyclotrons'16* (JACoW Publishing, Geneva), pp. 133–136. https://doi.org/10.18429/JACoW-Cyclotrons2016-TUA04. https://jacow.org/cyclotrons2016/papers/TUA04.pdf
38. G. Ising, Prinzip einer Methode zur Herstellung von Kanalstrahlen hoher Voltzahl. Ark. Mat. Astron. Fys. **18**(30), 1–4 (1924). https://cds.cern.ch/record/433984
39. R. Wideröe, Über ein neues Prinzip zur Herstellung hoher Spannungen. Archiv für Elektrotechnik **21**, 387–406 (1928). https://api.semanticscholar.org/CorpusID:109942448
40. L.W. Alvarez, The design of a proton linear accelerator. Phys. Rev **70**, 799 (1946)
41. T. Wangler, *RF Linear Accelerators*. Wiley Series in Beam Physics and Accelerator Technology (Wiley, Weinheim, 1998)
42. J.D. Cockcroft, E.T.S. Walton, Experiments with high velocity positive ions. R. Soc. Lond. **A129**, 477–489 (1930). https://doi.org/10.1098/rspa.1930.0169
43. D.H. Sloan, E.O. Lawrence, The production of heavy high speed ions without the use of high voltages. Phys. Rev. **38**, 2021–2032 (1931). https://doi.org/10.1103/PhysRev.38.2021
44. L.M. Bollinger, Low-β SC linacs: past, present and future, in *Proceedings of LINAC'98, Linear Accelerator Conference* (JACoW Publishing, Geneva), pp. 3–7. https://jacow.org/l98/papers/MO1001.pdf
45. J. Delayen, Superconducting accelerating structures for high-current ion beams, in *Proceedings of LINAC'88, Linear Accelerator Conference* (JACoW Publishing, Geneva), pp. 199–201. https://jacow.org/l88/papers/MO3-60.pdf
46. N.R. Holtkamp, EPS-AG 2008 Gersh Budker Prize presentation: the successful construction and commissioning of the Spallation Neutron Source, in *Proceedings of EPAC'08, European Particle Accelerator Conference*, no. 11 (JACoW Publishing, Geneva, 2008), pp. 2960–2962. https://jacow.org/e08/papers/THPPGM03.pdf

47. J. Wei et al., FRIB accelerator status and challenges, in *Proceedings of LINAC'12, Linear Accelerator Conference*, no. 26 (JACoW Publishing, Geneva), pp. 417–421. https://jacow.org/LINAC2012/papers/TU1A04.pdf
48. A. Saini, V.A. Lebedev, J.F. Ostiguy, N. Solyak, V.P. Yakovlev, Design of superconducting CW linac for PIP-II, in *Proceedings of IPAC'15* (JACoW Publishing, Geneva), pp. 565–567. https://doi.org/10.18429/JACoW-IPAC2015-MOPMA014. https://jacow.org/IPAC2015/papers/MOPMA014.pdf
49. Z. Li, P. Cheng, H. Geng, Z. Guo, Y. He, C. Meng, H. Ouyang, S. Pei, B. Sun, J. Sun, J. Tang, F. Yan, Y. Yang, C. Zhang, Z. Yang, Physics design of an accelerator for an accelerator driven subcritical system. Phys. Rev. ST Accel. Beams **16**, 080101 (2013). https://doi.org/10.1103/PhysRevSTAB.16.080101
50. J.L. Biarrotte, F. Bouly, J.P. Carneiro, D. Uriot, D. Vandeplassche, Design of the MYRRHA 17-600 MeV superconducting linac, in *Proceedings of SRF'13* (JACoW Publishing, Geneva), pp. 129–132. https://jacow.org/SRF2013/papers/MOP018.pdf
51. R. Garoby, A. Vergara, H. Danared, I. Alonso, E. Bargallo, B. Cheymol, C. Darve, M. Eshraqi, H. Hassanzadegan, A. Jansson et al., The European spallation source design. Phys. Scripta **93**(1), 014001 (2017)
52. H.J. Kim, H.J. Cha, M.O. Hyun, H.C. Jung, Y.J.K. Kim, M. Lee, Superconducting linac for RISP, in *Proceedings of LINAC'14, Linear Accelerator Conference*, no. 27 (JACoW Publishing, Geneva, 2014), pp. 245–247. https://jacow.org/LINAC2014/papers/MOPP082.pdf
53. R. Garoby, SPL at CERN, in *Proceedings of SRF'09* (JACoW Publishing, Geneva), pp. 930–933. https://jacow.org/SRF2009/papers/FROBAU04.pdf
54. I. Ben-Zvi, J. Brennan, The quarter wave resonator as a superconducting linac element. Nucl. Instrum. Methods Phys. Res. **212**(1), 73–79 (1983). https://doi.org/10.1016/0167-5087(83)90678-6. https://www.sciencedirect.com/science/article/pii/0167508783906786
55. A.W. Chao, M. Tigner, H. Weise, F. Zimmermann, *Handbook of Accelerator Physics and Engineering* (World scientific, Singapore, 2023)
56. F. Peauger et al., Status and first results of two high beta prototype elliptical cavities for ESS, in *Proceedings of IPAC'14* (JACoW Publishing, Geneva), pp. 2477–2479. https://doi.org/10.18429/JACoW-IPAC2014-WEPRI002. https://jacow.org/IPAC2014/papers/WEPRI002.pdf
57. H. Padamsee, J. Knobloch, T. Hays, *RF Superconductivity for Accelerators*. Wiley Series in Beam Physics and Accelerator Technology (Wiley, 1998). https://books.google.com/books?id=2leaQgAACAAJ
58. T. Ilg et al., Mechanical design of the drift-tube linac (DTL) for the Spallation Neutron Source, in *Proceedings of PAC'03* (JACoW Publishing, Geneva), pp. 2841–2843. https://jacow.org/p03/papers/RPAB051.pdf
59. A. Facco, Low- to medium-β cavities for heavy ion acceleration. Supercond. Sci. Technol. **30**(2), 023002 (2016). https://doi.org/10.1088/1361-6668/30/2/023002
60. M.P. Kelly et al., Low beta cavity development for an ATLAS intensity upgrade, in *Proceedings of SRF'13* (JACoW Publishing, Geneva), pp. 850–854. https://jacow.org/SRF2013/papers/THIOC01.pdf
61. A. Facco, V. Zvyagintsev, Beam steering in superconducting quarter-wave resonators: an analytical approach. Phys. Rev. Spec. Top. Accel. Beams **14**(7), 070101 (2011)
62. P.N. Ostroumov, K.W. Shepard, Correction of beam-steering effects in low-velocity superconducting quarter-wave cavities. Phys. Rev. ST Accel. Beams **4**, 110101 (2001). https://doi.org/10.1103/PhysRevSTAB.4.110101
63. A. Facco, Tutorial on low beta cavity design, in *Proceedings of SRF'05* (JACoW Publishing, Geneva), pp. 21–33. https://jacow.org/SRF2005/papers/SUA04.pdf
64. Z.A. Conway et al., Reduced-beta cavities for high-intensity compact accelerators, in *Proceedings of LINAC'12, Linear Accelerator Conference*, no. 26 (JACoW Publishing, Geneva), pp. 621–623. https://jacow.org/LINAC2012/papers/TUPB066.pdf
65. G. Olry et al., Recent developments on superconducting β 035 and β 015 spoke cavities at IPN for low and medium energy sections of proton linear accelerators, in *Proceedings*

of EPAC'04, European Particle Accelerator Conference, no. 9 (JACoW Publishing, Geneva, 2004), pp. 1003–1005. http://accelconf.web.cern.ch/e04/papers/TUPKF019.pdf
66. K.W. Shepard, P.N. Ostroumov, J.R. Delayen, High-energy ion linacs based on superconducting spoke cavities. Phys. Rev. ST Accel. Beams **6**, 080101 (2003). https://doi.org/10.1103/PhysRevSTAB.6.080101
67. J.R. Delayen, Applications of spoke cavities, in *Proceedings of LINAC'10, Linear Accelerator Conference*, no. 25 (JACoW Publishing, Geneva), pp. 377–381. https://jacow.org/LINAC2010/papers/TU302.pdf
68. M. Basten et al., Status of the first CH-Cavities for the new superconducting CW heavy ion linac@GSI, in *Proceedings of IPAC'16, International Particle Accelerator Conference*, no. 7 (JACoW Publishing, Geneva, 2016), pp. 886–888. https://doi.org/10.18429/JACoW-IPAC2016-MOPOY019. https://jacow.org/ipac2016/papers/MOPOY019.pdf
69. F. Dziuba et al., Development of superconducting crossbar-H-mode cavities for proton and ion accelerators. Phys. Rev. ST Accel. Beams **13**, 041302 (2010). https://doi.org/10.1103/PhysRevSTAB.13.041302
70. I.M. Kapchinskii, V.A. Teplyakov, A linear ion accelerator with spatially uniform hard focusing. Prib. Tekh. Eksp. **119**(2), 19–22 (1970)
71. R. DiGennaro et al., The design of a high current, high duty factor RFQ for the SNS, in *Proceedings of EPAC'00, European Particle Accelerator Conference*, no. 7 (JACoW Publishing, Geneva, 2000), pp. 495–497. https://jacow.org/e00/papers/THP6A07.pdf
72. N.K. Bultman et al., Design of the FRIB RFQ, in *Proceedings of IPAC'13* (JACoW Publishing, Geneva), pp. 2866–2868. https://jacow.org/IPAC2013/papers/WEPFI075.pdf
73. Y. Kondo, T. Morishita, Y. Sawabe, S. Yamazaki, T. Hori, A. Takagi, Beam test of a new RFQ for the J-PARC linac, in *Proceedings of LINAC'14, Linear Accelerator Conference*, no. 27 (JACoW Publishing, Geneva, 2014), pp. 267–269. https://jacow.org/LINAC2014/papers/MOPP091.pdf
74. A. Schempp, J.W. Goethe, *Properties of High Current RFQ Injectors* (1996)
75. K.R. Crandall, PARMTEQ with image charges. Technical Report, AT-1-92-213, Los Alamos National Laboratory (1991)
76. J. Staples, D. Oshatz, T. Saleh, Design of the SNS MEBT, in *Proceedings of LINAC'00, Linear Accelerator Conference*, no. 20 (JACoW Publishing, Geneva), pp. 250–252. https://jacow.org/l00/papers/MOD18.pdf
77. D. Oshatz, A. DeMello, L. Doolittle, P. Luft, J.W. Staples, A. Zachoszcz, Mechanical design of the SNS MEBT, in *Proceedings of PAC'01* (JACoW Publishing, Geneva), pp. 1574–1576. https://jacow.org/p01/papers/TPAH152.pdf
78. E. Pozdeyev et al., FRIB front end design status, in *26th International Linear Accelerator Conference* (2013), p. THPB097
79. G.W. Hill, On the part of the motion of the lunar perigee which is a function of the mean motions of the sun and moon. Acta Math. **8**, 1–36 (1886). https://doi.org/10.1007/BF02417081
80. S. Henderson et al., The Spallation Neutron Source accelerator system design. Nucl. Instrum. Methods Phys. Res., Sect. A **763**, 610–673 (2014). https://doi.org/10.1016/j.nima.2014.03.067. https://www.sciencedirect.com/science/article/pii/S0168900214003817
81. N. Bultman, J. Billen, Z. Chen, M. Collier, D. Richards, L.M. Young, Fabrication and tuning of the SNS CCL hot model, in *Proceedings of PAC'03* (JACoW Publishing, Geneva), pp. 2844–2846. https://jacow.org/p03/papers/RPAB052.pdf
82. H. Padamsee, *RF Superconductivity: Science, Technology, and Applications* (Wiley-VCH Verlag GmbH & Co. KGaA, Weinheim, 2009). https://doi.org/10.1002/9783527627172. First published: 18 March 2009, Copyright ©2009 Wiley-VCH Verlag GmbH & Co. KGaA, Weinheim
83. M. Plum, Beam loss in linacs, in *Proceedings of the Joint International Accelerator School: Beam Loss and Accelerator Protection* (CERN, Newport Beach, 2016), pp. 39–62. https://doi.org/10.5170/CERN-2016-002

84. T.P. Wangler et al., Design of a proton superconducting linac for a neutron spallation source, in *Proceedings of SRF'99* (JACoW Publishing, Geneva), pp. 336–344. https://jacow.org/SRF99/papers/WEA001.pdf
85. J. Wei et al., Design optimization and the path towards a 2 MW Spallation Neutron Source, in *Proceedings of PAC'01* (JACoW Publishing, Geneva), pp. 319–321. https://jacow.org/p01/papers/WOAA006.pdf
86. S. Henderson, Commissioning and initial operating experience with the SNS 1-GeV linac, in *Proceedings of LINAC'06, Linear Accelerator Conference*, no. 23 (JACoW Publishing, Geneva), pp. 1–5. https://jacow.org/l06/papers/MO1002.pdf
87. J.D. Galambos, D. Anderson, C. Barbier, K.B. Bekar, D. Bunch, J. Bullman Jr, M. Buchanan, G. Capps, M. Champion, A.E. Coleman et al., Final Design Report. Proton Power Upgrade Project. Technical report, Oak Ridge National Laboratory (ORNL), Oak Ridge (2020)
88. M.S. Champion et al., Progress on the Proton Power Upgrade project at the Spallation Neutron Source, in *Proceedings of LINAC'22, International Linear Accelerator Conference*, no. 31 (JACoW Publishing, Geneva, 2022), pp. 390–393. https://doi.org/10.18429/JACoW-LINAC2022-TUPOJO19
89. Y. Yamazaki, M. Mizumoto, Accelerator complex for the joint project of KEK/JHF and JAERI/NSP, in *Proceedings of PAC'99* (JACoW Publishing, Geneva), pp. 513–517 https://jacow.org/p99/papers/THDL1.pdf
90. K. Hasegawa, Commissioning of the J-PARC linac, in *2007 IEEE Particle Accelerator Conference (PAC)* (IEEE, 2007), pp. 2619–2623
91. A.M. Lombardi, Linac4: from initial design to final commissioning, in *Proceedings of IPAC'17* (JACoW Publishing, Geneva), pp. 1217–1222. https://doi.org/10.18429/JACoW-IPAC2017-TUYA1. https://jacow.org/ipac2017/papers/TUYA1.pdf
92. M. Vretenar, J. Vollaire, R. Scrivens, C. Rossi, F. Roncarolo, S. Ramberger, U. Raich, B. Puccio, D. Nisbet, R. Mompo, S. Mathot, C. Martin, L.A. Lopez-Hernandez, A. Lombardi, J. Lettry, J.B. Lallement, I. Kozsar, J. Hansen, F. Gerigk, A. Funken, J.F. Fuchs, N. Dos Santos, M. Calviani, M. Buzio, O. Brunner, Y. Body, P. Baudrenghien, J. Bauche, T. Zickler, *Linac4 Design Report*. CERN Yellow Reports: Monographs, vol. 6 (CERN, Geneva, 2020). https://doi.org/10.23731/CYRM-2020-006. https://cds.cern.ch/record/2736208
93. P.W. Lisowski, K.F. Schoenberg, The Los Alamos neutron science center. Nucl. Instrum. Methods Phys. Res., Sect. A **562**(2), 910–914 (2006)
94. T.P. Wangler et al., Basis for low beam loss in the high-current APT linac, in *Proceedings of LINAC'98, Linear Accelerator Conference* (JACoW Publishing, Geneva), pp. 657–659. https://jacow.org/l98/papers/TU4107.pdf
95. Q. Zhao et al., Beam dynamics studies for the FRIB driver linac, in *Proceedings of PAC'09, Particle Accelerator Conference*, no. 23 (JACoW Publishing, Geneva), pp. 3901–3903. https://jacow.org/PAC2009/papers/TH6PFP085.pdf
96. T. Xu et al., Completion of FRIB superconducting linac and phased beam commissioning, in *Proceedings of SRF'21, International Conference on RF Superconductivity*, no. 20 (JACoW Publishing, Geneva, 2022), p. 197. https://doi.org/10.18429/JACoW-SRF2021-MOOFAV10. https://jacow.org/srf2021/papers/MOOFAV10.pdf
97. P. Ostroumov, C. Contreras, A. Plastun, J. Rathke, T. Schultheiss, Taylor, A., J. Wei, M. Xu, T. Xu, Q. Zhao, et al.: Elliptical superconducting RF cavities for FRIB energy upgrade. Nucl. Instrum. Methods Phys. Res., Sect. A **888**, 53–63 (2018)
98. M. Xu, C. Compton, C. Contreras-Martinez, I. Gonin, W. Hartung, T. Khabiboulline, S.H. Kim, S. Miller, P. Ostroumov, A. Plastun et al., Design of β=0.65, 5 cells, 644 mhz elliptical cavity for FRIB upgrade. Fermi National Accelerator Laboratory (FNAL), Batavia (2018). https://doi.org/10.18429/JACoW-IPAC2018-THPAL035. https://www.osti.gov/biblio/1480483
99. D. Böhne, Status Report on the UNILAC, in *Proceedings of LINAC'79, Linear Accelerator Conference* (JACoW Publishing, Geneva), pp. 98–103. https://jacow.org/l79/papers/S2-7.pdf
100. W. Barth, A. Adonin, C.E. Düllmann, M. Heilmann, R. Hollinger, E. Jäger, O. Kester, J. Khuyagbaatar, J. Krier, E. Plechov, P. Scharrer, W. Vinzenz, H. Vormann, A. Yakushev, S.

Yaramyshev, High brilliance uranium beams for the GSI FAIR. Phys. Rev. Accel. Beams **20**, 050101 (2017). https://doi.org/10.1103/PhysRevAccelBeams.20.050101
101. M. Lindroos, S. Bousson, R. Calaga, H. Danared, G. Devanz, R. Duperrier, J. Eguia, M. Eshraqi, S. Gammino, H. Hahn et al., The European spallation source. Nucl. Instrum. Methods Phys. Res., Sect. B **269**(24), 3258–3260 (2011)
102. R.P. Stanek et al., PIP-II project overview and status, in *Proceedings of SRF'23, International Conference on RF Superconductivity*, no. 21 (JACoW Publishing, Geneva, 2023), pp. 19–24. https://doi.org/10.18429/JACoW-SRF2023-MOIXA02
103. D. Bernardi, A. Ibarra, F. Arbeiter, F. Arranz, M. Cappelli, P. Cara, J. Castellanos, H. Dzitko, A. García, J. Gutiérrez et al., The IFMIF-DONES project: design status and main achievements within the EUROfusion FP8 work programme. J. Fusion Energy **41**(2), 24 (2022)
104. J. Knaster, F. Arbeiter, P. Cara, S. Chel, A. Facco, R. Heidinger, A. Ibarra, A. Kasugai, H. Kondo, G. Micciche, K. Ochiai, S. O'hira, Y. Okumura, K. Sakamoto, E. Wakai, IFMIF, the European–Japanese efforts under the Broader Approach agreement towards a Li(d,xn) neutron source: current status and future options. Nucl. Mater. Energy **9**, 46–54 (2016). https://doi.org/10.1016/j.nme.2016.04.012. https://www.sciencedirect.com/science/article/pii/S2352179115301101
105. Y. He et al., Accelerator driven system—a solution to multiple problems of society, in *Proceedings of IPAC'23, International Particle Accelerator Conference*, no. 14 (JACoW Publishing, Geneva, 2023), pp. 5205–5209. https://doi.org/10.18429/JACoW-IPAC2023-FRYG2
106. H. Aït Abderrahim, P. Baeten, D. De Bruyn, R. Fernandez, MYRRHA–A multi-purpose fast spectrum research reactor. Energy Conversion Manag. **63**, 4–10 (2012)
107. D. Vandeplassche, J.L. Biarrotte, H. Klein, H. Podlech, The MYRRHA Linear Accelerator. Universitätsbibliothek Johann Christian Senckenberg (2011)
108. N. Added, B.E. Clifft, K.W. Shepard, Upgraded phase control system for superconducting low-velocity accelerating structures, in *Proceedings of LINAC'92, Linear Accelerator Conference* (JACoW Publishing, Geneva), pp. 181–183. https://jacow.org/l92/papers/MO4-49.pdf
109. A. Facco, Mechanical mode damping in superconducting low-β resonators. Part. Accel. **61**, 265–78 (1998)
110. S.N. Simrock, Lorentz force compensation of pulsed SRF cavities, in *Proceedings of LINAC'02, Linear Accelerator Conference*, no. 21 (JACoW Publishing, Geneva), pp. 554–558. https://jacow.org/l02/papers/WE204.pdf
111. R.E. Laxdal, Review of magnetic shielding designs of low-beta cryomodules, in *Proceedings of SRF'13* (JACoW Publishing, Geneva), pp. 800–807. https://jacow.org/SRF2013/papers/WEIOD01.pdf
112. M. Martinello et al., Magnetic flux expulsion in horizontally cooled cavities, in *Proceedings of SRF'15* (JACoW Publishing, Geneva), pp. 110–114. https://jacow.org/SRF2015/papers/MOPB014.pdf
113. M. Doleans et al., Plasma processing to improve SRF accelerating gradient (2015). Presented at SRF'15, Whistler, paper THBA01, Unpublished
114. R. Laxdal, M. Marchetto, The ISAC post-accelerator. Hyperfine Interact. **225**, 79–97 (2014). https://doi.org/10.1007/s10751-013-0884-8
115. L. Ge, C. Ko, Z. Li, J. Popielarski, Multipacting simulation and analysis for the FRIB = 0.085 quarter wave resonators using Track3P, in *Proceedings of IPAC'12* (JACoW Publishing, Geneva), pp. 2366–2368. https://jacow.org/IPAC2012/papers/WEPPC068.pdf
116. S. Kim, M. Doleans, D. Jeon, R. Sundelin, Higher order mode analysis of the sns superconducting linac, in *Proceedings of PAC'01* (JACoW Publishing, Geneva), pp. 1128–1130. https://jacow.org/p01/papers/MPPH149.pdf
117. M.P. Howell et al., History of cryomodule repairs at SNS, in *Proceedings of LINAC'14, Linear Accelerator Conference*, no. 27 (JACoW Publishing, Geneva, 2014), pp. 1108–1110. https://jacow.org/LINAC2014/papers/THPP109.pdf

118. A. Sukhanov, A. Lunin, V. Yakovlev, M. Awida, M. Champion, C. Ginsburg, I. Gonin, C. Grimm, T. Khabiboulline, T. Nicol, Y. Orlov, A. Saini, D. Sergatskov, N. Solyak, A. Vostrikov, High order modes in Project-X linac. Nucl. Instrum. Methods Phys. Res., Sect. A **734**, 9–22 (2014). https://doi.org/10.1016/j.nima.2013.06.113. https://www.sciencedirect.com/science/article/pii/S0168900213010796. Higher-Order-Mode Diagnostics and Suppression in Superconducting Cavities (HOMSC12)
119. H.J. Zheng, J. Gao, J.F. Chen, C. Pagani, HOM calculations for different cavities and beam induced HOM power analysis of ESS, in *Proceedings of SRF'15* (JACoW Publishing, Geneva), pp. 1061–1065 https://jacow.org/SRF2015/papers/THPB004.pdf
120. C.S. Hopper, J.R. Delayen, Higher order mode damping in superconducting spoke cavities, in *Proceedings of IPAC'14* (JACoW Publishing, Geneva), pp. 2669–2671. https://doi.org/10.18429/JACoW-IPAC2014-WEPRI076. https://jacow.org/IPAC2014/papers/WEPRI076.pdf
121. K. Papke, F. Gerigk, U. van Rienen, Progress of HOM couplers for CERN SPL cavities, in *Proceedings of IPAC'14* (JACoW Publishing, Geneva), pp. 2568–2570. https://doi.org/10.18429/JACoW-IPAC2014-WEPRI041. https://jacow.org/IPAC2014/papers/WEPRI041.pdf
122. I. Hofmann, Stability of anisotropic beams with space charge. Phys. Rev. E **57**, 4713–4724 (1998). https://doi.org/10.1103/PhysRevE.57.4713
123. V.A. Lebedev, N. Solyak, A.V. Aleksandrov, A.P. Shishlo, J.F. Ostiguy, Intrabeam stripping in H-linacs, in *Proceedings of LINAC'10, Linear Accelerator Conference*, no. 25 (JACoW Publishing, Geneva), pp. 929–931. https://jacow.org/LINAC2010/papers/THP080.pdf
124. A.J. Jason, D.W. Hudgings, O.B. van Dyck, Neutralization of H^- Beams by magnetic stripping. IEEE Trans. Nucl. Sci. **28**(3), 2703–2706 (1981). https://doi.org/10.1109/TNS.1981.4331890
125. S.D. Holmes, R. Alber, B. Chase, K. Gollwitzer, D. Johnson, M. Kaducak, A. Klebaner, I. Kourbanis, V. Lebedev, A. Leveling et al., Project X: accelerator reference design. arXiv preprint arXiv:1306.5022 (2013)
126. M. Oliphant, J. Gooden, G. Hide, The acceleration of charged particles to very high energies. Proc. Phys. Soc. **59**(4), 666 (1947)
127. E.M. McMillan, The synchrotron—a proposed high energy particle accelerator. Phys. Rev. **68**, 143–144 (1945). https://doi.org/10.1103/PhysRev.68.143
128. V.I. Veksler, New method for the acceleration of relativistic particles, *Doklady Akademii Nauk USSR*, vol. 43 (1944), pp. 346–348
129. N. Christofilos, Focussing system for ions and electrons (1956). https://ppubs.uspto.gov/dirsearch-public/print/downloadPdf/2736799. US Patent 2,736,799
130. E.D. Courant, M.S. Livingston, H.S. Snyder, The strong-focusing synchroton—a new high energy accelerator. Phys. Rev. **88**, 1190–1196 (1952). https://doi.org/10.1103/PhysRev.88.1190
131. J. Wei, D.T. Abell, J. Beebe-Wang, M. Blaskiewicz, P.R. Cameron, N. Catalan-Lasheras, G. Danby, A.V. Fedotov, C. Gardner, J. Jackson, Y.Y. Lee, H. Ludewig, N. Malitsky, W. Meng, Y. Papaphilippou, D. Raparia, N. Tsoupas, W.T. Weng, R.L. Witkover, S.Y. Zhang, Low-loss design for the high-intensity accumulator ring of the Spallation Neutron Source. Phys. Rev. ST Accel. Beams **3**, 080101 (2000). https://doi.org/10.1103/PhysRevSTAB.3.080101
132. M. Kinsho, Status of the J-PARC 3-GeV RCS, in *Proceedings of PAC'09, Particle Accelerator Conference*, no. 23 (JACoW Publishing, Geneva), pp. 1436–1438. https://jacow.org/PAC2009/papers/TU6PFP065.pdf
133. T. Koseki et al., Upgrade plan of J-PARC MR-toward 1.3 MW beam power, in *Proceedings of IPAC'18* (2018), pp. 966–969
134. J. Thomason, The ISIS Spallation Neutron and muon source—the first thirty-three years. Nucl. Instrum. Methods Phys. Res., Sect. A **917**, 61–67 (2019). https://doi.org/10.1016/j.nima.2018.11.129. https://www.sciencedirect.com/science/article/pii/S0168900218317820
135. S. Wang, S.X. Fang, S.N. Fu, W.B. Liu, H.F. Ouyang, Q. Qin, J.Y. Tang, J. Wei, Introduction to the overall physics design of CSNS accelerators. Chin. Phys. C **33**(S2), 1 (2009)
136. W.T. Weng, L. Ahrens, R. Damm, A.J. McNerney, Construction and early commissioning results of the AGS booster, in *Proceedings of PAC'91* (JACoW Publishing, Geneva), pp. 52–57

137. Y.Y. Lee et al., The AGS-Booster lattice, in *Proceedings of PAC'87* (JACoW Publishing, Geneva), pp. 865–867
138. M. Brennan, T. Roser, High intensity performance of the Brookhaven AGS, in *Proceedings of EPAC'96, European Particle Accelerator Conference*, no. 5 (JACoW Publishing, Geneva, 1996), pp. 530–532
139. S. Ozaki, T. Roser, Relativistic heavy ion collider, its construction and upgrade. Prog. Theor. Exp. Phys. **2015**(3), 03A102 (2015)
140. H. Hahn, E. Forsyth, H. Foelsche, M. Harrison, J. Kewisch, G. Parzen, S. Peggs, E. Raka, A. Ruggiero, A. Stevens, S. Tepikian, P. Thieberger, D. Trbojevic, J. Wei, E. Willen, S. Ozaki, S. Lee, The RHIC design overview. Nucl. Instrum. Methods Phys. Res., Sect. A **499**(2), 245–263 (2003). https://doi.org/10.1016/S0168-9002(02)01938-1. https://www.sciencedirect.com/science/article/pii/S0168900202019381. The Relativistic Heavy Ion Collider Project: RHIC and its Detectors
141. M. Harrison, S. Peggs, T. Roser, The RHIC accelerator. Annu. Rev. Nucl. Part. Sci. **52**(1), 425–469 (2002)
142. G. Lawrence, R. Hardekopf, A. Jason, P. Clout, G. Sawyer, Los Alamos high-current proton siorage ring; a status report. IEEE Trans. Nucl. Sci. **32**(5), 2662–2665 (2007)
143. R. Billinge, Booster reference design, part I. Technical report, Fermi National Accelerator Lab.(FNAL), Batavia (1968)
144. S.D. Holmes, R. Gerig, D.E. Johnson, The Fermilab main injector. Part. Accel. **26**(FERMILAB-CONF-89-183), 193–198 (1989)
145. C. Bovet, Present Performance of the CERN 800 MeV PS Booster (PSB), in *Proceedings of PAC'73, Particle Accelerator Conference*, no. 5 (JACoW Publishing, Geneva, 1973), pp. 221–226
146. R. Cappi et al., The CERN PS complex as part of the LHC injector chain, in *Proceedings of PAC'91* (JACoW Publishing, Geneva), pp. 171–174
147. E.J.N. Wilson, The lattice of the SPS, in *Proceedings of PAC'77* (JACoW Publishing, Geneva), pp. 1458–1461
148. D. Boussard, L. Evans, J. Gareyte, T. Linnecar, W. Mills, E.J.N. Wilson, Acceleration and storage of a dense single bunch in the CERN SPS, in *Proceedings of PAC'79* (JACoW Publishing, Geneva), pp. 3484–3487
149. L.R. Evans, LHC status, in *Proceedings of APAC'07* (JACoW Publishing, Geneva), pp. 46–50. https://jacow.org/a07/papers/TUXMA01.pdf
150. S. Ivanov, Accelerator complex U70 of IHEP-Protvino: status and upgrade plans, in *Proceedings of RuPAC'08* (JACoW Publishing, Geneva). https://jacow.org/r08/papers/TUBAU01.pdf
151. E. Troyanov, Status of the U-70 accelerator at the Institute of High-Energy Physics. Atomic Energy **93**(6), 923–929 (2002)
152. S. Ivanov et al., Accelerator complex U70 of IHEP: present status and recent upgrades, in *Proceedings of Conference* (2010), pp. 27–31
153. N. Angert, B. Franzke, The GSI heavy ion facility, in *Challenges and Goals for Accelerators in the XXI Century*, ed. by O. Brüning, S. Myers (World Scientific, Singapore, 2016), pp. 279–306
154. P.J. Spiller, K. Blasche, P. Hülsmann, A. Kraemer, H. Ramakers, H.R. Sprenger, High intensity uranium operation in SIS18, in *Proceedings of EPAC'04, European Particle Accelerator Conference*, no. 9 (JACoW Publishing, Geneva, 2004), pp. 1180–1182. http://accelconf.web.cern.ch/e04/papers/TUPLT020.pdf
155. H. Sasaki, The booster synchrotron utilization facility at KEK, in *Proceedings of the ICANS-IV Meeting* (1980)
156. K. Nakai, T. Ohshima, KEK-PS 1980s. Nippon Butsuri Gakkaishi (Proceedings of the Physical Society of Japan);(Japan) **47**(2) (1992)
157. J. Carpenter, D. Price, N. Swanson, IPNS-a national facility for condensed matter research. Technical report, CM-P00068627 (1978)
158. R.L. Kustom, Intense pulsed neutron sources. IEEE Trans. Nucl. Sci. **28**(3), 3115–3119 (2007)

159. H.T. Edwards, The Tevatron energy doubler: a superconducting accelerator. Annu. Rev. Nucl. Part. Sci. **35**(1), 605–660 (1985)
160. S.D. Holmes, V.A. Yarba et al., Fermilab Tevatron and upgrade status. Technical report, Fermi National Accelerator Lab., Batavia (1994)
161. H. Lengeler, The European spallation source study. Nucl. Instrum. Methods Phys. Res., Sect. B **139**(1–4), 82–90 (1998)
162. A.W. Chao, Accelerator physics studies for the SSC. EEE Trans. Nucl. Sci. Vol NS-**32**(5), 3442–3446 (1985). Particle accelerator conference. Accelerator engineering and technology Journal Page Range p. 3442–3446
163. S. Machida, Y. Mori, Y. Ishi, S. Shibuya, Design and beam dynamics issues of the JHF synchrotrons, in *Proceedings of APAC'98* (JACoW Publishing, Geneva), pp. 381–383
164. Y. Ishi, S. Machida, S. Shibuya, Y. Mori, Lattice design of JHF synchrotrons, in *Proceedings of APAC'98* (JACoW Publishing, Geneva), pp. 378–380
165. B. Boardman, Spallation neutron source: description of accelerator and target. Technical report, Science Research Council (1982)
166. G. Rees, Note on an RCS for a 2 MW spallation source at RAL. Private communication (2007)
167. J. Wei, Synchrotrons and accumulators for high-intensity proton beams. Rev. Mod. Phys. **75**, 1383–1432 (2003). https://doi.org/10.1103/RevModPhys.75.1383
168. Y. Cho, Y. Chae, H. Friedsam, D. Horan, R. Kustom, E. Lessner, W. McDowell, D. McGhee, H. Moe, R. Nielsen et al., *Proceedings of the 1996 European Particle Accelerator Conference*, Sitges (1996)
169. C. Warsop, Status and high intensity performance of ISIS, in *AIP Conference of Proceedings*, vol. 448 (American Institute of Physics, 1998), pp. 104–115
170. G. Rees, Important design issues of high output current proton rings, in *Proceedings of EPAC'94, European Particle Accelerator Conference*, no. 4 (JACoW Publishing, Geneva), pp. 241–246
171. C. Prior, G. Rees, Multiturn injection and lattice design for HIDIF. Nucl. Instrum. Methods Phys. Res., Sect. A **415**(1–2), 357–362 (1998)
172. J. Wei, Longitudinal Dynamics of the Nonadiabatic Regime on Alternating Gradient Synchrotrons. Ph.D. thesis, SUNY, Stony Brook (1990)
173. J. Wei, Y. Lee, A. Ruggiero, The effect of longitudinal space charge on multi-turn capture in the AGS booster. Part. Accel. **24**, 211–222 (1989)
174. Y. Irie, Y. Arakida, S. Igarashi, S. Muto, I. Sakai, K. Takayama, I. Yamane, H-Painting injection for the JHF 3-GeV synchrotron, in *Proceedings of EPAC'98, European Particle Accelerator Conference* (JACoW Publishing, Geneva, 1998), pp. 2103–2105
175. I. Sakai et al., H-painting injection system for the JKJ 3 GeV high intensity proton synchrotron, in *Proceedings of EPAC'02, European Particle Accelerator Conference* (JACoW Publishing, Geneva, 2002), pp. 1040–1042. https://jacow.org/e02/papers/THPLE007.pdf
176. J. Beebe-Wang, Y.Y. Lee, S. Machida, C.R. Prior, D. Raparia, J. Wei, Beam properties in the SNS accumulator ring due to transverse phase space painting, in *Proceedings of EPAC'00, European Particle Accelerator Conference*, no. 7 (JACoW Publishing, Geneva, 2000), pp. 1465–1467. https://jacow.org/e00/papers/MOP6A02.pdf
177. N. Christofilos, R. Hester, W. Lamb, D. Reagan, W. Sherwood, R. Wright, High current linear induction accelerator for electrons. Rev. Sci. Instrum. **35**(7), 886–890 (1964)
178. J. Griffin, Isolated bucket RF systems in the Fermilab antiproton facility. IEEE Trans. Nucl. Sci. **30**(4), 3502–3504 (1983)
179. K. Takayama, J. Kishiro, Induction synchrotron. Nucl. Instrum. Methods Phys. Res., Sect. A **451**(1), 304–317 (2000)
180. D. Boussard, The RF noise theory applied to the PS longitudinal controlled blow-up. Technical report (1985)
181. D. Raparia, Y.Y. Lee, J. Wei, W.T. Weng, Dependence of the SNS transfer lines and accumulator ring on linac energy, in *Proceedings of PAC'01* (JACoW Publishing, Geneva), pp. 3260–3262. https://jacow.org/p01/papers/RPPH018.pdf

182. A. Hofmann, S. Myers, Beam Dynamics in a double RF system (Birkhäuser Basel, Basel, 1980), pp. 610–614. https://doi.org/10.1007/978-3-0348-5540-2_83
183. J. Wei: Stachastic cooling with a double RF system, in *Proceedings of EPAC'92, European Particle Accelerator Conference*, no. 3 (JACoW Publishing, Geneva), pp. 833–836
184. P. Wilson, Beam loading in high-energy storage rings. Technical report, SLAC National Accelerator Lab., Menlo Park (1974)
185. F. Pedersen, Beam loading effects in the CERN PS booster, in *Proceedings of PAC'75* (JACoW Publishing, Geneva), pp. 1906–1910
186. D. Boussard, 2.4.4 Beam loading, in *Handbook of Accelerator Physics and Engineering*, ed. by A.W. Chao et al. (World Scientific, Singapore, 2022), pp. 160–164
187. R. Garoby, RF gymnastics in synchrotrons. arXiv preprint arXiv:1112.3232 (2011)
188. L.C. Teng, Design concepts for the beam scraper system of the main ring. FNAL Report FN-196/0400 (1969)
189. K. Yamamoto et al., Present status of beam collimation system of J-PARC RCS, in *Proceedings of EPAC'06, European Particle Accelerator Conference*, no. 10 (JACoW Publishing, Geneva, 2006), pp. 3200–3202
190. K. Yamamoto, M. Kinsho, F. Noda, Y. Irie, M. Shirakata, Beam collimator design for the 3 GeV synchrotron of the JAERI-KEK Joint Project, in *Proceedings of PAC'01* (JACoW Publishing, Geneva), pp. 3615–3617. https://jacow.org/p01/papers/RPPH305.pdf
191. N. Tsoupas et al., Beam extraction from the SNS ring and design of extraction kickers, in *Proceedings of EPAC'00, European Particle Accelerator Conference*, no. 7 (JACoW Publishing, Geneva, 2000), pp. 2270–2274. https://jacow.org/e00/papers/MOP6A12.pdf
192. G.D. Wait, M.J. Barnes, K.D. Metzmacher, L. Sermeus, The application of saturating inductors for improving the performance of the CERN PS kicker systems, in *Proceedings of PAC'97* (JACoW Publishing, Geneva), pp. 1328–1330
193. R. Cappi, M. Giovannozzi, Multiturn extraction: performance analysis of old and new approaches. Nucl. Instrum. Methods Phys. Res. Sect. A: Accel. Spectrom. Detect. Assoc. Equip. **519**(1–2), 442–452 (2004)
194. J. Borburgh, S. Damjanovic, S. Gilardoni, M. Giovannozzi, C. Hernalsteens, M. Hourican, A. Huschauer, K. Kahle, G. Le Godec, O. Michels, G. Sterbini, First implementation of transversely split proton beams in the cern proton synchrotron for the fixed-target physics programme. Europhys. Lett. **113**(3), 34001 (2016). https://doi.org/10.1209/0295-5075/113/34001
195. M. Vadai, A. Alomainy, H. Damerau, M. Giovannozzi, A. Huschauer, Barrier bucket gymnastics and transversely split proton beams: performance at the cern proton and super proton synchrotrons. Phys. Rev. Accel. Beams **25**, 050101 (2022). https://doi.org/10.1103/PhysRevAccelBeams.25.050101
196. A. Huschauer, H. Bartosik, S.C. Cave, M. Coly, D. Cotte, H. Damerau, G. Di Giovanni, S. Gilardoni, M. Giovannozzi, V. Kain et al., Advancing the cern proton synchrotron multiturn extraction towards the high-intensity proton beams frontier. Phys. Rev. Accel. Beams **22**(10), 104002 (2019)
197. S. Abernethy, A. Akroh, H. Bartosik, A. Blas, T. Bohl, S. Cettour-Cave, K. Cornelis, H. Damerau, S. Gilardoni, M. Giovannozzi, C. Hernalsteens, A. Huschauer, V. Kain, D. Manglunki, G. Métral, B. Mikulec, B. Salvant, J.L. Sanchez Alvarez, R. Steerenberg, G. Sterbini, Y. Wu, Operational performance of the cern injector complex with transversely split beams. Phys. Rev. Accel. Beams **20**, 014001 (2017). https://doi.org/10.1103/PhysRevAccelBeams.20.014001
198. H. Akbari, X. Altuna, S. Bardin, R. Bellazzini, V. Biryukov, A. Brez, M.P. Bussa, L. Busso, A. Calcaterra, G. Carboni et al., First results on proton extraction from the CERN-SPS with a bent crystal. Phys. Lett. B **313**(3–4), 491–497 (1993)
199. F.M. Velotti, L.S. Esposito, M.A. Fraser, V. Kain, S. Gilardoni, B. Goddard, M. Pari, J. Prieto, R. Rossi, W. Scandale, L.S. Stoel, F. Galluccio, M. Garattini, Y. Gavrikov, Septum shadowing by means of a bent crystal to reduce slow extraction beam loss. Phys. Rev. Accel. Beams **22**, 093502 (2019). https://doi.org/10.1103/PhysRevAccelBeams.22.093502

200. V.M. Biryukov, Y.A. Chesnokov, V.I. Kotov, *Crystal Channeling and Its Application at High-Energy Accelerators* (Springer, Berlin, 2013)
201. A.G. Afonin, V.T. Baranov, V.M. Biryukov, M.B.H. Breese, V.N. Chepegin, Y.A. Chesnokov, V. Guidi, Y.M. Ivanov, V.I. Kotov, G. Martinelli, W. Scandale, M. Stefancich, V.I. Terekhov, D. Trbojevic, E.F. Troyanov, D. Vincenzi, High-efficiency beam extraction and collimation using channeling in very short bent crystals. Phys. Rev. Lett. **87**, 094802 (2001). https://doi.org/10.1103/PhysRevLett.87.094802
202. C. Murphy, R. Carrigan, D. Chen, G. Jackson, N. Mokhov, H.J. Shih, B. Cox, V. Golovatyuk, A. McManus, A. Bogacz et al., First results from bent crystal extraction at the fermilab tevatron. Nucl. Instrum. Methods Phys. Res. Sect. B: Beam Interact. Mater. Atoms **119**(1–2), 231–238 (1996)
203. M. Garattini, D. Annucci, O. Blaco-Garcia, P. Gianotti, S. Guiducci, A. Liedl, M. Raggi, P. Valente et al., The sherpa experiment. POS Proc. Sci. **878**, 1–7 (2022)
204. M. Rose, Magnetic field corrections in the cyclotron. Phys. Rev. **53**(9), 715 (1938)
205. G. Danby, J. Jackson, C. Spataro, Precision magnetic elements for the SNS storage ring, in *Proceedings of PAC'99* (JACoW Publishing, Geneva), pp. 3333–3335. https://jacow.org/p99/papers/THP156.pdf
206. K. Halbach, Some eddy current effects in solid core magnets. Nucl. Instrum. Methods **107**(3), 529–540 (1973). https://doi.org/10.1016/0029-554X(73)90388-1. https://www.sciencedirect.com/science/article/pii/0029554X73903881
207. J.T. Tanabe, *Iron Dominated Electromagnets: Design, Fabrication, Assembly and Measurements* (World Scientific, Singapore, 2005)
208. D.A. Edwards, M.J. Syphers, *An Introduction to the Physics of High Energy Accelerators* (Wiley, Weinheim, 2008)
209. J. Wei, R. Talman, Theorem on magnet fringe field. Part. Accel. **55**, 93–98 (1996)
210. Y. Papaphilippou, J. Wei, R. Talman, Deflections in magnet fringe fields. Phys. Rev. E **67**, 046502 (2003). https://doi.org/10.1103/PhysRevE.67.046502
211. Y. Papaphilippou, Y.Y. Lee, W. Meng, Impact of magnetic field interference in the SNS ring, in *Proceedings of PAC'01* (JACoW Publishing, Geneva), pp. 1667–1669. https://jacow.org/p01/papers/TPPH007.pdf
212. D.A. Edwards, L.C. Teng, Parametrization of linear coupled motion in periodic systems. IEEE Trans. Nucl. Sci. **20**(3), 885–888 (1973). https://doi.org/10.1109/TNS.1973.4327279
213. L.C. Teng, Coupled transverse motion, in *Frontiers of Particle Beams; Observation, Diagnosis and Correction*, ed. by M. Month, S. Turner (Springer, Berlin/Heidelberg, 1989), pp. 133–149
214. S. Peggs, Coupling and decoupling in storage rings. IEEE Trans. Nucl. Sci. **30**(CERN-SPS-83-12-DI-MST), 2460–2462 (1983)
215. C.J. Gardner, Y.Y. Lee, Y. Papaphilippou, N. Tsoupas, J. Wei, Chromatic correction and optical compensation in the SNS accumulator ring using sextupoles, in *Proceedings of EPAC'00, European Particle Accelerator Conference*, no. 7 (JACoW Publishing, Geneva, 2000), pp. 1581–1583. URL https://jacow.org/e00/papers/MOP6A13.pdf
216. K. Schindl, Simultaneous dynamic compensation of stopbands and multipurpose location of correction lenses in the CERN PS booster. IEEE Trans. Nucl. Sci. **26**(3), 3562–3564 (1979). https://doi.org/10.1109/TNS.1979.4330100
217. C. Gardner et al., Analysis of resonance-driving imperfections in the AGS booster, in *Proceedings of EPAC'94, European Particle Accelerator Conference*, no. 4 (JACoW Publishing, Geneva), pp. 979–982

Chapter 3
Beam Dynamics: Intensity Limiting Mechanisms

Abstract This chapter reviews the key beam-dynamics issues relevant to high-power hadron accelerators, with emphasis on the dynamics governing the acceleration and transport of intense beams across different accelerator types, the mechanisms limiting beam intensity and stable operation, as well as the possible mitigation methods.

High-power hadron accelerators require careful consideration of beam dynamics, as even small imperfections in beam transport and acceleration can rapidly translate into uncontrolled beam loss, limits on achievable beam power, increased component activation, and reduced machine availability. In this chapter, several topics in beam dynamics are discussed, highlighting mechanisms that potentially limit the intensity of the beam in high-power hadron accelerators. Several important topics that may not be directly relevant to conventional high-power accelerators are discussed in the Appendices.

3.1 Beam Loss Mechanisms

The key to the design and operation of a high-power accelerator is to control beam loss. The beam loss is categorized into two classes: controlled and uncontrolled. Managing controlled beam loss and minimizing uncontrolled beam loss are crucial to maintaining radiologically sound operations with hands-on maintenance [177].

3.1.1 Controlled Beam Loss

Measures of loss control include beam collimation, beam dumping, targetry, and shielding for otherwise uncontrolled losses, such as charge stripping and charge selection. Controlled beam loss includes the beam that crosses the beam dump and

J. Wei et al., *High Power Hadron Accelerators*, Particle Acceleration and Detection,
https://doi.org/10.1007/978-3-032-17662-2_3

Table 3.1 Estimated FRIB controlled primary beam loss for typical 400 kW operation. MEBT: Medium Energy Beam Transport; F-Cup: Faraday Cup; LS: Linac Segment; FS: Folding Segment; RI: Rare Isotope [180]

Mechanism	Location	Fraction	Power [kW]
Front end beam tuning	MEBT F-Cup	2.5×10^{-4}	0.1
$\beta_0 = 0.041$ beam tuning	LS1 Nb plate	5×10^{-6}	0.002
LS1 beam tuning	FS1a dump	3.8×10^{-5}	0.015
LS1 beam tuning	FS1b dump	1.3×10^{-3}	0.5
Charge stripping loss	FS1 stripper	3.3×10^{-3}	1.3
Unwanted beam collection	FS1 selector	< 0.03	7–12
Unwanted charge state, ion contaminants collection	FS1 45° dipoles	$< 1\times10^{-3}$	0.1–0.4
Beam halo and ECR ion contaminants interception	Collimator 1–5	$< 4.3\times10^{-3}$	0.1–1.7
Charge exchange halo	Collimator 6–7	5×10^{-5}	0.02
LS1 beam tuning	FS1 F-Cup	7.5×10^{-5}	0.03
LS2 beam tuning	FS2 dump	3.5×10^{-4}	0.14
LS2 beam tuning	FS2 F-Cup	7.5×10^{-5}	0.03
Linac beam tuning	BDS dump	3.5×10^{-4}	0.14
Targetry protection	Collimators	2.5×10^{-4}	0.1
RI production	Target	< 0.2	< 80
Spent beam	Beam dump	< 0.8	< 320

Table 3.2 Estimated FRIB controlled secondary beam loss for typical 400 kW operation [180]

Mechanism	Location	Power [kW]
Target scattering	Post-target shield	< 15
RI beam cleaning	Fragment catchers	< 20
Post-target scattering	Thermal armors	< 17
Post-target scattering	Collimators	< 9
RI beam	Wedge	< 2
RI beam cleaning	Wedge slits	< 3
RI beam cleaning	Separator slits	< 5
RI beam cleaning	Focal plane slits	< 2

hits the collimators, both located in shielded regions. As an example, Tables 3.1 and 3.2 list the expected controlled beam loss in the FRIB complex. The leading causes of controlled primary beam loss are the spent beam downstream of the production target, the beam lost on the production target, the beam of unwanted charge states downstream of the charge stripper, loss upon charge stripping, and the lost beam collected at collimators and beam-tuning dumps. The leading causes of controlled secondary beams are the collection of unwanted isotope fragments, the loss of beam at fragment catchers and collimators, and the loss of fragment separation instrumentation, such as wedges and slits.

Table 3.3 lists the expected controlled primary beam loss in the SNS accumulator ring, the transports, and the target, including the tail of the beam and the residual from the linac, the H^- beam missing or escaping the stripping foil and reaching the

Table 3.3 Estimated controlled loss of a proton beam at 1 GeV in the SNS ring, linac-to-ring transport (HEBT), and ring-to-target transport (RTBT). Losses are given as a fraction of the total beam intensity. The total beam power is 2 MW [20, 177]

Mechanism	Location	Fraction	Power [kW]
HEBT:			
H^0 from linac	Linac dump	10^{-5}	0.02
Linac transverse tail	HEBT H/V collimator	10^{-3}	2
Energy jitter/spread from linac	HEBT L collimator	10^{-3}	2
Ring:			
Beam-in-gap	BIG kicker/collimator	10^{-4}	0.2
Excited H^0 at foil	Collimator	1.3×10^{-5}	0.026
Partial ionization at foil	Injection dump	10^{-2}	20
Foil miss	Injection dump	10^{-2}	20
Ring beam halo	Collimator	1.9×10^{-3}	3.8
Energy straggling at foil	Collimator	3×10^{-6}	0.006
RTBT:			
Kicker misfiring	RTBT collimator	10^{-5}	0.02
Target:			
Nuclear scattering at window	Target window	4×10^{-2}	80
Neutron production	Target	~ 0.94	$\sim 1.9\times10^3$

injection beam dump, the beam halo generated in the ring, energy straggling on the foil, and loss due to the kicker's misfiring [20].

The beam intercepting systems used for beam collection are described in Sects. 8.8, 8.9, and 8.10. The radiation shielding of such systems should be designed to allow maintenance access. Radiation-resistant measures must be implemented to ensure proper function in a high-radiation environment. Remote handling is usually mandatory (Sect. 8.9.3).

3.1.2 Uncontrolled and Accidental Beam Loss

Uncontrolled beam loss falls into several categories: those losses linearly proportional to the beam's intensity (e.g., Coulomb scattering and nuclear scattering at the injection foil, magnetic stripping, gas scattering, kicker malfunction, noises from the RF system, and the magnet power-supply system); those losses that grow progressively with intensity (e.g., space-charge-induced tune spread and coherent resonance crossing, instability, electron cloud effects); and those losses accidental to normal operations.

Table 3.4 lists some known mechanisms and the expected level of loss of the SNS ring [20]. As a design guide for hands-on maintenance, the average beam loss must be less than 1 W per meter. That translates into a level of radio-activation of

Table 3.4 Estimated uncontrolled loss of a proton beam at 1 GeV in the SNS ring, linac-to-ring transport (HEBT), and ring-to-target transport (RTBT). Losses are given as a fraction of the total beam intensity distributed in the specified machine length. The total beam power is 2 MW [20, 177]

Mechanism	Location	Fraction	Length (m)	Power (W/m)
HEBT:				
H^- magnetic stripping	All HEBT	1.7×10^{-6}	169	0.02
Collimator outscattering	HEBT achromat	7.5×10^{-6}	15	0.1
Ring:				
H^- magnetic stripping	Injection dipole	1.3×10^{-7}	1	0.3
Nuclear scattering at foil	Foil	3.7×10^{-5}	30	2.5
Collimation inefficiency	All ring	10−4	218	0.9

Table 3.5 Estimated FRIB uncontrolled beam losses for typical 400 kW operation [180]

Mechanism	Location	Power
Stripper scattering	Downstream of stripper	< 3 W/m at 17 MeV/u
Uncontrolled loss	All locations	< 1 W/m

Table 3.6 Accidental FRIB beam loss scenarios for typical 400 kW operation. MPS: Machine Protection System; PPS: Personnel Protection System [180]

Mechanism	Power	Response time	
		MPS	PPS
Damaged stripper	Small fraction of 50 kW	35 μs	N/A
Damaged target	Large fraction of 400 kW	200 μs	7 ms

approximately 1 mSv/h, measured at 30 cm from the surface, 4 h after the machine is shut down [108, 171].

Table 3.5 lists the expected level of uncontrolled loss with the FRIB accelerator complex. Table 3.6 lists some known mechanisms and the expected level of loss under accidental fault conditions. Again, uncontrolled losses must be kept below a level (∼1 W/m) to facilitate hands-on maintenance. The personnel protection system is designed to protect against radiation exposure in both normal and fault machine conditions (Sect. 1.4). Machine protection with a sufficiently short response time is designed to protect the accelerator against beam-induced damage (Sect. 8.2).

3.2 Phase-Space Matching

Phase space matching is the process of adjusting the properties of a particle beam upon transition (or at injection) to the next machine structure so that its distribution, usually represented by elliptical trajectories in phase space, is aligned with the intrinsic stable orbits of the machine. The goal is to maximize the beam's capture and acceleration efficiency while minimizing emittance growth and beam loss.

This section describes phase-space matching in linacs, transport beamlines, and synchrotron rings without taking into account space-charge effects. The phase-space matching techniques can be extended to include the linearized space-charge forces, as is done in the 3-D envelope matching codes such as TRACE3D [26] (Table 3.3). For future high-intensity accelerators, more advanced techniques for phase-space matching are being developed; see, for example, [83].

3.2.1 Linac Phase-Space Matching

The focusing of hadron beams in the transverse direction is provided by periodically located focusing elements along the beam propagation. The most popular focusing structure is the FODO, which is composed of focusing quadrupoles and accelerating gaps. The periodic focusing by SC solenoids has become popular in recently commissioned SC hadron linacs. The longitudinal focusing is provided by the proper selection of the synchronous phase. The transverse beam dynamics in accelerators with quadrupole focusing are described by Hill's equations [60, 81, 169]

$$u'' + K(s)\,u = 0\,, \tag{3.1}$$

where $u = x, y, z$; x and y are the horizontal and vertical deviations from the central axis, z is the deviation from the longitudinal position of the synchronous particle, and $K(s)$ is a periodic function with a focusing period of length L_c,

$$K(s) = \begin{cases} \dfrac{\pm qe\,G(s)}{Am_0c\,\beta_s\gamma_s} & \text{in a quadrupole magnet,} \\[2ex] \left(\dfrac{qeB_{\text{sol}}(s)}{2Am_0c\,\beta_s\gamma_s}\right)^2 & \text{in a solenoid,} \\[2ex] \dfrac{2\pi qe\,E_{\text{eff}}\sin(\phi)}{Am_0\,\beta_s^3\gamma_s^3\lambda} & \text{in an accelerating section,} \end{cases} \tag{3.2}$$

where $G(s)$ is the gradient of the quadrupoles, $B_{\text{sol}}(s)$ is the magnetic field strength of the solenoid, and E_{eff} is the accelerating gradient averaged over the focusing period, including the transit time factor. The synchronous phase is considered with respect to the maximum RF field; therefore, it is negative for stable acceleration in the standing wave. The equations can be applied to the axially symmetric beam in the solenoidal focusing channels. The more general case of the propagation of an asymmetric beam in the solenoidal channel is presented in [34].

Consider beam dynamics in the $(x,\, x')$ phase plane. The solution of Hill's equation can be expressed in phase-amplitude form, similar to that of a harmonic oscillator,

$$x(s) = \sqrt{2J_{0x}\beta_x(s)}\cos\left[\varphi_x(s) + \varphi_{0x}\right] , \tag{3.3}$$

where $\beta_x(s)$ is the horizontal amplitude function and $\varphi_x(s)$ is the corresponding phase function, and J_{0x} and φ_{0x} are the action and phase defined by the initial conditions, respectively. The trajectory satisfies the relation:

$$\gamma_x x^2 + 2\alpha_x x x' + \beta_x x'^2 = 2J_{0x} , \tag{3.4}$$

where $\alpha_x(s)$, $\beta_x(s)$, and $\gamma_x(s)$ are the Courant-Snyder (C-S) parameters [25] that characterize the machine lattice and $\gamma_x\beta_x - \alpha_x^2 = 1$. Similar equations are valid for the vertical phase space (y, y') and the longitudinal phase space (z, z'). The C-S parameters are functions of the longitudinal coordinate but are periodic with the focusing period. The phase advance of transverse oscillations per focusing period is defined as

$$\mu_{0x} = \int_{s0}^{s_0+L_c} \frac{ds}{\beta_x(s)} , \tag{3.5}$$

where L_c is the length of the focusing period. The geometry, accelerating, and focusing fields within the focusing period fully define the C-S parameters. The acceptance (or admittance) of the focusing channel is the area $A_x = 2\pi J_{0x,\max}$ of the largest ellipse that defines the boundary between transmitted and lost particles during propagation through the periodic focusing system.

In the general case, an ensemble of charged particles occupies a 6-D phase space with an arbitrary density distribution. The particle distribution in the phase space planes can be characterized by an ellipse, the parameters of which can be found from the second moments of the particle distribution in the phase space:

$$\gamma_{bx} x^2 + 2\alpha_{bx} x x' + \beta_{bx} x'^2 = \frac{\varepsilon_{bx}}{\pi}, \tag{3.6}$$

where $\alpha_{bx}(s)$, $\beta_{bx}(s)$, and $\gamma_{bx}(s)$ are the C-S parameters that characterize the beam distribution, along with $\gamma_{bx}\beta_{bx} - \alpha_{bx}^2 = 1$ and $\varepsilon_{bx} = \pi\overline{x^2}/\beta_{bx}$. ε_{bx} is known as a root mean square (rms) emittance. Similar values can be introduced in the phase planes (y, y') and (z, z').

The main goal of linac tuning is to match the beam rms ellipse parameters (Eq. 3.6) to the C-S parameters (Eq. 3.4) defined by the focusing lattice in the transitions between different focusing and accelerating structures. Such matching is required in multiple transitions from one to another focusing and accelerating lattice in a linac. The matched beam provides the smallest rms beam envelope size along the accelerator. In reality, there is some degree of mismatch due to many different factors; for example, the nonidentical properties of focusing periods due to acceleration; contribution of fringe fields; nonlinear components of focusing and accelerating fields; a dependence of the transverse phase advance per period on the particle's RF phase; the measurement errors; possible emittance growth, etc. The mismatch condition produces an effective emittance growth [81, 169] and a larger beam size along the focusing lattice. In reality, the beam distribution in phase space is close to a Gaussian distribution, and mismatch conditions can generate particle losses. The concept of beam rms parameters and beam matching to the following focusing channel can be well extended to include space-charge forces.

Normally, the accelerator is composed of several segments with different focusing and accelerating lattices. The accelerator tuning procedures are applied to match the beam parameters during the lattice transitions. The tuning procedure includes beam measurements to determine the beam C-S parameters in all phase-space planes and the identification of the settings of the accelerating and focusing devices to deliver the required beam rms parameters to the following segment of the accelerator.

There are many available beam dynamics computer codes that can be applied for the optimization and setting of focusing and accelerating devices to prepare the beam to match the following accelerator segment. These computer codes are based on transfer matrices of focusing and accelerating components, which is a very efficient method of solving Hill's equations [81]. The most popular design and optimization codes are TRACE3D [26], which is used primarily for the design and optimization of linear accelerators and transport systems, and MAD-X [21], which is used primarily for the tuning of circular accelerators. An optimization code, FLAME [58], was developed at FRIB to be applied for lattice design and optimization, including strong $x - y$ coupling and multiple charge-state beams. An arbitrary order beam dynamics simulation and analysis code, COSY INFINITY [100], is widely used for the study and tuning of accelerator lattices, beam transport systems, spectrographs, and isotope separators.

3.2.2 Ring Phase-Space Matching

In a typical hadron synchrotron ring, the frequency of the transverse betatron motion is much faster than that of the longitudinal synchrotron motion. Due to drastically different motion time scales, matching in the transverse and longitudinal phase spaces is often performed independently.

3.2.2.1 Transverse Hamiltonian and Matching

To minimize beam loss and preserve emittance, the transverse optical functions of the upstream and downstream injection or extraction lines must be matched to those at the injection or extraction kicker locations. Proper matching ensures a smooth transition of the beam envelope across the interface and prevents the excitation of mismatched oscillations and emittance growth due to nonlinear filamentation.

In transverse matching studies, the longitudinal motion is typically much slower and can be treated quasi-statically. As a result, the energy deviation is regarded as an external parameter. The matching should not only account for on-momentum particles through the continuity of the C-S parameters, the beta function β and its derivative (expressed via α), but also include off-momentum particles.

The transverse dynamics of a particle in an accelerator lattice can be described using a simplified Hamiltonian formalism under the paraxial approximation, in which transverse momenta are assumed to be small compared to the longitudinal momentum.

$$H\left(x, p_x, y, p_y, \delta, s\right) = -\left(1+\frac{x}{\rho}\right)\left(1+\delta-\frac{p_x^2}{2(1+\delta)}-\frac{p_y^2}{2(1+\delta)}\right)-e\frac{A_s(s)}{p_0}\left(1+\frac{x}{\rho}\right), \tag{3.7}$$

where A_s is the vector potential of the transverse magnetic field.

For an on-momentum particle ($\delta = 0$) in a linear lattice with bending only in the horizontal plane, the Hamiltonian yields the well-known Hill's equation.

$$x'' = -K_x(s)x = -\left(\frac{1}{\rho^2(s)}+k(s)\right)x \tag{3.8}$$

$$y'' = -K_y(s)y = k(s)y \tag{3.9}$$

where $k(s)$ is the normalized quadrupole strength normalized by rigidity, and $\rho(s)$ is the radius of bending with the periodic boundary conditions,

$$K_{x,y}(s) = K_{x,y}(s + C_s), \tag{3.10}$$

where C_s denotes the circumference of the ring. In this case, the Hamiltonian (Eq. 3.7) can be transformed using the action-angle variables of the transverse motion, $(J_x, \varphi_x, J_y, \varphi_y)$, as

$$H(J_x, \varphi_x, J_y, \varphi_y, s) = \frac{J_x}{\beta_x(s)} + \frac{J_y}{\beta_y(s)}, \tag{3.11}$$

where the transverse actions J_x and J_y are given by

$$J_x = \frac{1}{2\beta_x}\left[x^2+\left(\beta_x x'-\frac{\beta_x'}{2}x\right)^2\right], \quad J_y = \frac{1}{2\beta_y}\left[y^2+\left(\beta_y y'-\frac{\beta_y'}{2}y\right)^2\right], \tag{3.12}$$

where the transverse amplitude functions β_x and β_y satisfy the relations

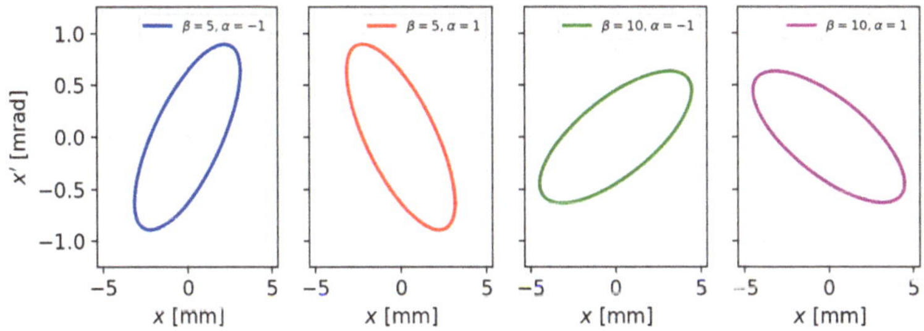

Fig. 3.1 Phase space trajectory for one particle with action 1 mm·mr with various C-S parameters

$$\frac{1}{2}\beta_x\beta_x'' - \frac{1}{4}\beta_x'^2 + K_x\beta_x^2 = 1\ , \quad \frac{1}{2}\beta_y\beta_y'' - \frac{1}{4}\beta_y'^2 + K_y\beta_y^2 = 1\ . \tag{3.13}$$

The solution to Hill's equation (Eq. 3.9) is given by

$$x(s) = \sqrt{2J_x\beta_x(s)}\cos\left[\varphi_x(s) + \varphi_{0x}\right]\ , \quad y(s) = \sqrt{2J_y\beta_y(s)}\cos\left[\varphi_y(s) + \varphi_{0y}\right]\ , \tag{3.14}$$

where φ_{0x} and φ_{0y} are the initial phases. As the Hamiltonian (Eq. 3.11) does not explicitly depend on the angle variable, both actions J_x and J_y are constants of motion, which also reveals that the particle trajectory is an ellipse in phase space, with the shape and orientation determined by the $\beta_{x,y}$ and $\alpha_{x,y} = -\beta'_{x,y}/2$ functions. Figure 3.1 shows an example of a particle phase space ellipse with different C-S parameters.

Therefore, for one-turn injection or fast extraction in a ring, the upstream and downstream beamlines should be matched in their Courant–Snyder (C-S) parameters. This ensures a smooth transition between the injection line and the accelerator ring, or between the ring and the extraction line, and prevents an artificial increase in the phase-space area due to mismatching. While matching is conventionally performed by aligning the C-S parameters, equivalent to matching the contours of the linearized Hamiltonian, in principle, the concept can be extended to matching with respect to the full nonlinear Hamiltonian, thereby accounting for nonlinear lattice effects and nonlinear interactions.

For off-momentum particles, the Hamiltonian leads to a non-homogeneous equation of motion:

$$x'' + \left(\frac{1}{\rho^2} + k(s)\right)x = \frac{\delta}{\rho}\ . \tag{3.15}$$

The particle trajectory can be separated as

$$x(s) = x_\beta(s) + D(s)\delta + O\left(\delta^2\right)\ , \tag{3.16}$$

where $x_\beta(s)$ represents the betatron oscillation of the on-momentum reference particle, which can be matched through the C-S parameter similarly to the matching procedures in Sect. 3.2.1, and the second term describes the first-order correction due to a fractional momentum deviation $\delta = \Delta p/p_0$. The coefficient $D(s)$, the dispersion function, follows a similar non-homogeneous equation:

$$D(s)'' + \left(\frac{1}{\rho^2} + k(s)\right) D(s) = \frac{1}{\rho}\,. \tag{3.17}$$

The dispersion function and its derivative must also be matched at the injection and extraction points of a ring accelerator. In a ring, the dispersion function is determined using periodic boundary conditions, which can be represented in matrix form:

$$\begin{pmatrix} D \\ D' \end{pmatrix}_{s=s_0} = M\left(s_0 + C, s_0\right) \begin{pmatrix} D \\ D' \end{pmatrix}_{s=s_0} + \begin{pmatrix} d \\ d' \end{pmatrix}, \tag{3.18}$$

where the matrix M represents the one-turn matrix for the betatron motion, and the vector (d, d') is the special solution of the dispersion equation (Eq. 3.17), which is related to the periodic solution as:

$$\begin{pmatrix} d \\ d' \end{pmatrix} = \left(I - M\left(s_0 + C_s, s_0\right)\right) \begin{pmatrix} D \\ D' \end{pmatrix}_{s=s_0}. \tag{3.19}$$

As a result, if the upstream beam line with a different dispersion $(\tilde{D}_{\text{inj}}, \tilde{D}'_{\text{inj}})^T$ is connected to the ring, the dispersion of later turns at the injection point will be given by:

$$\begin{pmatrix} \tilde{D} \\ \tilde{D}' \end{pmatrix}_{s=nC} = M^N \begin{pmatrix} D \\ D' \end{pmatrix} + \left(I - M^N\right) \begin{pmatrix} \tilde{D}_{\text{inj}} \\ \tilde{D}'_{\text{inj}} \end{pmatrix}. \tag{3.20}$$

In a hadron accelerator, the injection line is often a linear accelerator (linac). Since no bending magnets are typically present in the linac, the dispersion function is zero along the injection line. If the dispersion of the ring at the injection kicker is nonzero, the mismatch excites a dispersion wave in the ring. This oscillatory mismatch limits the maximum allowable momentum deviation of the injected particles. Specifically, if both the C-S parameter α and the dispersion slope D' are zero at the injection point, the amplitude of the dispersion wave at the injection point can increase to $2D$. As a result, the maximum tolerable energy deviation is reduced from A/D to $A/2D$, where A is the physical aperture at the injection point.

To avoid this limitation, dispersion at the injection point should ideally be matched. One straightforward solution is to design the lattice so that both D and D' vanish at the injection location. A common approach is to employ a dispersion suppressor, which uses special dipole arrangements to cancel the arc-generated dispersion. For example, in an arc composed of FODO cells with a phase advance Φ, the bending strengths of the dipoles in the two end cells can be scaled by the factors

$$\frac{1}{2(1-\cos\Phi)} \quad \text{and} \quad \frac{1-2\cos\Phi}{2(1-\cos\Phi)},$$

so that the dispersion is suppressed to zero in the straight section.

Alternatively, instead of modifying the ring design, dispersion can be introduced and controlled in the injection line itself. In this scheme, two dipoles are added to the injection line, and their bending angles are tuned, together with nearby matching quadrupoles, to generate the appropriate D and D' values. This approach allows for the simultaneous matching of both the dispersion and C-S parameters between the line and the ring. The same principles apply to beam extraction. A dispersion-matched extraction line can be realized with a similar dipole arrangement at the beginning of the line, ensuring that the beam is delivered to downstream systems with the desired dispersion properties.

3.2.2.2 Longitudinal Hamiltonian and Matching

The longitudinal motion of charged particles in a synchrotron may be described by a Hamiltonian [175]

$$H(\phi, W, t) = \frac{h_{\mathrm{RF}}^2\omega_s^2}{E_s\beta_s^2}\left[\frac{\eta_0}{2}W^2 + \frac{\eta_1}{3}\frac{h\omega_s}{E_s\beta_s^2}W^3 + O(W^4)\right] + \left(\dot{\phi}_s + \dot{\phi}_{\mathrm{RF}}\right)W$$
$$+\frac{qe\hat{V}}{2h_{\mathrm{RF}}}\left[\cos\phi - \cos\phi_s + (\phi-\phi_s)\sin\phi_s\right] - \int^{\phi}\dot{U}_{Z_\parallel}(\phi')d\phi', \tag{3.21}$$

where the subscript s denotes the synchronous value, ϕ is the phase of the RF field at the moment the particle passes the cavity, $W \equiv \Delta E/h\omega_s$ with energy deviation ΔE from the synchronous energy E_s, h_{RF} is the RF harmonic number, ω_s is the angular revolution frequency, $\hat{V}$ is the RF peak voltage, $\dot{\phi}_s$ and $\dot{\phi}_{\mathrm{RF}}$ are possibly ramping synchronous phase and RF phase, respectively, β_s is the synchronous velocity in units of speed of light c, and $U_{Z_\parallel}$ is the change in W due to the coupling impedance. The frequency slip factor η is given by

$$\eta \equiv \frac{1}{\delta}\frac{T-T_s}{T_s} = \eta_0 + \eta_1\delta + O(\delta^2),\quad \eta_0 = \alpha_0 - \frac{1}{\gamma_s^2},\quad \eta_1 = \alpha_0\alpha_1 - \frac{\eta_0}{\gamma_s^2} + \frac{3\beta_s^2}{2\gamma_s^2}, \tag{3.22}$$

where $T \equiv \frac{2\pi}{\omega}$, $\delta \equiv \Delta p/p_s = \frac{h_{\mathrm{RF}}\omega_s}{E_s\beta_s^2}W$, and the path length C is expressed in terms of the momentum compaction factors α_i, $i = 0, 1, \ldots$ as

$$\frac{C}{C_s} = 1 + \alpha_0\delta\left[1 + \alpha_1\delta + O(\delta^2)\right], \quad \gamma_T \equiv \frac{1}{\sqrt{\alpha_0}}, \tag{3.23}$$

γ_T is defined as the transition energy. Similarly to the motion in the transverse plane (Eqs. 3.7 and 3.13), the longitudinal motion governed by the Hamiltonian (Eq. 3.21) is best described in terms of the longitudinal amplitude function β_L [162, 175] that satisfies

$$\frac{1}{2}\beta_L\beta_L'' - \frac{1}{4}\beta_L'^2 + K_L\beta_L^2 = 1, \quad K_L = -\frac{2\pi h^3\omega_s^2\eta_0}{qe\hat{V}\cos\phi_s E_s\beta_s^2} \geq 0, \tag{3.24}$$

where K_L depends on the energy of the particle via η_0. When $\dot{\phi}_s$ $\dot{\phi}_{RF}$, and the higher-order terms are negligible, and when the non-resonant fields are small compared with the RF field, the system reduces to a harmonic oscillator in the limit of small amplitude where the "bunch area" $2\pi J_L$ is a constant of motion through the acceleration cycle, with the longitudinal action J_L given by

$$J_L = \frac{1}{2\beta_L}\left[W^2 + \left(\beta_L\Delta\phi + \frac{\beta_L'}{2}W\right)^2\right]. \tag{3.25}$$

Similarly to the transverse betatron motion, where the phase-space ellipse described by the transverse amplitude function $\beta_\perp$ tilts and evolves around the ring circumference, the longitudinal phase-space ellipse tilts and evolves along the acceleration cycle, as shown in Fig. 3.2.

Matching in the longitudinal phase space implies that one transfers a bunch of particles from one structure (e.g., HEBT beamline) or configuration to the next (e.g., a synchrotron) while preserving the phase space area occupied by the bunch without growth in longitudinal emittance and without beam loss. Usually, this process is carried out in the absence of acceleration when the longitudinal "RF bucket" is upright (that is, $\beta_L' = 0$), and when the area of the bunch is small compared to the area of the "RF bucket" corresponding to stable synchrotron oscillation. In this case, matching simply means choosing an RF voltage that provides a bucket ellipse with the same aspect ratio (between energy spread and phase spread) as the bunch in the longitudinal phase space.

The matching process can become complicated when the RF bucket is not stationary (e.g., during RF ramping when $\phi_s \neq 0$ and $\beta_L' \neq 0$) or when other forces effectively alter the focusing strength (e.g., the space-charge force). For example, the longitudinal space-charge force de-focuses the beam (widening the ellipse) below the transition energy γ_T and focuses the beam (narrowing the ellipse) above γ_T. This mismatch at the transition crossing cannot be easily avoided by manipulating the RF voltage (see Sect. 3.8).

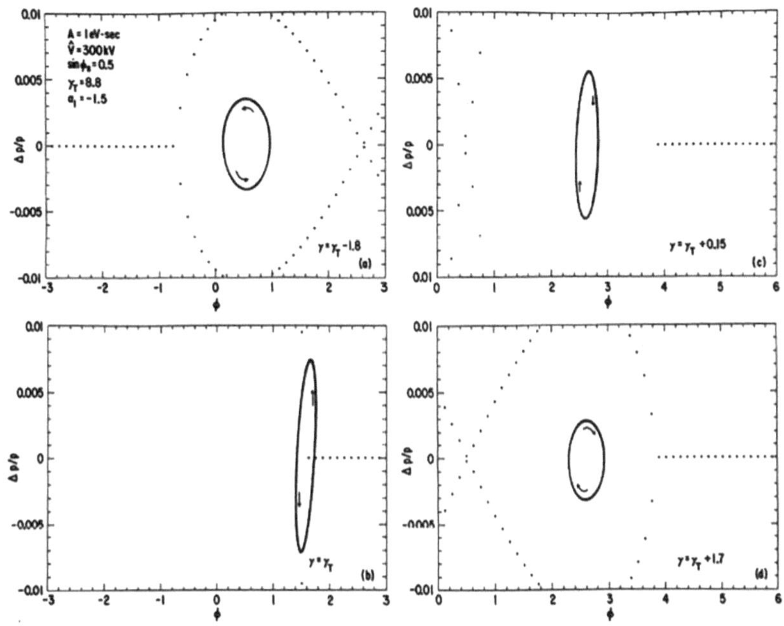

Fig. 3.2 Contours in the longitudinal phase space of a proton bunch upon crossing the transition energy. The simulation is performed with codes TIBETAN [175]

3.3 Resonance and Tuning

Resonance occurs when the driving frequency of the external periodic force exerted by the machine lattice matches the natural oscillatory frequency of a particle's motion, causing a large increase in the oscillation amplitude and likely leading to emittance growth and beam loss. In contrast, tuning is the active process of adjusting various accelerator parameters to control these resonant frequencies and optimize beam performance, avoiding unstable conditions.

3.3.1 Linac Resonances and Tuning

The stability of single-particle motion in a periodic focusing channel was analyzed by Gluckstern and Smith [155]. Their stability analysis of the general Hill's equation [60] resulted in the conclusion that $0° < \mu_{0x,y} < 180°$. This stability condition must be satisfied for all particles in the bunch once the phase advance depends on the particle phase with respect to the RF field. The equation of motion in the smooth approximation in all three directions can be written as

$$u'' + k_{0u}^2 u - qeE_{su} = 0, \tag{3.26}$$

where u represents the coordinate variables, i.e., $u = X, Y, Z$, $k_{0u} = \mu_{0u}/L_c$ is the zero-current phase advance per unit length, and $E_{\mathrm{sc}\,u}$ is the beam space charge field. The variables X, Y, and Z represent the slow components of the oscillations [81] in a smooth approximation. At the same time, the equations of motion are reduced to the equations of the harmonic oscillator by averaging fast-changing external forces over the focusing period. The expressions for μ_{0u} can be found in the literature; see, for example, [169]. For the focusing channel with quadrupoles,

$$\mu_{0x,y}^2 = \left(\frac{qeGl_{\mathrm{q}}L_c}{2Am_0c\beta_s\gamma_s}\right)^2 + \frac{1}{2}\mu_{0z}^2\frac{\sin\phi}{\sin(-\phi_s)}, \quad \mu_{0z}^2 = \frac{2\pi qeE_{\mathrm{eff}}\sin(-\phi_s)L_c^2}{Am_0c^2\lambda(\beta_s\gamma_s)^3}, \tag{3.27}$$

where G and l_{q} are the gradient and length of the quadrupole, respectively, ϕ is the particle RF phase, and ϕ_s is the synchronous RF phase. The frequency of the transverse oscillations of the particles is a function of the longitudinal position of the particle, as seen from (Eq. 3.27), even without considering the forces of the beam space charge. This coupling can result in the excitation of a parametric resonance of transverse oscillations if

$$\mu_{0x,y} = \frac{n}{2}\mu_{0z}, \tag{3.28}$$

where $n = 1, 2, \ldots$ [81]. This resonance can be easily avoided in normal-conducting linacs because the available accelerating gradients are not very high and μ_{0z} is low. However, in SC linacs, due to the high accelerating gradients available from SC cavities and the relatively long focusing periods, strong interactions between transverse and longitudinal motion may occur. Long focusing periods containing several cavities could be favorable for reducing the overall cost of the accelerator. Therefore, in some lattice designs, the transverse-longitudinal coupling can excite the parametric resonance of transverse oscillations [126].

Another possibility of a single-particle resonance is the parametric resonance of longitudinal motion. This resonance can be excited as a result of the properties of the accelerating lattice. The phase advance per period includes the averaged accelerating field E_{eff} over the period length. If the linac includes multiple accelerating structures interrupted by drift spaces for the installation of focusing devices, the linac lattice structure may induce a structure resonance. The structure resonance can be excited if the drift spaces are distributed with a "frequency" that is a multiple of the longitudinal oscillation frequency of the particles. The effect of parametric resonance of longitudinal motion on the stability area in the longitudinal phase space was studied in [35].

To investigate the effect of space charge effects in linear accelerators, it is convenient to approximate bunch with a 3-D ellipsoid with uniform distribution of particles inside the ellipsoid. This approximation results in linear space-charge fields in all three directions:

$$\begin{cases} E_{\mathrm{sc}\,x} = K_{\mathrm{sc}\,x} x\,, & K_{\mathrm{sc}\,x} = L\dfrac{3I\lambda(1-f)}{4\pi\varepsilon_0 c(r_x+r_y)r_z r_x}\,, \\ E_{\mathrm{sc}\,y} = K_{\mathrm{sc}\,y} y\,, & K_{\mathrm{sc}\,y} = \dfrac{3I\lambda(1-f)}{4\pi\varepsilon_0 c(r_x+r_y)r_z r_y}\,, \\ E_{\mathrm{sc}\,z} = K_{\mathrm{sc}\,z} z\,, & K_{\mathrm{sc}\,z} = \dfrac{3I\lambda f}{4\pi\varepsilon_0 c r_x r_y r_z}\,. \end{cases} \tag{3.29}$$

where $r_{x,y,z}$ are the semiaxes of the ellipsoid in the laboratory frame, $I = qeNc/\lambda$ is the average current over the RF period, N is the number of ions per bunch, the quantity f is the ellipsoid form factor [169]. The ellipsoid semiaxes are related to the rms beam sizes σ_u by $r_{x,y,z} = \sqrt{5}\sigma_{x,y,z}$. Using linear dependence of the space-charge fields in a bunch, the motion of a particle can be written as

$$u'' + \left(k_{0u}^2 - K_{\mathrm{sc}\,u}\right) u = 0 \tag{3.30}$$

The effect of space charge can be evaluated using the tune-depression ratio defined as

$$\frac{k_u}{k_{0u}} = \sqrt{1 - \frac{K_{\mathrm{sc}\,u}}{k_{0u}^2}} \tag{3.31}$$

The space charge forces reduce the phase advance per period, which is equivalent to weaker focusing conditions. The tune depression ratio shows a degree of the space-charge effects on particle motion. Once the parameter $K_{\mathrm{sc}\,u}$ includes the dependence from the beam size in all three directions, the space-charge force couples the motion in the transverse and longitudinal planes. Once transverse and longitudinal motions are coupled and particles perform oscillations similar to the coupled harmonic oscillator, one can assume that there could be many resonances depending on the relationship between transverse and longitudinal oscillation frequencies or phase advances per period.

One of the low-order resonances is the coupling between the single-particle motion and space-charge forces, which is modulated with the frequency of envelope oscillations. The 3-D rms envelope equations for bunched beams are

$$\begin{cases} \sigma_x'' + k_x(s)\sigma_x - \dfrac{\varepsilon_x^2}{\sigma_x^3} - \dfrac{3K_{\mathrm{sc,3D}}(1-f)}{(\sigma_x+\sigma_y)\sigma_z} = 0\,, \\ \sigma_y'' + k_y(s)\sigma_y - \dfrac{\varepsilon_y^2}{\sigma_y^3} - \dfrac{3K_{\mathrm{sc,3D}}(1-f)}{(\sigma_x+\sigma_y)\sigma_z} = 0\,, \\ \sigma_z'' + k_z(s)\sigma_z - \dfrac{\varepsilon_z^2}{\sigma_z^3} - \dfrac{3K_{\mathrm{sc,3D}} f}{\sigma_x\sigma_y} = 0\,, \end{cases} \tag{3.32}$$

where the 3-D space-charge parameter is defined as

$$K_{\mathrm{sc,3D}} = \frac{qeI\lambda}{20\sqrt{5}\pi\varepsilon_0 m_0 c^3 \beta_s^2 \gamma_s^3} \,. \tag{3.33}$$

The envelope of mismatched beam oscillates with the double frequency of single-particle oscillations [81]. If the zero current phase advance is selected to be above 90°, in the presence of space charge, the oscillations of some particles will satisfy the condition for parametric resonance with the envelope oscillations. It was found [68] that the stability area for transverse oscillations is $0° < \mu_{0x,y} < 90°$ for beams with notable space-charge forces. More recent work [76] suggests that a fourth-order resonance ($4\mu = 360°$) has a much stronger effect on 3-D beams and yields pronounced emittance growth and halo formation for zero-current phase advances around 90°. All three resonance types (particle-lattice, envelope-lattice, and fourth-order resonance) are avoided by keeping the zero-current phase advance per period in transverse planes below 90°, which is why this design rule is so fundamental [128, 169]. Similar conditions are applied in the longitudinal phase space as discussed theoretically in [66] and experimentally in [92] with the difference that the longitudinal period is twice the transverse period.

The second rule is to keep the phase advances per meter smooth across lattice transitions. This not only ensures a stable matching of transitions, but also results in a design that is matched for a wide range of beam currents. Together with keeping the phase advance smooth, all lattice transitions have to be matched for 6-D beams, including space charge.

The third rule can be derived from the core–core resonances, which are explained in [43]. It is basically imposed to keep the ratio between longitudinal and transverse phase advances within a certain range to avoid emittance exchange between non-equipartitioned planes of the beam [192].

The beam halo can be understood as a cloud of single particles traveling with the beam but having much larger distances from the beam center than the majority of beam particles. The most important mechanism for halo development is based on parametric 2:1 resonances [49, 119], which develop between the oscillations of a mismatched beam core and the movements of single particles. The core oscillations are generally excited by a beam mismatch, and the resonant interaction with single particles can be understood with a simple 1-D particle core model [170], which can be extrapolated to 3-D bunched beams.

Considering a single particle with amplitude x under the influence of the core space-charge field $E_{\mathrm{sc}\,x}$, one obtains

$$x'' + \left(k_{0x}^2 - K_{\mathrm{sc}\,x}\right) x = 0 \,. \tag{3.34}$$

The term $K_{\mathrm{sc}\,x}$ is inversely proportional to the size of the beam envelope a_x. If the beam is slightly mismatched with the periodic focusing channel and modulated with the amplitude δ_x, then,

$$\sigma_x(s) = \sigma_x \left[1 + \delta_x \cos\left(\frac{2\mu_x}{L_c}\right) s \right] . \tag{3.35}$$

The transverse oscillations appear to be parametrically coupled to the oscillations of the envelope. The core oscillates twice as fast as the single particle, demonstrating the mechanism of a parametric 2:1 resonance. The same principle of parametric resonances can be observed in realistic 3-D beams, when exciting one of the three beam eigenmodes [43].

Although the increase in halo-related emittance cannot be predicted by the particle–core model, as a constant beam emittance has been assumed, several important conclusions can still be drawn, which can be verified by 3-D beam simulations [43]:

- An rms mismatch results in a beam halo via parametric resonances.
- Single particles oscillate around certain fixed points in the phase space. The fixed-point–core distance is insensitive to the strength of the mismatch and the tune depression [170].
- The tune depression defines the time constant for the development of the halo [170]. A large tune spread increases the likelihood that single particles get trapped in 2:1 resonances, thus creating a halo, meaning that high tune depression yields higher halo density.
- An increased mismatch increases the number of particles affected by the mismatch and thereby increases the halo density.
- The maximum halo radius is limited. The 1-D particle core models predict a limit of ~ 7 times the rms radii, while the 3-D models predict more than 10 times the rms radii [43].

Statistical errors (e.g., lattice errors) also contribute to a beam halo through parametric resonances [43] and can lead to even larger single-particle radii. Furthermore, it should be mentioned that halo scraping removes halo particles but does not inhibit further halo development as long as the core is still oscillating due to a mismatch.

3.3.2 *Ring Tune Shifts, Resonances and Tuning*

In a typical hadron synchrotron ring, the period of longitudinal synchrotron oscillation is much longer than the revolution period of the particle around the ring. The longitudinal motion is not susceptible to resonances excited by the machine lattice. Also, coupled longitudinal-transverse resonance is usually not a concern, as the frequency of the longitudinal synchrotron motion is much smaller than that of the transverse betatron motion. Tuning in rings involves mainly avoiding dangerous resonance conditions in the transverse planes.

Table 3.7 Tune shift produced by various mechanisms on a 2 MW beam in the SNS ring with transverse emittance of 120 πmm·mr in each plane and momentum spread of ±0.7%

Mechanism	Maximum tune shift
Space charge (2 MW beam)	−0.2
Chromaticity (0.7% $\Delta p/p$)	±0.06
Kinematic nonlinearity (480 πmm·mr)	0.001
Fringe field (480 πmm·mr)	±0.025
Pre-compensation magnet error (480 πmm·mr)	±0.02
Post-compensation magnet error (480 πmm·mr)	±0.002
Fixed injection chicane	±0.004
Injection painting bump	±0.001
Electron cloud	∼0.04

3.3.2.1 Transverse Tune Shifts Overview

The tune in rings is defined as the number of oscillation cycles that a particle executes in one revolution around the circumference of the ring. The spread of the transverse tune shift in a beam is an important characterization that determines the non-linear resonance behavior and the stability of the beam. This section discusses the leading tune shift mechanisms and their mitigation and resonance correction schemes.

The amount of tune shift indicates its relative impact on the beam. In general, tune shifts from the ideal design value are due to changes in the effective strength of focusing on the particles [25]:

$$\Delta\nu = -\frac{1}{4\pi}\oint \beta_\perp \Delta K_\perp ds, \quad \Delta K_\perp = \frac{\partial B_y/\partial x}{B_0\rho}, \tag{3.36}$$

where $B_0\rho$ is the rigidity of the beam. When the perturbation in the amplitude function $\beta_\perp$ is relatively small, tune shifts due to various mechanisms may be superimposed according to their dependence on the oscillation amplitude and the deviation of the momentum of the particles. Table 3.7 lists the tune shifts produced by various mechanisms in the SNS accumulator ring, including space charge, natural chromaticity, and magnet imperfection.

3.3.2.2 Transverse Space-Charge Tune Shift

Space charge is a fundamental limitation in high-intensity circular accelerators. The amount of tune shift is linearly proportional to the product of the linear particle density and the circumference of the ring (that is, proportional to the total number of particles in the ring for a given bunching factor). The space-charge effect is drastically alleviated at higher beam energy when the transverse space-charge forces produced by the electric and magnetic forces tend to cancel each other. For a round beam of N_0 particles of charge state Z and atomic mass number A, the on-axis, direct, incoherent space-charge tune shift is given by

$$\nu_{sc} \approx -\frac{f_{sc} N_0 r_0}{4\pi B_f \epsilon_{rms} \beta^2 \gamma^3}, \quad r_0 = \frac{Z^2}{A} r_{p0} = \frac{Z^2 e^2}{4\pi \epsilon_0 A m_0 c^2}, \tag{3.37}$$

where r_{p0} is the classical radius of a proton, $\epsilon_{rms} = \sigma_\perp^2/\beta_\perp$, $\beta_\perp \approx R_0/\nu_0$, $2\pi R_0$ is the ring circumference, and σ_x and σ_y are the horizontal and vertical rms beam size. The bunch form factor f_{sc} is equal to $1/2$ for a uniform distribution and 1 for a Gaussian distribution. The bunching factor $B_f \leq 1$ is the ratio between the average and the peak beam density. Equation 3.37 shows the basic scaling of the space-charge tuning shift. The more comprehensive Laslett tune shift [94] contains contributions from the electric and magnetic images of the beam and the electron-cloud neutralization effects.

Rings are designed to avoid strong resonances, using a formula for the incoherent space-charge tune shift based on assuming a constant beam size. Taking into account the coherent-oscillation modes of the beam, the resonance condition becomes [3, 146, 154]

$$\frac{k}{l} = \nu_0 - C_l \Delta\nu_{sc} , \tag{3.38}$$

where ν_0 is the base tune, Δ_{sc} is the incoherent space-charge tune shift of an rms-equivalent uniform-density beam, k is the excited harmonic, l is the resonance order, and the deviation of the coefficient C_l from 1 represents the contribution of envelope oscillations. This contribution is most significant for low-order resonances [48, 63]. For a round beam near a half-integer resonance ($l = 2$), when the horizontal and vertical tunes are close to one another, $C_l = 1/2$ for the symmetric mode, and $C_l = 3/4$ for the antisymmetric mode. In the case of split tunes, $C_l = 5/8$. For the SNS ring at the split-tune working point with $\Delta\nu_{sc} \sim 0.15$, the effective tune shift is $\Delta\nu_{eff} = 5\Delta\nu_{sc}/8 = 0.09$. Therefore, the actual space-charge limit is less restrictive [39].

Equation 3.38 first derived [146] under the smooth approximation was generalized later to incorporate the periodic nature of the beam focusing lattice [125]. Based on the generalized formula, a semi-empirical resonance condition was recently proposed to predict coherent coupling resonances as well [121]. In contrast to the conventional understanding described above, the factor C_l in the semi-empirical formula depends only on the resonance order l; it does not change depending on the size of the tune split. The proposed resonance formula can be employed to construct a new type of beam stability chart applicable to both rings [121] and linacs [82]. The chart visualizes, in the conventional bare-tune space, dangerous operating regions where resonances may be excited coherently in the beam core and incoherently in the beam tail [89]. The S-POD trap system enables one to explore these space-charge-induced resonance effects systematically over much wider ranges of fundamental parameters (such as tunes and beam density) than in typical accelerator-based experiments [72, 73, 92, 109].

3.3.2.3 Transverse Chromatic Tune Shift

Variation of transverse focusing with beam energy generates the chromatic tune shift, which is generally expressed in terms of the chromaticity $\xi_\perp$ as

$$\nu_\perp = \xi_\perp \frac{\Delta p}{p}\,, \tag{3.39}$$

where $\Delta p/p$ is the relative momentum deviation of the particle. A linear lattice (consisting only of dipole and quadrupole magnets) with moderate focusing usually results in a negative, "natural" chromaticity of a value near the transverse tune [25]; negative chromaticity is necessary to damp instabilities below transition [23].

Existing rings, such as the ISIS and PSR, which accelerate or accumulate the beam solely below the transition energy, may operate without chromaticity control. For newly designed high-intensity rings, the chromatic tune spread often becomes significant when instability damping demands a large, sometimes intentionally broadened, beam-momentum spread. Then, sextupole magnets are used to adjust the chromaticity. The nonlinearity they introduce often causes a strong dependence of the machine chromaticity on the momentum deviation $\Delta p/p$. The SNS uses four-family chromatic sextupole magnets to control the chromatic tune shift throughout the momentum range [42].

3.3.2.4 Transverse Magnetic Imperfection

Quadrupole errors cause tune shifts and tune coupling for the entire beam. Sextupole and higher-order multipoles produce tune shifts that depend on the actions of the particles [182, 183]. The effects of even-order multipoles (quadrupole, octupole, etc.) can be directly assessed from the amount of tune shift produced, i.e. tune shift that is linearly proportional to the strength of the source. Multipoles of odd order (sextupole, decapole, etc.) do not produce first-order tune shifts, their impact may be evaluated through second-order tune shift analysis, dynamic aperture tracking, and resonance analysis.

In general, the 2-D magnetic field in a non-solenoidal magnet can be expressed in terms of multipole expansion when the relative impulse from the longitudinal field component on a particle's transverse momentum is small [133, 188] (Eq. 2.66 and Sect. 2.3.9.2),

$$B_y + jB_x = B_0 \sum_{n=1}^{\infty} (b_n + ja_a)(x + jy)^{n-1}, \tag{3.40}$$

where B_0 is the nominal field, b_n and a_n are the normal and skew multipole components, $n = 1, 2, \ldots$, respectively. To the first order of the error strengths a_n and b_n, retaining terms up to order 6 (12th-pole), the tune shifts are given by

$$\begin{aligned}\Delta\nu_x &= \left(-\frac{\Delta b_1}{2\rho} + \frac{b_1\delta}{2\rho} - C_1\right)\beta_x + 3C_2\beta_x^2\epsilon_x - 6C_2\beta_x\beta_y \\ &\quad + \frac{15}{2}C_3\beta_x^3\epsilon_x^2 - 45C_3\beta_x^2\beta_y\epsilon_x\epsilon_y + \frac{45}{2}C_3\beta_x\beta_y^2\epsilon_y^2\end{aligned} \tag{3.41}$$

$$\begin{aligned}\Delta\nu_y &= C_1\beta_y + 3C_2\beta_y^2\epsilon_y - 6C_2\beta_x\beta_y\epsilon_x - \frac{15}{2}C_3\beta_y^3\epsilon_y^2 \\ &\quad + 45C_3\beta_x\beta_y^2\epsilon_x\epsilon_y - \frac{45}{2}C_3\beta_x^2\beta_y\epsilon_x^2,\end{aligned} \tag{3.42}$$

where

$$\begin{aligned}C_1 &= \frac{1}{2}(\Delta b_2 - b_2\delta) + b_3\Delta_x - a_3\Delta_y + \frac{3}{2}(b_4\Delta_x^2 - a_4\Delta_y^2), \\ C_2 &= \frac{1}{2}\left(\frac{1}{4}b_4 + b_5\Delta_x - a_5\Delta_y\right) \\ C_3 &= \frac{1}{4}\left(\frac{1}{6}b_6 + b_7\Delta_x - a_7\Delta_y\right),\end{aligned} \tag{3.43}$$

and $\Delta_x = D_x\Delta p/p + x_c$, $\Delta_y = D_y\Delta p/p + y_c$ with x_c and y_c the horizontal and vertical closed orbit deviations from the axis of the magnet center.

3.3.2.5 Transverse Resonance Correction

The transverse resonance condition is given by the relation

$$m\nu_x + n\nu_y = k, \tag{3.44}$$

where ν_x and ν_y are the horizontal and vertical tunes, and m, n, and k are integers. If m and n have the same signs, the resonance is called a sum resonance, and if they have opposite signs, it is called a difference resonance. The order l of the resonance is $l = |m| + |n|$. The strength of the resonance is characterized by the quantity

$$\kappa = \frac{R_0}{\pi 2^l m!n!}\oint \beta_x^{|m/2|}\beta_y^{|n/2|}K_\perp^{(l)}\exp(i\varphi)d\theta, \tag{3.45}$$

where $K_\perp^{(l)}$ is the strength of the error field. It is given by

$$K_{\perp}^{(l)} = \frac{1}{B_0 \rho} \left(\frac{\partial^{(l-1)} B_i}{\partial x^{(l-1)}} \right), \tag{3.46}$$

with $B_i = B_y$ for a normal magnetic field (i.e., n even) and $B_i = B_x$ for a skew field (n odd). The integral is over the azimuthal angle $\theta = s/R_0$, and the phase φ satisfies $\varphi \approx k\theta$ when the tunes are near resonance. The resonance is excited by the kth harmonic of θ by the error component of $2l$-pole. In the vicinity of the resonance line, the motion of the particles becomes unstable. The size of the "stop band" is given by [54, 148]

$$\Delta \nu^{(m,n)} = \pm |\kappa| \epsilon_x^{|m|/2} \epsilon_y^{|n|/2} \left(\frac{m^2}{\epsilon_x} + \frac{n^2}{\epsilon_y} \right). \tag{3.47}$$

For systematic field errors, the major resonances are the so-called structure resonances that occur when k is an integral multiple of the lattice super-periodicity M. Resonance correction is based on minimizing κ (Eq. 3.45) using correctors of the appropriate order located at strategic positions of azimuth θ to generate the intended harmonics.

3.4 Collective Effects and Instabilities

Collective effects limit the beam current that may be transported through a high-power accelerator. Collective effects can arise either directly from the EM interaction of particles in the same beam (space-charge effect) or indirectly, as a particle beam generates an EM field while passing through discontinuities and variations in the vacuum chamber's cross-sectional shape and conducting properties.

Space-charge effects often set a fundamental limitation for low- and medium-energy high-power accelerators. Section 3.3.1 discussed the beam halo generated in linacs through parametric resonances of the core halo. Section 3.3.2.2 discussed space-charge-induced tune spread in synchrotron rings.

This section continues the discussion of collective effects in cyclotrons, linacs, and synchrotrons, with an emphasis on beam-coupling impedances and instabilities.

3.4.1 *Cyclotron Collective Effects*

The highest power proton beam was achieved in separated-turn isochronous cyclotrons in PSI [90, 193]. Space-charge forces in the isochronous cyclotron result in vortex motion in the radial-longitudinal plane, as Gordon noted [50]. This effect is well explained in a review paper by Baartman [5], which includes appropriate references to the original work by Gordon [50], Joho [79], Bertrand and

Ricaud [6], and Kleeven [85]. In the rotating frame, due to the Coriolis forces, the charged particles move perpendicular to the electric force along the equipotential of the space charge in the radial-longitudinal plane. As Baartman noted, the vortex motion results in circularly symmetric motion in the radial-longitudinal plane. The equipotential of the space charge is circular only if the density contours are circular. Bertrand and Ricaud created a simplified model [6], wherein the maximum bunch charge was coupled with the size of the sphere for non-relativistic motion in the cyclotron. In the following works [5, 85], 3-D envelope equations similar to Eq. 3.32 were applied to the bunched beam in the separated-turn isochronous cyclotron. Kleeven [86] recently published a comprehensive theory of 3-D envelope equations for cyclotrons [86]. The stationary or matched bunch size can be found from the envelope equation of the radial motion, assuming that the second-order derivative is equal to zero. Similarly to the KV envelope equations for linear accelerators, it results in the quartic equation for the matched envelope size of the bunch, $r_r = \sqrt{5}\, r_\sigma$

$$r_r^4 - C_r r_r - r_{r0}^4 = 0\,, \qquad C_r = \frac{q e Q_b g_r}{\pi \varepsilon_0 A m_0 \omega_c^2 \nu_r^2 \gamma_s^2}\,, \tag{3.48}$$

where $r_{r0} = \sqrt{2R_c\epsilon_r/\nu_r}$ is the matched beam radius for zero space charge, Q_b is the total charge in a bunch, g_r is the Carlson elliptic integral [5], which is equal to unity for the spherical shape of the bunch, ν_r is the radial tune, and R_c is the orbit radius. In the beam dominated by space charge, the maximum charge per bunch can be found from Eq. (3.48), assuming $r_{r0} = 0$,

$$Q_{b,\max} = \frac{\pi \varepsilon_0 A m_0 \omega_c^2 \nu_r^2 \gamma_s^2}{q e g_r} r_r^3\,. \tag{3.49}$$

The maximum bunch charge is proportional to the cube of the radial envelope size. The turn separation should be larger than the beam envelope by a factor of $\xi_{\rm gap} > 1$ to avoid unacceptable beam losses. As is well known, the turn separation is proportional to the beam energy gain, and the maximum bunch charge is proportional to the cube of the available RF voltage, $V_{\rm RF}$. The total charge per turn and, consequently, the maximum beam current can be increased by the factor of $h_{\rm RF}$, the RF harmonic number. Here, Baartman's estimate of the maximum beam current in the separated turn cyclotron is given by:

$$I_{\max} = \frac{h_{\rm RF}}{2 g_r \xi_{\rm gap}^3 \beta_s^3 \gamma_s \nu_r^4} \frac{V_{\rm RF}^3}{V_m^2 Z_0}\,, \qquad V_m = \frac{A m_0 c^2}{q e}\,, \tag{3.50}$$

where $Z_0 = \varepsilon_0 c)^{-1} = 377\ \Omega$ is the impedance of free space. According to equation Eq. 3.50, the maximum proton beam current in both the PSI injector 2 (72 MeV, $V_{\rm RF} = 0.86$ MV, $h_{\rm RF} = 10$) [90] and in the PSI Ring (590 MeV, $V_{\rm RF} = 4 \cdot 0.85$ MV,

$h_{\text{RF}} = 6)$ [51] is ~ 3.2 mA. This value is slightly higher than the operational beam current of 2.4 mA. Based on the expression (Eq. 3.50), Baartman concluded that

- The rate of nucleon delivery, $I_{\max} \cdot A/q \propto q/A$, is proportional to q/A. The highest rate of nucleon delivery can be achieved for protons only.
- A large radial tune at the extraction reduces the maximum beam current.
- Higher RF harmonics help but may cause difficulties in injection.

Detailed numerical studies of 3-D space-charge effects in PSI Injector 2 and the Ring have been reported in [90, 193]. The simulations show that the mismatched beam generates a significant beam halo. If the injected bunch is mismatched, the beam adjusts itself to the matched condition; however, a beam halo is generated. In the separated turn cyclotrons, the matched bunch length is short. If long bunches are injected, they break down into droplets [140].

The beam current limit has been derived, assuming zero emittance. Kolano et al. [90] discussed that the current limit can be higher with increased beam emittance. They concluded that the maximum beam current in PSI Injector 2 increases as $\sim r_r^4$ above 2 mA. In general, the current status of RF and cyclotron technology indicates that the limiting intensity in isochronous separated turn cyclotrons can be designed to deliver 10 mA proton beams.

3.4.2 Linac Collective Effects

In general, collective instabilities in accelerators refer to an intense charged-particle beam electromagnetically interacting with its vacuum chamber environment. As the beam interacts with its environment, it generates an EM field called the wakefield, and the wakefield acts back on the beam, disturbing its motion; if the perturbation is strong enough, the beam becomes unstable. This type of collective instability has not been observed in existing linacs. There were studies that, under some conditions, the collective instabilities could be excited at very high intensities of proton beams in specific accelerating structures. For example, the limiting current for the proton beam break-up instability was found to be 350 mA in the high-energy section of proton accelerators where acceleration is provided by the disk and washer structure [36].

Another class of collective effects in linacs is induced by the electromagnetic interactions between charged particles: the space charge effect. The influence of space charge can be quantified by a dimensionless parameter known as the generalized perveance, K_{sp}, defined as

$$K_{sp} = \frac{qeI}{2\pi\epsilon_0 A m_0 \beta^3 \gamma^3 c^3} \tag{3.51}$$

where I denotes the beam current, qe is the ion's charge, Am_0 is the atomic mass, β is the normalized velocity, γ is the relativistic Lorentz factor, and ϵ_0 is the vacuum permittivity. The perveance provides a convenient measure of the strength of space-charge forces, independent of beam geometry, allowing a direct comparison of space-charge effects across different accelerators and beam conditions.

The terminology of resonance, coherence, and instability due to space charge effects in accelerators is discussed in a recent book by Hofmann [64]. The coherent space-charge resonances were identified in 1979 for 2-D beams [61] and were then found to be responsible for emittance transfer between the different beam planes [65]. Mostly for this reason, linac designers in the past often applied equipartitioning [74] in their lattices for high-intensity machines, which demand that the product of phase advance per period (tune) times the rms emittance be equal in x, y and z planes:

$$\mu_x \epsilon_x = \mu_y \epsilon_y = \mu_z \epsilon_z \,. \tag{3.52}$$

In 2001, non-equipartitioned linac designs were systematically studied with 3-D simulations [44] and compared to the theoretical predictions [65] for 2-D beams. The emittance exchange between the longitudinal and the two transverse planes was found to be predicted by Hofmann's instability charts, which have since been used for the design of high-intensity linacs [43, 169]. An experimental proof of emittance exchange was conducted at GSI in 2009 [53].

As an example, an analysis [67] was performed with PARMILA [195] using the SCHEFF subroutine during the SNS linac design for the acceleration of the H^- beam to 1 GeV energy at a beam power of 2 MW (Sect. 10.1). The high beam power required careful control of emittance blow-up and beam halo, which were collimated in the HEBT line in both transverse directions and momentum space. The primary concern was transverse beam blow-up, which would result in ring injection foil miss, foil traversal and scattering increases, and an increase in radio-activation. To explore the sensitivity of the SNS design to space-charge resonances, an initially matched water-bag distribution was used, with the nominal emittances taken from the RFQ output (initially $\epsilon_z/\epsilon_x = 1.05$ and approaching an average of 1.2), compared to a doubled input longitudinal emittance. Figure 3.3 shows Hofmann's chart for the emittance ratio of 1.2, plotting the tune trajectory (red pluses) up to the end of the DTL at 87 MeV. The equipartitioning condition (Eq. 3.52) corresponds to the green vertical line.

As shown in Fig. 3.4, the fourth-order resonance stopband near the tune ratio of 1 is very weak, as the beam is almost equipartitioned there, with energy anisotropy of approximately 1.2. Therefore, crossing it in the first tank of the DTL does not affect emittance growth. In Fig. 3.3, it is seen that due to the reduced longitudinal focusing, the 1:2 third-order resonance (near tune ratio 1/2) is next crossed in the DTL. The potentially more dangerous 1:3 fourth-order resonance at a tune ratio of 1/3 is hardly reached at the end of the DTL, and no effect is seen in the simulation of Fig. 3.4.

The key to avoiding emittance exchange through coherent resonances in the beam core is to maintain the ratio between transverse and longitudinal phase advance within certain "stable" bands, which are given by Hofmann's charts. Nevertheless,

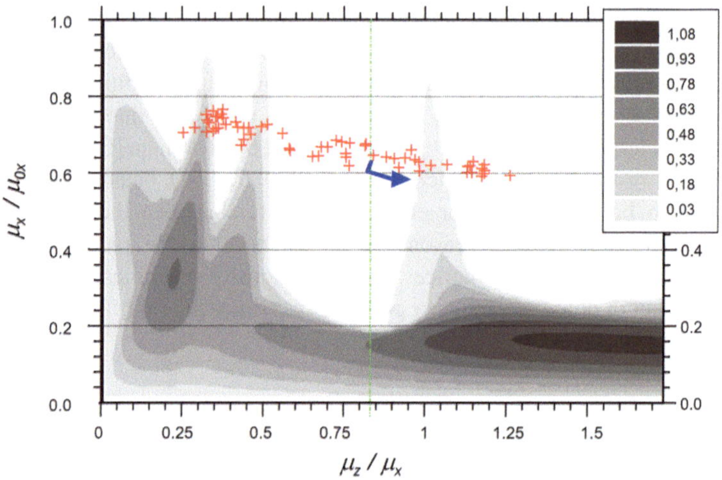

Fig. 3.3 Theoretical resonance chart for $\epsilon_z/\epsilon_x = 1.2$ with PARMILA tune trajectory through DTL for nominal SNS input emittances (at arrow). Grey scales indicate theoretical growth rates in units of inverse zero-current betatron periods per e-folding. The green vertical line indicates equipartition. [67] (Courtesy I. Hofmann)

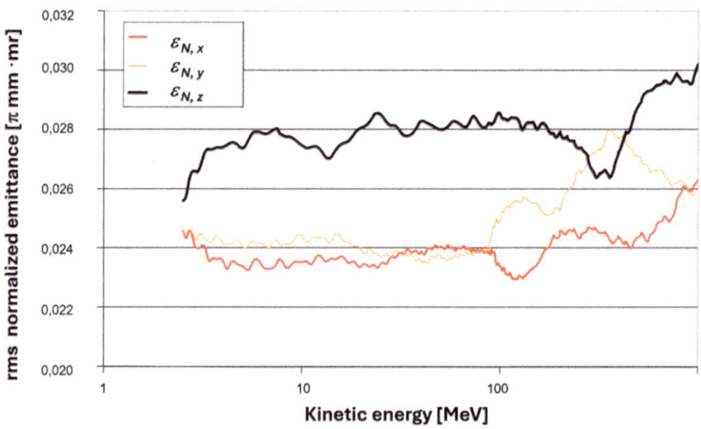

Fig. 3.4 PARMILA result to full energy showing normalized emittances of a 2 MW H^- beam in the SNS linac. The nominal input emittances from the SNS RFQ was used [67] (Courtesy I. Hofmann)

fast transitions of unstable bands are often possible without significantly affecting the emittances. Due to the continuous acceleration process in the linacs, the resonances can be crossed quickly and do not result in an exponential growth of oscillations or a sudden loss of the beam.

3.4.3 Ring Collective Effects and Instabilities

Collective effects in high-power rings usually refer to those whose impact is linearly proportional to the intensity of the beam. The collective force that the environment exerts on the beam may be described by a coupling impedance. This section describes such intensity-limiting effects in synchrotrons.

Effects in rings not covered in this section include intrabeam scattering (see Sect. 3.8.3), intrabeam stripping (see Sect. 3.7.4), and electron cloud (see Sect. 3.5). The effects of Coulomb scattering between particles (intrabeam scattering), which scales with q^4/A^2, are negligible during the short time of accumulation and acceleration.

3.4.3.1 Coupling Impedance

The coupling impedance describes the interaction between a beam and its environment [151, 200]. Assume a harmonic excitation of the beam current $\tilde{I}(\omega)$ at frequency ω that excites a harmonic field with a complex amplitude $\tilde{E}_z(\omega)$. The longitudinal coupling impedance is defined as

$$Z_{\parallel}(\omega) = -\frac{1}{\tilde{I}(\omega)} \oint \tilde{E}_z \exp\left(\frac{j\omega s}{\beta c}\right) ds. \tag{3.53}$$

The vertical and horizontal coupling impedances are defined as the integrals of the deflecting fields over one turn, normalized by the dipole moment of the excitation beam current,

$$\begin{aligned} Z_y(\omega) &= \frac{j}{\tilde{I}(\omega)\Delta y} \oint \left[\tilde{E}_y + \beta c \tilde{B}_x\right]\Big|_{x=0,y=\Delta y} \exp\left(\frac{j\omega s}{\beta c}\right) ds, \\ Z_x(\omega) &= \frac{j}{\tilde{I}(\omega)\Delta x} \oint \left[\tilde{E}_x - \beta c \tilde{B}_y\right]\Big|_{x=\Delta x,y=0} \exp\left(\frac{j\omega s}{\beta c}\right) ds, \end{aligned} \tag{3.54}$$

where $\tilde{E}_x$, $\tilde{E}_y$, $\tilde{B}_x$, and $\tilde{B}_y$ are the transverse components of the electric- and magnetic-field vectors in the frequency domain, and Δx and Δy are the horizontal and vertical offsets of the beam from the axis. With these definitions, Table 3.8 is an example of the coupling impedance "budget" estimated for the SNS accumulator ring at frequencies below 10 MHz. For simplicity, we use $Z_\perp$ to represent both Z_x and Z_y.

Minimizing impedance is an important measure to prevent collective instabilities. Below the transition energy, the reactive part of the impedance is generally dominated by space charge, whose value is given by

$$\frac{Z_{\parallel}^{\rm sc}(n\omega_s)}{n} = -\frac{g_0 Z_0}{2\beta\gamma^2}, \quad Z_{\perp}^{\rm sc}(n\omega_s) = -\frac{g_0 Z_0}{\beta^2\gamma^2}\left(\frac{1}{a^2} - \frac{1}{b^2}\right), \tag{3.55}$$

Table 3.8 Estimated beam coupling impedance of the SNS accumulator ring at frequency below 10 MHz. The beam revolution frequency is 1.058 MHz. The leading impedance source contributing to possible instability is the extraction kicker modules located inside the beam vacuum pipe

Device/Mechanism	$Z_\parallel/n$ (Ω)	$Z_\perp$ (kΩ/m)	Comment
Space charge	$-j196$	$j(-5.8+0.45)\times 10^3$	Incoherent and coherent part
Extraction kicker	$0.6n+j50$	$33+j125$	25 Ω termination at PFN
Injection kicker & Pipe	$0.5/n$	17.5	Pipe coated [b]
Injection foil assembly	$j0.05$	$j4.5$	MAFIA modeling
RF cavity	0.9[a]	18	To be damped
Resistive wall	$(j+1)0.71$ at ω_0	$(j+1)8.5$ at ω_0	
Broadband BPM	$j4$	$j18$	
Broadband bellows	$j1.1$	$j7$	Unscreened
Broadband steps	$j1.9$	$j16$	Tapered 1-to-3 ratio
Broadband ports	$j0.49$	$j4.4$	Screened
Broadband valves	$j0.15$	$j1.4$	Unscreened
Broadband collimator	$j0.22$	$j2.0$	

[a] At resonance peak
[b] Lowest tune at 200 Hz

where $n = \omega/\omega_s$, $Z_0 = (\epsilon_0 c)^{-1} = 377\,\Omega$, a is the effective beam radius, b is the radius of the vacuum pipe, and $g_0 \approx \frac{1}{2} + 2\ln\frac{b}{a}$ is for the geometric factor. The transverse and longitudinal space charge impedances can be reduced by allowing the contours of the conducting vacuum enclosure to follow the variations in the beam envelope. This practice was adopted in the ISIS synchrotron, using metal wires contoured to the beam envelope within the ceramic beam tube [12].

The growth of an instability is primarily associated with the resistive part of the impedance. For both the ISIS and J-PARC synchrotrons and the SNS accumulator ring, the leading resistive component of the coupling impedance comes from the extraction kicker modules located in the vacuum pipe. The coupling impedance can be minimized by the design of the pulse-forming network (PFN) power-supply circuits (low termination impedance). The frequency dependence of the coupling impedance further depends on the permeability of the ferrite material. Ferrite material with relatively low permeability and low resistive loss was found to be beneficial if the PFN was not terminated with low impedance [29]. Wrapping a special metal loop around the ferrite blocks was proposed as a method to influence the coupling response of the kicker module without changing the design kicker's magnetic field.

The impedance due to the resistive wall of the vacuum chamber can become important at low frequencies. When the thickness of the conducting chamber's wall is greater than the skin depth $\delta_s = \sqrt{2\rho_r/\mu\omega}$, where μ is the permeability of the conductor, and ρ_r is the resistivity of the wall material, the impedances are given by

$$\frac{Z_\parallel^{\rm rw}(n\omega_s)}{n} = (1+j)\frac{\beta_s Z_0 \delta_s}{2b}, \quad Z_\perp^{\rm rw}(\omega) = (1+j)\frac{R_0 Z_0 \delta_s}{b^3}. \tag{3.56}$$

Investigations have been conducted on the coupling impedance of thin resistive layers, such as the metallic coatings of ceramic vacuum chambers in kicker magnets.

Theoretical studies showed that the beam image current flows in the resistive layers, even in the low-frequency regime, when the depth of the skin δ_s is greater than the thickness of the layer, unless external structures offer alternative paths of lower impedance. At very low frequencies, where δ_s^2 exceeds the product of the radius of the chamber times the thickness of the layer, a significant reduction in the real part of the transverse impedance might be expected. These predictions were validated by preliminary bench measurements at CERN [136]. Two layers of coating were applied to the ceramic chamber of the SNS ring for injection kickers: a 1 μm thick copper layer for bypassing the image charge and a 0.1 μm thick titanium nitride (TiN) layer for low secondary electron yield, along with an exterior metal enclosure for bypassing DC current. This design allows the passage of the image current above a frequency of the lowest betatron sideband (∼200 kHz) without degrading the penetration of the magnetic field (a rise time of about 200 μs), eddy-current heating, and beam-induced heating. Improvements to the vacuum chamber bypass are also planned at the CERN PS Booster, and the shielding of the magnet septa and vacuum ports is underway at the CERN SPS.

Computer modeling is frequently used to predict the coupling impedance of a device. However, computation becomes inefficient when the geometry of the device is complex and asymmetric, and when the permeability of the material is high. Bench measurements are needed to determine the impedance of critical devices, e.g., the extraction kicker and its power-forming network (PFN) [29, 112]. After the machine is constructed, the total coupling impedance can be determined by measuring the beam transfer function, energy loss, and tune shifts [200].

3.4.3.2 Space Charge and Beam Halo

Space-charge effects often limit the performance of a high-intensity synchrotron if the injection energy of the beam is not sufficiently high. In contrast to linacs, where the longitudinal and transverse bunch sizes are similar, in synchrotrons the longitudinal size of the bunch is usually much larger than its transverse dimensions. Accordingly, transverse space charge tune shifts are orders of magnitude larger than longitudinal ones, so longitudinal and transverse space charge can usually be considered separately [80].

In the longitudinal direction, the space charge contributes a defocusing force below the transition energy. The longitudinal spread of the bunch can cause particles to escape from the RF bucket, which requires compensation by enhancing the focus of the RF field or introducing an inductive impedance into the ring [151].

The dominant effects of transverse space charge are the coupling of the transverse motions and the development of space charge resonances. The coherent tune of the beam, along with the associated collective oscillations, determines its resonant behavior. To prevent significant beam broadening and emittance growth, it is necessary to maintain coherent tunes of the beam away from low-order (integer and half-integer) imperfection and structure resonances (Sect. 3.3.2.2). The choice of working tune is based on a consideration of the space-charge resonances and the avoidance of higher-order resonances excited by lattice nonlinearity in the presence of space-charge-induced tune spread. When the horizontal and vertical tunes are close to each other, a significant exchange of emittance also occurs, generating a round beam in the transverse physical space. However, beam core parametric resonance induced by beam mismatch, which is the primary source of halos in a proton linac, is often unimportant, especially for a multi-turn injected ring when the halo is covered by the subsequently injected beam [37].

The most effective approach to solving space-charge problems is to increase the injection energy. The space charge can also be alleviated by longitudinal manipulation (e.g. dual harmonic RF system, barrier cavity) to improve the bunching factor and by transverse and longitudinal painting and controlled injection (Sect. 2.3.4.3). Longitudinal space charge compensation using inductive inserts has also been demonstrated [88, 116, 151].

3.4.3.3 Instabilities

Instabilities are common in proton rings [23, 95, 113]. Due to the short accumulation time (typically around one synchrotron-oscillation period), a proton beam in an accumulator is susceptible only to fast transverse instabilities, such as the electron-proton instability that limits the beam intensity in the PSR ring (Sect. 3.5; [99]). However, rapid-cycling and conventional synchrotrons are susceptible to an extended list of instabilities. Head-tail instabilities were observed near injection in the KEK PS, the CERN PS, and the BNL AGS and are suspected to be due to chromaticity changes caused by eddy-current-induced sextupole fields (proportional to $\dot{B}/B$ in the vacuum chamber under the dipole magnets. This type of instability can be cured by chamber correction windings (AGS Booster), chromaticity control, Landau damping with octupoles, and tune manipulation. Negative mass and microwave instabilities are observed at the CERN PS and SPS, the BNL AGS, and the KEK PS. They can be avoided by measures that reduce impedance (e.g., shielding vacuum ports and septa) and improve the bunching factor (e.g., using dual-harmonic or dilution cavities). Instability in coupled bunches was observed at the CERN PS Booster, PS, and SPS, and the BNL AGS and was damped by fast feedback systems and Landau damping systems. The ISIS programs the tunes in each cycle to accommodate natural variations in chromaticity, to depress space-charge tunes, and to avoid resistive wall head-tail instability [142].

Next-generation proton sources, such as the compressor ring for a neutrino factory or a muon collider, are expected to operate in the kiloampere regime of the peak beam current at low to medium energy [115]. In these machines, the intensity of the beam will be pushed close to its stability limits in the presence of strong space-

Table 3.9 Collective effects and estimated thresholds for a 2-MW beam power in the SNS accumulator ring. The dominant effects are expected to be the beam halo and the beam loss generated by space-charge-related resonance crossing. Other intensity-limiting mechanisms include electron-cloud effects and instabilities due to the extraction kicker coupling impedance

Mechanism	Threshold	Comment
Transverse space charge	$\Delta\nu_{sc} \approx -0.2$	Resonance crossing
Longitudinal space charge	15 kV induced RF voltage	60 kV RF voltage
Transverse microwave instability	$Z_\perp \approx 60\,\text{kV/m}$	Extraction kicker impedance
Longitudinal microwave instability	$\lvert Z_\parallel/n\rvert \approx 100\,\Omega$	Extraction kicker impedance
Resistive wall		Walls and kicker impedance
($\xi = 0$)	$Z_\perp \approx 1.3\,\text{k}\Omega\text{/m}$	At 200 kHz; rise time > 300 μs
($\xi = -3$)	$Z_\perp \approx 100\,\text{k}\Omega\text{/m}$	At 200 kHz
Electron cloud	Above 2 nC/m	Above 2% neutralization in beam

charge effects and nonlinear resonances (Table 3.9). A coasting-beam formalism may be applied to a bunched beam if the bunch length is much longer than the characteristic wavelength of the driving wake field, and if the instability growth rate is comparable to or faster than the synchrotron-oscillation frequency [168]. The criterion for longitudinal bunched-beam stability is the Keil-Schnell condition [15, 84, 117, 144]

$$\left|\frac{Z_\parallel}{n}\right| \le F_\parallel \frac{|\eta| B_f E_s}{e\beta^2 I_0} \left(\frac{\Delta E}{E_s}\right)^2_{\text{FWHM}}, \tag{3.57}$$

where I_0 is the average bunch current, η is the slip factor, and the form factor $F_\parallel \approx 1$ for a parabolic distribution. For transverse bunched-beam stability, this criterion is [147]

$$|Z_\perp| \le F_\perp \frac{4 B_f E_s}{e\beta\langle\beta_\perp\rangle I_0} \left(\frac{\Delta E}{E_s}\right)_{\text{FWHM}} |(n - \nu_\perp)\eta + \xi_\perp|, \tag{3.58}$$

where the form factor $F_\perp$ depends on the distribution of transverse particles, n is an integer, and $\xi_\perp$ is the chromaticity. Instability occurs only for slow waves with $n > \nu_{x,y}$. The lowest threshold obtained for n is just above the tune $\nu_{x,y}$. The chromaticity of the machine $\xi_\perp$ plays an important role in providing Landau damping [93]. Below the transition energy, $\eta < 0$, the natural chromaticity ($\xi_\perp < 0$) helps damp the transverse instability.

The observed threshold for longitudinal instability may not correspond to that estimated from the Keil-Schnell criterion, especially in the presence of a strong space charge. Examples include the ISIS synchrotron, the CERN PS, and the BNL AGS Booster. In the longitudinal direction, space charge typically reduces the growth rate of instability with its debunching force (below transition) and results in self-stabilization [62, 65, 91, 145]. Transversely, space charge lowers the instability threshold by altering the mode frequency and the mode coupling condition [9, 39]. Developments in computer simulation have made it possible to investigate the mechanism numerically [28, 69, 101].

Longitudinal and transverse coupled-bunch instabilities are primarily driven by parasitic resonances in the RF system. Such instability is often cured by passively damping such resonances or by installing a fast bunch-by-bunch damper. Feedback systems are frequently used to dampen instabilities. The long wake of resistive-wall impedance may cause a closed orbit drift during acceleration that can be damped by a slow-rate feedback system. However, fast-rate feedback systems are required to dampen electron-cloud instability and coupled-bunch instability.

3.5 Electron Cloud

Electron cloud effects that can severely limit the performance of high-intensity proton synchrotrons include trailing-edge tune shift and resonance crossing, electron-proton instability, emittance growth and beam loss, increases in vacuum pressure, heating of the vacuum pipe, and interference with beam diagnostics. The following are examples of hadron rings where electron-cloud effects were observed: the PSR at LANL, where a strong and fast transverse instability occurred both for coasting and for a stacked beam when a threshold intensity was exceeded [99]; the CERN PS and SPS, where a large number of electrons were produced by beam-induced multipacting when the machine parameters were configured for LHC injection [1, 110]; and the RHIC at BNL, where the vacuum pressure increased dramatically when beams were injected with halved nominal bunch spacing [198]. Electron-cloud instability occurred at the J-PARC Main Ring during the slow-extraction process [194]. It was also observed at the SNS accumulator ring during beam studies when the chromaticity was intentionally reduced, confirming theoretical predictions.

3.5.1 Electron Generation

Electron production is classified into the following categories: (1) electrons generated in the stripping foil; (2) electrons generated on the collimators and vacuum pipe surfaces due to the impact of lost protons; (3) electrons produced by multipacting from the vacuum chamber's wall; and (4) electrons produced around the ring from residual gas ionization.

3.5.1.1 Electron Generation at Injection

Figure 3.5 shows the distribution of the electron density flux measured in the PSR using electron detectors. The quantity f_e is defined as the ratio of the number of electrons that strike the vacuum pipe in one turn to the number of protons stored in the ring, scaled by the area of the detector surface. The electron density is high in the injection region, where the H^- beam is stripped of its electrons, and it is high in the extraction region due to the limited aperture. Near the injection stripping foil, a high concentration of electrons is expected, with a broad energy spectrum.

With an H^- beam, the stripped electrons carry twice the current of the injecting H^- beam and have a kinetic energy of $m_e c^2(\gamma - 1)$, where γ is the relativistic factor of the H^- beam. The injecting and circulating beams impacting the foil produce a secondary emission of electrons at low energy (tens of eV). Although the yield is low (0.006 for an 800-MeV proton beam incident on carbon material), the effect is proportional to the number of traversals of the foil. The injecting and circulating beams also produce knock-on electrons at high energy (up to several MeV). The stripping foil, operating at a high temperature of around 2000 K, emits low-energy thermionic electrons. All of these electrons can be back-scattered from the stripped electron collector and the surrounding surfaces (Table 3.10).

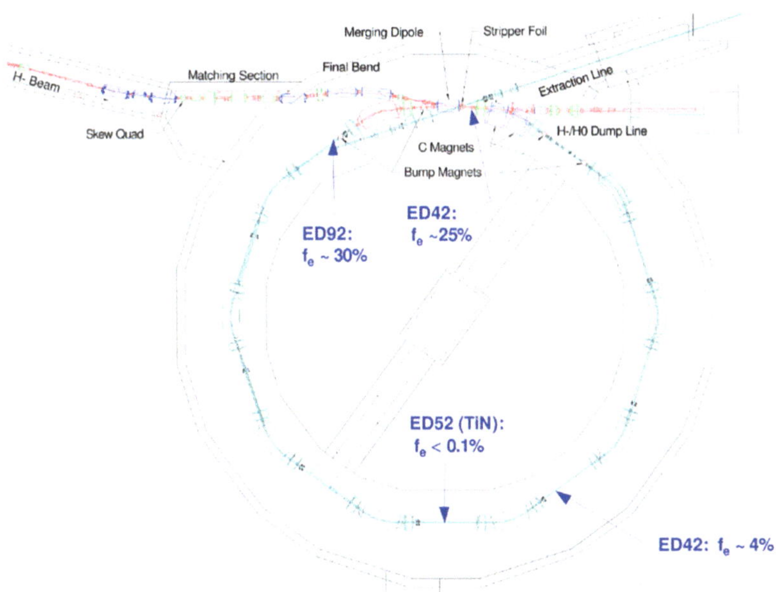

Fig. 3.5 Distribution of the electron flux measured on the wall of the vacuum pipe at the PSR [177] (Courtesy R. Macek, LANL). The circumference of the ring is 90.2 m. The kinetic energy of the proton beam is 800 MeV. The flux ratio f_e, which varies around the ring, is about 30% downstream of the extraction septum, about 25% downstream of the injection stripping foil, about 4% in section 4, and within the noise level in the TiN-coated section 5

Table 3.10 Estimated yield and kinetic energy of the electrons produced by the H^- beam at injection into the PSR. The yield is defined as the ratio of total number of electrons produced during the accumulation period per injected H^- particle. The average number of foil traversals is about 50. The kinetic energy of the injecting beam is 800 MeV. The average H^- beam current is 100 mA (Courtesy M. Plum)

Source	Yield	Kinetic energy
Stripped electron	2.0	430 keV
Secondary electron	1.0	Up to 20 eV
Knock-on electron	0.4	Up to 2.4 MeV
Thermionic electron	< 0.002	~0.24 eV
Residual-gas ionization	0.02	Up to 2.4 MeV

3.5.1.2 Electron Generation near Limiting Aperture

The region near the scrapers and collimators is susceptible to high beam loss and, potentially, is another location of high electron concentration. Protons incident on the collimator surfaces produce secondary electrons. Depending on the energy of the beam and the incident angle, the secondary-electron-to-proton yield can greatly exceed 1 when the incident beam is nearly parallel to the surface (i.e., the grazing angle $\theta_g \approx \pi/2$). The proton-induced yield Y_{ep} was shown to depend on the angle θ_g,

$$\frac{Y_{\text{ep}}}{Y_{\text{ep}}^{\text{max}}} = \frac{1.11\left\{1 - \exp\left[-2.3\left(\frac{E_k}{E_{k,\max}}\right)^{1.35}\right]\right\}}{\left(\frac{E_k}{E_{k,\max}}\right)^{0.35}\cos\theta_g} \tag{3.59}$$

where E_k is the kinetic energy of the primary proton, and the proton energy that corresponds to the maximum yield $E_{k,\max}$ is approximately 0.7 MeV. A serrated surface with triangular teeth may reduce the generation of secondary-emission electrons. The SNS ring uses a two-stage collimation system so that the beam halo is likely to be incident on the front edge of the secondary collimators.

3.5.1.3 Electron Generation Due to Gas Ionization

Due to gas ionization, the rate of increase in electron line density per unit length of circumference is given by the relation [199]

$$\frac{d^2\lambda_e}{dtds} = \frac{\rho_m \beta I \sigma_{\text{ion}} P}{133.3e}, \tag{3.60}$$

where I is the proton current, σ_{ion} is the ionization cross section, and P is in units of pascal. At a room temperature of 300 K, the molecular density ρ_m is about

$3.3 \times 10^{22}\,\mathrm{m}^{-3}$. The effect of photoemission is usually negligible for medium-energy protons because of the lack of synchrotron radiation.

3.5.1.4 Multi-bunch Multipacting

Beam-induced multipacting is believed to be the leading source of sustained electron production. The phenomenon of multibunch beam-induced multipacting was observed at the CERN PS and SPS. Multipacting occurs if the transit time of electrons crossing the vacuum pipe is comparable to the time between successive bunches and if the electrons gain enough energy to produce more than one secondary electron when they hit the vacuum pipe wall [52]. The multipacting parameter ζ_m is defined as the ratio between the transit time of electrons crossing the vacuum pipe and the time between successive bunches [52, 185],

$$\zeta_m = \frac{2b}{s_b}\frac{\beta}{\beta_e} \approx \frac{\beta^2 b^2}{r_e N_0 s_b}, \tag{3.61}$$

where s_b is the distance between the successive bunches, βc is the velocity of the proton beam, $\beta_e c$ is the average velocity of the electrons, and the second relation is valid if the electron motion is non-relativistic. Here, β_e is related to the energy gained by the electron from the passage of the proton bunch,

$$\Delta E_e = m_e c^2 \left[\sqrt{\left(\frac{2r_e N_0}{\beta b}\right)^2 + 1} - 1\right] \approx 2m_e c^2 \left(\frac{r_e N_0}{\beta b}\right)^2, \tag{3.62}$$

where $r_e = e^2/(4\pi\epsilon_0 m_e c^2$ is the classical radius of an electron. The condition for proper multi-bunch multipacting is given by $\zeta_m = 1$. The energy gained by an electron must be such that the electron-induced secondary emission yield satisfies $\alpha_e Y_{ee} > 1$, where $\alpha_e < 1$ is the electron survival rate in the bunch gap. Multi-bunch electron multipacting can occur for almost any value of ζ_m [199] (Fig. 3.6).

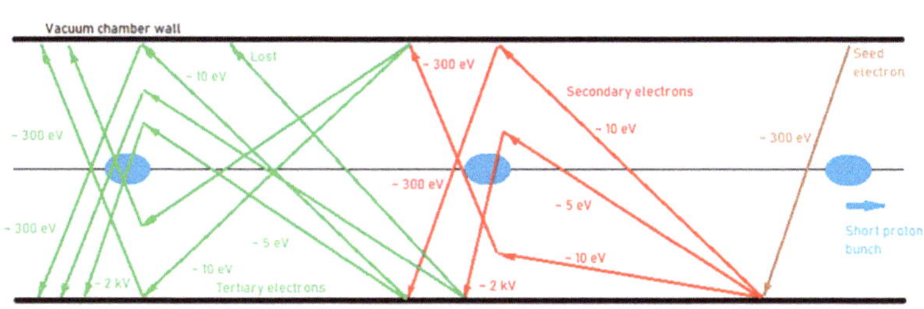

Fig. 3.6 An example of multi-bunch beam-induced electron multipacting. The spacing between the subsequent bunches is about 25 ns

3.5.1.5 Single-bunch Multipacting

Single-bunch, trailing-edge multipacting starts to dominate if the bunch length is long enough to sustain multiple passes of electrons. As shown in Fig. 3.7, electrons are attracted to the rising edge of the proton bunch. The motion is characterized by the electron bounce frequency,

$$\omega_e = \sqrt{2\pi r_e n_p}, \tag{3.63}$$

where n_p is the volume density of the proton beam. At the trailing edge of the proton bunch, electrons are released and accelerated by the bunch. The number of electrons increases dramatically at the trailing edge of the proton bunch, as observed at the PSR [99]. The electron cloud buildup due to this single-bunch mechanism is expected to have a weak dependence on bunch spacing, the vacuum pressure level, and the amount of residual protons in the beam gap. However, it critically depends on the length of the proton bunch and the variations in its longitudinal density. The single-bunch multipacting parameter ζ_s is defined as the ratio between the transit time of the electrons crossing the vacuum pipe and the passage time of half of the proton bunch [185],

$$\zeta_s = \frac{b}{s_b B_f}\frac{\beta}{\beta_e} \approx \frac{\beta b}{\sqrt{r_e N_0 s_b B_f}}. \tag{3.64}$$

The energy gained by an electron is approximately

$$\Delta E_e \approx 4 m_e c^2 \beta b \sqrt{\frac{r_e N_0}{s_b^3 B_f^3}}. \tag{3.65}$$

Single-bunch multipacting occurs if the condition $\zeta \ll 1$ is satisfied and if the energy gained by an electron is such that $Y_{ee} > 1$. Figure 3.7 shows an example of such single-bunch, trailing-edge multipacting with SNS parameters [135].

3.5.2 *Preventive and Mitigation Measures*

Mitigation of electron-cloud effects involves suppressing electron generation and enhancing Landau damping. The number of multipacting electrons can be effectively reduced by the surface treatment of the vacuum pipe. A surface coating of TiN has been shown to effectively suppress the electron flux by a factor of more than 100 at a coated section of the PSR. The thickness of the coating, typically about 100 nm, is chosen to withstand electron bombardment during the lifetime of machine operation. For critical elements, e.g. the ferrite of the extraction

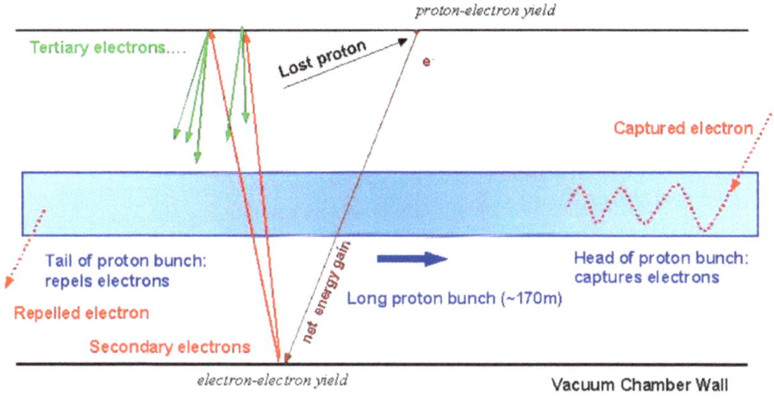

Fig. 3.7 An example of beam-induced electron multipacting at the trailing edge of a long proton bunch [177]

kicker inside the vacuum chamber at SNS, the pattern and thickness of the coating are selected to avoid eddy-current heating and to prevent changes in material properties.

Electrons in the injection region need to be guided to the collectors with a low back-scattering yield [97, 161], as shown in Fig. 8.40. A beam-in-gap kicker can ensure a clean beam gap [20, 191]. Vacuum ports can be screened, and steps in the vacuum pipe can be tapered to reduce the peaked electric fields that cause electron emission. A good vacuum can reduce electrons from gas ionization. Solenoids can be wound in straight sections to reduce multipacting [40]. Electrodes can be installed around the ring to remove the electron cloud and isolate areas of high electron concentration. Electron detectors need to be installed at locations susceptible to high electron concentrations to monitor the production of electrons.

The improvement of Landau damping starts with the design of the machine. A large vacuum-pipe aperture is needed, especially at locations of high dispersion, to further increase the momentum spread. A large RF voltage is required to provide sufficient momentum acceptance. Longitudinal painting can be used to expand the momentum spread of the injecting beam. Inductive inserts can be used to compensate for the space-charge effect, effectively increasing RF focusing [88, 116]. Landau-damping octupoles were shown in KEK PF and BEPC to raise the stability threshold. The lattice sextupole families (BEPC, SPS, KEKB, and SNS) can be used for chromatic adjustments, either to improve momentum acceptance or to enhance damping. Finally, a fast, wideband feedback system can be implemented to dampen instabilities.

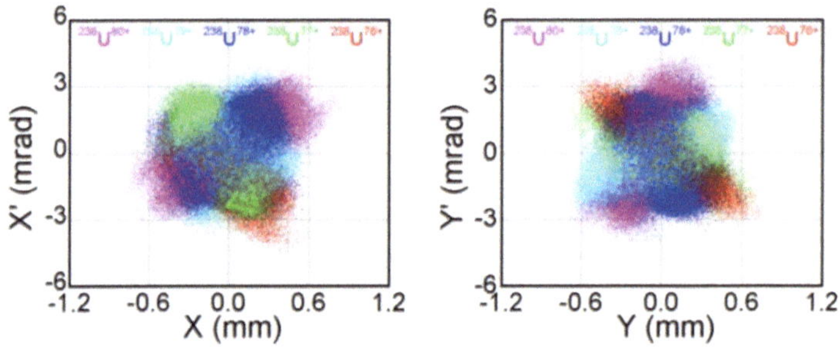

Fig. 3.8 Five charge states of the uranium beam designed to overlap at the FRIB target

3.6 Acceleration of Multiple Charge States

3.6.1 Beam Dynamics

To achieve a high beam intensity, simultaneous acceleration of heavy-ion beams with multiple charge states is often needed because of the broad charge spectrum resulting from stripping. The FRIB driver linac accelerates up to five charge states simultaneously, transversely overlapping at the charge stripper location and at the target (Fig. 3.8). An additional benefit of accelerating multiple charge states is the reduction in the dumped beam at the stripping points, which appreciably reduces the shielding requirements. Machine optics, diagnostics, and fault mitigation are designed in detail to meet performance goals.

The most abundant charge state of the uranium ions after stripping at 17 MeV/u in the FRIB driver linac is $q_0 = 78$. The simultaneous acceleration of neighboring charge states becomes possible because the high charge-to-mass ratio makes the required phase offsets small [129]. Note that different charge states of equal mass will have the same synchronous velocity profile along the linac if the condition

$$\frac{q}{A}\cos\phi_{s,q} = \frac{q_0}{A}\cos\phi_{s,q_0} \tag{3.66}$$

is satisfied. The simultaneous acceleration of ions with different charge states requires the injection of the beam with each charge state q at a synchronous phase:

$$\phi_{s,q} = \cos^{-1}\left(\frac{q_0}{q}\cos\phi_{s,q_0}\right). \tag{3.67}$$

Figure 3.9 shows the synchronous phase as a function of the charge state calculated for uranium ions in the FRIB driver linac. This particular example shows that if the linac phase is set for charge state $q_0 = 78$, it can accelerate a wide range of charge states. In fact, about 10 charge states of the uranium beam could be accepted and accelerated. For q ranging from 73 to 83, even in the worst-case scenario (charge

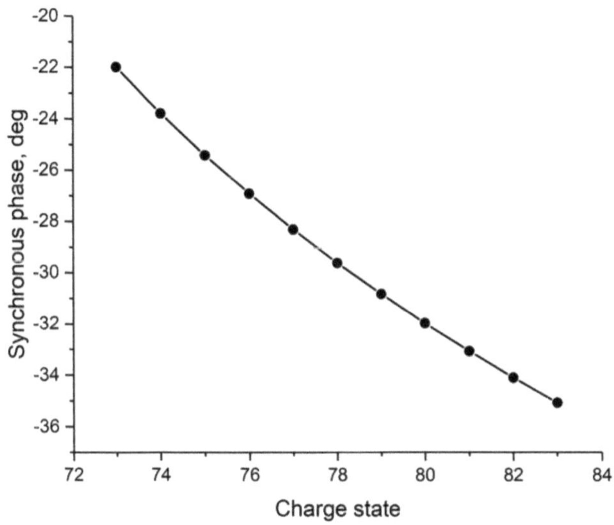

Fig. 3.9 The synchronous phase as a function of the uranium ion charge state. The designed synchronous phase for $q_0 = 78$ is $-30°$

state of 73), only a small change in the synchronous phase is required, from $-30°$ to 22°. At this phase, longitudinal acceptance is sufficient to accept the anticipated beam emittance. The design goal for the FRIB driver linac is to accept and accelerate 5 charge states: 76, 77, 78, 79, and 80.

The phase trajectories of the linear synchrotron oscillations are given by elliptical trajectories

$$\left(\frac{\Delta p/p}{(\Delta p/p)_m}\right)^2 + \left(\frac{\theta}{\theta_m}\right)^2 = 1 \tag{3.68}$$

where $\theta(q) = \phi - \phi_{s,q}$, and $\theta_m(q)$ are the amplitude of the phase oscillations; each particle with a different charge state q oscillates around its own synchronous phase with a different amplitude. The amplitude of the relative momentum oscillations is

$$(\Delta p/p)_m\,(q) = \sqrt{\frac{q}{A}\frac{\gamma_s e E_{\text{eff}} \sin|\varphi_{s,q}|}{2\pi\beta_s m_0 c^2}}\theta_m(q) \tag{3.69}$$

It would be entirely feasible to eliminate the relative oscillations of different charge states. If the linac has been tuned for the acceleration of a charge state q_0, then the particle bunches of different neighboring charge states could be injected into the linac at different neighboring RF phases to precisely match each charge to its own phase trajectory. One possible method of adjusting the phase of multiple charge states would be a magnetic system, such as a chicane, designed with appropriately varying path lengths for the various charge states. However, for the present application, such a system is not necessary since the acceleration of a multiple-charge-state beam is possible even without matching the different charge states to the proper synchronous phase. If all charge states are injected at the same time (at the same RF phase), then, as described above, each charge state bunch will perform coherent oscillations with respect to the tuned charge state q_0. One can view this as an increase in the total (effective) longitudinal emittance of the multiple-charge-state beam relative to the (partial) longitudinal emittance of the individual-charge-state bunches. Normally, for a heavy-ion SRF linac, the longitudinal rms emittance is determined by the multi-harmonic buncher located upstream of the RFQ and can be made as small as $\sim 0.15\ \pi\cdot$keV/u·ns for a single-charge-state beam. The SC linac provides ample headroom for the effective emittance growth introduced by the acceleration of multiple charge states. Note that if no phase-matching is done for different charge states, additional emittance growth will occur at frequency transitions in the linac. Heavy-ion linacs may have several transitions to permit efficient operation over the large velocity range required.

In the transverse phase planes, the main difference between single- and multiple-charge-state beams is that a mismatch in a single-charge-state beam is generally correctable and does not lead to transverse emittance growth. For multiple charge states, correction is more difficult and will generally induce growth in transverse emittance. For highly charged heavy-ion beams, the difference in C-S parameters for five charge states is sufficiently small that all charge states can be injected into the linac with the same transverse parameters. If these parameters are chosen to match, for example, the charge state $q_0 = 78$ for uranium after stripping at 17 MeV/u, the other charge states will only slightly mismatch. Multiple-charge-state beams are also more severely affected by misalignment errors.

The misalignments coherently deflect the beam. For a single charge-state beam, misalignment causes lateral displacement of the beam but no emittance growth, as long as the beam remains in the linear region of the focusing elements. With a beam containing multiple charge states, the differing betatron periods, as well as the differing displacements, cause growth in the transverse emittance. However, even for multiple charge-state beams, emittance growth can be substantially reduced by simple corrective steering procedures. The detailed simulation of the transverse dynamics of the beam of multiple charge states in the presence of misalignment and other errors in a driver linac is presented, for example, in [127].

3.6.2 *Tuning and Matching of Multiple Charge State Beams*

The FRIB linac has been designed to accelerate multiple charge-state beams after the stripper, ultimately achieving 400 kW beam power on the target. Additionally, multiple-charge-state acceleration dramatically reduces the power of unwanted charge states collimated in the Charge Selector Slits after the stripper.

The presence of the central charge state facilitates the linac tuning of the three (3-q) or five (5-q) charge state beams, such as ^{124}Xe and ^{238}U. The tuning of the dual

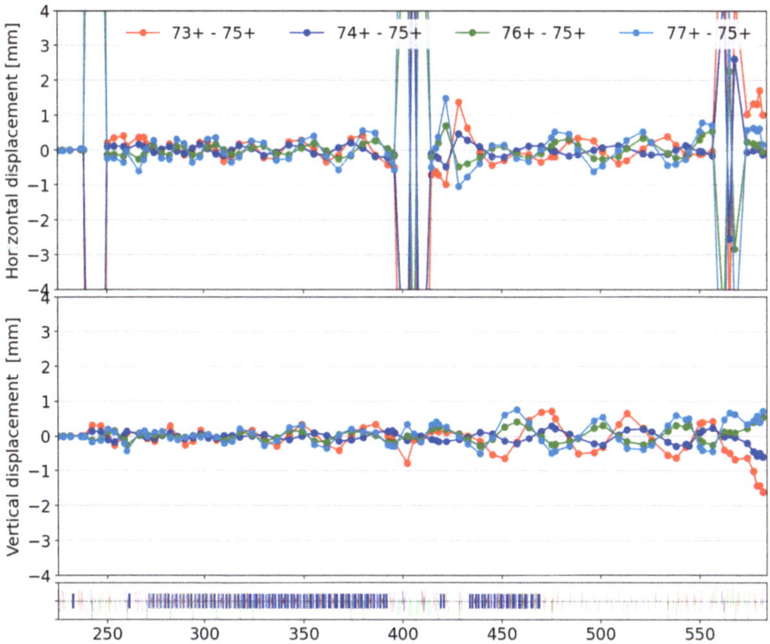

Fig. 3.10 Measured transverse beam position of each charge state of ^{238}U beam along the post-stripper linac (Courtesy T. Maruta, MSU). The horizontal plane is dispersive. The last beam position monitor is located right before the final focusing triplet

charge state beams, such as ^{48}Ca or ^{82}Se, is more complicated due to the absence of the central charge state, as reported in [130, 131]. An instant phase setting (IPS) application [138] is routinely used for phase and amplitude setting in all RF cavities in the linac. This method tremendously saves time in the development of the new velocity profile for the acceleration of new ion species, which is frequently required in FRIB to support the production of specific rare isotopes for science. For the no-loss acceleration of multiple charge state beams, the IPS is applied to the central charge state or lower charge state for 2-q ion beams. The dispersive sections of the linac are carefully tuned to combine multiple charge-state beams into the same phase space [132]. The lateral deviation of the beam centroid for the different charge states is tuned to be below 1 mm [103]. An example of the transverse position of each charge state along the FRIB linac is shown in Fig. 3.10.

To provide a low-loss beam tune, a simultaneous matching of the C-S parameters for each charge state is performed in several sections of the FRIB linac [130]. Figure 3.11 shows the envelopes of each charge state of the uranium beam matched to the target, with the rams beam size equal to 0.5 mm. Twelve quadrupoles are varied to provide the required beam size on the target, with constraints on the beam size along the beam delivery system. This optics has been utilized for the recent operations with a 20 kW uranium beam at FRIB.

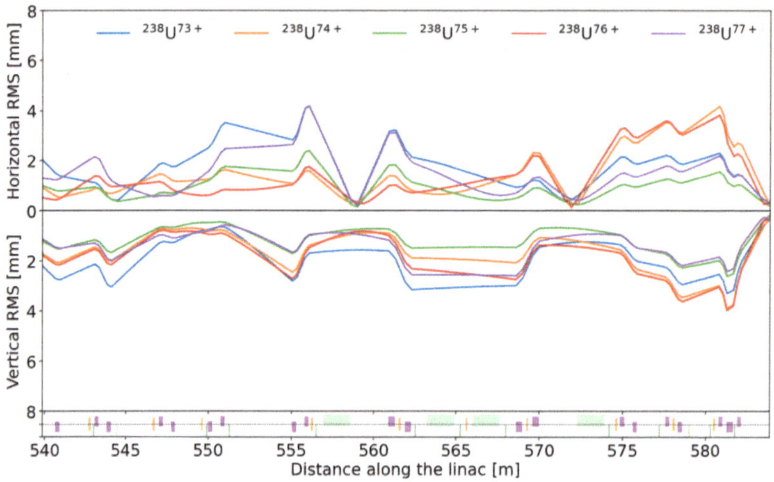

Fig. 3.11 Typical plots of five-charge-state ^{238}U beam matching into the target (Courtesy T. Maruta, MSU)

3.7 Uncontrolled Charge Stripping

This section discusses unwanted and uncontrolled charge stripping with the H^- beam in linacs, transport beamlines, cyclotrons, and synchrotron ring injection.

3.7.1 Gas Stripping

The H^- beam can be stripped by interaction with the residual gas in the beam pipe. The lifetime τ_m of an H^- ion in the presence of a residual gas with a single molecular species m is given by

$$\tau_m = \frac{1}{d_m \sigma_{sm} \beta c}, \tag{3.70}$$

where d_m and σ_{sm} are the molecular density and the interaction cross section for molecule m, respectively. Assuming several molecular species are present in the residual gas, the total lifetime of the H^- ion τ_i is given by

$$\frac{1}{\tau_i} = \sum_m \frac{1}{\tau_m}. \tag{3.71}$$

The experimental cross section of H^- beam stripping on various molecules of the residual gas is well known at energies $< 100\,\text{MeV}$, and theoretical cross sections for higher energies are given in the literature [46, 47]. The cross section for double stripping (H^- to H^+) is about 4% of the cross section for single stripping (H^- to H^0). The residual gas stripping is well known and observed in all H^- linacs [139]. At the typical residual gas pressure level of $\sim 2 \times 10^{-8}$ Torr, the contribution of residual gas stripping is much lower than the tolerable beam losses of 1 W/m in ~1 GeV linacs.

A new mechanism for possible losses in a high-power proton linac was recently identified in [77]. In the newly built CAFe linac, where the injection energy to the RFQ is low, it was found that protons can acquire two electrons due to their interaction with the residual gas in the LEBT, and the resulting H^- beam can be accelerated together with protons. The intensity of the H^- beam is more than three orders of magnitude lower than that of the proton beam intensity. However, the H^- beam can be accelerated and lost either during the transition to the higher-frequency section of the proton linac when the frequency ratio is an even number or in the bending system at full energy when the frequency ratio is odd.

3.7.2 *Stripping by Electromagnetic Field*

A strong electric field can remove one or even both electrons from an H^- ion. In reality, double-electron detachment of a hydrogen atom in its ground state is not achievable with the available accelerating fields. A very low rate of stripping of a single electron can take place in the transverse electric fields of high-gradient accelerating cavities. The electric field resulting from a Lorentz transformation of the magnetic field into the beam rest frame can become large enough to effectively strip H^- ions in medium-energy linacs and cyclotrons. For this reason, the magnetic fields in the transfer lines for H^- beams are kept at safe values to avoid stripping losses. In the highest energy, 500 MeV, H^- cyclotron at TRIUMF, stripping due to the magnetic field is the main contribution to total beam losses [4]. In fact, Lorentz stripping of the H^- beam is an energy limiting factor in isochronous cyclotrons [87].

The stripping in the magnetic field is rarely a problem in transfer lines, since the maximum allowable fields are readily calculable. The magnetic field B in the transport and injection region must be low enough so that beam loss due to magnetic stripping is acceptable. This condition becomes increasingly stringent for injection at higher energies. The scaling law is calculated from the requirement that the mean decay length λ_s due to Lorentz stripping be much longer than the path length in the laboratory frame [41],

$$\lambda_s = \frac{A_{s1}}{B} \exp\left(\frac{A_{s2}}{\beta\gamma B}\right), \tag{3.72}$$

where the coefficients $A_{s1} = (2.47 \pm 0.09) \times 10^{-6}$ Tm, and $A_{s2} = 15.0 \pm 0.03$ T are valid for $\beta\gamma B$ in the range of 0.62 to 2.34 T.

3.7.3 Blackbody Radiation Stripping

Photo-detachment of loosely coupled electrons in the H^- beam can be caused by blackbody radiation, but the stripping rate is minimal for today's H^- beam energies in linacs. The spectral density S of thermal photons per unit volume emitted by a beam pipe at a temperature T is given by the Planck formula [70]

$$S(\omega; T)d\omega = \frac{h_P}{\pi^2 c^3} \frac{\omega^3}{\exp\left(\frac{\hbar\omega}{k_B T} - 1\right)} d\omega\,, \tag{3.73}$$

where ω is the photon angular frequency, $\hbar = h_P/2\pi$, and k_B is the Boltzmann constant. Integration of Eq. 3.73 for a beam pipe radiating at 300 K leads to approximately 5.47×10^{14} photons per cubic meter. Figure 3.12 shows the photon spectral density for $T = 300$ K as a function of the photon energy in the laboratory frame and Doppler shifted into the rest frame of an 8 GeV beam. In the absence of an electric field, the photo-detachment cross section of H^- ions by thermal photons is given by [2]

$$\sigma(E) = 8\sigma_{\max} \frac{E_0^{3/2}(E - E_0)^{3/2}}{E^3}\,, \tag{3.74}$$

where $E = \hbar\omega$ is the photon energy, $\sigma_{\max} = 4.2 \times 10^{-21}$ m^2 and $E_0 = 0.7543$ eV, the lowest electron binding energy for H^-. The photo-detachment cross section as a function of the photon energy is also shown in Fig. 3.12. The overlap region between the photon spectral density distribution in the beam rest frame and the photo-detachment cross section is where the stripping of H^- ions by black-body radiation occurs. As can be seen in Fig. 3.12, the Doppler effect shifts the thermal photon energies upward and dramatically enhances the stripping probability. For example, at 8 GeV linac of Project X at FNAL, the stripping rate can reach 8×10^{-7} m^{-1} [78]. At this level of beam loss, photo-detachment becomes a serious concern, and mitigation methods, such as cooling the beam pipe to cryogenic temperatures, have been considered.

3.7.4 Intrabeam Stripping

The importance of intrabeam stripping of H^- to understand the mechanisms of beam loss in an H^- linac, like SNS, was established by Lebedev [96]. The relative beam intensity loss per unit length can be expressed as:

$$\frac{1}{N}\frac{dN}{ds} = F(\gamma\theta_x, \gamma\theta_y, \gamma\theta_s) \frac{N\sigma_{\text{strip}}\sqrt{\gamma^2\theta_x^2 + \gamma^2\theta_y^2 + \theta_s^2}}{8\pi^2\gamma^2\sigma_x\sigma_y\sigma_s}\,, \tag{3.75}$$

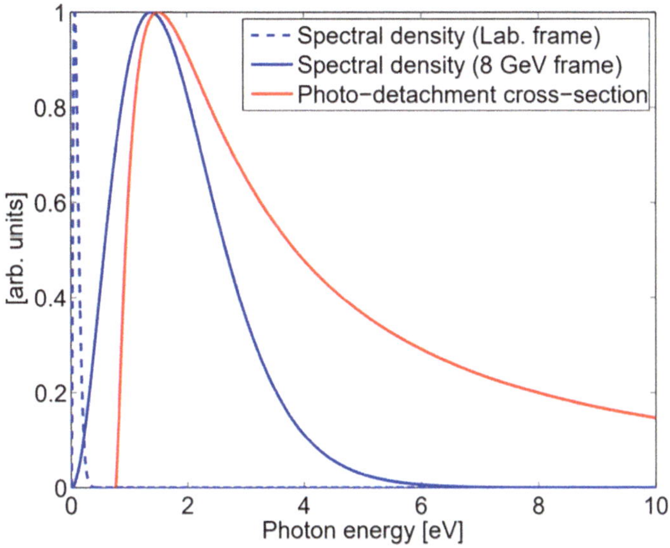

Fig. 3.12 Photon spectral density energy distribution at T = 300 K (in the laboratory frame and Doppler shifted to the 8 GeV beam frame) and H^- photo-detachment cross section [19] (Courtesy J.-P. Carneiro, FNAL)

where $\sigma_{x,y,s}$ are the transverse and longitudinal rms bunch sizes, $\theta_{x,y}$ are the transverse rms angular spreads, and the form factor F is a weak function of the beam parameters, ranging from 1 to 1.15. It appears [96] that the H^- stripping cross-section due to the reaction

$$H^- + H^- \longrightarrow H^- + H^0 + e \tag{3.76}$$

can be well approximated with $\sigma_{\text{strip}} \sim 4 \times 10^{-15}\,\text{cm}^2$ for the relative velocities of ions in the range $2 \times 10^{-4} < \beta_{\text{rel}} < 4 \times 10^{-3}$.

The cross section of intrabeam stripping was measured in the CERN Low Energy Antiproton Ring (LEAR) [22]. Comparison of beam losses by accelerating protons rather than H^- ions in the SNS and LAMPF showed that intrabeam stripping is by far the dominant loss mechanism. For the mitigation of beam losses due to intrabeam stripping, the designers of an H^- linac should consider reducing the bunch density in the high energy section of the linac. For the numerical multi-particle simulation of beam losses in an H^- linac, the beam dynamics code TRACK can be applied [18]. The code uses the Monte Carlo method to incorporate the above-mentioned mechanisms of H^- beam losses, including intrabeam stripping.

3.8 Transition Crossing in Synchrotrons

Among existing rings in which the beam must cross the transition energy (e.g., AGS, CERN PS and SPS, and FNAL Booster), beam loss and emittance growth are often observed due to chromatic nonlinearity, self-field mismatch, and instabilities ([157, 179]). At CERN SPS, the injection energy is increased to avoid transition crossing during intense single-bunch operation. Longitudinal head-tail instability caused by non-linear momentum compaction further complicates injection near transition. Among the newly built accelerators, J-PARC RCS and MR were designed with negative momentum compaction factors (i.e., imaginary transition energy) so that the beams do not cross the transition.

Near transition time $t = 0$, $\eta_0 \approx 2\hat{\gamma}_s t/\gamma_T^3$, the longitudinal motion (Sect. 3.2.2.2) becomes non-adiabatic with the characteristic non-adiabatic time T_c,

$$T_c = \left(\frac{\pi E_s \beta_s^2 \gamma_T^3}{qe\hat{V}\,|\cos\phi_s|\,\dot{\gamma}_s h \omega_s^2} \right)^{\frac{1}{3}} . \tag{3.77}$$

Complications at the transition energy are attributed to the nonadiabaticity of longitudinal motion. A high acceleration rate $\dot{\gamma}$ is essential for reducing complications during the transition.

3.8.1 *Single-Particle Effects at Transition*

Single-particle effects include mismatching with the accelerating RF bucket, coupling to transverse motion, and various types of mistiming in a duration comparable to T_c. Chromatic nonlinearities, as indicated by the η_1 term and beyond in Eqs. 3.21 and 3.22, cause particles of different momenta to cross the transition at different times, described by the time scale T_{nl},

$$T_{\text{nl}} = \left| \left(\alpha_1 + \frac{3\beta_s^2}{2} \right) \right| \frac{\sqrt{6}\sigma_p \gamma_T}{\dot{\gamma}_s} , \tag{3.78}$$

which is proportional to the rms momentum spread σ_p and inversely proportional to the acceleration rate $\dot{\gamma}_s$. As illustrated in Fig. 3.13, the effects include distortion of the phase-space ellipse, effective growth of the emittance, and eventual beam loss across transition. The growth in longitudinal emittance is proportional to the ratio T_{nl}/T_c.

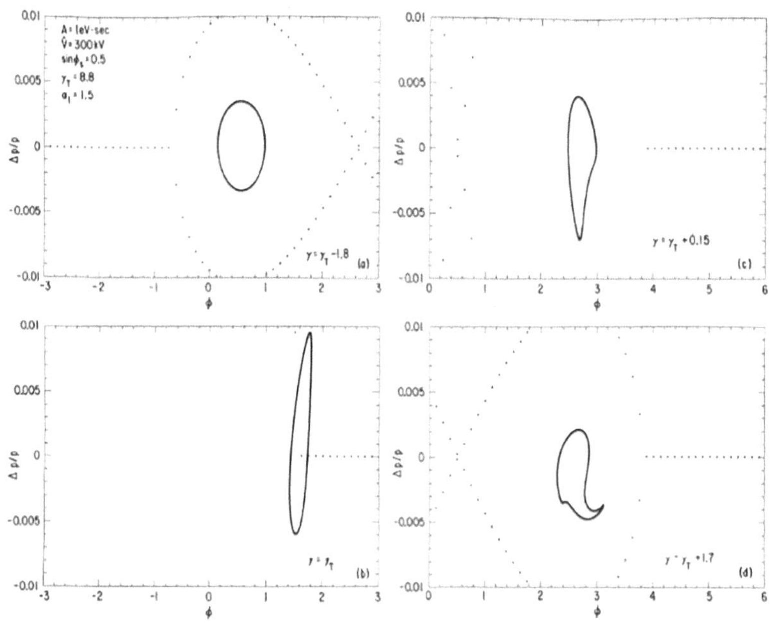

Fig. 3.13 Contours in the longitudinal phase space of a proton bunch upon transition crossing in the presence of chromatic nonlinearities. The simulation is performed with codes TIBETAN [175]

3.8.2 Multi-particle Effects at Transition

Multi-particle effects at transition include longitudinal bunch-to-bucket mismatches and microwave instabilities. A capacitive (or inductive) longitudinal coupling impedance $Z_{\|}$ near the frequency of the bunch defocuses (or focuses) it below transition, and focuses (or defocuses) it above. This change in focusing property causes a growth in longitudinal emittance proportional to the ratio of the beam-induced force to the RF accelerating force [158, 175], as shown in Fig. 3.14,

$$\frac{\Delta S}{S} \approx \frac{h\hat{I}\,|Z_{\|}/n|}{3\hat{V}\,|\cos\phi_s|\,\sigma_\phi^2}\,, \tag{3.79}$$

where $\hat{I}$ is the peak beam current at transition, σ_ϕ is the rms phase spread, and n is the frequency harmonic.

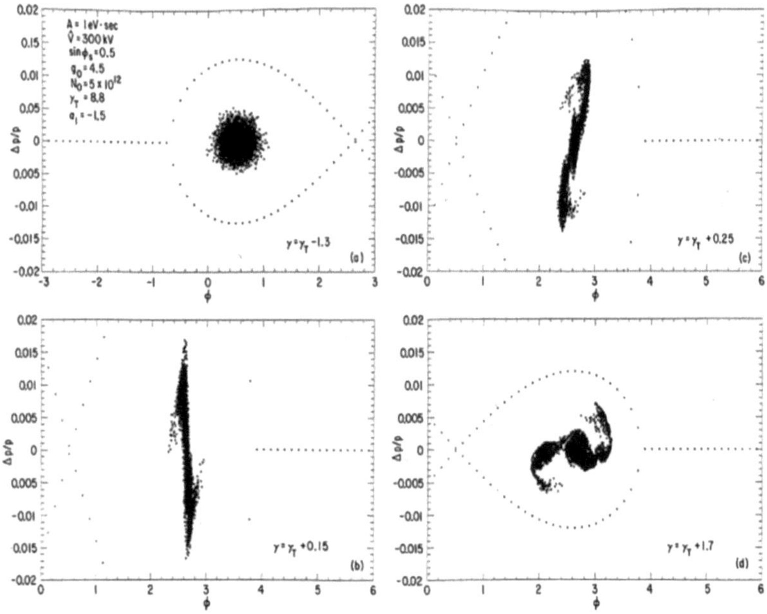

Fig. 3.14 Longitudinal phase space of a proton bunch upon crossing transition in the presence of a broad-band coupling impedance. The simulation is performed with codes TIBETAN [175]

3.8.3 *Mitigation*

A γ_T-jump has been demonstrated on many machines to improve crossing efficiency by effectively increasing the crossing rate. Without varying the tunes, a sudden change of γ_T is achieved by pulsing quadrupoles at locations of high dispersion. To minimize optical distortion and chromatic nonlinearity enhancement, "matched, first-order" schemes have been adopted for recently proposed accelerators incorporating two families of quadrupoles in regions of different dispersion.

With high-intensity synchrotrons, an important issue is to avoid crossing the transition during acceleration. A typical approach is to manipulate the lattice to achieve small or negative momentum compaction α_0 by enhancing a higher-order Fourier component of the momentum compaction (the harmonic approach) or by introducing bends in a region of negative dispersion (the modular approach [24, 25, 55, 163, 165]).

Appendix I: Intrabeam Scattering

Intrabeam scattering (IBS) refers to two-body Coulomb scattering processes that occur between particles within the same bunch [8, 102, 134, 137]. Energy exchange and increases in temperature in the beam rest frame manifest themselves as variations in beam emittance and momentum spread in the laboratory frame [176]. The rates of emittance exchange and growth are linearly proportional to the number of particles N_0 in the beam, strongly dependent ($\sim q^4/A^2$) on the charge state of the particles, and are approximately inversely proportional to the 6-D phase space area.

Appendix I: A. Mechanism

Intrabeam scattering mechanisms may be understood in the beam rest frame (Sect. 3.8.3) [181]. Energy exchanges and temperature increases in the particle rest frame manifest as variations in the beam's emittance and momentum spread in the laboratory frame.

Using the rest-frame (x, y, z, t) of the circulating synchronous particle, the rest-frame Hamiltonian (Sect. 3.8.3) in a simple system with bending dipoles of field B_0 and focusing quadrupoles of strength $n_1 = -(\rho/B_0)(\partial B_y/\partial x)$ is

$$H_R = \begin{cases} \dfrac{P_x^2 + P_y^2 + P_z^2}{2} + \dfrac{x^2}{2} - \gamma x P_z + V_C & \text{(bend)} \\[2ex] \dfrac{P_x^2 + P_y^2 + P_z^2}{2} - \dfrac{n_1(x^2 - y^2)}{2} + V_C + U_z & \text{(straight)} \end{cases} \tag{3.80}$$

where U_z is the potential provided by the RF system, and dimensions are measured in units of the characteristic distance ξ, with $\xi^3 = r_0\rho^2/\beta^2\gamma^2$, time in units of $\rho/\beta\gamma c$, and energy in units of $\beta^2\gamma^2 q^2 e^2/4\pi\epsilon_0\xi$, where $r_0 = q^2e^2/4\pi\epsilon_0 A m_0 c^2$ is the classical radius, βc and $\gamma A m_0 c^2$ are the velocity and energy of the reference particle, and ρ is the radius of curvature in the bending regions of the magnetic field B_0. The Coulomb potential is non-relativistic in the rest frame:

$$V_{Ci} = \sum_j \frac{1}{\sqrt{(x_j - x)^2 + (y_j - y)^2 + (z_j - z)^2}} . \tag{3.81}$$

In terms of the dispersion function D and betatron displacements x_β, this Hamiltonian can be transformed to

$$\bar{H}_R = \frac{P_{x\beta}^2 + P_y^2}{2} + \frac{1 - \gamma^2 G_z}{2} P_z^2 + V_C + U_z , \tag{3.82}$$

where

$$G_z = \begin{cases} D + DD'' + (D')^2 & \text{(bend)} \\ \\ DD'' + (D')^2 & \text{(straight)} \end{cases} \tag{3.83}$$

and

$$\langle G_z \rangle = \frac{1}{\gamma_T^2} \, . \tag{3.84}$$

Below the transition energy, $\gamma < \gamma_T$, particles are in a positive-mass regime. In an idealized case where the machine lattice is uniform along the ring circumference, the Hamiltonian in the rest frame is time-independent. The particle system is thus closed, and the total kinetic energy (or temperature) is bounded. Heat can be transferred from the high temperature to the low temperature direction. The system eventually reaches an equilibrium state when the temperature (i.e., rest-frame velocity) is the same in all directions.

In an actual alternating-gradient focusing ring, the beam sees a time-dependent potential modulated by the ring lattice frequency. The beam structure absorbs "phonons" and heats up [98, 186]. Intrabeam multiple scattering manifests as a mixture of thermal equalization and temperature growth, asymptotically approaching equal temperature in all directions in the rest frame.

Above the transition energy, $\gamma > \gamma_T$, the beam is in a negative-mass regime. The Hamiltonian (Eq. 3.82) indicates that even in the case of a uniform machine lattice, the beam temperature can grow simultaneously in the longitudinal and transverse directions.

Appendix I: B. Growth Rates and Beam Evolution

In the laboratory frame, the rate of emittance and momentum growth is obtained by assuming multiple small-angle scatterings among particles with a Gaussian distribution in the six-dimensional phase space. In the case that $D/\beta_x^{1/2}$ is nearly constant (e.g., for the FODO lattice), the growth rates of the rms beam sizes under IBS are given by [172]

$$\begin{bmatrix} \dfrac{1}{\sigma_p}\dfrac{d\sigma_p}{dt} \\ \\ \dfrac{1}{\sigma_x}\dfrac{d\sigma_x}{dt} \\ \\ \dfrac{1}{\sigma_y}\dfrac{d\sigma_y}{dt} \end{bmatrix} = \frac{q^4 N_0}{A^2}\frac{r_0^2\, m_0\, c^2 \mathcal{L}_C}{8\gamma \epsilon_{Nx}\epsilon_{Ny} S_{\text{rms}}} \mathcal{F}_{\text{IBS}}(\mathcal{X}) \begin{bmatrix} n_b(1-d_i^2) \\ \\ -\dfrac{a_i^2}{2} + d_i^2 \\ \\ -\dfrac{b_i^2}{2} \end{bmatrix} , \tag{3.85}$$

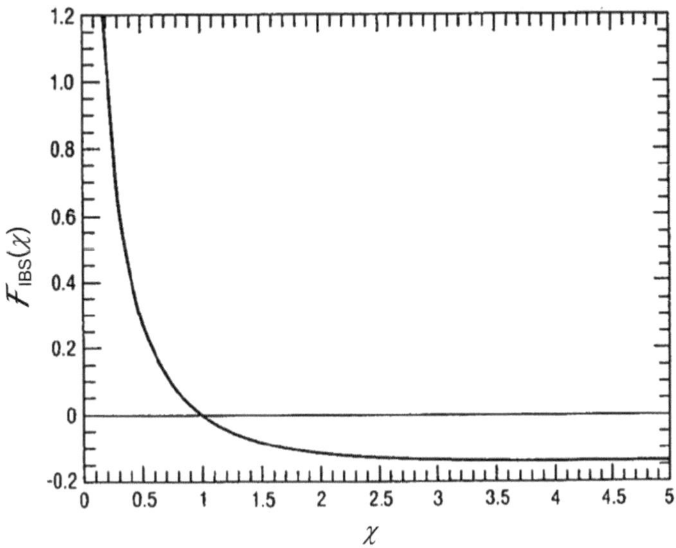

Fig. 3.15 Function $\mathcal{F}_{\text{IBS}}(\mathcal{X})$ with $0 < \mathcal{X} < \infty$ [172]

where $\mathcal{L}_C \approx 20$ is the Coulomb logarithm, $\epsilon_{Nx,y} = \beta\gamma\sigma_{x,y}^2/\beta_{x,y}$ is the normalized rms transverse emittance, $S_{\text{rms}} = \pi m_0 c^2 \beta\gamma\sigma_s\sigma_p/cA$ is the rms longitudinal bunch area, $\mathcal{X} = (a_i^2 + b_i^2)/2$, $d_i = D\sigma_p/\sqrt{\sigma_x^2 + D^2\sigma_p^2}$, $a_i = \beta_x d_i/D\gamma$, $b_i = (\beta_y\sigma_x/\beta_x\sigma_y)a_i$, n_b is equal to 1 if the beam is azimuthally bunched and equal to 2 if it is not. For azimuthally bunched beams, σ_s is the rms bunch length and N_0 is the number of particles per bunch; for unbunched beams, N_0 is the total number of particles and $\sigma_s = \sqrt{\pi}R_0$. In Eq. 3.85, $F_{\text{IBS}}(\mathcal{X})$ is an analytic function (Fig. 3.15) given by

$$\mathcal{F}_{\text{IBS}}(\mathcal{X}) = \frac{-3 + (1 + 2\mathcal{X})}{1 - \mathcal{X}} \mathcal{I}_{\text{IBS}}(\mathcal{X}), \tag{3.86}$$

where

$$\mathcal{I}_{\text{IBS}}(\mathcal{X}) = \begin{cases} \dfrac{1}{\sqrt{\mathcal{X}(\mathcal{X}-1)}} \text{artanh}\sqrt{\dfrac{\mathcal{X}-1}{\mathcal{X}}} & \mathcal{X} \geq 1, \\ \dfrac{1}{\sqrt{\mathcal{X}(1-\mathcal{X})}} \arctan\sqrt{\dfrac{1-\mathcal{X}}{\mathcal{X}}} & \mathcal{X} < 1. \end{cases} \tag{3.87}$$

The growth rates are linearly proportional to the bunch intensity N_0 and are strongly dependent ($\sim q^4/A^2$) on the charge state of the particle. Except for the form factors ξ, d, a, and b that depend on the ratio of the beam amplitudes in different dimensions, the rates are inversely proportional to the 6-D phase space area.

Under the IBS, the asymptotic beam distribution can be shown from Eq. 3.85 to be

$$\begin{cases} \langle\frac{\sigma_x}{\beta_x}\rangle \approx \langle\frac{\sigma_y}{\beta_y}\rangle \approx \frac{\sigma_p}{\gamma}\,, & \text{for } \gamma < \gamma_T\,, \\ \sqrt{n_b n_c}\,\langle\sigma_x\rangle \approx \langle D\rangle\sigma_p\,, & \text{for } \gamma > \gamma_T\,, \end{cases} \tag{3.88}$$

where n_c is equal to 1 if the horizontal and vertical motions are uncoupled, and is equal to 2 if they are fully coupled. Coupling between the horizontal and vertical motion helps reduce horizontal growth.

Appendix I: C. Beam Loss and Fokker-Planck Approach

The conventional theory based on an unbounded Gaussian distribution [8, 137] becomes inadequate in the presence of IBS-induced beam loss, as is the case with RHIC storage. Instead, an approach based on the Fokker-Planck equation was developed to describe the evolution of particle distribution in phase space. The general 6-D equation was simplified by the fact that the IBS growth time is typically much longer than the synchrotron-oscillation period, which is again much longer than the multiple collision relaxation time. For the case of RHIC, the leading source of beam loss is in the longitudinal direction due to the limited RF voltage and momentum aperture. The Fokker-Planck equation of the density function $\Psi(J)$ in the longitudinal direction, in terms of the action variable J, can be written as

$$\frac{\partial\Psi}{\partial t} = -\frac{\partial}{\partial J}(F_{\text{IBS}}\Psi) + \frac{1}{2}\frac{\partial}{\partial J}\left(D_{\text{IBS}}\frac{\partial\Psi}{\partial J}\right) \tag{3.89}$$

with the boundary conditions

$$\begin{cases} J = 0: & -F_{\text{IBS}}\Psi + \frac{D_{\text{IBS}}}{2}\frac{\partial\Psi}{\partial J}\,, \\ J = J_{\text{max}}: & \Psi = 0\,. \end{cases} \tag{3.90}$$

Here, the J-dependent $F_{\text{IBS}}(J)$ and $D_{\text{IBS}}(J)$ are the IBS coefficients of dynamic drift and diffusion, respectively [181].

Intrabeam scattering is the leading performance-limiting mechanism for the RHIC collider during the storage of the highly-stripped heavy-ion beams (Fig. 3.16) [181]. The Fokker-Planck approach appropriately describes the IBS mechanism, capable of evaluating not only emittance growth but also the evolution of density dis-

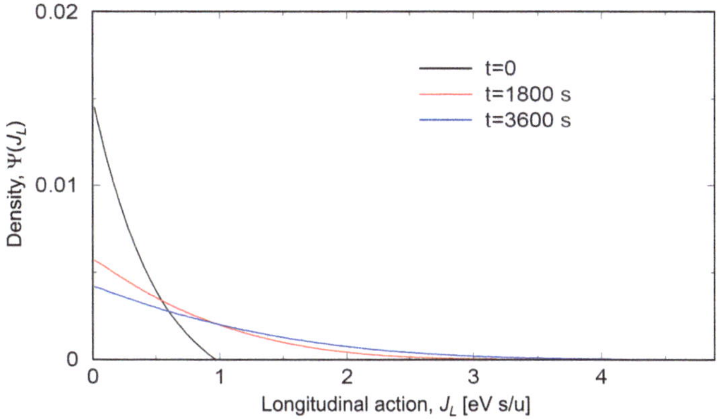

Fig. 3.16 Evolution of the longitudinal density distribution $\Psi(J_L)$ under intrabeam scattering during an 1 hour beam experiment at RHIC store. The simulation was performed with Fokker-Planck solver codes BBFP (Table 3.3 [187]). The edge of RF bucket corresponded to 4.9 eV·s/u. The distribution projected on the RF phase ϕ remained Gaussian-like [181]

tribution and beam loss. The approach was extended to include other mechanisms, such as stochastic cooling mitigation (Sect. 3.8.3), by introducing corresponding coefficients of dynamic drift F_{SC} and diffusion D_{SC} [187]. It would also be straightforward to include other slow mechanisms, like RF noise [33].

Appendix II: Beam Cooling and Crystallization

Beam cooling refers to the process of reducing the beam emittance in phase space. The characteristic time of the cooling process is typically comparable to beam storage in a ring and is much longer than the acceleration time in a hadron accelerator. . Therefore, beam cooling is not used for high-power hadron accelerators.

Appendix II: A. Beam Cooling and Experimental Results

This section introduces several beam cooling methods used in hadron storage rings.

Appendix II: A.1 Stochastic Cooling

Stochastic cooling, invented by van der Meer at CERN in 1972 [104, 166], is a negative feedback system in particle accelerators that reduces the emittance of a charged particle beam by detecting fluctuations in the sample beam's position or momentum and applying a correction pulse to the beam. A pickup sensor measures deviations in a sample of particles, sending an error signal to a kicker that applies an electric field to adjust the particles' momenta, thereby "cooling" the beam.

Stochastic cooling may be applied to both coasting and bunched beams, as well as to both the transverse and longitudinal phase spaces. It is, in particular, effective for lower-intensity beams with a high charge state q, using a cooling system with a wide frequency bandwidth at high frequencies. The technique has been successfully used in many accelerators, including CERN's ISR, AA, AD [59], and LEAR; FNAL's Debuncher and Accumulator rings; BNL's RHIC; GSI's ESR; BINP's NAP-M; and TARN at the University of Tokyo. Bunched-beam stochastic cooling fails for the proton beam stored in the Tevatron, as the coherent signal dominated the Schottky spectrum. On the other hand, bunched-beam stochastic cooling functioned as designed in the RHIC collider to mitigate the effects of intrabeam scattering at RHIC collision [10]. Bunched-beam stochastic cooling is expected to function well with a beam stored using a double-RF system [174].

Optical stochastic cooling (OSC) was proposed in 1993 to extend the cooling from the conventional microwave regime up to optical frequencies and bandwidths in the terahertz range [106], potentially increasing the achievable cooling rates by three to four orders of magnitude. The proof-of-principle experiment was successfully performed at the IOTA ring The OSC experiment insertion included a pickup undulator to generate light from the electron beam, a chicane to introduce a delay, and a kicker undulator to apply the amplified signal back onto the beam [75].

Appendix II: A.2 Electron Cooling

Electron cooling, invented by Budker at BINP in 1965 [17], is a process that reduces the emittance of a charged hadron beam by using a co-propagating beam of "cold" electrons. The electrons interact with the ions via Coulomb forces, absorbing their excess kinetic energy through a process similar to dynamic friction, thereby cooling the beam and increasing its phase-space density. A longitudinal solenoidal magnetic field may assist in passing an electron beam over distances without dilution, forming an intense electron beam with a small rms angular spread.

Electron cooling was successfully used at many accelerators including NAP-M [16]; FZJ's COSY [143]; CERN's AD, ELENA, and LEAR [164]; GSI's SIS-18; FNAL's Recycler ring; IUCF Cooler; and BNL's RHIC. The highest energy non-magnetized electron cooling was achieved with a 4.3 MeV electron beam for cooling 8 GeV antiprotons during Tevatron Run II [111]. Electron cooling system based on bunched-beam RF acceleration technology was achieved at RHIC to cool hadron beams at the injection energy [38].

Appendix II: A.3 Ionization Cooling

Ionization cooling is a technique that was proposed for colliders [114, 153] that reduces the emittance of a charged-particle beam by passing it through a material medium which causes the particles to lose energy and momentum through ionization. This energy loss occurs in all directions, while subsequent acceleration only restores longitudinal momentum. Repeated cycles of passing the beam through material absorbers and RF cavities can significantly reduce the beam's emittance, a necessary step for future high-luminosity colliders.

A working demonstration of the technique was achieved by the Muon Ionization Cooling Experiment (MICE) collaboration in 2020 [13, 31], where lithium hydride (LiH) disks and liquid hydrogen (LH_2) vessels were used as the energy-absorbing media through which muons passed to lose energy. This process of ionization cooling allows for the reduction of the muon beam's emittance and is a key step in building future muon colliders.

Appendix II: A.4 Radiation Damping

Synchrotron radiation damping is a passive phenomenon where energy loss from emitted radiation counteracts unwanted beam size growth. This effect, though weak for protons, is significant enough at the LHC's high energy (7 TeV) to require active countermeasures to prevent beam shrinkage.

Appendix II: A.6 Laser Cooling

The concept of laser cooling was proposed by Wineland and Dehmelt in 1975 for ions [189], a technique that uses the Doppler effect and light-matter interactions to reduce beam emittance. Laser cooling for ions in storage rings combines the Doppler cooling mechanism with the restoring force of an RF cavity. This technique reduces the longitudinal momentum spread of the ion beam, leading to a much higher phase space density than other cooling methods, potentially creating highly ordered or crystalline ion beams [56, 118, 149].

Appendix II: B. General Relativity Formalism

Analysis of particle interaction in the laboratory frame of storage rings is often complicated by the fact that particle motion is relativistic and that the reference particle trajectory is curved. The rotating rest frame of the reference particle is a convenient coordinate system to work with, within which particle motion is non-relativistic. In the rest frame, conventional techniques,like the molecular dynamics

(MD) method developed in condensed-matter physics, may be directly adopted to study the effects of intrabeam scattering, beam cooling, and beam crystallization.

The equations of motion in the rest frame can be rigorously derived using the general relativity formalism [107, 173]. Starting from the laboratory frame, the equations of motion were expressed in general tensor form. The Lorentz force experienced by the particle was constructed as a product of the EM field tensor and the four-velocity. Starting from the laboratory frame, the EM field tensor was expressed by terms of the components of the EM fields. Then, tensor algebra was used to transform this field tensor into the rest frame. With a similar transformation, the metric tensor of the rest frame was also obtained. The equations of motion were thus constructed in the rest frame, which includes centrifugal force, Coriolis force, time-dependent external EM forces, and electrostatic Coulomb forces.

Appendix II: C. Beam Rest-Frame Hamiltonian

The equations of motion described in Sect. 3.8.3 correspond to the Hamiltonian in the beam rest frame. Consider a general system of multi-species ions, using the rest frame (x, y, z, t) of a circulating reference particle with electric charge qe and atomic mass Am_0, and define for the ith species of ions with charge $q_i e$ and mass $A_i m_0$,

$$\bar{Z}_i \equiv q_i/q, \quad \text{and} \quad m_i \equiv A_i/A \tag{3.91}$$

Measure dimensions in units of the characteristic distance ξ with $\xi_B^3 = r_0\rho^2/\beta^2\gamma^2$, time in units of $\rho/\beta\gamma c$, and energy in units of $\beta^2\gamma^2 q^2 e^2/4\pi\epsilon_0\xi_B$, where $r_0 = q^2e^2/4\pi\epsilon_0 Am_0c^2$ is the classical radius, βc and γAm_0c^2 are the velocity and energy of the reference particle, and ρ is the radius of curvature in the bending regions of the magnetic field B_0. In a bending region with a pure dipole magnetic field, the Hamiltonian for particles of the ith species is [178]

$$\begin{aligned} H_{Ri} &= \frac{1}{2}\left(P_x^2 + P_y^2 + P_z^2\right) + \frac{1}{2}\left[x^2 + \left(\frac{\bar{Z}_i}{m_i} - 1\right)\gamma^2 z^2\right] + V_{Ci} \\ &\quad - \left(2 - \frac{\bar{Z}_i}{m_i}\right)\gamma x P_z - \left(2 - \frac{\bar{Z}_i}{m_i}\right)\gamma x P_z + \left(\frac{\bar{Z}_i}{m_i} - 1\right)\frac{\rho}{\xi_B} x \end{aligned} \tag{3.92}$$

where the Coulomb potential is

$$V_{Ci} = \sum_j \frac{\bar{Z}_i \bar{Z}_j / m_i}{\sqrt{(x_j - x)^2 + (y_j - y)^2 + (z_j - z)^2}} \tag{3.93}$$

The summation, j, is over all the other particles. In a non-bending region with a longitudinal electric field and non-dipole magnetic fields,

$$H_{Ri} = \frac{1}{2}\left(P_x^2+P_y^2+P_z^2\right) - \frac{n_1}{2}\frac{\bar{Z}_i}{m_i}(x^2-y^2) - n_{1s}\frac{\bar{Z}_i}{m_i}xy - \frac{n_2\xi_B}{6}\frac{\bar{Z}_i}{m_i}(x^3 - 3xy^2)$$

$$+V_{Ci} + U_z \tag{3.94}$$

where the quadrupole, skew quadrupole, and sextupole strengths are represented by $n_1 = -(\rho/B_0)(\partial B_y/\partial x)$, $n_{1s} = -(\rho/B_0)(\partial B_y/\partial y)$, and $n_2 = -(\rho/B_0)(\partial^2 B_y/\partial x^2)$, respectively, the electrical force F_s is expressed in terms of the electric field E_s in the laboratory frame,

$$F_z \equiv -\frac{\partial U_z}{\partial z} = \frac{\bar{Z}_i}{m_i}\frac{q_0 e E_s \xi_B}{A m_0 c^2}\left(\frac{\rho}{\xi_B \beta\gamma}\right)^2 \tag{3.95}$$

In the usual case of a single species of ion, $\bar{Z}_i = m_i = 1$.

Appendix II: D. Conditions for Crystalline Beams

A crystalline beam is an ordered state of charged particles circulating in a storage ring, where the particles arrange themselves into a stable, crystal-like structure via Coulomb force by removing their random motion through cooling techniques like tapered cooling [124]. This transition from a chaotic "warm" beam to a highly ordered "cold" beam results in a state of unprecedented brilliance, offering unique research opportunities beyond standard accelerator physics.

There are two necessary conditions [184] to form and maintain a multidimensional crystalline beam: (i) The ring is alternating-gradient (AG) focusing operating below transition, and (ii) The ring lattice periodicity is at least $2\sqrt{2}$ as high as the maximum betatron tune. Condition (i) arises from the criterion of stable kinematic motion under Coulomb interaction when particles are subjected to bending in a storage ring. Condition (ii) arises from the criterion that there is no linear resonance between the phonon modes of the crystalline structure and the machine lattice periodicity. A ring lattice periodicity as high as four times the maximum betatron tune is preferred so that linear resonance stopbands are not crossed during the cooling process as the beam density is increased. Condition (ii) is relaxed for the ordering of 1-D states in the azimuthal direction.

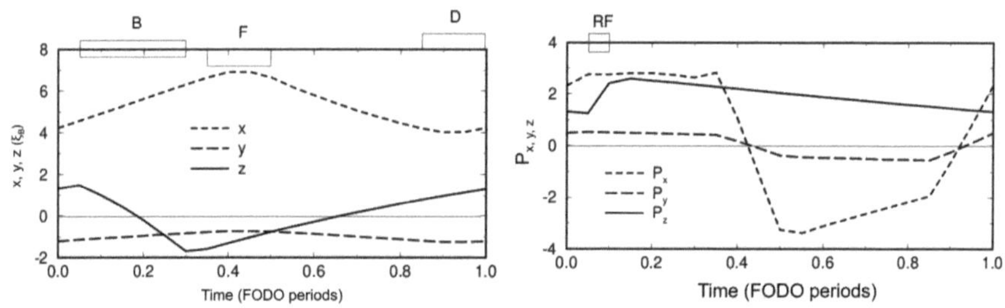

Fig. 3.17 Particle trajectory of a bunched crystalline beam. The machine consists of 10 FODO cells with $\nu_x = 2.8$, $\nu_y = 2.1$, and $\gamma = 1.4$. Lattice components in each cell are displayed on the figure: B is a bending section, F and D are focusing and de-focusing quadrupoles, and RF is the bunching RF cavity [178]

Appendix II: E. Molecular Dynamics and Numerical Methods

Numerical studies of the crystalline state have been performed using the molecular dynamics (MD) method [57, 141, 184] to iterate the equations of motion derived from the Hamiltonian. Ewald-type summation is performed in the azimuthal direction to evaluate the long-range Coulomb forces among particles and their image charges modeled in periodic "supercells" for computational efficiency.

Appendix II: E.1 Crystalline Beam Ground State

In a crystalline ground state, the motion of the circulating particles is periodic in time [184] with the period of the machine lattice. As shown in Fig. 3.17, the particle trajectory in the transverse direction conforms to AG focusing (breathing), and in the longitudinal direction, it conforms to the change in bending radius (shear). In the presence of a longitudinal electric field, momentum P_z also varies periodically, conforming to the energy gain at the cavity.

The ground state structure is a 1-D chain when the beam line density is low [57, 184]. The structure becomes 2-D, lying in the plane of weaker transverse focusing if the normalized line density λ_B in the machine is

$$\lambda_B > 0.62\gamma\xi_B^{-1}\left[\min(\nu_y^2, \nu_x^2 - \gamma^2)\right]^{1/3} \tag{3.96}$$

where $\nu_{x,y}$ are the transverse tunes. For even higher density, the particles arrange themselves into 3-D crystals, becoming helices and then helices within helices. Figure 3.18 shows a multi-shell structure at a defocusing location of the lattice. The average ratio between the horizontal and vertical particle spacing is about $[\nu_y^2/(\nu_x^2 - \gamma^2)]^{1/3}$. The maximum spatial density in the laboratory frame is approximately $\gamma\nu_y\sqrt{\nu_x^2 - \gamma^2}/(2\xi_B^3)$. If a sinusoidal electric field is present, the crystalline structure can be bunched azimuthally.

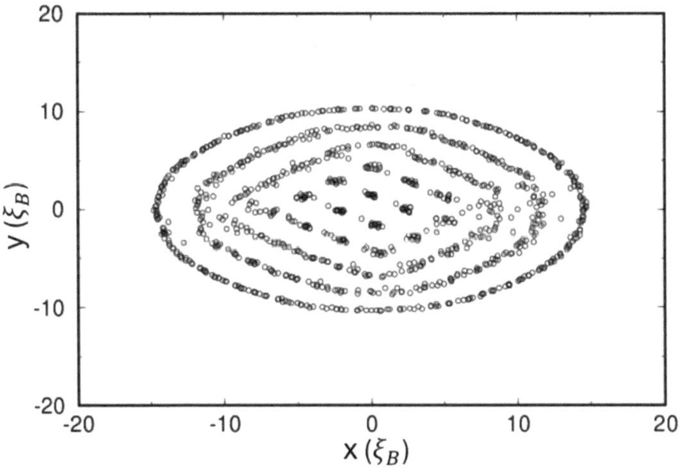

Fig. 3.18 A multi-shell structure with particle positions projected into the x-y plane ($\lambda_B = 25\gamma\xi_B^{-1}$) [178]

Appendix II: E.2 Lattice Heating

A crystalline beam in its ground state, despite breathing and shear motion, remains in the zero-temperature state [184]. At any nonzero temperature, the crystalline beam absorbs energy and heats up under time-dependent external forces caused by variations in lattice focusing and bending. In the high temperature limit, this intrabeam scattering (Sect. 3.8.3) results in a growth rate $\propto \lambda_B T_N^{-5/2}$ for coasting beams and $\propto \lambda_B T_N^{-3}$ for bunched beams, where the normalized temperature T_N (with its components defined as the deviation of P_x, P_y, and P_z from their ground-state values, squared and averaged over particles) is related to the conventional beam temperature T_B at high temperature by

$$T_N \approx \frac{2k_B\rho^2}{\beta^2\gamma^2 A m_0 c^2 \xi_B^2} T_B \tag{3.97}$$

with k_B the Boltzmann constant. The peak heating rate occurs at a temperature of about $T_N \approx 1$ when the ordering starts to occur, as shown in Fig. 3.19 [184].

Typically, strong spatial correlation appears in all directions when the temperature $T_N \lesssim 0.05$. Lattice shear and AG focusing have similar effects on beam heating. Heating behavior is similar for both bunched and coasting beams. The effects of machine lattice imperfection, ion neutralization, and envelope instability have been studied.

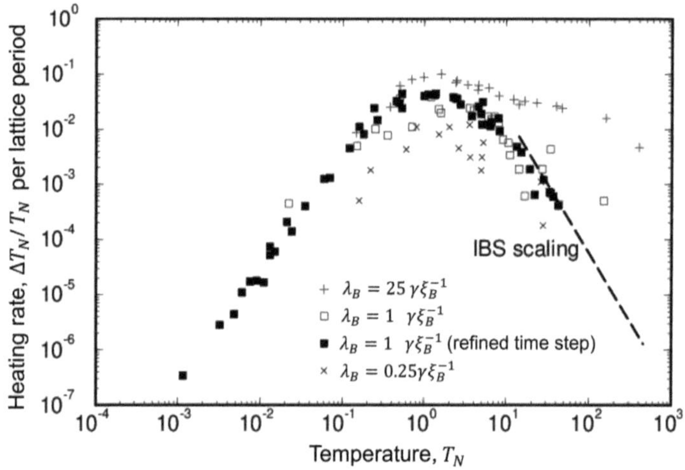

Fig. 3.19 Typical heating rates as functions of temperature obtained by MD simulation at various line density λ_B [178]

Appendix II: F. Cooling Methods and Experimental Approaches

To attain a crystalline state, the beam must be effectively cooled in three dimensions at a sufficient speed to overcome the heating. Both electron (Sect. 3.8.3) and laser (Sect. 3.8.3) cooling provide high cooling efficiency in the longitudinal direction, reaching a beam temperature of less than 1 K, but not in the transverse direction (approximately 100 K) [149]. “Sympathetic cooling” due to intrabeam scattering does produce transverse cooling [105], but the heat exchange becomes ineffective as the beam approaches an ordered state. Coupling cavities operating on a synchro-betatron resonance [123] or regular RF cavities [120] in a dispersive region can provide effective 3-D cooling. MD simulation shows that laser cooling in the longitudinal direction, together with such synchrotron-betatron resonance coupling, can achieve ordering at low beam intensity [196]. Realization of multidimensional crystalline beams may require cooling that provides the ions with a constant angular velocity, rather than a constant linear velocity (so-called tapered cooling) [124].

Experimentally, 3-D ion crystallization has been observed in RF traps [32, 190], static traps [45], and RF quadrupole rings [7] using laser cooling (Sect. 3.8.3). Ion traps were used to experimentally simulate features of an AG-focusing storage ring [122]. “Moving” crystals were generated in a ring-shaped Paul trap system [150]. Figure 3.20 shows the laser-induced fluorescence images of Coulomb crystals formed in the S-POD using the Doppler cooling technique.

Evidence of an anomaly in the Schottky signal was observed on an electron-cooled (Sect. 3.8.3) proton beam at NAP-M [30]. 1-D ordering was reported using electron cooling with both protons and heavier ions at storage rings ESR [160], CRYRING [27], and S-LSR [152].

Fig. 3.20 Laser-induced fluorescence images of Coulomb crystals formed in the S-POD with the Doppler cooling technique (Courtesy H. Okamoto, Hiroshima University). Each bright dot represents a single $^{40}Ca^{+}$ ion

In storage rings of sizable circumference with beams of significantly high energy, 3-D crystallization has not been successful despite considerable experimental efforts made in Europe [56, 149] and in Japan [159]. Practical conditions peculiar to a typical storage ring make 3-D laser cooling far more troublesome than in a compact RF trap [197]. For instance, extra effort is required to maintain the lattice periodicity (in other words, to compensate for various error fields) to meet the maintenance condition stated in Sect. 3.8.3. Furthermore, the existence of strong momentum dispersion induced inevitably by bending magnets gives rise to serious shear forces that can easily destroy 2-D and 3-D crystalline ordered structures. The tapered cooling must then be carried out as discussed below.

Appendix III: AI/ML Application in Beam Tuning

Artificial Intelligence (AI) and Machine Learning (ML) techniques have been increasingly applied to the optimization and control of particle accelerators since the early 2010s. Rapid developments in this area have shown that AI/ML approaches offer distinct advantages in specific aspects of beam diagnostics, dynamics modeling, and beam control. Using the vast amounts of data generated by accelerators, along with synthetic data from numerical simulations, AI/ML methods have enabled the identification of hidden patterns, fault prediction, and automated tuning processes.

In this section, recent progress is highlighted in applying AI/ML to enhance beam tuning in high-intensity hadron accelerators.

Appendix III: A. Virtual Diagnostics

In hadron accelerators, significant effort is devoted to achieving beam matching during acceleration or transport to minimize emittance growth and beam loss. Beam tuning typically relies on diagnostic devices that monitor key beam properties, such as beam position monitors (BPMs), profile monitors, beam current monitors (BCMs), and beam loss monitors (BLMs). However, some diagnostics are interceptive and may cause beam loss during measurement. Furthermore, in some scenarios, beam properties must be inferred in regions where direct diagnostics are unavailable; as it is not practical to include diagnostic devices in any places of interest.

Machine learning-based virtual diagnostics (VDs) offer a promising solution to these challenges. Using archived or real-time data from non-invasive sensors, such as BPMs, BLMs, and non-intercepting profile monitors. Predictive models may be trained as VDs to infer beam characteristics. These models can then support various control strategies and optimization processes without interfering with the beam.

VDs have been actively developed in high-power hadron accelerators. Improvement and advancement are updated very frequently. In SNS, machine learning-based errant beam prediction [11] was developed to prevent RF cavity damage and reduce device downtime. FRIB developed a neuron network model that can predict the bunch length from Beam Position Monitor (BPM) readings [167], as well as perform the prediction of beam quadrupole moments, i.e., the difference in horizontal and vertical beam variance, from BPM readings and an efficient quadrupole magnet scan for CS reconstruction [71].

Accurate knowledge of the beam distribution is particularly important in high-power hadron accelerators to achieve optimal matching conditions, mitigate emittance growth, and minimize beam loss. Since the direct measurement of the full high-dimensional phase space is not feasible, it must be reconstructed from a series of lower-dimensional diagnostics, such as wire scanners, camera images, and profile monitors. This process is commonly referred to as beam phase-space tomography. Recently, generative machine learning models have shown promise in reconstructing high-dimensional phase-space distributions from multiple low-dimensional data sources, including 2-D profile images and 1-D wire scan results. These reconstructions can be used to improve beam matching and reduce beam loss. However, further developments in machine learning methods and diagnostic instruments are still needed to accurately model the beam halo.

Appendix III: B. Machine Learning based Beam Tuning

Hadron accelerators usually require frequent beam tuning due to changes in operation mode, switching between different ion species, fluctuations in the beam distribution from the source, and slow drift under ambient conditions. Therefore, there is a strong motivation to develop fast and efficient tuning methods, particularly those that use machine learning techniques.

Since the early 2010s, Bayesian optimization (BO) based on Gaussian processes (GP) has been widely adopted by the accelerator community for automatic beam tuning. Beam tuning can be formulated as an optimization problem, where a set of pre-defined tuning knobs $\mathbf{x}$ is adjusted to maximize an objective function $f(\mathbf{x})$. The goal is to identify the optimal machine setting $\mathbf{x}^*$ such that:

$$\mathbf{x}^* = \arg\max(f(\mathbf{x})) \tag{3.98}$$

In most practical cases, the objective function $f(\mathbf{x})$ cannot be modeled analytically. Instead, it is approximated by a GP-based surrogate model that is iteratively updated using experimental data. GP is a non-parametric stochastic model of distribution over functions. GP takes the set of m settings and their evaluated values or measurement data and assumes that they are sampled from a Gaussian distribution $GP(\mu(m), \Sigma(m, m'))$, with μ as the mean function and Σ as the kernel function (also known as the covariance function), which characterizes the "similarity" between two function evaluations or measurements.

Through Bayes theorem, the GP model proposes new sampling points that are likely to yield improvement by using an acquisition function. This process involves an automatic trade-off between exploration (searching new regions of the parameter space) and exploitation (refining knowledge of the most promising areas). Popular acquisition functions include expected improvement (EI), maximum probability of improvement (MPI), and upper confidence bound (UCB) [156].

The BO procedure updates the surrogate model iteratively as follows:

- Find the next sampling point by optimizing the acquisition function over the surrogate model.
- Obtain a possibly noisy sample from the objective function f.
- Add the sample to the previous samples and update the surrogate model.

Figure 3.21 shows an example of a Gaussian-process-based surrogate model for a mathematical function with nine existing observations and the next suggested sampling point based on the EI acquisition function.

Recently, there has been a growing number of applications of BO with GP modeling in accelerator operation, with various enhancements such as adding physical or domain information, leveraging prior means, and including machine-specific customizations [14, 71].

However, until now, success in realizing BO in one optimization problem does not easily transfer to other problems, which increases the challenge of applying BO in a new accelerator facility.

Problems

3.1 (Negative Hydrogen Beam Magnetic Stripping) An H^- beam of 1 GeV kinetic energy is transported through an achromat of 90° bend before injection into the ring. Suppose that the magnetic stripping loss criterion is for the fractional beam loss to be less than 10^{-7} per meter. Use the following mean decay path length (in meters in the laboratory frame)

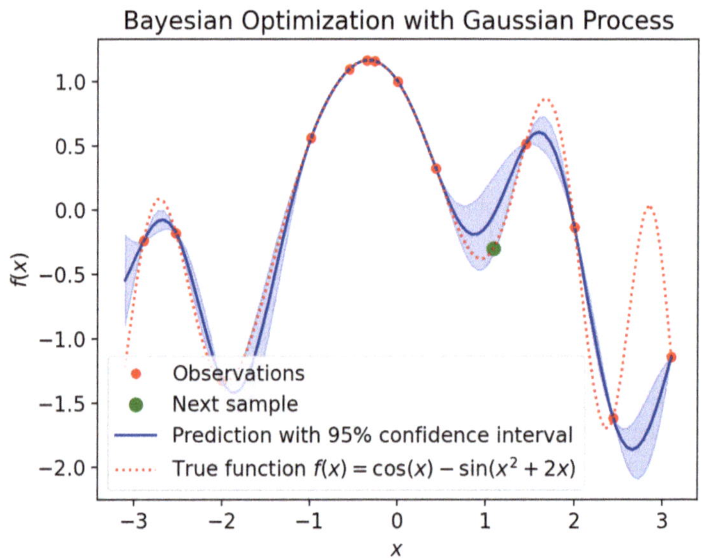

Fig. 3.21 Example of Bayesian Optimization based on Gaussian Process to model a function. The red line dotted line represents the true function, while the blue line and shaded area indicate the surrogate model with uncertainty. The next sampling point (the green dot) is suggested by the Expected Improvement acquisition function

$$\lambda_s = \frac{A_{s1}}{B} \exp \frac{A_{s2}}{\beta \gamma B}, \quad A_{s1} = (2.47 \pm 0.09) \times 10^{-6}\,\text{Tm}, \quad A_{s2} = 15.0 \pm 0.03\,\text{T} \tag{3.99}$$

where B is the magnetic field, and β and α are relativistic factors corresponding to the beam velocity.

1. Estimate the maximum magnetic field that can be used to transport the beam under the loss criteria
2. The achromat consists of 4 FODO cells, each containing 2 dipoles and 2 quadrupoles. What is the minimum length of the dipole?
3. The beam trajectory has a maximum transverse orbit deviation of 2 cm from the magnet center in the quadrupoles. What is the maximum gradient of the quadrupole that can be used under the loss criteria?
4. Estimate the maximum dipole field when the loss criterion is instead 10^{-6} per meter.
5. Estimate the minimum dipole length when the H^- beam is injected with 8 GeV kinetic energy, and the loss criterion is 10^{-5} per meter.

References

1. G. Arduini et al., Electron cloud: observations with LHC-type beams in the SPS, in *Proceedings of EPAC'00* (2000), pp. 939–941. https://jacow.org/e00/papers/THP1B16.pdf
2. B.H. Armstrong, Empirical analysis of the H- photodetachment cross section. Phys. Rev. **131**(3), 1132 (1963)
3. R. Baartman, Betatron resonances with space charge, AIP Conf. Proc. **448**(1), 56–72 (1998). https://doi.org/10.1063/1.56781
4. R. Baartman et al., Beam developments at TRIUMF, in *Proceedings of PAC'81* (1981), pp. 2879–2882
5. R.A. Baartman, Space charge limit in separated turn cyclotrons, in *Proceedings of Cyclotrons'13* (2013), pp. 305–309. https://jacow.org/CYCLOTRONS2013/papers/WE2PB01.pdf
6. P. Bertrand, C. Ricaud, Specific cyclotron correlations under space charge effects in the case of a spherical beam, in *Proceedings of Cyclotrons'01* (2001)
7. G. Birkl, S. Kassner, H. Walther, Multiple-shell structures of laser-cooled 24 mg+ ions in a quadrupole storage ring. Nature **357**(6376), 310–313 (1992)
8. J.D. Bjorken, S.K. Mtingwa, Intrabeam scattering. Part. Accel. **13**, 115–143 (1982)
9. M. Blaskiewicz, Fast head-tail instability with space charge. Phys. Rev. Spec. Top. Accel Beams **1**(4), 044201 (1998)
10. M. Blaskiewicz, J.M. Brennan, J. Wei, Stochastic Cooling Studies in RHIC, II, in *Proceedings of EPAC'04* (2004), pp. 2861–2863. http://accelconf.web.cern.ch/e04/papers/THPLT171.pdf
11. W. Blokland, K. Rajput, M. Schram, T. Jeske, P. Ramuhalli, C. Peters, Y. Yucesan, A. Zhukov, Uncertainty aware anomaly detection to predict errant beam pulses in the Oak Ridge Spallation Neutron Source accelerator. Phys. Rev. Accel. Beams **25**, 122802 (2022). https://doi.org/10.1103/PhysRevAccelBeams.25.122802
12. B. Boardman, *Spallation neutron source: description of accelerator and target*. Tech. rep., Science Research Council (1982)
13. M. Bogomilov, Y. Karadzhov, D. Kolev, I. Russinov, R. Tsenov, G. Vankova-Kirilova, L. Wang, F. Xu, S. Zheng, R. Bertoni et al., The MICE Muon Beam on ISIS and the beam-line instrumentation of the Muon Ionization Cooling Experiment. J. Instrum. **7**(05), P05009 (2012)
14. T. Boltz, J.L. Martinez, C. Xu, K.R.L. Baker, Z. Zhu, J. Morgan, R. Roussel, D. Ratner, B. Mustapha, A.L. Edelen, Leveraging prior mean models for faster bayesian optimization of particle accelerators. Sci. Rep. **15**(1) (2025). https://doi.org/10.1038/s41598-025-95297-z
15. D. Boussard, *Observation of microwave longitudinal instabilities in the CPS*. Tech. rep., CERN (1975). https://cds.cern.ch/record/872559/files/cer-002556319.pdf
16. G. Budker, N. Dikansky, V. Kudelainen, I. Meshkov, V. Parchomchuk, D. Pestrikov, A. Skrinsky, B. Sukhina, Experimental studies of electron cooling. Part. Accel. **7**, 197–211 (1976)
17. G.I. Budker, An effective method of damping particle oscillations in proton and antiproton storage rings. Sov. At. Energy **22**(5), 438–440 (1967)
18. J.P. Carneiro, B. Mustapha, P. Ostroumov, Numerical simulations of stripping effects in high-intensity hydrogen ion linacs. Phys. Rev. Spec. Top. Accel Beams **12**(4), 040102 (2009)
19. J.P. Carneiro, B. Mustapha, P.N. Ostroumov, Beam dynamics studies for multi-GeV proton and H-minus linacs, in *Proceedings of LINAC'10* (2010). https://proceedings.jacow.org/LINAC2010/html/sessi0n.htm
20. N. Catalan-Lasheras, Y.Y. Lee, H. Ludewig, N. Simos, J. Wei, Optimization of the collimation system for the Spallation Neutron Source accumulator ring. Phys. Rev. ST Accel. Beams **4**, 010101 (2001). https://doi.org/10.1103/PhysRevSTAB.4.010101
21. CERN, MAD-X (2002). http://mad.web.cern.ch/mad/
22. M. Chanel, R. Giannini, P. Lefévre, R. Ley, D. Manglunki, D. Möhl, Measurements of H^- intra-beam stripping cross section by observing a stored beam in LEAR. Phys. Lett. B **192**(3), 475–477 (1987). https://doi.org/10.1016/0370-2693(87)90141-9

23. A.W. Chao, *Physics of collective beam instabilities in high energy accelerators* (Wiley, Hoboken, 1993)
24. E.D. Courant, A. Garren, U. Wienands, Low momentrum compaction lattice study for the SSC low energy booster, in *Proceedings of PAC'91*, 2829–2832 (1991)
25. E.D. Courant, H.S. Snyder, Theory of the alternating-gradient synchrotron. Ann. Phys. **3**(1), 1–48 (1958)
26. K. Crandall, D. Rusthoi, *TRACE 3-D documentation*. Tech. rep., Los Alamos National Laboratory (1997)
27. H. Danared, A. Källberg, K.G. Rensfelt, A. Simonsson, Model and observations of schottky-noise suppression in a cold heavy-ion beam. Phys. Rev. Lett. **88**(17), 174801 (2002)
28. V. Danilov, J.D. Galambos, J.A. Holmes, Transverse impedance implementation in ORBIT, in *Proceedings of PAC'01*, 1752–1754 (2001). https://jacow.org/p01/papers/TPPH040.pdf
29. D. Davino, H. Hahn, Y.Y. Lee, Measurements of the coupling impedance of the SNS extraction kickers, in *Proceedings of EPAC'02*, 1467–1469 (2002). https://jacow.org/e02/papers/WEPRI036.pdf
30. E. Dement'ev, N. Dikanskii, A. Medvedko, V. Parkhomchuk, D. Pestrikov, Measuring proton beam thermal noises on the nap-m storage ring. Zh. Tekh. Fiz. **50**(8) (1980)
31. Demonstration of cooling by the muon ionization cooling experiment. Nature **578**(7793), 53–59 (2020)
32. F. Diedrich, E. Peik, J. Chen, W. Quint, H. Walther, Observation of a phase transition of stored laser-cooled ions. Phys. Rev. Lett. **59**(26), 2931 (1987)
33. G. Dôme, *Theory of rf acceleration and rf noise*. Tech. rep., CERN (1984)
34. T. Ebisawa, K. Takayama, 3d beam-envelope approach for an asymmetric beam in a high intensity linac with a solenoid channel. AIP Adv. **13**(4), 045316 (2023). https://doi.org/10.1063/5.0146071
35. A. Fateev, P. Ostroumov, The effect of drift spaces on the longitudinal motion in an ion linac. Nucl. Instrum. Methods Phys. Res. **222**(3), 420–427 (1984)
36. A.P. Fateev, S.K. Esin, L.V. Kravchuk, P.N. Ostroumov, V.V. Paramonov, Beam breakup in a multi-sectional ion linac, in *Proceedings of LINAC'84*, 392–393 (1984). https://jacow.org/l84/papers/THP0024.pdf
37. A.V. Fedotov, D.T. Abell, Y.Y. Lee, N. Malitsky, Y. Papaphilippou, J. Wei, Effect of non-linearities on beam dynamics in the SNS accumulator ring, in *Proceedings of EPAC'00*, 1492–1494 (2000). https://jacow.org/e00/papers/MOP6A05.pdf
38. A.V. Fedotov et al., Experimental demonstration of hadron beam cooling using radio-frequency accelerated electron bunches. Phys. Rev. Lett. **124**, 084801 (2020). https://doi.org/10.1103/PhysRevLett.124.084801
39. A.V. Fedotov, I. Hofmann, Half-integer resonance crossing in high-intensity rings. Phys. Rev. ST Accel. Beams **5**, 024202 (2002). https://doi.org/10.1103/PhysRevSTAB.5.024202
40. Y. Funakoshi et al., KEKB performance, in *Proceedings of EPAC'00*, 28–32 (2000). https://jacow.org/e00/papers/MOYE01.pdf
41. M. Furman, 7.1.8 Lorentz Stripping of H^- Ions, in *Handbook of Accelerator Physics and Engineering*, ed. by A.W. Chao et al., vol. 689 (World Scientific, Singapore, 2022)
42. C.J. Gardner, Y.Y. Lee, Y. Papaphilippou, N. Tsoupas, J. Wei, Chromatic correction and optical compensation in the SNS accumulator ring using sextupoles, in *Proceedings of EPAC'00*, 1581–1583 (2000). https://jacow.org/e00/papers/MOP6A13.pdf
43. F. Gerigk, Beam halo in high-intensity hadron linacs. Ph.D. thesis, Tech. U. Berlin (2006)
44. F. Gerigk, I. Hofmann, Beam dynamics of non-equipartitioned beams in the case of the SPL project at CERN, in *Proceedings of PAC'01*, 2872–2874 (2001). https://jacow.org/p01/papers/RPAH014.pdf
45. S.L. Gilbert, J. Bollinger, D. Wineland, Shell-structure phase of magnetically confined strongly coupled plasmas. Phys. Rev. Lett. **60**(20), 2022 (1988)
46. G.H. Gillespie, Double closure calculation of the electron-loss cross section for H- in high-energy collisions with H and He. Phys. Rev. A **15**(2), 563 (1977)

47. G.H. Gillespie, High-energy cross sections for H- ions incident on intermediate and high-Z atoms. Phys. Rev. A **16**(3), 943 (1977)
48. R.L. Gluckstern, Oscillation modes in two dimensional beams, in *Proceedings of LINAC'70*, 811–822 (1970). https://jacow.org/l70/papers/H-5.pdf
49. R.L. Gluckstern, Analytic model for halo formation in high current ion linacs. Phys. Rev. Lett. **73**(9), 1247 (1994)
50. M.M. Gordon, The longitudinal space charge effect and energy resolution, in *Proceedings of Cyclotrons'69* (1969). https://jacow.org/c69/papers/CYC69D04.pdf
51. J. Grillenberge, Is a megawatt accelerator sustainable?, in *Sustainable Accelerators Workshop*, Daresbury (2022). https://indico.stfc.ac.uk/event/646/contributions/4148/subcontributions/134/attachments/1462/2596/SusWorkshop2022.pdf
52. O. Gröbner, *Proceedings of 10th Int. Conf. High-Energy Accel.*, Protvino (1977)
53. L. Groening et al., Experimental evidence of space charge driven emittance coupling in high intensity linear accelerators. Phys. Rev. Lett. **103**, 224801 (2009). https://doi.org/10.1103/PhysRevLett.103.224801
54. G. Guignard, *The general theory of all sum and difference resonances in a three-dimensional magnetic field in a synchrotron*. Tech. rep., CERN (1975)
55. R. Gupta, J. Botman, M.K. Craddock, High transition energy magnet lattices. IEEE Trans. Nucl. Sci. **32**(5), 2308–2310 (1985)
56. J. Hangst, M. Kristensen, J. Nielsen, O. Poulsen, J. Schiffer, P. Shi, Laser cooling of a stored ion beam to 1 mK. Phys. Rev. Lett. **67**(10), 1238 (1991)
57. R.W. Hasse, J. Schiffer, The structure of the cylindrically confined coulomb lattice. Ann. Phys. **203**(2), 419–448 (1990)
58. Z. He, Y. Zhang, J. Wei, Z. Liu, R. Talman, Linear envelope model for multicharge state linac. Phys. Rev. Spec. Top. Accel Beams **17**(3), 034001 (2014)
59. W. Höfle et al., Recommissioning of the CERN AD Stochastic Cooling System in 2021 after Long Shutdown 2, in *Proceedings of COOL'21*, 20–25 (2021)
60. G.W. Hill, On the part of the motion of the lunar perigee which is a function of the mean motions of the sun and moon. Acta Math. **8**, 1–36 (1886). https://doi.org/10.1007/BF02417081
61. I. Hofmann, Coherent space charge instability of a two-dimensional beam, in *Proceedings of HIF Workshop*, 388 (1979)
62. I. Hofmann, Suppression of microwave instabilities. Laser Part. **3**(1), 1–8 (1985)
63. I. Hofmann, Stability of anisotropic beams with space charge. Phys. Rev. E **57**, 4713–4724 (1998). https://doi.org/10.1103/PhysRevE.57.4713
64. I. Hofmann, *Space Charge Physics for Particle Accelerators*. Lecture Notes in Physics (Springer, New York, 2017). https://doi.org/10.1007/978-3-319-62157-9
65. I. Hofmann, O. Boine-Frankenheim, Resonant emittance transfer driven by space charge. Phys. Rev. Lett. **87**, 034802 (2001). https://doi.org/10.1103/PhysRevLett.87.034802
66. I. Hofmann, O. Boine-Frankenheim, Revisiting the longitudinal 90° limit in high intensity linear accelerators. Phys. Rev. Lett. **118**, 114803 (2017). https://doi.org/10.1103/PhysRevLett.118.114803
67. I. Hofmann et al., Emittance coupling in high intensity beams applied to the SNS linac, in *Proceedings of PAC'01*, 2902–2904 (2001). https://jacow.org/p01/papers/RPAH025.pdf
68. I. Hofmann, L.J. Laslett, I. Haber, Stability of the Kapchinskij-Vladimirskij (KV) distribution in long periodic transport systems. Part. Accel. **13**, 145 (1983)
69. J.A. Holmes, V. Danilov, J.D. Galambos, K. Woody, Longitudinal impedance simulations in ORBIT: benchmarking and application to the SNS extracton kicker, in *Proceedings of PAC'01*, 2057–2059 (2001). https://jacow.org/p01/papers/TPPH311.pdf
70. J. Honerkamp, Estimation of parameters. in *Statistical Physics: An Advanced Approach with Applications*, 309–337 (2002)
71. K. Hwang, A. Plastun, K. Fukushima, P. Ostroumov, Q. Zhao, T. Maruta, T. Zhang, Machine-learning-assisted beam tuning at FRIB, in *JACoW LINAC2024*, THXA004 (2024). https://doi.org/10.18429/JACoW-LINAC2024-THXA004

72. T. Ikeda, K. Ito, H. Okamoto, Novel tabletop experiment demonstrating the nonlinear resonance excitation observed at the cern proton synchrotron. Jpn. J. Appl. Phys. **60**(7), 070901 (2021)
73. K. Ito, H. Okamoto, Y. Tokashiki, K. Fukushima, Coherent resonance stop bands in alternating gradient beam transport. Phys. Rev. Accel. Beams **20**(6), 064201 (2017)
74. R.A. Jameson, Beam-intensity limitations in linear accelerators. IEEE Trans. Nucl. Sci. **28**(3), 2408–2412 (1981). https://doi.org/10.1109/TNS.1981.4331708
75. J. Jarvis, V. Lebedev, A. Romanov, D. Broemmelsiek, K. Carlson, S. Chattopadhyay, A. Dick, D. Edstrom, I. Lobach, S. Nagaitsev et al., Experimental demonstration of optical stochastic cooling. Nature **608**(7922), 287–292 (2022)
76. D. Jeon, L. Groening, G. Franchetti, Fourth order resonance of a high intensity linear accelerator. Phys. Rev. ST Accel. Beams **12**, 054204 (2009). https://doi.org/10.1103/PhysRevSTAB.12.054204
77. D. Jia, Z. Wang, H. Jia, B. Liu, M. Yi, D. Guo, H. Cai, W. Chen, S. Liu, T. Wang et al., Investigation of beam loss mechanism by parasitic h-in high-power proton linac. Phys. Rev. Accel. Beams **28**(9), 090101 (2025)
78. D.E. Johnson, Challenges sssociated with 8 GeV H- transport and injection for Fermilab Project-X, in *Proceedings of HB'08*, 290–299 (2008). https://jacow.org/HB2008/papers/WGC07.pdf
79. W. Joho, High intensity problems in cyclotrons, in *Proceedings of Cyclotrons'81*, 337–347 (1981). https://jacow.org/c81/papers/EI-03.pdf
80. F.W. Jones, H.O. Schönauer, G. MacKenzie, ACCSIM: a program to simulate the accumulation of intense proton beams. Part. Accel. **31**, 199–204 (1989)
81. I. Kapchinskiy, *Theory of linear resonance accelerators* (1985)
82. A. Kasagaki, H. Okamoto, Simple design criterion for high-intensity hadron linacs. Prog. Theor. Exp. Phys. **2024**(7), 073G01 (2024)
83. K. Kato, H. Okamoto, Theoretical study of mismatch-induced halo formation from an intense hadron bunch in a uniform focusing channel. Prog. Theor. Exp. Phys. **2025**(3), 033G01 (2025). https://doi.org/10.1093/ptep/ptaf031
84. E. Keil, W. Schnell, *Concerning longitudinal stability in the ISR*. Tech. rep., CERN (1969). https://cds.cern.ch/record/1229157/files/ISR-69-48.pdf
85. W. Kleeven, Theory of Accelerated Orbits and Space Charge Effects in an AVF Cyclotron. Ph.D. thesis, Eindhoven Tech. U. (1988)
86. W. Kleeven, An analytical envelope model for space charge in cyclotrons. J. Instrum. **18**(03), P03018 (2023). https://doi.org/10.1088/1748-0221/18/03/P03018
87. H.W. Koay, Lorentz dissociation of hydrogen ions in a cyclotron. J. Instrum. **18**(03), P03027 (2023)
88. K. Koba et al., Longitudinal impedance tuner using new material FINEMET. Rev. Sci. Instrum. **70**(7), 2988–2992 (1999). https://doi.org/10.1063/1.1149859
89. K. Kojima, H. Okamoto, Characterization of overlapping betatron resonances above the phase advance of 90° per cell. Phys. Rev. Accel. Beams **25**(2), 024201 (2022)
90. A. Kolano, A. Adelmann, R. Barlow, C. Baumgarten, Intensity limits of the PSI Injector II cyclotron. Nucl. Instrum. Methods Phys. Res., Sect. A **885**, 54–59 (2018). https://doi.org/10.1016/j.nima.2017.12.045
91. S. Koscielniak, New longitudinal space charge algorithm, in *Proceedings of PAC'99*, 1839–1841 (1999). https://jacow.org/p99/papers/TUP147.pdf
92. M. Kuroda, A. Kasagaki, H. Okamoto, K. Ito, Table-top ion-trap experiment on the stability of intense short bunches in linear hadron accelerators. Phys. Rev. Accel. Beams **27**(5), 054201 (2024)
93. L.D. Landau, *Oscillations of an Electron Plasma* (US Atomic Energy Commission, Germantown, 1946)
94. L. Laslett, *The brookhaven national laboratory informal report no. BNL-7534* (1963)
95. L.J. Laslett, V.K. Neil, A.M. Sessler, Coherent electromagnetic effects in high current particle accelerators: III. Rev. Sci. Instrum. **32**(3), 276–279 (1961)

96. V.A. Lebedev, N. Solyak, A.V. Aleksandrov, A.P. Shishlo, J.F. Ostiguy, Intrabeam stripping in H- linacs, in *Proceedings of LINAC'10*, 929–931 (2010). https://jacow.org/LINAC2010/papers/THP080.pdf
97. Y.Y. Lee, G.J. Mahler, W. Meng, D. Raparia, L. Wang, J. Wei, SNS ring injection stripped electron collection: design analysis and technical issues, in *Proceedings of PAC'05*, 2384–2386 (2005). https://jacow.org/p05/papers/WPAE035.pdf
98. X.P. Li, H. Enokizono, H. Okamoto, Y. Yuri, A.M. Sessler, J. Wei, Phonon spectrum and the maintenance condition of crystalline beams. Phys. Rev. Spec. Top. Accel. Beams **9**(3), 034201 (2006)
99. R.J. Macek et al., Electron proton two-stream instability at the PSR, in *Proceedings of PAC'01*, 688–692 (2001). https://jacow.org/p01/papers/FOAB007.pdf
100. K. Makino, M. Berz, Cosy infinity version 9. Nucl. Instrum. Methods Phys. Res., Sect. A **558**(1), 346–350 (2006)
101. N. Malitsky, J. Smith, J. Wei, R. Talman, UAL-Based simulation environment for Spallation Neutron Source ring, in *Proceedings of PAC'99*, 2713–2715 (1999). https://jacow.org/p99/papers/THA40.pdf
102. M. Martini, CERN Report PS/84-9 (AA), 1984. Part. Accel **17**, 1 (1985)
103. T. Maruta et al., Primary beam development for FRIB experiments, in *Proceedings of HIAT2025*, 112–117 (2025). https://doi.org/10.18429/JACoW-HIAT2025-TUZ02
104. S. van der Meer, *Stochastic damping of betatron oscillations in the ISR*. Tech. rep., CERN (1972). https://cds.cern.ch/record/312939
105. H.J. Miesner, R. Grimm, M. Grieser, D. Habs, D. Schwalm, B. Wanner, A. Wolf, Efficient, indirect transverse laser cooling of a fast stored ion beam. Phys. Rev. Lett. **77**(4), 623 (1996)
106. A. Mikhailichenko, M. Zolotorev, Optical stochastic cooling. Phys. Rev. Lett. **71**(25), 4146 (1993)
107. C. Møller, *The Theory of Relativity* (Oxford University Press, Oxford, 1952)
108. N.V. Mohkov, W. Chou (eds.), *Beam halo and scraping*. Proceedings of 7th ICFA Mini-Workshop (2000)
109. K. Moriya, M. Ota, K. Fukushima, M. Yamaguchi, K. Ito, H. Okamoto, Double stop-band structure near half-integer tunes in high-intensity rings. Phys. Rev. Accel. Beams **19**(11), 114201 (2016)
110. E. Mètral, R. Cappi, M. Giovannozzi, G. Mètral, F. Zimmermann, Electron cloud effects in the CERN PS, in *Proceedings of PAC'01*, 682–684 (2001). https://jacow.org/p01/papers/FOAB005.pdf
111. S. Nagaitsev et al., Experimental demonstration of relativistic electron cooling. Phys. Rev. Lett. **96**(4), 044801 (2006)
112. G. Nassibian, F. Sacherer, Methods for measuring transverse coupling impedances in circular accelerators. Nucl. Instrum. Methods **159**(1), 21–27 (1979)
113. V.K. Neil, A.M. Sessler, Longitudinal resistive instabilities of intense coasting beams in particle accelerators. Rev. Sci. Instrum. **36**(4), 429–436 (1965)
114. D.V. Neuffer, Principles and applications of muon cooling. Part. Accel. **14**, 75 (1983)
115. K.Y. Ng, B. Zotter, Space charge and beam stability of the proposed Fermilab proton driver in phase I, in *Proceedings of PAC'01*, 2887–2889 (2001). https://jacow.org/p01/papers/RPAH020.pdf
116. K.Y. Ng et al., Recent experience with inductive insert at PSR, in *Proceedings of PAC'01*, 2890–2892 (2001). https://jacow.org/p01/papers/RPAH021.pdf
117. C. Nielsen, A. Sessler, K. Symon, *High energy accelerators and instrumentation* (1959)
118. A. Noda et al., Laser cooling for 3-D crystalline state at S-LSR, in *AIP Conf. Proceedings of* **821**, 491–500 (2006)
119. J. O'Connell, T. Wangler, R. Mills, K. Crandall, Beam halo formation from space-charge dominated beams in uniform focusing channels, in *Proceedings of PAC'93*, 3657–3659 (1993)
120. H. Okamoto, Transverse laser cooling induced through dispersion at an RF cavity. Phys. Rev. E **50**(6), 4982 (1994)

121. H. Okamoto, M. Aoki, C. Ichikawa, K. Kojima, T. Kurauchi, Y. Yamane, Coherent and incoherent space-charge effects in high-intensity hadron rings. J. Instrum. **15**(07), P07017 (2020)
122. H. Okamoto et al., Experimental simulation of beam propagation over long path lengths using radio-frequency and magnetic traps. Nucl. Instrum. Methods Phys. Res., Sect. A **733**, 119–128 (2014)
123. H. Okamoto, A.M. Sessler, D. Möhl, Three-dimensional laser cooling of stored and circulating ion beams by means of a coupling cavity. Phys. Rev. Lett. **72**(25), 3977 (1994)
124. H. Okamoto, J. Wei, Theory of tapered cooling. Phys. Rev. E **58**, 3817–3825 (1998). https://doi.org/10.1103/PhysRevE.58.3817
125. H. Okamoto, K. Yokoya, Parametric resonances in intense one-dimensional beams propagating through a periodic focusing channel. Nucl. Instrum. Methods Phys. Res., Sect. A **482**(1–2), 51–64 (2002)
126. P. Ostroumov, *Design features of high-intensity medium-energy superconducting heavy-ion linac*. Tech. rep., Argonne National Lab (2002)
I have formatted the bibliography according to your specific requirements: Author names (Initial. Surname) followed by a comma, Article titles in regular text, Book titles in *Title Case*, and Publisher details (along with the year) in parentheses. LaTeX
127. P. Ostroumov, V. Aseev, B. Mustapha, Beam loss studies in high-intensity heavy-ion linacs. Phys. Rev. Spec. Top. Accel. Beams **7**(9), 090101 (2004)
128. P. Ostroumov, F. Gerigk, Superconducting hadron linacs. Rev. Accel. Sci. Technol. **6**, 171–196 (2013)
129. P. Ostroumov, R. Pardo, G. Zinkann, K. Shepard, J. Nolen, Simultaneous acceleration of multiply charged ions through a superconducting linac. Phys. Rev. Lett. **86**, 2798–801 (2001). https://doi.org/10.1103/PhysRevLett.86.2798
130. P. Ostroumov, J. Wei, First two years of FRIB operation, in *Proceedings of LINAC'24*, no. 32 in Linear Accelerator Conference (JACoW Publishing, Geneva, 2024), pp. 278–283. https://doi.org/10.18429/JACoW-LINAC2024-TUXA004
131. P. Ostroumov et al., Accelerator and beam physics challenges in support of FRIB experiments, in *Proceedings of IPAC'23*, no. 14 in International Particle Accelerator Conference (JACoW Publishing, Geneva, Switzerland, 2023), pp. 1729–1732. https://doi.org/10.18429/JACoW-IPAC2023-TUPA180
132. P.N. Ostroumov, K. Fukushima, T. Maruta, A.S. Plastun, J. Wei, T. Zhang, Q. Zhao, First simultaneous acceleration of multiple charge states of heavy ion beams in a large-scale superconducting linear accelerator. Phys. Rev. Lett. **126**, 114801 (2021). https://doi.org/10.1103/PhysRevLett.126.114801
133. Y. Papaphilippou, J. Wei, R. Talman, Deflections in magnet fringe fields. Phys. Rev. E **67**, 046502 (2003). https://doi.org/10.1103/PhysRevE.67.046502
134. G. Parzen, Intrabeam scattering at high energies. Nucl. Instrum. Methods Phys. Res., Sect. A **256**(2), 231–240 (1987)
135. M.T.F. Pivi, M.A. Furman, Electron cloud development in the proton storage ring and in the spallation neutron source. Phys. Rev. ST Accel. Beams **6**, 034201 (2003). https://doi.org/10.1103/PhysRevSTAB.6.034201
136. A. Piwinski, Penetration of the field of a bunched beam through a ceramic vacuum chamber with metallic coating. IEEE Trans. Nucl. Sci. **24**(3), 1364–1366 (1977)
137. A. Piwinski, Beam losses and lifetime, in *CAS - CERN Accelerator School: General Accelerator Physics*, no. CERN-85-19-V-2, ed. by P.J. Bryant, S. Turner (CERN, Geneva, 1985), pp. 432–462. https://doi.org/10.5170/CERN-1985-019-V-2.432
138. A.S. Plastun, P.N. Ostroumov, Instant phase setting in a large superconducting linac, in *Proceedings of NAPAC'22*, North American Particle Accelerator Conference (JACoW Publishing, Geneva, 2022), pp. 885–890. https://doi.org/10.18429/JACoW-NAPAC2022-THZD1
139. M.A. Plum, Beam loss in linacs, in *Proceedings of the Joint International Accelerator School: Beam Loss and Accelerator Protection*, vol. 2, ed. by R. Schmidt (CERN, Geneva, 2016), pp. 39–62. https://doi.org/10.5170/CERN-2016-002.39

140. E. Pozdeyev, *CYCO and SIR: New Tools for Numerical and Experimental Studies of Space Charge Effects in the Isochronous Regime*, Ph.D. thesis (MSU, Michigan, 2003)
141. A. Rahman, J. Schiffer, Structure of a one-component plasma in an external field: a molecular-dynamics study of particle arrangement in a heavy-ion storage ring. Phys. Rev. Lett. **57**(9), 1133 (1986)
142. G. Rees, Important design issues of high output current proton rings, in *Proceedings of EPAC'94*, no. 4 in European Particle Accelerator Conference (JACoW Publishing, Geneva, 1994), pp. 241–246
143. V.B. Reva, M.I. Bryzgunov, A.J. Halama, V. Kamerdzhiev, P.J. Niedermayer, V.V. Parkhomchuk, COSY experience of electron cooling, in *Proceedings of COOL'19* (JACoW Publishing, Geneva, Switzerland, 2019), pp. 1–5. https://doi.org/10.18429/JACoW-COOL2019-MOX01
144. A. Ruggiero, V. Vaccaro, Effect of conducting plates on coherent space charge phenomena, Tech. rep. (CERN, Geneva, 1968). https://cds.cern.ch/record/1229159/files/ISR-69-51.pdf
145. G. Rumolo, U. Oeftiger, I. Hofmann, R. Bär, K. Beckert, M. Steck, Measurement and interpretation of longitudinal instabilities in the ESR. Nucl. Instrum. Methods Phys. Res., Sect. A **415**(1), 363–369 (1998). https://doi.org/10.1016/S0168-9002(98)00407-0
146. F.J. Sacherer, *Transverse Space-Charge Effects in Circular Accelerators*, Ph.D. thesis (University of California, Lawrence Radiation Lab., Berkeley, 1968)
147. F.J. Sacherer, Transverse - part ii: Bunched beams, in *Theoretical Aspects of the Behaviour of Beams in Accelerators and Storage Rings*, no. CERN-77-13, ed. by M.H. Blewett (CERN, Geneva, 1977), pp. 198–218 https://doi.org/10.5170/CERN-1977-013.198
148. A. Schoch, Cern report. CERN **57**, 23 (1958)
149. S. Schröder, R. Klein, N. Boos, M. Gerhard, R. Grieser, G. Huber, A. Karafillidis, M. Krieg, N. Schmidt, T. Kühl et al., First laser cooling of relativistic ions in a storage ring. Phys. Rev. Lett. **64**(24), 2901 (1990)
150. T. Schätz, D. Habs, C. Podlech, J. Wei, U. Schramm, Towards crystalline ion beams–the PALLAS ring trap. AIP Conf. Proc. **457**(1), 269–273 (1999). https://doi.org/10.1063/1.57465
151. A.M. Sessler, V.G. Vaccaro, Longitudinal instabilities of azimuthally uniform beams in circular vacuum chambers with walls of arbitrary electrical properties, Tech. rep. (CERN, Geneva, 1967)
152. T. Shirai, M. Ikegami, S. Fujimoto, H. Souda, M. Tanabe, H. Tongu, A. Noda, K. Noda, T. Fujimoto, S. Iwata et al., One-dimensional beam ordering of protons in a storage ring. Phys. Rev. Lett. **98**(20), 204801 (2007)
153. A. Skrinskii, V. Parkhomchuk, Methods of cooling beams of charged particles. Sov. J. Part. Nucl. **12**(3), 223–247 (1981)
154. L. Smith, Effect of gradient errors in the presence of space charge forces, in *4th International Conference on High-Energy Accelerators* (1963), pp. 1232–1234
155. L. Smith, R.L. Gluckstern, Focusing in linear ion accelerators. Rev. Sci. Instrum. **26**(2), 220–228 (1955)
156. J. Snoek, H. Larochelle, R.P. Adams, *Practical Bayesian Optimization of Machine Learning Algorithms* (2012). https://arxiv.org/abs/1206.2944
157. A. Sørensen, Crossing the phase transition in strong focusing proton synchrotrons. Part. Accel. **6**, 141–165 (1975)
158. A. Sorenssen, Longitudinal space-charge forces at transition, in *Proceedings of the 6th International Conference on High Energy Accelerators (HEACC 67)* (Cambridge, Massachusetts, USA, 1967), p. 474
159. H. Souda, M. Nakao, H. Tongu, K. Jimbo, K. Osaki, H. Okamoto, Y. Yuri, H. Zhengqi, M. Grieser, A. Noda, Efficiency enhancement of indirect transverse laser cooling with synchrobetatron resonant coupling by suppression of beam intensity. Jpn. J. Appl. Phys. **52**(3R), 030202 (2013)
160. M. Steck, K. Beckert, H. Eickhoff, B. Franzke, F. Nolden, H. Reich, B. Schlitt, T. Winkler, Anomalous temperature reduction of electron-cooled heavy ion beams in the storage ring ESR. Phys. Rev. Lett. **77**(18), 3803 (1996)

161. T. Tabata, R. Ito, S. Okabe, An empirical equation for the backscattering coefficient of electrons. Nucl. Instrum. Methods **94**(3), 509–513 (1971)
162. K. Takayama, Phase dynamics near transition energy in the Fermilab main ring. Part. Accel. **14**, 201–225 (1984)
163. L. Teng, Infinite transition energy synchrotron lattice using Pi straight sections. Part. Accel. **4**, 81–85 (1972)
164. G. Tranquille et al., Comissioning of the LEIR electron cooler with Pb+54 ions, in *Proceedings of RuPAC'06* (JACoW Publishing, Geneva, 2006). https://jacow.org/r06/papers/WEBO01.pdf
165. D. Trbojevic, D. Finley, R. Gerig, S. Holmes, Design method for high energy accelerator without transition energy, in *Proceedings of EPAC'90*, no. 2 in European Particle Accelerator Conference (JACoW Publishing, Geneva, 1990), pp. 1536–1539
166. S. Van Der Meer, Stochastic cooling and the accumulation of antiprotons. Rev. Mod. Phys. **57**(3), 689 (1985)
167. J. Wan, A. Plastun, P. Ostroumov, Machine learning-based non-destructive measurement of bunch length at FRIB, in *JACoW LINAC2024*, TUPB006 (2024). https://doi.org/10.18429/JACoW-LINAC2024-TUPB006
168. J.M. Wang, C. Pellegrini, On the condition for a single bunch high frequency fast blow-up, in *Single Bunch Effects in Accelerators* (Birkhäuser Basel, Basel, 1980), pp. 554–561. https://doi.org/10.1007/978-3-0348-5540-2_72
169. T. Wangler, *RF Linear Accelerators* (Wiley, Hoboken, 2008)
170. T. Wangler, K. Crandall, R. Ryne, T. Wang, Particle-core model for transverse dynamics of beam halo. Phys. Rev. Spec. Top. Accel. Beams **1**(8), 084201 (1998)
171. T.P. Wangler et al., Basis for low beam loss in the high-current APT linac, in *Proceedings of LINAC'98*, Linear Accelerator Conference (JACoW Publishing, Geneva, Switzerland, 1998), pp. 657–659. https://jacow.org/l98/papers/TU4107.pdf
172. J. Wei, Evolution of hadron beams under intrabeam scattering, in *Proceedings of PAC'93* (JACoW Publishing, Geneva, 1993), pp. 3651–3654
173. J. Wei, General relativity derivation of beam rest-frame Hamiltonian, in *Proceedings of PAC'01* (JACoW Publishing, Geneva, Switzerland, 2001), pp. 1678–1680. https://jacow.org/p01/papers/TPPH011.pdf
174. J. Wei, Stochastic cooling with a double RF system, in *Proceedings of EPAC'92*, no. 3 in European Particle Accelerator Conference (JACoW Publishing, Geneva, Switzerland, 1992), pp. 833–836
175. J. Wei, *Longitudinal Dynamics of the Nonadiabatic Regime on Alternating Gradient Synchrotrons*, Ph.D. thesis (SUNY, Stony Brook, 1990)
176. J. Wei, Intensity dependent effects in RHIC, in *AIP Conference Proceedings*, vol. 496 (American Institute of Physics, 1999), pp. 197–212
177. J. Wei, Synchrotrons and accumulators for high-intensity proton beams. Rev. Mod. Phys. **75**, 1383–1432 (2003). https://doi.org/10.1103/RevModPhys.75.1383
178. J. Wei, Crystalline Beam, in *Handbook of Accelerator Physics and Engineering*, ed. by A.W. Chao et al. (World Scientific, Singapore, 2022), pp. 250–253
179. J. Wei, Transition Crossing, in *Handbook of Accelerator Physics and Engineering*, ed. by A.W. Chao et al. (World Scientific, 2022), pp. 453–455
180. J. Wei, C. Alleman, H. Ao, B. Arend, D. Barofsky, S. Beher, G. Bollen, N. Bultman, F. Casagrande, W. Chang et al., Technological developments and accelerator improvements for the FRIB beam power ramp-up. J. Instrum. **19**(05), T05011 (2024)
181. J. Wei, A. Fedotov, W. Fischer, N. Malitsky, G. Parzen, J. Qiang, Intra-beam scattering theory and RHIC experiments, in *AIP Conf. Proc.*, vol. 773 (American Institute of Physics, College Park, 2005), pp. 389–393
182. J. Wei, M. Harrison, Tune spread due to magnetic multipoles in RHIC. Int. J. Mod. Phys. Proc. Suppl. A **2B**, 1031–1033 (1993)
183. J. Wei, M. Harrison, The RHIC project–design, status, challenges, and perspectives, in *Proceedings of 16th RCNP Osaka International Symposium* (1997), pp. 198–206

184. J. Wei, X.P. Li, A.M. Sessler, Low-energy states of circulating stored ion beams: Crystalline beams. Phys. Rev. Lett. **73**(23), 3089 (1994)
185. J. Wei, R. Macek, *Electron Cloud Effects in High-Intensity Proton Accelerators* (2002)
186. J. Wei, H. Okamoto, A.M. Sessler, Necessary conditions for attaining a crystalline beam. Phys. Rev. Lett. **80**(12), 2606 (1998)
187. J. Wei, A.G. Ruggiero, Beam life-time with intrabeam scattering and stochastic cooling, in *Proceedings of PAC'91* (JACoW Publishing, Geneva, 1991), pp. 1869–1872
188. J. Wei, R. Talman, Theorem on magnet fringe field. Part. Accel. **55**, 93–98 (1996)
189. D. Wineland, H. Dehmelt, Proposed laser fluorescence spectroscopy on Tl+ mono-ion oscillator III (side band cooling). Bull. Am. Phys. Soc **20**(4), 637 (1975)
190. D.J. Wineland, J. Bergquist, W.M. Itano, J. Bollinger, C. Manney, Atomic-ion Coulomb clusters in an ion trap. Phys. Rev. Lett. **59**(26), 2935 (1987)
191. R.L. Witkover, P.R. Cameron, T.J. Shea, R.C. Connolly, M. Kesselman, Beam instrumentation for the Spallation Neutron Source ring, in *Proceedings of PAC'99* (JACoW Publishing, Geneva, 1999), pp. 2250–2252. https://jacow.org/p99/papers/WEA169.pdf
192. Y. Yamane, H. Okamoto, K. Kojima, Excitation and suppression of synchrobetatron resonances in high-intensity hadron linacs. Phys. Rev. Accel. Beams **24**, 084201 (2021). https://doi.org/10.1103/PhysRevAccelBeams.24.084201
193. J.J. Yang, A. Adelmann, M. Humbel, M. Seidel, T.J. Zhang, Beam dynamics in high intensity cyclotrons including neighboring bunch effects: Model, implementation, and application. Phys. Rev. ST Accel. Beams **13**, 064201 (2010). https://doi.org/10.1103/PhysRevSTAB.13.064201
194. B. Yee-Rendón, H. Kuboki, R. Muto, K. Satou, M. Tomizawa, T. Toyama, M. Uota et al., Electron cloud measurements at J-PARC main ring, in *Proceedings of the 7th International Particle Accelerator Conference*, Busan, 2016, pp. 4175–4177
195. L. Young, J. Billen, The particle tracking code PARMELA, in *Proceedings of PAC'03*, vol. 5 (2003), pp. 3521–3523
196. Y. Yuri, H. Okamoto, Generating ultralow-emittance ion beams in a storage ring. Phys. Rev. Lett. **93**(20), 204801 (2004)
197. Y. Yuri, H. Okamoto, Feasibility of beam crystallization in a cooler storage ring. Phys. Rev. Spec. Top. Accel. Beams **8**(11), 114201 (2005)
198. S.Y. Zhang et al., RHIC pressure rise and electron cloud, in *Proceedings of PAC'03* (JACoW Publishing, Geneva, 2003), pp. 54–56. https://jacow.org/p03/papers/MOPA010.pdf
199. F. Zimmermann, The electron cloud Instability: summary of measurements and understanding, in *Proceedings of PAC'01* (JACoW Publishing, Geneva, 2001), pp. 666–670. https://jacow.org/p01/papers/FOAB001.pdf
200. B.W. Zotter, S.A. Kheifets, *Impedances and Wakes in High-Energy Particle Accelerators* (World Scientific, Singapore, 1998)

Chapter 4
Cryogenic Refrigeration

Abstract This chapter presents a comprehensive treatment of large-scale cryogenic systems with an emphasis on helium refrigerators/liquefiers used in modern high-power particle accelerators. It introduces the fundamental thermodynamic concepts, including energy quality and exergy, and explains how these principles guide system design and performance evaluation. Major cryogenic system components such as compressors, cold boxes and distribution are discussed, followed by an in-depth discussion of optimum system design under realistic, variable operating conditions. Finally, successful application of floating pressure technology in large facilities such as FRIB and JLab are demonstrated, providing a practical system level perspective.

Beginning with this chapter, several major systems essential to modern high-power hadron accelerators are introduced, starting with the cryogenic refrigeration system.

Cryogenic helium refrigerators are a critical enabling technology for present and future particle physics facilities that employ superconducting (SC) magnets and superconducting radio-frequency (SRF) systems. These refrigeration plants are inherently complex and highly energy-intensive, operating near fundamental thermodynamic limits. Historically, many of the key components used in cryogenic helium refrigeration plants have been adapted from other industrial applications. Consequently, effective system design requires a detailed understanding of individual components, their integration into functional subsystems, and the interactions among subsystems within the overall cryogenic plant. The primary objective is to achieve a process design that is efficient, reliable, maintainable, and capable of stable operation over a wide range of operating conditions.

A large-scale cryogenic helium refrigeration plant utilized in particle accelerators consists of several major subsystems, including the compressor system, the 4.5 K cold box, the sub-atmospheric (2 K) cold box, and the cryogenic distribution system. These are supported by additional auxiliary systems such as oil removal, gas management and storage, purification, liquid helium storage, and quench handling. The strong coupling among these subsystems necessitates a system-level design approach, in which the operational strategy and control philosophy are considered alongside component-level performance.

J. Wei et al., *High Power Hadron Accelerators*, Particle Acceleration and Detection,
https://doi.org/10.1007/978-3-032-17662-2_4

Many of the examples presented in this chapter are drawn from the Facility for Rare Isotope Beams (FRIB) cryogenic system. This system has successfully supported accelerator commissioning and ongoing operation at near-peak Carnot efficiency over a wide operating range (approximately 30–100% of cryogenic load), enabled by Ganni-cycle floating-pressure technology.

4.1 Background

Refrigeration systems transfer heat energy from a low temperature to a high temperature. The input power required for these systems depends mainly on:

1. Quality of energy, which is based on the source and sink thermal reservoir temperatures; i.e., the ratio of the ambient temperature to the low temperature
2. Actual process cycle [1] and how it is implemented; e.g.,
   (a) Brayton cycle
   (b) Claude cycle
   (c) Collins cycle
   (d) Ganni cycle [2]
3. "Carnot Step" [3] selection (i.e., temperature and pressure level-selection concerning the load characteristics)
4. Efficiency and effectiveness of the selected components
   (a) Compressors (i.e., isothermal and volumetric efficiency)
   (b) Expanders (i.e., adiabatic efficiency)
   (c) Heat exchangers (i.e., net thermal rating (UA) and number of transfer units (NTUs) effectiveness)
5. Real component characteristics under actual operating conditions
6. System control
7. Optimization goals

4.2 Quality of Energy

Thermal systems are governed by the laws of thermodynamics, which define the quality of energy. For a reversible system, the quality of energy is defined by the Clausius equality [4]. This is the basis for the definition of the "entropy" property of the intensive fluid, which is key to understanding the quality of energy. The terms "quality" (of energy), "availability," or "exergy" are synonymous. The equality of Clausius is *inequality* if the cycle is not reversible. On the other hand, for a reversible process,

$$\frac{\Delta Q_L}{T_L} = \frac{\Delta Q_H}{T_H}, \tag{4.1}$$

where ΔQ_L and ΔQ_H are heat loads at low and high temperatures T_L and T_H, respectively.

Equation 4.1 is a statement of thermal energy quality equivalence.
Eq. 4.1 implies that

$$\frac{300\,\text{W}}{300\,\text{K}} = \frac{4\,\text{W}}{4\,\text{K}} = \frac{2\,\text{W}}{2\,\text{K}}, \tag{4.2}$$

i.e., $Q_L = 1\,\text{W}$ at $T_L = 4.22\,\text{K}$ is equivalent in quality to $Q_H = 70\,\text{W}$ at $T_H = 300\,\text{K}$.

4.3 Carnot Refrigeration System

A Carnot refrigeration system is a reversible cycle (that is, there is no lost work) and defines the minimum input work required for a given rate of thermal energy transfer between two thermal reservoirs. For cryogenic systems, the inverse performance coefficient (COP_{INV}), which is the input work required per unit of cooling provided, is typically used for evaluation, as shown in Fig. 4.1. For an actual refrigerator, the input power required is greater than that required for the Carnot refrigerator, so COP_{INV} is higher than the values shown in Fig. 4.1.

The typical input power required for an actual helium refrigerator that supports loads at different temperatures (ranging from thermal shield loads to liquid helium) against the general exergetic efficiency of a given refrigerator is shown in Fig. 4.2. These values represent a reasonably well designed refrigerator and illustrate the effect of the overall exergetic efficiency of the refrigerator system on the specific input power required (COP_{INV}), to support the load at the required load temperature [5]. The typical COP_{INV} in Fig. 4.2 is on a logarithmic scale and can vary by an order of magnitude depending on the required operating temperature.

4.4 Carnot Step

Typically, in helium refrigeration systems, there are multiples of some similar non-simple process steps to accomplish a given overall process; e.g., warm compression stages and expansion stages in a cold box. The Carnot step is defined (by the author) as the distribution (or step "spacing") of a given number of similar process steps that produce the minimum irreversibility [3]. It is important to note that the Carnot step is not necessarily a reversible "step", since it depends on whether the process and/or components are reversible. A typical helium refrigeration system consists of

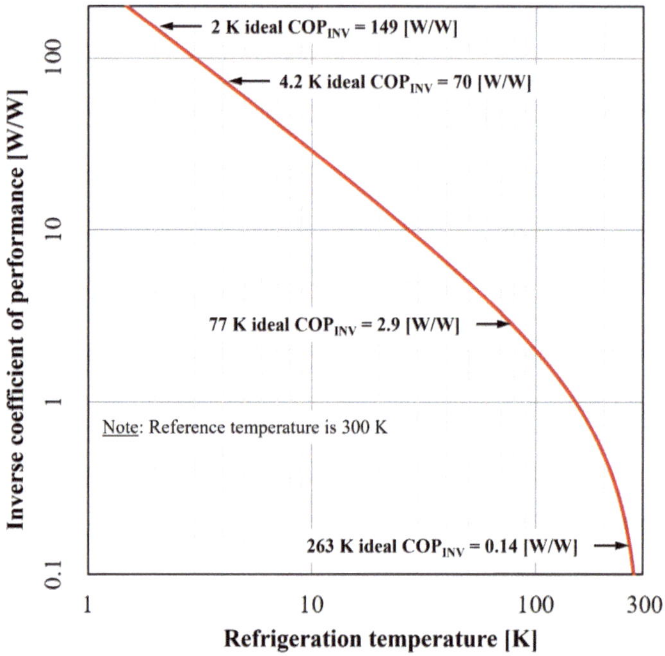

Fig. 4.1 The variation of COP_{INV} for various load temperatures [W/W]

three main parts: (1) load, (2) refrigerator cold box, and (3) compressor. Clearly, an efficient overall system depends on how these main parts match each other and on the efficient design of each part. Each part may require a number of Carnot steps for minimal losses (i.e., maximum exergy utilization).

4.4.1 The Load

A helium refrigeration system is designed to support specific external load(s); it can be a refrigeration load, a liquefaction load, or any combination of the two. The heat exchange between the load and the helium (refrigerant) can occur at a constant (isothermal) or varying (non-isothermal) temperature. Examples of isothermal loads, at least approximately so, include pool boiling, thermosiphon cooling with phase separation, superfluid cooling, and heat transfer into supercritical or subcooled helium with a modest temperature rise. Examples of non-isothermal loads include thermal shield cooling (e.g., 35 to 55 K), cold compressor load (e.g., 4.5 to 30 K), and current lead flow cooling (e.g., 4.5 to 300 K). When the non-isothermal load extends from the load temperature to the ambient temperature (e.g., 4.5 to 300 K), it is usually referred to as a liquefaction load.

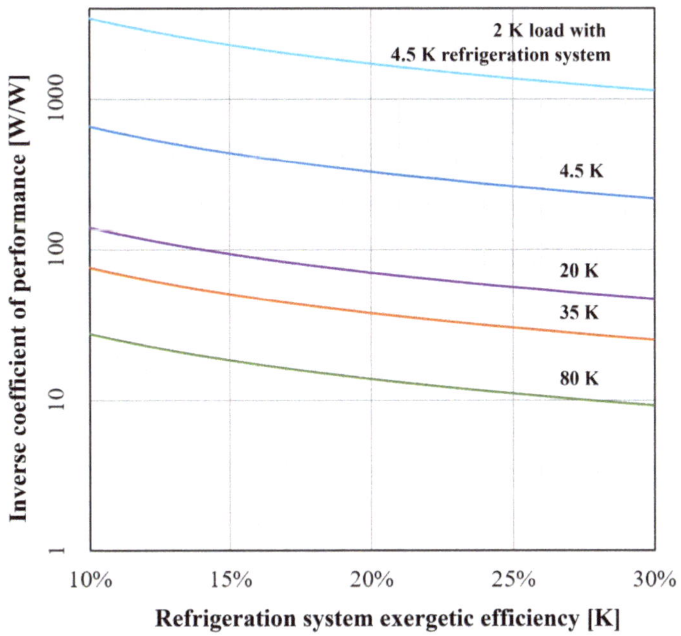

Fig. 4.2 COP_{INV} as a function of the refrigeration system exergetic efficiency

The heat energy from the loads is absorbed in the minimum entropy (liquid) helium, thereby increasing its entropy. Every attempt should be made at the load-temperature level to minimize the increase in entropy of the helium and to recover the exergy of the load return flow while satisfying the load requirements. Any wasted refrigeration, as well as losses introduced by a distribution system, becomes a load on the refrigeration system as well.

Some load temperatures and their characteristics are fixed by the experimental system requirements, whereas some loads can be traded from liquefaction to refrigeration and to shield loads. As an example, in an attempt to minimize load losses, consider the temperature selection of a single thermal shield to be used between 300 K and 4 K to minimize the heat input to the primary (colder) load. Assuming equal conductance on both sides of the shield, the optimum choice for the shield temperature to minimize the total reversible input power is found by equating the temperature ratios and is approximately 35 K. Therefore, a careful tradeoff analysis of the various types of loads (e.g., refrigeration, liquefaction, shield) that minimizes the total exergy required by all the loads is known as the load Carnot step analysis and will lead to a system with lower capital and operating costs.

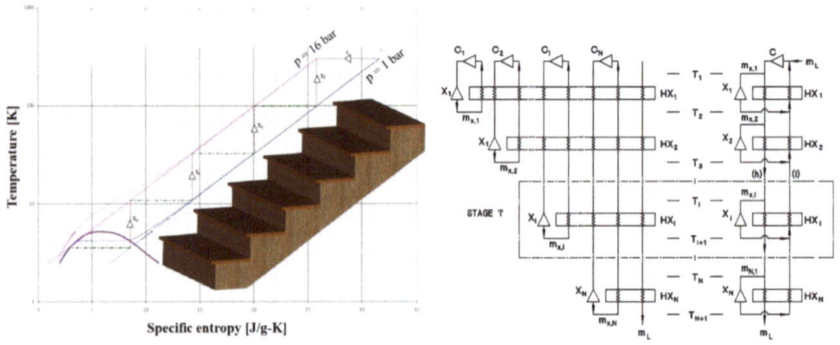

Fig. 4.3 An example of the cold box Carnot step [3]

4.4.2 *The Cold Box*

The cold box bridges the temperature difference from the load temperature to the ambient conditions by transferring the increase in the entropy of the load to the ambient-temperature compressors. The cold box does not receive input power and can only utilize the availability (i.e., exergy) supplied by the compressor(s). It is important for the cold box to minimize its use of exergy supplied by the compressors to pass as much as possible to the load. The cold box provides a process path analogous to carrying a load from a deep basement floor (4.2 K) to the ground floor (300 K) by walking up stairs (Fig. 4.3). So, given the "height" difference between the "basement" and "ground floor" (e.g., 4.2 K to 300 K), the question would be on the optimal spacing of a given number of steps that yield a minimum irreversibility [5]. Excluding liquid nitrogen (LN) pre-cooling and Joule-Thompson (JT) effects, similar non-simple process steps in a cold box include the expansion stages, which provide refrigeration by extracting work. Figure 4.3 shows a very simple arrangement of *N* expansion stages. Although an expander is shown for each stage in Fig. 4.3, there may be more than one in series for each expansion stage (as long as there is no heat exchange between them). The number of expansion stages is equal to the number of cold box Carnot steps and provides a means of evaluating the efficiency of a given cold box system design by establishing the ideal temperature "step" distribution for the expanders.

4.4.3 *The Compressor System*

The compressor system uses the input energy (which is usually electrical energy) to increase the availability (i.e., exergy) of the helium gas supplied to the cold box. It compresses the helium gas from (nominally) atmospheric pressure to the

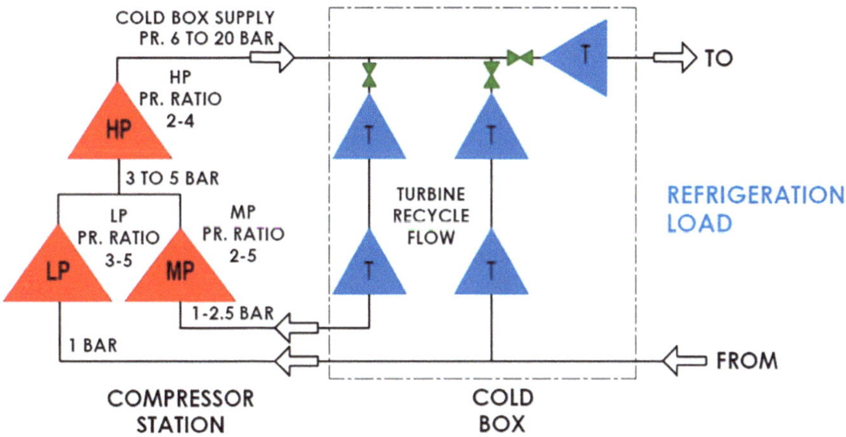

Fig. 4.4 A typical helium refrigeration compressor system

high pressure required by the cold-box design. In practice, this compression is accomplished using several stages of compressors.

For a multistage polytropic compression process, an equal pressure ratio among each of the equal-efficiency stages yields the minimum specific input work. So, for the compressor system, the Carnot step is an equal pressure ratio for each compression stage of equal efficiency. Figure 4.4 shows a typical configuration for a helium refrigeration compressor system. Figure 4.5 shows the typical volumetric and isothermal efficiency of oil-injected twin rotary screw compressors in helium service [6]. As can be seen, the efficiency of these machines is primarily dependent on the pressure ratio and the built-in volume ratio (BVR), and it is also dependent on the stage type (i.e., high-, medium-, and low-pressure stage). The Carnot step provides a means to evaluate the efficiency of a given compressor system design by establishing the ideal pressure "step" distribution for compressor stages.

4.5 Main Differences between a Refrigerator and a Liquefier

A highly efficient liquefier requires many stages of expanders, including an expander at the coldest temperature level (usually known as a "load", "wet", or "JT" expander). The output power for these expanders is higher than that of a refrigerator for the same load exergy flux. In addition, heat exchangers (HXs) inherently have an unbalanced flow, resulting in a smaller HX than would be needed in a refrigerator. The term "unbalanced" flow means that the ratio of stream capacity rates (i.e., mass flow rate multiplied by the isobaric specific heat) is not close to unity; while the term "balanced" flow indicates that this ratio is close to unity. A highly efficient refrigerator requires much larger HXs, as they have a more balanced flow than a liquefier. However, there are fewer expanders and less output power

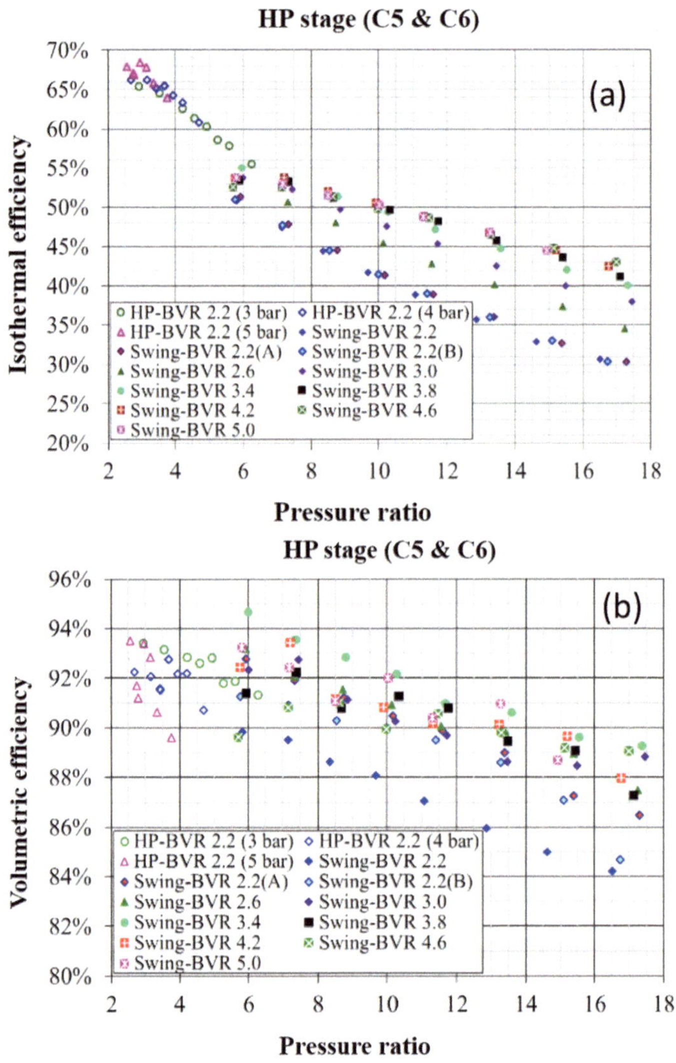

Fig. 4.5 Test results for FRIB high pressure (HP) stage compressor (**a**) isothermal efficiencies, and (**b**) volumetric efficiencies [6]

from these expanders than from the liquefier. For an ideal system, 100 W at 4.5 K refrigeration capacity requires the same amount of exergy as liquefying 1 g/s from room temperature and is shown in Fig. 4.6 for an "ideal system". The "size" of the HX can be qualitatively described using the net thermal rating (UA) and the number of transfer units (NTU). The net thermal rating can be thought of as describing the HX volume or cross-sectional area, and the NTUs can be thought of as describing the HX length.

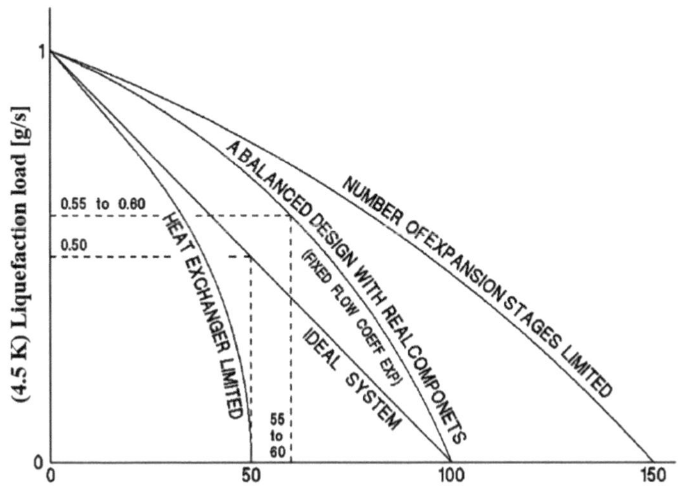

Fig. 4.6 Effect of components on the system load capacity

An example of how component limitations affect the performance of the system, ranging from a refrigerator to a liquefier, is presented in Fig. 4.6.

4.6 Optimum Refrigeration System Design

Traditionally, a design that provides maximum efficiency (minimum operating cost) at a single design point is termed the optimum system design. However, this design has two major underlying assumptions. One is that the loads are well-characterized and known. The second is that the as-built system is essentially the same as the intended design. In actuality, neither of these assumptions is correct. Additionally, various margins allocated during the design phase make the system operate under off-design conditions, even if the first two assumptions are correct. In light of these realities, a good question could be as follows: Can a cycle be envisioned, considering the optimization goals and using real components, that is capable of operating close to maximum efficiency for a load varying from maximum to minimum capacity and from full refrigeration to full liquefaction mode, or in any partial load combinations? The Ganni floating pressure cycle approaches these optimization goals for a real system design [3]. This process maintains high overall operational efficiencies at full and reduced plant capacities, allowing system pressures to follow load by varying the system charge, which determines the pressure levels and mass flow in the system. This is naturally accomplished through control intervention only to protect the equipment from shutdowns and damage.

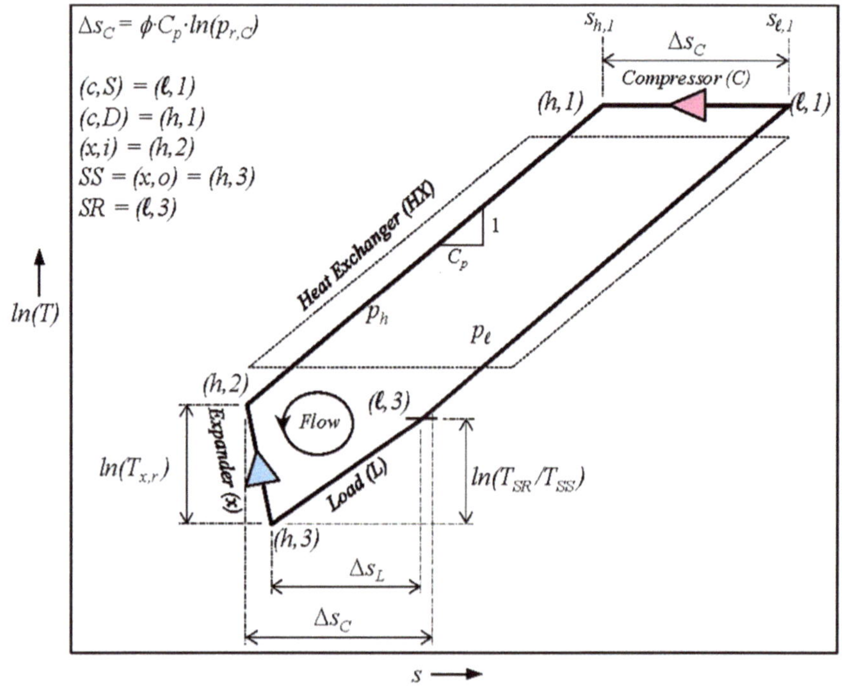

Fig. 4.7 Temperature (T) – entropy (S) diagram of a shield refrigerator [2]

4.6.1 *Floating Pressure Ganni Cycle*

Typical helium refrigeration systems have multiples of some similar process steps to perform a given overall process efficiently, as stated in Sect. 4.4. Therefore, a simple shield refrigeration system operating on a modified Brayton cycle based on fixed pressure process paths is shown in Fig. 4.7 below, and a more complicated system will have multiple similar cycles nested in one system, as shown in Sect. 4.4.

The Ganni floating pressure cycle [2] operates on a constant pressure ratio process, where the fixed displacement compressor sets the low pressure and the fixed flow coefficient expander sets the high pressure of the system; which are typical components in large capacity helium refrigeration systems. Once the compressor and expander flow coefficients are fixed, the cycle pressure ratio is also fixed. This pressure ratio sets the specific efficiency of the process cycle. The demand for refrigeration load sets the pressure level, which makes the efficiency of the system independent of the capacity as long as the constant pressure ratio is maintained. This process is applicable to existing and new systems [3]. This floating pressure design was well demonstrated by the NASA-JSC Chamber-A 20 K helium refrigeration system to support loads from 2 to 20 kW at maximum efficiency, and performance is presented in [2, 7].

4.6.2 *Example: FRIB/JLab 18 kW 4.5 K System Design*

Figure 4.8 presents the process flow diagram for the FRIB 4.5 K refrigeration system, which has a capacity of 18 kW of equivalent 4.5 K refrigeration on an equal exergy basis [2]. A similar refrigerator for the JLab 12 GeV upgrade has been successfully supporting 2 K linac operations since 2012 [8].

The actual test performance of the 4.5 K refrigeration systems used for the FRIB and JLab 12 GeV upgrade is shown in Fig. 4.9 [8, 9].

A comparison of the required load exergy in each mode (1 to 6) for the two plants is shown in Fig. 4.10. This figure illustrates the flexibility of the system to adapt to various types of loads and is very close to the maximum efficiency of 30% to 100% of the exergetic capacity of the system.

4.7 2 K Systems

For large 2 K systems, the volume flow and pressure ratios are too high for efficient compression of the sub-atmospheric helium gas at room temperature. Therefore, multistage cryogenic centrifugal compressors (cold compressors) are used to compress the helium vapor from approximately 0.030 bar to 1.2 bar. Systems that use cold compressors (CC) have become the norm for present-day large 2 K systems that provide cooling to accelerators that employ SRF cavities. The efficient and reliable operation of these systems for more than three decades at JLAB [11] and two decades at SNS has established their performance and reliability records. The subatmospheric system at FRIB employs five cold compressor stages for 2 K operation, which are housed in a vacuum vessel (cold box). A simplified schematic of the FRIB 2 K system and the FRIB subatmospheric cold box is provided in Fig. 4.11.

The design, fabrication, commissioning, and testing of the FRIB 180 g/s 2 K helium cold-compressor system was presented in [13]. Figure 4.12 shows the results of the performance testing and the operational-range mapping. During this test, the outlet pressure remained nearly constant at around 1.15 bar, and the inlet pressure was varied to achieve different operating pressure ratios. As observed in the figure, this system has a considerable operating range and a turndown capacity of nearly 2 to 1 near the nominal operating condition (i.e., a total or overall pressure ratio of approximately 40). The operation of these cold compressors has been found to be very stable, and the inverse performance coefficient of the FRIB 2 K system under its nominal operating conditions is less than 800 W/W.

Research efforts at FRIB have aided in reducing the dependence on empirical correlations used for the selection of relative speed ratios during transient processes (such as pumping down from 1 bar to the operating condition of 0.030 bar) and normal operation. These empirical correlations were developed over many years of operation through trial-and-error methodologies. The research at FRIB resulted in:

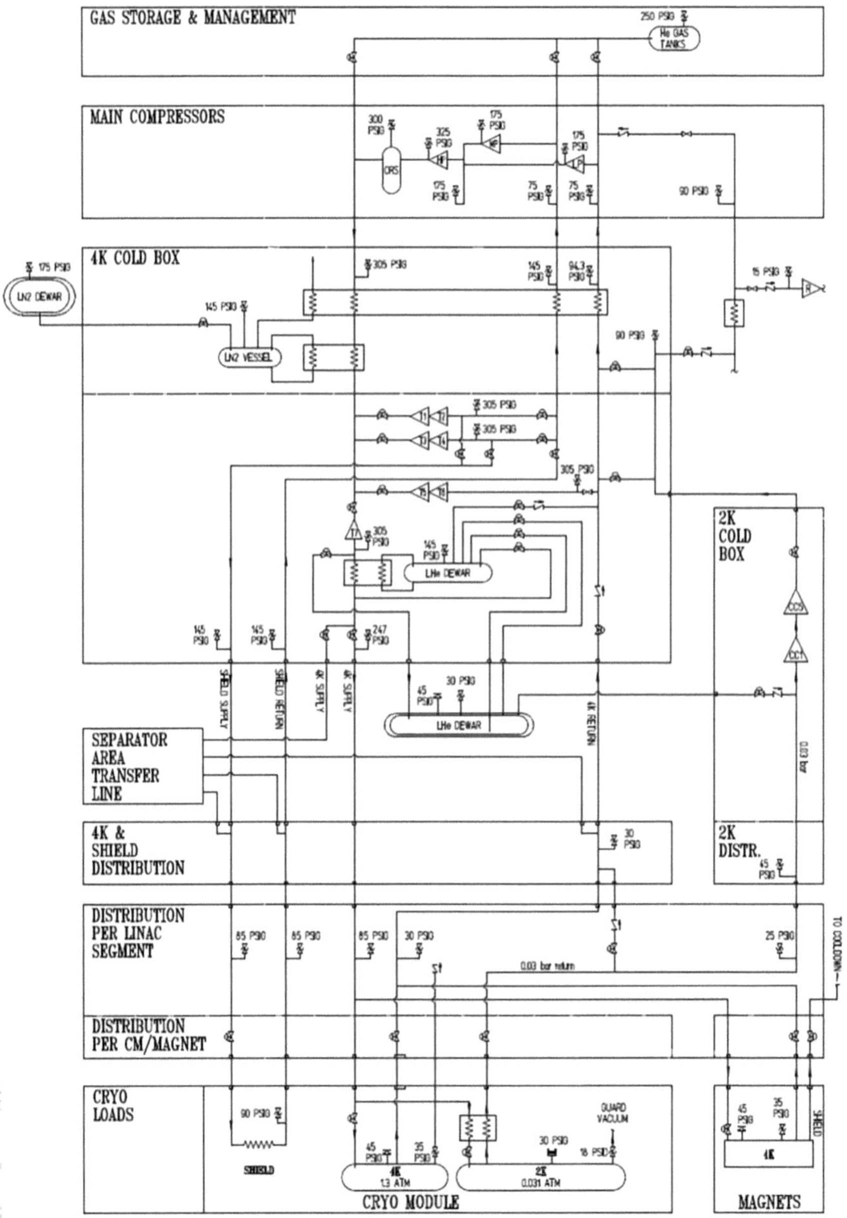

Fig. 4.8 Process flow diagram for the FRIB cryogenic system

(1) the development of a mean-line centrifugal compressor model that characterized the cold compressor train and aided in optimal steady-state speed ratio selection, (2) the development of a sub-atmospheric system pressure model which estimated

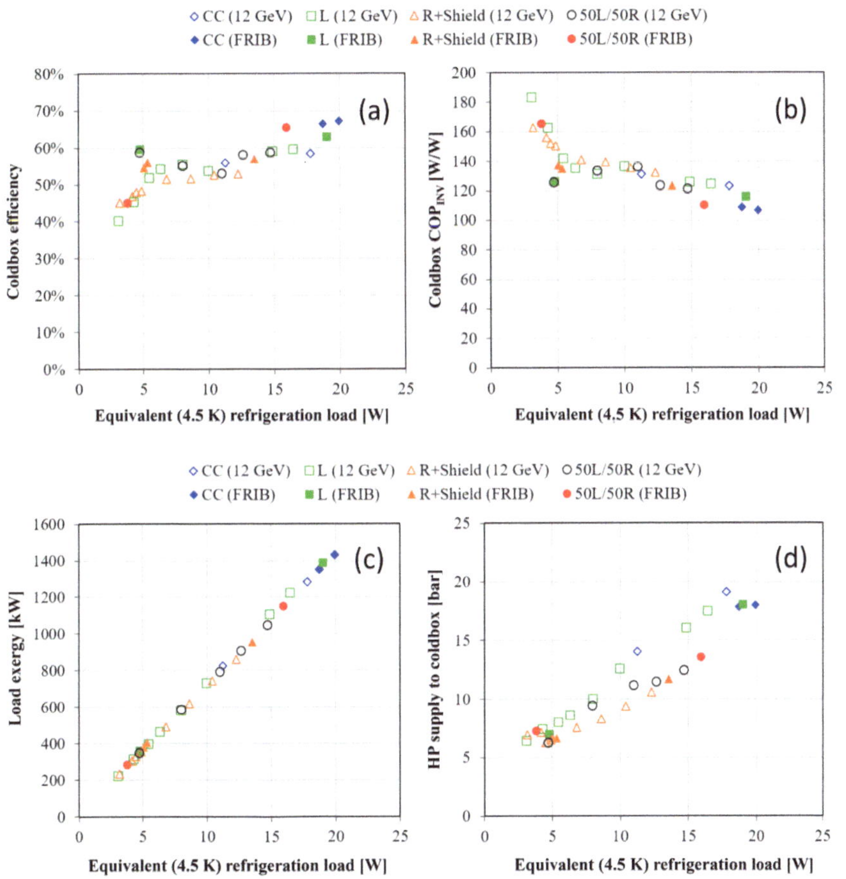

Fig. 4.9 (**a**) Carnot efficiency, (**b**) cold box COP_{INV}, (**c**) load exergy, and (**d**) HP supply pressure as a function of the equivalent 4.5 K refrigeration load [9]

the required steady-state cold compressor mass flow and enabled the prediction of the system pressure during transient operation (i.e. "pump-down"), and (3) the selection of a simplified transient process path which produced a stable and reliable pump-down sequence, including the development of a method for transient mass flow selection.

The stability of the multi-stage compressor train is critically important during transient operation and steady-state operation near minimum or maximum mass flow conditions. From the research studies conducted at FRIB, substantial progress was made in understanding the fundamental behavior of the sub-atmospheric system and the multi-stage cold compressor train. Figure 4.13 shows the results of the predictive model, comparing the predicted subatmospheric system pressure and the measured pressure during a pump-down sequence. The models developed can be

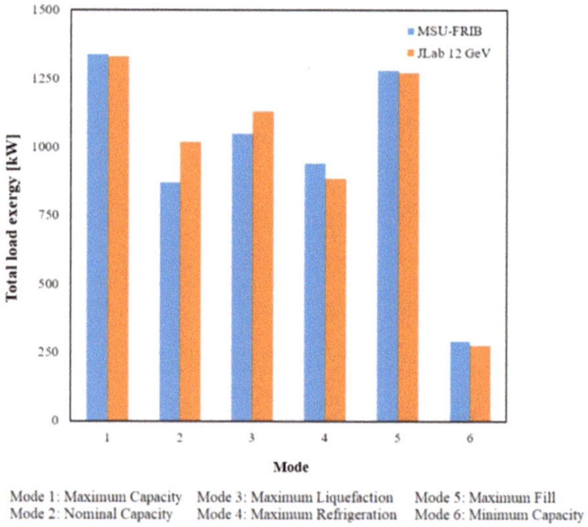

Fig. 4.10 Comparison of required load exergy at different design modes for the FRIB and JLab 12 GeV 4.5 K cryogenic refrigeration system [10]

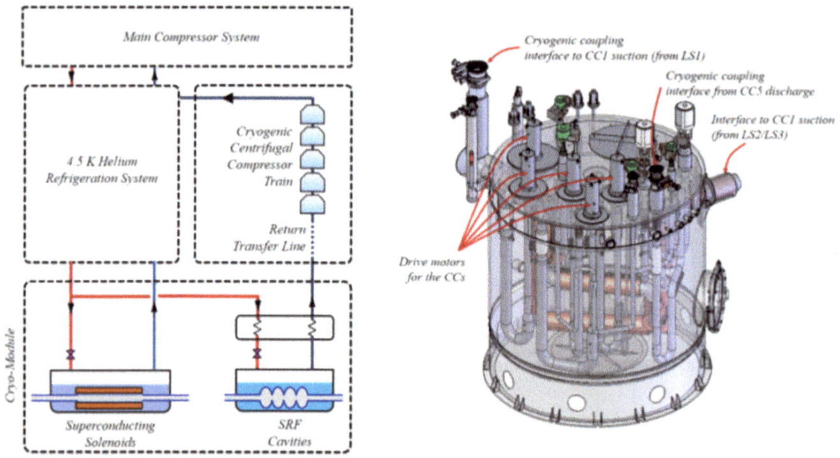

Fig. 4.11 Left: simplified schematic of the FRIB sub-atmospheric system. Right: a 3-D model of the FRIB 2 K coldbox [12]

applied to other large-scale subatmospheric systems that utilize multi-stage cold compressor technology, as the analysis applied minimal amounts of empirically developed correlations. More details on the development of predictive methodologies (mean-line compressor performance, subatmospheric system pressure, and pump-down paths), the selection of relative speed ratios and transient mass flow rate profiles, and additional case studies experimentally tested in FRIB can be found in [12, 14, 15].

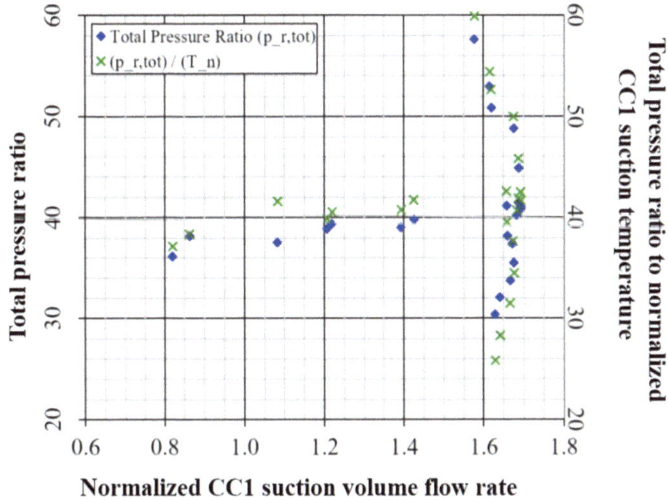

Fig. 4.12 Variation of the FRIB cold-compressor train total pressure ratio as a function of the normalized CC1 volumetric flow rate [13]

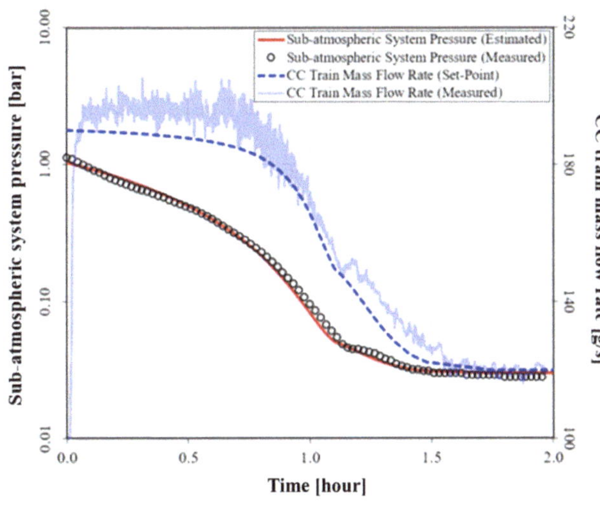

Fig. 4.13 Transient variation of FRIB linac cryomodule pressure and corresponding CC mass flow rate (model estimation versus as measured) [12]

4.8 Cryogenic Distribution

For large-scale systems, the transmission of refrigeration (cryogen) to/from the load is equally challenging as generating the refrigeration. The liquid helium required to support the refrigeration or liquefaction load at the end user is usually supplied through vacuum-jacketed transfer lines with multi-layer insulation (MLI) inside the annular space. The vacuum and MLI combination provides roughly three orders of magnitude lower heat flux compared to conventional thermal insulation systems, such as perlite powder, polymeric foams, or fiberglass, operating under ambient pressure conditions.

The challenge of providing vacuum jacketing with carefully executed MLI at every point along the distribution system is a main driver of the mechanical design of the overall system. These transfer lines come in many shapes, sizes, and forms, including small coaxial lines to very large and complex transfer line distribution systems used at locations such as CERN, RHIC, FNAL, CEBAF, and FRIB. The design aspects and considerations for these complex transfer lines are outlined in various publications (e.g., [16], including a detailed description of the FRIB transfer line designs [17, 18]). The transfer-line designs at CEBAF and SNS were based on concepts originally developed at FNAL. The interface connections from the refrigerator and the loads to the transfer lines were accomplished in various ways. They included permanent connections with isolation valves, joints (sometimes called "field joints") that involved both the vacuum jacket and the process line [16], or cryogenic couplings (also referred to as "bayonets"). Some of these bayonet-style couplings were designed for easy disconnection at positive pressures and at cryogenic temperatures [16], such as those used for CEBAF and SNS cryogenic distribution (which were scaled-up and adopted from smaller FNAL bayonet couplings). Other bayonet-coupling designs, typically provided by industry, can only be disconnected when the inner process line is at ambient temperature.

In general, it is observed that most of the heat in-leak in the distribution system is contributed by components such as valves and bayonet couplings, and to a lesser extent, by the insulated piping sections of the cryogenic transfer-line. Although relatively unpublished, it is not uncommon for the heat in-leak of the distribution system to be underestimated, sometimes significantly. To reduce the effect of heat in-leak, the cold temperature lines are thermally shielded (from radiation) by lines carrying higher temperature (but still cryogenic) fluid, and the valves and bayonets are thermally heat intercepted.

Long transfer lines can pose cool-down and stability issues in the operation of refrigeration systems, especially if the vacuums degrade. Since MLI requires a high vacuum (i.e., below about 10^{-5} mbar) to operate properly, vacuum degradation can be a serious issue for all elements of the distribution system. The technology of bake-out processes, getters, adsorbents, evacuation ports, safety relief devices, and vacuum pumping has been successfully developed in the last 50 years to protect against most vacuum failures. However, the long life and high reliability of high-vacuum systems remain a present challenge.

4.8.1 Example: FRIB Cryogenic Distribution System

The cryogenic distribution of the FRIB accelerator was divided into three segments, and an independent transfer connection was designed for each segment from the refrigerator to facilitate step-wise commissioning, troubleshooting, and maintenance [17, 18]. The transfer-lines and the cryomodules were connected with CEBAF / SNS style bayonet couplings to facilitate commissioning, operations, and maintenance of the cryomodules. The linac cryogenic transfer lines were composed of 46 standard sections to interface with the cryomodules. There were three shaft transfer lines that connected the above ground refrigerator to the three segments in the underground accelerator. The main parts of the FRIB cryogenic distribution system are shown in Figs. 4.14, 4.15 and 4.16.

Although the cryogenic loads from the FRIB experimental system were part of the same accelerator beamline, the corresponding cryogenic distribution system was separated into a fourth segment (with subsequent branches) following their geometrical arrangement along the accelerator beamline leading to the target, separator, and experimental halls (spectrograph, etc.) to simplify the installation, commissioning, and troubleshooting efforts and to meet a very aggressive project schedule. All four distribution segments (linac and experimental system) were independently connected to the FRIB cryogenic refrigeration system, and any one of them could be disconnected or operated under different process conditions.

The experimental system (target, pre-separator, and separator) SC magnets, supported by the fourth distribution segment, was significantly different in design and operating requirements. The cross section of the cryogenic transfer line and the layout of this distribution system are shown in Fig. 4.17. The cryogenic distribution system consists of six cryogenic process piping headers – two each (supply and return) for primary (4.5 K), thermal shield (helium, 50 K), and cool-down, enclosed in a single vacuum insulated pipe (vacuum jacket). The cool-down return header was also used during any upset condition (magnet quench) to segregate the quench flow from a specific magnet to the rest of the magnets – thereby ensuring stable 4.5 K operation for the remainder. The pressure pulse (and energy) due to a magnet quench was absorbed in the 10,000 liter liquid helium Dewar via the cool-down return header (Fig. 4.17). Bayonet-style couplings were used for each of the process headers at the cryo-plant, magnet, liquid-helium Dewar, and future expansion extension interfaces. Based on experience gathered from the design and operation of the cryogenic distribution system in CEBAF, NSCL, and SNS, this provided great flexibility in the commissioning and maintenance of the system. Several utility piping headers that ran along this distribution system for warm helium supply, cool-down/contaminated return from magnets, magnet lead flow, etc., were also incorporated into the distribution. The design, fabrication, and installation of most elements of the cryogenic distribution system were carried out from the ground up and in-house at FRIB. More details of the design of this distribution system can be found in [19, 20].

Fig. 4.14 Different sections of FRIB cryogenic distribution system – (**a**) straight section of the transfer line with (**b**) cryomodule interface boxes at intermediate locations, and (**c**) horizontal cross-over and tee branch for Linac Segment 3 cryogenic transfer line [17]

Several branches of the experimental system at FRIB had been reconfigured and integrated from legacy systems from the NSCL era [21]. The design of the cryogenic distribution for each of these legacy systems followed the same concept as discussed above. However, several additional components, such as liquid nitrogen thermosiphons for thermal shield flow management and flange-style cryogenic couplings, were designed and incorporated. The flange-style cryogenic couplings were used to meet the space constraints within the existing legacy experimental areas while maintaining a means of positive isolation between operating and non-operating cryostats. The geometrical layout of such a segment (reconfigured

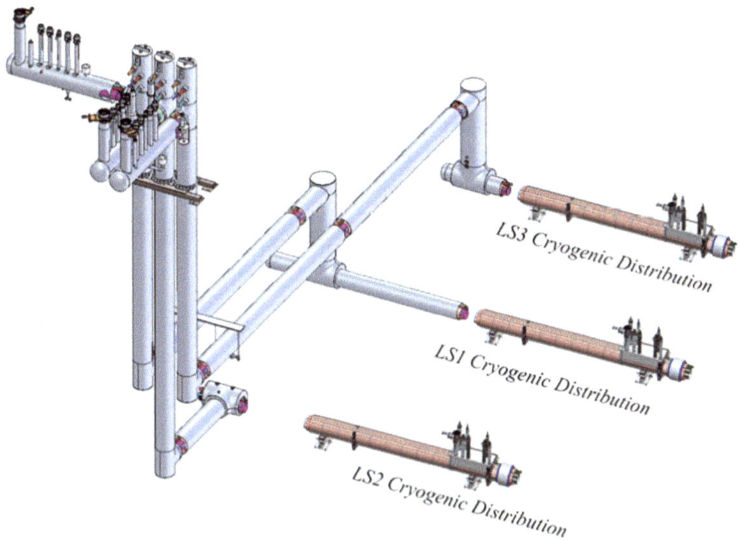

Fig. 4.15 Layout of the FRIB cryogenic distribution system for three linac segments

fragment separator A1900), together with magnet valve boxes and flange-style cryogenic couplings, is shown in Fig. 4.18.

4.9 Auxiliary and Utility Systems

Helium refrigeration systems require many auxiliary systems to support their operation and loads. They include helium gas storage, liquid cryogen (typically helium and nitrogen) storage Dewars, and helium recovery and purification systems. The size requirements for these systems depend on the cryogenic load and many other operational requirements of the facility.

As helium is becoming scarce, the operation of cryogenic systems is subject to uncertainties regarding the availability and wide price fluctuations of this valuable commodity. To support the uninterrupted and sustained operation of these systems, it is critical to preserve and recover the helium inventory whenever possible. In large-scale accelerator facilities, one of the major sources of helium loss is through cryostat pressure relief valves due to quenching in a superconducting magnet. Quenching results in a loss of both the cryogenic coolant, which is typically released into the atmosphere, and the stored refrigeration. At FRIB's experimental system, the thermal energy and boil-off flow released during the quench in the SC magnets were handled using a unique system [20, 22, 23], along with a concept that was originally developed and successfully demonstrated at the Superconducting

Fig. 4.16 Major components of the FRIB LINAC cryogenic distribution system – (**a**) FRIB cryogenic plant to the linac distribution interfaces, (**b**) JLab-SNS style cryogenic couplings between the FRIB linac cryogenic distribution and the cryomodules, and (**c**) a panoramic view of the linac cryogenic distribution segment 1 [18]

Super-Collider Laboratory (SSCL) for the magnet string test in 1993. This quench management system developed at FRIB consisted of a 10,000 liter liquid helium Dewar (buffer vessel) and an associated cryogenic distribution system. A 3-D model of the FRIB quench recovery Dewar (QRD) and the associated pressurization due to the experimental system magnet quench is shown in Fig. 4.19.

In the same context of helium preservation, the recovery and subsequent purification of helium to very low levels (less than 1 ppmv of contaminants) is important for cryogenic applications. Historically, contaminated helium from sources such as vapor-cooled magnet leads and cryostat cool-downs was not recovered. However, with the present helium economy, helium recovery and purification systems are becoming essential auxiliary systems in small- to large-scale accelerator facilities [25].

Helium refrigeration systems also require many conventional utility systems to support their operation and loads. They include electric power (both primary and standby), cooling water, and instrument air (Sect. 8.13).

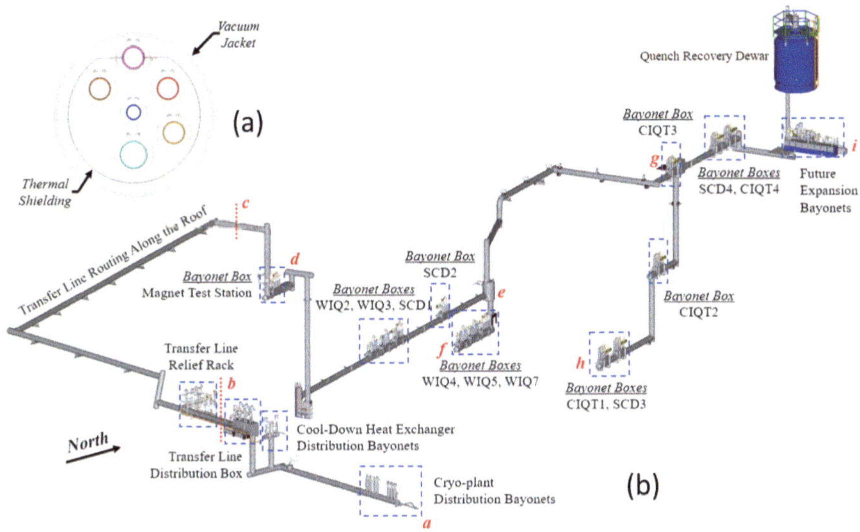

Fig. 4.17 (**a**) Cryogenic transfer line cross-section, and (**b**) layout of the FRIB target and pre-separator cryogenic distribution system [20]

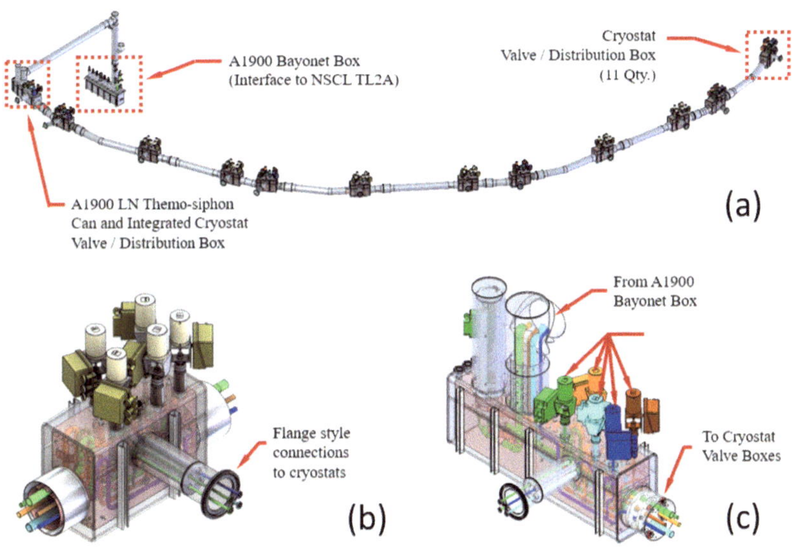

Fig. 4.18 Major components of FRIB A1900 fragment separator cryogenic distribution system – (**a**) cryogenic transfer line layout with magnet interfaces, (**b**) magnet valve boxes with flange-style couplings, and (**c**) unified liquid nitrogen thermo-siphon vessel with a magnet valve box [21]

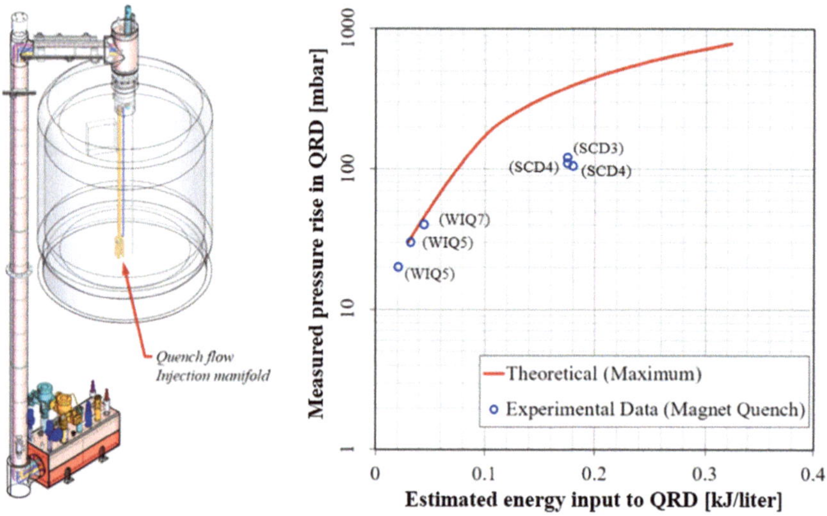

Fig. 4.19 Left: 3-D model of the FRIB quench recovery Dewar. Right: calculated and measured pressure rise in this Dewar due to quench energy addition [23, 24]

4.10 Project Planning

The design basis and sizing requirements depend on present loads, uncertainties, future projected expansions, and system design goals (efficiency, flexibility, reliability, cost, and schedule). The size, shape, and foundation requirements of the equipment, as well as the utility and auxiliary equipment, and the environmental conditions of the equipment space should be carefully analyzed. For large-scale accelerator facilities, appropriately sizing the cryogenic distribution system while considering potential future expansion is critical. Upgrading the cryogenic distribution system is significantly more complex than upgrading the cryogenic refrigeration capacities.

All of the cryogenic systems and subsystems discussed require space and conventional utilities (power, cooling water, etc.). These have substantial cost and schedule drivers and should be carefully considered at the beginning of project planning.

References

1. R. Barron, *Cryogenic Systems*. Monographs on Cryogenics (Oxford University Press, Oxford, 1985). https://books.google.com/books?id=SjjTwAEACAAJ
2. V. Ganni, P. Knudsen, Optimal design and operation of helium refrigeration systems using the Ganni cycle. AIP Conf. Proc. **1218**(1), 1057–1071 (2010). https://doi.org/10.1063/1.3422267

3. P. Knudsen, V. Ganni, Simplified helium refrigerator cycle analysis using the ‘Carnot step’ (2006). https://www.osti.gov/biblio/885354, https://www.osti.gov/servlets/purl/885354
4. R. Clausius, *The Mechanical Theory of Heat* (Macmillan, London, 1879)
5. V. Ganni, P. Knudsen, Helium refrigeration considerations for cryomodule design (2014). https://doi.org/10.1063/1.4860928, https://www.osti.gov/biblio/1995419, https://www.osti.gov/servlets/purl/1995419
6. P. Knudsen, V. Ganni, N. Hasan, M. Wright, F. Casagrande, G. Vargas, N. Joseph, FRIB helium refrigeration system commissioning and performance test results. IOP Conf. Ser.: Mater. Sci. Eng. **755**(1), 012090 (2020). https://doi.org/10.1088/1757-899X/755/1/012090
7. J. Homan, M. Montz, V. Ganni, A. Sidi-Yekhlef, P. Knudsen, S. Garcia, J. Garza, Commissioning of the liquid nitrogen thermo-siphon system for NASA-JSC Chamber-A. AIP Conf. Proc. **1573**(1), 345–351 (2014). https://doi.org/10.1063/1.4860721
8. P. Knudsen, V. Ganni, N. Hasan, K. Dixon, R. Norton, J. Creel, Modifications to JLab 12 GeV refrigerator and wide range mix mode performance testing results. IOP Conf. Ser.: Mater. Sci. Eng. **171**(1), 012015 (2017). https://doi.org/10.1088/1757-899X/171/1/012015
9. P. Knudsen, V. Ganni, N. Hasan, M. Wright, F. Casagrande, G. Vargas, N. Joseph, FRIB helium refrigeration system commissioning and performance test results. IOP Conf. Ser.: Mater. Sci. Eng. **755**, 012090 (2020). https://doi.org/10.1088/1757-899x/755/1/012090
10. V. Ganni, P. Knudsen, D. Arenius, F. Casagrande, Application of JLab 12 GeV helium refrigeration system for the FRIB accelerator at MSU. Adv. Cryo. Eng. **59**, 323 (2014)
11. V. Ganni, D.M. Arenius, B.S. Bevins, W.C. Chronis, J.D. Creel, J.D. Wilson, Design, fabrication, commissioning, and testing of a 250 g/s, 2-K helium cold compressor system. AIP Conf. Proc. **613**(1), 288–304 (2002). https://doi.org/10.1063/1.1472033
12. J.W. Howard, Characterization of a cryogenic centrifugal compressor train for reliable wide-range operation of helium sub-atmospheric systems. Ph.D. thesis (2024)
13. P. Knudsen, V. Ganni, F. Casagrande, A. Fila, N. Hasan, M. Wright, G. Vargas, N. Joseph, Design, fabrication, commissioning and testing of FRIB 2 K cold compressor system. IOP Conf. Ser.: Mater. Sci. Eng. **755**(1), 012092 (2020). https://doi.org/10.1088/1757-899x/755/1/012092
14. J. Howard, A. Engeda, P. Knudsen, N. Hasan, V. Ganni, Characterization of centrifugal compressors operating in FRIB’s sub-atmospheric compression system. IOP Conf. Ser.: Mater. Sci. Eng. **1301**(1), 012105 (2024). https://doi.org/10.1088/1757-899X/1301/1/012105
15. J. Howard, P. Knudsen, A. Engeda, Empirical loss model optimization for the prediction of centrifugal compressor off-design performance (2023). https://doi.org/10.1115/GT2023-103013
16. V. Ganni, J. Fesmire, Cryogenics for superconductors: Refrigeration, delivery, and preservation of the cold. AIP Conf. Proc. **1434**(1), 15–27 (2012). https://doi.org/10.1063/1.4706901
17. V. Ganni, K. Dixon, N. Laverdure, S. Yang, T. Nellis, S. Jones, F. Casagrande, FRIB cryogenic distribution system and status. IOP Conf. Ser.: Mater. Sci. Eng. **101**, 012034 (2015). https://doi.org/10.1088/1757-899x/101/1/012034
18. V. Ganni, P. Knudsen, F. Casagrande, S. Jones, Cryogenic design of FRIB cryomodule and distribution system and the present status. IOP Conf. Ser.: Mater. Sci. Eng. **755**, 012091 (2020). https://doi.org/10.1088/1757-899x/755/1/012091
19. N. Hasan, V. Ganni, A. Fila, F. Casagrande, Design of cryogenic heat exchangers and associated sub-systems for controlled cool-down and testing of superconducting magnets at FRIB. IOP Conf. Ser.: Mater. Sci. Eng. **756** (2021)
20. N. Hasan, M. Wright, V. Ganni, F. Casagrande, S. Jones, C. Nguyen, A. Fila, N. Joseph, Design, fabrication, and installation of the cryogenic distribution system for FRIB target and fragment pre-separator superconducting magnets. IOP Conf. Ser.: Mater. Sci. Eng. **1240**, 012083 (2022)
21. N. Hasan, M. Wright, V. Ganni, F. Casagrande, S. Jones, B. Laumer, C. Nguyen, A. Fila, N. Joseph, Design, fabrication and installation of the cryogenic distribution system for re-configured FRIB A1900 fragment separator. in *Proceedings of ICEC28-ICMC 2022*, ed. by L. Qiu, K. Wang, Y. Ma (Springer Nature Singapore, 2023), pp. 233–239

22. N. Hasan, V. Ganni, P. Knudsen, Analysis and management of thermal energy release during quench in a superconducting magnet, in *Proceedings of the 2022 ASME International Mechanical Engineering Congress and Exposition (Proceedings of IMECE2022)*, vol. 86700 (2021). https://doi.org/10.1115/IMECE2022-95762
23. N. Hasan, V. Ganni, P. Knudsen, F. Casagrande, B. Laumer, S. Bott, Commissioning and operational experience from FRIB target and fragment pre-separator superconducting magnet quench management system, in *Proceedings of ICEC28-ICMC 2022* (2023). https://doi.org/10.1007/978-981-99-6128-3_30
24. N. Hasan, V. Ganni, P. Knudsen, Analysis and management of thermal energy release during quench in a superconducting magnet, in *Proceedings of ASME IMECE 2022* (2022)
25. D. Kroll, N. Hasan, V. Ganni, A. Fila, A. Engeda, Freeze-out purifier for helium refrigeration system applications. IOP Conf. Ser.: Mater. Sci. Eng. **755**, 012056 (2020). https://doi.org/10.1088/1757-899x/755/1/012056

Chapter 5
Superconducting Radiofrequency System

Abstract In this chapter, we introduce the high-power hadron accelerator SRF technology state of the art, including cavities, RF couplers, tuners, and cryomodules, with focus on specific problems related with high beam current. Solutions adopted in FRIB and several other examples are presented, giving an overview of the work done in this field in most of the existing and planned facilities.

Two critical aspects strongly influence the SRF technology of high-power hadron accelerators (HPHAs). The first is the need to limit the beam losses to negligible fractions of the beam current. The second is the need to couple very high RF power to the resonators to cope with high beam loading. Both aspects influence the cavity and cryomodule (CM) design, requiring large beam apertures and large forward power couplers. Especially in low-β cavities, a large aperture affects the cavity shunt impedance and its maximum achievable operating gradient. However, for beam dynamics reasons, a smooth acceleration is required at low-β in order to preserve good emittance, and the use of a very high gradient in low-β cavities is often impossible in HPHAs. Most of the beam power is usually delivered more efficiently by high beta sections, which are made of large cavities coupled to very large RF sources. For multi-cell elliptical ones, a large beam aperture is usually not an issue. In both low-β and high-β cases, difficulties are mainly related to the need to handle high RF power and to accommodate RF couplers in CMs with sizes comparable, in some cases, to the cavity size itself.

5.1 Superconducting Resonators Technology

5.1.1 *Bulk Nb Resonators*

SC resonators used in HPHAs (see Sect. 2.2) are typically built with high-purity niobium (RRR > 250) [1]. The parts are formed and machined starting from sheets and rods. After cleaning, they are joined together by means of electron beam

J. Wei et al., *High Power Hadron Accelerators*, Particle Acceleration and Detection,
https://doi.org/10.1007/978-3-032-17662-2_5

welding under high vacuum to preserve metal purity. The electromagnetic design (EM) of the resonator aims to minimize E_p/E_a and B_p/E_a while keeping R_{sh} as high as possible. The shape is also optimized to minimize multipacting or to keep multipacting levels far from the operating voltage. The mechanical design aims to minimize the sensitivity of the cavity frequency to pressure, df/dP, and the Lorentz force detuning coefficient, $df/dE_a{}^2$, while moving the frequencies of resonant mechanical modes as high as possible. The helium vessel surrounding the cavity, usually made of stainless steel or titanium, can be shaped and welded to the cavity walls to reinforce the entire structure. The cryogenic design aims to cool the cavity efficiently, especially in areas of high RF power dissipation, while guaranteeing He gas exhaust and maintaining stable cryogenic and thermal flow.

Although there are many variations, and considering that the technology of resonator processing is in continuous evolution, a somewhat "present standard" procedure for the treatment of Nb resonators, performed not only in the laboratory environment but also by industry as part of the production of SC cavities, can be briefly summarized. After fabrication, cavities undergo surface treatment – chemical polishing (CP) or electropolishing (EP) [1, 2] – that removes approximately 100–200 μm of material from the RF surface to eliminate damaged layers created during forming. This treatment also helps remove contaminants introduced during fabrication and improves surface roughness. To prevent Q-disease (see Sect. 2.2.4.4), hydrogen is removed from the bulk by heat treatment at 600–800 °C in a high-vacuum furnace for several hours. The cavities are then subjected to a light chemistry, either CP or EP (5–10 μm), followed by a high pressure water rinse (HPR) for several hours to remove particulate contamination from the surface. In some cases, a final low temperature baking at ∼120 °C in vacuum is performed to reduce high-field Q slope [2]. After processing, the cavities are ready for clean assembly in their CM (Fig. 5.1).

5.1.2 *Alternate Resonator Technology: Nb Sputtering on Cu in QWR*

Elliptical cavities, as well as QWRs with a large aperture closed by a removable bottom plate ("open" QWRs), allow the introduction of large electrodes. This makes them suitable for the deposition of a thin (∼1 μm) layer of superconducting material like Nb on their RF surface using sputtering technology [4]. There are several possible advantages of Nb sputtering over bulk Nb. The SC material substrate can be made of OFHC copper, which is characterized by a thermal conductivity much higher than that of Nb. This improves the cooling of the SRF surface and allows the use of thick walls in resonators, reducing their pressure and Lorentz force detuning coefficients df/dP and $df/dE_a{}^2$. The lower cost of the resonator material allows for the construction of cavities machined from a single piece of copper, with a minimum number of welds and reduced construction costs. Moreover, sputtered

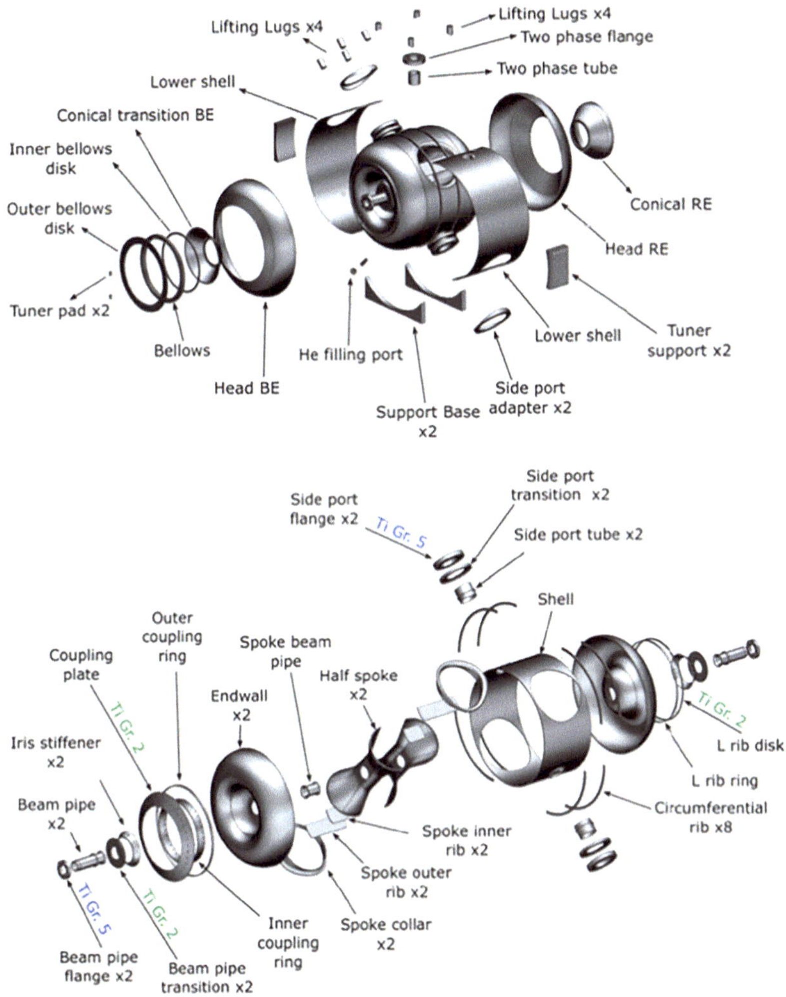

Fig. 5.1 HPHA low-β resonator examples with the 325 MHz, $\beta_0 = 0.475$ single spoke resonator SSR2 developed at FNAL for the PIP-II project with (top) exploded view of the jacketed cavity and (bottom) exploded view of the bare cavity, showing parts to be EB welded in high vacuum. The components with no indication about the material are made from RRR grade Nb [3]. (Courtesy FNAL, USA)

Nb cavities have shown a reduced sensitivity of SRF performance to an external magnetic field and a substantial absence of Q-disease. However, except for the case of low frequency QWRs, sputtered Nb cavities have shown significantly lower SRF performance than bulk Nb ones, with a strong Q-slope and, in general, a lower operating gradient. Nb sputtering technology is not yet as mature and reproducible as bulk Nb technology, and it is presently limited to resonators with simple shapes and large apertures. However, it is still evolving and remains very promising. Nb sputtering was first used for 352 MHz elliptical cavities of LEP at CERN [5] and, later, in two low current, low-β QWR linacs [6, 7]. This technology is definitely suitable for HPHAs where a large aperture and moderate gradient are required, and sputtered QWRs for high-power applications have already been proposed [8].

5.1.3 *Frontiers in SC Resonators Technology*

There are presently two main fronts that are rapidly progressing toward substantial improvements in cavity performance; both are related to bulk Nb cavities. The first is nitrogen doping [9]; the second is Nb_3Sn coating [10].

5.1.3.1 N-Doping

It consists of a recently discovered surface treatment of Nb that is able to raise the quality factor of the cavities beyond the value that could be calculated for bulk Nb using the standard BCS formula, which was previously considered the upper possible limit. N-doping consists of the controlled inclusion of doping elements (in principle, N atoms, but other contaminants might also play an important role) in a thin layer of Nb on the RF surface of the resonator. The procedure is relatively simple, based on exposure to a low pressure of N gas for a limited time when a resonator is heated in a high-vacuum furnace to an elevated temperature. R&D is still ongoing to find the optimal N-doping parameters, with continuous improvements in performance. Doubts about the long-term stability of this treatment have been raised, which will only be fully confirmed after many years of operation. Although understanding the physics behind this remarkable increase in Q is not yet complete, the technology developed so far in elliptical cavities is already sufficiently advanced for mass production and is relatively inexpensive. The technology of N-doped cavities has been transferred to industry for electron accelerator projects [11] and could also be used profitably to reduce the cryogenic load in large HPHAs.

5.1.3.2 Nb_3Sn Cavity

This material is a superconductor with a higher critical temperature, a higher critical magnetic field, and a lower $R_s(T)$ compared to pure Nb, potentially allowing high gradient operation at 4.2 K with high frequency cavities instead of 2 K. At present, Nb_3Sn coated prototype cavities, realized by post-processing bulk Nb cavities, can already reach very high Q at 4.2 K with little or no Q slope, but they cannot yet achieve a very high gradient and tend to quench above $B_p \sim 70\,\text{mT}$ surface magnetic field [12, 13]. This technology might find useful applications, especially in elliptical cavities for medium gradient CW HPHAs, where operation at 4.2 K could considerably reduce the complexity and operational cost of the cryogenic system.

5.2 Superconducting Resonators Ancillaries

5.2.1 *High Power Couplers*

Fundamental power couplers (FPC) have the function of efficiently transmitting the RF power coming from a waveguide fed by an RF power source to the resonator (see Sects. 2.2.3.4 and 8.5). Because RF waveguides are in air at room temperature and the interior of the SRF cavity is in a vacuum at cryogenic temperatures, FPCs must also provide vacuum and thermal separation. The transfer efficiency of RF power is particularly critical in SC cavities: the challenge of high power (up to $\sim$1 MW) is increased by the requirement of a low thermal load to the cryogenic system (typically aiming at $<$1 W per FPC delivered to liquid helium).

Minimizing reflected RF power is, of course, highly desirable to minimize wall plug power consumption, and optimum coupling is determined by beam loading. However, if the optimum loaded resonator RF bandwidth is insufficient to achieve reliable cavity control (e.g., in the presence of mechanical noise typical of an accelerator environment), some over-coupling is required to make this bandwidth broader, at the cost of increasing the reflected power and the RF losses in the FPC antenna.

5.2.1.1 RF Transmission Lines

RF power is transported from the RF source to the cavities by RF transmission lines, either of rectangular or coaxial types. Rectangular waveguides, used in the TE10 mode, are very efficient and easy to cool, but they have a low-frequency cutoff $f_c \sim c/\lambda_c = c/2a_w$ that depends only on their longer-side width a_w [14]. The resulting constraint $a_w > \lambda/2$ makes them quite large and impractical below 350 MHz RF frequency. In contrast, coaxial waveguides used in the TEM mode have no low-frequency cutoff. Although coaxial lines exhibit higher RF power losses than rectangular waveguides of similar size, they can be chosen with a diameter well below $\lambda/2$, making them ideal for low frequencies. In typical 50 Ω coaxial lines, about 2/3 of RF losses are concentrated in the inner conductor (IC), which is more difficult to access and cool. Air and low-loss solid dielectric materials in coaxial lines of smaller diameter provide heat removal and prevent multipacting.

5.2.1.2 RF Coupling

FPCs are either of the waveguide or coaxial type. Waveguide coupling consists of a fixed aperture between the waveguide and the cavity, which allows for proper superposition of their EM modes. Coaxial couplers consist of coaxial lines terminated inside the cavity and coupling either to its magnetic field, by means of a conductive loop (inductive coupling), or to its electric field by means of a rod-shaped antenna (capacitive coupling). Compared to waveguide couplers, coaxial couplers are smaller in size and provide greater thermal resistance between the ambient temperature environment and the cold cavity. Moreover, their coupling strength β can be easily adjusted by varying the penetration of the antenna in the cavity. The outer conductor (OC) of the FPC must be grounded and in perfect contact with the resonator RF port to avoid RF losses and EM field irradiation. This creates thermal continuity between the cold SC cavity and the outside world, which is becoming a major source of cryogenic load and a significant constraint in the FPC design.

5.2.1.3 RF Windows

RF couplers include one or two RF windows, which provide separation between the cavity interior, which is under high vacuum, and the RF transmission line, which is under atmospheric pressure. In the single-window case, the window itself must necessarily be at room temperature – one side is in air – and must be kept far enough from the SC cavity to limit thermal flow. Windows can be made with a planar shape (the most common in SRF), a cylindrical shape, or a conical shape. Windows are usually made of high purity Alumina (Al_2O_3), a dielectric material with good mechanical properties, low RF losses ($\tan \delta_{RF} \sim 10^{-4}$) and good thermal conductivity in the relevant temperature range. RF windows must be brazed to metal flanges, which often include flexible parts to compensate for differential thermal expansion/contraction between metal and ceramic.

Because the secondary electron emission coefficient (SEY) of alumina is well above unity, the risk of multipacting is enhanced on RF windows. This can be mitigated by the deposition of a very thin layer (a few nm) of low SEY material (typically TiN) on the alumina surface, in addition to careful EM design of the surrounding volume.

A vacuum failure in a single-window FPC would, of course, directly expose the cavity and its beamline to contamination from outside.

In a double-window system, a second window (usually called the cold window and kept at cryogenic temperatures) is located between the cavity vacuum and the CM insulation vacuum. The RF transmission line section in between is located in the insulating vacuum volume (see Sect. 5.4.6) and is usually pumped through slots in its outer conductor. The cold window reduces thermal radiation and outgassing toward the cavity, and it reduces the risk of beamline contamination in the event of a vacuum failure of one of the windows. This allows more options for the penetration of the FPC inside the vacuum vessel and makes its assembly easier. The cooling of the cold window can be achieved through a thermal link to the FPC outer conductor, which can be directly accessed by cryogenic fluids. In capacitive couplers, the antenna protrudes into the cavity vacuum without touching its walls, and even when kept at a much higher temperature (e.g., at room temperature), it delivers a quite

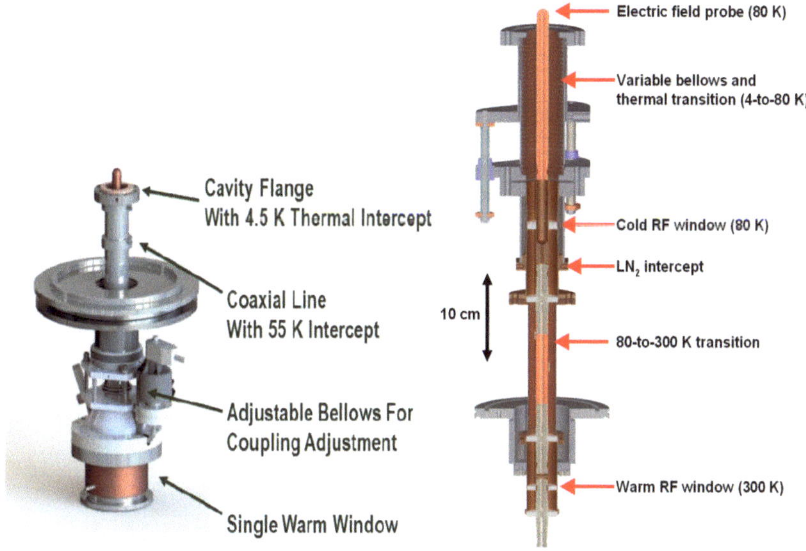

Fig. 5.2 Examples of (left) single-window and (right) double-window coaxial couplers fed by coaxial RF lines [15, 16]

acceptable heat load to the cavity through thermal radiation. Depending on its RF losses, the antenna can be cooled either simply by thermal conduction through the RF window or by flowing gas or liquid coolant (e.g. water) inside it. In inductive couplers, the cooling fluid can be brought into the inner conductor through the loop side connected to the outer conductor.

Two-window couplers are usually made of two parts that can be mechanically separated. The "cold" part with its window can be installed in the cavity RF port aperture inside the clean room, thus protecting the cavity RF surface from contamination during its subsequent installation in the CM. This simplifies the overall mechanical design of the CM and its assembly procedure at the cost of increased complexity and expense of the FPC design (Fig. 5.2).

In SRF HPHAs, the RF frequency is generally well below 1 GHz, and FPCs are of the coaxial capacitive type in most cases.

When external waveguides are of the rectangular type, a rectangular-to-coaxial transition section is required (see Fig. 5.3). To optimize RF matching, the diameter of the initial part of the coaxial inner conductor can be increased ("doorknob"). The transition can also be designed as an impedance transformer to reduce multipacting in the coaxial antenna in a vacuum, as described later in this section.

In addition to the RF current required for coupling, parasitic currents may be induced by the cavity magnetic field on the antenna surface – usually made of copper – causing additional RF heating, which strongly depends on the antenna shape and location. Losses are higher in inductive couplers, located in a high B volume of the cavity, than in capacitive couplers working in a high E, low B volume.

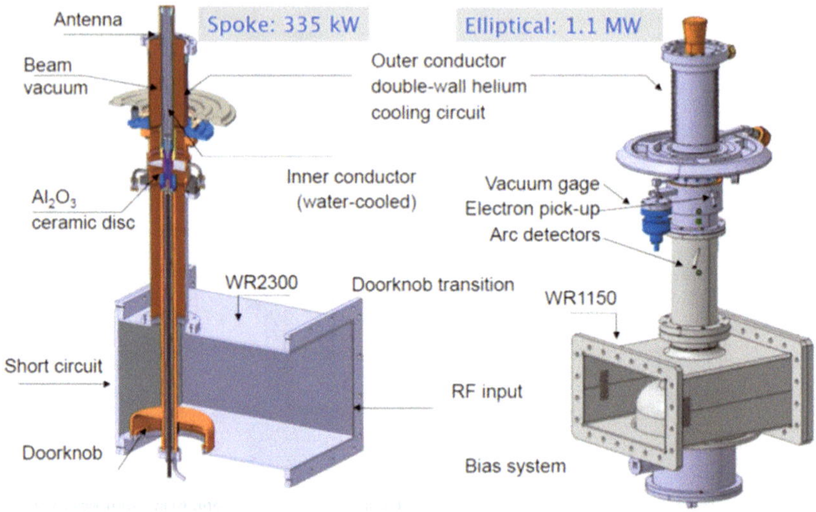

Fig. 5.3 ESS single-window, water cooled FPCs for SC spoke (352 MHz) and elliptical (704 MHz) cavities. Both include a "door-knob" rectangular to coaxial waveguide transition [17]. (Courtesy C. Darve and G. Olry, ESS)

FPC antennas can distort the cavity field. In elliptical cavities, to avoid surface peak field enhancement and the risk of sparks, antennas are located in the beam pipe, in the cut-off region outside the cells where the EM field gradually decays to zero.

5.2.1.4 Fixed, Variable and Adjustable FPCs

In fixed couplers, the coupling strength is permanently fixed by design. In variable couplers (rare in HPHAs), the coupling strength can be modified remotely during operation by means of actuators, which can vary the penetration into the cavity. This can be very useful when changing the operating parameters of the resonator. For example, the reflected RF power can be minimized for different beam current settings. Moreover, to address problems of cavity detuning due to mechanical instabilities, the RF cavity bandwidth can be increased by enhancing the coupling. Among several virtues, the drawback of variable couplers is their increased mechanical complexity, which is hardly affordable in very large power FPCs. An intermediate configuration, which can usually be accomplished with moderate cost and increased risk, is one of the "adjustable" couplers designed with the possibility of manually adjusting their coupling strength during CM maintenance periods without breaking the cavity vacuum.

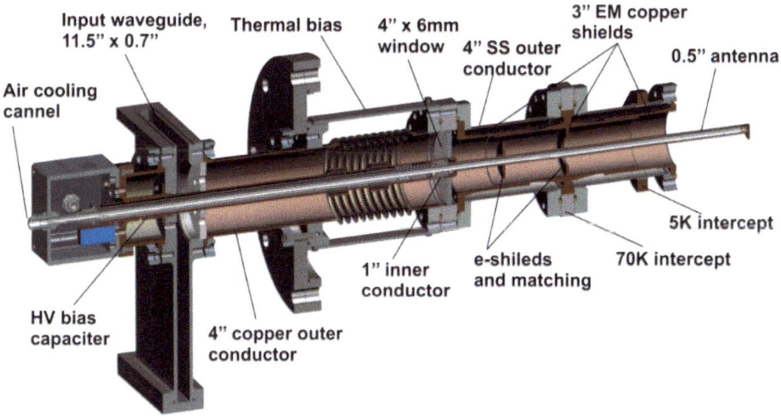

Fig. 5.4 Cut-away of the FNAL PIP-II FPC for 650 MHz elliptical cavities [18] (Courtesy O. Pronitchev, FNAL). To be noted: (1) rectangular to coaxial waveguide transformer section; (2) high voltage bias; (3) air cooling of the room temperature IC; (4) warm ceramic window; (5) OC's 70 K and 5 K thermal intercepts; (6) bellows for coupling adjustment; and (7) shields for protection of the cold window from backstreaming electrons

5.2.1.5 Mechanical Design Issues

To minimize thermal load from the room temperature environment to the superconducting cavity, the coaxial FPC outer conductor line includes a section made of a thin, non-magnetic stainless steel tube with high thermal resistance (e.g., SS316L). The RF surface is, in most cases, coated with Cu to a thickness of several times the RF skin depth $\delta_s(\omega)$, representing a compromise between minimizing RF losses and increasing conduction heat leakage. To accommodate differential thermal contraction during cool-down/warm-up, bellows must be included in the structure. In some cases, they allow for sufficient displacement for coupling adjustment. An FPC for SRF cavities, with all accessories and high power capability, may result in a rather complex and expensive mechanical piece (see Fig. 5.4).

5.2.1.6 Cooling

Cooling and correct temperature distribution in FPCs are provided by thermally linking their outer conductor to one or more cold parts inside the CM and, when needed, by active cooling with helium (liquid or gas) or even liquid nitrogen when present. The inner conductor, which is the main source of RF losses, is more difficult to access and can be cooled either by conduction through the Alumina windows (transferring the heat to the outer conductor) or directly, with gas or even water injected into the inner conductor from outside the CM. In this case, the cooling line to the inner conductor must first pass through the outer conductor without creating an electrical short between the two. If a rectangular-to-coaxial transition is present,

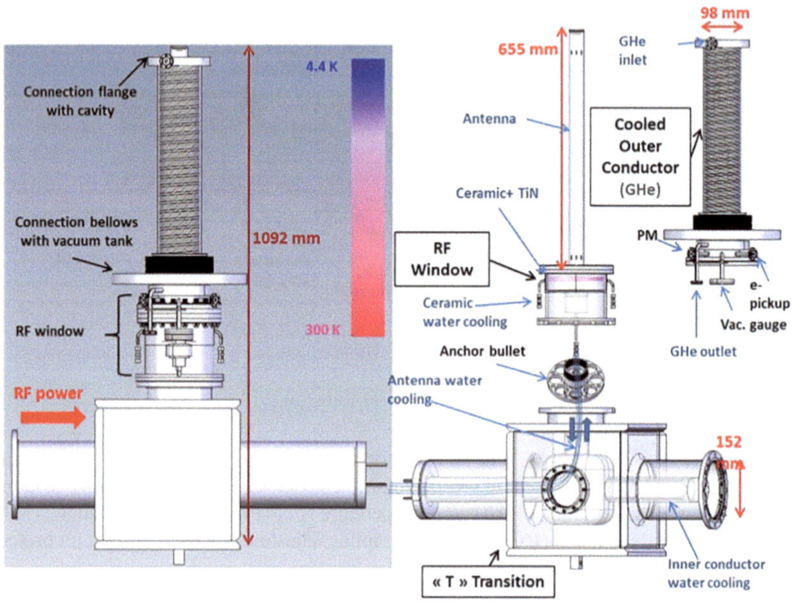

Fig. 5.5 The 100 kW FPC for the IFMIF-EVEDA, 175 MHz HWR [19]. Water-cooling is supplied to the antenna through a 175 MHz quarter-wave stub (note that in the two figures the coupler is observed from two opposite sides. (Courtesy H. Jenhani, CEA/F4E)

the IC entrance on the wall of the rectangular waveguide can be used (Fig. 5.5). In a fully coaxial line system, this can be achieved using a dedicated $\lambda/4$ stub, tuned to the cavity RF frequency, and thus unable to transmit the RF power of this frequency to the ground.

5.2.1.7 Electromagnetic Design Issues

The EM design must provide proper RF coupling and matching, tolerable multipacting levels, and RF power transfer with a minimal cryogenic load. In addition, it must avoid peak field enhancement in the cavity and locate the dielectric RF windows far from the high-voltage nodes (RF losses in dielectric materials are caused by the electric field). The direct line of sight to the windows should be avoided because electrons generated in the cavity could be accelerated toward the coupler interior and damage the alumina. The high dielectric constant of alumina may require some local design changes in the coaxial line to provide good impedance matching. The overall FPC design can result in a rather complex structure and requires accurate EM, mechanical, and thermal simulations by means of multi-physics codes. The design must take into account not only steady operation but also several transient situations, such as RF conditioning (e.g., high pulsed power processing requiring gradients well above the operational values) and operation with low beam loading

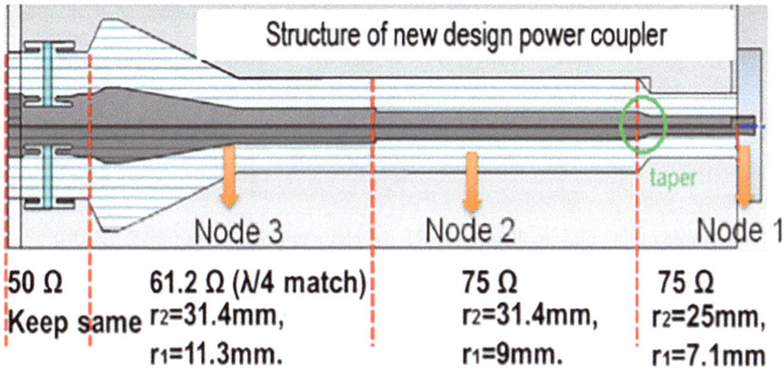

Fig. 5.6 Example of impedance transformation through a λ/4 section along the FPC antenna to reduce MP: the FRIB HWRs coupler [15]. (Courtesy Z.H. Zheng)

(generating reflected power and standing waves with high voltage and high current nodes in the FPC), etc.

5.2.1.8 Multipacting in FPC

Multipacting (MP) can completely jeopardize an FPC design if it cannot be completely overcome. Calculations show that, in many cases, it is nearly impossible to remove all MP levels in a coaxial geometry. However, by a proper choice of the line diameter and impedance, MP bands can be moved to a very low field, thus facilitating conditioning with low and harmless RF power, and to a very high field above the operating conditions. Therefore, although starting with the standard 50 Ω impedance at their warm end, FPCs can include sections with higher impedance in the vacuum part (typically between 50 Ω and 100 Ω) as shown in Fig. 5.6.

An effective way to shift the MP above the operational fields in capacitive couplers is DC-bias, i.e., the supply of a DC voltage to the antenna of up to a few kV (the energy range in which the SEY is significant). This bias modifies the energy of the electrons that interact with the RF surfaces and prevents their capture in resonant trajectories. DC bias can be accomplished by electrically isolating the antenna from the RF line inner conductor by means of a capacitor in series, which is able to transmit the RF signal but not the DC components, and connecting it to a DC voltage source via a cable shielded from RF (see Fig. 5.7). As for water cooling, this can be done directly in rectangular-to-coaxial transitions where one side of the antenna is accessible (see Fig. 5.7) or, in the case of coaxial systems, inside a resonant quarter-wave stub tuned to the resonator frequency or even through an inductive line. With DC bias, any water cooling of the antenna requires an electrically isolated local water circuit at the same DC potential to avoid shorts.

Before the installation of high-power FPCs in SC cavities, preliminary MP conditioning is usually performed in a dedicated normal conducting cavity, characterized

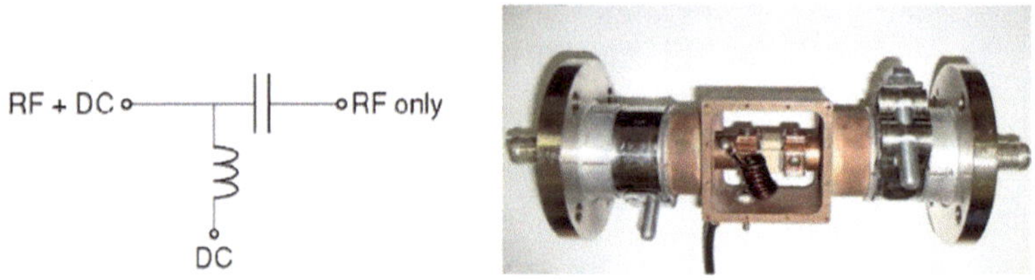

Fig. 5.7 Example of DC-bias application in RF coupler's coaxial lines of SARAF [20]

Fig. 5.8 The FRIB FPC coupling cavity system for MP conditioning of two couplers [21]. (Courtesy MSU, USA)

by an electromagnetic field geometry around the coupler that is similar to the final one in operation. This allows for coupler conditioning with no risk of damaging the SC cavity and without the use of cryogenic power. As shown in Fig. 5.8, the conditioning cavity hosts two couplers, one connected to the main RF source and the second to an RF load. Changing the RF frequency and the load impedance allows one to excite and condition all possible MP modes that are expected in the cavity during operation. Strong and identical coupling of both FPCs to the cavity makes RF transmission from one FPC to another nearly 100%. The memory of this preliminary conditioning is, to some extent, preserved after the clean installation of the FPC in the final cavity, considerably reducing the final conditioning time.

5.2.1.9 FPC Diagnostics

Strong MP and sparks can cause overheating, metal sputtering, and RF windows to break, which, in the worst cases, may require the replacement of the FPC and the removal and reprocessing of all cavities of a CM. Electron probes (small antennas), arc/spark detectors (optical fibers looking inside the EM volume to detect light when, e.g., RF breakdown occurs), and vacuum gauges are usually installed in high-power FPCs to promptly detect electron activity, combined with anomalies in the RF signals. This information is used by the RF control and the machine protection systems to stop or limit RF power when necessary to prevent possible damage to resonators, FPCs, and other systems during operation.

5.2.2 *SRF Cavity Tuners*

Frequency tuning in SC resonators must be performed without contamination of the cavity interior or enhancement of the peak fields and multipacting. Thus, movable plungers with sliding RF contacts, which are common tools in NC cavities, cannot be used. SRF cavity tuners usually act by deforming the cavity walls using a mechanical device applied to an area with low RF power density, avoiding a reduction in cooling capability at critical points but with a sufficiently high electric field to achieve adequate tuning sensitivity. The tuning works according to the Slater perturbation theorem [14] for small deformations:

$$\frac{\Delta f}{f} = \frac{\mu H^2 - \varepsilon E^2}{4U} \Delta V_{\mathrm{RF}}$$

where ΔV_{RF} is the amount of RF volume removed by this deformation, ε is the permittivity, μ is the permeability, H and E are the magnetic and electric field strengths in that volume, respectively, and U is the RF stored energy in the resonator.

High-Q SC cavities with narrow RF bandwidths require a tuning step on the order of a few Hz; therefore, their tuners provide micrometric displacements that must be precise, stable, reproducible, and backlash-free. In most cases, the required force is rather large, as the cavities must be made intrinsically stiff to minimize unwanted detuning from the helium pressure and the Lorentz force.

Many different types of mechanical tuners have been developed for SC resonators, satisfying a wide variety of cavity shapes, sizes, and performance requirements [22, 23].

In elliptical cavities, tuning is always performed by varying the cavity length. The tuner squeezes or stretches the two cavity ends along the cavity axis in a spring compression-like action. This primarily varies the cavity volume occupied by a high electric field, as shown in Fig. 5.9.

In TEM cavities, deformation is usually applied in the beam port region, where $\mu H^2 \ll \varepsilon E^2$ and the RF surface power density is low. Fine-tuning can also be

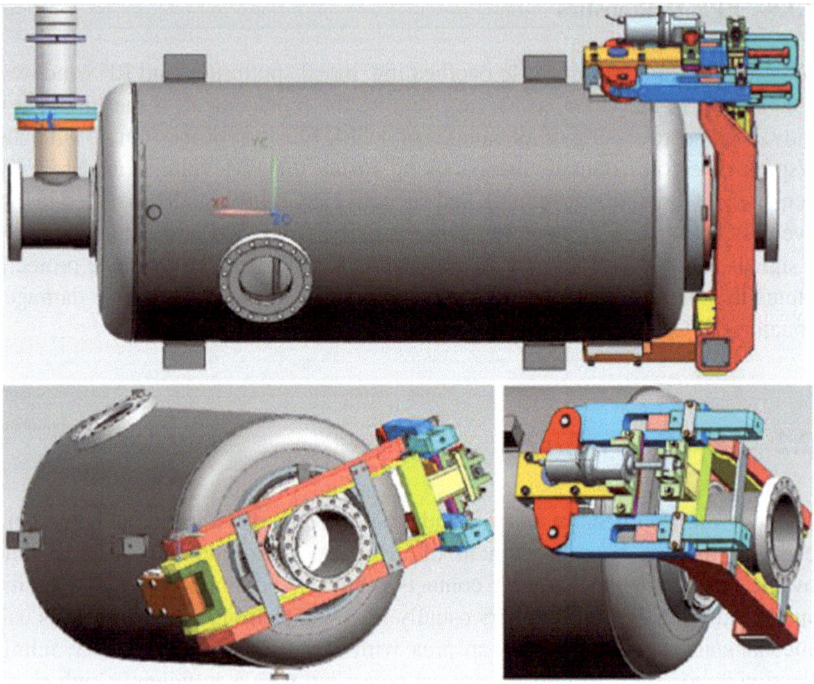

Fig. 5.9 Example of tuner for elliptical cavities: design of the lever tuner for the PIP-II at FNAL mounted on an elliptical cavity with helium jacket [24]

performed in open QWRs by deforming the thin bottom plate, as shown in Fig. 5.10. In some special cases of cavities with very high stiffness or space limitations, deformable bellow plungers have been developed in high field volumes of E, and movable plungers have been inserted in a high volume of B, with acceptable losses in RF.

Possible issues in an SRF tuner design may result from excessive cryogenic load, the presence of poorly shielded magnetic components, lack of accessibility for maintenance and repair, excessive backlash, insufficient resolution, and the excitation of mechanical modes in cavities due to rough tuner actuation.

Slow tuners are actuated in most cases by stepper motors or gas pressure (pneumatic tuners). Stepper motors are located either outside (warm) or inside cryostats (cold). Stepper motors, combined with mechanical reducers, are generally slower compared to other types of actuators, but they can produce a huge force and maintain the tuner position even in the absence of a driving signal. They are well suited for cavities with large bandwidths, which require frequency adjustment only from time to time. In pulsed or in high loaded-Q machines, they are often combined with piezoelectric tuners that can provide additional fast tuning within a small frequency range (see Fig. 5.11).

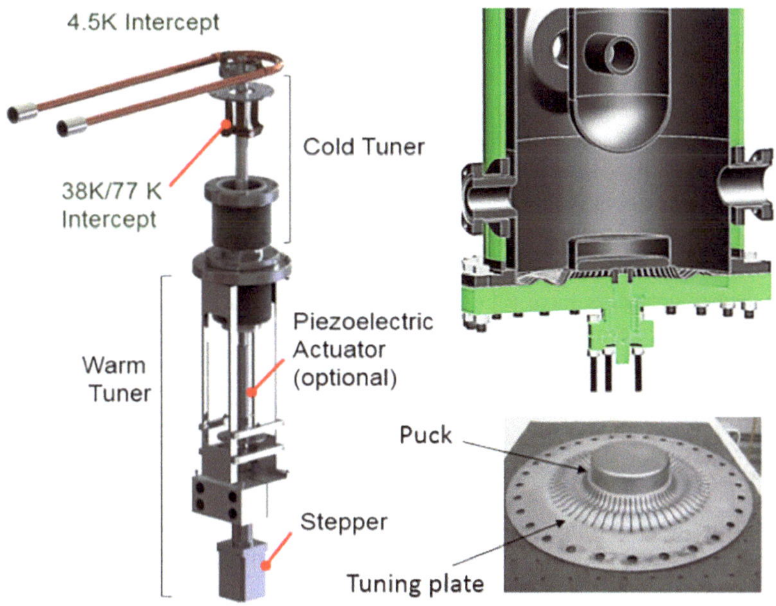

Fig. 5.10 Example of tuner for open QWRs with warm motor acting on a thin tuning plate: the ReA QWR tuner [25]

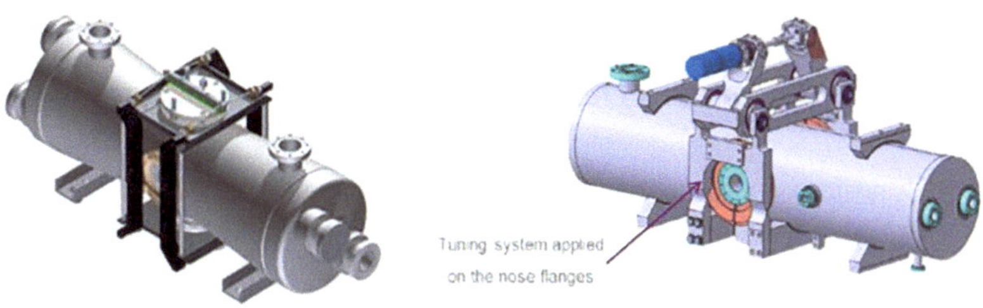

Fig. 5.11 Examples of tuners on HWRs. Left: pneumatic [26]; right: with cold stepper motor [27]

Pneumatic tuners [26] are based on pressure-controlled actuating bellows. They can exert a very large force at a relatively high speed, with a large stroke in one direction only, and make use of the elastic restoring force of the cavity to move in the opposite direction. Consequently, the tuning speed in the two directions may differ significantly and strongly depends on the cavity pre-load. All mechanical parts and the actuating bellows are located inside the CM, but the gas pressure controller is at room temperature. In comparison to stepper motor-based tuners, their frequency tracking is dynamic, and the gas pressure must be controlled continuously. They are particularly suitable for SC cavities with narrow RF bandwidths, which require frequent or continuous retuning for frequency tracking. One general benefit of this type of tuner is that it does not generate a mechanical impulse like a stepper motor, which may introduce additional vibration unless carefully designed.

Fast tuners have the function of tracking the cavity frequency in the presence of fast detuning, e.g., that induced by resonant mechanical modes or the fast LFD caused by EM field transients in pulsed accelerators. In HPHA cavities, the systems are based on piezoelectric actuators; however, other types have been developed and used in the past. Fast tuners are installed mainly in elliptical cavities operated in pulsed mode; however, they have also been developed for TEM cavities and are planned for low- and medium-β cavities of future pulsed SRF machines. Piezoelectric actuators have a large pushing force and a rather fast response, but a very small stroke on the order of 100 μm. When they are present, they always complement mechanical tuners. They are usually mounted inside the cryostat, close to the cavity, and at cryogenic temperatures, which reduces their stroke by 70∼90%. This can be mitigated by a special thermal design to maintain their temperature higher than the cavity temperature. Piezoelectric actuators must be properly cooled during operation to avoid rupture.

Because they are subject to wear, care must be taken to make these parts easily accessible and maintainable through additional ports in the cryostat.

5.3 HPHA SRF Resonator Example: FRIB Resonators and Ancillaries

5.3.1 FRIB Resonators Design and Production Strategy

The FRIB project aimed to build, on time and on budget, a large HPHA user facility with high availability and reliability. A main constraint was the need for a design that allowed for the industrial production of nearly 400 SRF cavities of four different types, with limited costs and minimal risk of delays or under-performance. Resonator technology and shapes were chosen to facilitate production and to be cost-effective, even accepting a slight reduction in the maximum SRF performance. The standard bulk Nb technology was chosen, and the resonator parts to be electron-beam welded together were made mostly of cylinders, toroids, and semi-spheres produced by standard mechanical techniques well mastered by the industry and adapted for ultrapure niobium. These simple and symmetric shapes could withstand high pressure with a minimal amount of material and allow resonator tuning in a relatively straightforward manner (see below). Separate beam line vacuum (see Sect. 5.4.3) was chosen to better preserve the cavity interior in the long term, although the best-performing QWR SRF linac in operation at that time (ISAC-II at TRIUMF) was still working with a common vacuum. Optimization of the system determined the choice to operate for the first time a low- and medium-beta heavy-ion SRF linac at 2 K temperature (superfluid helium), with increased performance and pressure stability. The system can also work at 4 K, which is very convenient during linac operation setup and conditioning, although with reduced performance in the HWR sections at 322 MHz. The most widely used configuration for user operation is currently 4 K for the 80.5 MHz QWR section and 2 K for the 332 MHz HWR section.

The resonator production was preceded by a prototyping phase, which included several steps of validation, ending with an integrated test of the fully equipped cavity in operation for 24 hours, locked at its final operating conditions (gradient, frequency/phase, and power within design intervals as in FRIB), with no interruptions. After the successful test, a pre-production run of eight cavities was launched and validated before starting large-scale production. FRIB personnel transferred to the industry the know-how for cavity construction and assisted in critical construction steps, such as cavity tuning and the planning of critical welds. Two or more manufacturers were always in charge of the construction of each cavity type to expedite parallel production and to avoid the risk of significant delays in the event of the withdrawal or inability of one of them. The niobium material was purchased directly from different vendors by FRIB personnel, who tested and validated the quality of the material before delivering it to the industry. Each cavity was subjected to mechanical and geometric checks, chemical and thermal treatments, high-pressure water rinsing, clean room assembly, and SRF testing at the FRIB SRF facility. Cavities that did not fulfill the design goals were returned to the previous step and re-processed or rejected in the case of irrecoverable flaws (a very rare event). Ten percent more spare cavities were built to account for possible under-performing ones, for future R&D, and to build spare CMs ready for operation, if a fast replacement becomes necessary in the linac. It should be noted that FRIB includes a post-accelerator for rare isotope beams (ReA) that shares the same $\beta_0 = 0.085$ QWR CMs as FRIB.

5.3.2 Choice of the Resonator Geometry

The FRIB QWRs [28] were cylindrical resonators with open ends (see Fig. 5.12). The open structure allowed for easy access to the resonator interior without the need for additional ports for RF surface processing. The bottom plate was made of a thin sheet of Nb, shaped to be easily deformable, and used to fine-tune the cavity in operation by means of a commercial actuator. In addition, a "puck" on that plate was adapted in size to tune the cavity frequency with a few kilohertz precision during construction (Fig. 5.10). Each QWR was surrounded by a titanium-made helium vessel, which was welded to it at the ports, in the bottom aperture, and in the niobium top plate of the cavity through a ring of reinforcement called "crown" (Fig. 5.12), which increased the stiffness of the cavity and reduced deformation due to variations in helium pressure. The "crown"connection, in particular, increases the frequency of the inner conductor mechanical oscillation modes and reduces their sensitivity to low-frequency environmental noise. To reduce the quality factor of these modes and

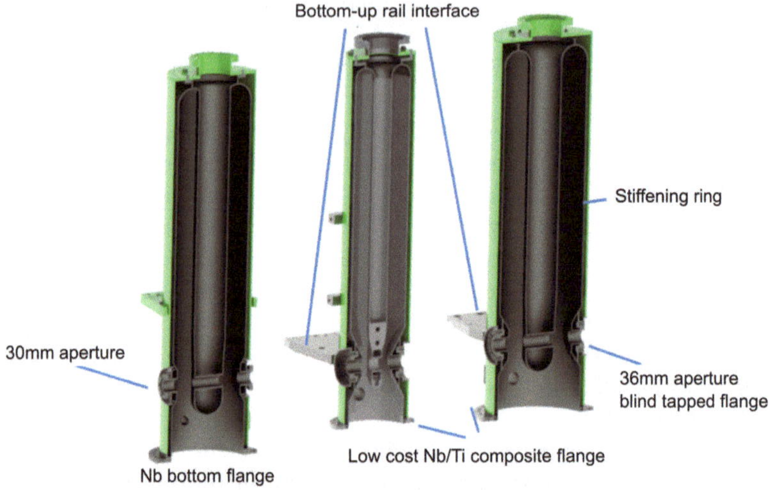

Fig. 5.12 MSU QWR designs for $\beta_0 = 0.085$: (**a**) first generation (RIA prototype); (**b**) second generation (original ReA3 design); and (**c**) third generation (final design)

prevent large oscillations, a mechanical damper capable of working in liquid helium was used [29, 30] (see Fig. 5.12).

Titanium was chosen for the helium vessel instead of stainless steel or normal grade niobium because it presented excellent mechanical characteristics and is non-magnetic. It had a thermal contraction rate similar to that of Nb; it could be welded directly to it, and it had a considerably lower cost. To further save material costs, the bottom flange of the cavity was also made primarily of titanium, with a small helium-cooled niobium surface for connection to the bottom plate to preserve the SRF continuity (see Fig. 5.12).

A titanium flange covered the bottom plate and sealed the vacuum of the cavity interior through indium wires. Although the RF geometry was rather similar to that of ALPI/PIAVE [23] and ISAC-II QWR [31] with a common vacuum, FRIB innovations allowed for a significant cost reduction and several advantages in large-scale production and long-term operational performance.

The FRIB HWRs were of the cylindrical coaxial type (see Fig. 5.13) [32]. Compared to the more widespread spoke geometry, this shape was easier to build, tune, and condition, at the cost of a very limited reduction in the achievable SRF parameters. In particular, these HWRs contain moderate multipacting bands, which allow for full conditioning within approximately 30 minutes in linac cryomodules. In addition to the RF coupler and beam ports, two rinsing ports were placed at each of the shorting plates to facilitate cavity RF surface processing and inspection at every point of the inner surface. In addition, these rinse ports allow for the electro-polishing of this coaxial cavity. The RF antenna was finally placed in one of these ports. The helium vessel was made of titanium, welded to the niobium at the seven

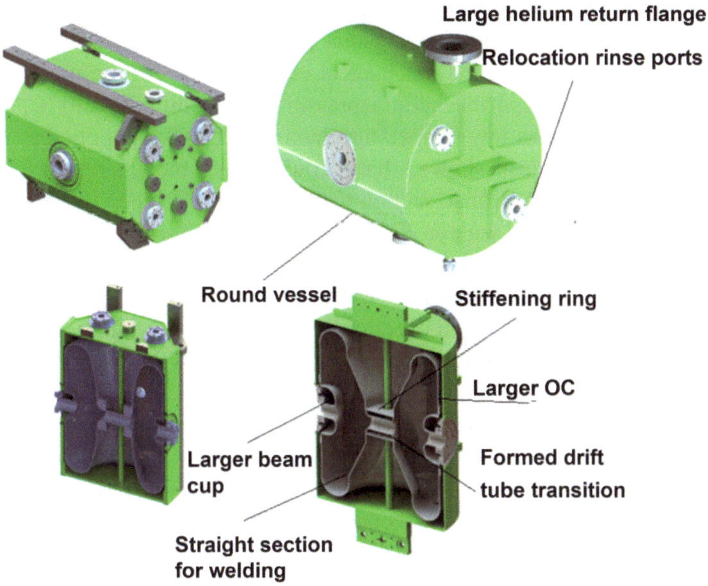

Fig. 5.13 MSU HWR designs for $\beta_0 = 0.53$: (left) prototype HWR and (right) final production HWR

ports, and it contributed to the overall stiffness of the cavity. The tuning was obtained by elastic deformation of the cylinder employing a pneumatic tuner, which pushed the beam ports closer to the inner conductor when the pressure was increased or allowed them to elastically return when the pressure was decreased. This system was reliable, cost-effective, and easier to install than traditional mechanical tuners [33, 34]. Its compact design allowed cavities to be placed very close to each other in the cryomodules.

5.3.3 Cavity Processing

In view of the large production, the standard chemical polishing for niobium was chosen, with 120 μm removal of the material, followed by thermal treatment under high vacuum at 650 °C. A light etching (20 μm or variable, depending on the frequency) and robotic high-pressure deionized water rinse of the entire RF surface for more than 3 hours finalized the surface treatment before the qualification test. Chemical polishing was chosen because it is well known, reliable, finely controllable, and could be implemented in the FRIB SRF infrastructure as well as in industry. FRIB cavities were designed to allow EP in the case of a possible future upgrade, if needed. The temperature of 650 °C was chosen for the high-temperature vacuum baking of all cavities after CP because it would remove the hydrogen absorbed by niobium during that process (which causes Q-disease; see Chap. 2) without modifying the mechanical characteristics of the material, which must be preserved to hold the helium pressure. An additional 120 °C, 48 hours of baking, also known as low temperature baking (LTB), was applied to a limited number of QWRs, and tests showed that it could reduce the Q slope at 4 K but not at 2 K. However, the QWR linac section was originally designed to operate at 2 K, so LTB was not performed with production cavities for linac construction. Skipping that treatment allowed for savings in cost and production time, with an acceptable increase in thermal load at 4 K. It is worth noting that this choice was not easy at that time. Currently, the QWRs section is successfully operating at 4 K without microphonics issues; thus, only spare and refurbished QWRs experience in-situ or furnace LTB.

Treatments have been carried out on the FRIB SRF infrastructure. Cavities that did not achieve performance must undergo treatment again after a visual inspection. After qualification, the cavities were moved to the clean room and assembled in the cold mass together with SC solenoids and diagnostic elements. The cold mass was then vacuum checked and, after being sealed from the outside world, was transferred to the cryomodule assembly area outside the clean room. The FRIB resonator performance statistics in the single cavity ("vertical") test are shown in Fig. 5.14. In operation, the FRIB cavity gradients were kept within the design values to guaranty good availability and at a rather low field emission radiation level below 300 μSv/h to minimize cavity degradation in the long term. The change in the performance of FRIB cavities' over the first few years of operation is shown in Fig. 5.15.

5.3.4 Choice of the Design Gradient

The design gradient (see Table 5.1) was chosen based on the experience gained from the operation of low-β linacs and prototype cavities of similar frequency and β, particularly ISAC-II at TRIUMF, which was the best performing at that time, with an average of approximately $E_p = 30$ MV/m during operation under common vacuum. The design gradient $E_p < 40$ MV/m and $B_p < 70$ mG were chosen based on the implementation of separate vacuum and the use in FRIB of the most recent handling and assembly techniques successfully developed for high-β SRF linacs. Although low- and medium-β cavities, due to their less favorable geometry, could not truly compare with elliptical ones in terms of gradient, the maximum peak surface electric and magnetic fields could be quite similar and independent of the frequency. Once the operational fields were defined, it was decided to accept only cavities that could achieve a 10% higher field in the qualification tests.

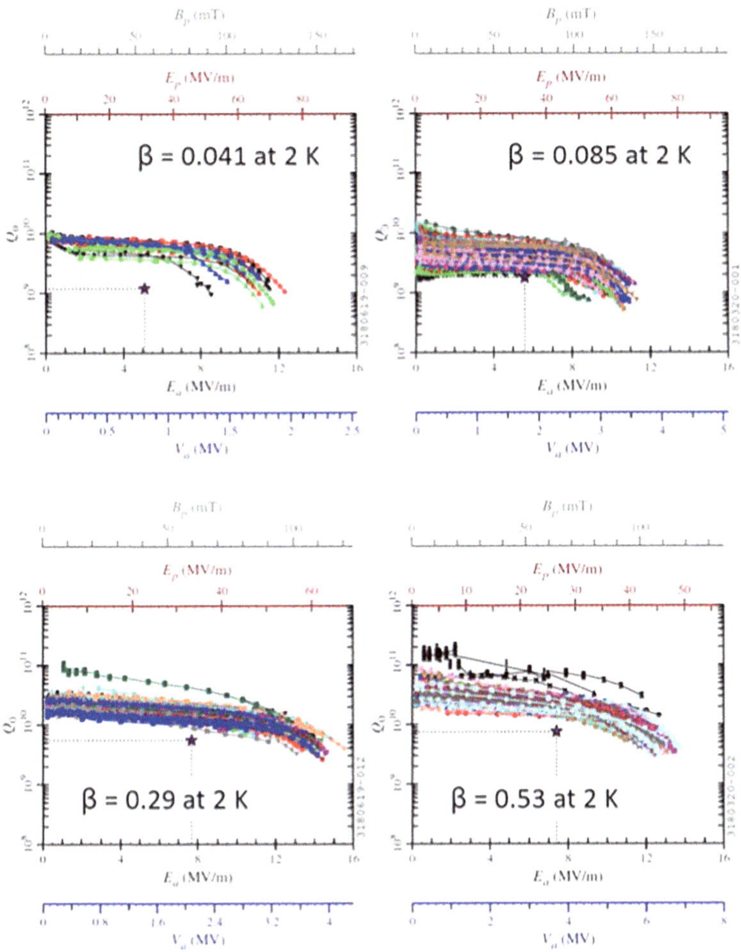

Fig. 5.14 FRIB cavities test results by types [35]

5.3.5 *Choice of the Design RF Bandwidth*

The choice of the RF bandwidth in SC cavities, determined by the RF coupling factor and thus by the positioning of the fundamental power coupler (FPC, see Sect. 2.2.3.4), was driven by two conflicting requirements: the minimum possible installed RF power cost and sufficient RF power to guaranty reliable RF locking of the cavities in the presence of the beam. Figure 5.16 shows the design bandwidth versus the measured bandwidth in the CMs installed in the accelerator tunnel. The difference, due to the unavoidable error in positioning the coupler at room temperature, was within acceptable values. To assess the minimum reliable bandwidth requirements, the environmental mechanical noise was measured at the linac

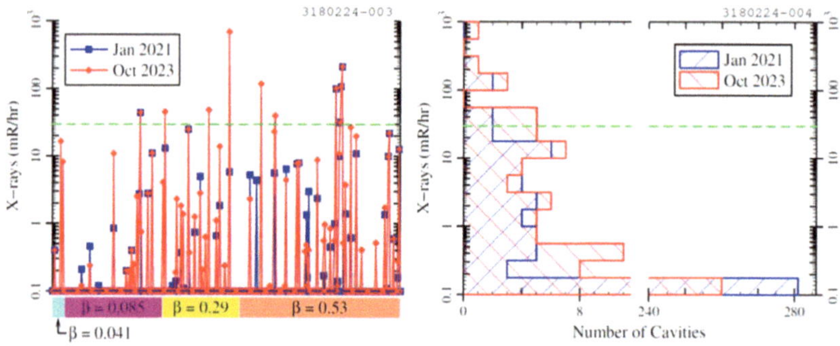

Fig. 5.15 Measured field emission X-rays at the design gradients (or administrative limit) for cavities in the FRIB linac, measured before (blue) and after (red) the start of user operations. Left: X-rays as a function of position along the linac. Right: X-ray histogram. The X-ray sensors' background level was approximately 1 μSv/h (blue dashed line). The administrative field emission limit was 300 μSv/h (green dash line). (Courtsey W. Hartung, MSU)

Table 5.1 SRF design parameters for FRIB cavities. The intrinsic (unloaded) Q at the design gradient was the requirement for the acceptance Dewar test

Cavity type	QWR	QWR	HWR	HWR
β_0	0.043	0.086	0.29	0.54
f [MHz]	80.5	80.5	322	322
V_a [MV]	0.81	1.78	2.09	3.70
E_a [MV/m]	5.1	5.6	7.7	7.4
E_p/E_a	6.1	6.0	4.3	3.6
B_p/E_a [mT/(MV/m)]	10.8	12.4	7.7	8.6
R/Q [Ω]	401	455.4	224.4	229.5
G [Ω]	15.3	22.3	77.9	107.4
Aperture [m]	0.036	0.036	0.040	0.040
T [K]	2.0	2.0	2.0	2.0
Intrinsic Q [$\times 10^9$]	1.4	2.0	6.7	9.2

location, along with the cavity phase and frequency error distribution during 24-hour periods when operated with this environmental noise at the required gradient in open loop. Twelve times the root mean square of that distribution was taken as a realistic expected maximum peak-to-peak detuning; that is, the minimum bandwidth that should be managed by the RF system in operation. Moreover, the expected minimum resolution and maximum backlash of the tuners were added to the required bandwidth, as well as the contribution of the expected helium pressure fluctuations, which can be represented by $df/dP \times \Delta P$ (where df/dP is the resonator frequency sensitivity to pressure, and ΔP is the maximum expected value of pressure fluctuations in the cryogenic system). Slow pressure fluctuations were tracked by the mechanical tuner and did not enter that calculation. An additional safety factor of 2 was included in the specifications, resulting in the final estimated minimum "safe" requirements of approximately 40 Hz for QWRs and 30 Hz for

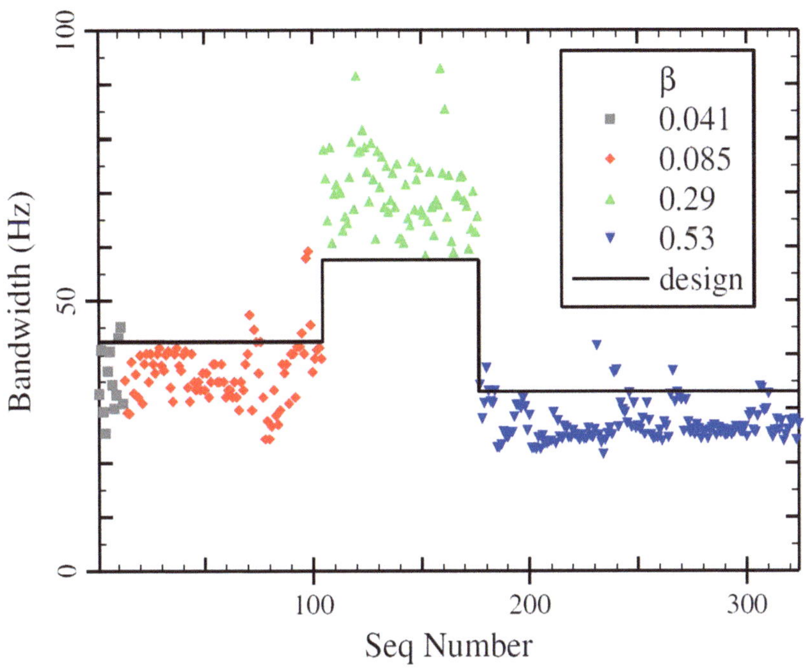

Fig. 5.16 RF bandwidth of the FRIB SRF resonators. The measured average bandwidths were (35.2, 36.6, 70.2, 27.2) Hz that correspond to $Q_L = (2.28, 2.20, 4.58, 11.9) \times 10^6$

Table 5.2 Bandwidth for FRIB SRF resonators

Parameter	QWR041	QWR085	HWR290	HWR530
β_0	0.041	0.085	0.29	0.53
P_b (max) [W][a]	590	690	1526	2373
Design bandwidth [Hz]	43	41.5	57	33.3

[a]For uranium beam operation

HWRs concerning microphonics. For the $\beta_0 = 0.29$ HWRs, the coupling needed for beam loading has resulted in a bandwidth that already exceeds the specifications. For the remaining three types of cavities, the beam loading requirements were insufficient for that scope, and extra over-coupling had to be added. Table 5.2 shows the effective RF bandwidth implemented in the FRIB driver linac.

5.3.6 Tuning Procedure During Fabrication

Tuning cavities to their design center frequency required several steps of RF measurement, machining, and welding, each of which improved the precision of the tuning. Every weld in the cavity and in the helium vessel would cause a partially predictable detuning that must be taken into account. The use of cylindrical geometries made these operations simpler compared to more complex shapes.

The first step consisted of adjusting the cavity length, testing the frequency, and welding the Nb cylinders (which already included the beam ports) to the shorting plate(s) to obtain roughly ±50 kHz precision. The subsequent steps included adjusting the length of the tuning plate puck in the QWRs, virtual welding of the HWR and its Ti vessel [36], differential chemical polishing inside the cavities, and plastic deformation of the cavity in the beam port region (HWRs) or in the tuning plate (QWRs). The process ended when the frequency reached the central part of the tuning range provided by the mechanical tuner, which is approximately 0.025% of the cavity frequency.

5.3.7 Choice of RF Couplers

As usual, for RF power in the kW range or higher for SC cavities, capacitive couplers were chosen for all FRIB cavities. The coupling could be manually adjusted without breaking the cavity vacuum. In QWRs, the RF port was located with the cold window on the outer conductor side, quite close to the tuning plate. That position, with a high E-field but a low B-field, minimized the RF losses on normal-conducting surfaces of the coupler while providing the required coupling strength. To bring the RF connectors to the warm window at the bottom of the CM, while guaranteeing a static thermal load below 1 W and the required flexibility to account for thermal contractions, a short 90° RF line was used inside the CM, as shown in Fig. 5.17.

In HWRs, the coupler was located in the cavity equator, and its RF connector extended directly outside the bottom of the CM, utilizing the single RF window design as shown in Fig. 5.18. The inner conductor was shaped with two steps in diameter to suppress multipacting [15]. That shape allowed for the transform of the coupler shaft impedance while preserving 50 Ω at the RF connector. The coupler was equipped with two thermal intercepts to minimize the heat load to the 2 K circuit. The cold cathode gauge, spark detector, and e-probe were used to monitor the operation of the FPC and to protect it from damage caused by electron activities.

5.3.8 Choice of RF Tuners

The QWR tuner was a commercial stepper motor that moved the tuning plate. It was mounted outside the CM, which was easily accessible and replaceable [32]. It could provide a maximum force of 200 N at a speed of 1 μm/s and a resolution of 0.2 μm. The original choice of the stepper motor used a Teflon lead screw and required current to hold its position. After the first couple of years of operation and

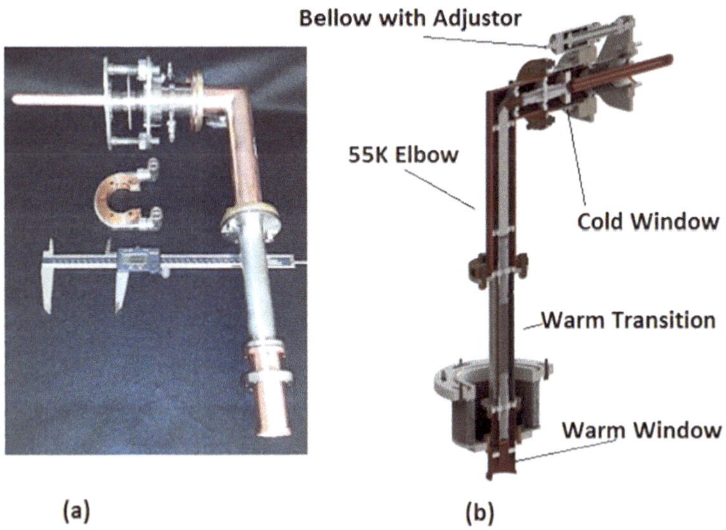

Fig. 5.17 FRIB QWR FPC: (**a**) a prototype built by ANL (Courtesy ANL, USA), and (**b**) the production design model [32]

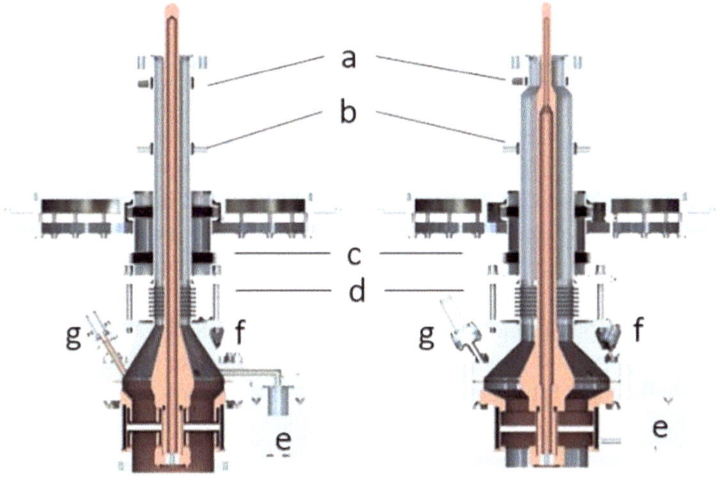

Fig. 5.18 FRIB HWR FPC: (left) the first prototype design, and (right) the production design minimizing MP with (**a**) 4 K cooling line; (**b**) 55 K cooling line; (**c**) vessel interface; (**d**) coupling adjustor; (**e**) cold cathode gauge; (**f**) spark detector; and (**g**) e-probe [32]

several motor failures, and after observing that a few tuners required a tuning force right at the edge of their maximum force, they were replaced with different motors made of radiation-hard material and with gears that allow one to increase their force and self-lock to the position without a holding current. That significantly improved the lifetime of the tuners and the reliability of cavity operation.

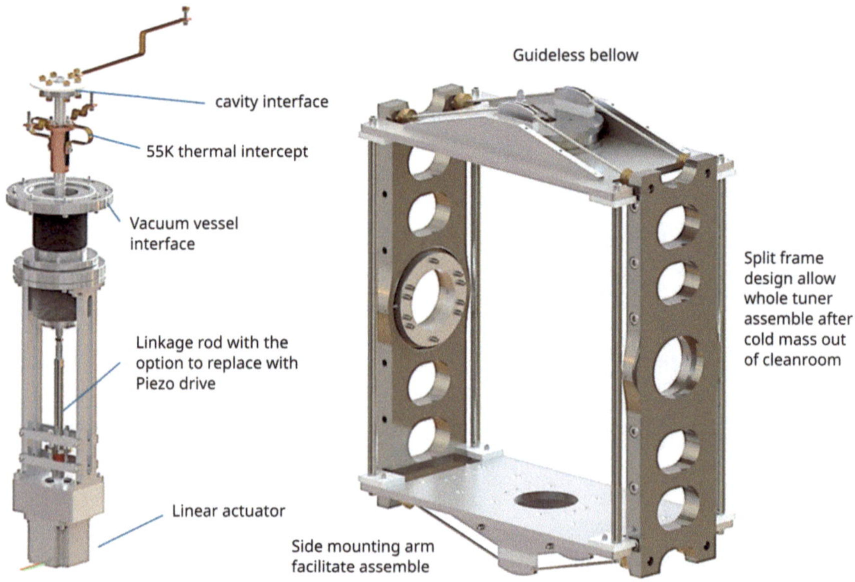

Fig. 5.19 FRIB tuners: (left) QWR tuner and (right) HWR tuner [32]

The pneumatic HWR tuner, as shown in Fig. 5.19 [33, 34, 37], was specially developed for FRIB in collaboration with ANL, which owned the original design. This type of tuner, compared to the more widely used screw-type ones, presents many advantages in terms of cost, weight, operation speed, and easy mounting/dismounting. Moreover, the frequency could be set in a very short time at the beginning of operation, and the force could be released immediately at the end of operation, preventing accidental plastic deformations of the cavity during the warm-up of the linac.

The measured phase and voltage stability of the FRIB cavities are shown in Fig. 5.20. It is important to note that FRIB cavity tuners are of the slow tuner type. The data confirm that the HPHA stability requirements for low-β structures in CW operation can be fulfilled without the need for piezoelectric fast tuners.

5.4 Cryomodule

5.4.1 Cryomodule Basic Concepts

Cryomodules are accelerator units devoted to maintaining and operating SC resonators and SC solenoids at cryogenic temperatures (typically set in the range of 1.8–4.5 K). Their design (see, e.g., [38]) includes mechanical solutions that minimize thermal flow from their outer vessel to the cold mass, namely: mechanical links with high thermal resistance to reduce thermal conduction; high vacuum

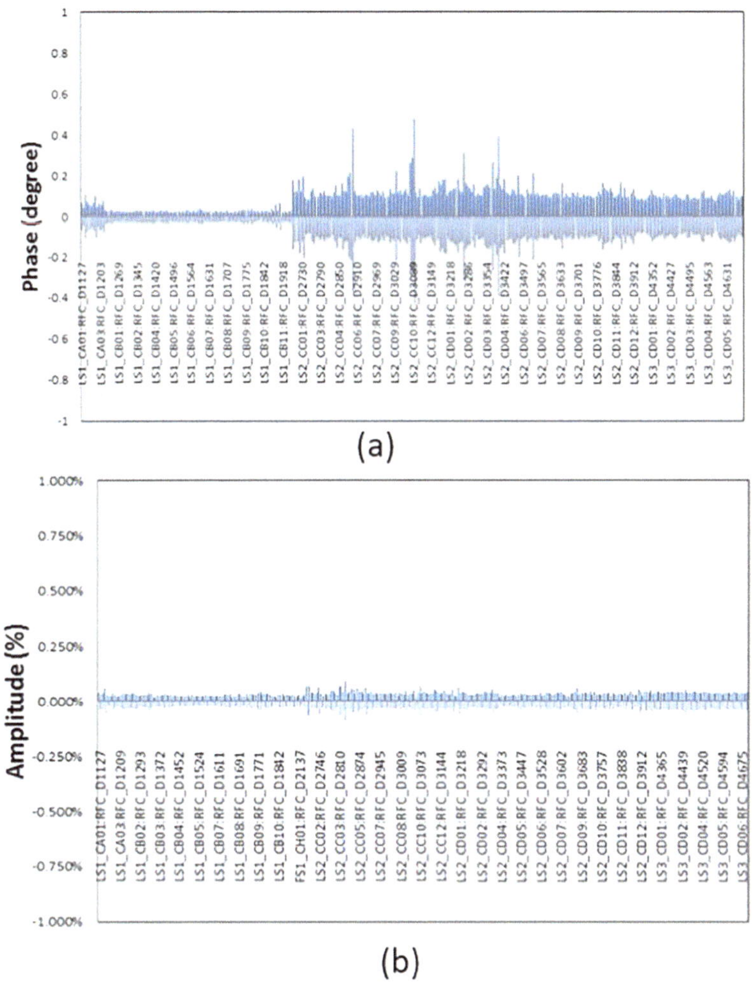

Fig. 5.20 FRIB cavity resonance control performance at the design gradient during linac operation: (**a**) the phase peak-to-peak error and (**b**) the amplitude peak-to-peak error

to eliminate thermal convection; thermal shields at intermediate temperatures to reduce thermal radiation (e.g., actively cooled shields and Multi-Layer Insulation (MLI)); and optimized thermal paths and temperature distribution inside the CM. Thermal conductivity and other mechanical characteristics depend on temperature in different ways across various materials. This has a strong impact on the mechanical and thermal design. In a vacuum, heat transfer is generated by radiation and conduction through the residual gas. The contribution from radiation can be calculated using the Stefan-Boltzmann equation: between two surfaces A_1 and A_2, with emissivities ε_1 and ε_2 and temperatures T_1 and T_2, the radiative heat flow rate can be expressed as

$$\dot{Q} = F_e F_{12} \sigma_{SB} A_1 (T_2^4 - T_1^4)$$

where $\sigma_{SB} = 5.67 \times 10^{-8}$ W/(m^2 K^4) is the Stefan-Boltzmann constant, and F_e and F_{12} are factors that depend on the thermal emissivity values of the two materials and the geometrical configuration (e.g., in the common case of large flat surfaces with the same emissivity $\varepsilon \le 1$, facing each other at a small distance, $F_e \sim \varepsilon/(2-\varepsilon)$ and $F_{12} \sim 1$) [39]. The fourth power dependence on temperature can cause a large heat flow to the cold mass at 2 K or 4.5 K if it is directly exposed to the CM vacuum vessel at room temperature. This can be substantially reduced by the use of intermediate thermal shields and MLI. A shield that is actively cooled at a temperature T_i, with $T_1 < T_i \ll T_2$, intercepts the heat flow from room temperature and delivers a much lower thermal load to the cold mass ($T_i^4 - T_1^4 \ll 300^4 - T_1^4$). Moreover, insertion between two surfaces of $N \gg 1$ layers of thermally isolated material, thus floating in temperature, creates a multilayer insulation (MLI) that reduces the emissivity factor F_e in the equation and, consequently, the heat flow by approximately $1/N$.

The heat flow from conduction through residual gas, under high vacuum conditions where the mean free path of the gas particles is larger than the distance between the two surfaces at different temperatures, is proportional to $p(T_2 - T_1)$, where p is the gas pressure. This contribution to the cryogenic heat load can generally be negligible if the residual pressure is kept below 10^{-5} mbar.

When possible (with a few exceptions, such as fundamental power couplers and sometimes tuner actuators), cables and pipes connecting the outside world to the CM interior are made to enter through the same port (sometimes called the "chimney"); thus reducing the number of apertures and facilitating the design of an optimum thermal distribution for a minimum heat load.

In typical linac CMs, at least two temperatures are actively maintained by direct cooling: the cold mass, typically 2/4.5 K, and the intermediate thermal shield (typically in the 35–80 K range, cooled by gaseous helium or liquid nitrogen). In the case of a cold mass operated at 2 K (~1.8–2.1 K), cooled by superfluid helium, an additional layer actively cooled to 4–20 K is sometimes present. Helium is typically delivered to CMs at supercritical conditions (3–5 bar, 4.5 K) or as a saturated liquid/gas in a two-phase condition, depending on the design of the cryogenic system [40].

In CMs for cavities that operate near 4.2 K, helium is collected in a reservoir (also called a phase separator, since the liquid collects at its bottom) and is then delivered to cavities and SC solenoids. In 2 K systems, the incoming supercritical helium is first sent through a heat exchanger to be further cooled by return vapor and then through a JT (Joule–Thomson valve) to saturated pressure before being delivered to a phase separator that remains at the saturation pressure of superfluid helium (31.4 mbar for 2 K) by pumps located in the cryogenic system He return circuit [40]. The pumps are usually placed on the cold end to maximize thermal efficiency for a large system, while in small systems they are placed at room temperature to minimize capital cost and system complexity.

Typically, the pressure of the liquid helium on the cavity walls is kept below ~2 bars to prevent the risk of plastic deformation or collapse, and this is guaranteed by safety pressure relief valves and burst disks. Cryomodules include not only cryogenic and vacuum equipment but also all accessories and diagnostics for the operation of SC resonators and beam transport. Warm-to-cold transitions between cold parts and the outside world (RF power lines, cabling, actuators, and the supporting structure of the cold mass) are designed to maintain a minimum thermal load. In addition, the cold-mass support structure also serves to guaranty the precise position and alignment of all beam components during transport, cool-down/warm-up, and under operational conditions, as well as to provide attenuation of external mechanical vibrations down to negligible amplitudes at the cavity side.

The materials used in CM construction must have good mechanical characteristics at cryogenic temperatures, good vacuum characteristics with modest outgassing, must be non-magnetic to avoid trapped flux in resonators, and must be able to sustain high radiation levels without losing their properties. The most widely used materials are stainless steels, typically AISI 304 and AISI 316, which are relatively inexpensive and characterized by excellent mechanical and vacuum properties, as well as low thermal conductivity. Carbon steel, which is ferromagnetic, is sometimes used for vacuum vessels and outside the magnetic shields. In addition to many other metals and alloys, Aluminum, Copper, and Titanium alloys, special materials characterized by a very low thermal contraction coefficient (e.g., INVAR) are sometimes used for alignment rails and rail holders. Alloys with high magnetic permeability ($\mu_r > 100{,}000$) are used for the magnetic shielding of the resonators [41]. Thermally insulating materials are also needed. Good mechanical properties are important, especially when they serve structural functions.

Cryomodules contain large amounts of cryogenic fluids that, in the case of a vacuum failure, can be rapidly heated and transformed into a high-pressure gas with explosive potential. Therefore, they must comply with the safety codes for high pressure vessels (for example, ASME Boiler and Pressure Vessel Code, BPVC) enforced in the country where they are installed and operated [13].

Assembly and installation of all components, while preserving the extremely high level of cleanliness required by SRF, is one of the main challenges: even small errors can jeopardize the achievement of the linac design performance. For this reason, even if large infrastructural investments are required, in most SC linac projects, this fundamental step is not assigned to industry but is still kept at the laboratory level, where highly specialized SRF personnel are available and quality assurance and control can be better guaranteed. This is expected to change in the near future, making way for greater participation from the industry (Fig. 5.21).

5.4.2 *Clean Assembly*

The clean components that enclose the beam line in a CM, namely, resonators with their RF power couplers, SC solenoids, and dipole correctors, and, in some cases, beam diagnostic components, are assembled and aligned in a string (the so-called cold mass), terminated at both ends by vacuum valves that keep its interior clean

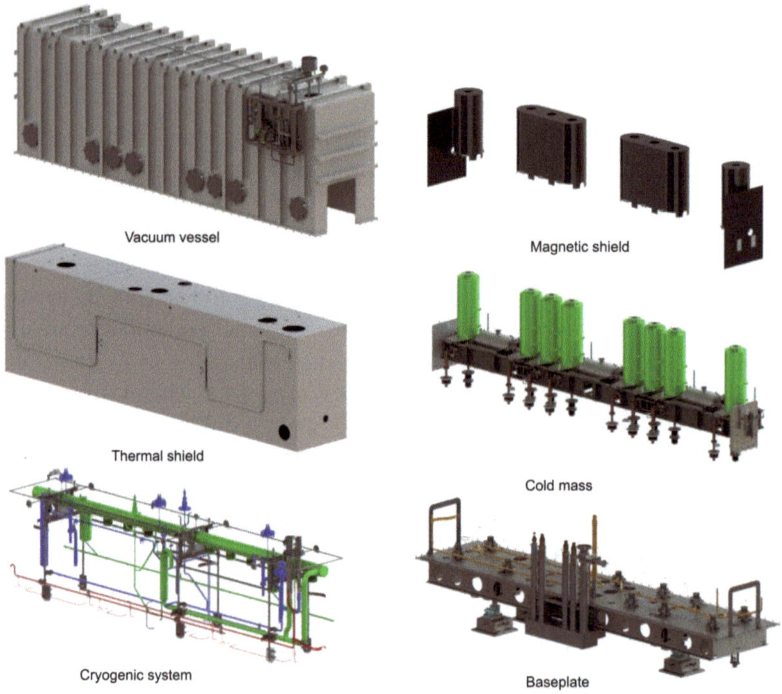

Fig. 5.21 A low-β CM example: main components of the FRIB $\beta_0 = 0.085$ unit, which hosts 8 QWRs and 3 SC solenoids (bottom-up type CM with local magnetic shielding) [25]

and isolated from the outside world. This string sits on special rails and includes all components that must be operated at liquid helium temperature. The beamline elements are interleaved by vacuum bellows to cope with thermal contractions. The supporting rail must be machined with high precision to allow reliable alignment of the beamline components and must respond in a predictable manner to the temperature changes inside the CM. The cold mass is assembled in clean rooms of class 10/100 (ISO4/ISO5) to prevent contamination of surfaces facing the electromagnetic field and, in general, of the entire beam vacuum side. The resulting sub-assembly is inserted into the CM, its rail is connected to the vacuum vessel by means of mechanical links with high thermal resistance, and it is aligned along the beam axis. The cavities are equipped with frequency tuners [22], connected to cryogenic equipment for cavity cooling, heating, and temperature monitoring, linked to RF and diagnostic cables, and finally enclosed in magnetic shields and surrounded by MLI, according to the design of the CM. The CM vacuum vessel is then closed, tested to specifications, and installed in the linac. Of course, the details of the CM assembly depend on its particular design, which can vary significantly for different machines, beam energies, and applications (Fig. 5.22).

Cavities of different shapes and sizes (see Sects. 2.2 and 5.1.1) used in HPHAs result in different types of CM, which can be roughly grouped according to some

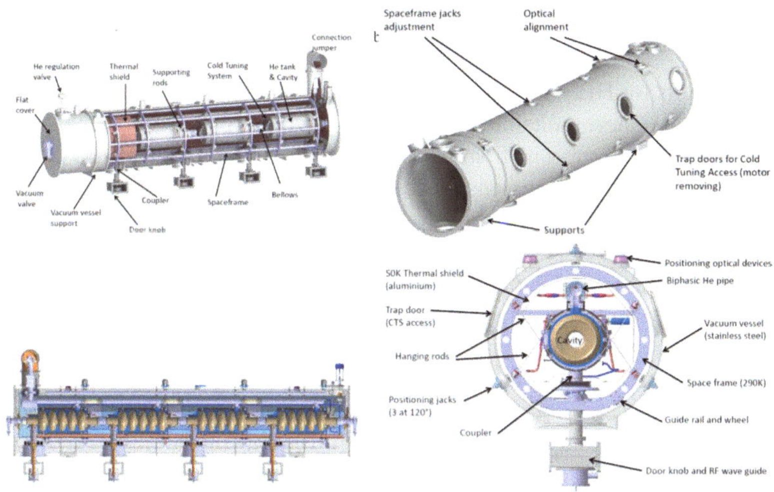

Fig. 5.22 High-β CM example: different views of the ESS CM for 704 MHz elliptical cavities [42], with cylindrical vacuum vessel and internal supporting frame for the cavities. (Courtesy ESS, Sweden)

general characteristics, such as operational temperature, vacuum configuration, top-down or bottom-up construction, and vacuum vessel shape.

5.4.3 Vacuum Considerations

Inside the CMs for HPHAs, the vacuum volume of the beam (that is, the ultra-high vacuum volume inside the beam tubes and cavities, in the range $\sim 10^{-8}$–10^{-10} mbar) is separated from the insulation vacuum (that is, the high vacuum outside the beam space and required for thermal insulation, in the range $\sim 10^{-5}$–10^{-7} mbar). The two volumes are evacuated independently by different pumping systems. This separation contributes to maintaining a high level of cleanliness within the resonators during installation, operation, and maintenance. Separate vacuum is widely considered the best way to achieve the highest gradients and long-term reliability. This configuration is straightforward in CM for elliptical cavities, which can be seen as corrugated sections of the beam pipe. In comparison, in low-β CMs containing cavities such as QWRs and HWRs that extend most of their volume far from the beam axis, vacuum separation requires a large number of vacuum-tight connections and bellows, which increase the CM length and assembly costs. It should be mentioned that, for low beam current applications at 4.2 K, low-β CMs configured with beam and insulation vacuum in common have been routinely used for decades, in some cases with rather good performance despite their considerably lower complexity and cost [43].

5.4.4 Top-Down and Bottom-Up Construction

Vacuum vessels for HPHA CMs are made in the shape of cylinders with removable caps, which are more suitable for elliptical cavities, or in the shape of boxes with a removable cover, which are especially well suited for TEM low-β cavities with significant transverse extension from the beam line. In both cases, the cold mass and its rail can be suspended from the top of the vacuum CM vessel ("top-down"configuration) or supported from the bottom ("bottom-up" configuration). In "top-down" CMs, cold masses and their rails hang from the removable top cover, suspended by flexible systems that accommodate thermal contractions without creating dangerous structural tensions and without losing internal alignment. Suspension rods create a long mechanical and thermal path from the ground to the resonators and contribute to the mechanical filtering of environmental noise. With low-β cavities and SC solenoids, careful design of the suspension system can minimize the displacement of the beamline after cooling; however, achieving predictable and reproducible alignment is always challenging. Top-down CMs that host only cavities without focusing magnets usually have less demanding alignment specifications and can be built with simpler holding systems.

An alternate, more recent configuration is the bottom-up design, where everything is supported by the CM base plate, which hosts all interfaces, while the upper part of the vacuum vessel is used as a removable cover. The cold mass is supported by short pillars of thermally insulating material (e.g., G10), which provide a short mechanical path between the ground and the resonators, minimizing thermal displacements. Linear bearings on top of the pillars allow the contraction of the above cold mass rails along a well-established mechanical path to a fixed point. This approach facilitates good alignment of the parts with the minimum required adjustments and has shown very good predictability and reproducibility of the final cavities' positions after cooldown [25]. Bottom-up designs have previously been successfully used for different applications, such as superconducting magnets (Fig. 5.23).

5.4.5 Local and Global Magnetic Shielding

Magnetic shielding [41] is obtained by surrounding resonators with materials of high magnetic permeability. Cryomodule vessels made of carbon steel provide a rather modest shielding of the earth's magnetic field. To reach typical target residual fields on the cavity surface on the order of 1 μT (10 mG) or lower, it is necessary to shield it with high-μ_r materials, which are rather expensive and can easily lose their characteristics if not handled properly. Magnetic properties depend on temperature, and shielding materials must be chosen according to their position in the CM. Magnetic shielding must also take into account the shape, size, and location of the cavity and can be optimized in different ways [45]. Global magnetic shielding

Fig. 5.23 Examples of (left) top-down [44] (Courtesy Z. Conway) and (right) bottom-up [25] low-β CMs during assembly. (Courtesy S. Miller, MSU)

consists of enclosing the entire cold mass of the CM in the shield, whereas local shielding encloses single cavities or small groups of neighboring cavities. Apertures in the shield, needed to fit cavity interfaces and accessories (RF couplers, cavity tuners, cooling pipes, cables, etc.), can cause leakage of the external magnetic field into the shield; to minimize degradation of shielding performance, extended tubes are often attached to the aperture. In large CMs, the use of global shielding causes a large magnetic flux concentration and requires an increased thickness of the shielding material to prevent saturation. Local shielding is often preferred, with the additional advantage of protecting cavities from the fringe field of neighboring SC solenoids as well. Global shielding and SC solenoids can co-exist in the CM if proper degaussing procedures are applied before cooling down cavities to the SC state. Global and local shielding can be implemented in the same CM as well.

5.4.6 TEM Cavity Cryomodule Examples

Low-β HPHA CM technology derives largely from that previously developed for low current heavy-ion linacs, and its origin can be dated back more than four decades [46]. Many types of CMs have been developed for QWRs and HWRs in HPHA projects, some of which are currently in the construction phase. Some of them will be briefly described in this paragraph, focusing more on recent developments. The aim is not to provide an exhaustive list but rather some significant examples of the wide variety of solutions that have been adopted for different types of beams, which are in continuous evolution.

Short CMs, which house 1–4 cavities operating at $\sim$4.2 K and without SC solenoids, are used in low- and medium-β sections with NC quadrupole focusing [47–49]. The SPIRAL2 linac, with 40 MeV and 5 mA CW for the deuteron

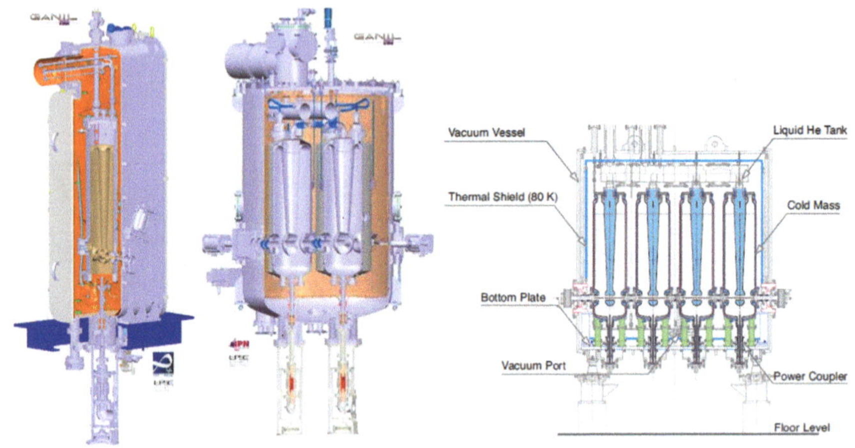

Fig. 5.24 Left: SPIRAL2 $\beta_0 = 0.07$ and $\beta_0 = 0.12$ QWR cryomodules [47] (Courtesy C. Marchand, GANIL). Right: SRILAC cryomodule [48]. (Courtesy N. Sakamoto, RIKEN)

beam, and 8 MeV/u for $A/q \leq 1/6$, was built with rather short top-down-type CMs containing either one $\beta_0 = 0.07$ (box-type CM with global shielding) or two $\beta_0 = 0.12$ QWRs (vertical cylinder CMs with local shielding; see Fig. 5.24 [47]. The $\beta_0 = 0.07$ cavities were fed by 12 kW forward power couplers inserted from the bottom of the CM and tuned by deformation of the outer conductor at the beam port level. The $\beta_0 = 0.12$ cavities used Nb rod insertion in the magnetic region.

Cryomodules working at 2 K for HWR and spoke cavities without SC solenoids can also be found [48, 50, 51]. The ESS linac, a 5 MW, 2 GeV pulsed proton accelerator for neutron production currently under construction, includes a $\beta_0 = 0.5$ section made of 26, 352 MHz double-spoke resonators. CMs were cylindrical and hosted only two cavities (see Fig. 5.25) [50]. The CM received 2 K helium from an external dedicated cryo-distribution box containing a JT valve and a heat exchanger. The coaxial RF couplers were mounted from the bottom of the CM and fed by rectangular guidelines via a door-knob type transformer. Each resonator delivered about 350 kW of peak power to the pulsed beam. A similar design [51] was also used for the single-spoke resonators section, equipped with 20 kW CW RF couplers, of the 650 MeV, 4 mA CW proton linac project MYRRHA in Leuven (Belgium) aiming at 2.4 MW average beam power for ADS applications. RAON CMs [52] included up to six spoke cavities.

The first example of a long CM with HWRs and SC solenoids built for HPHA was the top-down CM of SARAF Phase 1, as shown in Fig. 5.26 [53], hosting six 175 MHz HWRs with vertical orientation working at 4.2 K and operated with proton and deuteron beams above 1 mA CW (upgraded Phase 2 CMs are currently being developed for 5 mA CW operation [54]). Several years later, the CiADS HWRs $\beta_0 = 0.12$ CM [55], with six cavities (lying horizontally instead of vertically, as in the previous example), reached 2.7 mA with protons, shortly followed by one more $\beta_0 = 0.12$ and a $\beta_0 = 0.15$ CM with five vertical HWRs.

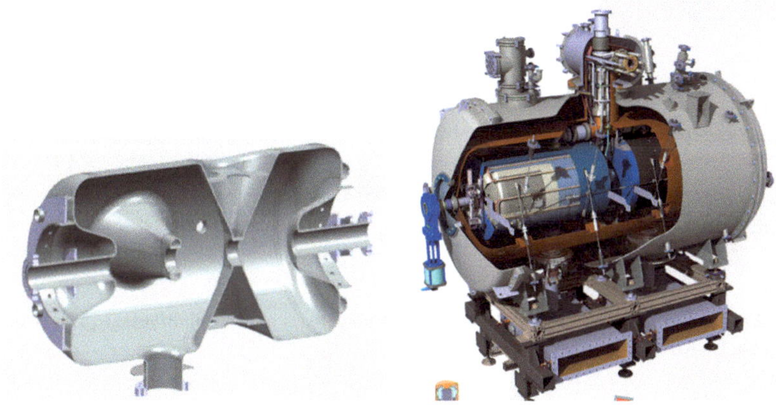

Fig. 5.25 Cutaway of the ESS $\beta_0 = 0.5$ DSR resonator and its cryomodule (not to scale) [17]. (Courtesy ESS, Sweden)

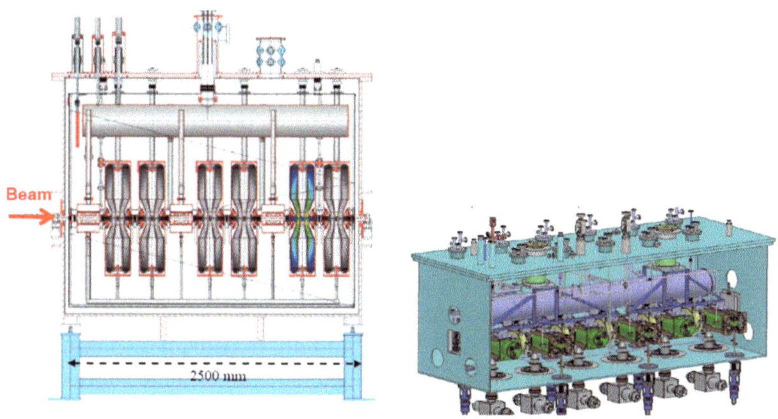

Fig. 5.26 SARAF Phase 1 HWR (left) [56] (Courtesy Z. Conway) and CiADS Injector-II cryomodules [55]. (Courtesy Z.J. Wang, IMP)

A CM of the same type was designed to accelerate deuteron beams with a record intensity of 125 mA CW. The LIPAc low-β CM (IFMIF-EVEDA) [57], as the demonstrator of the IFMIF linac, aims for a 40 MeV beam with a final average beam power of 5 MW [58]. This CM hosted eight 175 MHz, $\beta_0 = 0.09$ HWRs and eight 5 T SC solenoids with integrated steering correctors. The vacuum vessel was a square-shaped cross-section cylinder, closed at both ends by removable walls, where a top-down frame held the cold mass. HWRs are oriented horizontally.

The cold mass components were allowed to slide longitudinally on the contracting rails during cooldown, maintained in place by an INVAR rod. This prevented tensions between the cavities and the FPCs, which were connected to the stiff, heavy high-power RF lines located at room temperature. One of the main challenges in

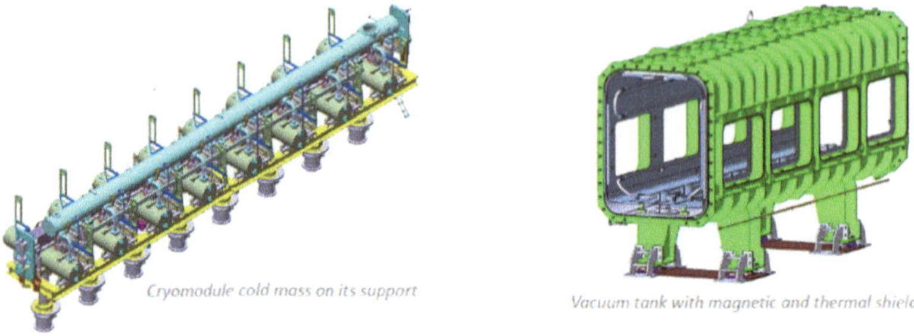

Fig. 5.27 View of the LIPAc low-β cryomodule cold mass and vacuum vessel [57]. (Courtesy H. Dzitko, CEA/F4E)

this CM was represented by the $\sim$105 kW CW RF couplers, which are of a size comparable to the cavity, vertically mounted through the bottom wall of the CM vessel (see Fig. 5.27). Magnetic shielding was global. The mechanical structure of the CM was designed to withstand severe seismic events.

The first long CMs for TEM cavities working at 2 K have been built for the FRIB SC linac, which accelerates ions of all masses (from protons to uranium) to above 200 MeV/u and almost 400 kW CW final beam power with 324 SC cavities [59]. FRIB CMs, of six different types, host 80.5 MHz QWRs and 322 MHz HWRs, working at 2 K, together with 8 T SC solenoids with integrated steering correctors working at 4.5 K [25, 30]. A 2–4 K heat exchanger is placed in each CM to maximize the 2 K capacity. The vacuum vessel, made of carbon steel, is box-shaped, with a strong base plate to which the cryogenic system interfaces are connected. The cold mass rail is mounted in a bottom-up configuration, supported by short posts made of thermally insulating material, G-10. To limit maximum displacement after thermal contraction, the cold mass rail is divided into two or three shorter sections, each holding a maximum of four elements (cavities or solenoids). The cold mass is built with sliding supports that make each section converge to a fixed point, with very limited mechanical tensions and rather small alignment errors under operating conditions. The large size of the cold masses required local magnetic shielding, with boxes enclosing up to three cavities each. The fundamental power couplers for the HWRs are single-window couplers rated for CW operation at 10 kW and extending out of the bottom of the CM. In QWRs, where the thin bottom plate is used for tuning, 4 kW couplers with a cold window are located on the cavity side and are entirely contained in the CM; 90° coaxial RF lines connect them to their warm windows on the bottom side of the CM. The rather long 2 K phase separator, spanning up to eight cavities and connected to them through large bellows, is also supported by the bottom plate during assembly. However, it is finally connected to the strong top cover of the vacuum vessel to decouple its possible mechanical vibrations from the resonators. This modular CM design scheme can be extended to the elliptical cavities of the FRIB upgrade (Fig. 5.28).

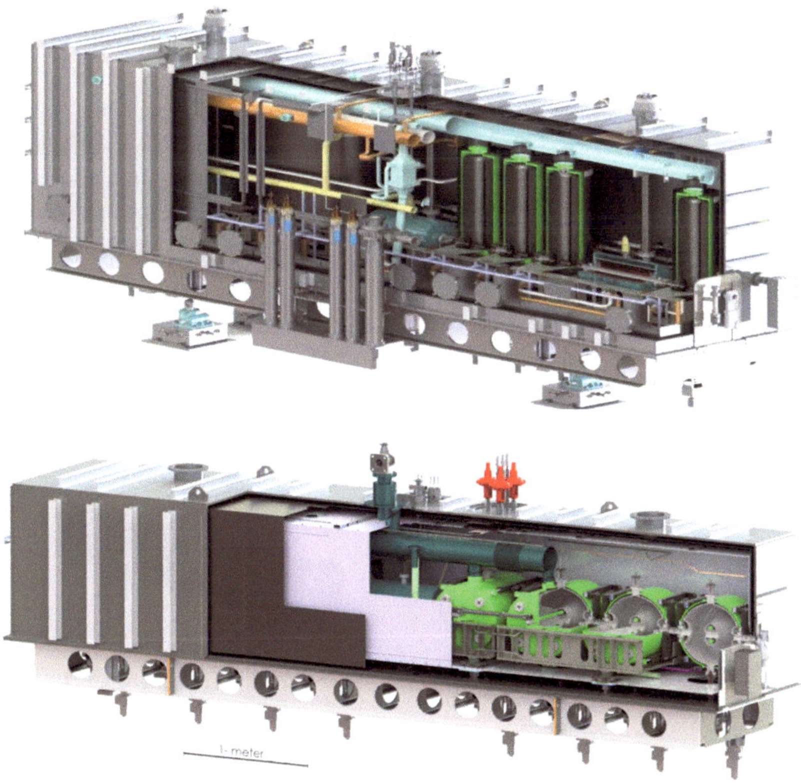

Fig. 5.28 Cutaway of the FRIB $\beta_0 = 0.085$ QWR and $\beta_0 = 0.53$ HWR cryomodules [30]

Long 2 K CMs have also been developed for the PIP-II linac, currently in its technical construction phase, for 5 mA, 800 MeV pulsed proton beams [60] with a CW upgrade plan. It included five different types of CMs hosting HWRs, spoke resonators, and elliptical cavities. The first section consisted of a box-type CM for eight 162 MHz, $\beta_0 = 0.11$ biconical HWRs and eight SC solenoids [56]. It had a top-down structure with global magnetic shielding, and cavities were horizontally mounted on a long rail that hung from the CM top plate. 10 kW CW RF couplers were inserted vertically from the bottom. 2 K helium was produced within the CM (Fig. 5.29).

The second section of PIP-II at FNAL included spoke cavities of two different types, hosted in nine CMs. Unlike the HWR one, the CMs for the much more compact 324 MHz, $\beta_0 = 0.22$ and $\beta_0 = 0.47$ single-spoke resonators had a completely different bottom-up design [12] (Fig. 5.30). A long strongback, kept at room temperature to eliminate longitudinal displacement of cavities during cooldown, sustained cavities and solenoids, each sitting on a single post with low thermal conductance. The entire cold mass was inserted into a cylindrical CM vessel with global magnetic shielding. 17 kW RF couplers were inserted from the side with

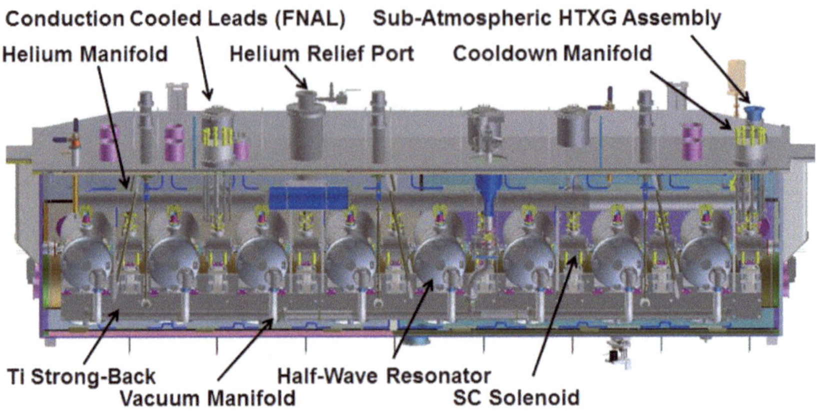

Fig. 5.29 PIP-II $\beta_0 = 0.11$ HWR cryomodule [56]. (Courtesy Z. Conway)

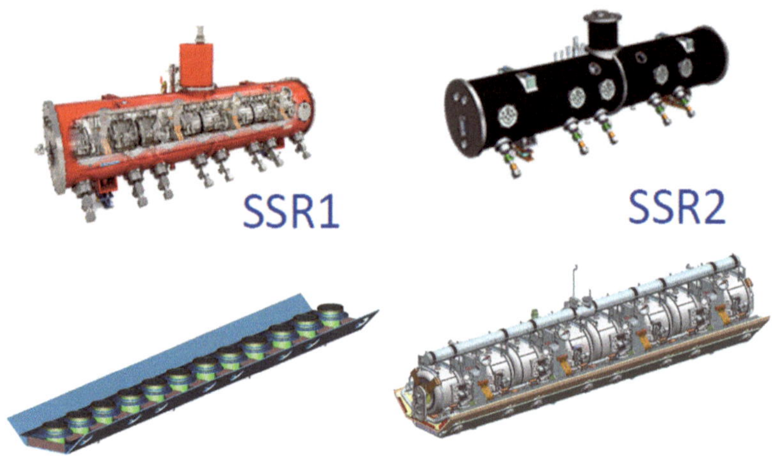

Fig. 5.30 PIP-II, SSR1 and SSR2 cryomodules for spoke cavities [12] (Courtesy T. Nicol, FNAL). Bottom: the strongback with posts and coldmass be-fore insertion in the vessel, and cutaway view of the assembled cryomodule

~45° inclination. 2 K helium was produced within the CM. A similar structure was also used in PIP-II CMs containing elliptical cavities and no solenoids.

A fairly similar bottom-up design philosophy, with a room temperature strong-back and composite posts, was developed and already installed for the CiADS Injector-1 in Lanzhou [60]. Two such CMs, hosting seven $\beta_0 = 0.12$ spoke resonators and seven SC solenoids each, accelerated a 10 mA pulsed proton beam in 2016 and possessed full CW operation capability (Fig. 5.31).

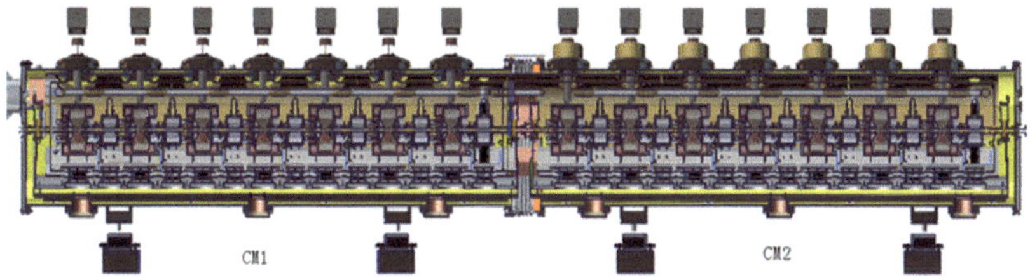

Fig. 5.31 CiADS $\beta_0 = 0.12$ spoke cryomodules [61]. (Courtesy R. Han, IHEP)

5.4.7 Elliptical Cavity Cryomodule Examples

Cryomodules for elliptical cavities in HPHAs differ significantly less among them in comparison with those of the TEM cavities. This is due to the similarity in the shapes and sizes of the cavities and to the fact that most CMs do not include SC solenoids and are somehow derived from the SNS design [62]. However, there are important differences related to RF power coupler requirements, beam current values (average and peak), resonator gradients, and operation modes (pulsed or CW). The elliptical cavities of HPHAs are in the frequency range 644–805 MHz, and their CMs are designed to operate at a temperature of 2 K. However, in some cases (for example, in SNS), they can be conveniently operated at 4.5 K at a reduced gradient. Some relevant examples are given below.

SNS [63] (Sect. 10.1) was the first operating HPHA based on SC elliptical resonators and strongly influenced the design of subsequent ones, particularly concerning CMs (see Sect. 2.2). The SNS linac H^- beam design parameters, at the injection of the associated accumulation ring, were 1 GeV beam energy and 38 mA peak current pulsed at 60 Hz, with a 6% duty cycle (1.6 mA average current). The 1 MW beam power was reached and reliably delivered after several years of operation [64], and the proton power upgrade project (Sect. 10.1.5) to 1.3 GeV and 2.8 MW was also completed [65, 66]. SNS CMs [62] were designed based on the experience gained from the operation of electron linacs (such as CEBAF in JLab, the laboratory where all SNS CMs have been developed). They hosted three $\beta_0 = 0.61$ or four $\beta_0 = 0.81$ 6-cell elliptical cavities working at 804 MHz. A cylindrical space frame structure was used to hold the cavities suspended by means of INVAR rods. This frame was meant to provide robust support for the ready-to-use CM transportation from JLab to ORNL, but it also allowed for reasonably simple insertion of rather long cavity strings in cylindrical vacuum vessels. The cavity helium vessel is made of titanium. The tuners [67] were of the TESLA type and also included a piezoelectric fast tuner for pulsed operation. The single-window, coaxial 50 Ω RF couplers are rated for 550 kW peak power and 45 kW average, and their inner conductor is equipped with a 2.5 kV DC bias voltage to eliminate multipacting. HOM couplers, initially implemented in all cavities, were found to be unnecessary and were affecting cavity performance; they were later removed. The 2.1 K helium was produced in the CM by JT valves, starting from 4.5 K He at 3 bars delivered

Fig. 5.32 Cutaway of the SNS high- and medium-β cryomodules [62]

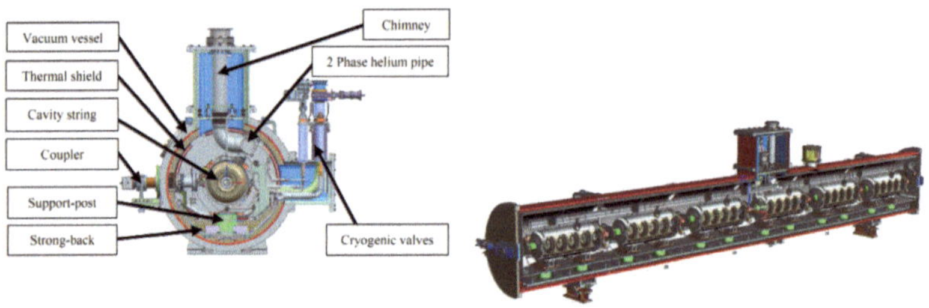

Fig. 5.33 Cross-sections of the FNAL PIP-II high-β 650 MHz cryomodule [24]

by the cryogenic system, which was also used to cool the RF couplers. CMs are connected to the cryogenic system by means of bayonet-type removable transfer tubes for the easy mounting or dismounting of the CM without heating up the entire system. 35 K He gas at 4 bars is used for intermediate thermal shield cooling. "End cans" close the cylindrical CM vessel and provide the function of "chimneys" for helium pipes entering the CM. Magnetic shielding includes two layers: one global, outside the space-frame, and one at the cavity helium vessel (Fig. 5.32).

The SNS CM has been successfully operated for many years. Its design underwent several revisions but preserved most of its original characteristics and became a reference for all subsequent HPHA CMs.

ESS [68], in addition to the spoke CMs described above, included 30 CMs, each containing four 704.42 MHz cavities with a geometrical beta of 0.67 (6-cell) or 0.86 (5-cell) [42, 69]. These cavities for pulsed operation received a peak RF power of up to 1.1 MW through single-window FPCs. The CM design with the space-frame resembled the SNS one. ESS chose to have all of its 120 elliptical cavities of the same length, despite different β_0, to have only one type of CM design. Only one layer of cold global magnetic shielding was used. Cavities were equipped with mechanical and piezoelectric tuners.

A similar ESS-type design [70] is planned for the MYRRHA elliptical cavity section, which will operate in CW mode at 4 mA.

The PIP-II 650 MHz elliptical cavity CM at FNAL [24] (in the production stage) differs from the previous ones due to its bottom-up construction, which is based on a strongback rail inside the vacuum vessel that holds the cavities through cylindrical support posts, similar to the spoke CM mentioned above for the same project (Fig. 5.33).

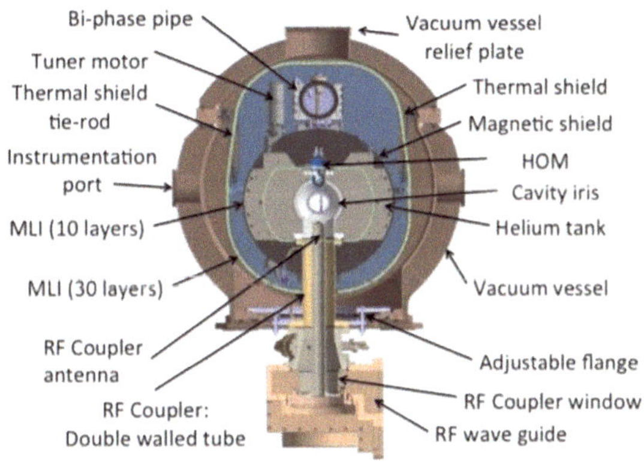

Fig. 5.34 Cross-section of the SPL cryomodule for 704 MHz elliptical cavities [71]. (Courtesy V. Parma, CERN)

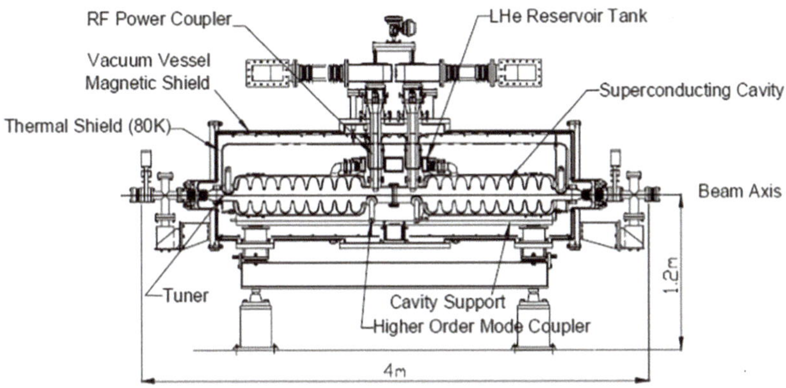

Fig. 5.35 Cross-section of the J-PARC upgrade cryomodule [72]

A rather innovative bottom-up design, which aimed at minimizing the thermal flow to the resonators and considerably simplifying the internal design, was proposed for the SPL H^- linac project at CERN [71], which is still under development. The supporting posts were eliminated, and cavities sit directly on their single-window couplers located underneath at room temperature. Cavities had sliding connections and bellows to the neighboring ones to create a system that is insensitive to thermal contractions (Fig. 5.34).

The J-PARC (Sect. 10.2) upgrade prototype CM [72] hosted two nine-cell $\beta_0 = 0.725$, 972 MHz cavities powered by a single RF source for pulsed operation with 50 mA H^- beam. 2 K helium was produced internally, and both magnetic and intermediate temperature thermal shields were positioned at 80 K. The cavities were equipped with HOM couplers and supported from the bottom (Fig. 5.35).

Finally, it should be noted that cavities for hadron and electron beams become quite similar at relativistic beam energies. Future high-energy hadron linacs may take advantage of the advanced CM technology already developed for electron linear colliders and SC linac light sources.

References

1. H. Padamsee, J. Knobloch, T. Hays, *RF Superconductivity for Accelerators*. Wiley Series in Beam Physics and Accelerator Technology (Wiley, New York, 1998)
2. H. Padamsee, *RF Superconductivity: Science, Technology, and Applications* (Wiley-VCH Verlag GmbH & Co. KGaA, Weinheim, 2009). https://doi.org/10.1002/9783527627172. First published: 18 March 2009, Copyright ©2009 Wiley-VCH Verlag GmbH & Co. KGaA, Weinheim
3. M. Parise, P. Duchesne, D. Longuevergne, D. Passarelli, D. Reynet, F. Ruiu, Mechanical design and fabrication aspects of prototype SSR2 jacketed cavities, in *Proceedings of the SRF'19* (JACoW Publishing, Geneva), pp. 424–429. https://doi.org/10.18429/JACoW-SRF2019-TUP014. https://jacow.org/srf2019/papers/TUP014.pdf
4. S. Stark, et al., Niobium sputter-coated QWRs, in *Proceedings of the SRF'97* (JACoW Publishing, Geneva), pp. 1156–1163. https://jacow.org/SRF97/papers/SRF97D37.pdf
5. C. Benvenuti, Superconducting coatings for accelerating RF cavities: past, present, future, in *Proceedings of the SRF'91* (JACoW Publishing, Geneva). https://jacow.org/SRF91/papers/SRF91C04.pdf
6. A.M. Porcellato, V. Palmieri, S. Stark, F. Stivanello, Niobium sputtered QWRs, in *Proceedings of the SRF'05* (JACoW Publishing, Geneva), pp. 118–122. https://jacow.org/SRF2005/papers/MOP04.pdf
7. W.V. Delsolaro, et al., Nb sputtered quarter wave resonators for the HIE-ISOLDE, in *Proceedings of the SRF'13* (JACoW Publishing, Geneva), pp. 767–772. https://jacow.org/SRF2013/papers/WEIOA03.pdf
8. T. Tan, Nb/Cu cavity development at IMP for the stable operation of high current CW beam. Presentation at the 2018 TTC Meeting, Milano (2018)
9. A. Grassellino, N doping: progress in development and understanding, in *Proceedings of the SRF'15* (JACoW Publishing, Geneva), pp. 48–54. https://jacow.org/SRF2015/papers/MOBA06.pdf
10. S. Posen, M. Liepe, Nb3Sn – present status and potential as an alternative SRF material, in *Proceedings of the LINAC'14, No. 27 in Linear Accelerator Conference* (JACoW Publishing, Geneva, 2014), pp. 431–436. https://jacow.org/LINAC2014/papers/TUIOC03.pdf
11. C.E. Reece, F. Marhauser, A.D. Palczewski, The transfer of improved cavity processing protocols to industry for LCLS-II: N-doping and electropolishing, in *Proceedings of the SRF'15* (JACoW Publishing, Geneva), pp. 418–422. https://jacow.org/SRF2015/papers/MOPB110.pdf
12. T.H. Nicol, et al., SSR1 cryomodule design for PXIE, in *Proceedings of the NAPAC'13* (JACoW Publishing, Geneva), pp. 1373–1375. https://jacow.org/PAC2013/papers/THPMA09.pdf
13. M.P. Howell, et al., The first ASME code stamped cryomodule at SNS, in *Proceedings of the IPAC'12* (JACoW Publishing, Geneva), pp. 2465–2467. https://jacow.org/IPAC2012/papers/WEPPC108.pdf
14. S. Ramo, J. Whinnery, T. Van Duzer, *Fields and Waves in Communication Electronics* (Wiley, 1984). https://books.google.com/books?id=xMVoQgAACAAJ
15. Z. Zheng, A. Facco, Z. Liu, J. Popielarski, K. Saito, J. Wei, Y. Zhang, Prediction and suppression of two-point 1st order multipacting. Nucl. Instrum. Methods Phys. Res. Sect. A **735**, 596–601 (2014). https://doi.org/10.1016/j.nima.2013.09.052. https://www.sciencedirect.com/science/article/pii/S0168900213012862
16. P.N. Ostroumov, M.P. Kelly, S.M. Gerbick, M. Kedzie, S.I. Sharamentov, A new fast tuning system for ATLAS intensity upgrade cryomodule, in *Proceedings of the LINAC'10, No. 25 in Linear Accelerator Conference* (JACoW Publishing, Geneva), pp. 884–886. https://jacow.org/LINAC2010/papers/THP057.pdf
17. C. Darve, The SRF linac components for the European spallation source: first test results, in *Proceedings of the LINAC'16, No. 28 in Linear Accelerator Conference* (JACoW Publishing, Geneva, 2017). https://proceedings.jacow.org/linac2016/talks/we1a03_talk.pdf

18. O.V. Pronitchev, S. Kazakov, Design of main coupler for 650 MHz SC cavities of PIP-II project, in *Proceedings of the NAPAC'16* (JACoW Publishing, Geneva), pp. 121–123. https://doi.org/10.18429/JACoW-NAPAC2016-MOPOB24. https://jacow.org/napac2016/papers/MOPOB24.pdf
19. H. Jenhani, N. Bazin, N. Berton, G. Devanz, P. Hardy, V.M. Hennion, Manufacturing of the IFMIF series power couplers, in *Proceedings of the IPAC'16, No. 7 in International Particle Accelerator Conference* (JACoW Publishing, Geneva, 2016), pp. 2122–2124. https://doi.org/10.18429/JACoW-IPAC2016-WEPMB005. https://jacow.org/ipac2016/papers/WEPMB005.pdf
20. B. Kaizer, Y.B. Aliz, I. Fishman, J. Rodnizki, L. Weissman, Warming rate reduction of the SARAF RF couplers by application of a high voltage DC bias, in *Proceedings of the LINAC'14, No. 27 in Linear Accelerator Conference* (JACoW Publishing, Geneva, 2014), pp. 1168–1170. https://jacow.org/LINAC2014/papers/THPP132.pdf
21. J. Popielarski, P. Glennon, M. Hodek, J. Wlodarczak, Development and testing of prototype fundamental power couplers for FRIB half wave resonators, in *Proceedings of the SRF'11* (JACoW Publishing, Geneva), pp. 353–355. https://jacow.org/SRF2011/papers/TUPO004.pdf
22. R. Paparella, Overview of recent tuner development on elliptical and low-beta cavities, in *Proceedings of the SRF'15* (JACoW Publishing, Geneva), pp. 1425–1431. https://jacow.org/SRF2015/papers/FRAA01.pdf
23. A. Facco, Low- to medium-β cavities for heavy ion acceleration. Supercond. Sci. Technol. **30**(2), 023002 (2016). https://doi.org/10.1088/1361-6668/30/2/023002
24. V. Roger, T.H. Nicol, Y.O. Orlov, Design of the high beta 650 MHz cryomodule – PIP II, in *Proceedings of the NAPAC'16* (JACoW Publishing, Geneva), pp. 149–151. https://doi.org/10.18429/JACoW-NAPAC2016-MOPOB36. https://jacow.org/napac2016/papers/MOPOB36.pdf
25. S.J. Miller, et al., Construction and performance of FRIB quarter wave prototype cryomodule, in *Proceedings of the SRF'15* (JACoW Publishing, Geneva), pp. 1446–1452. https://jacow.org/SRF2015/papers/FRAA06.pdf
26. Z.A. Conway, M.P. Kelly, P.N. Ostroumov, K.W. Shepard, Innovative tuner designs for low beta SRF cavities, in *Proceedings of the SRF'11* (JACoW Publishing, Geneva), pp. 943–945. https://jacow.org/SRF2011/papers/FRIOA02.pdf
27. N. Bazin, et al., Cavity development for the linear IFMIF prototype accelerator, in *Proceedings of the SRF'13* (JACoW Publishing, Geneva), pp. 878–883. https://jacow.org/SRF2013/papers/THIOD03.pdf
28. A. Facco, et al., Superconducting resonators development for the FRIB and ReA linacs at MSU: recent achievements and future goals, in *Proceedings of the IPAC'12* (JACoW Publishing, Geneva), pp. 61–63. https://jacow.org/IPAC2012/papers/MOOAC03.pdf
29. A. Facco, Mechanical mode damping in superconducting low-β resonators. Part. Accel. **61**, 265–278 (1998)
30. T. Xu, et al., FRIB cryomodule design and production, in *Proceedings of the LINAC'16, No. 28 in Linear Accelerator Conference* (JACoW Publishing, Geneva, 2017), pp. 673–678. https://doi.org/10.18429/JACoW-LINAC2016-WE2A02. https://jacow.org/linac2016/papers/WE2A02.pdf
31. V. Zvyagintsev, et al., Status of superconducting ISAC-II and eLinac accelerators, and SRF activities at TRIUMF, in *Proceedings of the RuPAC'16* (JACoW Publishing, Geneva, 2016), pp. 134–138. https://doi.org/10.18429/JACoW-RUPAC2016-THZMH02. https://jacow.org/rupac2016/papers/THZMH02.pdf
32. T. Xu, et al., SRF development and cryomodule production for the FRIB linac, in *Proceedings of the NAPAC'16* (JACoW Publishing, Geneva), pp. 847–852. https://doi.org/10.18429/JACoW-NAPAC2016-WEB3IO01. https://jacow.org/napac2016/papers/WEB3IO01.pdf
33. S. Stark, et al., FRIB HWR tuner development, in *Proceedings of the LINAC'16, No. 28 in Linear Accelerator Conference* (JACoW Publishing, Geneva, 2017), pp. 535–537. https://doi.org/10.18429/JACoW-LINAC2016-TUPLR029. https://jacow.org/linac2016/papers/TUPLR029.pdf

34. J.T. Popielarski, et al., FRIB tuner performance and improvement, in *Proceedings of the NAPAC'19* (JACoW Publishing, Geneva), pp. 755–758. https://doi.org/10.18429/JACoW-NAPAC2019-WEPLM70. https://jacow.org/napac2019/papers/WEPLM70.pdf
35. C. Zhang, W. Hartung, J. Popielarski, K. Saito, S.H. Kim, W. Chang, T. Xu, Certification testing of production superconducting quarter-wave and half-wave resonators for frib. Nucl. Instrum. Methods Phys. Res., Sect. A **1014**, 165675 (2021). https://doi.org/10.1016/j.nima.2021.165675. https://www.sciencedirect.com/science/article/pii/S0168900221006604
36. A. Facco, C. Compton, J. Popielarski, G.J. Velianoff, Virtual welding as a tool for superconducting cavity coarse tuning, in *Proceedings of the IPAC'15* (JACoW Publishing, Geneva), pp. 3412–3414. https://doi.org/10.18429/JACoW-IPAC2015-WEPTY060. https://jacow.org/IPAC2015/papers/WEPTY060.pdf
37. W. Chang, et al., Enhanced pneumatic tuner control for FRIB half-wave resonators, in *Proceedings of the SRF'21, No. 20 in International Conference on RF Superconductivity* (JACoW Publishing, Geneva, 2022), p. 829. https://doi.org/10.18429/JACoW-SRF2021-THPTEV002. https://jacow.org/srf2021/papers/THPTEV002.pdf
38. J.G. Weisend II (ed.), *Cryostat Design: Case Studies, Principles and Engineering*. International Cryogenics Monograph Series (Springer, Cham, 2016)
39. R. Barron, *Cryogenic Systems*. Monographs on Cryogenics (Oxford University Press, Oxford, 1985). https://books.google.com/books?id=SjjTwAEACAAJ
40. P. Knudsen, V. Ganni, Process options for nominal 2-K helium refrigeration system designs. AIP Conf. Proc. **1434**(1), 800–813 (2012). https://doi.org/10.1063/1.4706993. https://doi.org/10.1063/1.4706993
41. R.E. Laxdal, Review of magnetic shielding designs of low-beta cryomodules, in *Proceedings of the SRF'13* (JACoW Publishing, Geneva), pp. 800–807. https://jacow.org/SRF2013/papers/WEIOD01.pdf
42. G. Olivier, J.P. Thermeau, P. Bosland, C. Darve, ESS cryomodules for elliptical cavities, in *Proceedings of the SRF'13* (JACoW Publishing, Geneva), pp. 341–344. https://jacow.org/SRF2013/papers/MOP084.pdf
43. R. Laxdal, M. Marchetto, The ISAC post-accelerator. Hyperfine Interact. **225**, 79–97 (2014). https://doi.org/10.1007/s10751-013-0884-8
44. Z.A. Conway, et al., Cryogenic performance of a new 72 MHz quarter-wave resonator cryomodule, in *Proceedings of the LINAC'14, No. 27 in Linear Accelerator Conference* (JACoW Publishing, Geneva, 2014), pp. 437–439. https://jacow.org/LINAC2014/papers/TUPP001.pdf
45. Z. Zheng, J.T. Popielarski, K. Saito, T. Xu, Cost reduction for FRIB magnetic shielding, in *Proceedings of the LINAC'16, No. 28 in Linear Accelerator Conference* (JACoW Publishing, Geneva, 2017), pp. 818–820. https://doi.org/10.18429/JACoW-LINAC2016-THPRC023. https://jacow.org/linac2016/papers/THPRC023.pdf
46. J. Delayen, Superconducting accelerating structures for high-current ion beams, in *Proceedings of the LINAC'88, Linear Accelerator Conference* (JACoW Publishing, Geneva), pp. 199–201. https://jacow.org/l88/papers/MO3-60.pdf
47. C. Marchand, et al., Performances of Spiral2 low and high beta cryomodules, in *Proceedings of the SRF'15* (JACoW Publishing, Geneva), pp. 967–972. https://jacow.org/SRF2015/papers/WEBA04.pdf
48. N. Sakamoto, et al., Development of superconducting quarter-wave resonator and cryomodule for low-beta ion accelerators at RIKEN Radioactive Isotope Beam Factory, in *Proceedings of the SRF'19* (JACoW Publishing, Geneva), pp. 750–757. https://doi.org/10.18429/JACoW-SRF2019-WETEB1. https://jacow.org/srf2019/papers/WETEB1.pdf
49. H. Jung, H. Jang, Y. Kim, Y. Kwon, SRF status of the RAON heavy ion accelerator project. Presented at SRF2019 in Dresden, Germany, unpublished (2019)
50. P. Duthil, et al., Design and prototyping of the spoke cyromodule for ESS, in *Proceedings of the HB'16* (JACoW Publishing, Geneva), pp. 416–421. https://doi.org/10.18429/JACoW-HB2016-WEAM4Y01. https://jacow.org/hb2016/papers/WEAM4Y01.pdf

51. J.L. Biarrotte, F. Bouly, J.P. Carneiro, D. Uriot, D. Vandeplassche, Design of the MYRRHA 17–600 MeV superconducting linac, in *Proceedings of the SRF'13* (JACoW Publishing, Geneva), pp. 129–132. https://jacow.org/SRF2013/papers/MOP018.pdf
52. W. Kim, M. Lee, H. Kim, Y. Kim, G.T. Park, H. Kim, D.O. Jeon, SSR1 and SSR2 cryomodules for the heavy ion accelerator raon. Phys. Proc. **67**, 853–856 (2015)
53. D. Berkovits, et al., Operational experience and future goals of the SARAF linac at SOREQ, in *Proceedings of the LINAC'12, No. 26 in Linear Accelerator Conference* (JACoW Publishing, Geneva), pp. 100–104. https://jacow.org/LINAC2012/papers/MO1A01.pdf
54. C. Madec, et al., CEA cryomodules design for SARAF phase 2, in *Proceedings of the SRF'17* (JACoW Publishing, Geneva), pp. 70–73. https://doi.org/10.18429/JACoW-SRF2017-MOPB011. https://jacow.org/srf2017/papers/MOPB011.pdf
55. Z.J. Wang, et al., Beam commissioning for a superconducting proton linac. Phys. Rev. Accel. Beams **19**, 120101 (2016). https://doi.org/10.1103/PhysRevAccelBeams.19.120101. https://link.aps.org/doi/10.1103/PhysRevAccelBeams.19.120101
56. Z.A. Conway, et al., Progress towards a 2.0 K half-wave resonator cryomodule for Fermilab's PIP-II project, in *Proceedings of the LINAC'16, No. 28 in Linear Accelerator Conference* (JACoW Publishing, Geneva, 2017), pp. 906–909. https://doi.org/10.18429/JACoW-LINAC2016-THPLR027. https://jacow.org/linac2016/papers/THPLR027.pdf
57. H. Dzitko, et al., Design and manufacturing status of the IFMIF-LIPAC SRF linac, in *Proceedings of the IPAC'15* (JACoW Publishing, Geneva), pp. 3686–3688. https://doi.org/10.18429/JACoW-IPAC2015-THPF006. https://jacow.org/IPAC2015/papers/THPF006.pdf
58. J. Knaster, Y. Okumura, P. Cara, A. Kasughai, M. Sugimoto, Challenges of the high current prototype accelerator of IFMIF/EVEDA, in *Proceedings of the IPAC'16, No. 7 in International Particle Accelerator Conference* (JACoW Publishing, Geneva, 2016), pp. 52–57. https://doi.org/10.18429/JACoW-IPAC2016-MOZB02. https://jacow.org/ipac2016/papers/MOZB02.pdf
59. J. Wei, et al., The FRIB superconducting linac – status and plans, in *Proceedings of the LINAC'16, No. 28 in Linear Accelerator Conference* (JACoW Publishing, Geneva, 2017). https://doi.org/10.18429/JACoW-LINAC2016-MO1A01. https://jacow.org/linac2016/papers/MO1A01.pdf
60. G. Wu, SRF cryomodules for PIP-2 at fermilab. Presented at SRF2019 in Dresden, Germany, unpublished (2019). https://proceedings.jacow.org/srf2019/talks/wetea2_talk.pdf
61. R. Han, et al., Design of C-ADS injector-I cryomodule for 325 MHz cavities, in *Proceedings of the SRF'17* (JACoW Publishing, Geneva), pp. 294–296. https://doi.org/10.18429/JACoW-SRF2017-MOPB099. https://jacow.org/srf2017/papers/MOPB099.pdf
62. W.J. Schneider, et al., Design of the SNS cryomodule, in *Proceedings of the PAC'01* (JACoW Publishing, Geneva), pp. 1160–1162. https://jacow.org/p01/papers/MPPH162.pdf
63. S. Henderson, Commissioning and initial operating experience with the SNS 1-GeV linac, in *Proceedings of the LINAC'06, No. 23 in Linear Accelerator Conference* (JACoW Publishing, Geneva), pp. 1–5. https://jacow.org/l06/papers/MO1002.pdf
64. J. Galambos, SNS high power operation – expectations and experience, in *Proceedings of the HB'10* (JACoW Publishing, Geneva), pp. 11–15. https://jacow.org/HB2010/papers/MOIB01.pdf
65. S.H. Kim, M. Doleans, J. Galambos, M.P. Howell, J.D. Mammosser, Superconducting linac upgrade plan for the second target station project at SNS, in *Proceedings of the SRF'15* (JACoW Publishing, Geneva), pp. 268–271. https://jacow.org/SRF2015/papers/MOPB069.pdf
66. S.H. Kim, Status of the spallation neutron source proton power upgrade. Presented at LINAC'24, Chicago, IL, Aug. 2024, paper TUXA001, unpublished (2024)
67. J. Preble, et al., SNS cryomodule performance, in *Proceedings of the PAC'03* (JACoW Publishing, Geneva), pp. 457–461. https://jacow.org/p03/papers/ROAA001.pdf
68. M. Lindroos, H. Danared, C. Darve, D.P. McGinnis, S. Molloy, The ESS linac design, in *Proceedings of the LINAC'12, No. 26 in Linear Accelerator Conference* (JACoW Publishing, Geneva), pp. 768–772. https://jacow.org/LINAC2012/papers/TH2A01.pdf

69. F. Peauger, et al., Progress in the elliptical cavities and cryomodule demonstrators for the ESS linac, in *Proceedings of the SRF'15* (JACoW Publishing, Geneva), pp. 544–548. https://jacow.org/SRF2015/papers/TUPB007.pdf
70. F. Bouly, et al., Developments and tests of a 700 MHz cryomodule for the superconducting linac of MYRRHA, in *Proceedings of the SRF'13* (JACoW Publishing, Geneva), pp. 250–254. https://jacow.org/SRF2013/papers/MOP057.pdf
71. V. Parma, et al., Status of the Superconducting Proton Linac (SPL) cryomodule, in *Proceedings of the SRF'13* (JACoW Publishing, Geneva), pp. 345–348. https://jacow.org/SRF2013/papers/MOP085.pdf
72. N. Ouchi, J-PARC upgrade, in *Proceedings of the SRF'09* (JACoW Publishing, Geneva), pp. 934–940. https://jacow.org/SRF2009/papers/FROBAU05.pdf

Chapter 6
Magnets

Abstract This chapter begins with an overview of magnets used in high-power hadron accelerator facilities, categorized by their function and type for both normal-conducting and superconducting magnets. Key design considerations including achievable field strength and beam aperture, radiation and beam-induced heating, as well as construction and operational costs, are discussed. A general magnet design process is presented to provide a complete view of magnet design that extends beyond electromagnetic analysis, including mechanical, thermal, and operational aspects. Finally, a large-scale superferric dipole magnet developed for FRIB is presented as a case study to illustrate the important design elements and the associated engineering decision-making process during design and construction.

Magnets are optical devices for transporting and focusing beams through particle accelerators. Both normal-conducting (NC) and superconducting (SC) magnets are widely used in high-power hadron accelerator facilities. In this chapter, we introduce magnets for various applications, followed by a general design process and a design example of an SC dipole magnet.

Magnets confine charged particles in accelerators for acceleration. A magnet with the field direction along the beam axis z having fields in its body,

$$\begin{cases} B_x = B_y = 0 \\ B_z = B_s \end{cases}, \tag{6.1}$$

provide coupled focusing in the transverse (horizontal x and vertical y) directions. Such solenoidal magnets are often used near the ion source or in heavy-ion linacs that accelerate multiple species and charge states of ions. Most accelerator magnets are separate-function, non-solenoidal magnets, with fields in the body taking the form

$$\begin{cases} B_y + iB_x = B_0 \sum\limits_{n=1}^{\infty} (b_n + ia_n)(x + iy)^{n-1} \\ B_z = 0 \end{cases}, \tag{6.2}$$

J. Wei et al., *High Power Hadron Accelerators*, Particle Acceleration and Detection,
https://doi.org/10.1007/978-3-032-17662-2_6

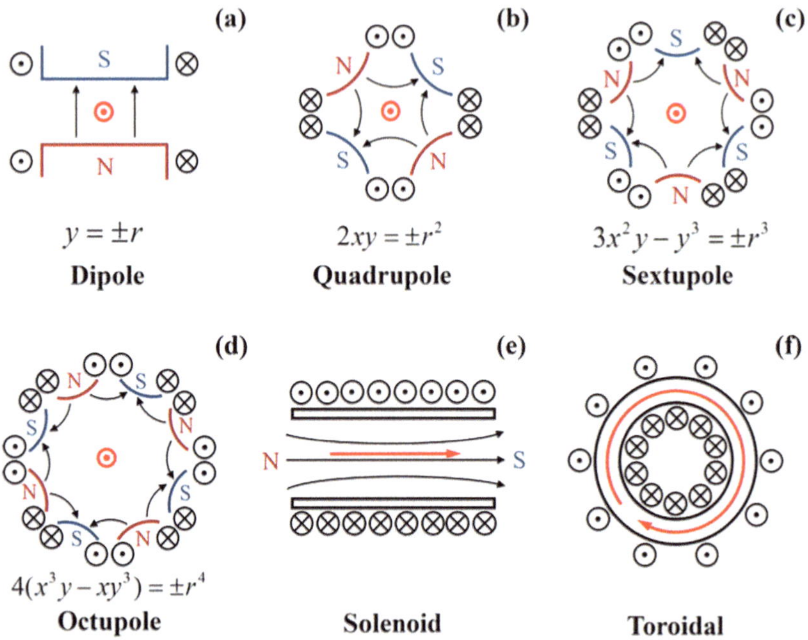

Fig. 6.1 Magnet schematics of (**a**)–(**d**) normal multipoles, (**e**) solenoid magnets, and (**f**) toroidal magnets. Equations under schematics of (**a**)–(**d**) are the ideal pole profiles in 2-D with respect to bore radius r. x and y are the horizontal and vertical axis with origin at the center of the magnet. A skew multiple a_n may be obtained by rotating around the beam (z) axis the normal multipole b_n by $\pi/2n$ with $n = 1, 2, 3, 4, \ldots$ for dipole, quadrupole, sextupole, octupole, ... respectively. Red color indicates the beam direction

where n is the multipole order, b_n are the normal multipoles, and a_n are the skew multipoles. The normal dipole b_1 and the skew dipole a_1 bend the beam horizontally and vertically, respectively. The normal quadrupole b_2 provides focusing in the transverse directions. The skew quadrupole a_2 is often used for transverse coupling. The normal sextupole b_3 is used to vary chromaticity. Nonlinear $n \geq 3$ multipoles are used for resonance corrections. Figure 6.1 illustrates the layout of the magnetic pole tip and the field directions of these magnets. Combined-function magnets that combine dipole b_1 and quadrupole b_2 field components were used in some early-generation accelerators.

Realistic magnets contain allowed multipoles along the magnet body and fringe fields near the ends of the magnet. The leading multipoles are those that are allowed geometrically by the symmetry and manufacturing design of the magnet. For a magnet b_n of $2n$ poles ($n = 1$ for a dipole, $n = 2$ for a quadrupole magnet, and so on) the allowed multipoles n' are

$$n' = n(2k + 1), \qquad k = 0, 1, 2, \ldots \tag{6.3}$$

For a dipole magnet b_1, the allowed multipoles include sextupole b_3, decapole b_5, etc. For a quadrupole magnet b_2, the allowed harmonics are at the 12th pole b_6, 20th pole b_{10}, etc. Tolerance to magnetic field error is an important factor that dictates the feasibility of engineering design and the cost of fabrication. Typically, the tolerable fractional field error in the "good field region" occupied by the beam is of the order of 10^{-4} for main magnets in rings where the beam passes for many hundreds of revolutions; 10^{-3} for main dipole and quadrupole magnets in linacs where the beam is a single pass, and for chromatic sextupoles in rings; and 10^{-2} for correction magnets.

6.1 Magnet Applications in Particle Accelerators

This section introduces several types of magnets for various applications in hadron accelerators.

6.1.1 Confinement Devices for Low Energy Beams

Electrostatic devices are often used at low beam energy, particularly near the ion source and in the area of low-energy beam transport. As a particle of charge q under EM experiences the Lorentz force

$$\mathbf{F}_L = q\,\mathbf{E} + q\,\mathbf{v} \times \mathbf{B} \tag{6.4}$$

favors the electric field when the speed of the particle $|\mathbf{v}|$ is sufficiently low [11]. Furthermore, electrostatic devices have the advantage that the deflection and focusing strength provided by the electric field is independent of the charge state of the particles [56]. Thus, such devices, in particular, are suitable for heavy-ion facilities that accelerate multiple charge states of ions at low energies [44].

For reliable operations, the operating electric voltage is typically kept below half the breakdown voltage, which depends on the material, design geometry, surface finish, contaminants, and conditioning. The corresponding operating electric field in vacuum is typically limited to about 10^7 V/m.

Figure 6.2 shows part of the 40 m long LEBT of the FRIB driver linac, consisting mostly of an electrostatic deflector and electrostatic quadrupoles [34]. Stable ions of different charge states (for example, Ar^{8+} and Ar^{9+}) are transported simultaneously through the LEBT at an energy of 0.5 MeV/u. Beyond this energy, electromagnetic magnets are used to steer and focus the beam to 200 MeV/u and beyond.

Fig. 6.2 The low-energy beam transport (LEBT) at the FRIB driver linac (Courtesy MSU, USA). The 40 m long LEBT consists of upper (left), vertical, and lower (right) sections containing electrostatic deflectors, electrostatic quadrupoles and EM solenoids to guide and focus the heavy ion beams

6.1.2 Permanent Magnets

Permanent magnets have the advantage of providing a high magnetic field in a compact size and operate in the absence of a power supply and water-cooling complexities [26]. They are also immersible in other fields without perturbing them. The main disadvantages of permanent magnets are that the fields cannot be adjusted and that they lose their permanent magnetic properties when the temperature exceeds the Curie temperature or because of radiation damage.

Figure 6.3 shows the permanent magnet quadrupoles (PMQ) used in the SNS DTL. Their compact sizes allowed them to be used in drift tubes operating at the relatively high radio frequency of 402.5 MHz. However, a complication occurred during DTL fabrication due to imperfect shielding, as the magnetic field of the PMQ distorted the electron beam during e-beam welding and damaged the drift tube enclosure. Repeated welding corrections at an excessive temperature above the Curie point further compromised the magnetic properties of the PMQ. Significant efforts were made to repair and replace damaged PMQs in DTLs, mitigate issues, and restore performance to the design level. Such repair efforts were successful, as demonstrated by the results of linac commissioning [1].

6.1.3 Resistive Magnets

Electromagnets are most often used in high-power accelerators because of their operational flexibility. Such resistive magnets operating at room temperature offer relatively low engineering and integration complexity, as well as fabrication costs.

Resistive magnets consist mainly of coils of normal conductors (NC), such as copper, aluminum, and iron poles, along with a yoke of high permeability. The conceptual design can be made using Ampère's law, assuming $\mu = \infty$ and thus

Fig. 6.3 Permanent magnet quadrupole used in the SNS drift tube linac. The material is Samarium Cobalt (SM_2CO_{17}) The integrated gradient is around 1.3 T (Courtesy SNS, US)

$H_{iron} = 0$ in the iron in a 2-D configuration with infinite extent in the beam direction z,

$$\oint \mathbf{H} \cdot d\mathbf{l} = \frac{B(2a)}{\mu_0} + H_{iron} l_{iron} = I \ , \tag{6.5}$$

where **B** is the magnetic field in the gap, $2a$ is the height of the gap, and μ_0 is the permeability of the vacuum. As illustrated in Fig. 6.4, for an H-frame dipole magnet with N_{coil} turns of coil wound around both the top and bottom poles, with current I_{coil} powering them, the magnetic field in the gap is given by

$$B\,[\mathrm{T}] = \frac{4\pi\, 10^{-7} 2 N_{coil} I_{coil}}{2a}\ [\mathrm{A} \cdot \mathrm{turn} \cdot \mathrm{m}^{-1}] \, . \tag{6.6}$$

Similarly, for a quadrupole magnet with pole radius a and N_{coil} turns of coil wound around each of the four poles with current I_{coil} powering them, the magnetic field gradient is given by

$$B'\,[\mathrm{T/m}] = \frac{4\pi\, 10^{-7} 2 N_{coil} I_{coil}}{a^2}\ [\mathrm{A} \cdot \mathrm{turn} \cdot \mathrm{m}^{-2}] \, . \tag{6.7}$$

With a finite extent of magnet length, realistic yoke and pole geometry, and finite material permeability, non-fundamental multipoles are introduced. An H-dipole magnet usually contains a sextupole component as an allowed harmonic; a C-dipole magnet contains a quadrupole component; a quadrupole magnet contains a dodecapole component, etc. Longitudinally, the field usually extends outside the magnet edge for a distance comparable to the pole aperture.

The construction of the core (or yoke), in particular the pole geometry and the yoke material composition, largely determines the field quality of an iron-

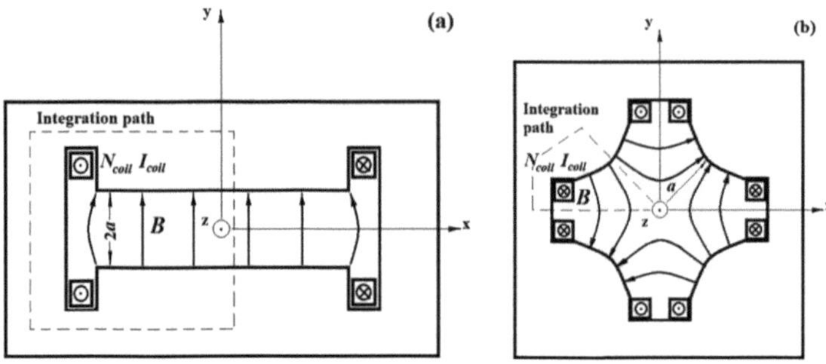

Fig. 6.4 Cross sections of (left) H frame dipole and (right) quadrupole magnets with integration path of the Ampere law indicated

dominated magnet. Pole shaping and pole end-profile compensation are often used to improve field quality and reduce unwanted multipole harmonics. As the amplitude functions β_x and β_y do not vary significantly across the magnet (unlike interaction region magnets in high-energy colliders), the goal of magnetic field optimization, assisted by 3-D modeling codes, is often to minimize the leading integral multipole harmonics across the magnet.

The magnetic properties of the core material (permeability, saturation, and coercivity) depend on both the chemistry and the heat treatment process. Low carbon steels are often chosen for their overall performance, cost, and ease of fabrication. A B-H curve, also known as a magnetization or hysteresis curve, describes the relationship between the strength (H) of the external magnetic field applied to the material and the resulting magnetic flux density (B) within a magnetic material such as steel. The B-H curves of several types of commonly used magnetic steels are shown in Fig. 6.5. The magnetic property can vary with the chemistry of the material for the same type; thus, testing material samples for each lot is important. For critical applications, it is preferred to use the same lot of material for the magnetic poles of each magnet.

The construction of the coil, magnet steel, and power supply usually dominates the cost and reliability of the magnet. With the exception of low-current correctors, most electromagnets are water-cooled, often with the coil made of a hollow copper conductor through which chilled water flows to increase current density. The flow speed, pressure drop, and temperature drop are usually configured so that the water flow is turbulent to promote heat transfer while keeping internal stress and erosion to copper minimal. The conductor is often wrapped with polyimide or fiberglass tape and impregnated with epoxy for insulation, mechanical stability, and radiation resistance. The performance of the epoxy material, which is chosen for its optimized dielectric properties, flexibility, strength, and endurance to temperature variation and radiation damage, is the key to the reliability of the magnet.

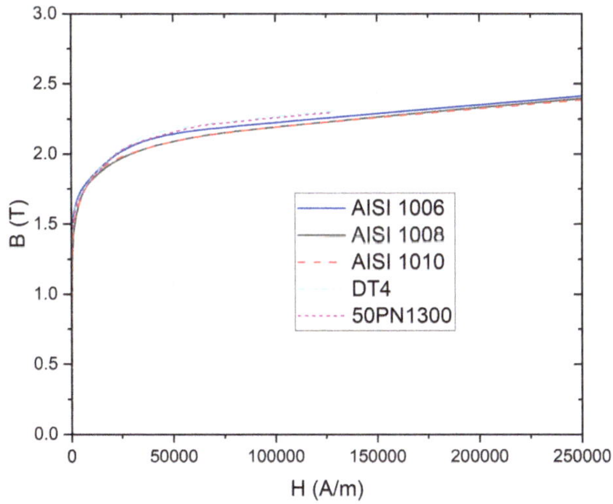

Fig. 6.5 B-H curves of several types of commonly used magnetic steels. Data source for the plot is from the following reference [8, 15, 57]

6.1.3.1 DC Magnet

Linacs and accumulator rings mostly use DC magnets for beam confinement. Because of the lack of ramp-induced eddy currents, either solid-steel or laminated-steel cores can be used. For magnets whose field quality is less stringent, such as those in a linac, solid steel is often used for its low fabrication cost. The maximum field is typically chosen to be below 1.8 T to avoid over-saturation.

Magnets in rings are grouped into families and powered by common power supplies for both cost savings and magnet-to-magnet field tracking. Field quality requirements apply to the magnet transfer function (defined as the ratio between the magnetic field integrated along the path of the particle and the magnet excitation current), the magnet center deviation, and the magnetic multipoles. Laminated steel cores are sometimes chosen for their good mechanical accuracy, high uniformity of material and field, and cost savings in high-volume production. The laminations produced in different production periods are often randomized to minimize variation in material properties.

Figure 6.6 (left) shows the main dipole magnet of the SNS accumulator ring with a large gap of $2a = 170\,\text{mm}$. To ensure high field quality, the maximum design field was chosen modestly as 0.77 T for the nominal operational beam energy of 1 GeV and 0.93 T for the upgraded beam energy of 1.3 GeV. Solid steel was selected for the magnet cores to reduce costs. Although individually the field quality ($\sim 10^{-4}$ relative error at full acceptance) was satisfactory, excessive (up to 0.25%) magnet-to-magnet variation was found in the dipole transfer function and its current dependence, as shown in Fig. 6.6 (right). Thin (tens of micrometers) layers of shims

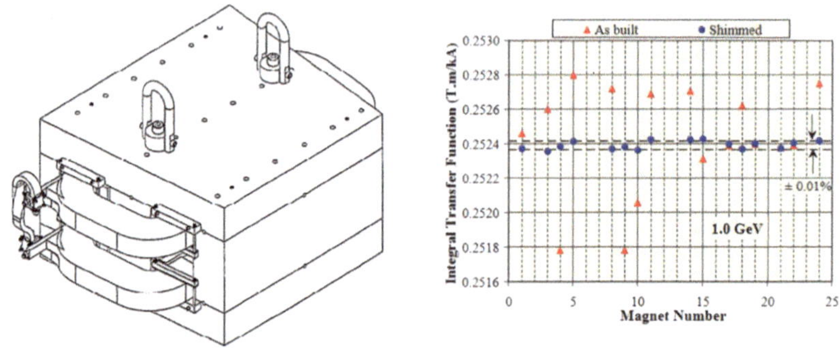

Fig. 6.6 (Left) Schematics of the main dipole of the SNS accumulator ring. (Right) Integral transfer function before and after shimming at 1 GeV [50] (Courtesy P. Wanderer, BNL)

Table 6.1 Major parameters of the main dipole magnet of the SNS accumulator ring [50]

Parameters	Values
Effective length	1.44 m
Gap	170 mm
Pole width	450 mm
B_0 at 1.0 GeV (1.3 GeV)	0.77 (0.925) T
I_{op} at 1.0 GeV (1.3 GeV)	4398 (5408) A
$\int B \cdot dl$ at 1.0 GeV (1.3 GeV)	1.11 (1.33) T·m

were inserted at the base of the pole or the back leg of the magnets to reduce the variation in the transfer function below 10^{-4} for operation at nominal energy [50].

Table 6.1 lists the main parameters of the magnet. Table 6.2 lists the integral harmonics of a representative magnet measured by a rotating coil 0.92 m long and 136.8 mm in diameter. The harmonics are given at a reference radius of 80 mm in "units", which are 10^{-4} of the fundamental field. Most of the harmonics are smaller than 1 unit. The exceptions are the normal quadrupole component of -105 units (-1%) caused by the magnet taper by design and the correspondingly affected normal octupole of 2 units. The harmonics are essentially unchanged by the shimming. Typical harmonic changes in the measured magnets are less than 0.1 units.

6.1.3.2 Ramping Magnet

Ramping magnets are needed in many accelerators, including rapid-cycling synchrotrons, injection and extraction kickers, and septum magnets. The key challenges are engineering complications caused by eddy current heating and magnetic field deterioration due to time-dependent effects, hysteresis effects and eddy current effects as discussed in Sect. 3.4.3.1. The peak magnetic field and peak ramp rate of fast ramping magnets used in high-power rings are typically limited to below 1.5 T and 50 T/s, respectively, which limits the repetition rate to below 50 Hz.

Table 6.2 Normal and skew multipole harmonics of the SNS dipole magnet measured by a rotating coil. The harmonics are given at a reference radius of 80 mm in "units," which are 10^{-4} of the fundamental field [50]

Harmonic	Normal	Skew
Dipole	10,000.00	0.00
Quadrupole	− 105.21	0.10
Sextupole	0.56	− 0.50
Octupole	2.06	0.01
Dccapole	0.97	− 0.04
Dodecapole	0.18	− 0.06
14-pole	0.28	0.04
16-pole	0.05	0.05
18-pole	− 0.06	− 0.02
20-pole	0.01	− 0.04
22-pole	− 0.22	0.01

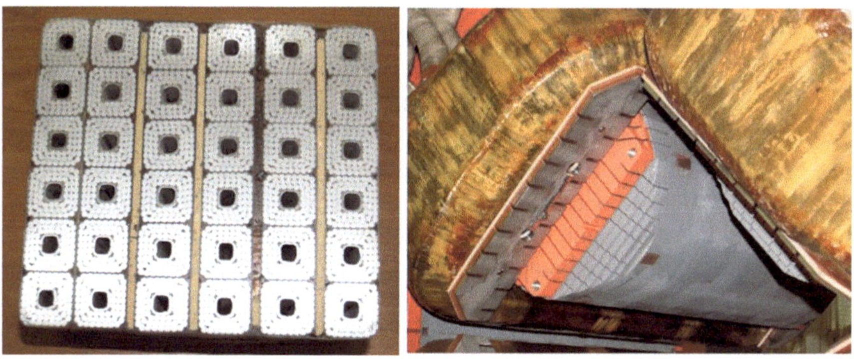

Fig. 6.7 Conductors made of stranded aluminum wire surrounding a water-cooling channel (left) and laminated steel core with cuts near the ends (right) developed for the main magnets of the J-PARC rapid-cycling synchrotron [47, 48] (courtesy N. Tani, J-PARC)

Various engineering measures need to be implemented to mitigate eddy-current effects. The magnet core should be made of laminated steel. In addition, cuts are made near the ends of the magnet core, where the longitudinal component of the magnetic field and its change are not negligible, as shown in Fig. 6.7.

Figure 6.7 shows the main dipole magnet of the J-PARC 3 GeV RCS, which operates at a repetition rate of 25 Hz with a peak field of the dipole magnet of 1.4 T. Due to the combination of a relatively high field and a high repetition rate, J-PARC developed a water-cooled stranded aluminum conductor for the main magnet coils to mitigate large eddy-current losses [47, 48]. Table 6.3 lists the major parameters of the dipole magnet. Table 6.4 lists the calculated eddy current loss and mean temperature compared to the measured temperature.

Rapid-cycling synchrotrons with a repetition rate of 25 Hz and higher have been using sinusoidal waveforms to power magnets [47]. Ceramic vacuum chambers are typically used to mitigate the effects of eddy currents induced by magnets on the walls of the vacuum chamber. During the early stage of the SNS construction, a

Table 6.3 Major parameters of the main dipole magnet of the J-PARC rapid-cycling synchrotron [47]

Parameters	Values
Field	0.27 T (for 400 MeV)
	1.1 T (for 3 GeV)
Useful horizontal aperture	± 120 mm
Gap height	210 mm
Core length	2770 mm
AC current	970 A
DC current	1600 A

Table 6.4 Calculated eddy current loss and mean temperature in comparison with the measured temperature [47]

2-D Model	Eddy loss	Mean temperature	Measured temperature
Without slits	0.971 kW	144°C	146°C
With slits	0.572 kW	91°C	105°C

study was performed on a possible RCS design as an alternative to the baseline accumulator ring design [52] (Sect. 10.1). The RCS design adopted programmable magnet power supplies for a 30 Hz double ring configuration to minimize eddy currents and maintain good field quality. It maintains a moderate ramp rate below 30 T/s with a linear up-ramp of 24 ms and a down-ramp of 8 ms. This RCS design was abandoned due to economic challenges, particularly those associated with the programmable magnet waveform.

The AGS Booster was designed as a versatile synchrotron that serves as an injector for the AGS at BNL, supporting both heavy-ion RHIC operations and high-power proton acceleration. The power supply for the main magnets of the Booster must accommodate a wide range of cycles and parameters, ranging from storage for several seconds to rapid cycling at 7.5 Hz [53]. To accommodate fast ramping, Inconel vacuum chambers were used to contain the beam. Passive eddy-current correction windings were used in the vacuum chamber beneath the main dipole magnet to compensate for the effects of the induced sextupole harmonic error.

6.1.3.3 Kicker Magnets

Kickers are fast-ramping dipole magnets in accelerators that are often used in the injection and extraction processes, which rapidly switch a particle beam between two paths. Conceptually similar to a railroad switch in function, a kicker magnet must switch on very rapidly and then maintain a stable magnetic field for a minimum period. The switch-off time can be less important if the subsequent beam does not follow immediately.

The dipole bump magnets for the SNS ring injection painting were powered by programmable power supplies. These bump magnets operated at a repetition rate of 60 Hz, covering an accumulation time 1 ms. Ferrite was used as the core material for its high-frequency response and low loss (Fig. 6.8). The coils were

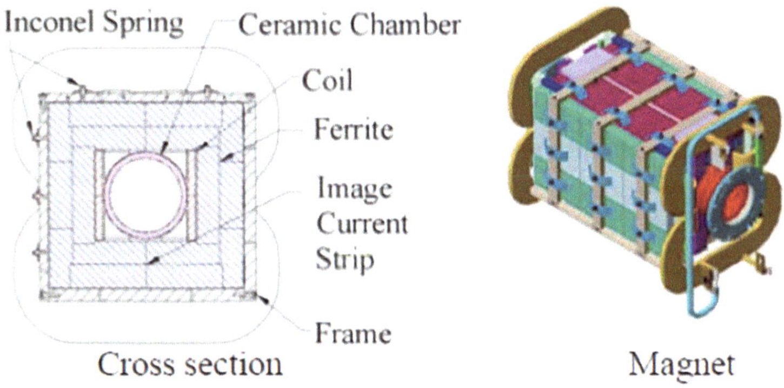

Fig. 6.8 SNS accumulator ring injection kicker magnet for the transverse phase-space painting [33] (Courtesy C. Pai, BNL)

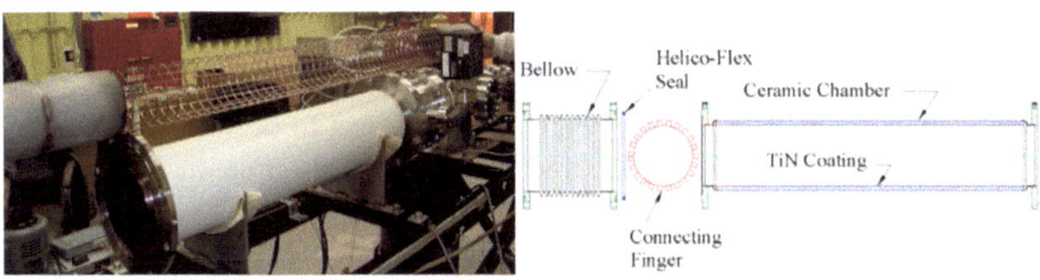

Fig. 6.9 Vacuum chamber of the SNS accumulator ring injection kicker magnet [33] (Courtesy C. Pai, BNL)

wound with copper conductors and were water-cooled. The vacuum chamber was made of alumina ceramic to avoid eddy currents induced by the pulsed magnetic field. The inner surface of the ceramic chamber was coated with titanium nitride (TiN) to carry beam-induced image current and to reduce the secondary electron yield (Fig. 6.9).

Kickers with a relatively fast rise and fall time but a weak deflecting field usually work together with the septa that operate at DC or with a slow rise and fall time but a strong field (Sect. 6.1.3.4). A group of individually powered kickers is often needed to adequately deflect the beam. Each individual kicker magnet is powered by a high voltage source (usually in the range of tens of kV) called a power modulator (Sect. 8.6.3), which uses a pulse-forming network to produce a short pulse of current (usually in the range of a few ns to a μs and several kA in amplitude) to power the kicker magnet.

Figure 6.10 shows the lumped kicker magnet used in the SNS ring to deflect the beam vertically. The coil was made of a single-turn copper conductor. The yoke was made of a high-permeability ferrite material to enhance magnetic deflection. The ferrite yoke of the kicker had a window frame design but was separated at the mid-plane by a copper ground plane, which extended for the full length of the unit. This arrangement provided a low longitudinal coupling impedance for

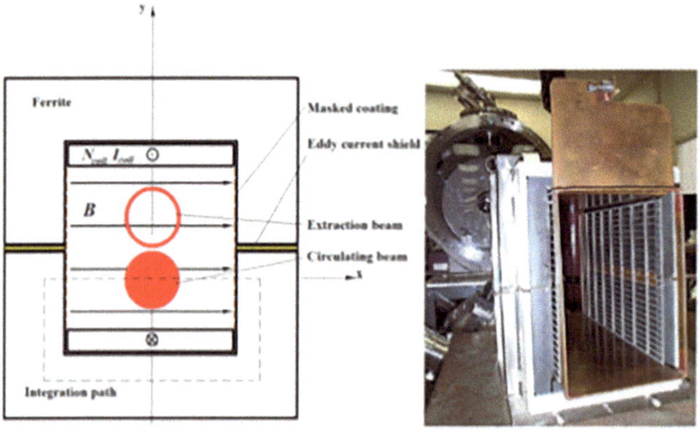

Fig. 6.10 SNS extraction kicker: (left) cross-section of a lumped kicker magnet and (right) kicker magnet contained inside a bakeable vacuum vessel [32] (Courtesy C. Pai, BNL)

the circulating beam. The magnet height was designed to be sufficiently large not only to contain the circulating and extracted beams but also to reduce the transverse coupling impedance [17].

The kicker assembly was designed to be baked to 250 °C so that the porous ferrite material could adequately release gas when reaching the required vacuum level of about 10^{-8} Torr. As the ferrite surface had a high secondary electron yield upon bombardment by the stray beam, the exposed ferrite surfaces were coated with a thin layer of TiN film of 0.1 μ m thickness to suppress the secondary emission. Because the TiN film was conductive, gaps in the coating pattern were used to prevent high voltage breakdown against the magnet coil and to mitigate eddy-current heating induced by the fast-varying magnetic field.

6.1.3.4 Septa

A septum is a partition boundary that provides different deflecting fields on each side of the boundary. It is typically used together with kicker magnets (Sect. 6.1.3.3) in a two-step process, either during injection into a synchrotron, with the beam injected through the septum being deflected by the downstream kickers for single-turn injection, or bump magnets for multi-turn injection, or during extraction from a synchrotron, with kickers deflecting the beam into the extraction channel of the septum (Chap. 3).

There are two main types of septa: the electrostatic septa and the magnetic septa. A magnetic septum is either a pulsed or DC dipole magnet with a thin (typically $\sim$ 2 mm) separation between the zero-field and high-field regions. An electric septum is a pulsed or DC electrostatic device with a very thin (typically $\sim$ 0.1 mm) separation between the zero-field and high-field regions. A challenge in the septum

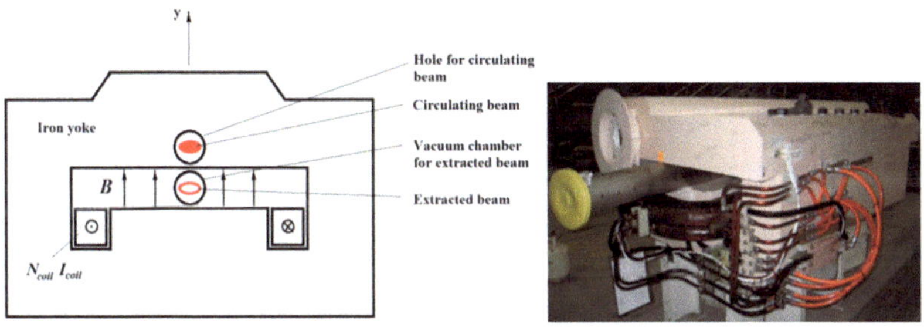

Fig. 6.11 SNS Lambertson magnetic septum for the accumulator ring extraction [35] (Courtesy J. Rank, BNL)

design is to ensure a low leakage field affecting the circulating beam and to provide high field homogeneity in the region for the deflecting beam.

Two types of magnetic septa are commonly used in high-power accelerators: Lambertson iron-septum dipoles and current-carrying septum dipoles [36]. Septa design involves the analysis of field uniformity, end fields, leakage fields, vacuum properties, cooling, corrosion, insulation, radiation resistance, forces, expansion, clamping, and electrical power. In the case of a pulsed septum, the effects of the eddy current must also be assessed.

Fast extraction of high-power beams requires that a bunched beam be extracted in a single passage without beam loss after the beam is energized to the extraction energy. Although the goal of extraction efficiency is 100%, a possible kicker misfire usually results in a stray beam hitting the septum partition, causing damage or radio-activation. The Lambertson septum is thus favored.

Figure 6.11 shows the Lambertson septum built for the SNS accumulator ring. The circulating beam is contained in the field-free upper circular channel in the steel. Upon extraction, the beam is deflected vertically by the kicker magnets to enter the extraction channel underneath on the subsequent turn. The DC magnetic field bends the beam horizontally into the downstream transfer line and to the neutron target. A major advantage of the Lambertson septum is that the conductors are enclosed in the steel yoke at a distance far from the circulating beam, the extracted beam, and the possible stray beam caused by partially fired kickers that are likely to strike the steel separating the two beam channels. The SNS Lambertson septum is also designed to be installed at a small ($\sim$ 4%) "roll angle" to create a vertical bending angle, which is countered by a downstream skew dipole magnet that accommodates the different beam heights as determined by the design of the accumulator ring and the target.

Electrostatic septa are used when a thin boundary partition is required. During slow extraction, when a stream of the DC beam is continuously extracted from the circulating beam (3), the extraction efficiency must be high (typically about 99%) so that the amount of uncontrolled beam loss is low and the radio-activation is tolerable. Such high efficiency can only be achieved if the ripple of the magnet (e.g. resonance sextupole magnet) power supply is extremely low and if the thickness of the first septum is extremely thin (typically $\sim$ 0.1 mm). This can be achieved

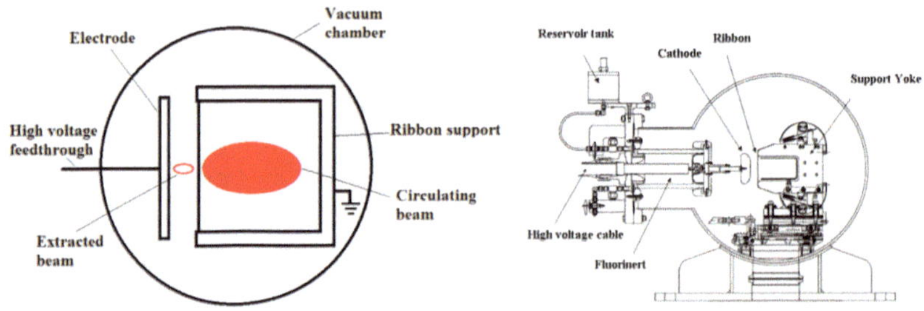

Fig. 6.12 Illustration of the electrostatic septum (left) and the unit developed for the slow extraction of the J-PARC Main Ring (right) [6] (Courtesy Y. Arakaki, J-PARC)

with an electrostatic septum: The beam is deflected by an electric field in a direction perpendicular to the initial beam direction, as shown in Fig. 6.12 (left). The partition can be formed by a grounded array of tungsten wires with a thickness of 0.05 mm and a spacing of 1 mm. Figure 6.12 (right) shows the electrostatic septum developed for the slow extraction of the 30 GeV proton beam of the J-PARC Main Ring [6].

6.1.4 Superconducting Magnets

SC magnets, widely used in hadron accelerators, are usually cooled by liquid helium to cryogenic temperatures for operation. SC wires are chosen for their ability to carry a high current density so that the magnet can achieve a higher field and/or a more compact size. In its SC state, the wire can carry a high current without DC dissipation. The trade-off is the initially high construction cost of the wire and cryostat, as well as the static and dynamic heat load that is now deposited at cryogenic temperatures. There can often be general savings in power consumption. SC magnets are typically used when the magnetic field is above 2 T or the bore size is large.

6.1.4.1 Superferric Magnets

Superferric magnets refer to iron-dominated magnets with coils of SC wires. The magnet iron core can operate either at room temperature or at cryogenic temperatures, submerged in liquid helium. Currently, most superferric magnets operate at a maximum magnetic field of around 2 T. At this field level, the iron geometry largely determines the quality of the magnetic field in the static state. The coil dimensions and winding quality can be relaxed, having a minimal impact on field quality. Compared with resistive magnets operating at room temperature, the technical complexities are primarily associated with cryogenic interfaces and factors that may result in premature quenching of the magnets.

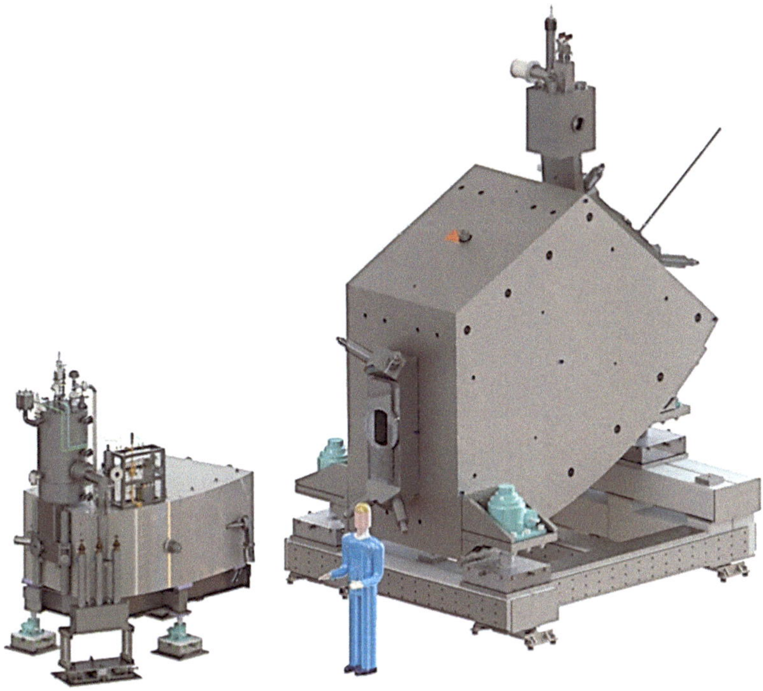

Fig. 6.13 FRIB superferric dipole magnets at the folding segment 2 of its driver linac (left) and at the fragment separator region downstream of the production target (right) (Courtesy MSU, USA)

At Michigan State University, superferric magnets have long been used in the beamlines of the NSCL cyclotron facility [30, 31, 41]. FRIB uses superferric magnets in the folding segment 2 of its driver linac and in the fragment separator region downstream of the production target. These magnets individually weigh approximately 30 tons to 180 tons, as shown in Fig. 6.13. The magnet iron is kept at room temperature, while the coils are housed in a cryostat. Although field quality requirements for such static, single-pass beams are less stringent compared with what is required for ramping, multi-pass rings, significant challenges exist with magnet construction, installation, and maintenance. These challenges include temperature uniformity requirements during cool-down and warm-up (typically within a 50 K temperature range), interlacing installation and maintenance logistics associated with congested spaces and external vacuum vessel containment, and handling in areas of potentially high radio-activation. The case study in Sect. 6.3 provides some details of the challenges and solutions.

A similar superferric design was also applied to quadrupole magnets. Figure 6.14 shows one of the cold iron triplets designed and built at FRIB. The steel yoke and poles were integrated with SC coils and fully immersed in a helium bath. This was one of the disadvantages of the design: the cooldown thermal mass was huge, and the helium inventory was large. One of the challenges was the different thermal

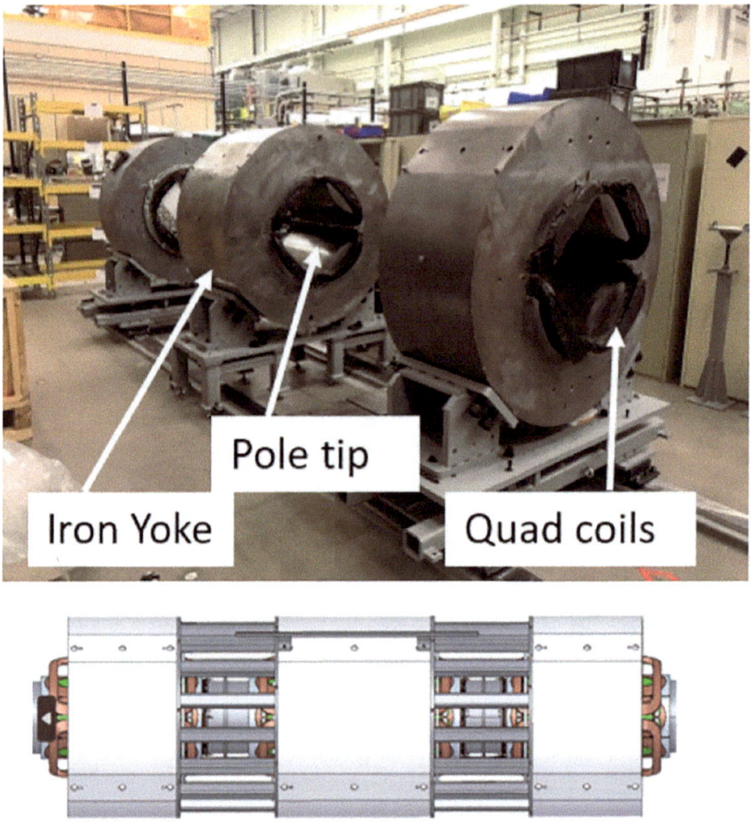

Fig. 6.14 FRIB cold iron quad triplet magnet at the fragment separator region downstream of the production target (Courtesy to MSU, US)

contractions between the steel and the coils. The poles are saturated at the maximum design field and slightly above 2 T, as shown in Fig. 6.15. The operating current I_{op} is typically less than 30% of the critical current I_C of the conductor of the random-wound coil. It typically takes 10 to 20 training quenches before reaching the full field.

6.1.4.2 Current-dominated Magnets

Magnets operating at fields much higher than 2 T are usually coil-dominated, where the placement of the coil conductor dictates the magnet's field quality and properties. In ideal conditions, the $\cos\theta$ current distribution produces a pure dipole, while the $\cos 2\theta$ current distribution produces a quadrupole field. The so-called $\cos\theta$ dipole magnets are used in many collider storage rings that require a high magnetic field. Figure 6.16 shows RHIC arc dipole magnets with an 80 mm aperture

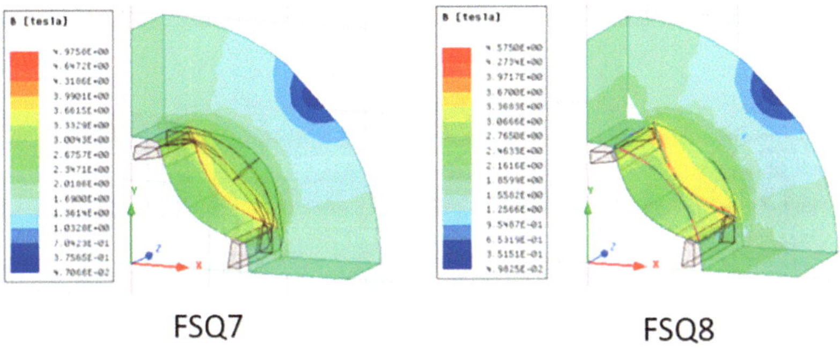

Fig. 6.15 Magnetic field distribution at the maximum design fields of the FRIB cold iron quadrupoles FSQ7 and FSQ8 (Courtesy MSU, USA)

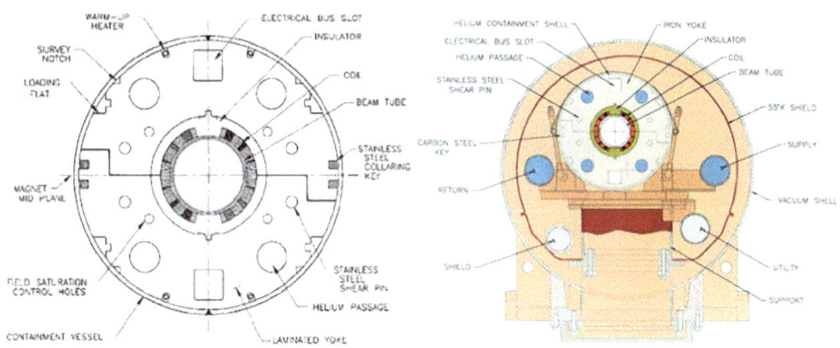

Fig. 6.16 RHIC arc dipole, cross-section view of coldmass (left) and magnet (right) [3] (Courtesy BNL, USA)

and a length of 9.7 m that operate at 3.45 T. The single-layer coil used an NbTi Rutherford cable. Cold iron was used for both the yoke and collar. The magnets operate in forced-flow helium at a nominal temperature of 4.6 K.

Figure 6.17 shows the cross-section of main arc dipole magnets of some collider storage rings: HERA [39], Tevatron [55], SSC [7], RHIC [18], and LHC [10]. The LHC dipole cold mass houses vacuum chambers for both counter-rotating beams in the 56 mm coil aperture [37]. The magnet of the double layer coil of the NbTi conductor operates in 1.9 K superfluid helium, reaching an 8.3 T field with a nominal operating current of about 11,850 A. Each magnet, with a length of 16.5 m, stores 7 MJ of energy. Once or twice a day, as the protons accelerate from 450 GeV to 7 TeV, the field of the SC bending magnets increases from 0.54 T to 8.3 T. In a recent LHC upgrade, 11 T dipole magnets made by Nb_3Sn technology were installed in the dispersion suppressor regions [38].

High-temperature superconducting (HTS) conductors, such as $Bi_2Sr_2CaCu_2O_{8+x}$ (Bi-2212) round wire, $Bi_2Sr_2Ca_2Cu_3O_{10+x}$ (Bi-2223) tapes, and rare earth barium copper oxide (REBCO) tape, are attractive for future hadron accelerators in three distinct applications: (1) high radiation tolerance for use in the target region; (2)

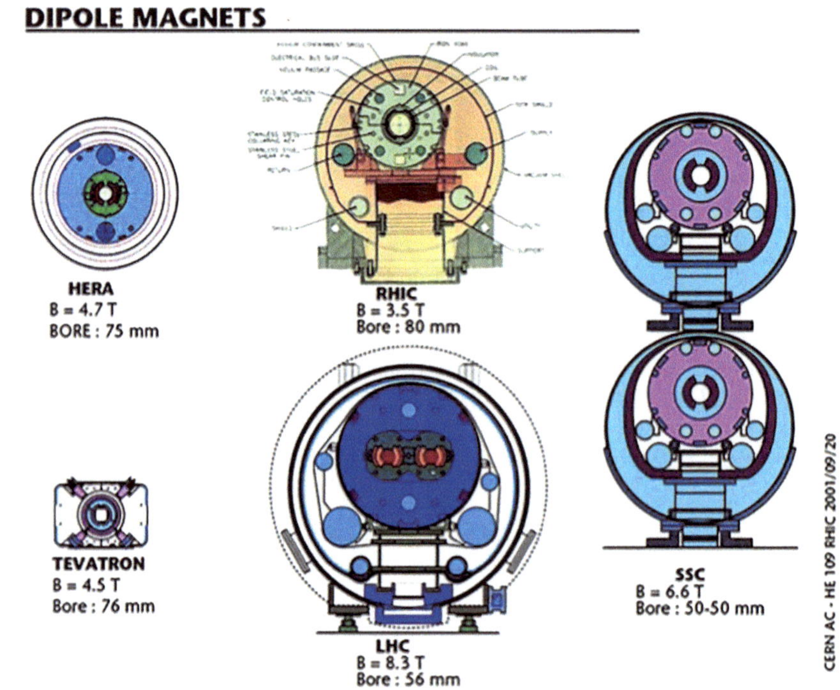

Fig. 6.17 Comparison of main arc dipole magnets of some collider storage rings [40]

higher magnetic fields beyond Nb_3Sn for next-generation ring dipole magnets (above 20 T) [51] and quadrupole magnets to allow for a more compact design; and (3) higher operating temperatures in the 35–55 K range to facilitate the utilization of the capacity of large-scale cryogenic systems after the first turbine string or the cryo-cooler 1st stage to save operational costs and reduce helium inventory.

REBCO conductors have a high I_C in a broad temperature range and a particularly high in-field current density due to their high pinning force at the microscopic level. They also have significant mechanical properties as a result of the Hastelloy substrate. These advantages allow them to have a wide range of application markets, including the HTS motor, transformer, magnetic resonance imaging (MRI), fusion reactor, and particle accelerator. Many companies have actively produced long-length conductors, with prices varying on the basis of the manufacturing process and critical current density.

REBCO conductors are good choices under ionizing irradiation. Different research groups have characterized HTS conductors under the influence of ionizing radiation. For example, γ-ray irradiation was found to have a minimal impact on the critical temperature T_C of the YBCO conductor at 50 MGy and result in an improvement in I_C in the absence of a background field [27, 28]. The impact of fast neutron irradiation on a set of commercially available REBCO conductors was studied, showing a 3% degradation in T_C per $10^{18}cm^{-2}$ particle flux, and that a conductor with artificial pinning centers exhibited less degradation [13]. The feasibility of using an HTS conductor in a fusion magnet was verified, demonstrating that a YBCO conductor was stable under $2 \times 10^{18}n\ cm^{-2}$ neutron flux irradiation [14]. Therefore, an HTS conductor is expected to be suitable for use in the target region of high-power hadron accelerators without concerns regarding radiation lifetime. Another advantage is that REBCO coils operating at low or liquid helium temperatures can have a substantial temperature margin against beam heating deposited to the coils, which is a major concern when SC magnets are used in the target region. Although REBCO has high stability, it also has very slow quench propagation, which results in challenging quench detection and protection in HTS-based magnets. Recently, non-insulation winding technology has been developed, demonstrating outstanding self-protection characteristics [19, 20].

During the FRIB design phase, an HTS dipole was proposed for the FRIB fragment separator region [43]. A set of sub-scale curved REBCO testing coils was wound and tested [4]. The design was not selected due to concerns about the overall cost and compatibility with the project schedule. The design is worth revisiting as the conductor technology matures and the cost becomes more affordable as a result of fusion magnet applications.

BNL [16] designed and fabricated a full-size prototype HTS quadrupole magnet for the FRIB target region to address concerns about radiation dose and beam heating. The magnet had an iron-dominated design, with the iron pole and return yoke outside the cryostat. The coils were co-wound in a double pancake configuration using a computer-controlled coil winder with $\sim$ 12 mm wide REBCO tapes and slightly wider stainless steel tape for turn-to-turn insulation, as shown in Fig. 6.18. All coils were individually tested at liquid nitrogen temperature and then tested as a quadrupole using cold helium gas to the full design field.

Fig. 6.18 HTS quad prototype for FRIB Target area by BNL: (left) racetrack coil winding using REBCO tapes, and (right) fully assembled FRIB quadrupole with iron yoke for dunk testing [16] (Courtesy R. Gupta, BNL)

Table 6.5 Radiation lifetime of some organic and inorganic materials used in magnet fabrication [22, 25]

Material	γ-dose tolerance (Gy)
Organic material	
Teflon (PTFE; FEP; in vacuum)	10^2–10^6
Polyamide (nylon)	10^3–10^4
Phenolic resin (unfilled; asbestos filled; glass laminate)	10^5–10^8
Epoxy resin (aromatic-type curing agent)	10^7–10^8
Polyimide (Kapton; in vacuum)	10^7–10^8
Cyanate ester resin	$\sim 10^8$
Inorganic material	
Concrete; mica; glass	10^8–10^9
Oxide mineral (beryllium oxide; magnesium oxide; aluminum oxide)	10^{10}–10^{12}
Quartz	10^{12}–10^{13}
NbTi filament	$> 10^{15}$
Steel; copper	10^{15}–10^{16}

6.1.5 Radiation Resistant Magnet

Radiation-resistant magnets are often needed in high-power target and dump regions where radiation can be extremely high. For example, magnets in the FRIB target area can receive a dose level of 1–10 MGy per year at full beam power. The radiation tolerance of organic materials, often defined as its tensile strength being significantly reduced, typically lies between 10^5 and 10^7 Gy. Polyimide, like Kapton, can serve as primary insulation with acceptable radiation resistance.

In the AGS and later in the SNS accumulator ring, epoxy/glass insulation was used in the vacuum impregnation process for magnet coils in areas of high radiation. Radiation protection was provided by wrapping the cables with glass tape and a Kapton layer [49]. This material provided the radiation protection of Kapton with the bonding ability of glass fiber and epoxy (Table 6.5).

Similarly, KEK and J-PARC used pre-impregnated glass cloth made of polyimide resin to insulate the magnet coil. The radiation lifetime of the insulator was tested with the KEK-PS proton beam that survived up to 4×10^8 Gy radiation dose without suffering serious damage [45].

When possible, NC magnets are designed with low current density ($< 3.5\,\mathrm{A/mm^2}$) copper coils. The low current density reduces the resistive energy loss in the coils and keeps the voltage drop across the magnets low, thus reducing the possibility of turn-to-turn shorts occurring in the coil should the insulation be damaged by radiation. For radiation resistance and reliability, metal fittings and tubing are used to carry the cooling water to and from the coils. The coil water lines are mounted on insulated stand-offs and routed to a common water manifold on the magnet stand, which has an insulated ceramic break.

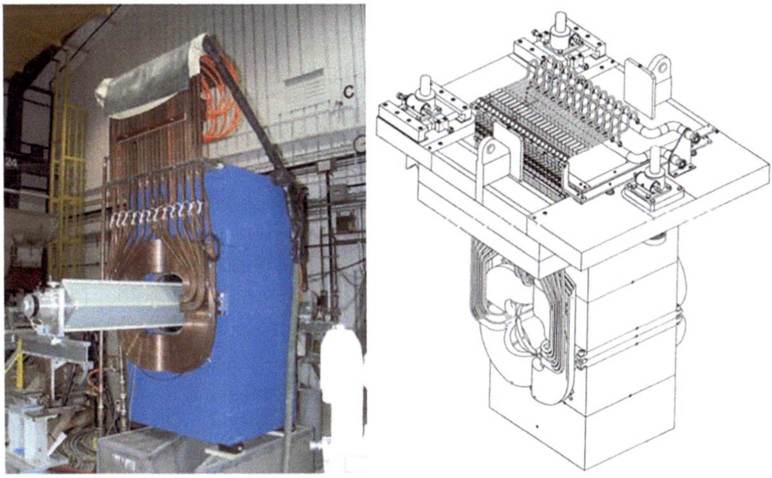

Fig. 6.19 (Left) Radiation resistant quadrupole magnet under measurement at BNL for the SNS RTBT. (Right) The magnet is designed with top shielding with the lifting fixture allowing vertical handling (Courtesy C. Pearson, BNL)

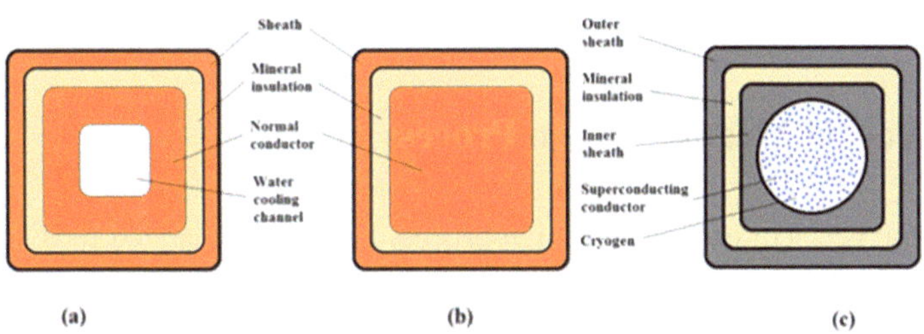

Fig. 6.20 Cross-sections of mineral insulated (**a**) NC hollow-core wire, (**b**) NC solid-core wire, and (**c**) SC cable-in-conduit wire

Magnets built with only inorganic materials are used in an environment where the radiation dose is expected to be between 10^8 and 10^{11} Gy. Inorganic mineral insulation performs much better after exposure to high levels of radiation doses than organic materials, where the formation of gaseous byproducts results in the loss of material and, ultimately, polymer chain fragmentation. Such magnets are often used in areas near the production target of secondary beams in high-power facilities.

Figure 6.19 shows one of the four radiation-resistant quadrupole magnets later installed in the ring-to-target beam transport (RTBT) line immediately upstream of the SNS neutron target. As shown in Fig. 6.20a, the copper conductor wire insulated with mineral powder of magnesium oxide contains a hollow channel for the cooling of water. The rigid conductor was carefully wound to avoid sharp turns. The conductor came with a grounded sheath that was terminated with a brazed ceramic feed-through to keep the wire insulation sealed. The fabricated magnet was physically attached to a block of concrete mounted from the top to shield against radio-activation during maintenance handling when an overhead crane was used to lift the shielded magnet vertically. In addition, remote handling fittings were used for water connections, and quick disconnect flanges were used for vacuum chamber connections.

Figure 6.20b shows an alternative mineral-insulated wire with a solid copper core. In the absence of the central cooling channel, the cross-section of the wire can be reduced to facilitate bending and coil winding. Separate water cooling channels must be designed and implemented to provide indirect cooling. The magnet coil developed at KEK used a solid core conductor insulated with magnesium oxide powder, assembled with a copper cooling tube, and potted with solder [46].

Similarly to resistive magnets operating at room temperature, the components of SC magnets that are most susceptible to prompt radiation damage include the organic content of the insulation material. SC filaments, such as NbTi material, can tolerate an equivalent radiation dose of up to 10^{11} Gy with modest degradation of the critical current density [23]. Depending on radiation-resistant requirements, several design approaches have been pursued. Cyanate ester resin was shown to meet the magnet performance requirements of the ITER fusion facility after exposure to more than 10^8 Gy (more than ten times the anticipated radiation exposure for the coils of the ITER test facility) [24].

6.2 General Magnet Design Process

The general accelerator magnet design process can be summarized by a flow chart shown in Fig. 6.21. As a general rule, the maximum design field of the iron pole is limited to $\sim$ 1.6 T for NC magnets and $\sim$ 2 T for superferric magnets, credited to the lack of DC losses, respectively. Typically, NC magnets are preferred over SC magnets for manufacturing and operational simplicity. NC magnets also have advantages in cases where beam loss and AC loss are high due to the physics requirements.

EM design must consider space for non-field-generating components such as the cryostat, thermal shield, and helium vessel for an SC magnet. The design should meet the required field strength and quality, minimize the peak magnetic field and fringing fields outside the magnet, and keep the overall magnet size compact as required. EM analysis involves analyzing the magnet's performance, protection, and force input for the mechanical design, as well as key aspects that cover the entire operating range, including:

- Field distribution along the beam path, including central and fringe fields;
- Integrated field strength and magnetic effective length;
- Field uniformity within the good field region;
- Inductance and stored energy;
- Quench characteristics for SC magnets;
- Lorentz force calculations at $1.07 \times I_{\mathrm{op}}$;

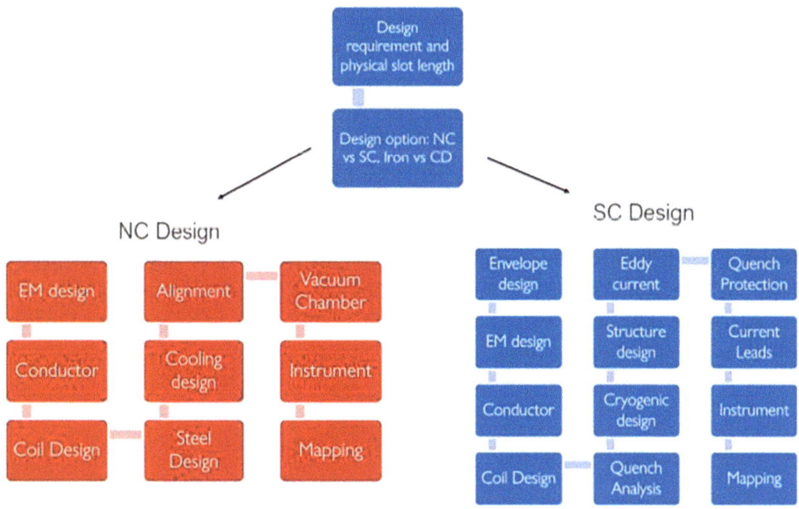

Fig. 6.21 Flow diagram of general magnet design

- Transient forces due to induced eddy currents during a quench;
- Impact of coil/magnet misalignment on magnet performance.

In EM analysis, the magnetic vector potential **A** is often used when there is current density and a time-varying electric field,

$$\mathbf{B} = \nabla \times \mathbf{A}, \tag{6.8}$$

while the magnet scalar potential ψ is used in a space free of current and without transients,

$$\mathbf{H} = -\nabla\psi, \quad \mathbf{B} = -\nabla\phi, \quad \mu_0\psi = \phi. \tag{6.9}$$

The ideal 2-D pole shapes for some common magnets are shown in Fig. 6.1. Magnet strengths are specified by the value of k_n (normalized to the beam rigidity $B_0\rho$) in most lattice codes:

$$B(x) = B_0\rho \sum_{n=1}^{\infty} \frac{k_n x^{n-1}}{(n-1)!} \tag{6.10}$$

where $n = 1$ denotes the dipole component, $n = 2$ denotes the quadrupole component, etc. Commonly used EM software codes include (Chap. A):

- OPERA [12]: It offers predefined coils and conductors, allowing for different current distributions. It enables field calculations in the absence of iron and includes modules for quench and transient eddy current analysis.

- Ansys Maxwell [5]: It integrates with CAD software like Solidworks [42], but it lacks predefined coils, conductors, and a quench model.
- COMSOL Multiphysics® [8]: It is a multi-physics platform that enables coupled analysis and accepts CAD models. The AC/DC module is needed for magnet analysis, but it lacks a quench model.
- Poisson/Superfish [21, 29]: It is a 2-D tool suitable for conceptual design.

Because of the stored energy, the stress generated by the EM force is a major consideration in the design of electromagnets, particularly at high fields. Material selection is often determined by the stress conditions in the coil winding. The tensile stress σ_m of the conductor or structural support material can be evaluated using the Virial Theorem for the first order.

$$\sigma_m \geq \frac{\rho_m U_m}{M_m}\,, \tag{6.11}$$

where ρ_m is the density of the material, U_m is the stored energy of the magnet, and M_m is the mass of the magnet. Equation 6.11 indicates that the magnet mass (conductor and structural material) is scaled with the stored energy U_m to keep the stress in the magnet below the allowable stress level, which is below the yield stress $\sigma_{m,y}$ of the magnet material with a safety margin. The radial Lorentz force in the winding will create the tensile force T_m in the conductor, resulting in the Hoop stress, with the relation:

$$\sigma_m = \frac{T_m}{A_m} = \frac{B \cdot I \cdot R}{A_m}\,, \tag{6.12}$$

where A_m is the cross-section of the conductor, B is the magnetic flux density, I is the electrical current in the conductor, and R is the bending radius, as shown in Fig. 6.22.

The mechanical strength of the conductor limits the design current due to the Lorentz force or requires additional structural support, which is often the case for superconducting magnets. In addition, the forces caused by thermal contraction and pressure in the cooling circuit should also be taken into account.

The choice of material is an important aspect of magnet design. As discussed in Sect. 6.1, copper and aluminum alloys are common options for normal conducting magnet conductors. Copper and its alloys are preferred because of their low resistivity, high thermal conductivity, ease of fabrication, strength, bonding capabilities, and moderate annealing temperature. Copper is also an ingredient in most regular SC (or low-temperature superconducting, LTS) conductors. It forms well with NbTi and Nb_3Sn alloys and serves as a stabilizer due to its low resistivity. It is worth mentioning that the purity of copper affects not only the resistivity at low temperatures but also significantly impacts thermal conductivity. A residual resistivity ratio (RRR) of 100 and above is typically specified in the fabrication of LTS conductors.

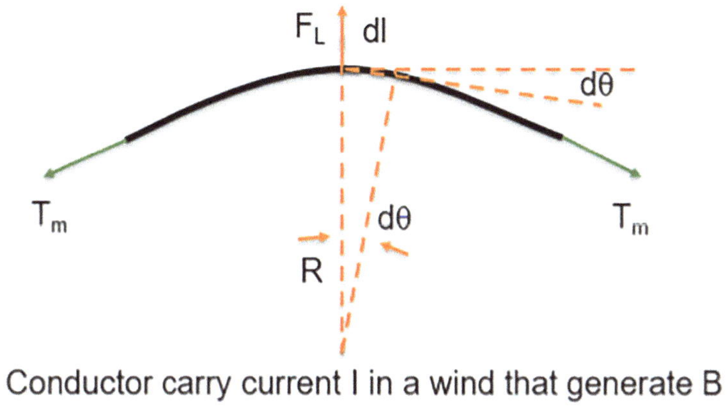

Fig. 6.22 Illustration of the Lorentz force F_L on a magnet winding that carries current I that generates a magnetic flux density B

Another aspect that needs attention is the material heat capacity in the operating temperature range. The design of DC water-cooled magnets is not concerned with the heat capacity of the magnetic materials. The situation is different for pulse and SC magnets. Pulse magnets produce high fields for short durations ($\sim 10\,\text{ms}$) with energy supplied by a capacitor bank or another pulsed power supply. To limit the maximum temperature, many pulsed magnets are initially cooled to liquid nitrogen temperatures before pulsing. For SC magnets, the heat capacity of the material can determine the thermal mass required for cool down, the transient stability of the conductor, and the protection of the coil during a quench.

Structural analysis and pressure safety are additional major design tasks in conjunction with the EM design. A detailed analysis must be performed to identify the worst case scenario of a force combination that includes electromagnetic, pressure, and thermal forces. Structure analysis must also consider transportation loads and seismic requirements. In the USA, the American Society of Mechanical Engineers (ASME) Boiler Pressure Vessel Code (BPVC) [2] Section VIII Div. 2 Par 5 design by analysis is usually used for structural analysis. Most allowable stresses for materials are well established and available in ASME BPVC Section II Part D. This topic is further discussed in Sect. 8.1.6.

Cooling is a limiting factor for the choice of magnet current density. For NC magnets, the heat generated by thermal heating from conductor resistance needs to be carried away by air or water. For SC magnets, cryogen needs to carry away the AC loss, the load on current leads, and the structural heat leak.

The main considerations for resistive magnet cooling are (1) a steady temperature increase in the water, (2) a local temperature increase of the coil due to heat transfer, (3) hydraulic resistance in the water cooling path, and (4) the overall effect on magnet performance. The water cooling is by low conductivity water (Sect. 8.13.2.3) and the total temperature rise in the water is usually limited to 30C.

Table 6.6 shows the resistive magnets used in the FRIB driver linac. The current density is around 10 A/mm^2 for water-cooled magnets and 1 A/mm^2 for air-cooled magnets, respectively. Air-cooled magnets are limited to small correctors due to the low current density. The power supply voltage needs to account for the DC voltage drop and inductive voltage to ramp up the magnet. The copper conductor resistivity is sensitive to temperature, and the voltage drop of the coil needs to be calculated at the design temperature instead of room temperature. The total pressure drop in the cooling loop is constrained by the pressure head of the water pump. Practically, the total pressure drop is kept below 50 psi (0.35 MPa) for the efficiency and cost of the water system. The heat transfer coefficient h_w is given by

$$h_w \sim \frac{1.4\, v_w^{0.8}}{D_w^{0.2}} \sim 30 \tag{6.13}$$

where v_w is the velocity of water and D_w is the characteristic dimension of the cooling channel. The water pressure drop ΔP_w is given by

$$\Delta P_w \propto \frac{L_w\, v_w^2}{D_w} \tag{6.14}$$

where L_w is the total length of the flow path. The optimization is to maximize h_w while minimizing ΔP_w. The parallel cooling scheme is commonly used to reduce the total pressure drop. Equations 6.13 and 6.14 are applicable to magnets with a modest magnetic field. The cooling requirements for high-field magnets can be quite different.

SC magnets are typically cooled by immersion in liquid helium. The high magnet inductance limits the ramping rate. AC loss is usually not a concern. The coils are typically impregnated with epoxy resin. The cooling neither contributes to the quench protection nor impacts the stability of the coils. For cable-in-conduit-conductor (CICC) magnets, cooling is achieved by the forced flow of supercritical helium using sensible heat. Helium contributes to the stability of the conductor and protection against quenching [9]. The conduction cooling scheme has been used more often recently due to (1) the compact design, (2) the advancements in cryo-cooler technology, and (3) the increasing cost of helium. The thermal link between the coil and the cooling source is critical. High RRR OFHC copper and aluminum (5N, 6N-Al, and 6N8-Al) are popular choices for thermal link materials. Thermal linkage material is often added between layers of large coils to enhance cooling and limit the temperature difference.

Table 6.7 shows the current density of the FRIB warm iron SC magnets. The current density is about one magnitude higher than that of the NC magnets (Table 6.6); however, when the cryogenic vessel used for the magnet is included, the current density is similar to or lower than that of the NC magnets. Therefore, in a heavy iron particle accelerator, a superferric SC magnet does not provide a more compact solution but allows for a higher field at the pole ($\sim$ 2 T) without excessive

Table 6.6 FRIB driver linac RT magnet parameters

Magnet type	Current [A]	Voltage [V]	Wire size [mm]	Cooling channel [mm]	Wire type	Current density [A/mm^2]	ΔT [°C]	Flow rate [LPM]
Q1	307	20.3	6.48	3.81	Square	10.01	16	5.9
Q2	320	55.4	6.48	3.81	Square	10.46	16.5	20.2
Q3	320	63.4	6.48	3.81	Square	10.46	20.6	21.5
Q4	311	56.4	6.48	3.81	Square	10.16	16.3	18.9
Q5	311	63.8	6.48	3.81	Square	10.16	19.8	17.7
Q6	252	33.3	6.48	3.81	Square	8.24	16.3	11.9
D1	307	20.3	6.48	3.81	Square	10.03	15.4	10.3
D2	214	113	6.48	3.81	Square	6.99	25.1	14.6
D4	135	160	6.48	3.81	Square	4.41	23.1	14.2
C1	124	1.62	6.48	3.81	Square	4.05	1.78	3.4
C3	186	4	6.48	3.81	Square	6.08	8.96	2.5
H1	4.25	26	2.14	0	Square	1.18	N/A	N/A
H2	4.25	45.6	2.14	0	Square	1.18	N/A	N/A
H3	4.38	34.7	2.14	0	Square	1.22	N/A	N/A

Table 6.7 FRIB warm iron SC magnets current density

Magnet type	Current density in coil [A/mm^2]	Current density including cryogenic vessel [A/mm^2]
FS2D3 dipole	134	4
SCD1 30° dipole	88	7
SCD3 50° dipole	106	9
FSQ2 quadrupole	121	23
FSQ5 quadrupole	112	27

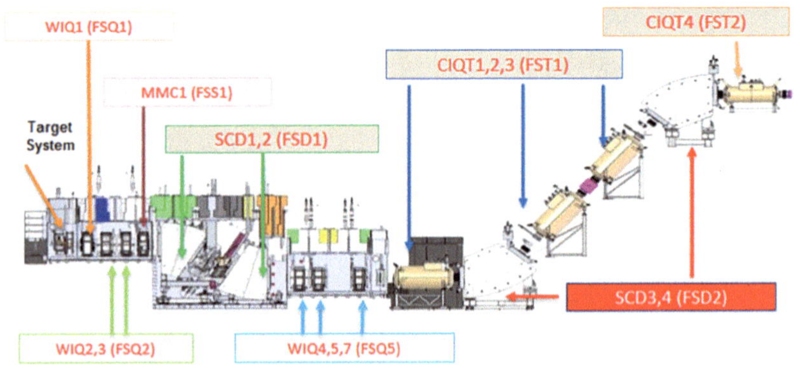

Fig. 6.23 FRIB target and vertical pre-separator layout

penalties on power consumption. This also provides a rough estimate of the size of the SC magnets in the conceptual design stage.

6.3 A Case Study with FRIB SC Dipole Magnet

This section uses the 150 ton FRIB SC 50° dipole (SCD3) as an example to demonstrate the design and optimization process for a large-scale magnet in high-power hadron accelerators.

As shown in Fig. 6.23, SCD3 is the first vertical-bending dipole downstream of the FRIB target area. It pairs with a similar 50° dipole SCD4 to transport the secondary beam to ground level; it is also part of the separator used to filter impurities in the fragment. Table 6.8 lists the general functional requirements of the magnet. The main challenges of this dipole are (1) EM design with a large aperture to meet field quality requirements, (2) quench detection and protection for large stored energy, (3) mechanical structure support to address large EM forces and sizable thermal contraction due to size, (4) cryogenic considerations to accommodate different utility modes and quench, and (5) assembly and installation aspects due to the weight and size.

Table 6.8 System requirement parameters for the FRIB SCD3 magnet

Parameter	Value
Bending radius	4 m
Bending angle	50°
Entrance pole face rotation, θ_i	0
Exit pole face rotation, θ_e	0
Full gap (perpendicular to bend plane)	0.2 m
Full aperture (perpendicular to bend plane)	≥ 0.24 m
Half width aperture (in bend plane) to outside	≥ 0.24 m
Half width aperture (in bend plane) to inside	≥ 0.24 m
Lower bound of operating range	≥ 0.5 T
Upper bound of operating range	≤ 2.0 T
Field index	0.2
Dipole field uniformity deviation $\Delta B/B_0$ within full aperture	≤ 0.004
Dipole good field region, over operating range	≥ 0.46 m
Field instability (fraction of actual field over operating range)	$\leq 1.2 \times 10^{-4}$

6.3.1 *EM Design*

SCD3 is a typical superferric H-shaped dipole magnet comprising a warm iron yoke, D-shaped warm iron poles, and a cryostat with SC coils that follow the shape of the poles. Figure 6.24 shows the mechanical CAD model compared to the simplified model for EM design. Mechanical details have been removed, and only iron and coils are used in the EM model. A good example of EM optimization is shown below; although the EM designs before and after the optimization both meet the functional requirements for the magnet.

Figure 6.25a shows the 3-D shape of the filter to shift the saturation. Purcell filters are used on the poles to achieve better field quality near the maximum field. Figure 6.25b shows the results of the integrated field quality at different current/field levels. One design iteration of this magnet is the iron yoke design. An additional 30 ton of iron was added as two additional slabs after the initial design. The comparison before and after the additional steel is shown in Table 6.9. The additional steel reduced the peak field in the coil, the stored energy, the current needed to meet the field strength, and the fringe field. The optimization also included balancing the EM force in the radial direction caused by the D-shaped magnet. Figure 6.5 shows the B-H curve of the used steel of AISI 1006.

6.3.1.1 Conductor

The coil and conductor dimensions are listed in Table 6.10. The packing factor was set to 83% based on practical experience. The coil design had an even number of layers so that the lead in and lead out were both at the bottom. Figure 6.26 shows

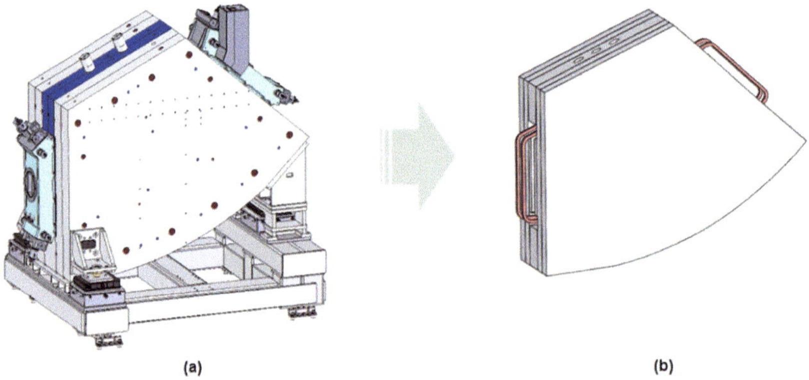

Fig. 6.24 SCD3 design 3-D model: (**a**) CAD model in Solidworks [42], and (**b**) simplified EM model in Ansys Maxwell [5]

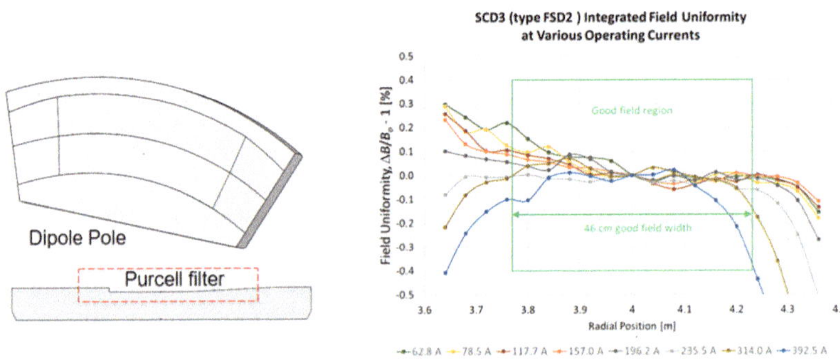

Fig. 6.25 SCD3 dipole pole 3-D model with the Purcell filter (left) and the field uniformity plot (right)

Table 6.9 Comparison of SCD3 preliminary and final design parameters

	Preliminary design	Final design
Yoke height [m]	2.93	2.93
Yoke width [m]	1.68	2.09
Yoke mass [kg]	101,705	141,986
Operating current [A]	471	392.5
Center field [T]	1.99	1.98
Peak field in coil [T]	3.13	2.91
Integrated field strength [T-m]	6.907	6.939
Integrated field uniformity	+0.24%/−0.24%	+0.2%/−0.4%
Stored energy [MJ]	2.2	1.99

the inductance and the stored energy as functions of the operating current. The conductor was Nb/Ti filament in a copper matrix. The rectangular wire had a Cu to SC material ratio of 4.6, with a filament size of 56 . The stability of the conductor

Table 6.10 Specifications of the SC wire and coil for SCD3

SC wire	
Material	NbTi/Cu
Bare conductor	1.17 mm × 2.40 mm
Cu:SC ratio	4.6
Filament diameter	56
I_{op}/I_C	19.50%
T_C at I_{op}	8.1 K
Coil	
Cross-section width	51.3 mm
Cross-section thickness	46.1 mm
Turn/coil	637

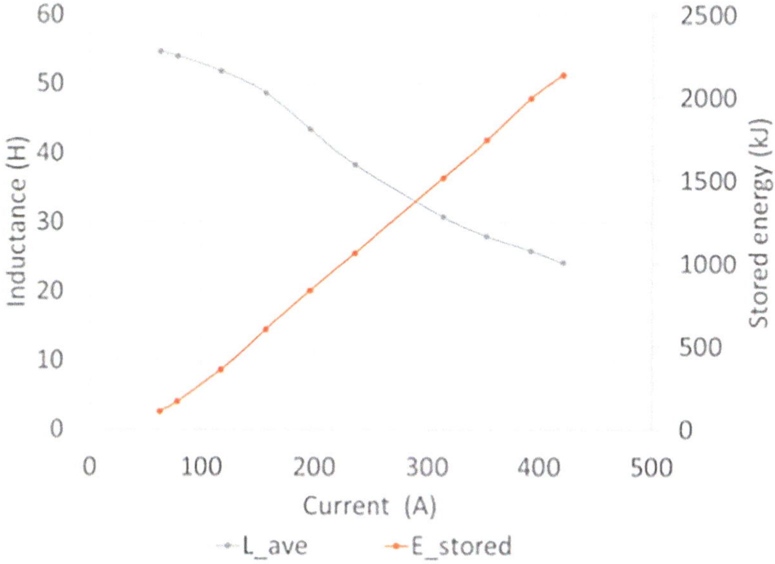

Fig. 6.26 SCD3 inductance and store energy as functions of the operating current

was verified against the previous dipole magnets designed and operated at MSU (Table 6.11). The ratio I_{op}/I_C of 19.5%, chosen based on the relatively low cost of the SC wire compared to the total cost of the magnet, provided a large margin against conductor quench. The coil was designed using the wet-wind method with StyCast^{TM} 2850.

6.3.1.2 Quench Protection

As elaborated in Sect. 8.3.1, quench detection and protection are critical for SC magnet operations. The quench analysis of the SCD3 magnet was conducted using

Table 6.11 SCD3 conductor parameters in comparison with other FRIB dipole magnets

	A1900 dipole	XFP dipole	FS2D3 rev	SCD3
Conductor				
I_C	500 A @ 4.2 K, 2 T	315 A @ 4.2 K, 3 T	500 A @ 4.2 K, 2 T	2029 A @ 4.2 K, 2.9 T
No. of filaments	87	30	60	198
Conductor size	1.898 mm × 0.898 mm	0.838 mm diameter	1.06 mm diameter	2.40 mm × 1.17 mm
Cu:SC ratio	9	4.35	7.02	4.6
Coil				
I_{op}	171 A	180 A	200 A	392.5 A
B_{max} in coil	1.2 T	1.2 T	1.2 T	2.9 T
T_{op}	4.2 K	4.2 K	4.2 K	4.2 K
MQE[a]	10 mJ	0.66 mJ	2.2 mJ	3.7 mJ
Max MIITS[b] (at 150 K)	0.225 MA·s	0.062 MA·s	0.062 MA·s	1.63 MA·s
Stekley factor[c]	0.226	0.603	0.53	0.49

Note, [a] MQE is the minium quench energy, [b] MIITS stands for Mega–Ampere^2–Seconds and a measurement for energy deposition into conductor during quench, [c] Stekley factor is a dimentionless parameter for cryo-stability

the Wilson codes [54] to simulate one of the two dipole coils, as the two dipole coils were separated in the helium vessel, and the thermal link between them was poor with respect to the time scale for the quench event. The transverse to longitudinal quench velocity ratio g_q was set to 0.1 for the epoxy (StyCast^{TM} 2850) used for coil impregnation.

Two configurations were analyzed: (1) without the dump resistor and (2) with a 1 Ω dump resistor and 120 ms quench detection and switch open time, as shown in Fig. 6.27. The results of both cases showed that even when protected by the fast dump circuit, the quench voltage was above the desired condition < 600 V. Thus, quench heaters were added to help propagate the quench throughout both coils. Figure 6.28 shows the electrical diagram for quench protection with heater wiring. A small resistor was connected in series with a fast dump resistor to provide power to the quench heaters upon the detection of a quench. A TRIAC was used to regulate the current that would trigger the quench heaters to help mitigate quench heater trips when the magnet was commanded to a slow dump, with the fast-dump switch remaining closed. Figure 6.29 shows the peak voltage and temperature when the quench heaters forced both coils to quench with the fast dump resistor.

6.3.2 Mechanical Supporting Structure

As shown in Fig. 6.30, the cold mass of the SCD3 magnet is supported by 5 compression links and 12 tension links. The primary functions of the tension links

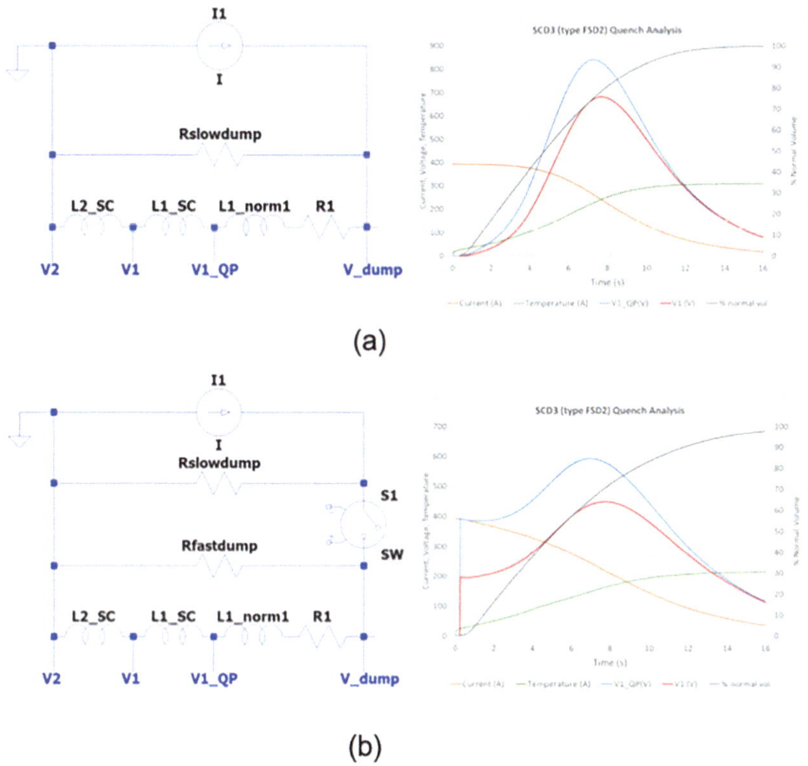

Fig. 6.27 SCD3 quench analysis (**a**) without quench detection and (**b**) with quench detection and fast dump resistor of 1 Ω. (Please re-plot: need to use subscripts $R_{\text{slow dump}}$ and $R_{\text{fast dump}}$)

were to carry the weight and to allow for the adjustment of the cold mass position. The compression links were used to limit the coil deformation due to the Lorentz force. Three different types of link designs were used according to load condition and function (Fig. 6.31). The vertical links were designed with a carbon fiber belt that has low heat leakage and is more compliant under load due to its lower stiffness in comparison to the metal links. The horizontal links, designed with Ti alloy, were used to stabilize the cold mass only; the load rating was only 6,000 lbs (27 kN), and the Ti link kept the design simple and compact. The compression links with the highest load rating (~ 35,000 lbs, 156 kN) were made of G10 tubes.

An important design aspect is that the design must consider the misalignment of the coil with respect to the steel, as the EM force is sensitive to the coil position. The misalignment of up to 7 mm in different directions was analyzed. The force ranges were fed back to the mechanical analysis to ensure that the links were safe under all operating conditions.

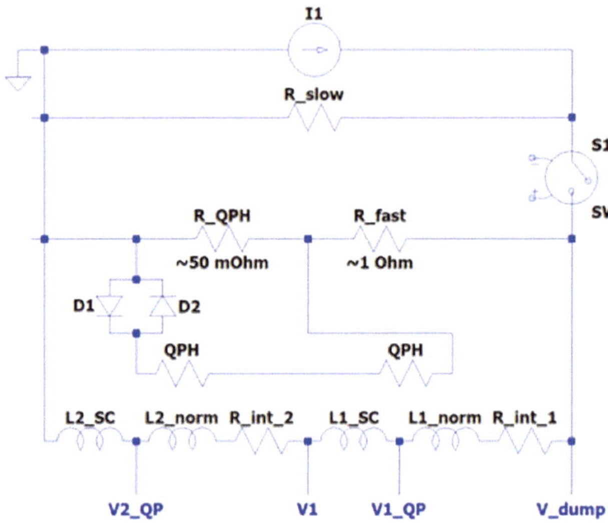

Fig. 6.28 SCD3 quench protection diagram with quench heaters

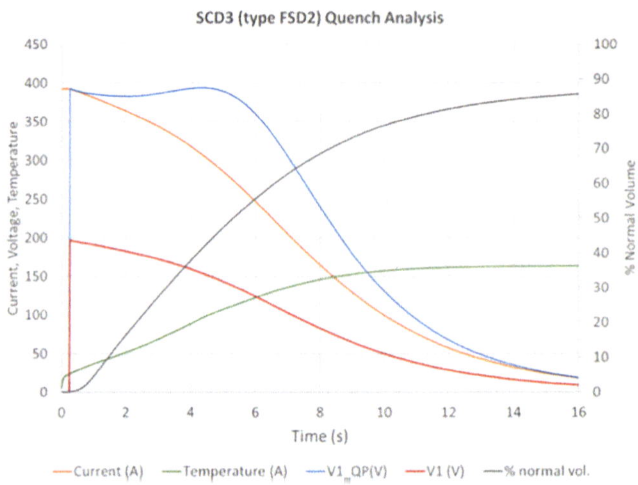

Fig. 6.29 SCD3 quench analysis result with configuration shown in Fig. 6.28

The helium vessel was analyzed under thermal contraction, helium pressure, and electromagnetic force loading. The total deformation was limited to 2 mm, as shown in Fig. 6.32. As the SCD3 coils were not subdivided while installed in the helium vessel after impregnation, the stress was in compression in all scenarios.

Tensile stress should generally be avoided for epoxy-impregnated coils. Tensile stress above 18 MPa can result in epoxy cracks and immature coil quenching.

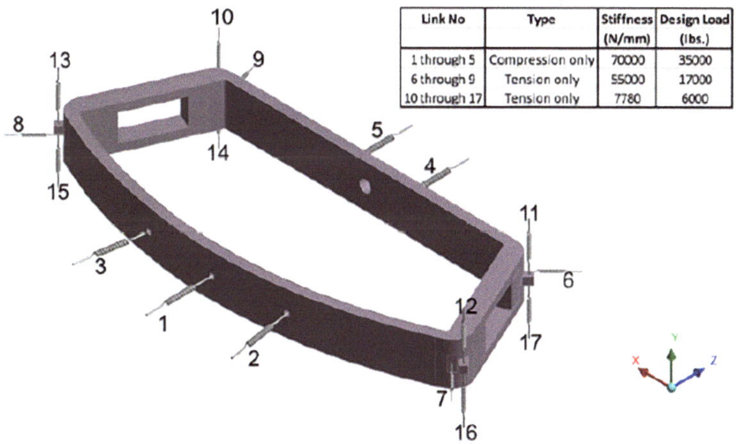

Link No	Type	Stiffness (N/mm)	Design Load (lbs.)
1 through 5	Compression only	70000	35000
6 through 9	Tension only	55000	17000
10 through 17	Tension only	7780	6000

Fig. 6.30 SCD3 coldmass support system with 5 compression links and 12 tension links

Fig. 6.31 SCD3 link design of (**a**) the compression link, (**b**) the carbon fiber vertical link, and (**c**) the titanium alloy horizontal link

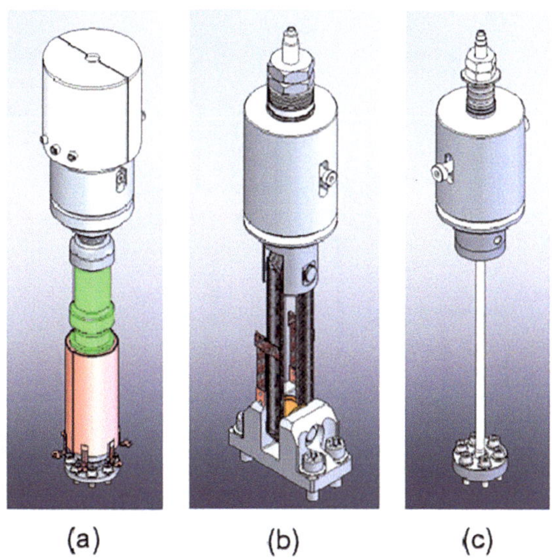

6.3.3 *Cryogenic Design*

As shown in Fig. 6.33, the SCD3 magnet is interfaced with the cryogenic system through 5 U-tube connections that allow for independent cool-down and warm-up. To address the challenge of the large 4 K cold mass ($\sim$ 3.4 ton) and cool-down energy (300 MJ), a dedicated cool-down line was utilized to force the supply gas to the lowest point of the magnet, ensuring that the cooling was uniform and efficient. The temperature difference between the supply and the return gas was limited to 50 K to minimize the transient thermal stress. A minimum cool-down time of 4

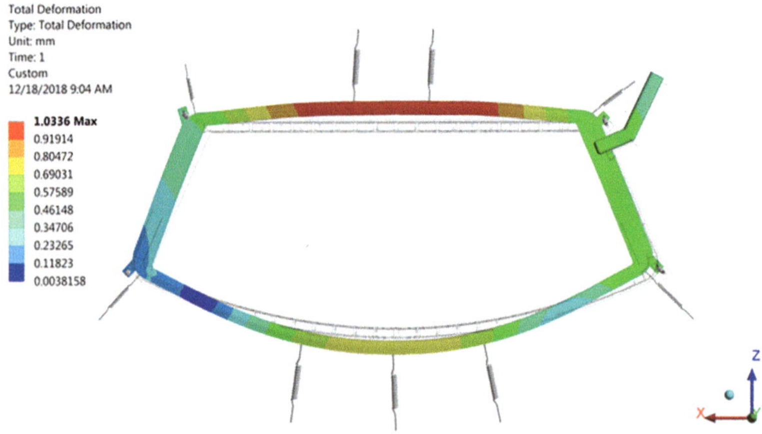

Fig. 6.32 SCD3 helium vessel total deformation plot with pressure and maximum EM force in the finite element analysis (FEA)

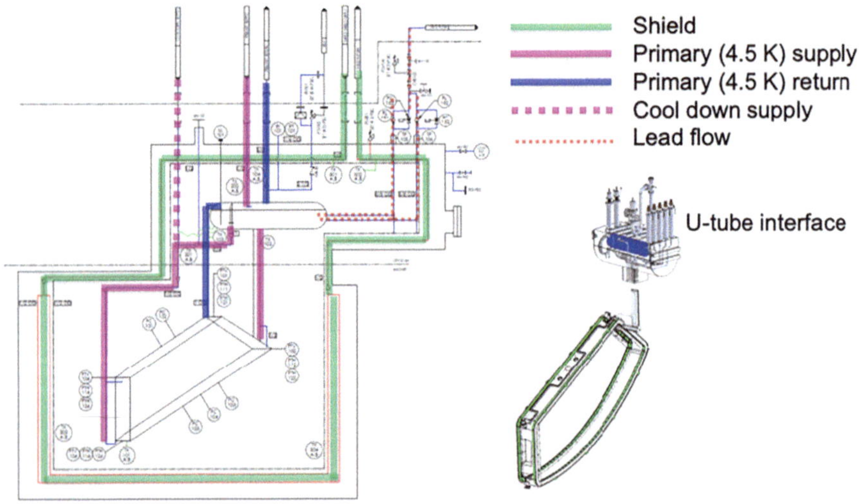

Fig. 6.33 SCD3 magnet cryogenic flow diagram

days was estimated based on this constraint. After the liquid accumulated in the magnet, the cool-down supply line was isolated. The liquid was supplied to the helium reservoir on the top instead, and the rest of the magnet was cooled using a thermal siphon loop. The thermal shield was designed with alternating copper and stainless steel panels to mitigate eddy currents during the magnet fast dump.

In addition to ensuring that the magnet can cool down and warm up properly, as well as confirming that there are no unstable conditions under normal operational circumstances, safety design and mitigation against various fault scenarios must be considered in the cryogenic design. As shown in Table 6.12, one category was the

Table 6.12 SCD3 magnet pressure relief under various fault scenarios

Scenario	Description	Type	Set point	Mass flow	Max. (KA) required[a]	Temperature @ max. (KA)
Loss of vacuum	Sudden (air)	Relief valve	35 psig	1.25 kg/s	3.6 cm^2	6.33 K
Loss of vacuum	Sudden (helium)	Relief valve	35 psig	1.33 kg/s	3.9 cm^2	6.32 K
Loss of vacuum	1/2″ opening	Relief valve	35 psig	0.82 kg/s	2.4 cm^2	6.31 K
Quench	Local hotspot temperature	Relief valve	35 psig	0.56 kg/s	2.3 cm^2	8.95 K
Quench	Uniform energy dissipation	Burst disc	45 psig	5.46 kg/s	17.2 cm^2	7.75 K

Note, KA is the effective relief area. K is the discharge coefficient and A is the relief area.

loss of vacuum, which caused a large heat flux to the cold mass; the other was the stored energy that was released into liquid helium in a short period. Two different pressure relief methods were chosen: the primary method was equipped with a resealable relief valve at 35 psig (0.24 MPa gauge pressure), and the secondary method was equipped with a rupture disk at 45 psig (0.31 MPa gauge pressure). Most relief cases were covered by primary relief, and the worst case of quench was covered by the rupture disk, as shown in the table.

6.3.4 Assembly and Testing

The assembly, installation, and testing processes should also be included in the design. A list of tooling must be designed to facilitate fabrication, as shown in Fig. 6.34, since the construction of helium and vacuum vessels included heavy welding. Sufficient tolerance must be included to accommodate welding distortions. The welding fixture was used to ensure the critical dimension, and the $\frac{1}{8}''$ (3 mm) tolerance was applied for the SCD3 vessel and the steel pole fitting. Multiple handling and rotation tooling were designed for the welding. Alignment features were also added to various stages of assembly to track the coil position.

A list of tests was used to ensure the quality of the construction, including tests for dimensions, pressure, and electrical components throughout the system. To validate electrical insulation, high-pot tests were performed at different stages of assembly to eliminate magnet ground faults during operation, starting at the highest voltage in the earliest assembly steps and stepping down towards the final assembly. The high-pot voltage was designed to cover the worst quench voltage with a safety margin. The final high-pot test from terminal to terminal was rated at the maximum fast-dump voltage. The pressure test was performed according to the ASME requirements regarding the maximum design pressure. All links were tested under design load, and destructive testing was also performed for each type to ensure

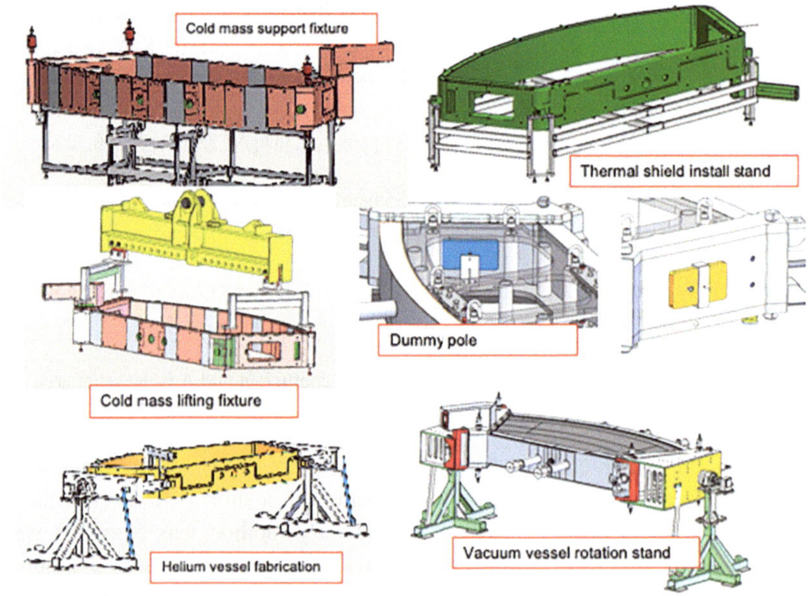

Fig. 6.34 SCD3 fabrication tooling

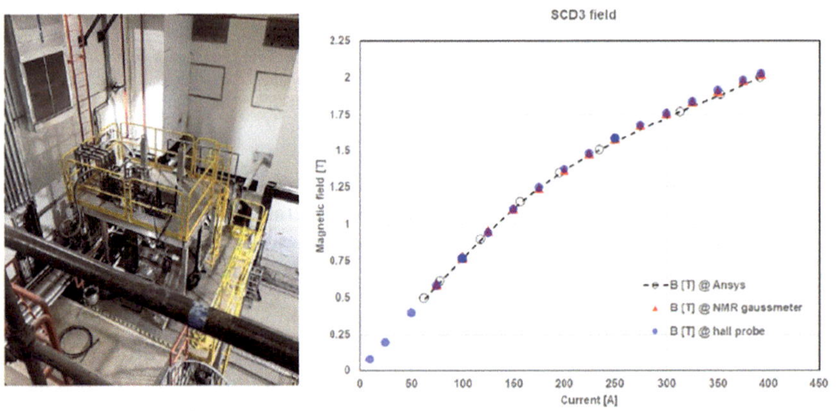

Fig. 6.35 SCD3 magnet cold test (Courtesy MSU, USA). Left: SCD3 installed at the FRIB vertical pre-separator area. Right: measured center field as a function of the current

the safety margin. A dedicated mapping system was also designed and fabricated to measure the magnetic field in the good-field region at different currents.

The final cold test was performed at the installation location, and the results are shown in Fig. 6.35. SCD3 was able to ramp up to full current without a training quench; the low I_{op}/I_C ratio and well-designed support were the keys. The test of the quench protection system was also performed by triggering a fast dump at full current to activate quench heaters to force the magnet to quench.

The design and optimization case presented in this section provides an overview of how a large-scale magnet is designed, following the process flow outlined in Fig. 6.21. The main purpose is to demonstrate that the design involves complex system-level work that extends well beyond basic electromagnetic considerations.

Problems

6.1 (Allowed Magnetic Multipoles in a Quadrupole Magnet) Prove that for a quadrupole magnet, the magnetic errors that are allowed by the quadrupole symmetry are quadrupole, 12-pole, 20-pole, etc.

6.2 (Iron Dominated Normal-Conducting Magnet) A C-dipole magnet is designed to operate at a field of 1 T. The magnet has a width of 0.8 m, height 0.6 m. The iron material has a relative permeability of 2500. The gap height is 18 cm. The maximum current is 5000 A. How many coil turns are needed for the top and bottom poles?

6.3 (Conductor Hoop Stress under Lorenz Force) Calculate the Hoop stress of a conductor of 2 mm^2 wire cross-section under 2 T background magnetic flux density, when the current in the conductor is 1 kA, and the bending radius is 2 m? Will the conductor survive if its yield stress is 50 MPa? What cross-section should be chosen for the conductor to be self-supporting?

References

1. A. Aleksandrov, S. Assadi, I. Campisi, P. Chu, S. Cousineau, V. Danilov, B. Dodson, J. Galambos, D. Jeon, S. Henderson, et al., Commissioning experience for the sns linac. *AIP Conf. Proc.* **773**, 73–78 (2005)
2. American Society of Mechanical Engineers (ASME), *ASME Boiler and Pressure Vessel Code (BPVC)* (2023)
3. M. Anerella, J. Cottingham, J. Cozzolino, P. Dahl, Y. Elisman, J. Escallier, H. Foelsche, G. Ganetis, M. Garber, A. Ghosh, C. Goodzeit, A. Greene, R. Gupta, M. Harrison, J. Herrera, A. Jain, S. Kahn, E. Kelly, E. Killian, M. Lindner, W. Louie, A. Marone, G. Morgan, A. Morgillo, S. Mulhall, J. Muratore, S. Plate, A. Prodell, M. Rehak, E. Rohrer, W. Sampson, J. Schmalzle, W. Schneider, R. Shutt, G. Sintchak, J. Skaritka, R. Thomas, P. Thompson, P. Wanderer, E. Willen, The rhic magnet system. Nucl. Instrum. Methods Phys. Res., Sect. A **499**(2), 280–315 (2003). https://doi.org/10.1016/S0168-9002(02)01940-X
4. M. Anerella, R.C. Gupta, H. Hocker, R.P. Johnson, S.A. Kahn, W. Sampson, J. Schmalzle, Construction and testing of curved ReBCO coils. IEEE Trans. Appl. Supercond. **28**(3), 4601505 (2018)
5. Ansys, Ansys Mechanical. https://www.ansys.com/products/structures/ansys-mechanical
6. Y. Arakaki, et al., Electrostatic septum for 50 GeV proton synchrotron in J-PARC. *Proc. IPAC'10*, 3900–3902 (2010). http://accelconf.web.cern.ch/IPAC10/papers/THPEB010.pdf
7. A. Chao, Accelerator physics studies for the ssc. IEEE Trans. Nucl. Sci. **32**(5) (1985)
8. COMSOL Multiphysics® v. 6.3, COMSOL AB, Stockholm, Sweden. https://www.comsol.com
9. L. Dresner, *Stability of Superconductors* (Springer Science & Business Media, 2001). https://doi.org/10.1007/978-1-4615-1583-1

10. L.R. Evans, LHC status. *Proc. APAC'07*, 46–50 (2007). https://jacow.org/a07/papers/TUXMA01.pdf
11. A. Faltens, 7.2.11 Electrostatic Lenses. in *Handbook of Accelerator Physics and Engineering*, ed. by A.W. Chao, et al., 734–735 (World Scientific, 2022)
12. V. Fields, *Opera-3d Reference Manual*. Vector Fields Limited, England **3010**(10) (2004)
13. D. Fischer, R. Prokopec, J. Emhofer, M. Eisterer, The effect of fast neutron irradiation on the superconducting properties of REBCO coated conductors with and without artificial pinning centers. Supercond. Sci. Technol. **31**(4) (2018)
14. R. Fuger, M. Eisterer, H.W. Weber, YBCO coated conductors for fusion magnets. IEEE Trans. Appl. Supercond. **19**(3) (2009)
15. G2ELab, *BHmag User Manual*. G2ELab, Grenoble, France, version 2 edn. (2013)
16. R. Gupta, M. Anerella, J. Cozzolino, P. Joshi, W. Sampson, P. Wanderer, A. Zeller, Hts quadrupole for frib—design, construction and test results. IEEE Trans. Appl. Supercond. **25**(3), 1–6 (2015). https://doi.org/10.1109/TASC.2014.2365560
17. H. Hahn, Impedance measurements of the spallation neutron source extraction kicker system. Phys. Rev. ST Accel. Beams **7**, 103501 (2004). https://doi.org/10.1103/PhysRevSTAB.7.103501
18. H. Hahn, E. Forsyth, H. Foelsche, M. Harrison, J. Kewisch, G. Parzen, S. Peggs, E. Raka, A. Ruggiero, A. Stevens, S. Tepikian, P. Thieberger, D. Trbojevic, J. Wei, E. Willen, S. Ozaki, S. Lee, The RHIC design overview. Nucl. Instrum. Methods Phys. Res., Sect. A **499**(2), 245–263 (2003). https://doi.org/10.1016/S0168-9002(02)01938-1
19. S. Hahn, K. Kim, K. Kim, X. Hu, T. Painter, I. Dixon, S. Kim, K.R. Bhattarai, S. Noguchi, J. Jaroszynski, D.C. Larbalestier, 45.5-tesla direct-current magnetic field generated with a high-temperature superconducting magnet. Nature **570**, 496–499 (2020)
20. S. Hahn, D.K. Park, J. Bascunan, Y. Iwasa, HTS pancake coils without turn-to-turn insulatio. IEEE Trans. Appl. Supercond. **21**, 1592–1595 (2011)
21. K. Halbach, Superfish-A computer program for evaluation of RF cavities with cylindrical symmetry (1976)
22. C. Hanks, D. Hamman, *Radiation Effects Design Handbook* (NASA, Washington, DC, 1971)
23. C. Henning, et al., Tiber II—An upgraded Tokamak ignition/burn experimental reactor. Fusion Technol. **10**(3P2B), 1079–1087 (1986)
24. M.W. Hooker, et al., Industrialization of radiation-resistant cyanate ester magnet insulation. IEEE Trans. Appl. Supercond. **19**(3), 2367–2370 (2009)
25. R.L. Keizer, M. Mottier, Radiation-resistant magnets. Tech. rep., CERN (1982)
26. R. Kraus, The overview and history of permanent magnet devices in accelerator technology. IEEE Trans. Magn. **30**(4), 1547–1554 (2002)
27. T. Kutsukake, H. Somei, Y. Ohki, K. Nagasawa, F. Kaneko, Gamma-irradiation effect on high-Tc superconductor YBa2Cu3O7. Jpn. J. Appl. Phys. **28**(8), 1393–1394 (1989)
28. L. Luo, Y. Zhang, S. Hu, W.H. Liu, G. Zhang, W. Hu, Gamma radiation effects on some properties of YBCO. Physica C **178**, 11–14 (1991)
29. M. Menzel, H.K. Stokes, *Users Guide for the Poisson/Superfish Group of Codes*. Tech. rep., Los Alamos National Lab (1987)
30. D. Morrissey, The coupled cyclotron project at the nscl. Nuclear Physics A **616**(1), 45–55 (1997). https://doi.org/10.1016/S0375-9474(97)00073-0
31. D. Morrissey, B. Sherrill, M. Steiner, A. Stolz, I. Wiedenhoever, Commissioning the a1900 projectile fragment separator. Nucl. Instrum. Methods Phys. Res., Sect. B **204**, 90–96 (2003). https://doi.org/10.1016/S0168-583X(02)01895-5
32. C. Pai, et al., Mechanical design of fast extraction kicker and PFN for SNS accumulator ring. *Proc. PAC'03*, 2147–2149 (2003). https://jacow.org/p03/papers/WPAB049.pdf
33. C. Pai, et al., Mechanical design of a ferrite-based injection kicker for SNS accumulator ring. *Proc. PAC'03*, 2144–2146 (2003). https://jacow.org/p03/papers/WPAB048.pdf
34. E. Pozdeyev, et al., FRIB Front End Design Status. *Proc. LINAC'12*, 1047–1049 (2012). https://jacow.org/LINAC2012/papers/THPB097.pdf

35. J. Rank, et al., The Lambertson septum magnet of the spallation neutron source. *Proc. PAC'05*, 3847–3849 (2005). https://jacow.org/p05/papers/MPPT071.pdf
36. G.H. Rees, Extraction. in *CAS - CERN Accelerator School*, 5th General Accelerator Physics Course (CERN, 1992)
37. L. Rossi, The LHC superconducting magnets. *Proc. PAC'03*, 141–145 (2003). https://jacow.org/p03/papers/TOAB001.pdf
38. F. Savary, et al., Progress on the development of the Nb3Sn 11T dipole for the high luminosity upgrade of LHC. IEEE Trans. Appl. Supercond. **27**(4), 1–5 (2017). https://doi.org/10.1109/TASC.2017.2666142
39. P. Schmüser, F. Willeke, 5 the electron-proton collider hera: The largest accelerators and colliders of their time. in *Accelerators and Colliders*, 428–443 (Springer, Berlin, 2013)
40. R.G. Sharma, Superconducting magnets in accelerators. in *Superconductivity: Basics and Applications to Magnets*, 257–303 (Springer International Publishing, 2015). https://doi.org/10.1007/978-3-319-13713-1_8
41. B. Sherrill, D. Morrissey, J. Nolen Jr, J. Winger, The A1200 projectile fragment separator. Nucl. Instrum. Methods Phys. Res., Sect. B **56**, 1106–1110 (1991)
42. SolidWorks, Dassault systèmes software. https://www.solidworks.com/
43. H. Song, et al., Design and engineering of an HTS dipole in the FRIB fragment separator. IEEE Trans. Appl. Supercond. **25**(3), 4602506 (2015)
44. J.W. Stetson, et al., A Comparison of electrostatic and magnetic focusing of mixed species heavy ion beams at NSCL/MSU. *Proc. PAC'05*, 2281–2283 (2005). https://jacow.org/p05/papers/TPPE033.pdf
45. K.H. Tanaka, E. Hirose, H. Takahashi, K. Agari, M. Ieiri, Y. Katoh, M. Minakawa, R. Muto, Y. Sato, S. Sawada, M. Takasaki, A. Toyoda, H. Watanabe, Y. Yamanoi, K. Kato, Y. Saitoh, H. Satoh, Y. Satoh, K. Yahata, Radiation-resistant magnets for J-PARC. IEEE Trans. Appl. Supercond. **26**(4), 4000804 (2016)
46. K.H. Tanaka, et al., Development of radiation-resistant magnets for the JHF project. IEEE Trans. Appl. Supercond. **10**(1), 206–209 (2000)
47. N. Tani, T. Adachi, H. Someya, Y. Watanabe, H. Sato, J. Kishiro, Eddy current effect of magnets for J-PARC 3-GeV synchrotron. IEEE Trans. Appl. Supercond. **14**(2), 421–424 (2004)
48. N. Tani, Y. Watanabe, H. Suzuki, T. Adachi, H. Someya, S. Igarashi, K. Nakamura, T. Watanabe, Design and Fabrication of an RCS Magnet Coil Using a Stranded Conductor of J-PARC 3-GeV Synchrotron. IEEE Trans. Appl. Supercond. **18**(2), 314–317 (2008). https://doi.org/10.1109/TASC.2008.920829
49. J. Tuozzolo, J. Brodowski, G. Danby, J. Jackson, W. Meng, Magnets for the National Spallation Neutron Source accumulator ring. *Proc. PAC'97*, 3233–3235 (1997)
50. P. Wanderer, et al., The SNS ring dipole magnetic field quality. *Proc. EPAC'02*, 1317–1319 (2002). https://jacow.org/e02/papers/WEPLE055.pdf
51. X. Wang, S.A. Gourlay, S.O. Prestemon, Dipole magnets above 20 tesla: Research needs for a path via high-temperature superconducting REBCO conductors. Instruments **3**(4) (2019). https://doi.org/10.3390/instruments3040062
52. J. Wei, Y.Y. Lee, D. Raparia, J. Sandberg, J. Tuozzolo, W.T. Weng, Spallation Neutron Source ring – status, challenges, issues, and perspectives. *Proc. PAC'03*, 571–575 (2003). https://jacow.org/p03/papers/ROPA001.pdf
53. W.T. Weng, Status of the AGS Booster Project. *Proc. EPAC'88*, 436–439 (1988)
54. M.N. Wilson, *Superconducting Magnets* (Plenum Press, 1983)
55. R.R. Wilson, The tevatron. Phys. Today **30**(10), 23–30 (1977)
56. H. Wollnik, *Optics of Charged Particles* (Academic Press, 2021)
57. X. Zhou, *Electrician Technology Handbook* (Chinese Electric Power Publish Company, Beijing, China, 2004)

Chapter 7
Controls

Abstract This chapter presents the control systems used in high-power accelerators, along with the design features that drive the software and hardware architecture. The control system provides critical global capabilities that are essential to the accelerator for all phases of the project. The standard software architecture for Supervisory Control and Data Acquisition (SCADA). The standard hardware architecture consists of several networks for high speed, low latency, and precise timing. In addition to the architectural aspects, there are a few application-specific hardware, firmware, and software architectures, there are a few facility-specific requirements that require special implementations.

7.1 Accelerator Control Systems Introduction

An accelerator control system is needed to coordinate data acquisition and control for the facility that allows scientists and engineers to test, commission, operate, and automate the facility in a safe and reliable manner. An accelerator control system includes traditional industrial control I/O and operator applications. It also requires high-performance timing, networks, and instrumentation that have sampling rates in the hundreds of MHz, latency measured in μs, jitter measured in picoseconds (ps), and data rates in gigabits per second (Gbps). Accelerator controls also require applications and software architecture that support physics applications for beam commissioning and automation [1].

7.1.1 Integrate the Instrumentation

The most basic function of the control system is to provide read/write access to all control system equipment. In modern accelerator user facilities, there are approximately 1,000,000 physical and computed process variables (PVs), 200,000 physical signals, and some 10,000 device instances, which include about 250 unique device types.

J. Wei et al., *High Power Hadron Accelerators*, Particle Acceleration and Detection,
https://doi.org/10.1007/978-3-032-17662-2_7

7.1.1.1 Industrial Control Equipment

Industrial control equipment includes instruments that can be read in tens of milliseconds to many seconds. These include vacuum, temperature, and power supplies, such as those used for electromagnets, high power RF, or vacuum gauges. These instruments are traditionally integrated through programmable logic controllers (PLCs). More recently, many of these instruments have been integrated directly over Ethernet. PLCs are also typically integrated using protocols implemented over Ethernet. A large facility may have on the order of 8,000 devices with 200,000 signals to integrate [2].

7.1.1.2 Instrumentation Control Equipment

Instrument control encompasses the high-speed, high-performance instrumentation required for accelerator control. These instruments are required to read analog signals at more than 100 MHz. These instruments are critical for understanding the operation of the accelerator. High-speed data are provided by these instruments for analyzing machine trips. 100 kHz frequency data may be used for fast feedback systems to control the beam trajectory. These data must be transmitted over dedicated communication links (1–10 Gbps) to the computational nodes that compute corrections and change set points. These instruments also produce lower-frequency data for operator viewing and long-term archiving. Most of these instruments are now implemented in field programmable gate arrays (FPGAs), and have embedded triggering, data buffering, and time stamping functionality to synchronize and correlate data. These instruments are critical for understanding and operating accelerators. A user facility typically has two thousand device instances with 10,000 signals [3].

7.1.2 Provide Tools to Monitor and Control Instrumentation

Foremost, a control system provides read and write access to all instrumentation. In industrial control system parlance, supervisory control and data acquisition (SCADA) systems collect data from a distributed control system to an operator console. From the operator console, the accelerator complex can be monitored, and commands can be sent to equipment facility-wide. In an accelerator facility, there can be one million PVs. Operator user interfaces must provide graphical updates at 5–60 Hz, provide display hierarchies, display call-up times on the order of a second, and provide response to operator commands in less than a second. More recent user interfaces provide an integrated environment that supports operator workflow, such as Control System Studio (CS-Studio) [4].

7.1.3 Provide Tools to Archive Data

A data archiver in an accelerator facility needs to be provided to support trending in normal operations. It is not uncommon to archive on the order of one million PVs. These data arc typically archived at 1 Hz and are maintained for years. The time stamp is used to correlate data from signals that are distributed as widely as kilometers. The data is time stamped as near to the time it is read or written so that communication and task latency do not hinder the correlation of potential causal relationships. Time-series data are typically retrieved for a small number of PVs over the last work shift. The return of data over one shift for 20 signals is a typical request and should be retrieved and plotted within 1 s [5].

7.1.4 Provide Tools to Support Fault Analysis

Good tools for fault data collection and analysis greatly reduce the time required to identify and address system issues. High-speed sampling of instrumentation control equipment is sent to an operator or data archive only when there is a fault or a request. A period on the order of 1–10 s is kept in local circular buffers and latched on demand or when a fault occurs. At 1 MHz, this is one million samples. There are hundreds of such transducers in an accelerator. The trigger for these buffers is typically delivered through hardware, causing the event to be time stamped and the buffers to freeze until they are downloaded. These events are not monotonic. They can occur very frequently during commissioning or during a period of change or equipment failure, and then not occur again for a long period. The collection, viewing, access, and management of these data sets provide machine physicists and scientists with high-resolution correlated data that help identify source issues to reduce the time required to resolve them and minimize operational downtime.

7.1.5 Provide Alarm Management

With more than 1 million PVs, the notification on exception provides operators with notifications when something is off normal. The management of alarms can significantly improve operations and maintenance. Care is needed to determine which alarm the operations team needs to respond to. Alarms can be used to notify plant engineers that some equipment is off-normal. The distinction between a process alarm that requires the operator to take action and those that inform an engineer that something is incorrect, but not significant or actionable by the operations team is important. Significant time must be spent distinguishing alarms required for operations.

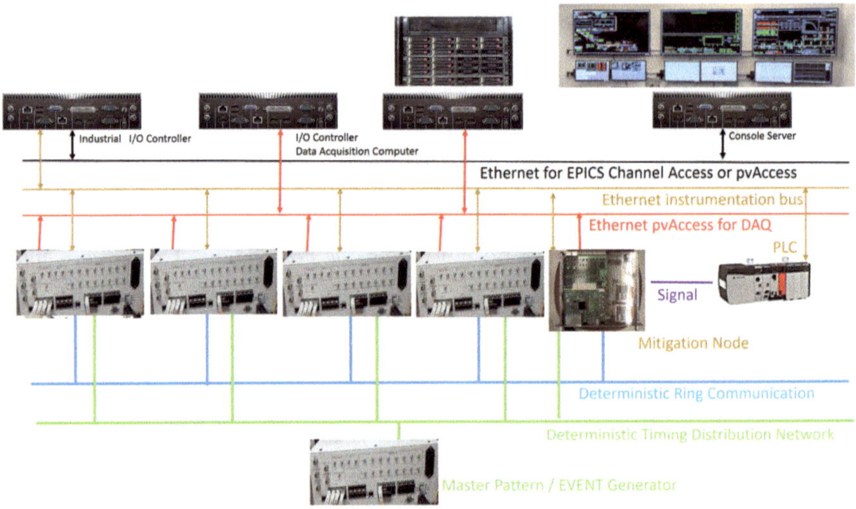

Fig. 7.1 Hardware interconnect for control system components

7.1.6 Provide Tools to Support Commissioning

Applications can be very specific to a facility. However, services that enable the rapid development of thin commissioning clients significantly improve commissioning. Some tools that greatly improve commissioning include a directory service that allows reference to subsystems or devices, an online model that simulates the machine, a scan engine that supports the study of control space, and the ability to save, manage, and restore set points.

7.2 Central Instrumentation Subsystems

Central instrumentation subsystems are those that are common to many of the subsystem-specific applications. They provide all connectivity for data transmission and synchronization. These systems communicate over fast serial interfaces using different protocols. These networks include: SCADA network, instrumentation network, timing network, machine protection network, and fast feedback network. Figure 7.1 shows the hardware interconnect for the control system components.

7.2.1 SCADA Network

The SCADA network provides communication between remote I/O interfaces and client applications to provide facility-wide monitoring and control. This network typically uses standard Ethernet protocols, such as user datagram protocol (UDP) or the transmission control protocol/internet protocol (TCP/IP). The network typically has hundreds of remote I/O interface controllers (IOCs) that pass data to tens of middle-layer servers or operator computers. In the SCADA network, the data are referenced by PV names. These names represent the system, device, and signals. These PVs are located within the hundreds of IOCs. Once they are located, operations on this network allow clients to get, put, or subscribe to change notifications. On a user-facility, it is common for this network layer to support more than 1 million notifications per second, which are sent to the tens of middle-layer servers.

7.2.2 Instrumentation Network

The instrumentation network is used by the IOC to communicate with various sources of input and output signals. It is increasingly common for this connection to be established over dedicated serial interfaces running some level of the Ethernet protocol. This remote I/O integrates both the industrial I/O and the instrumentation I/O. There are hundreds of I/O boxes that are housed in a number of formats: PLCs with proprietary backplanes, I/O housed in μTCA, ATCA, and custom crates, or standalone devices housed in 1u rack mount crates. A typical I/O box supports from 1 device to tens of devices, with fewer than 500 signals. There are thousands of I/O boxes in a typical user facility.

7.2.3 Event and Timing

Timing provides two critical capabilities for accelerator control systems: synchronized control of the equipment and global information that is used to timestamp and correlate data for analysis. The events and data are transmitted from a single master (event generator, EVG) over a unidirectional network to every node that requires high-precision event triggering for control or data acquisition. The distribution of the timing bus is achieved by fanning out the timing stream to each node.

7.2.3.1 Global Information

The event system typically transmits global data to IOCs and instruments with embedded event receivers (EVR). This data includes the time of day, beam parameters, machine mode, beam path, and pulse structure information that can be used for various feedback and protection systems. This data is not changed frequently, perhaps 1 or 10 Hz.

The time of day is obtained from a global positioning system (GPS) receiver connected to the EVG, and the frequency is synchronized to a high-precision RF frequency generator.

The machine mode, beam path, and pulse structure are all set by operators or high-level applications that define the currently approved operational envelope. The broadcast of these parameters to all controllers and high-performance devices ensures that all critical equipment protection devices are synchronized with the current operating mode of the machine and that they will respond when they detect excursions. These modes are also used to set soft limits that provide additional protection by limiting the values that can be sent to the controllers.

7.2.3.2 Synchronized Control

All accelerators require synchronized control and the acquisition of devices. This supports the control of the phase, amplitude, and position at the appropriate granularity. Some machines require pulses of tens of femtoseconds at a 1 MHz rate/resolution to be created, compressed, accelerated, and transported from different sources to different destinations. In synchrotrons, global position control requires diagnostics to be read and transmitted to support the calculation and settling time of new set points for the fast corrector at 10 kHz over 1 km. Triggers that are sent over large distances require the calibration of offsets based on distance and fiber type. Hardware is required to synchronize instruments at this accuracy and distance. Examples of this hardware include the EVG and the EVR developed at SLS with Micro Research Finland (MRF) [6], which were widely sold throughout the community in the 2000s & 2010s. The features include 8 ns trigger resolution, less than 10 ps pulse-to-pulse jitter, and less than 15 ps long-term drift.

All accelerators benefit from the ability to correlate data read into the control system. The ability to time-stamp data with a globally synchronized identification allows machine physicists to study the effects on beam quality and the causes of beam loss by making it possible to observe the causes and effects of various conditions in different subsystems. The more care is taken to provide the best possible time stamp, the more useful the correlation data becomes. For example, many beam instruments now have an EVR embedded in the FPGA so that a time stamp is taken at the 8 ns pulse when the value was read or the device was controlled. This time stamp must be communicated through the control system hardware and software stack along with the measurement as an atomic value. In the I/O controller (IOC) of the Experimental Physics and Industrial Control System

(EPICS) [7–9], great care is taken to allow the time stamps to originate anywhere in the HW/SW stack and to be preserved through all layers. The absence of hardware correlation does not mean that there is no chance of correlation. Using a GPS in conjunction with the network time protocol (NTP) allows control engineers to provide millisecond-level time stamps for signals that are read through hardware that does not communicate with the accelerator timing system. In these cases, the latency or frequency with which values are read into the control system is the largest source of time misalignment. A PLC typically scans the data using the local control configured in the PLC. The EPICS IOC then reads the value from a memory area at a specified frequency. The sum of these two periods creates the worst-case offset that may occur in the timing system. The time difference in this case can easily be managed to be in the 10–100 ms range, which is helpful in many situations where temperature, pressure, or flow can affect beam quality. The ability to correlate data for analysis greatly reduces the time required for initial commissioning and performance improvements with data that can be analyzed to discover sources of beam perturbations.

7.2.3.3 Machine Protection Systems

The primary function of machine protection is to protect the accelerator components from beam damage. The beamline components typically take a long time to repair or are very expensive to replace. The machine protection system consists of several distinct and important components: the beam permit system, parameter management, fast machine protection [10], fault data, and fault analysis.

Beam Permit System

The beam permit system, or the run permit system (RPS, Sect. 8.2.3), is needed to limit the approved and controlled modes of operation. This system has two parts: (1) the monitoring of key parameters that permit a defined state and (2) the monitoring of those parameters to determine if the current state is allowed. This is typically done in software and is typically in the 100 ms response range. Multiple states may be permitted at the same time, but the machine is in one and only one state. State transitions may be manual or automated, such as an operator or physicist requesting a certain state or the ramping up or down of a repetition rate as conditions allow. The definition of the machine states is required by the beam permit system from the machine physicists and engineers to ensure that well-planned operations are in place to avoid unexpected beam loss that may result in radiation outside the planned areas or damage to critical components, which can cause large costs or schedule delays to a project.

Parameter Management

The parameter management is required to give administrative control to those with the authority to approve the operating range of the machine in a given state. These machine modes, specified by machine physicists and engineers, have clear limits of operation. This section identifies the signals that constitute the permit and the range in which the operation is safe. In certain modes of the machine, for instance, the beam may not be able to enter different areas of the accelerator. In this mode, only areas that see the beam need to be included in the fast protection logic. In different modes of the machine, different ranges may be allowed for control parameters such as current or beam rate. These parameters must be closely managed, approved, loaded for operations, and confirmed to be the appropriate version.

Fast Protection

The fast protection system (FPS) is the hardware and firmware that integrate all the monitored values and the mitigation devices to provide the fastest possible response time to limit damage to the equipment or radiation from beam loss (Sect. 8.2). The primary data path for FPS is from the monitors that detect loss to the mitigation devices that stop loss. These systems may also be used to create fast, reliable paths between slow devices that change the beam permits, such as a device moving off a limit switch and into the beam, or an interlock from the beam steering that could signal the fast protection system when the beam steering is off, but prior to significant loss outside the beam pipe. The FPS must be designed to be fail-safe. It is not as critical as a safety system like PPS, but it is typically treated similarly in the verification of its operation during any new start-up period. The primary difference between MPS and PPS is that MPS does not require regulatory compliance.

Fault Data

Different machines require different time periods to be saved that match the types of problems that accelerator physicists expect. These known cases are 1 s of the data at 1 MHz or 10 s of the data at 100 kHz. This amounts to approximately one million (1M) samples per signal. There are hundreds of signals that are used to monitor trip conditions. The collection of hundreds of 1M sample buffers can be time-consuming. It may be useful to consider double buffering of fault data so that if a single fault occurs, the collection of the buffers does not limit the time it takes to restore operation. On the other hand, machine protection trips are not normal; they typically stop operation and require repairs to certain devices. During commissioning or ramp-up, this is more likely to occur. In those cases, double buffering would be more useful. Figure 7.2 shows an example of BPM traces after an MPS fault. Fault mitigation during commissioning frequently includes increasing the trip threshold, reducing the beam rate, or reducing the beam intensity.

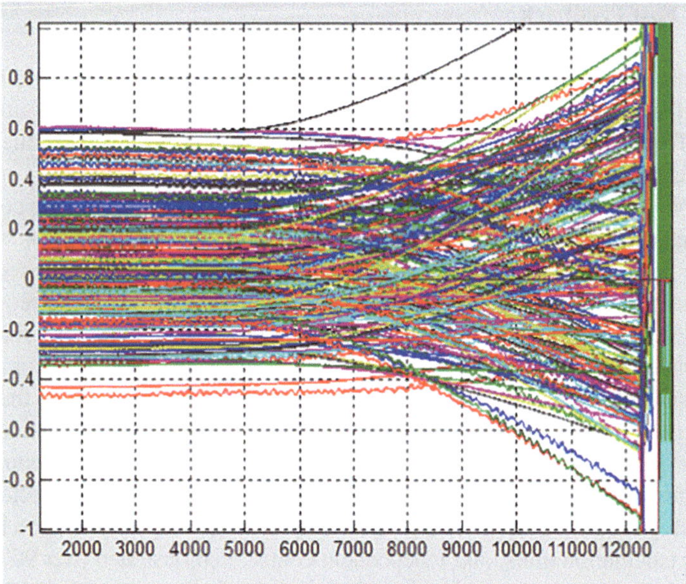

Fig. 7.2 An example of beam position monitor (BPM) traces after an MPS fault

Fault Analysis

As machine protection occurs on the microsecond time scale. It is important to have the ability to determine the cause of the trip on this time scale. Many of the interfaces to the MPS are driven by this requirement. It is typical for an MPS trip to send a notification to all devices that can cause the trip. This signal locks high-speed circular buffers that can be downloaded and analyzed to identify the cause of a trip. The causes of a trip can range from bad cabling or malfunctioning beam loss monitors to beam loss that exceeds the limits for a given area. The time required to download and analyze the data may affect the commissioning or operational time. The amount of data available can be significant with modern instrumentation that can keep 300 MHz of raw data. The identification of what data will be needed for fault analysis is important early in the project, as this usually imposes requirements on beam instrumentation that would not otherwise exist. A common scenario is for the MPS to turn off the beam at the source (or a bending magnet to a beam dump) and also to feed a trip indicator into the timing system that is propagated to every device synchronized with the global timing system. This signal, through the timing system, is decoded in the FPGA of each fast device (and perhaps into software-managed circular buffers) to lock the fast buffers until they are downloaded, and the Fast Protection System is rearmed, allowing the beam permit system to give the operator the ability to restart the beam.

7.2.3.4 Fast Feedback

Fast feedback is an automation implemented to control the beam parameters to improve the beam size. These control loops are used to suppress noise sources up to 10 kHz. To provide global fast feedback, hardware must be designed to collect inputs from their sources, communicate the data for the computation of current errors and new set points for correction, and communicate the new set points to the correctors. This time critical system must be designed into the sensor and corrector hardware. The time budget must be carefully managed.

An example of fast feedback is the NSLS II beam steering [11]. The requirement was to position the beam to 0.5 μm. The time budget was divided by the latency of the analog front end, the transmission of the data, the global calculation, the transmission of the new set points, the setting of the new values, and the latency for that new setting to affect the position of the beam. NSLS II demonstrated a 10 kHz feedback, with 240 BPMs in a storage ring of 792 m circumference, with 90 fast correctors (3 dB bandwidth of 5 kHz) and 180 slow correctors (3 dB bandwidth of 10 Hz), 30 cell controllers acting as two-tier communication hubs and feedback calculation units, and independent control demonstrated over 90 different eigen modes in both horizontal and vertical planes. The latency of the BPM analog electronics and the settling time of the magnetic field used most of the 100 μs time budget. The digital electronics used 14 μs for global BPM data transfer, 24 μs for full eigen-mode calculations, and 7 μs for fast corrector data transfer and settings.

7.3 Instrumentation Architecture

7.3.1 Industrial Instrumentation Systems

Industrial instrumentation is typically integrated into a PLC system. There are many commercially available industrial PLCs, such as Allen Bradley [12], Beckhoff [13], Siemens [14], and Yokogawa [15]. They provide a variety of hardened I/O modules that can be distributed throughout a plant with little need for cooling. These I/O systems are chosen for their cost, ease of maintenance, reliability, ease of installation, and the variety of signal conditioning provided. These I/O products are typically supported for the life of a facility, 25 years or longer.

Until recently, PLCs had no precision time synchronization. This limited the ability to correlate the data collected facility-wide to within hundreds of milliseconds. In recent years, these platforms have also started to support NTP or Precision Time Protocol (PTP), which allows PLC signals to be time stamped and correlated with other signals in less than a millisecond.

Motion control is a specialized application of industrial systems that is now performed using specialized controllers. These controllers handle limits, velocity, acceleration, coordinate transformations, and position readback.

Subsystems that fall into this category are high-power RF, vacuum, cooling, EPICS facility, mirror control, and safety.

7.3.2 High Performance Instrumentation Systems

High performance instrumentation is typically developed on FPGA based platforms. There is an embedded EVR, gigabits of random access memory (RAM), an Ethernet interface to communicate with the SCADA system, and fast serial ports to allow peer-to-peer communication. The lifetime of an FPGA is limited by the development chain support from the FPGA manufacturer; this typically lasts around seven years. A good design that isolates the digital portion of the instrument from the digital and analog I/O helps limit the impact of updating FPGAs.

The timing system provides events for data collection, beam control, and RF control. For example, the MRF [6] system delivers 8 ns events with less than 10 ps jitter facility-wide. This system can be used to replicate the RF clock. The system receives 16 bits every 8 ns, where 8 bits are used to deliver events and the other 8 bits are used to transmit data. Using the data capability, the event system transmits changing conditions in the accelerator so that high performance systems can respond within 10 s of nanoseconds.

RAM can be used to store and forward a variety of data. In typical applications, RAM is divided into configuration registers, monitor registers, and deep circular buffers. The configuration registers are used to set parameters for the device, such as conversion coefficients, deep buffer size, and control constants for feedback systems. The monitor registers are used to hold values that are sent to different applications that consume data at much lower rates, such as slow global feedback, operator monitoring, and data archiving. The deep circular buffers are used to capture either raw data or high-speed data, which can be stopped through the event system and collected to study the frequency performance of accelerator components.

The SCADA system typically sets and reads the FPGA memory registers over an Ethernet connection between 100 Mbps and 10 Gbps. For most of the data being integrated, any rate is adequate, as this is a dedicated serial line. When the large circular buffers are transmitted, higher speed connections are preferred, particularly in the case of an MPS fault. MPS faults cannot be cleared until the fault data are secured. To minimize this time, MPS data can be double buffered in the FPGA or SCADA system.

The dedicated fast serial interfaces are used to develop 100 kHz feedback and 10 μs fast mitigation solutions. These system-wide timing requirements are spread among analog front-end latency, data transmission, computing, analog output latency, and system response times. The architecture of the FPGA communication network is critical to minimizing the data transmission component of this budget. Many fast protection networks use local marshalling of fault signals to a small set of nodes local to the mitigation devices, with many dedicated serial links to minimize transmission delays. For fast feedback, the distribution and execution of the feedback algorithm must also be considered.

Systems that fall into this category include fast orbit feedback, LLRF, diagnostics, and magnet power supply.

7.4 Software and Firmware Architecture

7.4.1 FPGA Based Programs

FPGA based instruments are used for accelerator instruments such as BPMs, BLMs, fast orbit feedback, and LLRF field control. They communicate with the control system over Ethernet using an assortment of Ethernet layers. In the best case, they have an EVR that can directly synchronize time stamps, decode data broadcast over the timing system, such as machine mode and time of day, and decode and respond to any number of timing events. FPGAs sample data at speeds of up to hundreds of MHz. They provide data to the subsystems at appropriate rates, such as machine protection response rates of 1 MHz, feedback rates of 100 kHz, and operator rates of 60 Hz or less. They capture raw data and data from the fast subsystem into onboard RAM that is used for analysis. Typical buffers include enough raw data to study the hardware performance of the device and system-wide data for about 1 s to analyze the performance of the accelerator.

Figure 7.3 shows an example of FPGA communications with a star timing distribution and ring communication for a fast orbit feedback and machine protection system. The blue network transfers data for fast feedback and machine protection between FPGA-based instruments over 6.25 Gbps communication links. The green network provides the event system data to all FPGA based devices. This is an event stream used to synchronize the instruments with the accelerator at 8 ns with jitter in the ps range. The event stream also delivers a data stream that contains the clock time and global machine data, such as the beam path, pulse identification (ID), and machine mode.

7.4.2 Instrumentation Integration

All instrumentation must be integrated into a SCADA system to provide real-time instrumentation data for commissioning, machine studies, and operations. At this layer, the SCADA system collects information from FPGA based instruments, industrial PLCs, crate based I/O such as VME or μTCA, and other intelligent devices that expose a communication protocol.

As shown in Fig. 7.4, Ethernet is the primary instrumentation bus for integration into the control system. The brown network is used to configure the FPGAs as well

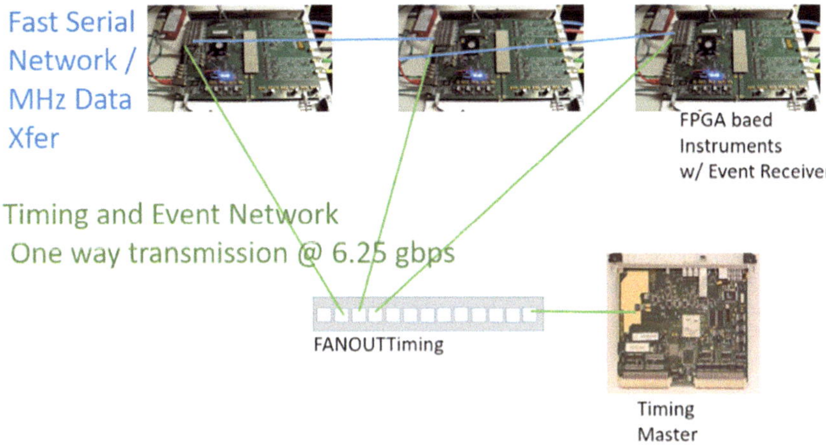

Fig. 7.3 FPGA communications with a star distribution of timing and a ring communication for fast orbit feedback and machine protection system

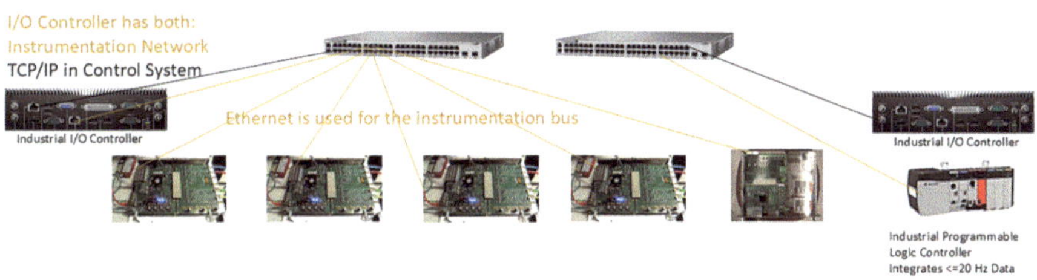

Fig. 7.4 Instrumentation integration with Ethernet as the primary instrumentation bus

as to provide access to the operator and memory buffers for raw and processed synchronous data.

7.4.3 *Distribution of Instrumentation Data*

Once the data are read into the SCADA system, they are available through a client/server protocol. This protocol is layered on UDP and/or TCP/IP. In the case of one of the most commonly used control system toolkits, EPICS [7, 8], the protocol provides access in several ways: get, put/get, and monitor or publish/subscribe. Each of the IOCs has a local directory that responds to UDP broadcast searches for PVs. Once the connection is made over TCP/IP, the get, put/get, and subscribe commands are supported from the client to the service. This communication is conducted among many clients that support operations, commissioning, automation, and optimization. Operations are dependent on this network layer for all monitoring and supervisory control of the facility. This network is typically redundant to some

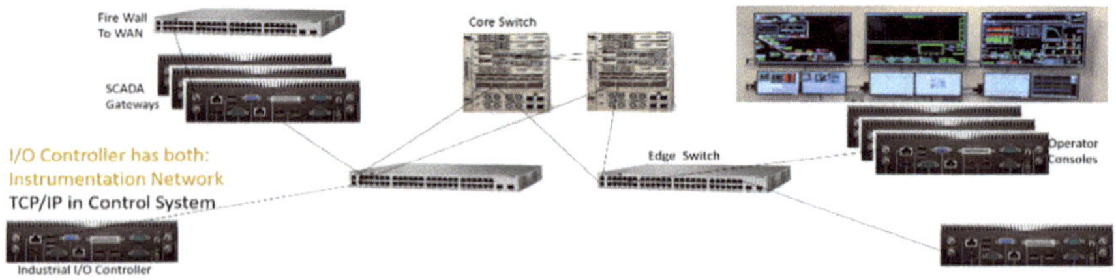

Fig. 7.5 Control system communication with Ethernet where PVs are sent with time and status metadata

degree. Additionally, this network typically uses a gateway to allow managed access to the control network for external computers. A firewall is typically provided to offer further protection to the control network.

Figure 7.5 shows the communication of the control system with Ethernet, where PVs are sent along with time and status metadata. The black network illustrates the connection between IOCs, intelligent routers (which are also redundant in some installations), redundant core switches, a set of gateway processors that connect to a wide-area network (WAN) through a firewall, and a set of control room computers that provide synoptic displays, alarm viewers, data archivers, and many services that can support efficient operations.

7.4.4 *Middle-Layer Service*

SCADA systems primarily provide an interface to the instrumentation, which is the main service of a control system. As systems reach higher power, electronics achieve higher performance, and instrumentation becomes more ubiquitous, the control system has grown significantly in complexity. There are many configuration and aggregation functions that are provided as standalone utilities. A partial list of these services includes: directory, alarm, archive, synchronous data acquisition, data indexing, experimental data, electronic log book, and save/retrieve. These middle-layer services enable the development of powerful operator tools and custom high level applications (HLAs) that are lightweight. The tools can be written in a scripting language, such as Java [16] or C++ [17]. Figure 7.6 shows the software architecture of the control system, including clients, I/O servers, and middle-layer services.

7.4.4.1 Directory Service [18]

In an accelerator facility, it is common to have hundreds of thousands of physical I/O points and hundreds of thousands of computed variables, which result in over a million named PVs. The directory service allows engineers to assign properties and tags to the PVs. An existing design supports up to 120 properties per PV. The

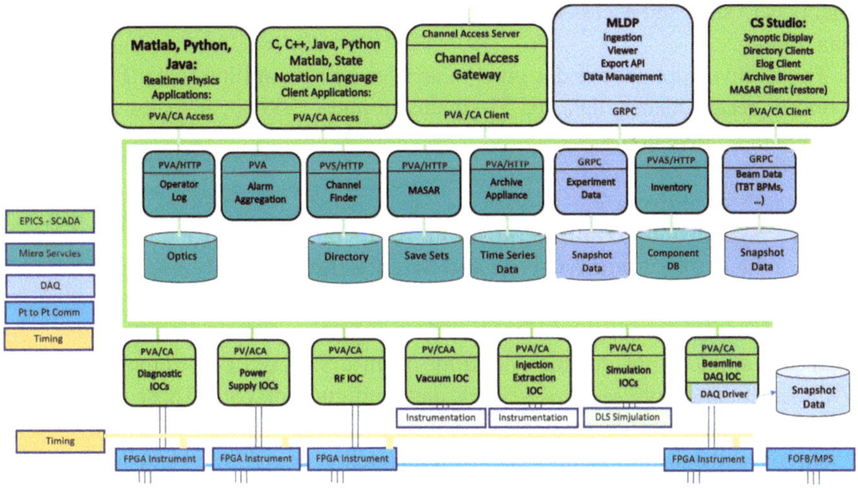

Fig. 7.6 Control system software architecture—clients, I/O servers and middle-layer services

properties can include the physical position of the device, the internet address of the service that contains this device, the device type and instance, and the subsystem name. These properties can be used to find all PVs that are X and Y position monitors for an accelerator and return them in position order without knowing the unique name of the PV. Tags can be used to assign PVs to different systems. For instance, a position monitor could belong to the injection linac and also to the warm linac. With tags and properties, a request can be made for all X- or Y-position readbacks in a portion of the linac. With a directory service in place, a physics application can be written to perform slow orbit feedback that will require no modifications when new beam position monitors or magnet power supplies are added to the accelerator.

7.4.4.2 Alarm Data Service [19]

The alarm data service provides grouping, notification, and masking of alarm states of signals in the control system. Signals can be configured into more than one group. The group alarm status is supported. In the group, the most severe alarm conditions are the state of the group. Notification can be made at the signal or group level.

7.4.4.3 Archive Data Service

The data archiving capability is designed to save time series data and provide it to users for studying machine operations. Time-series data are used most frequently to review recent operations. It also provides data for engineers, physicists, and scientists to save and access their time series data to perform system analysis using data mining.

The archive appliance is designed to archive up to 1 million scalar PVs at 1 Hz. It returns recent history data for 32 signals in under 500 ms. It scales through the addition of archive nodes.

7.4.4.4 Synchronous Data Acquisition, Artificial Intelligence and Machine Learning, Fault Analysis and Science Data [20, 21]

Synchronous data acquisition (DAQ) and management collect data that are correlated at a single point in time and come from a variety of instruments: images, N-dimensional arrays, vectors, scalars, and configuration data. These data include high-speed accelerator instruments, scientific data from experiments, and fault data.

High-speed data may be used for artificial intelligence (AI) and machine learning (ML) or for scientific data that results from an experiment. These data may come from a variety of instruments and configuration databases. These data are consumed by different analysis codes and require different formats to use these codes. Data sets created from this analysis must be stored along with the raw data to support others in recreating the results of the analysis. It also supports re-analysis of the raw data if a source of error is discovered, such as mis-calibration or an unrecorded configuration change. As these data have a long life cycle; indexing, annotating, and managing the data are required.

Beam synchronous data involve the collection, management, and retrieval of data gathered from distributed instruments at high speeds into local buffers that are latched by distributed triggers, typically through a facility-wide event system. Fault data are a special case of this that is triggered by the MPS to study the sequence of events that lead to a beam spill. Data are typically collected from fast diagnostics and LLRF. In a user facility, this involves up to 1,000 instruments collecting 1 MHz data for 1 s. Other PVs may be stored in this set, such as the beam current. User interfaces and programming interfaces will provide the ability to visualize and analyze this data. The analysis data and additional metadata must be attached to the data set as a result of the analysis.

7.4.4.5 Data Index Service

Fault data, beam synchronous data, and experimental data are all sets of data from a single event or a tightly related series of events. The data are stored in a single file. The files also contain configuration data or independent variables that are controlled at the time the data are collected. These files are some sort of self-describing data, such as HDF5 [22], XML [23], or more colloquial data, such as SDDS [24]. To support data mining, searches are used to find data sets whose metadata matches certain criteria. A data index service is used to provide high-speed searches. The metadata for these data files will be placed into the data index service, along with the file location information, to provide fast search implementations without opening and parsing self-describing data files.

7.4.4.6 Electronic Log Book Service [25]

The electronic log book service (ELS), or OLog, provides a service to manage operator log entries. Entries are submitted and retrieved from a backend database. It has features to connect images, signal lists, tags, and additional metadata. Entries can be edited, with previous versions kept and accessible.

7.4.4.7 Data Save and Retrieve Service [26]

The save and retrieve service allows applications to define groups of PVs and to store and retrieve them as named, tagged, and described sets. This tool has been extensively used to manage set points for machine commissioning.

Data sets include scalar and vector values. Sets of up to 1,000 PVs have been demonstrated to support all magnet power supplies in an accelerator facility. The retrieval and display of a given save set is specified to be completed in 500 ms.

7.4.5 Operator Tools

There are CS-Studio widgets that support the configuration and viewing of middle-layer services. The CS-Studio environment enables the use of other services. From a CS-Studio panel, a query can be made to the channel finder service, which returns a list of PVs. This list of variables can then be used to bring up a list of PVs in the data browser. The list of PVs can be dropped into the MASAR [27] configuration tool. Any panel can be placed in the OLog to report operations.

7.4.5.1 Save and Retrieve

The save and retrieve tool allows CS-Studio and Python [28] applications to compare, scale, set, and manage these settings. Figure 7.7 shows an example of save sets of configurations that are saved, compared, and sent to PVs.

7.4.5.2 Data Browser and CS-Studio

The data browser application runs in CS-Studio. It allows users to retrieve data from the archive appliance and plot time-series data. The data browser application

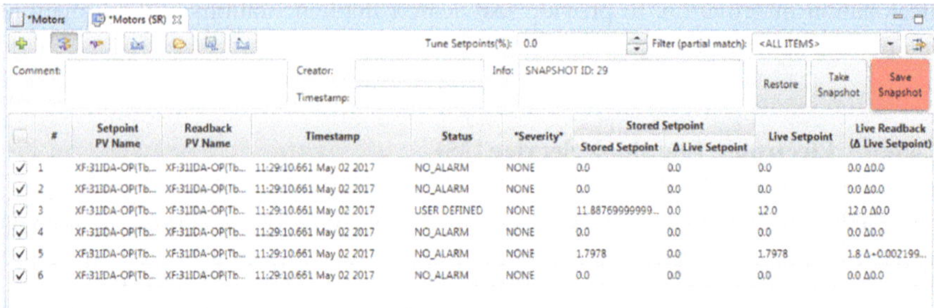

Fig. 7.7 An example of save sets of configuration

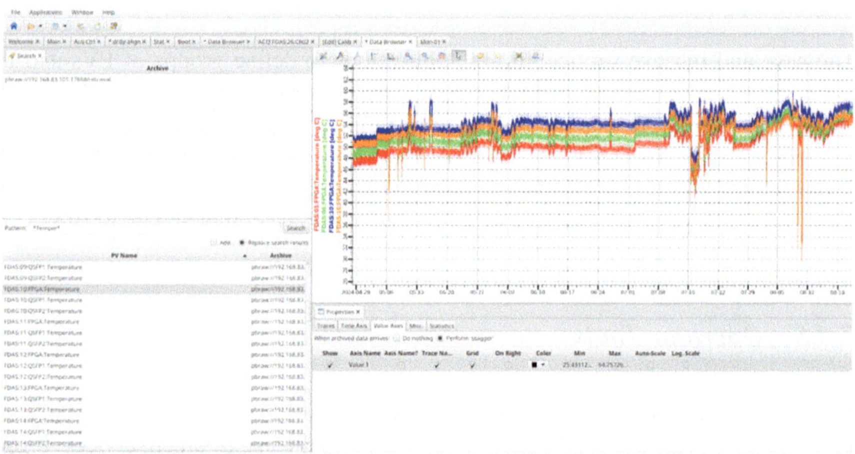

Fig. 7.8 An example of the time series data retrieved from the archive servers

plots real-time data when the time range includes the present. Figure 7.8 shows an example of the time series data retrieved from the archive servers.

7.4.5.3 Alarm Viewer

Each signal can potentially be under alarm conditions. PVs can be configured in the IOC to represent the sum of alarm signals for a given device. These sums can also be performed at the middle layer. An alarm viewer for BEAST [29] is available in CS-Studio. Figure 7.9 shows an example of the alarm data servers that collect PV alarms and aggregate the status in configured hierarchies.

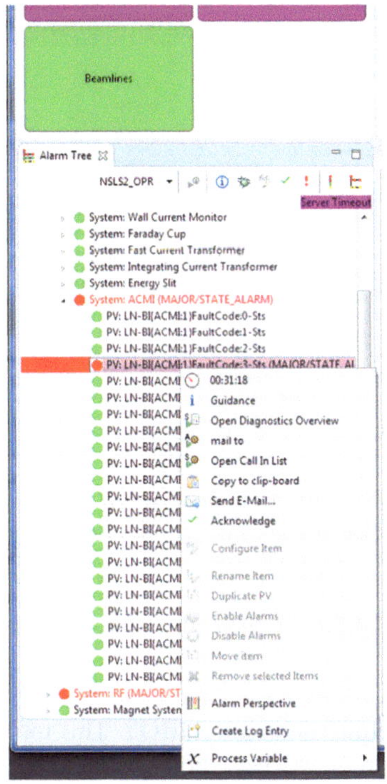

Fig. 7.9 An example of the alarm data servers collecting PV alarms and aggregating the status in configured hierarchies

7.4.5.4 Directory Configuration and Use

The directory service has a CS-Studio configuration parameter that allows users to add properties and tags to sets of PVs. These properties and tags allow subsequent users to access a list of PVs based on the values of the properties or tags. This can be used to find all tags for a particular device, all devices in a geographic area or subsystem, or all steering devices, as examples.

There are existing widgets in CS-Studio for viewing and editing the channel finder service. The channel finder service can be used directly with many of the CS-Studio widgets. Figure 7.10 shows an example of the directory services that are used to provide PV properties that allow searches by device, location, and/or subsystem.

7.4.5.5 Electronic Log Viewer

The electronic log viewer is a CS-Studio widget for viewing operator entries. The viewer allows for pursuing one line entries or the full entry. The viewer also allows for edits to the entries and views of previous versions.

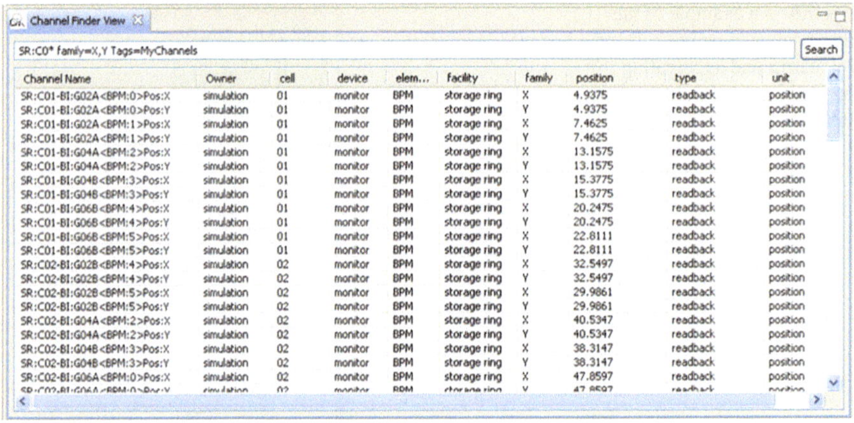

Channel Name	Owner	cell	device	elem...	facility	family	position	type	unit
SR:C01-BI:G02A<BPM:0>Pos:X	simulation	01	monitor	BPM	storage ring	X	4.9375	readback	position
SR:C01-BI:G02A<BPM:0>Pos:Y	simulation	01	monitor	BPM	storage ring	Y	4.9375	readback	position
SR:C01-BI:G02A<BPM:1>Pos:X	simulation	01	monitor	BPM	storage ring	X	7.4625	readback	position
SR:C01-BI:G02A<BPM:1>Pos:Y	simulation	01	monitor	BPM	storage ring	Y	7.4625	readback	position
SR:C01-BI:G04A<BPM:2>Pos:X	simulation	01	monitor	BPM	storage ring	X	13.1575	readback	position
SR:C01-BI:G04A<BPM:2>Pos:Y	simulation	01	monitor	BPM	storage ring	Y	13.1575	readback	position
SR:C01-BI:G04B<BPM:3>Pos:X	simulation	01	monitor	BPM	storage ring	X	15.3775	readback	position
SR:C01-BI:G04B<BPM:3>Pos:Y	simulation	01	monitor	BPM	storage ring	Y	15.3775	readback	position
SR:C01-BI:G06B<BPM:4>Pos:X	simulation	01	monitor	BPM	storage ring	X	20.2475	readback	position
SR:C01-BI:G06B<BPM:4>Pos:Y	simulation	01	monitor	BPM	storage ring	Y	20.2475	readback	position
SR:C01-BI:G06B<BPM:5>Pos:X	simulation	01	monitor	BPM	storage ring	X	22.8111	readback	position
SR:C01-BI:G06B<BPM:5>Pos:Y	simulation	01	monitor	BPM	storage ring	Y	22.8111	readback	position
SR:C02-BI:G02B<BPM:4>Pos:X	simulation	02	monitor	BPM	storage ring	X	32.5497	readback	position
SR:C02-BI:G02B<BPM:4>Pos:Y	simulation	02	monitor	BPM	storage ring	Y	32.5497	readback	position
SR:C02-BI:G02B<BPM:5>Pos:X	simulation	02	monitor	BPM	storage ring	X	29.9861	readback	position
SR:C02-BI:G02B<BPM:5>Pos:Y	simulation	02	monitor	BPM	storage ring	Y	29.9861	readback	position
SR:C02-BI:G04A<BPM:2>Pos:X	simulation	02	monitor	BPM	storage ring	X	40.5347	readback	position
SR:C02-BI:G04A<BPM:2>Pos:Y	simulation	02	monitor	BPM	storage ring	Y	40.5347	readback	position
SR:C02-BI:G04B<BPM:3>Pos:X	simulation	02	monitor	BPM	storage ring	X	38.3147	readback	position
SR:C02-BI:G04B<BPM:3>Pos:Y	simulation	02	monitor	BPM	storage ring	Y	38.3147	readback	position
SR:C02-BI:G06A<BPM:0>Pos:X	simulation	02	monitor	BPM	storage ring	X	47.8597	readback	position

Fig. 7.10 An example of the directory services that are used to provide properties to PVs that allow search by device, location, and/or subsystem

7.4.5.6 CS-Studio Synoptic Display Viewer

CS-Studio has many components. The primary component is the synoptic display editor and viewer, BOY. This tool is being replaced by a new editor; this work is in progress. The ability to handle pvAccess is provided primarily by the data layer below CS-Studio (DIIRT). The goal is to use the most recent developments for this demonstration. In either case, the data communication layer that connects to pvAccess is demonstrated. There are also a number of additional applications and widgets that are specific to MASAR, DataBrowser, OLog, and Channel Finder. All of these will further demonstrate the ability to build integrated services for operators, physicists, and experimenters.

References

1. L.R. Dalesio, M.E. Thuot, Control system architecture: the standard and non-standard models, in *Proceedings of PAC'93* (JACoW Publishing, Geneva), pp. 1806–1811 (1993)
2. D. Piso, S. Birch, A. Nordt, T. Gahl, P. Arnold, J. Weisend II, T. Korhonen, ESS PLC controls strategy, in *Proceedings of IPAC'15* (2015), pp. 3066–3068
3. M.G. Minty, F. Zimmermann, *Measurement and Control of Charged Particle Beams* (Springer, Berlin, 2003)
4. Git repository for control system studio, https://www.controlsystemstudio.org/
5. Git repository for the archive appliance, https://github.com/archiver-appliance/epicsarchiverap
6. MRF Timing System, Micro-Research Finland Oy. https://www.mrf.fi/
7. EPICS, Experimental Physics and Industrial Control System. https://epics-controls.org/
8. L.R. Dalesio, J.O. Hill, M. Kraimer, S. Lewis, D. Murray, S. Hunt, W. Watson, M. Clausen, J. Dalesio, The experimental physics and industrial control system architecture: past, present, and future. Nucl. Instrum. Methods Phys. Res. Sect. A: Accel. Spectrom. Detect. Assoc.

Equip. **352**(1), 179–184 (1994). https://doi.org/10.1016/0168-9002(94)91493-1. https://www.sciencedirect.com/science/article/pii/0168900294914931
9. A. Kozubal, D. Kerstiens, J. Hill, L. Dalesio, Run-time environment and application tools for the ground test accelerator control system. Nucl. Instrum. Methods Phys. Res., Sect. A **293**(1–2), 288–291 (1990)
10. K. Ha et al., NSLS-II active interlock system for fast machine protection, in *Proceedings of ICALEPCS'15* (JACoW Publishing, Geneva), pp. 554–557. https://doi.org/10.18429/JACoW-ICALEPCS2015-TUC3O05. https://jacow.org/ICALEPCS2015/papers/TUC3O05.pdf
11. Y. Tian et al., NSLS-II fast orbit feedback system, in *Proceedings of ICALEPCS'15* (JACoW Publishing, Geneva), pp. 34–37. https://doi.org/10.18429/JACoW-ICALEPCS2015-MOC3O05. https://jacow.org/ICALEPCS2015/papers/MOC3O05.pdf
12. Allen Bradley PLC Controllers, Rockwell Automation. https://www.rockwellautomation.com/
13. Beckhoff, New Automation Technology. https://www.beckhoff.com/
14. Siemens, Industrial Automation. https://www.siemens.com/
15. Yokogawa, Control Devices. https://www.yokogawa.com/
16. Java programming landuage, Oracle Java. https://www.java.com/
17. B. Stroustrup, A history of c++ 1979-1991, in *History of Programming Languages-II* (1996), ACM Press, New York, pp. 699–769
18. Git repository for the channel finder service, https://github.com/ChannelFinder/ChannelFinderService
19. GIT Repository for BEAST Alarm Service, https://github.com/ControlSystemStudio/phoebus/tree/master/services/alarm-server
20. C. Allen et al., A data science and machine learning platform supporting large particle accelerator control and diagnostics applications, in *Proceedings of IPAC'24, IPAC'24—15th International Particle Accelerator Conference*, no. 15 (JACoW Publishing, Geneva, 2024), pp. 1843–1845. https://doi.org/10.18429/JACoW-IPAC2024-TUPS71. https://indico.jacow.org/event/63/contributions/4032
21. C. McChesney et al., The data platform: an independent system for management of heterogeneous, time-series data to enable data science applications, in *Proceedings of IPAC'24, IPAC'24—15th International Particle Accelerator Conference*, no. 15 (JACoW Publishing, Geneva, 2024), pp. 1839–1842. https://doi.org/10.18429/JACoW-IPAC2024-TUPS70. https://indico.jacow.org/event/63/contributions/4052
22. HDF5, Hierarchical Data Format. https://www.hdfgroup.org/solutions/hdf5/
23. T. Bray, J. Paoli, C.M. Sperberg-McQueen, E. Maler, F. Yergeau, J. Cowan, Extensible markup language (xml) 1.0 (2000)
24. M. Borland, L. Emery, H. Shang, R. Soliday, User's guide for SDDS toolkit version 5.0. Advanced Photon Source (2020)
25. Git repository for olog, https://github.com/Olog/phoebus-olog
26. Git repository for save/restore, https://github.com/ControlSystemStudio/phoebus/tree/master/app/save-and-restore
27. G. Shen, Y. Hu, M.R. Kraimer, K. Shroff, D. Dezman, NSLS II middlelayer services, in *Proceedings of ICALEPCS'13* (JACoW Publishing, Geneva), pp. 467–470. https://jacow.org/ICALEPCS2013/papers/MOPPC155.pdf
28. G. Van Rossum, A brief timeline of python. The history of Python (2009)
29. K.U. Kasemir, X.H. Chen, E. Danilova, The best ever alarm system toolkit, in *Proceedings of ICALEPCS'09* (JACoW Publishing, Geneva), pp. 46–48. https://jacow.org/icalepcs2009/papers/TUA001.pdf

Chapter 8
Other Technologies for High Power Hadron Accelerators

Abstract Following the previous chapters on cryogenics, superconducting radiofrequency, magnets, and controls systems, this chapter discusses the remaining technical systems essential to high-power hadron accelerators: personnel, machine, and equipment protection systems, ion sources, radio frequency and power supply systems, vacuum system, various types of beam-interception system, beam diagnostics, survey and alignment, and conventional utilities. Aspects on radiation impact mitigation are also highlighted.

Several major technical systems for high-power hadron accelerators have been discussed in previous chapters: cryogenic refrigeration (Chap. 4), superconducting radiofrequency system (Chap. 5), magnets (Chap. 6), and controls (Chap. 7). In this chapter, other major technical systems are introduced, focusing on their practical application in high-power hadron accelerators.

The chapter starts with protection systems that are essential for the operation of high-power hadron accelerators: personnel protection (Sect. 8.1), machine protection (Sect. 8.2), and equipment protection (Sect. 8.3). Then, ion sources that produce protons, H^-, heavy ions, and rare isotope beams are discussed in Sect. 8.4. Sections 8.5 and 8.6 discuss the radio frequency and power supply systems that are used to energize accelerator devices. Section 8.7 covers the vacuum system. Sections 8.8, 8.9, and 8.10 describe various types of beam-interception systems that often become performance bottlenecks, limiting high-power accelerator performance. Section 8.11 covers beam diagnostics. Section 8.12 covers the survey and alignment. Conventional utility systems, including electricity, water, cryogen, and HVAC, are described in Sect. 8.13. Section 8.14 highlights the design aspects of radiation impact mitigation.

J. Wei et al., *High Power Hadron Accelerators*, Particle Acceleration and Detection,
https://doi.org/10.1007/978-3-032-17662-2_8

8.1 Personnel Protection

The personnel protection system is designed to protect accelerator workers, the public, and the environment from hazards associated with various forms of stored energy within the accelerator system. Hazards may be classified into two categories: standard industrial and laboratory hazards, such as electrical, mechanical, chemical, and fire hazards, and accelerator-specific hazards. Standard industrial and laboratory hazards are usually controlled and mitigated in accordance with the corresponding standards-based management system. An example is to use lock-out-tag-out (LOTO) procedures for hazard mitigation. This section focuses on accelerator-specific hazards.

8.1.1 Hazard Analysis, Safety Assessment, and Credited Controls

A hazard analysis (HA) forms the basis for the design of personnel protection systems. The analysis must exhaustively include all "maximum credible events" (MCEs) associated with the entire life-cycle of the accelerator facility. An MCE used for hazard analysis and risk assessment refers to a hypothetical worst-case scenario that is realistically possible.

Based on the HA, a safety assessment document (SAD) is prepared, containing the mitigation strategies for each identified hazard. Hazard mitigation often uses credited design features, credited engineered controls (CECs), and credited administrative controls (CACs). In general, such mitigation favors reliance on passive rather than active design features and favors engineered over administrative controls.

The attributes of the credited controls often include the following:

- Competence: Only competent individuals design and perform work;
- Configuration management: The configuration basis is identified and controlled, and changes follow a defined process;
- Testing and verification: Controls must be validated initially and periodically to ensure that they function as intended;
- Record keeping: Design documentation, validation results, and records of procedures performed are maintained;
- Responsibility: A responsible owner must be identified, with suitable experience to address the system hazards;
- Specificity: The elements and boundaries of the control must be clearly identified;
- Monitoring: The state of the control should be monitored;
- Fail-safe: Failures in the control or its energy sources should result in a safe state;
- Independence: Functional requirements, configuration, and work authorization are controlled by an independent party.

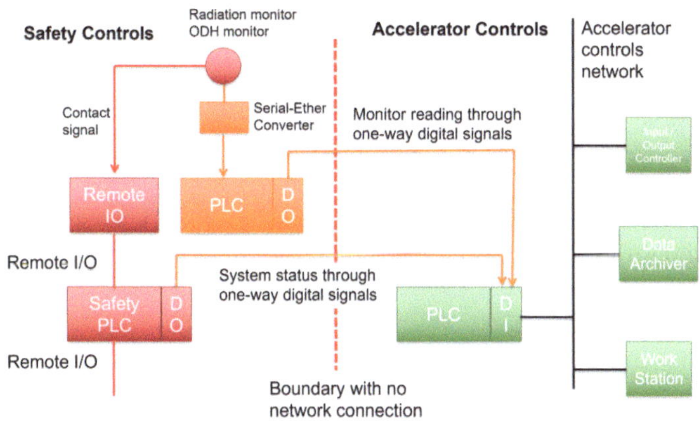

Fig. 8.1 Architecture of the FRIB PPS control network and the accelerator control network (D: digital, I: input, O: output). (Courtesy M. Ikegami and S. Beher)

The control network of the personnel protection system (PPS) is usually "air gapped" from the general control network for security reasons. Figure 8.1 shows the architecture of the dedicated FRIB PPS control network, which contains the dedicated programmable logic controller (PLC) that is "air gapped" from the general control network for security. The PPS control system is developed by qualified engineers and is maintained under strict configuration control. The PPS console is used for monitoring and operating in the control room. The PLC is locked in its own rack cabinet, with connecting cables running through its own conduits and cable tray.

Redundancy is mandatory in the development and implementation of PPS controls. For example, the IFMIF/EVEDA PPS system is configured such that the signal paths of all equipment (console, key boxes, door control boxes, emergency stop buttons, monitoring cameras, etc.) are centrally connected by hard-wire to the two separately configured and independently operated programmable logic controller (PLC) systems [1], as shown in Fig. 8.2. The dual PLC system permits certain actions only when both PLCs make the same judgment. Similarly, FRIB uses a single safety PLC with dual processors for its PPS system. In the event that a specified control is not functional, compensatory measures with equal functionality must be deployed to allow accelerator operations.

8.1.2 Radiation Safety

For high-power hadron accelerators, the foremost hazard is due to ionizing radiation generated by the beam and the electromagnetic fields produced by the accelerator system. For nearly all high-power accelerator facilities, shielding is the primary mitigation for both prompt radiation and induced radio-activation. The building that

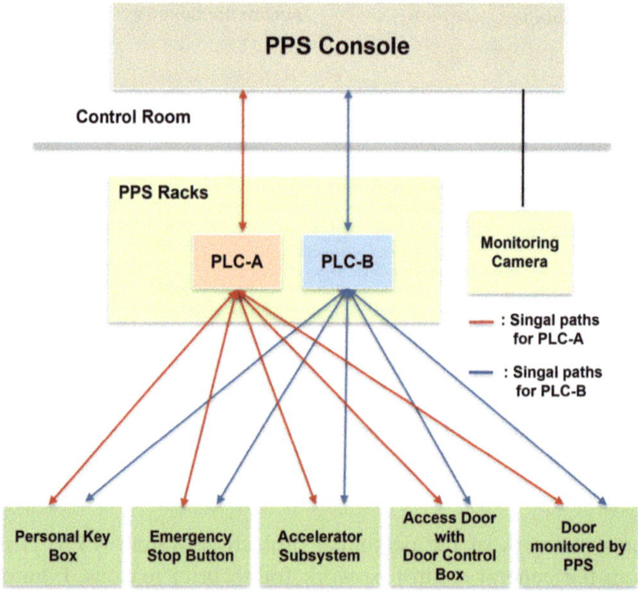

Fig. 8.2 The IFMIF/EVEDA personnel protection controls system with redundant programmable logic controller [1]. (Courtesy T. Kojima)

houses the main accelerator body is usually constructed underground so that the earth is used as part of the shielding. The access control system prevents personnel access to hazardous areas deemed unsafe due to prompt radiation (*that is, it keeps people away from radiation hazards*). On the other hand, the radiation control system prevents personnel outside the shielded enclosure from being exposed to radiation hazards (*that is, it keeps radiation hazards away from people*).

8.1.2.1 Access Control System

Access control refers to the management of personnel entry to the shielding enclosure of the accelerator facility. It consists of both engineered and administrative implementations. The engineered implementation consists of sensors that monitor access status, a control system that processes the signals from the sensors, triggers responses, and implements hazard mitigation devices. The administrative implementation includes the search-and-evict process that removes people from the shielding enclosure, as well as response training for emergencies and abnormal conditions.

To meet the credited control requirements, the shielding enclosure, including the tunnel/vault wall and the entry gates, needs to be verified for construction quality and then placed under configuration control so that alterations are only made following established review processes. Figure 8.3 shows the layout of the

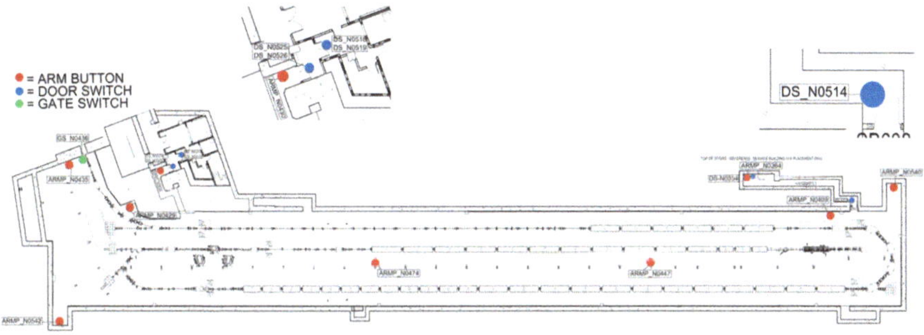

Fig. 8.3 FRIB driver linac tunnel with access control and radiation control indicated. (Courtesy MSU, USA)

tunnel that houses the FRIB driver accelerator. The tunnel is located underground inside a restricted access building. The tunnel is secured by two entrance gates that allow for independent egress. Although both the east and west entrance gates have two contactors, the west gate also has double entrance doors that can accommodate personnel screening upon entry. The access status is displayed at each gate and shows one of the three states: "secure tunnel" (red), "sweep/survey in progress" (yellow), and "unsecured tunnel" (green). Warning lights are installed both inside the tunnel and outside the gates. The emergency stop buttons are distributed on the wall inside the tunnel and in the control room. Activation of any of the gate contactors or emergency stop buttons brings the facility to the "unsecured" state. To transition from the "tunnel unsecured" state to the "tunnel secured" state, a search-and-evict must be carried out with two operators sweeping the tunnel, following a predefined procedure and sequentially activating the "sweep arm button" along the defined path within a time limit. To transition from the "tunnel secured" state to the "tunnel unsecured" state without triggering a fault, a radiation survey must be conducted to identify and secure a possibly activated area.

8.1.2.2 Radiation Control System

The radiation control system is designed to limit personnel exposure to ionizing radiation for both accelerator workers and the general public. The US Nuclear Regulatory Commission (NRC) requires that the occupational dose to individual adults be controlled, except for planned special exposures, to an annual limit of the total effective dose equivalent of 0.05 Sv. Like many other accelerator facilities, FRIB follows the As Low As Reasonably Achievable (ALARA) principle [2] with a radiation safety envelope well below the annual NRC limit [3]:

- Determined dose of 50 mSv/year to a worker;
- Determined dose of 0.02 mSv in any 1 hour to a member of the public;
- Determined dose of 1 mSv/year to a member of the public.

The design of the radiation control system must address both the prompt and residual radiation aspects. With prompt radiation, conditions for both normal operations and abnormal events must be addressed. The residual radiation must be low enough to allow hands-on maintenance, with a dose level below $\sim$1 mSv/h measured 30 cm from the device surface after 4 hours of cooling following month-long accelerator operations.

The FRIB radiation control system consists of radiation monitors placed outside the shielding enclosure at strategic locations, including shafts for HVAC and cryogenic transfer lines, as well as an unsealed hatch opening for equipment transport. The system response time (including radiation sensoring time, control signal processing time, and beam inhibiting time) is designed to react to foreseeable abnormal events and fault conditions without personnel exposure above the limit. These monitors are interlocked with the beam inhibit devices to stop the source of prompt radiation within the specified response time. Inhibitors include an AC contactor for the high voltage power supply of the ion source, the fail-safe beam plug that physically blocks the beam at the LEBT, and the AC/DC contactor for the power supply to the electrostatic bending device. The system response time of $\sim$1 ms is designed to mitigate the worst abnormal events at the design beam power.

8.1.3 Oxygen Deficiency Hazard Mitigation

A common hazard associated with SC magnets and SRF resonators is due to oxygen deficiency caused by cryogenic incidents. Examples include liquid helium used in SC devices, liquid nitrogen used as thermal shields, and liquid hydrogen and methane used as neutron moderators. The hazard occurs when inert gasses, such as nitrogen or helium, displace room air, thus lowering the oxygen concentration below that required for human life. In reference to the oxygen concentration of 20.9% in the normal ambient atmosphere, the US Occupational Safety and Health Administration (OSHA) defines an area as oxygen deficient if the percentage of oxygen is less than 19.5% by volume. Oxygen deficiency is a particular hazard in cryogenics because of the large volume change between a cryogen in the liquid state and the resultant gas at room temperature and pressure. The volumetric ratio is about 700 for liquid helium, transitioning from liquid at 4.2 K to gas at 300 K at 1 bar pressure. Relatively small amounts of cryogen, if converted to gas, may produce inert atmospheres that are unable to support life.

8.1.3.1 ODH Design Engineering

The design engineering of the accelerator facility is the primary line of defense to minimize oxygen deficiency hazards. High-power accelerators located in underground tunnels or vaults typically avoid liquid nitrogen because of the excessive risk associated with nitrogen gas filling the space in the event of a cryogen release incident. Instead, SNS and FRIB both used helium gas at a temperature of 35 ~55 K as a shield for the liquid helium cryogen at 2 ~4.5 K. Compared with nitrogen gas, helium gas is much lighter than air and tends to stay near the ceiling.

Segmentation effectively reduces the cryogen inventory in the event of a catastrophic release. The FRIB cryogenic refrigeration system supplies cryogen to the driver linac, which contains approximately 50 cryomodules distributed across three segmented distribution lines. Each cryomodule is individually connected to the cryogenic distribution line with U-tubes and is supplied with individual control valves, pressure relief, and safety interlocks.

The design of civil engineering must be compatible with the mitigation of hazards. Access to the work space must be designed so that at least one egress path to a "safe area" can be reached on foot under all foreseeable conditions. The safe area must be free of oxygen deficiency hazards and remain so during all plausible equipment failures. If the safe area relies on ventilation, emergency power must be provided to its ventilation systems. The path to the safe area must be adequately marked, illuminated, and free of obstructions during plausible ODH incidents.

8.1.3.2 Oxygen Deficiency Hazard Control System

The ODH control system is designed to alert workers to evacuate spaces that are susceptible to oxygen deficiency hazards. The ODH control system consists of ODH monitors that are set to alarm when the measured oxygen level falls below a preset value between 18% and 19.5%. These monitors are strategically mounted as part of the cryogenic system design so that a trained worker can escape to the safe area at a typical walking pace in any foreseeable catastrophic condition.

As a safety system with credited control for personnel protection, the FRIB ODH control system was designed with redundant monitors. The safety programmable logic controller (PLC) contains dual processors. The cables connecting the monitors and the control system were run through dedicated conduits and cable trays. Both hardware and software have been under configuration control and have required periodic revalidation. Any worker entering the tunnel must be trained to respond properly to ODH alarms.

8.1.4 Lock-Out-Tag-Out

Hazards in accelerator facilities are associated with multiple forms of stored energy, including mechanical (e.g., hydraulic and pneumatic pressure), electromagnetic, and chemical. Lock-out-tag-out (LOTO) is a safety procedure commonly used when working on hazardous equipment. LOTO requires that hazardous energy sources be isolated and rendered inoperative before work is started on the equipment in question. The isolated power sources are then locked, and a tag is placed on the lock identifying the worker who placed it. The worker then holds the key for the lock, ensuring that only he or she can remove the lock and start the machine. This prevents the accidental startup of a machine while it is in a hazardous state or while a worker is in direct contact with it.

The LOTO policy may differ by country and institution. In the United States, a lock selected by color, shape, or size, such as a red padlock, is used to designate a standard safety device for locking and securing hazardous energy. No two keys or locks should ever be the same. A person's lock and tag must only be removed by the individual who installed the lock and tag unless removal is accomplished under the direction of the employer. Worker procedures and training for this removal must have been developed, documented, and incorporated into the employer's energy control program. The LOTO policy may also differ by discipline. The safety procedure before working on electrical equipment may comprise the following five steps: disconnect completely; secure against re-connection; verify that the installation is dead; carry out earthing and short-circuiting; and provide protection against adjacent live parts. Group lockout is often practiced when two or more people are working on different parts of a larger system.

Engineered trapped key interlocking utilizes locks and keys for the sequential control of equipment to ensure safe operations. A safe sequence of operations is enabled through the transfer of keys that are either trapped or released in a predetermined order. Figure 8.4 shows the cage that contains the electron cyclotron resonance (ECR) ion source, which resides on the high voltage platform. A trapped-key interlock system is used to secure the entrance gate. This system ensures that an authorized worker must follow the engineered sequence to shut down the power supply and apply grounding before entering the high-voltage enclosure. A similar

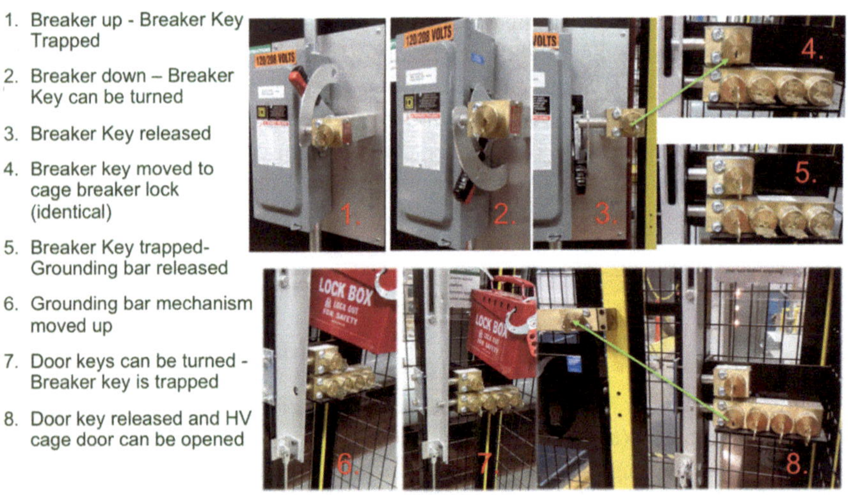

Fig. 8.4 FRIB ECR ion source trapped lock system ensuring that an authorized worker must follow the engineered sequence to shut down the power supply and apply grounding before entering the high voltage enclosure. (Courtesy G. Machicoane)

trapped key interlock system is used inside the cage to ensure that the ion source is de-energized and that the prompt X-ray radiation hazard is mitigated before the lead shielding panels can be moved to allow for maintenance of the ion source.

8.1.5 Electrical Safety Conformance

An electrical safety program must be in place to provide rules, standards, and guidelines for the safe design, installation, inspection, operation, alteration, and maintenance of electrical and electronic systems and equipment in accordance with regulations. For example, the FRIB electrical safety program served as a means of complying with Michigan Occupational Safety and Health Administration (MIOSHA) regulations [4], the National Electric Code (NEC®) [5] and other established safety standards to reduce or eliminate the dangers associated with the use of electrical energy. Such a program enforces procedures such as LOTO (Sect. 8.1.4) to protect personnel during electrical work. It also enforces standards to protect equipment from electrical hazards (Sect. 8.3.5).

Lessons learned from an electrical shock incident in an accelerator facility typically include the need to improve on the following aspects:

- Proper work planning with qualified electrical workers, certified and authorized to execute work;
- Thorough risk assessments and clear communication;
- Strict adherence to safety protocols, including proper lockout/tagout procedures and personal protective equipment (PPE); and
- A culture of safety in which workers are empowered to report potential hazards is essential to prevent similar incidents from occurring in the future.

8.1.6 Pressure Vessel Safety Conformance

Pressure safety is an important aspect addressed through the design and operation of large modern particle accelerators. The introduction of SC devices and an associated large cryogenic system is a significant part of the challenge. Liquid helium expands 700 times when it warms to room temperature. In any fault condition, this could cause catastrophic failure if the system is not designed and protected correctly. Equipment damage and possible personnel injuries can lead to significant downtime of accelerator operations.

Proper design, fabrication, and protection of cryogenic devices can be critical. The industry pressure vessel code, which was developed for the same reason of ensuring the safety of industrial equipment in the late nineteenth and early twentieth centuries, has been widely adopted by research complexes to standardize the safety margin, fabrication, and inspection of accelerator designs. The ASME boiler and pressure vessel code (BPVC) [6] and the B31.3 process piping code [7] are commonly used in US laboratories. Division I is widely used for cryogenic systems; while division II (design by analysis) is more fitting for more complicated SRF and SC magnet devices. The safety factor can be affected by the material properties, and in general, the allowable stress is based on the lesser of 1/3.5 of the tensile strength and 2/3 of the yield strength.

The maximum design pressure should be chosen collectively throughout the accelerator. First, the device design pressure must accommodate the operational parameters of the cryogenic plant. For example, if the thermal shield circuit is cooled directly by the high-pressure stream of the cryogenic plant, the thermal shield circuit of the devices must be rated for the maximum pressure of the cryogenic plant. Second, the strategy is to minimize cryogen relief in the tunnel to avoid a dangerous ODH situation. The relief device set points shall be staged in a way that the cryogenic distribution is slightly higher than the corresponding devices. Third, special considerations. The cost of niobium cavities scales with the cost of the material. The thicker the way, the cost will not only increase, but the thermal feedback effect will also degrade Q and limit the maximum gradient. It is typical not to exceed 4 mm for niobium cavities for this purpose and to have a maximum pressure limit at room temperature of 2 atm.

One topic worth mentioning is the fact that there is no established industrial standard for the allowable stress of high purity niobium. The mechanical data on high-purity niobium have a large spread, and high-temperature treatment for hydrogen degassing can further complicate the situation. Data show that any heat treatment at 800 °C or above can reduce the yield stress of the material. The design needs to consider surface processing recipes.

Sizing of the relief device is another important aspect of safety design. The superconducting magnet quench can release a large amount of stored energy $\sim$100 kJ to 10 MJ in less than one minute, and fault conditions, such as loss of insulating vacuum, can apply a high transition peak heat flux of 4 kW/m^2 or much higher to the helium system. A thorough analysis of all fault conditions and consequent load cases is used to size the relief piping and safety devices. Primary relief devices are usually resealable, and they can be sized for room-temperature over-pressurization cases as well as for less severe over-pressure cases. Secondary relief devices are usually non-resealable rupture disks and can be used as a final defense against the most severe cases of overpressure.

The last but not least is the LOTO practice of operation; LOTO must be applied to the pressure source for maintenance jobs, particularly when working on helium compressors, as well as air and water systems.

8.1.7 Toxic Liquid Metal Hazard Mitigation

Liquid metals are used in an increasing number of high-power accelerators to facilitate heat dissipation in beam-intercepting devices, such as the liquid lithium stripper for charge stripping (Sect. 8.8.2.2) and the liquid mercury target for secondary beam production (Sect. 8.9.1).

Hazards associated with the mercury target at J-PARC and SNS include the radiological and toxicological impacts. In SNS, approximately 20 tons of mercury are continuously pumped through an enclosed piping system. As the primary proton beam strikes, the chemically toxic mercury also becomes radioactive, along with the airborne mercury dispersion. The radionuclides of the spallation and activation products include ^{203}Hg, ^{197}Hg, ^{194}Hg, ^{189}Hg, ^{148}Gd and ^{178}Ta. The confinement of the target mercury is a primary design requirement. The target service bay is lined with stainless steel and is enclosed by massive shielding designed to withstand natural phenomena, including severe seismic events [8]. Radiological and toxicological impacts are evaluated based on the amount of mercury released during various accidents. Controls chosen to mitigate radiological impacts also mitigate toxicological impacts.

8.1.8 Chemical Hazard Mitigation

Similarly to electrical safety (Sect. 8.1.5), a chemical safety program must be in place to provide rules, standards, and guidelines for safe chemical processing and handling according to regulations in places such as the SRF surface processing facilities at FNAL, JLab, and FRIB. For example, FRIB chemical safety followed the MSU chemical hygiene plan and complied with MIOSHA regulations [4], the Clean Air Act (CAA) [9] and the National Ambient Air Quality Standards (NAAQS) [10] of the EPA, and other established safety standards to reduce or eliminate chemical hazards.

Surface treatment of superconducting components using chemical polishing and electro-polishing techniques involves hazardous chemicals and produces toxic by-products. Typical buffered chemical polishing (BCP) uses a 2:1:1 (volume ratio) mixture of concentrated 85% (w/w) phosphoric, 70% (w/w) nitric, and 48% (w/w) hydrofluoric acids. The etching process produces NO_x gas (like nitric oxide, NO, and nitrogen dioxide, NO_2), and hydrogen fluoride is emitted through evaporation. The chemical reactions of the BCP process show the following [11, 12]:

$$6\,Nb + 10\,HNO_3 \rightarrow 3\,Nb_2O_5 + 10\,NO + 5\,H_2O$$

$$Nb_2O_5 + 10\,HF \rightarrow 2\,NbF_5 + 5\,H_2O$$

Typical electro-polishing (EP) uses a 10:1 mixture of concentrated 98% (w/w) sulfuric and 48% (w/w) hydrofluoric acids. Electro-polishing reactions generate ppm level sulfur dioxide and hydrogen sulfide [13]. Hydrogen fluoride is emitted through evaporation directly from the electrolyte solution. A small amount of hydrogen is also produced through the reaction and needs to be purged out. The chemical reaction of the EP process shows the following [12]:

$$2\,Nb + 5\,SO_4^{2-} + 5\,H_2O \rightarrow Nb_2O_5 + 10\,H^+ + 5\,SO_4^{2-} + 10\,e^-$$

$$Nb_2O_5 + 6\,HF \rightarrow H_2NbOF_5 + NbO_2F\,0.5\,HO + 1.5\,H_2O$$

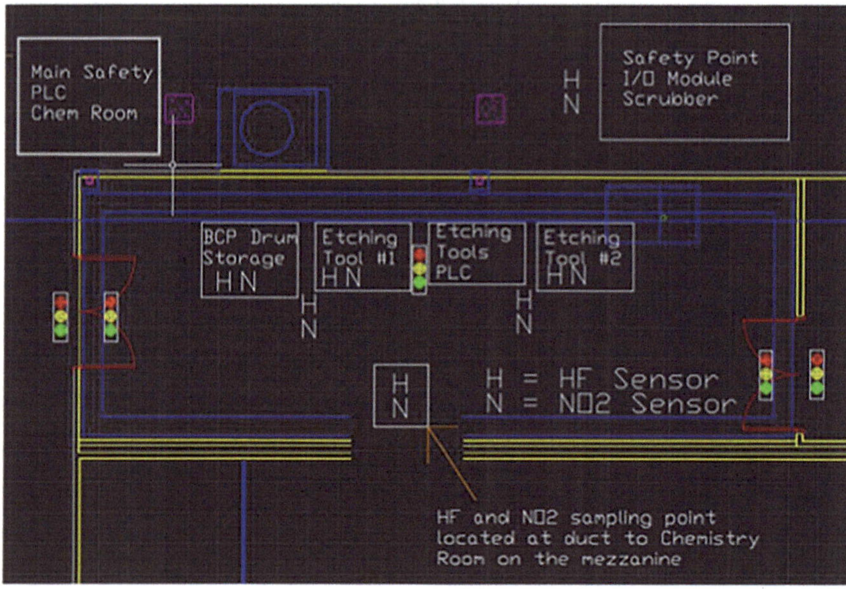

Fig. 8.5 Schematic layout of the FRIB SRF building chemistry room. (Courtesy L. Popielarski)

$$NbO_2F \cdot 0.5\,H_2O + 4\,HF \rightarrow H_2NbOF_5 + 1.5\,H_2O$$

Figure 8.5 shows the schematic layout of the FRIB SRF chemistry room as an example. The exposure limit for HF and NO_2 in the workplace is defined by OSHA as the permissible exposure limit (PEL). For example, the 2025 OSHA PEL for HF is 3 ppm for an 8 hour time-weighted average. HF and NO_2 gas monitoring sensors are installed in the chemical hood, inside and around the chemical room, with redundancy and are periodically calibrated. The safety PLC that handles the gas monitoring system is isolated from other control systems with a proper rating of safety integrity level (SIL) [14].

The chemical scrubber is an essential part of chemical safety. It pulls contaminated air from the etching hood, removes most of the toxic gas using sodium hydroxide, and sends the gas through the stack. Another key function of the scrubber is to keep the chemical room under negative pressure and maintain adequate face velocity for the workers to operate in front of the hood. The scrubber should be equipped with a backup blower, along with a backup generator, in case of a power outage.

Other safety measures include the mandate for personal protective equipment (PPE), spill containment and waste handling procedures, and emergency response. For example, secondary containment is often included in system design. Safety showers with eyewash stations are located nearby, and first aid, including calcium gluconate, is readily available, along with postings of safety response protocols.

8.1.9 Environmental Protection

Control of environmental radiation emission (air emissions and groundwater activity) and waste disposal must be part of the facility's environment, safety, and health (ES&H) program. The accelerator tunnel enclosure usually operates under negative pressure to control the flow of activated air. Radiation monitors are installed on the exhaust stack to monitor air emissions and to interlock operations if necessary. Wells located at strategic locations near the facility boundary are used to monitor the groundwater activity of suspected isotopes and a possible correlation with machine operations above the measurable background.

The US NRC requires the implementation of ALARA [2] for air emissions of radioactive material into the environment, such that an individual member of the public is not expected to receive a total effective dose equivalent in excess of 0.1 mSv per year from such emissions [15].

8.2 Machine Protection

Machine and equipment protection systems are designed to protect the accelerator facility from catastrophic incidents that interrupt accelerator operations. In this section, machine protection related to beam-induced malfunction events is discussed. Equipment protection pertaining to events not related to the accelerated beam is discussed in Sect. 8.3.

8.2.1 Requirements and Architecture

High-power hadron accelerators create, accelerate, and transport beams that carry megawatt (MW) scale power or store gigajoule (GJ) scale energy. The machine protection systems (MPS) must guard against both acute, catastrophic failures with a fast response time, mitigating errant beam events on time scales as short as several microseconds, and chronic, long-term performance degradation with high detection sensitivity.

8.2.1.1 Acute Beam Loss

Acute beam loss typically refers to fault scenarios when the beam is accidentally mis-steered and intercepted by surrounding materials. The required MPS response time usually corresponds to the duration before the impacted material yields to thermal stress.

$$t_{\text{mps}} \sim \frac{4\pi}{\sqrt{3}} \frac{\sigma_x \sigma_y \rho_T c_{v,T} \sigma_m}{\hat{I} S_{p,\max} \alpha_l E_y} \tag{8.1}$$

where σ_x and σ_y are the horizontal and vertical rms radii of the projectile beam, respectively; $\hat{I}$ is the beam peak current, $S_{p,\max}$ is the maximum stopping power, ρ_T is the absorber material density, $c_{v,T}$ is the material's specific heat, α_l is the linear expansion coefficient, E_y is Young's modulus, and σ_m is the material yield stress. The response time is inversely proportional to the stopping power, or the energy loss per unit length in the material. The stopping power as a function of the evolving projectile velocity βc upon entering the intercepting material is given approximately by a simplified version of the Bethe-Bloch formula [16],

$$\begin{aligned} -\frac{S_p}{\rho_T} &= \frac{1}{\rho_T}\left(\frac{dE}{dx}\right) \\ &= \frac{0.313 \times 10^{-3} Z_T}{\beta^2 A_T} Z_{P,\text{eff}}^2 \left[\ln\left(\frac{\beta^2}{I_S(1-\beta^2)}\right) - \beta^2\right] \left[\text{MeV/mg} \cdot \text{cm}^{-2}\right], \end{aligned} \tag{8.2}$$

where the subscript P indicates the projectile, the subscript T indicates the absorber, Z_P and Z_T are the atomic numbers of the projectile and the absorber, respectively, and A_P and A_T are the atomic mass numbers of the projectile and the absorber, respectively, and

$$I_S = \begin{cases} (11.2 + 11.7 Z_T) \times 10^{-6}\ \text{MeV} & \text{for } Z_T \le 13 \\ (52.8 + 8.71 Z_T) \times 10^{-6}\ \text{MeV} & \text{for } Z_T > 13 \end{cases}$$

$$Z_{P,\text{eff}} = Z_P \left[1 - C(Z_P)\exp(-0.879\,\beta_R)\right], \quad \beta_R = \frac{137.0\,\beta}{Z_P^{0.65}}$$

with

$$C(Z_P) = 1.035 - 0.4\exp(-0.16\,Z_P)$$

Note that for a given absorber, the stopping power

$$S_p \propto Z_{P,\text{eff}}^2 \tag{8.3}$$

is highly enhanced by the atomic number of the beam. The corresponding stopping range is given by

$$R \propto \frac{E_{K,P}^{1.74}}{A_P^{0.76} Z_{P,\text{eff}}^2} \tag{8.4}$$

where $E_{K,P}$ is the kinetic energy per nucleon of the projectile beam. Compared with a proton beam of comparable velocity and current, an intercepted heavy-ion beam of high atomic number Z_P has a far shorter stopping range (Eq. 8.4) and higher stopping power (Eq. 8.2), thus demanding a more stringent response time for machine protection (Eq. 8.1),

$$t_{\mathrm{mps}} \propto Z_{P,\mathrm{eff}}^{-2}. \tag{8.5}$$

Software codes are available to evaluate the stopping power and range of ions in matter [17]. Figure 8.41 shows examples of the energy loss dE/dx of U, Xe, Ca, and H when traversing a carbon stripper as a function of ion energy.

8.2.1.2 Chronic Beam Loss

Chronic beam loss typically refers to normal operating conditions when the surrounding environment intercepts the beam halo and contaminants that cause longer-term performance degradation. An example is the performance degradation of SRF resonators located in areas susceptible to excessive chronic beam loss. In situ restoration processing and renovation maintenance must be performed to recover performance. Another example is the impact of stray beam irradiation, which causes material fatigue and electronics failure. Sensitive electronics must be placed away from exposure to the beam.

MPS sensors of high sensitivity, though slow in response time, need to be in place to detect chronic beam loss.

8.2.1.3 Multi-Layer, Multi-Time-Scale Machine Protection

Due to the complexity of machine protection requirements, a multilayer, multi-time-scale machine protection system is often needed for high-power operations. A multilayer system refers to both preemptive and passive controls. Preemptive controls detect abnormal signals in accelerator devices and stop the beam to prevent loss. It is also called a beam (or run) permit system (RPS). Passive controls detect signals of beam loss and stop the beam before causing significant damage. It is also called a fast protection system (FPS) (Sect. 7.2.3.3).

FRIB adopted such approaches: (1) the fast protection system (FPS) is designed to prevent damage from acute beam loss by quickly activating the beam inhibiting device; (2) the run permit system of different time scales (RPS-1, RPS-2, and RPS-3) continuously queries the machine state and provides permission to operate with the beam. For example, the slow but highly sensitive thermo-sensor prevents the slow degradation of the SRF system under small beam loss (Table 8.1).

Table 8.1 Multi-layer, multi-time-scale machine protection for the FRIB accelerator complex

Mode	Time	Detection	Mitigation
FPS	~35 μs	Differential BCM Ion chamber monitor Halo monitor ring [18] Fast neutron detector	LEBT bend electro-static deflector chopper
RPS-1		LLRF controller Dipole current monitor	
RPS-2	~100 ms	Vacuum status Cryomodule status Non-dipole PS status Quench signal Vacuum excursion signal Motor rotation status Beam dump position status	As above; Ion source HV
RPS-3	>1 s	Thermo-sensor Cryo. heater power	As above

8.2.2 *Sensors, Controls, and Mitigation*

The machine protection system consists of sensors, controls, and mitigation [19]. The total duration from fault detection to mitigation must be within the required response time.

Sensors for preemptive (or beam permit, or run permit) controls include LLRF controller signals, current monitoring of the power supply for main magnets, and thermal coupler signals from beam intercepting devices. For example, when the LLRF controller detects a deviation in the RF phase in an accelerating resonator, or when the dedicated current monitoring module detects a deviation of the read-back from the setting of the main magnet power supply, "Not OK" signals are sent to the controls processor, activating beam inhibition.

Sensors for passive (or fast protection) controls include beam diagnostic devices such as an ion-chamber-based beam loss monitor, a neutron monitor, a differential beam current monitor, and a beam halo monitor. Often, a combination of both fast (yet with a larger threshold of deviation to minimize false trips) and high-sensitivity (yet slower to allow statistical averaging to minimize false trips) controls is used, providing "Not OK" signals for beam inhibition when excessive stray beam is detected.

The control network and the processor for machine protection are described in Sect. 7.2.3.3.

MPS mitigation almost always corresponds to inhibiting the beam at low energy when the impact is minimized. The mitigation devices can be high voltage switches to the ion source and to the LEBT deflectors. Note that the total duration of beam mitigation includes both the time of activating the mitigation device and the time before the remaining beam in the accelerator is emptied. For low-energy heavy-ion machines, the mitigation time can be significant.

8.2.3 *Run Permit System*

The run permit system (RPS), or the beam permit system (Sect. 7.2.3.3), examines whether the conditions are met for beam acceleration and issues the beam permit based on input from configuration management databases, machine and personnel protection systems, and beam characteristics and destination. The RPS prohibits beam operation when beam-inhibiting interlocks are activated. Such interlocks include both non-beam-induced events, such as accelerator setting violations against predetermined logic, and beam-induced slow events, such as vacuum pressure excursions, quench signals, motor rotation speeds, critical device positions, thermo-sensor temperature rises, and cryogenic heat-load increases (Table 8.1).

The FRIB RPS was designed to operate in concert with the personnel protection system, the machine protection system, and the global timing system [20]. Two important concepts of configuration management are machine modes, which define the geographic scope of permissible beam propagation, and beam modes, which define the permissible range of ion species and charge states, beam energy, beam intensity, timing structure, and ramping strategies [21]. The RPS then (1) determines if conditions are met for beam acceleration based on configuration management systems indicating approval of hardware for use; critical signal thresholds are distributed (for example, power supply operating ranges); and the absence of PPS, MPS, and critical signal alarms or faults; (2) indicates to dependent systems that beam acceleration may commence or continue by issuing the run permit signal; (3) determines if conditions are such that beam operations may continue in concert with PPS and MPS mechanisms: the failure of critical signals revokes the run permit; and (4) prevents modifications to critical signals during beam operations.

8.3 Equipment Protection

Equipment protection systems are designed to protect the accelerator facility from catastrophic incidents during energization. This subsection lists several examples common to high-power accelerators, particularly those using superconducting technology.

8.3.1 *Quench Protection*

A quench is an abnormal termination of the operation of a SC device that occurs when part of the SC material enters the normal (resistive) state. A quench may occur when the temperature, magnetic field, or current density is raised above their critical values. Quench protection is different for SC magnets and SRF resonators. SC magnets typically have a stored energy range of 100 kJ to 10 MJ; quench protection means properly dissipating this energy without destroying the device. On the other hand, SRF resonators have a stored energy on the order of 10 J; quench protection is more about flux trapping through the normal zone and preventing further driving of RF power into the cavity after the quench. Quench events of SC magnets are always associated with large transient heat flux and can cause helium inventory loss if the recovery design is not well-considered.

8.3.1.1 Magnet Quench Protection

Practically, quenches may start (1) intrinsically due to alternating current (AC) losses and conductor instability with respect to flux jumps; (2) mechanically due to the motion of the conductor under the Lorentz force and delamination of impregnation epoxy; and (3) thermally due to excess heating in splices or current leads, external heat leaks, and local heating induced by stray beams. A quench can cause damage if the coil winding temperature is above the design value or if the voltage is too high, leading to insulation failure when stored energy is discharged. Section 6.3 gave a detailed example of the quench analysis and protection scheme of the FRIB large dipole magnet.

The challenge of magnet quench protection is the high current density and large stored energy in the SC coils, and it is often a major topic of magnet design. Typical protection strategies are: (1) extracting the energy to the external resistor at room temperature; (2) helping to spread the normal zone in the coil to increase the thermal mass to absorb the energy; and (3) subdividing coils to limit the stored energy in each section. Typical quench protection design can be categorized into passive protection, which does not require quench detection, and active protection, which requires fast quench detection within a few milliseconds.

Quench detection is based on the measurement of resistive voltages. Inductive voltages are eliminated either by subtraction or by imbalance bridge circuits that detect resistive voltage in any branch of the coil winding by comparing the potential of a pre-selected voltage tap to that provided by a resistive divider. Non-voltage-quench diagnostics are primarily used as secondary tools. The example given in Sect. 6.3 uses the comparison of the voltages of two dipole coils as a detection signal and a quench heater to help spread the normal zone.

Figure 8.6a shows the passive quench protection schemes adopted in colliders such as RHIC and LHC, where many SC magnets are powered in series. The cold diode connected across each magnet prevents unquenched coils in series from

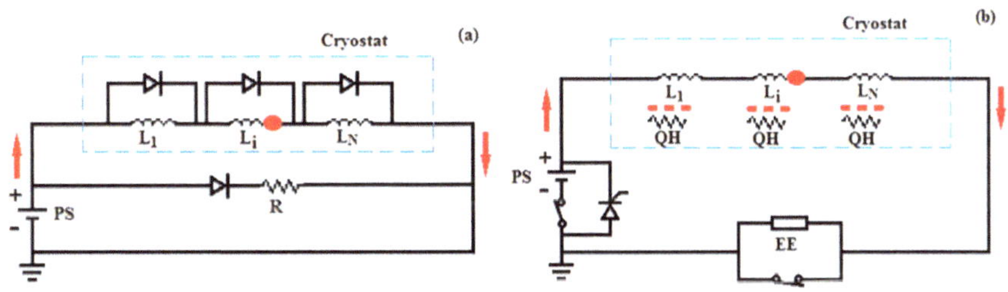

Fig. 8.6 Schematic diagram of (**a**) passive and (**b**) active quench protection of superconducting magnets

dumping their energy into the quenched magnet. When one coil quenches, a sizable voltage develops that shorts the current through the diode. The other magnets can continue to carry current that, because it does not flow through the quenched coil, does not dump their energy in the quenched coil. The warm diode in series with the warm external resistor provides the next level of protection. When quenching occurs, the increasing resistance of the coil generates enough voltage to allow the warm diode to conduct the current, thus reducing the temperature rise in the coil.

Figure 8.6b shows the active quench protection schemes using quench heaters (as adopted by Tevatron) and energy extractors (external resistors). When a limited quench is detected, the heaters are activated to induce more general quenches, avoiding energy dumping from unquenched magnets in series without using cold diodes. This process causes a quench in a larger region with rapid propagation, reducing local heating. An active scheme using a fast switch to disconnect the power supply and connect the external resistor provides ultimate protection. The resistors are chosen to be high enough to promptly reduce the current to avoid excessive local heating, yet low enough to prevent excessive voltages that may cause electrical breakdown.

8.3.1.2 Resonator Quench Protection

SRF resonator (Chap. 5) quenches can affect accelerator operation and damage equipment in various ways. During a quench, part of the cavity RF surface is turned to the NC state. The NC area may remain relatively small and stabilized (soft quench) or extend to most or all RF surfaces (hard quench). This causes a drop in the Q of the cavity by several orders of magnitude, and correspondingly, an increase in the input RF power is needed to keep the cavity locked at the operating gradient. In high-power hadron accelerators, the large amount of RF power available and the strong RF coupling allow the quenching cavity to be locked for an extended period, increasing the heat load to liquid helium by orders of magnitude. This suddenly generates large helium pressure fluctuations, which can unlock other cavities, heavily interfere with accelerator and cryogenic system operations, and damage equipment. In addition, if the quench is associated with field emission sustained for a long time by the large amount of available RF power, the electron discharge can seriously damage the RF surface of the resonator, spray the Nb material around, and irreversibly affect cavity performance. To avoid such occurrences, the individual quench gradient is preliminarily determined for all cavities, and then each is operated below that value (with a 1 ~2 MV/m safety margin). However, a quench may also occur at a lower gradient (e.g., induced by beam losses, multipacting, or field emission), and a quench protection algorithm is usually implemented in the LLRF resonators that stops the RF power within a time on the order of 1 ms or less in the case of a quench.

For fast quench detection, a rapid analysis of the forward, reflected, and pickup power RF signals is performed, as changes in cavity temperature and helium pressure signals, although strong, are too slow. To avoid false detection, this analysis must distinguish between quench and other conditions affecting the behavior of the anomalous RF power signal (for example, the detuning of the cavity due to mechanical vibrations) [22]. The switch-off thresholds and conditions depend on the particular resonator parameters in operation and may vary across different machines. For example, in pulsed machines, the cavity field decay time may be a good quench indicator [23], while in CW machines, it is necessary to analyze the dynamic behavior of all RF signals [24].

In the case of a resonator quench, depending on the position and function of the resonator, the beam dynamics conditions may change enough to lead to severe beam losses. Especially in high-power hadron accelerators, the detection of a cavity quench typically activates the beam permit system to inhibit the beam at the source (Sect. 8.2.3).

8.3.2 Cryogen Inventory Protection and Electricity Backup

Many incidents of loss of cryogen inventory are caused by failures in the supply of electricity to the cryogenic refrigeration system, particularly if the storage is inadequate to save the cryogen in the gaseous state. Strategically planned power supply backup is key to investment protection and facility availability.

As illustrated in Fig. 8.7, FRIB has relied on three tiers of power backup in the event of a power outage, together with a redundant electrical system featuring two sources of power, multiple service feeders, and double-ended substations.

- Uninterruptible Power Supply (UPS) located on site. The redundant 500 kW UPS system is constantly on standby to protect critical safety and control systems from any power interruption with a UPS system that provides no interruption in service; these include the PPS, the fire protection system, and the computer and communication network system, including the controls of the cryogenic system.
- Backup generators located on site. Three 800 kW diesel engine standby generators connected in a shared load bus configuration power other systems, including

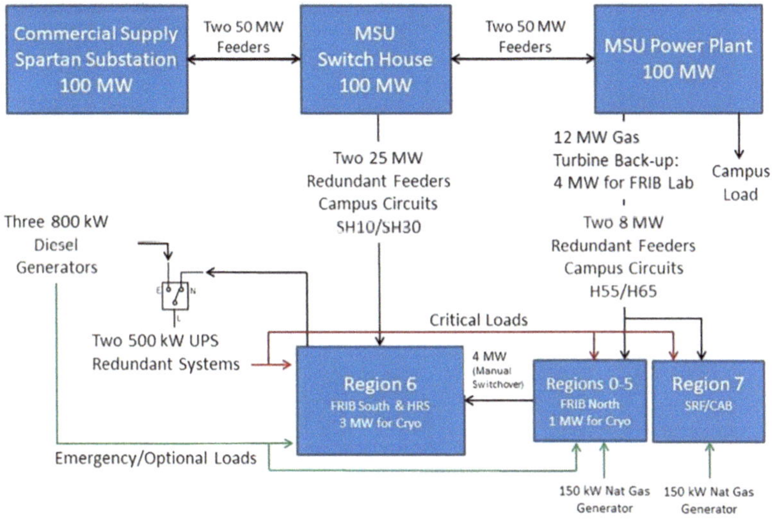

Fig. 8.7 Electricity power schematics of the FRIB complex. The cryogenic refrigeration plant is housed in Region 6. (Courtesy MSU, USA)

emergency lighting, elevator power, confinement systems, and required utilities. Power is transferred within 10 seconds of a power outage.

- Gas turbine generator located at the MSU campus power plant. The 4 MW gas turbine generator provides cryogenic inventory protection, supplying power to cryogenic systems to maintain the valuable liquid helium inventory and to avoid the warm-up of SC cavities and magnets. This power is provided by the university power plant to the Laboratory within 30 minutes of a power outage.

Periodic system testing and revalidation are required to ensure that the electrical backup system works as designed.

8.3.3 Vacuum Venting Protection for SRF Linacs

SRF linacs operate at fields much higher than those of NC RF linacs (Chap. 5). Particle-free cleaning and high vacuum are crucial for avoiding field emission, which affects performance. Protecting it from contamination is an important topic not only in operation but also throughout design, construction, and installation.

Accidental venting of beamline space to air is one of the failure scenarios threatening SRF cryomodules because air is considered dirty in the absence of a clean room equipped with proper filters and flow rate control. The result can be devastating and costly. Most of the time, the entire contaminated structure must be disassembled, and the cavities must be reprocessed, certified, and assembled.

Fig. 8.8 FRIB $\beta_0 = 0.085$ QWR coldmass SCM005 venting incident in 2017. (Courtesy MSU, USA). (Left) View of the bottom cold tuner assembly, and (Right) tuner bellow that was attached to the tuning plate damaged during assembly

Accidental venting can occur during construction, transport, and installation, or during operation due to beam-induced damage, fatigue-induced component failure, and operational errors. The root cause of venting can be design flaws, human factors, or both. Several examples of incidents are provided as follows:

- Venting during construction and assembly. During FRIB cryomodule production, one $\beta_0 = 0.085$ QWR cold mass was vented during cryomodule assembly outside the clean room. As shown in Fig. 8.8, when a technician tried to remove a galled fastener from the tuner, the bellows were damaged, resulting in venting the entire cold mass to the atmosphere. This incident resulted in the total rebuilding of the cold mass and impacted the schedule by six weeks. The root causes were: (1) insufficient design and planning; special tooling should have been designed for this difficult-to-reach location, and work instructions should have contained warnings about the high risk at the beamline vacuum boundary; (2) human factors; brute force should have been avoided, and the production manager should have communicated better with the technician. As lessons learned, (1) new tooling was designed and fabricated; (2) work instructions were reviewed, and all steps that might carry a risk of damaging the beamline were marked as high risk; (3) task forces were formed with assessments made on-site for high risk repair and rework tasks.

- Venting during transportation. Several incidents of beamline venting occurred during the production, testing, and transport of LCLS-II cryomodules [25]. There were incidents during the assembly with a similar root cause as the FRIB incident discussed above. Yet, the most valuable experience was related to transportation. During transportation from FNAL to SLAC in January 2018, the F1.3-06 cryomodule was shipped under vacuum and was found to be vented upon arrival. Several issues contribute to the root cause: (1) as a result of using outdated BPM design drawings from XFEL, bolts made of a different titanium material were used, resulting in loosened fasteners after thermal cycles; (2) the shipping fixture failed to provide sufficient support, allowing excessive motion of the cold mass inside the vacuum vessel during transportation, which damaged the FPC cold bellows. Both issues had the common root cause that changes to the original XFEL design resulted in unexpected impacts. A list of necessary improvements was identified to correct the mistakes and enhance the performance of the shipping fixture.
- Venting during operation. TRIUMF reported a venting incident due to an operational error at the beamline downstream of the ISAC-II SRF linac [26]. The interlock on the beamline gate valve was bypassed due to a request for alignment for an experiment. Subsequently, a wrong valve was opened, followed by a malfunctioning fast valve, resulting in the contamination of 20 SRF cavities. Field emission onset was degraded to 1.5 MV/m for the worst case, with the performance degradation correlated to the distance to the venting source. Thermal cycling did help to partially restore the performance. As lessons learned, the protection system logic was upgraded, including arming the fast valve and interlocking isolation valves. Other examples of accidental venting during operations include bellows failure due to fatigue, turbo pump and foreline pump failure, premature activation of disk ruptures, ceramic window failure, and vacuum enclosure punctured by intense beams.

Mitigation of accidental venting includes:

- Segmented vacuum with safety interlocks. The vacuum space should be segmented (Sect. 8.7) with isolation gate valves that can be interlocked with pressure gauges. When not operating with a beam, the valves should be closed to reduce the risk of venting.
- Implementation of fast gate valves. The cold section of the linac is separated from the warm section by fast valves that can be activated in $\sim$10 ms. The pressure sensor must be placed at least 3.4 m away from the fast valve to activate the valve in time against the sonic air flow at a speed of $\sim$340 m/s due to venting. Once the fast valve trips, the MPS can inhibit the beam in $\sim 10\,\mu$s. For example, FRIB deployed seven fast gate valves surrounding the linac cold segments and one for the target against the downstream beamline. All these fast valves require periodic testing and re-validation.
- Protective covers on fragile beamline components, such as bellows, windows, and rupture disks, to avoid accidental hits.
- Administrative control to prevent unnecessary crossings under the beamline.

- Proactive maintenance includes: (1) bellows that are replaced before reaching their specified lifetime; (2) pumps that are on a maintenance schedule; (3) turbo pumps that are interlocked based on the bearing temperature; and (4) foreline pumps that are equipped with anti-suckback valves to protect against loss of power.
- Properly sized relief against beamline venting to mitigate sudden pressure increases in the cold linac segment due to cryogen expansion.
- Operator training and emergency procedures are in place with clear protocols for managing a vacuum breach, including shutting down the accelerator and isolating the affected area.
- Design and engineering considerations to prevent venting during construction, repair, transportation, and installation.

8.3.4 Liquid Metal Fire Hazard Mitigation

Alkali metals, such as sodium and lithium, in liquid form are used in reactors and accelerators because of their favorable thermodynamic properties. When used as a neutron producing target (Sect. 8.9.1), liquid lithium reacts with the impinging proton beam through the reaction $^{7}\mathrm{Li}\,(\mathrm{p, n})\,^{7}\mathrm{Be}$ while also serving as a beam dump, dissipating the thermal power by the forced flow of the liquid metal. When used as a charge stripper (Sect. 8.8), a thin film of liquid lithium is maintained to remove electrons from passing ions while dissipating the heat generated by the impinging beam, taking advantage of the solid-like density, thermal capacity, and conductivity, as well as the low vapor pressure of liquid lithium.

Liquid lithium reacts forcefully with both air and moisture, imposing a fire hazard; however, it does so with noticeably less vigor than liquid sodium. Severe incidents have occurred when heated liquid lithium under pressure in a vacuum was suddenly sprayed into the air, causing fire and an explosion. Mitigation of fire and chemical hazards induced by liquid lithium is largely associated with preventing the lithium from being exposed to air. The safety system consists of sensors that detect lithium leaks and air leaks, a control system that processes the signals and activates the responses, and mitigation measures, including de-energizing the lithium system.

The FRIB liquid lithium charge stripper (Sects. 8.8 and 10.3.3.3) used double containment vessels with inert argon gas filled in between. The credited engineered controls included the secondary vessel that maintained specified argon gas pressure and the vacuum gate valves that isolate the stripper system from the beamline vacuum in the event of an incident.

8.3.5 Electrical Fire Hazard Mitigation

Electrical fires are often attributed to factors such as the overheating of cables or components due to high current loads, insulation degradation, faulty connections, short circuits, arcing, power surges, improper wiring practices, or malfunctioning power supplies; all of which can generate enough heat to ignite nearby combustible materials. Mitigation strategies often include: (1) proper design and installation with adequate capacity and proper grounding; (2) regular maintenance checks (for example, with infrared scanning); (3) implementing fire detection and automatic fire suppression with appropriate extinguishing agents; (4) compartmentalization of high-risk areas; (5) using fire-resistant materials; and (6) training personnel on fire safety procedures; all while considering the environment of the accelerator tunnel and service buildings, including ventilation and potential radiation exposure.

In operating accelerator facilities, electrical fires may occur, particularly under two conditions:

- Newly developed or custom-made equipment is susceptible to fire hazards when tests are insufficient, particularly in the complex accelerator environment. For example, electromagnetic noise from the surroundings can interfere with countermeasures against electrical discharge; conducting construction debris, such as aluminum foil, may be transported through ventilation to unintended areas, causing arcing.
- Aging legacy equipment is susceptible to fire hazards due to worn insulation, deterioration from radiation exposure, and aging materials. Proactively planning the replacement of aging components is mandatory.

In general, it is of primary importance to ensure that potential fires are isolated, self-extinguishing, and do not disperse radiative material that may harm personnel or endanger the environment.

8.4 Ion Source

Ion sources are devices that produce charged particles for subsequent acceleration. For high-power hadron accelerators, the ion sources must deliver the required ion species with appropriate charge states, design intensity, duty factor, and emittances. In addition, they must meet the design requirements for output energy, outlet pressure, and beam purity, along with material efficiency, stability, and reliability. Ion sources are often located in accessible places for frequent maintenance while the rest of the accelerator is secured for operations.

8.4.1 Proton Source

Proton sources are often favored for cyclotrons (such as the PSI cyclotron) and linacs (such as the LANSCE and ESS linacs) [27]. The most popular proton source in many pulsed accelerators (CERN, BNL, LANL, GSI, etc.) has been a plasmatron frequently referred to as the duoplasmatron. The prefix "duo" indicates the constriction of the discharge (plasma) by a simple aperture and a shaped magnetic field [28]. Energetic thermal electrons emitted from the cathode ionize a support gas and create plasma. The original duoplasmatron proton source was developed by Ardenne in 1948 [28, 29]. The LANSCE facility has used a duoplasmatron as a proton source for the past 55 years, and several upgrades have been made to produce $\sim$20 mA pulsed current with the rms emittance $<0.2\ \pi$mm·mr and a lifetime $\sim$30,000 hours [29, 30]. Similarly, CERN has operated a duoplasmatron [31] for decades to produce a pulsed beam current of up to 300 mA with $\sim 1.6\ \pi$mm·mr rms emittance. A plasmatron ion source capable of producing 200 mA DC protons has been developed in Frankfurt, Germany [32]. The rms emittance measured at this source for a 75 mA proton beam was 0.15 πmm·mr. The ion beam extracted from a plasmatron typically contains about 85%–90% protons.

For cyclotrons and continuous-wave (CW) linear accelerators, the requirement for high currents and high reliability makes the electron-cyclotron resonance (ECR) proton source the preferred choice. Direct current proton sources using 2.45 GHz ECR plasma heating, with high production efficiency and no cathode, deliver stable beams with high availability and small emittances. They are available as injectors for accelerators with proton currents greater than 100 mA in several facilities, including LEDA [27, 33] and IFMIF [34]. ECR sources are also chosen for accelerator-driven subcritical facilities such as CiADS [35]. As reported in [34], the ECR proton source can produce a beam of high brightness with the rms emittance $\sim 0.1\ \pi$mm·mr at 100 mA.

8.4.2 H^- Source

Despite the more complicated technologies required for the ion source, the H^- ion beam is often favored over the proton beam due to the charge-exchange process that facilitates high-power beam acceleration. Charge exchange is used in cyclotrons to allow efficient beam extraction (such as in the TRIUMF cyclotron), and in rapid-cycling synchrotrons (such as in ISIS, J-PARC, and CSNS) and accumulators (such as in PSR and SNS) to stack high-intensity proton beams.

Table 8.2 shows the characteristics of some typical H^- sources. Sources for high-power accelerators must provide the required beam current at the design duty with sufficiently low emittance compatible with the acceleration scheme. They must also operate reliably, with lifetimes compatible with the operating schedule.

Table 8.2 Features of typical H^- ion sources [36, 37]: magnetron source, Penning source, multicusp surface, volume, and hybrid sources

Type	Peak current [mA]	Duty [%]	Pulse length [ms]	Emittance[a] [πmm·mr]	Reference
Magnetron	100	<1	~100	0.3–0.4	[38]
Penning (with Cs)	35 (100)	1.2	–	0.05	[39–42]
Multicusp, surface	20	12	1	0.2	[43–45]
Multicusp, volume (with Cs)[b]	35 (120)	0.1–7	0.1–1	0.1–0.2	[46–48]
CW	60	100	–	0.1	[49, 50]

[a]Normalized rms emittance
[b]Volume production with surface enhancement by cesium seeding

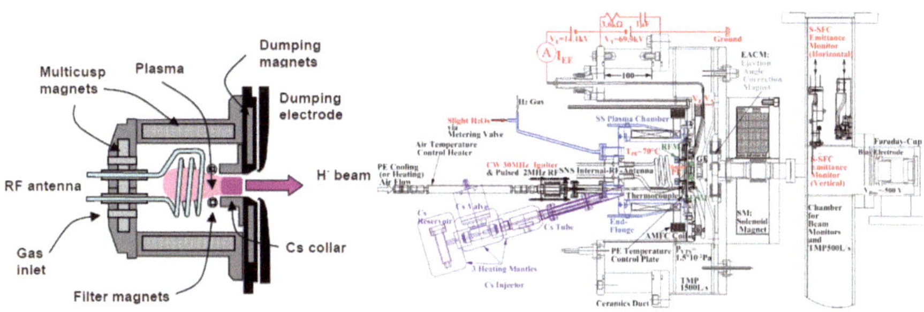

Fig. 8.9 Examples of the Cs-enhanced, RF-driven, and internal-antenna H^- ion sources: (Left) SNS ion source [51]. (Courtesy R. Keller) and (Right) J-PARC ion source [52]. (Courtesy A. Ueno)

An H^- ion is formed by a neutral hydrogen atom that captures an extra electron with an affinity of 0.7542 eV, which is low compared to the electron binding energy of neutral hydrogen, which is 13.6 eV. According to the production mechanism, H^- sources can be classified as surface, volume, or hybrid sources. Sources dominated by surface production generally involve the introduction of cesium to reduce the surface work function and facilitate electron capture.

SNS operated with the accumulator delivering short (~1 μs), intense (~35 A) pulses of protons bombarding the target at 60 Hz with an average beam power of 1.7 MW. To feed the 1 GeV, 1.7 MW H^- linac, the ion source provided a peak current of 50-60 mA H^- beam with a duty factor of 6%. Figure 8.9 (left) shows its Cs-enhanced, multicusp, RF-driven internal-antenna ion source that has delivered ~1 ms long beam pulses, extracted from electrostatic LEBT at 65 kV energy [51].

The J-PARC ion source supplied a peak current of 72 mA H^- beam at 52.5 kV to the linac. The linac operated at a 25 Hz repetition rate and a 1.8% duty factor,

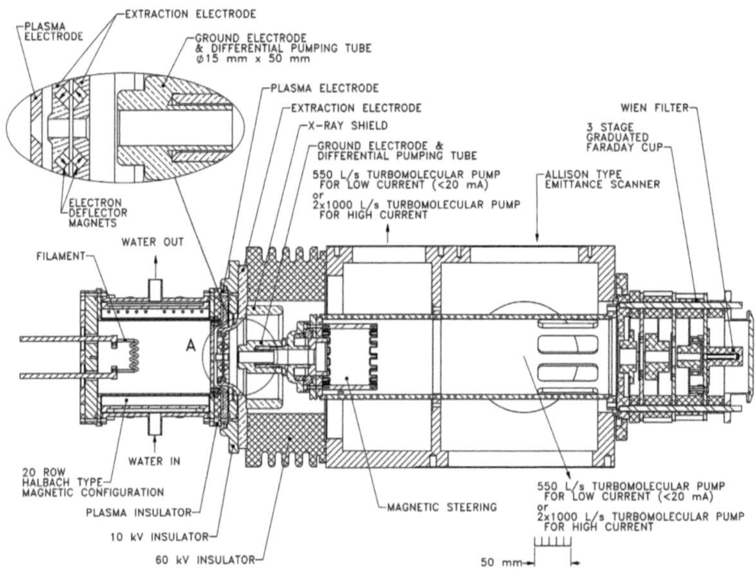

Fig. 8.10 Schematic view of the CW H^- ion source developed at TRIUMF [50]. (Courtesy K. Jayamanna, TRIUMF)

accelerating the H^- beam to 400 MeV for injection into the subsequent rings [52]. As shown in Fig. 8.9 (right), the ion source was also enhanced with cesium and RF driven by an internal RF antenna.

TRIUMF uses an H^- beam to facilitate efficient extraction from the cyclotron. Limited by space charge effects (Sect. 3.4.1), up to 1 mA of the CW H^- beam was injected for acceleration. During development, the high-current volume production multicusp-type ion source delivered over 60 mA DC beam at a normalized rms emittance of 0.11 πmm·mr (Fig. 8.10, [50]). Such a high-intensity CW beam can be important for the development of ADS programs (Sect. 1.6.3). Similar CW H^- sources are now commercially produced with technologies licensed from TRIUMF [53].

8.4.3 *Multi-charged Heavy Ion Source*

Multi-charged sources refer to sources of positively charged ions, up to the heaviest stable element, uranium. Higher ion charge states are preferred for attaining high acceleration efficiency of ions with a high charge-to-mass ratio. However, there is generally a trade-off between intensity and charge state from a source. To achieve a high intensity and charge state, the ions must interact with high-density, high-energy electrons in a plasma for a time long enough to reach the desired charge state.

Table 8.3 Features of typical multi-charged heavy ion sources [54]: electron-cyclotron-resonance ion source (ECR), electron beam ion source (EBIS), laser on source (LIS), Penning ionization gauge (PIG), and metal vapor vacuum arc ion source (MEVVA)

Type	Substances	Ion type	Duty	Current	Reference
ECR	Gases	Xe^{30+}	DC	117 μA	[55–57]
	Solids	U^{35+}	DC	300 μA	
PIG	Gases, volatile compounds,	Xe^{15+}	Pulsed	55 μA	[58, 59]
	solids	U^{10+}	Pulsed	350 μA	
EBIS	Solids, gases	Au^{32+}	Pulsed	mA's[a]	[60, 61]
LIS	Solids	Pb^{27+}	Pulsed	mA's[b]	[62–64]
MEVVA	Solid conductors	U^{4+}	Pulsed	10s–100s mA	[65, 66]

[a] ~10 μs, 5 Hz
[b]Few mA for few μs, ≤1 Hz

Table 8.3 shows the features of some typical multi-charged heavy ion sources. Sources for high-power accelerators must provide preferred ion species of preferred charge states at the required beam current and design duty, with sufficiently low emittance compatible with the design. In addition, lifetime, purity, reliability, stability, and ease of change of species are also to be considered.

Among the sources listed in Table 8.3, an ECR source is essentially the only option for high-intensity and high-charge-state CW beams [67]. An ECR source uses electron-cyclotron resonance to ionize a plasma. Microwaves are injected into a volume at the frequency corresponding to the electron-cyclotron resonance, which is defined by the magnetic field applied to a region inside the volume. ECR sources continue to move to higher RF frequencies and higher confining magnetic fields. High power ECR sources operate at frequencies up to 28 GHz [68] and RF power of ~15 kW. The required SC sextupole and solenoid magnets push the state-of-the-art in SC technology (Chap. 6).

Figure 8.11 shows the design layout of the FRIB front end area housing three high-voltage platforms, each containing an ECR source. Such an arrangement allows for the prompt change of ion species for high operational efficiency.

Figure 8.12 shows the charge state distribution of the FRIB SC-ECR ion source when supplied with RF power of 3.8 kW at 28 GHz and 1.0 kW at 18 GHz, and tuned for uranium with a 35+ charge state. A high temperature micro-oven was used [69] to heat uranium dioxide to 2100 °C, injecting the uranium vapor into the plasma. Plasma-confining magnetic fields were at 4 T (injection axial), 0.5 T (middle axial), 1.9 T (extraction axial), and 2 T (radial), respectively. The normalized horizontal and vertical rms emittances measured were 0.097 and 0.095 πmm·mr, respectively.

Two charge states are designed to be simultaneously accelerated downstream of the ion source until the charge stripper. Charge stripping (Sect. 8.8.2) at an energy of up to 20 MeV/u raises the average charge state from 35+ to 78+ without excessive degradation of beam quality, allowing efficient acceleration by the linac of up to five charge states (76+, 77+, 78+, 79+, and 80+) simultaneously to the design

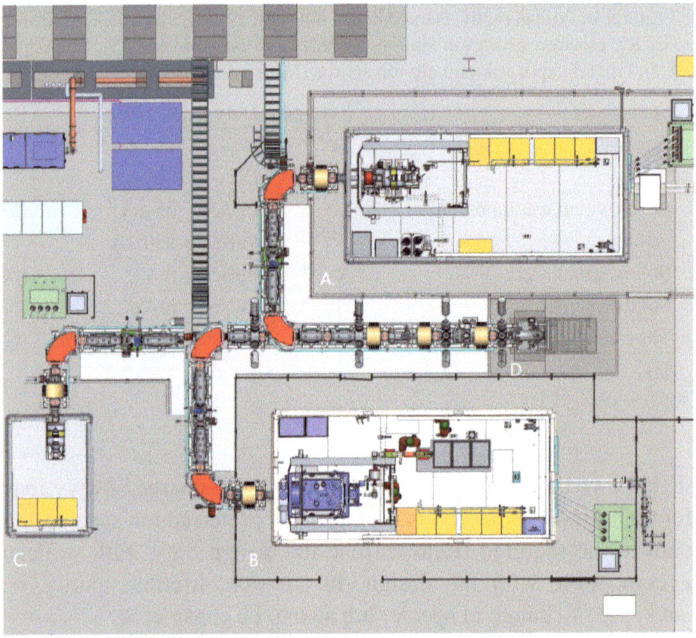

Fig. 8.11 Layout of the FRIB front end including the high-voltage platforms for two operational ECR ion sources (A: NC-ECR source with 14 GHz RF supply; B: SC-ECR source with 28 GHz and 18 GHz RF supplies) and a third ECR ion source (C) to be added later. All of the sources deliver ion beams to the LEBT where they are bent by 90° into the vertical beam line down to the linac tunnel (D). (Courtesy G. Machicoane, MSU)

energy, which meets the requirements of maximizing beam power and minimizing both controlled and uncontrolled beam losses. The beam currents in charge states 34+ and 35+ shown in Fig. 8.12 would be adequate to deliver 400 kW beam power to the FRIB production target.

8.4.4 Rare Isotope Sources

Rare or radioactive isotope (RI) beams can be created with a charge breeder or an accelerator. A charge breeder is a device that increases the charge state of ions, particularly RI, to allow for efficient acceleration in subsequent stages of an accelerator system. RI charge breeding was demonstrated at REX-ISOLDE/CERN with the use of an electron beam ion source (EBIS) and at several facilities using ECR ion sources [70]. EBIS and electron beam ion traps (EBIT) are capable of delivering clean, highly charged beams within a short transformation time, both for low-energy experiments and for post-acceleration at RI facilities, such as

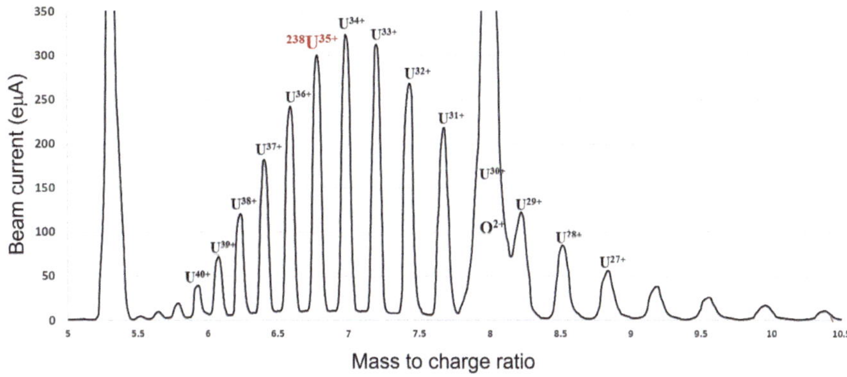

Fig. 8.12 Charge state distribution of the FRIB SC-ECR ion source when supplied by the RF power of 3.8 kW at 28 GHz and 1.0 kW at 18 GHz, and tuned for uranium with 35+ charge state. (Courtesy J. Guo and G. Machicoane, MSU)

the MSU re-accelerator (ReA3, ReA6). Another example is the CARIBU facility (Californium Rare Isotope Breeder Upgrade) that served as an injector into the ATLAS accelerator after charge breeding [71].

RI beams (Fig. 8.13) can be produced by an accelerator mainly by two methods [73, 74]: the isotope separation on line (ISOL) [75, 76] and in-flight fragmentation separation (IF) [77–80] (Sect. 1.6.1.2). As illustrated in Fig. 8.14a, the ISOL method uses high-energy light projectiles (e.g., protons, deuterons, or neutron beams) to bombard a thick target (e.g., uranium carbides; or a catcher). The nuclear reaction (e.g., spallation, fission) products are stopped and then transported to an ion source that provides element separation through chemical selection and ionization. Upon extraction, the desired isotopes are obtained through electromagnetic separation before being re-accelerated (Fig. 8.15, top). Currently, the main ISOL facilities [81] operating in the world include ISOLDE at CERN [82], SPIRAL at GANIL [83], and TRIUMF (Fig. 1.12) [84].

The IF method illustrated in Fig. 8.14b uses high-energy heavy projectiles (e.g., uranium beams) to bombard a thin target made of a relatively light element (e.g., graphite). The reaction (e.g., fragmentation and fission) products retain high velocities determined by the reaction kinematics. The wanted fragments typically exhibit a narrow momentum distribution, and a substantial part of them may be collected directly through a fragment separator. A major advantage of an in-flight separator is that the limits on half-lives for the separated rare isotopes are determined only by the time of flight through the optical system. Currently, the main operating IF facilities in the world include RIBF at RIKEN [86] and FRIB at MSU [87]. As shown in Fig. 1.14, upon production in the target hall, the FRIB rare isotope beams can be transported through various beamlines, providing fast, stopped, or re-accelerated beams to different user vaults for experiments [85].

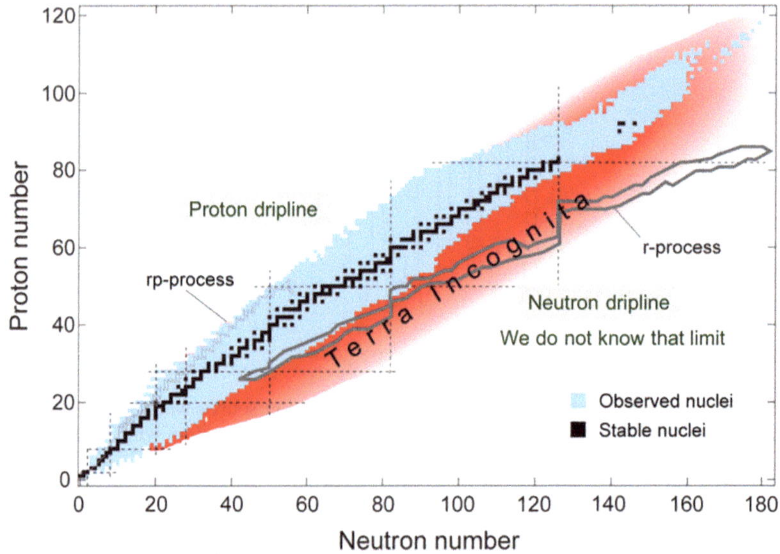

Fig. 8.13 Chart of elements and isotopes showing about 300 stable nuclei found in natural environment (black), more than 3000 nuclei discovered in the laboratories (blue), and unknown number in the unexplored territory (red) [72]

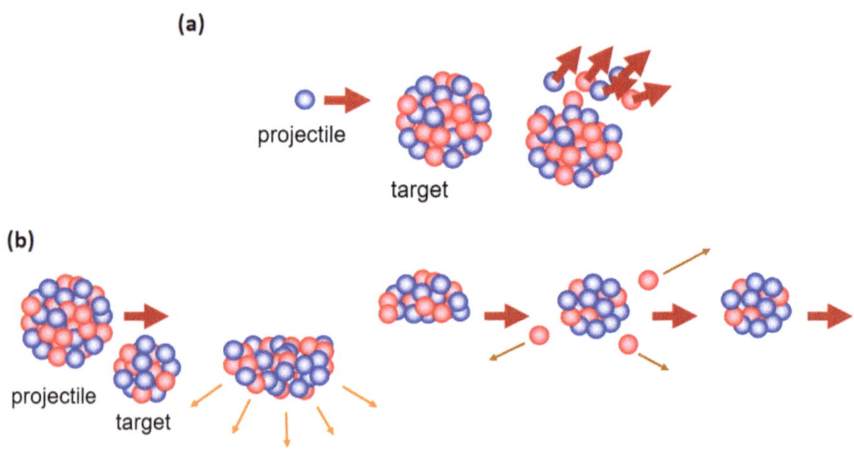

Fig. 8.14 Sketch of (**a**) ISOL and (**b**) IF reaction mechanisms [85]

The figures of merit for RI facilities are high efficiency, selectivity, sensitivity, and short delay times [73]. The final aim is the production of intense and pure rare isotope beams with good optical quality, proper timing characteristics, and energies essentially varying from rest (~meV/u) to the highest attainable energies (~GeV/u).

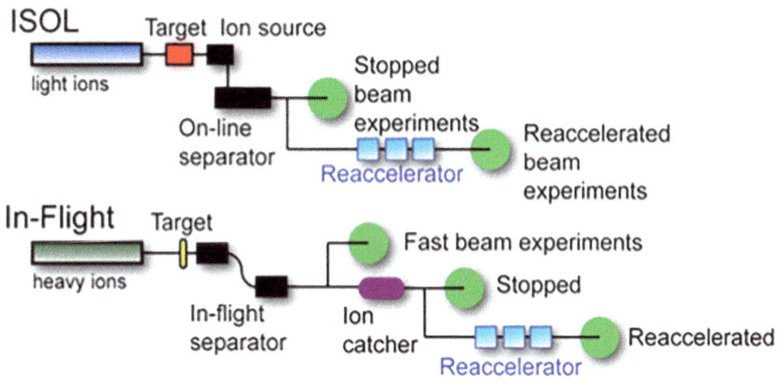

Fig. 8.15 Illustration of (top) ISOL and (bottom) IF acceleration and separation layout [85]

8.5 Radio-Frequency System

Except for the DC acceleration at the low-energy front end, the acceleration of high-quality, high-power hadron beams in accelerators is performed by RF electric fields. The RF system includes a master oscillator, RF amplifiers, accelerating structures, and low-level RF (LLRF) controls to synchronize distributed accelerating structures with a high phasing accuracy of $\sim 0.1^\circ$ at the local frequency. Independently of the accelerator type, the RF systems require RF amplifiers and DC or pulsed power supplies. The latter is known as modulators.

In NC linacs, as the power of the amplifiers is typically at the megawatt level or higher to support the high accelerating gradients and power transfer to the beam in NC linacs and cyclotrons, the RF system is often the biggest cost driver in construction and operation. The choice of RF power amplifiers and accelerating structures is important in accelerator design.

The acceleration processes for cyclotrons, linacs, and rings have been discussed in Sects. 2.1.3, 2.2, and 2.3.5, respectively. SRF accelerating structures for linacs have been extensively discussed in Sect. 2.2 and Chap. 5. Sections 8.5.1, 8.5.2, and 8.5.3 will briefly cover RF systems for cyclotrons, NC linacs, and rings, respectively.

8.5.1 Cyclotron RF Systems

Cyclotrons are highly efficient in producing high-power CW beams, as discussed in Sect. 2.1.3. For example, the acceleration in the isochronous cyclotron at PSI [88] is provided by four single-gap NC resonators with the parameters shown in Table 8.4 [89]. The PSI cyclotron produces 1.4 MW beam power with 0.85 MV RF voltage in the cavity. Potentially, these cavities can support the production of 1.8 MW beam

Table 8.4 Parameters of the single-gap resonator in the isochronous cyclotron at PSI

Parameter	Value	Unit
Frequency	50.6	MHz
Quality factor	4×10^4	
Maximum voltage	1.2	MV
Power to the beam	400	kW
Wall deformation due to air pressure	20	mm
Accuracy of hydraulic tuning devices	10	µm

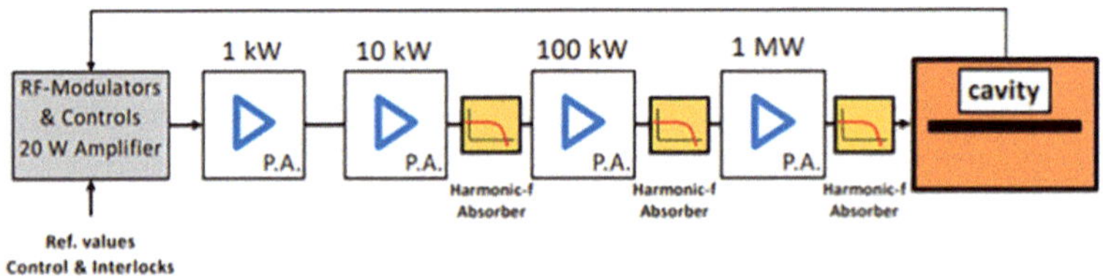

Fig. 8.16 Schematic layout of the RF system at the PSI cyclotron. (Courtesy PSI, Switzerland)

power. The cavity (Fig. 2.7) requires 1 MW CW RF power generated by a four-stage triode amplifier. The schematic layout of the RF system is shown in Fig. 8.16 [90].

8.5.2 *Linac RF Systems*

There are three different types of linacs to produce high-power hadron beams: (1) The highest-power RF amplifiers and modulators are required for pulsed NC linacs, while the largest fraction of the RF power in pulsed NC linacs is converted to RF power loss on structures, which is necessary to create an accelerating field; (2) Pulsed SRF linacs require a relatively high-power pulsed RF system to support beam acceleration; (3) The CW SRF linacs require low-power RF amplifiers to feed each SC cavity, and the power is only needed for the low-current CW beam acceleration. The SRF accelerating cavities are detailed in Chap. 5. This section discusses only NC structures along with the linac RF systems.

8.5.2.1 NC Accelerating Structures

Several types of NC accelerating structures are required for the economical acceleration of hadron beams from low velocities after the DC front end to typically up to a velocity of $\sim$80% of the speed of light. The transition between the different accelerating structures is dictated by cost minimization and typically requires different RF frequencies. The front end of all modern hadron linacs includes an RFQ. The development and application of RFQ for the acceleration of high-current beams has resulted in two main breakthroughs: (1) the capability to form

Fig. 8.17 SNS RFQ operating at 402.5 MHz frequency. (Courtesy ORNL, USA)

a compact longitudinal emittance with a substantially lower halo compared to a front end based on external bunchers and DTLs, and (2) the transverse emittances remaining unchanged despite the bunched structure of the beam exiting the RFQ. As an example, Fig. 8.17 shows the four-vane RFQ of the SNS linac operating at 402.5 MHz and delivering an H^- beam at 2.5 MeV [91]. Its internal view with the removed end plate is depicted in Fig. 8.18.

In a pulsed NC hadron linac, an RFQ is usually followed by DTLs that accelerate protons or H^- ions up to 50 to 100 MeV (Sect. 2.2). The focusing EM or permanent-magnet quadrupoles are installed inside the drift tubes. A significant challenge in the design and construction of DTL resonators for high-duty-cycle linacs of 5–15% is the management of RF power losses in the walls.

Figure 8.19 shows the internal view of the 324 MHz J-PARC DTL [92]. The transition to a higher RF frequency is required to maintain a high shunt impedance at beam energies above 50–100 MeV. For higher-energy linacs, DTLs are typically followed by CCLs that operate at a higher frequency (Sect. 2.2). Three types of multi-cell CCL are used in operations: (1) side-coupled accelerating structures at LAMPF [93], and SNS [94, 95]; (2) annular-ring coupled structures (ACS) at J-PARC [96]; and (3) disk–and–washer CCL at MMF [97]. Figure 8.20 shows the side cross-section view of the J-PARC ACS and the accelerating cell. Figure 8.21 shows the ACS section of the J-PARC linac.

Fig. 8.18 Internal view of the SNS 4-vane RFQ. (Courtesy ORNL, USA). The RFQ accelerates the H^- beam from 65 keV to 2.5 MeV energy

Fig. 8.19 Internal view of the J-PARC DTL. (Courtesy J-PARC, Japan). The DTL operating at 325 MHz RF frequency accelerates the H^- beam from 3 to 50 MeV energy

8.5.2.2 Klystrons and Tube Power Amplifiers

Historically, DTLs for pulsed proton acceleration below 100 MeV operate at ~200 MHz [97, 98] with a peak RF power of up to 3–5 MW, typically using triode [99] and tetrode [100] grid tubes available from industry. The grid tube amplifiers generate high peak and average RF power in a small volume, and the probability of failure is notable.

With an increase in the duty cycle of pulsed accelerators, klystrons were selected as RF generators [101]. Klystrons have been used in linacs at a high frequency of

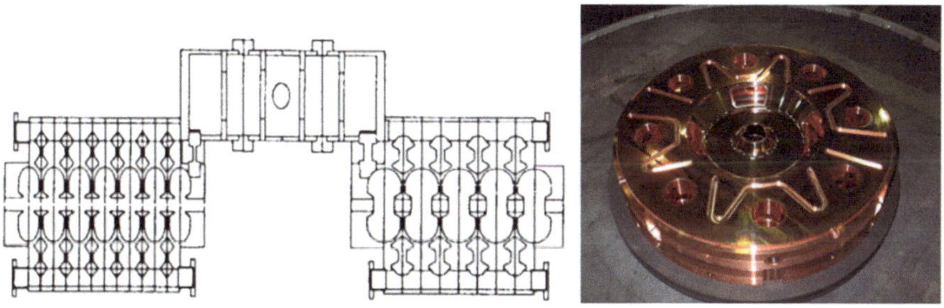

Fig. 8.20 (Left) Side cross-section view of the J-PARC ACS, and (Right) the ACS accelerating cell. (Courtesy J-PARC, Japan)

Fig. 8.21 The ACS section of the J-PARC linac. (Courtesy H. Ao). The ACS accelerates the H^- beam from 181 to 400 MeV energy

805 MHz, since the 1960s [98]. A klystron is a mini accelerator that consists of an electron gun, RF cavities, and beam-focusing devices. Klystrons can generate RF power in both CW and pulsed modes. Klystrons operating in pulsed mode are equipped with a modulation anode, which plays a role similar to that of the grid in triodes and tetrodes. Solenoids or permanent magnets provide beam focusing. Electron beams emitted from the electron gun (cathode) are accelerated by a static electric voltage of $\sim$100 kV or higher, modulated at the first cavity externally driven to form bunched beams, and further shaped by the subsequent cavities, one or more of which have resonant frequencies of the second harmonics. The first cavity is thus referred to as an input cavity. The bunched electron beam generates EM waves in the final cavity, losing energy and being dumped at the collector. The EM waves

generated in the final cavity are extracted by a coupler (or couplers) installed in the cavity. Because klystrons are very similar to electron accelerators, the design and manufacturing of klystrons utilize almost the same technologies as those used for NC accelerating cavities.

Klystrons are typically developed for RF frequencies above 300 MHz. The typical RF power generated by a long pulse ($\geq$ ms) and/or high duty ($\geq$ a few percent) klystron is several MW. The power generated by a CW klystron is typically limited to about 1 MW. The cooling requirements of the klystron collectors usually limit the highest possible power for CW or high-duty klystrons. The discharge limit tends to decrease as the pulse is elongated. Pulsed and high peak power klystrons typically operate with power conversion efficiencies of about 40–55%.

Compared with other RF amplifiers operating at frequencies greater than 300 MHz, such as vacuum tubes (triodes and tetrodes), inductive output tubes (IOT), and semiconductors, klystrons have the highest gain and generate the highest power. Even in the case of the medium power that vacuum tubes can generate, the high gain of klystrons can minimize the number of amplification stages, significantly increasing the reliability and availability of the systems. Another important choice is whether to use cathode modulation or anode modulation for pulsed operations. Cathode modulation is typically chosen for the short pulse ($\leq$10 μs), high peak power ($\geq$10 MW) cases, while anode modulation is typically chosen for the long pulse ($\geq$100 μs), medium peak power ($\leq$1 MW) cases. The long-pulse, high-power modulator is significantly more expensive than the short-pulse modulator. In particular, a pulse-forming network (PFN) commonly used for the short-pulse modulator becomes oversized for the long-pulse design.

The J-PARC (Sect. 10.2) linac uses anode modulation for their klystrons with a peak power of 3 MW and a pulse length of 600 μs [102] (see Fig. 8.22). On the other hand, both the ESS linac and the SNS [103] (Sect. 10.1) linac use cathode modulation for their klystrons. With the recent SNS proton power upgrade project (Sect. 10.1.5), 28 new 700 kW, 805 MHz klystrons have been installed to power the 28 SRF cavities for an upgraded section of the SNS linac, as shown in Fig. 8.23 [104]. The RF systems are largely the same as the existing RF systems, with some improvements based on operational experience since the SNS construction was completed in 2006.

The modulator system is one of the most significant cost drivers for linacs using tube amplifiers. The reliability of the linacs is dependent on the modulator performance. The SNS linac uses high-voltage converter modulators, as shown in Fig. 8.24. Numerous improvements have been implemented over many years, even after the start of SNS operations in 2006 [105].

While klystrons are used for RF frequencies above 300 MHz, devices such as tetrodes and inductive output tubes (IOTs) are commonly used for RF frequencies below 300 MHz. A single-stage tetrode can generate up to 100 kW RF power with an efficiency of about 60%.

The IOT, often called the "klystrode", is partly a klystron and partly like a tetrode. Like a klystron, an IOT magnetically focuses electron beams and is equipped with an output cavity and a collector. However, the electron beam current is modulated

Fig. 8.22 The 324 MHz klystron for the J-PARC linac. (Courtesy J-PARC, Japan)

Fig. 8.23 The 805 MHz klystron gallery at the SNS. (Courtesy ORNL, USA)

by a control grid closely placed to the cathode. Unlike a klystron, the DC current is modulated in the IOT. For this reason, the power conversion efficiency can be higher ($\sim$75%). As beam-shaping cavities are no longer necessary, an IOT's length can be much shorter than that of a klystron's of similar frequency. However, the transit-time effect limits the frequency to $\sim$1.5 GHz (high enough for hadron linacs) and

Fig. 8.24 RF high voltage modulator undergoing heat test at the SNS. (Courtesy ORNL, USA)

the gain to around 25 dB. The latter is a serious shortcoming compared to the typical klystron gain of about $\sim$50 dB. The typical output powers of IOTs are several tens of kW CW and several hundreds of kW (peak) pulsed, respectively.

8.5.2.3 Fabrication Technology of NC RF Cavities and Klystrons

This Section provides manufacturing details of klystrons and NC cavities. The klystrons and NC cavities are typically made nearly entirely of pure copper because of its high electrical conductivity. The mechanical rigidity of the RF cavities can be enhanced by using stainless steel parts in the non-RF surface area. Copper-plated steel is one example that maintains high electrical conductivity and mechanical strength.

Electron beam welding (EBW) and/or furnace brazing with silver-gold alloys are commonly used for manufacturing klystrons and NC cavities. As a general rule of thumb, EBW is the most reliable method for vacuum-tight joining; gold brazing in a well-temperature-controlled furnace is the second best and, at worst, silver brazing with a melting point higher than 800 °C. Low-temperature soldering is not an option for vacuum-tight joining because long-term reliability cannot be guaranteed. The cavity and klystron parts assembly for brazing and EBW should be performed in clean rooms with a class better than 1,000.

Frequently, brazing is performed in stages using alloys with different melting temperatures. The melting points can be increased by using higher ratios of gold to silver in the brazing alloys. Both vacuum and hydrogen furnaces are commonly used. The hydrogen furnace facilitates temperature uniformity during brazing, since the hydrogen gas tends to equalize the parts' temperatures. However, hydrogen can reduce the discharge limit. The choice between hydrogen and vacuum furnaces is thus a matter of cost-performance optimization and risk mitigation. Ideally, the brazing furnace should not be used for purposes other than providing clean brazing of the RF structures. The inner surfaces of furnaces are often contaminated by the heat treatment of dirty materials, which can stain the RF structures during their brazing.

The vacuum tightness of the water cooling channels should be highly reliable, as vacuum leaks from the water cooling channels are difficult to locate. The leak may be temporarily stopped by icing the leaking water through its adiabatic expansion, but powering the cavities melts the ice. For reliability, the brazing of the water-vacuum joints should be minimized. In addition, the copper thickness between vacuum and air, or water, should be at least 5 mm.

There are two typical ways of placing brazing alloys: (1) place thin films of the alloys between the brazing surfaces, and (2) place brazing wires in grooves near the brazing surfaces. The melted alloy permeates through the micro-space between the contacting brazing surfaces by capillarity.

Oxygen-free copper (OFC) must be used for reliable brazing and EBW. Otherwise, oxygen emitted from copper seriously impairs brazing reliability. The brazing surface should be reasonably clean and oil-free. High-quality OFC or well-controlled copper plating is necessary due to the required high Q value and the discharge limit. For example, well-controlled copper plating was used for J-PARC DTL and SDTL and has achieved an almost ideal Q value.

The deterioration of Q values arises from impurities and damage to the RF surface. In particular, oxidization developed in a humid atmosphere is a common cause of deterioration, especially when the bulk copper cavities' RF surface becomes particularly active for oxidization after the RF conditioning. To minimize the damage or imperfections of the copper surface, a high-quality diamond cutter can be used with an ultra-high-accuracy lathe for finishing machining. Polishing the copper surface with sandpaper is prohibited, as the process rubs impurities into the copper surface layer and mechanically damages its structure, reducing the Q value and the discharge limits. The furnace brazing process can recover surface perfection through heat treatment. In addition, the hydrogen furnace can deoxidize the oxidized surface. Due to these reasons, the Q value of a brazed structure often increases. However, brazing cannot remove the impurities that are rubbed into the surface layer by sandpaper.

8.5.2.4 Solid-State Power Amplifiers

Recent rapid progress in power semiconductor technology and digital LLRF control systems is making the solid-state power amplifier (SSPA) system increasingly promising. For example, an SSPA unit that includes a field effect transistor (FET) with a driver module, impedance matchers, distortion compensation circuits, and other control devices can generate an RF power of 2 kW CW. An SSPA system

Fig. 8.25 LLRF modules (the left rack) and three 8 kW solid-state amplifiers (right) installed at the FRIB. (Courtesy D. Morris, MSU)

can combine several SSPA units in multiple stages to increase the output power. Figure 8.25 shows the LLRF modules (left rack) and three 8 kW solid-state amplifiers (right) installed at FRIB [106]. By combining 4 drawers of 2 kW SSPA units in one rack, the 8 kW amplifier powers an HWR of $\beta_0 = 0.53$. With Class C operation, the efficiency of an SSPA is greater than 75%, and development efforts are underway to achieve higher efficiency.

Table 8.5 shows the parameters of the FRIB RF system. Except for the ion sources and RFQ at the front end, all other RF structures of the FRIB driver linac, including 324 SRF cavities from the low-velocity ($\beta_0 = 0.041$) QWRs to medium-velocity ($\beta_0 = 0.53$) HWRs, are powered by SSPA systems. Figure 8.26 shows a typical RF block diagram for a $\beta_0 = 0.085$ QWR at an 80.5 MHz frequency. The $\beta_0 = 0.65$ elliptical cavities developed for the FRIB400 energy upgrade (Sect. 10.3.8) will be powered by SSPA systems operating at a frequency of 644 MHz.

Table 8.5 FRIB driver linac RF system parameters [106]

Device name	Frequency [MHz]	Device type	RF power [kW]	Amplifier type	Tuner type	Quantity
ARTEMIS ECR IS	14×10^3	NC ECR	2	Klystron	–	1
HP ECR IS	18×10^3	SC ECR	2	Klystron	–	1
HP ECR IS	28×10^3	SC ECR	5	Gyrotron	–	1
LEBT multi-harmonic buncher (MHB)	40.25–120.75	NC	0.1	SSPA	–	3
RFQ driver	80.5	NC	8	SSPA	–	1
RFQ tetrode	80.5	NC	100	Tetrode	Servo (water)	1
RFQ[a]	80.5	NC	100	SSPA	Servo (water)	1
MEBT buncher	80.5	NC	4	SSPA	Stepper	2
QWR[b] ($\beta_0 = 0.041$)	80.5	SC	0.7	SSPA	Stepper	12
QWR[c] ($\beta_0 = 0.085$)	80.5	SC	2.5	SSPA	Stepper	92
IH multi-gap buncher	161	NC	30	SSPA	Stepper	2
HWR[d] ($\beta_0 = 0.29$)	322	SC	3.0	SSPA	Pneumatic	72
HWR[e] ($\beta_0 = 0.53$)	322	SC	5.0	SSPA	Pneumatic	148
Elliptical[f] ($\beta_0 = 0.65$)	644	SC	15	SSPA	Stepper/piezo	55

[a]Established in 2025
[b]For 12 cavities in 4 accelerating cryomodules
[c]For 88 cavities in 11 accelerating cryomodules and 4 cavities in 1 matching cryomodule
[d]For 72 cavities in 12 accelerating cryomodules
[e]For 144 cavities in 18 accelerating cryomodules and 4 cavities in 1 matching cryomodule
[f]For 55 cavities in 11 accelerating cryomodules under development for the FRIB400 project

In 2025, an SSPA system was developed in preparation to replace the tetrode amplifier for the RFQ to increase the system's availability. By combining 6 drawers of 2 kW SSPA units in one rack, each rack can provide 12 – 15 kW RF power, depending on the type of load cavity. For the NC cavity with a matched load and minimum reflected power, the output power is up to 15 kW per rack. As shown in Fig. 8.27, eight of these racks are combined in three stages, reaching the required power of 100 kW by the RFQ.

8.5.2.5 Low-Level RF Control

The electric accelerating field in each cavity must be precisely controlled in amplitude and phase, typically 0.1% and 0.1°, respectively. Figure 8.28 shows a schematic layout of the resonator operation together with the LLRF control system. It is a common practice to extract an RF signal from a cavity through a pick-up antenna or a magnetic loop. The LLRF compares the phase and amplitude of the resonator field with the master oscillator reference signal and adjusts the phase and amplitude of the pre-amplifier input power.

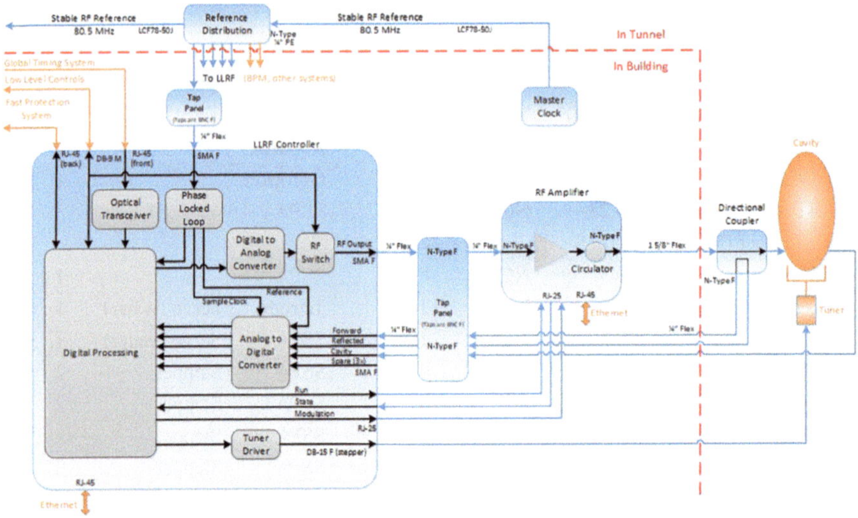

Fig. 8.26 RF block diagram for a $\beta_0 = 0.085$ SRF cavity of the FRIB driver linac [106]. (Courtesy D. Morris and S. Zhao, MSU)

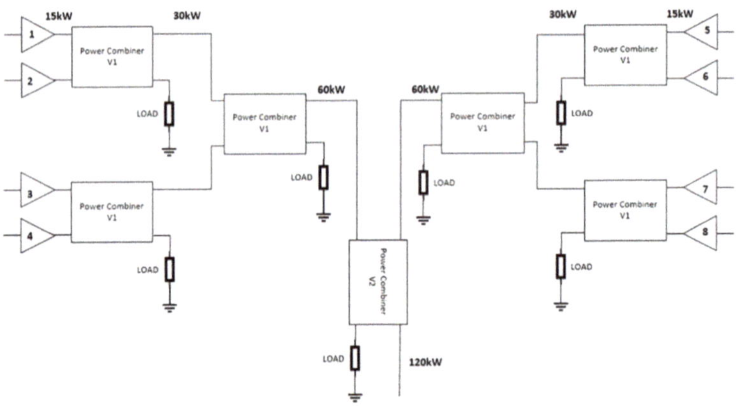

Fig. 8.27 Schematics of three-stage power combining with hybrid couplers for the FRIB RFQ SSPA system. (Courtesy S. Zhao, MSU)

8.5.3 *Ring RF Systems*

Unlike the linac RF system, the frequency of the ring RF system must increase as the beams are accelerated. The RF cavities, therefore, have to quickly tune their resonant frequencies. The variable frequency RF acceleration system traditionally uses shorted, ferrite loaded coaxial transmission lines as inductance to resonate with the accelerating gap or drift tube capacitance, reaching an accelerating gradient approaching 20 kV/m in a frequency range of 1 to 20 MHz, as shown in Fig. 8.29. Ferrite serves dual purposes of enhancing the magnetic field, allowing a small cavity

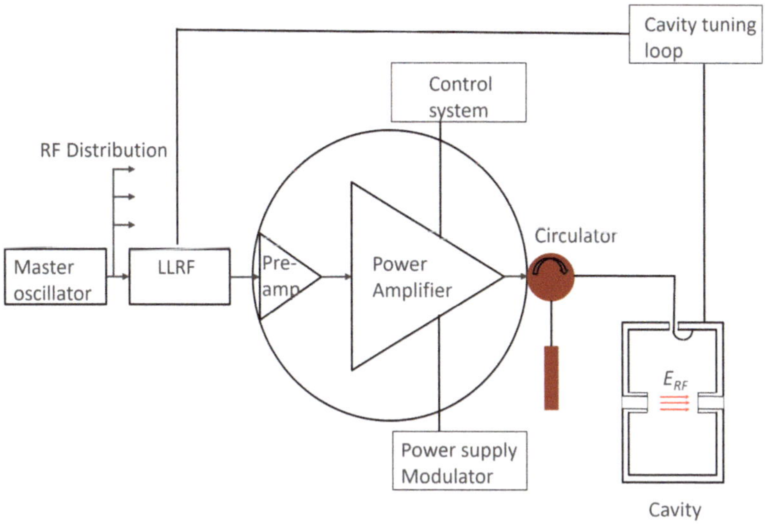

Fig. 8.28 Functional diagram of the RF resonator operation

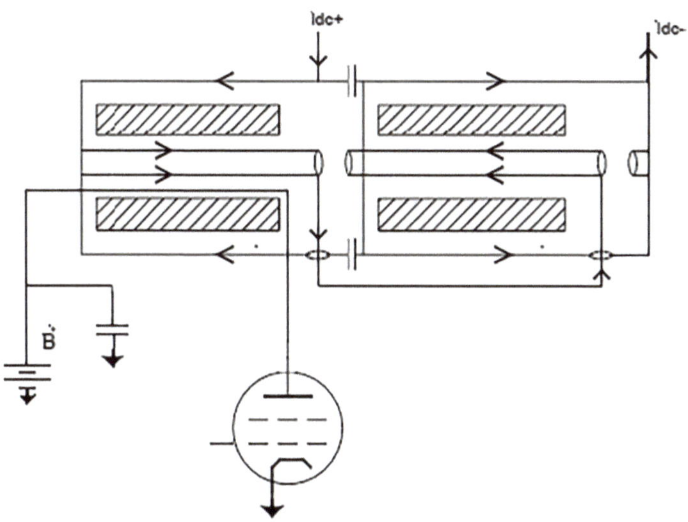

Fig. 8.29 Functional diagram of the ferrite-loaded RF cavity and the power amplifier system of the SNS accumulator ring [107]. (Courtesy A. Zaltsman, BNL). The two sets of ferrite rings (shaded) are DC biased whereas the outer vessel, beam pipe, and the gap-connecting bus-work form a single turn bias winding

size, and enabling dynamic frequency tuning of the cavity using the bias current. Because the (magnetic) permeability of ferrite is significantly dependent on the static bias magnetic field, the inductance can be appropriately lowered by the bias field to increase the resonant frequency. The high permittivity also lowers the Q

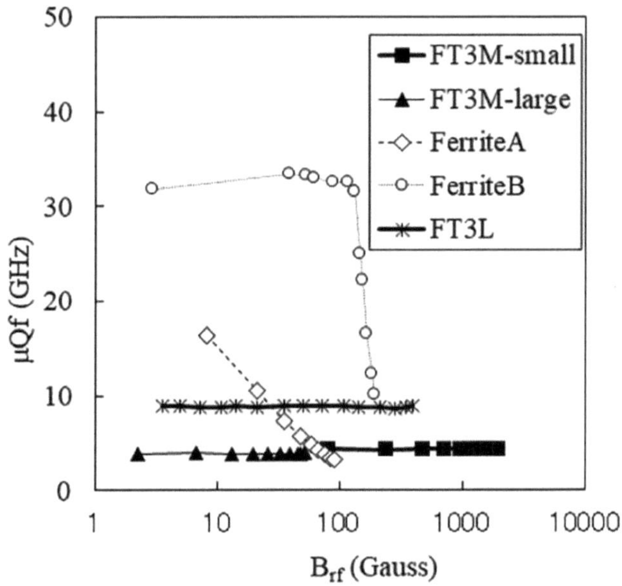

Fig. 8.30 $\mu Q f_{RF}$ as a function of the peak RF magnetic induction field $B_{R,\max}$ for some ferrite (Ferrite A and B) and MA (FT3L, FT3M) material [109]. (Courtesy C. Ohmori, J-PARC)

value. The LLRF system must incorporate the bias field control system, significantly complicating the RF system.

The shunt impedance R_{sh} of an RF accelerator determines how effectively the accelerator can convert the supplied RF power to the accelerating gradient (Sect. 2.2.1). The shunt impedance can be expressed by the relation [108]

$$R_{sh} = (\mu Q f_{RF}) 2\pi L_0 \tag{8.6}$$

where μ, Q and f_{RF} are the permeability of the loaded magnetic material, the cavity quality factor, and the RF frequency, respectively, and L_0 is the air-cored inductance.

The quantity $\mu Q f_{RF}$ is often used as a figure of merit for the loaded material in a given cavity structure. Figure 8.30 shows the dependence of $\mu Q f_{RF}$ as a function of the peak RF magnetic induction field $B_{RF,\max}$ for various ferrite materials [109]. When $B_{RF,\max}$ exceeds about 100 G, which is approximately equivalent to a field gradient of 10 kV/m, $\mu Q f_{RF}$ decreases rapidly. In other words, it is difficult for ferrite-loaded cavities to generate a field gradient above 10 kV/m.

After many years of persistent research, J-PARC RF cavities loaded with magnetic alloy material (Finemet®) were developed and used successfully in both the J-PARC RCS and MR (Sect. 10.2.4). As shown in Fig. 8.31, with saturation magnetization far exceeding that of ferrite, the cavities loaded with the magnetic alloy material reached approximately twice the accelerating gradient of the ferrite-loaded cavities [110]. However, because of the lossy nature of the material, power dissipation and consumption are significant.

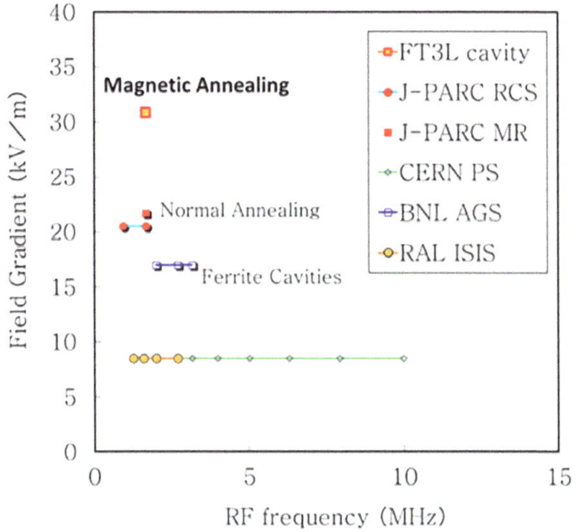

Fig. 8.31 RF system operating gradient and frequency of some hadron synchrotrons [109]. (Courtesy C. Ohmori, J-PARC)

High-power accumulator rings use RF cavities for injection capture, bunch confinement, and beam stability. The SNS accumulator ring used ferrite-loaded cavities operating in both $h_{\mathrm{RF}} = 1$ and $h_{\mathrm{RF}} = 2$ harmonics to create a "flat" longitudinal beam density profile that minimized peak density, increased the beam instability threshold, and retained a clean beam gap for loss-free extraction.

8.6 Power Supply

The power supply system converts energy from the electrical grid to energize beamline devices such as magnets and RF power amplifiers. The power supplies used to energize the RF devices have been discussed in Sect. 8.5. This section discusses power supplies that energize magnets and pulsed kickers.

8.6.1 Power Conversion Requirements

A magnet power supply converts the input alternating-current (AC) electrical voltage to a variable direct-current (DC) voltage. The magnet, which consists of a magnetic core and winding coils, corresponds to an equivalent circuit of an inductance in series with a resistor [111].

The power supply requirement includes the design current and voltage necessary to power the device throughout the acceleration cycle. The operating quadrant must also be identified. As shown in Fig. 8.32, in quadrants 1 and 3, the device is passive,

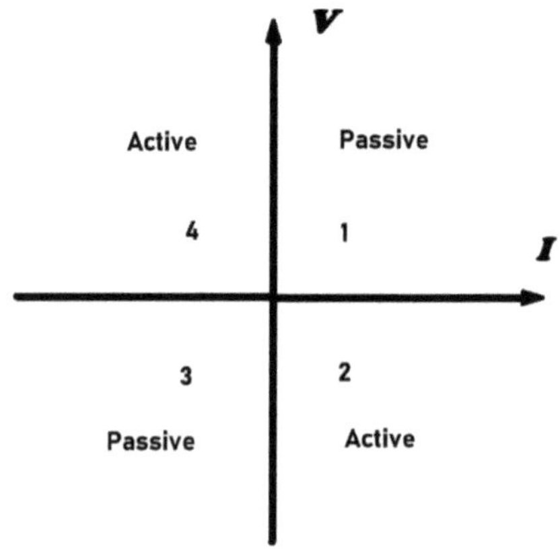

Fig. 8.32 Four-quadrant voltage-current diagram with the device in passive or active mode

consuming power from the supply circuit. In quadrants 2 and 4, the device is active, returning power to the circuit. The main bending dipole magnets in a rapid-cycling synchrotron usually operate in quadrants 1 and 2, consuming power during the magnetic field ramp up and returning power during the ramp down. Ramping corrector magnets often operate in all four quadrants (bipolar power supply). Such four-quadrant power supplies can provide positive or negative output voltage and can source or sink current.

Power ripple is the residual periodic variation of the DC voltage within a power supply that has been derived from an AC source. The ripple, which is possibly caused by the AC source, the switching frequency, charging and discharging, or transient ringing, must be controlled within an acceptable level, typically an order of magnitude lower than the tolerable magnetic field error. For example, if the required magnetic error $\Delta B/B$ is within $\pm 10^{-4}$, the power ripple needs to be controlled to the order of 10^{-5} of the maximum value. The requirement is more stringent for sensitive cases, such as slow beam extraction in synchrotrons.

Long term stability is another key requirement of accelerator power supply systems. Power supply output variations can be caused by long term drift in current measurement electronics or by various external factors, such as changes in ambient air temperature, cooling water temperature, AC input, and load resistance. The most dramatic output variations are typically attributed to changes in ambient air temperature; therefore, the HVAC systems in the power supply rack areas should be designed accordingly, or some accelerators should use temperature-controlled power supply racks.

Personnel protection (Sect. 8.1), machine protection (Sect. 8.2), and equipment protection (Sect. 8.3) are important aspects of the high-power accelerator power supply system. Isolation transformers are typically used to isolate magnets from the electrical grid. Magnets must be properly connected to the ground, with ground current monitored to inhibit power in the event of an earth fault in the DC circuit (Sect. 8.13.6). Precision current transducers, such as direct current current transformers (DCCT), are often used to measure the output current of the power supply and inhibit the beam when a fault is detected. Magnets must be equipped with interlocking systems that inhibit power in the event of faults, such as magnetic resistance temperature detectors (RTD) and cooling water flow meters for normal-conducting magnets, and quench protection and cryogenic monitoring for superconducting magnets. For superconducting magnets, dump resistors are needed as safety components that protect the magnet by dissipating the stored energy outside of the cryostat.

8.6.2 Magnet Power Supplies

During the past several decades, power electronics has evolved rapidly due to the development of power semiconductor technology (from tubes to insulated gate bipolar transistors, IGBT) and power converter typologies (thyristor-controlled rectifiers, switch-mode power converters, and discharged power converters) along with high-precision current measurement and regulation theory [112].

8.6.2.1 Power Supply for DC Magnets

Most power supplies in cyclotrons, linacs, and accumulator rings operate in a static state. A typical linac, like the FRIB driver linac, requires a large number of power supplies with different capabilities. For example, 728 power supplies were required for the FRIB linac through the vertical pre-separator [113]. All FRIB power supplies are solid state, switch mode power supplies, using conventional switch mode circuits such as the H-bridge to convert and regulate power. Switch mode power supplies have greater efficiency, which reduces electricity costs compared to other standard power supply topologies, such as linear regulators. Historically, linear regulators were commonly used to meet the demanding stability requirements at particle accelerators and also to reduce electrical noise. However, switch mode power supplies have taken the forefront with modern digital controllers, along with the proper design of the switching device, snubber circuit, output filter, and high-frequency grounding. The FRIB power supplies can be divided into 5 families, as shown in Fig. 8.33, and described below:

1. High voltage power supplies, quantity 74, are used in the FRIB front end, where the beam energy is low, as described in Sect. 6.1.1 and in electrostatic devices. High voltage, low current power supplies are used with electrostatic quadrupoles and dipoles to create electric fields for focusing and steering the beam. To account for beams of electrons striking the electrostatic device, the FRIB electrostatic high voltage power supplies require 2-quadrant operation (quadrants 1 and 4 or 2 and 3). The FRIB ion sources required several 1-quadrant HV power supplies in the 1 to 100 kV range, with up to 2 kW power; for example, to set the initial beam energy from the ion source with a 100 kV high voltage power supply that can

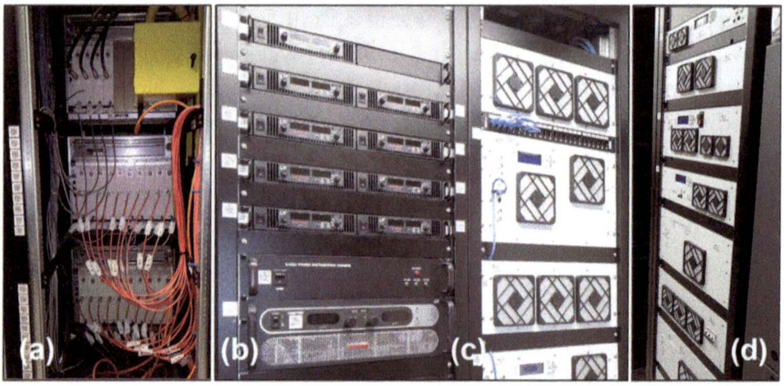

Fig. 8.33 FRIB power supply types installed at the FRIB: (**a**) electrostatic high voltage power supplies, (**b**) NC magnet power supplies for quadrupole (bottom) and corrector/cryogenic heater (top), (**c**) NC dipole modular power supply, and (**d**) modular SC magnet power supply. (Courtesy K.Holland, MSU) [113]

regulate voltage within 100 ppm. After the beam is accelerated, higher voltages will be required to use electrostatic devices; therefore, it is no longer practical to use these devices due to limitations related to high voltage and the voltage withstanding capabilities of materials.

2. NC magnet power supplies, quantity 312, are used at FRIB through the vertical pre-separator. As described in Sect. 6.1.1, resistive (room temperature) magnets are used to create magnetic fields to focus or steer the beam. The requirement to regulate multipole magnet current within 500 ppm or higher allowed for the use of commercially available power supplies. A parallel modular configuration was selected to meet the magnet requirements, where higher power is needed. A custom module added a free-wheeling diode to protect the magnet and power supply from inductive kickback and ground fault detection to protect smaller gauge control wiring during fault conditions. Additionally, polarity reversal switches were used to allow the single quadrant power supplies to operate in quadrants 1 and 3 for use with corrector steering magnets.
3. Sixty cryomodule heater power supplies are used to maintain the pressure in the cryomodule cryogenic system. 1000 ppm stability class allowed the use of commercially available power supplies. The heater resistance was optimized to use one of the room temperature magnet power supply models, minimizing the number of power supply types.
4. NC dipole power supplies, quantity: 9. Although these are a type of resistive magnet, a separate power supply family was created due to special requirements, such as 100 ppm stability class and complex auxiliary components. A modular series configuration was chosen to meet the higher power magnet requirements and higher voltage, where dipole magnets are connected in series.
5. SC magnet power supplies: quantity 273. Where the highest magnetic fields are required, superconducting magnets, as described in Sect. 6.1.4, are used to focus or steer the beam. The FRIB superconducting magnet power supplies are all true

bipolar 4-quadrant systems, which allow for a controlled ramp-up and ramp-down during beam tuning and enable the degaussing of the magnets. Although the superconducting dipole magnets only require current in one polarity and therefore require only 2-quadrant operation (quadrants 1 and 2), FRIB uses 4-quadrant dipole power supplies to reduce the number of power supply types. One key feature of the FRIB SC magnet power supplies is the capability to safely discharge the stored energy in the magnet inductance. Since the SC magnet has practically zero resistance to dissipate stored energy, an energy dump resistor is incorporated into the power supply. The concept is similar to dissipating the stored energy in an electric train; in fact, the FRIB dump resistors are energy braking resistors most commonly used in trains. The dump resistor limits the peak voltage if the power supply shuts off for any reason, such as loss of AC power, an operator command, power supply failure, or magnet interlock. Most FRIB magnets are self-protecting and require only a slow dump resistor that limits the voltage to a low value. However, some magnets require additional fast dump circuitry to expedite the dissipation of stored energy. The fast dump circuit is typically accomplished by a fast dump switch that opens very quickly and an additional resistor with a higher resistance that allows a higher peak voltage and a faster exponential decay. The stability classes of these power supplies range from 20 ppm for dipole magnets to 1000 ppm for solenoids.

8.6.2.2 Power Supply for Ramping Magnets

For high-power synchrotrons, the magnet power supply for rapid-cycling or fast ramping is challenging, as precise power tracking is especially required for beam loss mitigation.

High-power hadron rings use large aperture magnets to reduce uncontrolled beam loss. The electrical current required to produce the magnetic field B_0 over a full aperture of height g is

$$I = \mu_0^{-1} B_0 g \tag{8.7}$$

where μ_0 is the permeability of the vacuum. The associated electrical energy is thus proportional to the square of the magnet aperture.

Although magnets can be powered either individually or in series, a common practice in ring design is to connect magnets of the same type in series. This practice helps maintain tracking between the magnets and the super-symmetry of the optical lattice. In addition, trim magnets and correctors are usually powered individually for flexible adjustments.

Rapid-cycling synchrotrons require a large amount of electrical power (typically tens of megawatts) for magnet and RF systems. The higher the repetition rate, the larger the peak power drawn from the power grid, and the more likely the need for local energy storage. Traditionally, flywheel energy storage has been used by accelerating or decelerating a rotor (flywheel) at very high speeds to store or retrieve

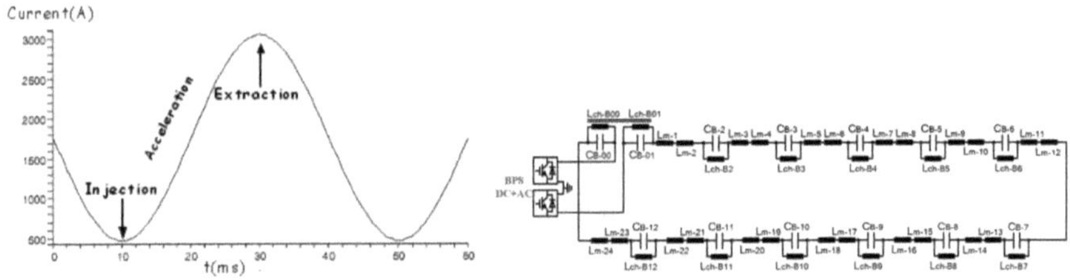

Fig. 8.34 CSNS DC biased resonant waveform and the power supply network feeding the main dipole magnets at 25 Hz repetition rate [114–116]. (Courtesy CSNS, China)

energy in terms of rotational energy. Recent accelerator facilities, such as the J-PARC RCS, store energy in their power supply using large-capacity capacitors and resonance circuits.

Most RCS with a repetition rate greater than 20 Hz use resonant power systems (Fig. 8.34) [117]. With a resonant power system, the strength of the magnetic field varies sinusoidally along with the rigidity of the beam. The consequences are the lack of a "flat bottom" for a static injection condition and the demand for high-voltage amplitude by the RF system to sustain the largest ramping slope of the sinusoidal waveform. A proposed variation is a dual harmonic ramp with a slower slope during the up ramp (acceleration) and a faster slope during the down ramp [118].

The magnet power system for non-sinusoidal, programmable ramping includes bridge rectifiers directly connected to the power grid (e.g., BNL AGS Booster) and bridge rectifiers with local energy storage in the generator set (e.g., BNL AGS). To satisfy the increasing need to accurately control injection "flat bottom" and extraction "flat top" and to have a flexible program of magnet current, high-frequency AC bridge rectifier systems are being developed, such as those based on IGBT (insulated gate bipolar transistor) rectifiers. A programmable ramp using this technology provides maximum flexibility in manipulating the profile of the rise and fall, thus reducing the peak ramp rate and current-induced magnetic imperfections.

Tracking various types of magnets to the design ramping rate is crucial in minimizing undesired effects, including closed-orbit deviation and focusing-amplitude mismatching. Efforts are made to group magnets into families, with magnets in the family sharing a common power supply, and to minimize the number of families to facilitate tracking.

8.6.3 *Pulsed Power Supplies*

Pulsed power supplies are used to activate devices that deflect the beam during a short period of the acceleration cycle; for example, they activate the chopper system

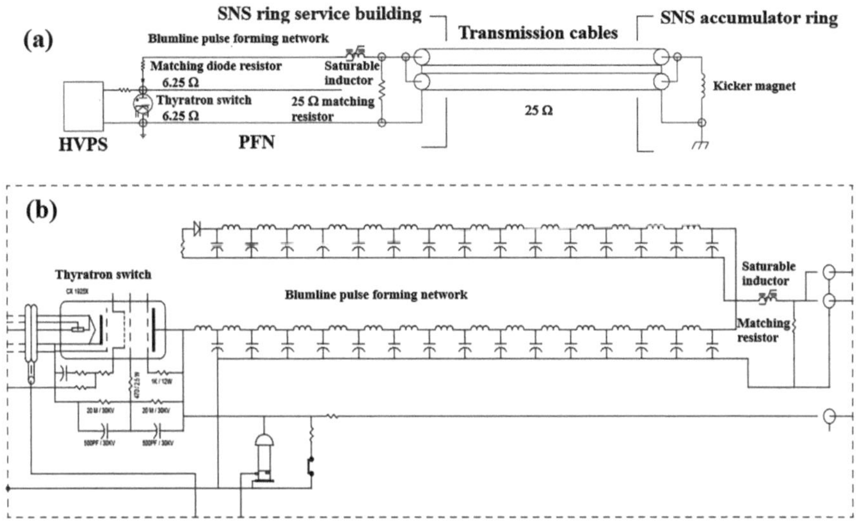

Fig. 8.35 (**a**) Simplified schematics of the SNS extraction kicker system containing the Blumlein pulse-forming network [119, 120]. (**b**) Blumlein pulse forming network modulator schematic diagram [121]. (Courtesy J.-L. Mi, BNL)

at the front end of the linac (Sect. 8.10.1) and energize the kicker magnets during the injection and extraction of the beams in the rings (Sects. 2.3.4 and 2.3.7).

High-power rings are more susceptible to uncontrolled beam losses incurred during the injection and extraction of the beams. The pulse power supplies must provide a high voltage amplitude, a fast rise/fall time, and a steady flat top. Figure 8.35 shows the pulse-forming network that is used to power the 14 extraction kickers of the SNS accumulator ring (Sect. 2.3.7.1) [119, 120]. The system consisted of fourteen high voltage modulators (Fig. 8.36), their local control and auxiliary systems, and fourteen window frame magnet sections. High-voltage transmission cables were used for pulse transmission from modulators to their corresponding magnet sections. The high voltage and high current pulse had a rise time of 200 ns, a pulse duration of $\sim$700 ns, a fall time of 17 ns, and a current amplitude of $>$2500 A at $>$35 kV per modulator. The total output of the system reached several GW peak power at a 60 Hz repetition rate. The modulator design featured a Blumlein pulse-forming network with ultra-low inductance, high-energy storage pulsed capacitors, an ultra-fast high-power hollow anode thyratron, a parallel resistor stack for low beam-impedance termination, and a saturable inductor stack for beam current isolation and pulse rise time sharpening. This design was driven by considerations of low beam loss, high maintainability, and reliability. All high voltage modulators were placed outside the ring tunnel.

Fig. 8.36 Pulse-forming network modulator for the SNS extraction kicker system [121]. (Courtesy J.-L. Mi, BNL)

8.7 Vacuum

The vacuum system serves the primary purposes of preserving beam quality in the accelerator and protecting the beam enclosure surfaces from contamination. It preserves beam quality by limiting the amount of interaction between the beam particles and the gas molecules in the vacuum chamber, mitigating machine performance limitations, including beam lifetime reduction, emittance growth, pressure-induced instabilities, and electron-induced instabilities. It protects beam-enclosure surfaces from contamination that causes surface sparking, field emission, and outgassing, affecting the performance of electro-magnetic devices.

8.7.1 Partition and Protection

The vacuum requirements differ in various accelerator segments. Table 8.6 shows the design vacuum pressure levels across the SNS accelerator segments: front end (IS, LEBT, RFQ, and MEBT), NC linac (DTL and CCL), SC linac (SCL), high energy beam transport (HEBT), accumulator ring, and ring-to-target beam transport (RTBT). Ultra-high vacuum (UHV) is needed for the segments of the SCL and for the ring where the beam circulates for up to one thousand turns. For example, the beam of charged ions needs to be contained in an ultrahigh

Table 8.6 Vacuum design requirements across the SNS accelerator segments

Segment	IS/LEBT	RFQ	MEBT	DTL	CCL	SCL	HEBT	Ring	RTBT
Pressure [Torr]	10^{-4}	10^{-6}	5×10^{-7}	1×10^{-7}	1×10^{-7}	10^{-9}	5×10^{-8}	10^{-8}	10^{-7}
Energy [MeV]	0.065	2.5	2.5	87	186	10^3	10^3	10^3	10^3

vacuum to suppress pressure instability caused by, e.g., ion-induced desorption and electron-cloud effects. Gate valves are commonly used to isolate vacuum spaces in large-scale accelerator systems, providing flexibility in accelerator maintenance, repair, and upgrades.

For accelerator systems under UHV conditions and for sensitive devices such as SRF cryomodules, which are susceptible to performance degradation caused by surface contamination, fast gate valves play an essential role in protecting the beam vacuum enclosure from accidental atmospheric exposure and contamination resulting from vacuum fault incidents from both the accelerator system and the user stations. The design of a fast gate valve system must consider the speed of the system response. As discussed in Sect. 8.3.3, a sufficient distance is needed between the valve sensor and the valve body. For example, a commercially available 100 mm opening fast gate valve can close in about 10 ms [122] from the moment its hardwired cold-cathode gauge sensor detects a vacuum event. Assuming that the air shock wave travels at the speed of sound of about 340 m/s along the vacuum chamber, the fast gate valve should be located at least 3.4 m from the nearest potential vacuum incident location, where the sensors are installed.

As fast gate valves are designed to temporarily deter vacuum deterioration caused by catastrophic events, the gain in fast response time is usually at the expense of valve lifetime (e.g. 2,000 actuations), compromised leak rate, and elastomer gate seal. Subsequently, regular sectoring gate valves are used to isolate the segment after the fast gate valves have been activated. Fast gate valves may not be applicable in a reactive gas environment, such as lithium vapor, due to the elastomer gate seal. For critical protection, they should be placed under the credited control program that requires periodic maintenance and performance verification.

For systems like the SRF cryomodule or liquid lithium charge stripper (Sect. 8.8.2.2), the beam vacuum can be separated from and protected against the secondary insulating vacuum that ensures device performance.

8.7.2 Vacuum Chamber and Shielding

The vacuum chamber through which the circulating beam passes provides an environment intended to reduce the effects of residual gas scattering and to lower unintended charge stripping. The high beam current can drive gas ions into the walls of the chamber, leading to positive feedback and pressure influx, unless the ion-induced desorption coefficient is kept below a threshold value [123]. In addition, the enclosure offers a conducting surface to carry the induced image current so that the coupling impedance is minimized.

For stand-alone devices like ion sources, RFQ, accelerating RF cavities, SC cryomodules, SC magnets, injection and extraction kickers, and special instrumentation stations, the beam vacuum enclosure is an integral part of the device. For areas exposed to beam-induced or field-induced heating, vacuum chambers often require custom design or development. Fabrication of such complex chambers usually needs to follow defined specifications, including machining, cleaning methods, joint techniques (e.g., welding, brazing), allowable leak rates, and shipping and packaging precautions. For more conventional areas, such as beam transfer lines or regular connections, commercially available vacuum chambers may be used. Fabrication of such chambers follows the standard industrial UHV-specifications and practices. Welded or brazed vacuum joints are common weak spots in a vacuum enclosure. Design considerations are often necessary to avoid high heating power deposition on the welded or brazed joints. Welding can sometimes be choreographed to minimize or correct structural distortion.

In time-varying magnetic fields, the vacuum enclosure must accommodate several factors: (1) thermal heating from the eddy current generated by the varying magnetic field in a resistive material; (2) the delay in the rise time of the magnetic field inside the chamber; and (3) the continuous passage of the beam image current. Conventionally, the vacuum chamber is made of either a metallic pipe (stainless steel or aluminum) or is directly formed from the magnet pole faces. For rapid-cycling synchrotrons, the vacuum chambers need to be RF shielded to provide a high impedance to the eddy current while maintaining a low impedance to the induced image current. Consider a pipe of thickness d_w and volume resistivity ρ_r, as an example, for a resistive loop of width w_r and height h_r penetrated by a magnetic field B, the instantaneous power per unit length is given by

$$\frac{dP_{inst}}{ds} = \frac{\dot{B}^2 w_r^2 h_r d_w}{2\rho_r} . \tag{8.8}$$

For a circular pipe of radius b, the average power is proportional to the repetition rate, the field variation rate, the magnetic-field amplitude squared, and the pipe's radius b cubed. It is also inversely proportional to the sheet resistivity ρ_r/d_w.

ISIS adopted ceramic chambers with supported internal stainless steel wires, interrupted by ceramic capacitors to allow only the passage of the beam image charge at high frequency [124]. The wires were contoured to the design beam envelope, reducing the impedance of the beam coupling. For newly proposed synchrotrons with much higher beam power, the main concern with this approach is the failure of the wires and the consequent unavailability of the machine. Other candidates include a ceramic chamber with printed internal silver wires, external shielding, and internal coating, or a thin Inconel chamber of high mechanical strength, such as those in the 7.5 Hz AGS Booster, with six pairs of self-correcting coils mounted on the dipole chambers, which cancel the eddy current field [125]. Figure 8.37 shows the circular vacuum chamber used in the straight section of the 3 GeV synchrotron of the J-PARC project [126]. For both rapid-cycling

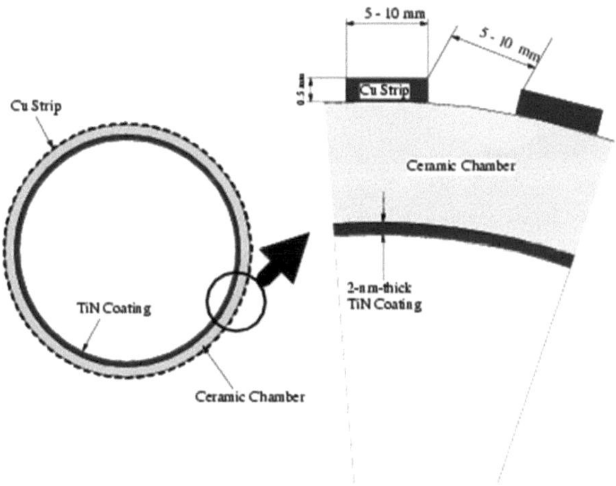

Fig. 8.37 Schematic view of the ceramic vacuum chamber and the RF shield of the 3 GeV synchrotron of the J-PARC project. (Courtesy M. Kinsho, J-PARC). The 5 to 6 mm thick chamber wall is made of 99.8% high-purity alumina [126]

synchrotrons and accumulators, ceramic vacuum chambers are often used to allow the fast ramping of trim quadrupole and dipole fields. Ceramic chambers are also used in the injection region to accommodate the fast-ramping of bump magnet fields for phase-space painting. Figure 8.38 shows the titanium alloy-lines of the thin-walled vacuum chamber used for the HIAF facility [127].

8.7.3 Instrumentation and Practices

8.7.3.1 Vacuum Gauges and Residual-Gas Analyzers

Vacuum pressure is monitored by vacuum gauges that can measure well-defined physical properties that are functions of gas density, such as thermal conductivity and ionization potential. The sensitivities and resulting measurements of most vacuum gauges are highly gas species dependent; therefore, careful calibration and correction are necessary to ensure the accuracy of the readings. Pirani gauges are commonly used in the range from 0.5 to 10^{-4} Torr and are linear in the range from 10^{-2} to 10^{-4} Torr. The operating principle is based on the variation of the thermal conductivity of the gasses as a function of pressure. A resistor under vacuum is heated to a constant temperature. The heating current required to keep the temperature constant is a measure of pressure. Convection enhanced Pirani gauges can extend the upper measuring range to a few hundred Torr [128].

Penning type cold cathode gauges are commonly used in the range of 10^{-4} to 10^{-9} Torr for monitoring and interlocking purposes. The operating principle is based on the measurement of the discharge current in a Penning cell, which

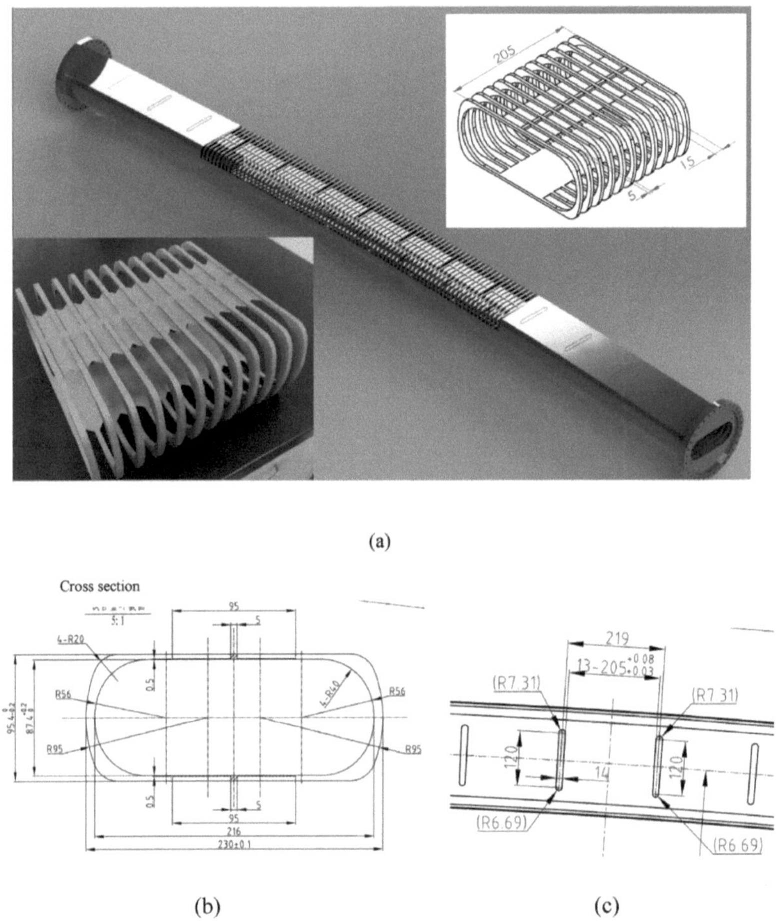

Fig. 8.38 The titanium alloy-lined vacuum chamber with wall thickness 0.3 mm of the Heavy Ion Accelerator Facility (HIAF): (**a**) a 3-D structure diagram; (**b**) the cross section of the titanium alloy ring; and (**c**) the enlarged view of the concave surface. (Courtesy IMP, China)

is a function of pressure. Electrons are produced by field emission and perform oscillations in the crossed electric and magnetic fields. Along the path length, molecules are ionized, and ions are collected onto the cathode. Inverted magnetron type cold cathode gauges extend the measuring range down to 10^{-12} Torr and have since replaced the Penning type gauges [129].

Residual gas analyzers (RGA) are used in the vacuum range from 10^{-4} to 10^{-12} Torr for gas content analysis. A filament produces electrons that ionize the residual gas inside a grid cage. A mass filter is used between the grid and the ion collector to sweep through the selected mass ranges. The ion current can be measured in the Faraday mode or in the secondary electron multiplier mode. RGA is often used to identify contaminants and leaks by tracing gas contents such as hydrocarbons, argon, and nitrogen. It can also be used to interlock, together with the machine protection system, for safe operation.

8.7.3.2 Vacuum Pumps

Gas molecules are removed from the vacuum chambers by vacuum pumps. There are three types of vacuum pumps commonly used in accelerators: positive gas displacement pumps, kinetic pumps, and gas-binding pumps. The first two types of pumps are also called gas transfer pumps; they can, in principle, operate continuously. Since these pumps usually contain movable parts that are subject to mechanical failure, which contaminates the vacuum environment, they are typically used only for initial pump down and are not preferred for steady accelerator operations. Gas-binding or capture pumps are the dominant pumps in the ultrahigh- and high-vacuum (UHV and HV) regions of accelerator applications [130], particularly for steady-state operations. They have a limited gas absorption capacity and usually need to be regenerated at process-dependent intervals.

Positive gas displacement pumps use a mechanical rotary mechanism to physically "sweep" the gas molecules from the vacuum system. These pumps are non-discriminatory towards gas species and have large gas handling capacities; however, they cannot achieve high vacuum. They are used for two main purposes: to remove ("rough") the bulk of the air from a system that is initially at atmospheric pressure and, once this is achieved, to "back" kinetic pumps, which cannot exhaust against atmospheric pressure. Commonly used positive displacement pumps include diaphragm pumps, piston pumps, scroll pumps, and roots pumps. These pumps typically compress the gas to atmospheric pressure before being discharged at the outlet. The lubricant-free dry scroll pump dominates this application.

Kinetic pumps function by impacting and transferring momentum to the gas molecules, accelerating them in the pumping direction either through a mechanical drive system or through a directed vapor stream that is condensed at the end of the pumping section. Two popular kinetic pumps are the turbo-molecular pump and the diffusion pump. With a turbo-molecular pump, gas molecules obtain momentum in a desired direction through repeated collisions with a moving solid surface. A rapidly spinning rotor "hits" gas molecules from the inlet of the pump towards the exhaust to create and maintain a vacuum. The pump usually has multiple stages of rotating rotor blades and stationary stator blade pairs. The gas captured by the upper stages is pushed into the lower stages and successively compressed to the next stage, where it again collides with the rotor surface, eventually leading outward through the exhaust. As the pump blades rotate at high frequency, catastrophic mechanical failure may occur, leading to contamination of the beamline vacuum. Consequently, turbo-molecular pumps are often used only for initial or temporal pumping solutions. Often, it is installed with its backing pump on a mobile cart so it can be removed after valving-off. Automated interlocking does reduce catastrophic failures of turbo-molecular pumps by monitoring the key pump parameters (such as the draw current) and then isolating it from the vacuum system with a gate valve.

More conservative methods (like a sputter ion pump) are usually used for stable operating conditions.

In a diffusion pump, a high-speed vapor jet generated by heating the fluid directs gas molecules from the pump throat down into the bottom of the pump and out towards the exhaust. Most modern diffusion pumps use low-vapor-pressure fluids, such as silicone oil or polyphenyl. Often, several jets are used in series to enhance the pumping action. The outside of the diffusion pump is cooled by a water-cooled jacket for fluid condensation and recovery. The gasses pumped out of the diffusion pump outlet are compressed to ambient pressure by a mechanical forepump and exhausted. Unlike turbo-molecular pumps, diffusion pumps have no moving parts and are thus more reliable. One major disadvantage of diffusion pumps is the tendency to backstream oil into the vacuum chamber, causing contamination. Sometimes, cold traps and baffles are used at the inlet to minimize back streaming. However, modern turbo-molecular pumps have largely replaced diffusion pumps in accelerator vacuum systems.

Gas-binding pumps or capture pumps retain gas molecules through chemical or physical interactions on their surfaces, thereby eliminating the complicated valve interlocks and moving parts found in turbo-molecular pumps. Capture pumps either bind the gas to an active substrate or condense the gas at a suitable temperature. Three popular capture pumps are a sputter ion pump, a non-evaporative getter pump, and a cryogenic pump. With a sputter ion pump, a cloud of electrons produced inside a Penning trap ionizes the gas molecules. The resultant ions are accelerated to strike a chemically active titanium cathode. Upon impact, the accelerated ions will either become buried within the cathode or sputter cathode material onto the walls of the pump. The freshly sputtered, chemically active cathode material acts as a getter. Gasses are trapped by a combination of chemical reactions with the surface of the highly reactive sputtered material or physically trapped beneath that material. The discharge current of the sputter ion pumps is proportional to the vacuum levels, thus providing a very reliable measurement of the system pressures.

With a non-evaporable getter pump (NEG), gas molecules penetrate into the bulk of the getter material and form stable chemical compounds. The getter materials, mostly porous alloys or powder mixtures of Al, Zr, Ti, V and Fe, can be used in either coated or bulk form. With a coated getter, the nonvolatile metal powder is sintered or sputtered into a porous coating onto the inner surface of vacuum vessels. This coating can be applied even to narrow spaces with limited conductance, thus providing effective distributed pumping. The activation of the coated getter does require heating the vacuum vessels to elevated temperatures and needs to be included in the vessel design. With a bulk getter, sheets, strips, or pellets of gas-absorbing metals are mounted together with heating elements in a separate housing [131]. NEG pumps are often activated by heating under vacuum conditions. They can be reactivated to restore their initial pumping efficiency upon saturation, when the hydrogen content approaches the embrittlement limit, or if the NEG material is exposed to atmospheric pressure.

Cryogenic pumps, or cryopumps, trap certain types of gasses and vapors by condensing them on a cold surface. The effectiveness depends on the vapor pressures of the gasses at the cold surface temperature. Trapping helium and hydrogen requires extremely low temperatures (<10 K). Even at this temperature, helium and hydrogen have low trapping efficiency [132] and are the predominant molecules in ultrahigh vacuum systems. Cryopumps are sometimes used to block particular contaminants, for example, in front of a diffusion pump to trap backstream oil. Most cryopumps can be valved off from the vacuum systems so the cryopumping surfaces can be "regenerated" by warming up and removing the condensed gasses with mechanical pumps. Superconducting RF cavity and magnet surfaces operating at liquid helium temperature effectively serve as non-detachable cryopumps, leading to low vacuum pressure in such cold sections of an accelerator.

8.7.3.3 Operational Practices

Outgassing and vacuum leakage are the main sources of degradation in vacuum systems. While outgassing causes the pressure to rise to the level of the saturated vapor pressure of the involved molecules. Without pumping, leakage causes the pressure to rise linearly with time. The leak may also indicate defects in a vacuum joint that can develop into long-term operational problems.

Leaks may occur in detachable connections, such as flanged joints, in permanent joints, such as welded or brazed seams, and due to material porosity, particularly following mechanical deformation or thermal processing of cast components. Leaks may occur only at elevated temperatures or under load when the defects open up; or only at low temperatures due to superfluid helium flow through the leak passage [133], thus complicating the effectiveness of the leak check. The leak check is essential to ensure the achievable low design pressure of a clean vacuum system, both for the preservation of beam quality and for the protection of the devices. For a complex system, every vacuum component must be leak-checked prior to assembly and installation, particularly for those components that are difficult to access in the field. The leak check may adopt negative pressure ("outside-in") or positive pressure ("inside-out") methods. Negative pressure methods evacuate the vessel and utilize techniques such as acoustic scanning, a mass spectrometry leak detector, and a residual gas analyzer. Positive pressure methods rely on pressurizing the vessel while detecting leaks by "bubbling," foam spray, or "sniffing" with a leak detector using a tracer gas such as helium. The sensitivity of positive methods is usually a few decades worse than that of the negative methods. Certain types of leaks may occur only under negative pressure.

Outgassing can be effectively mitigated by pre-installation or in-situ bake out. Whenever possible, vacuum chambers and components should be baked before installation using a specifically designed and controlled thermal cycle, minimizing thermal distortion and ensuring material compatibility. Baking may also expose leaks that have been "plugged" during production. In situ bake-out allows for vacuum restoration after extended operations or atmospheric exposure. It can be necessary when ultrahigh vacuum conditions are required, when the surface is sensitive to the contamination by particular gas molecules, or when heat activation is required for NEG coating. In-situ bake-out does put substantial requirements and thermal stress onto adjacent devices; therefore, it needs to be implemented from design through assembly and commissioning.

8.8 Charge Stripping

The charge stripper removes one or more electrons from an ion and changes its charge state to facilitate acceleration or manipulation. Traditional charge strippers in accelerators are designed using thin films of solid materials; in many cases, carbon. Key design considerations include the charge distribution of the stripped beam, stripping efficiency, foil lifetime, impact on the quality of the stripped beam (energy loss, emittance growth), and the effects of beam scattering and beam loss.

8.8.1 H^- Charge Stripping

Injection into synchrotrons or accumulator rings for pulsed high-power operations requires high-efficiency charge stripping of the H^- beam. The density of the foil, typically $200 - 400\,\mu g/cm^2$, is a compromise between the stress and heating of the foil and the stripping efficiency (Fig. 8.39).

Injection of intense H^- beams requires sophisticated charge stripping designs. Figure 8.40 shows the SNS charge stripping at the injection of the ring. For easy placement, the foil, minimized in size to reduce the number of traversals of the beam, was hung from a frame supported by thin fiber and mounted on quick-exchange devices. Foils made of diamond material show excellent heat resistance and an improved lifetime. The injection occurred in a straight region of zero dispersion, thereby allowing independent control of the phase-space painting both transversely and longitudinally. The injection process was more tolerant of deviations in linac energy. A long, uninterrupted straight section was preferred to contain the entire injection chicane so that the effects of lattice tuning could be minimized. Special magnets with tapered poles were used to steer the stripped electrons to a precisely located heat-resistant catcher. The magnetic field of the magnets was carefully designed to minimize the uncontrolled beam loss associated with H^0, H^-, and scattered H^+ during stripping.

The stripping of the two electrons from an H^- circulating beam in a cyclotron has interesting consequences. The power of the stripped electrons is equal to the power of the extracted protons, scaled by a factor of 2 times the mass ratio of electrons to protons. In the TRIUMF case with the proton beam at 500 MeV energy and $100\,\mu A$ current, 54 W of power was dissipated into the foil if the stripper is in a magnetic field, as it unavoidably is in the TRIUMF case, since the electron orbits have radii in the millimeter range, which are smaller than the foil dimensions. The foil accumulates damage from high temperatures, compromising its lifetime.

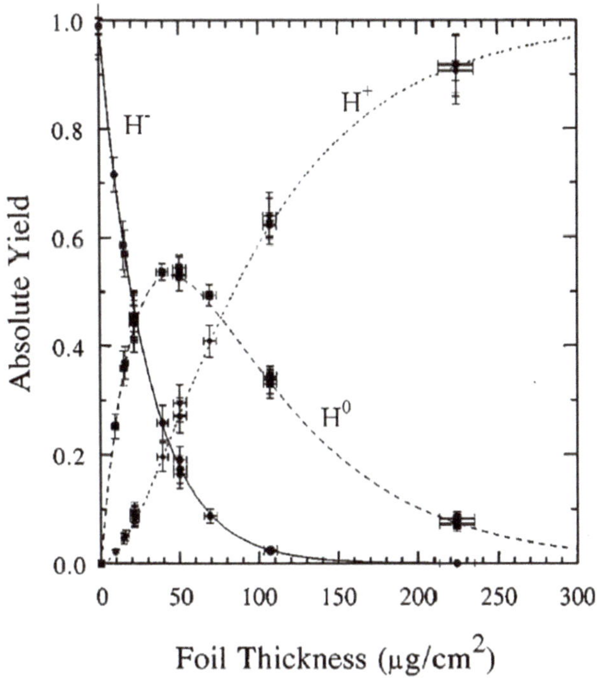

Fig. 8.39 Measured yield during H^- stripping by carbon foil at 800 MeV [134]

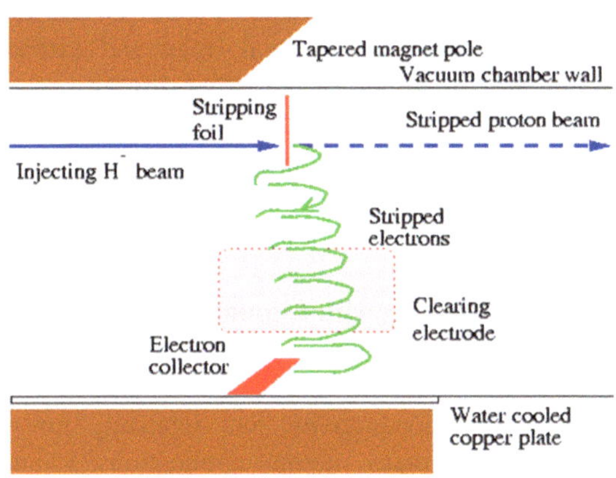

Fig. 8.40 Schematic layout of the SNS H^- charge stripping at the injection region of the accumulator ring [135]

For a 5 mg/cm^2 pyrolytic graphite foil, all the electron energy is deposited (in ~16 traversals of the foil in the TRIUMF case) near the beam spot on the target. However, if the foil is thinner, the scattering of the electrons allows them to escape the foil by performing a random walk along the magnetic field lines that intersect the foil. Thus, if the foil is too thick, all the electron energy is deposited, and the lifetime is decreased. Conversely, if it is too thin, too much neutral beam survives, resulting in an optimum thickness. Another technique is to direct electrons away from the foil by tilting it (see [136]).

8.8.2 *Heavy Ion Charge Stripping*

Charge stripping is required for heavy-ion acceleration at one or more stages of beam energy to raise the charge state for efficient acceleration. The choice of the number of striping stages, stripping energy, and stripper density thickness is an optimization process that maximizes the output beam energy and intensity. FRIB used single-stage stripping at a beam energy of 16–20 MeV/u to accelerate various heavy-ion beams to 200–300 MeV/u. The stripping efficiency of the uranium beam was up to 83% when up to five charge states of the beams were simultaneously accelerated. RIKEN used a two-stage stripping scheme to accelerate the beams to 345 MeV/u energy.

8.8.2.1 Stripping with Solid Media

Compared to H^- stripping, the energy loss per ion per unit length is much higher for heavy ions during the stripping process (Eq. 8.2 and Fig. 8.41). Both radiation damage and thermal effects must be considered. At the energies of the FRIB stripper, the uranium beam would deposit $\sim$30 MW/cm^3 power densities in the carbon foil.

Rotation alleviates thermal stress and extends the life of the stripping foil. Figure 8.42 shows the FRIB charge stripping system, with both rotating carbon foil and liquid-lithium film devices located next to each other. The rotating foil stripper was only used for lighter ions with relatively low beam power (less than 1 kW/u on the foil).

8.8.2.2 Stripping with Liquid Lithium

Liquid metals have superior heat transport properties, including low vapor pressure, high thermal conductivity, and relatively low viscosity, making them natural candidates for many thermal applications. Liquid lithium offers a viable option for charge stripping of high-power beams because it can produce a charge state as high as that of a solid charge stripper and is self-replenishing, mitigating radiation and thermal impact [137, 138]. However, since liquid lithium reacts forcefully with air or water, precautions must be taken to mitigate this fire hazard (Sect. 8.3.4). Furthermore, corrosion and contamination aspects must be handled carefully, including the avoidance of surrounding material that may easily react with lithium (such as copper) and the containment of lithium vapor with cold traps to prevent contamination of superconducting surfaces in particular.

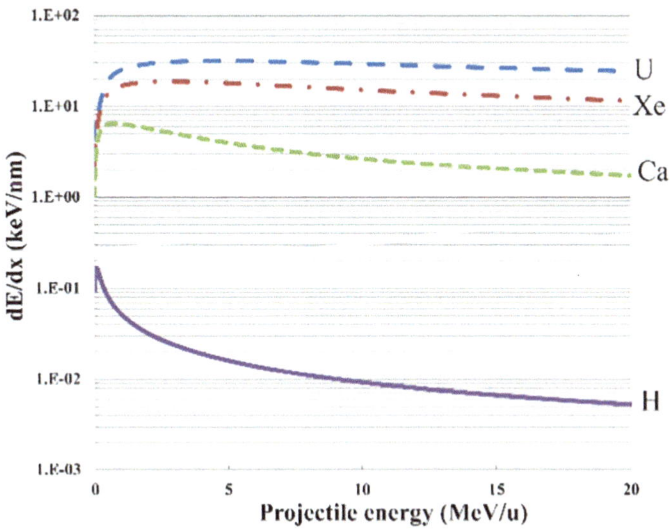

Fig. 8.41 Comparison of the energy loss per unit length in carbon foils for some heavy ions and protons. An uranium ion deposits three orders of magnitude higher power than a proton of the same velocity [137]. (Courtesy J. Nolen and F. Marti)

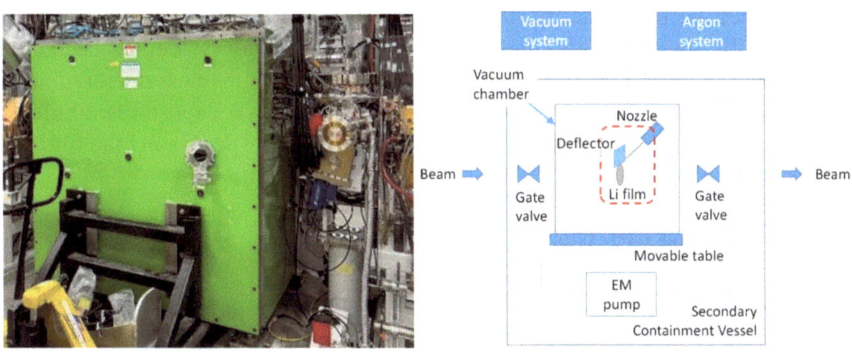

Fig. 8.42 (Left) FRIB charge stripper systems: the secondary vessel containing the primary liquid-lithium chamber (green box) located upstream of the chamber containing the carbon foil rotating at up to 500 rpm (on the post); (Right) FRIB lithium charge stripper schematics with the area marked by the red dashed line shown in Fig. 8.43. (Courtesy T. Kanemura and F. Marti, MSU)

The liquid lithium film in the FRIB charge stripper was produced by a lithium jet moving at ~50 m/s speed and striking a deflector (Fig. 8.43). Lithium was heated by electric heaters to a liquid state and circulated by an electromagnetic pump (Fig. 8.42) [139]. The lithium loop was contained in the secondary vessel filled with argon gas. The uniformity of the film thickness could be determined with either a diagnostic electron beam or directly with the stripped ion beam. Feedback controls operating on both 0.1 s timescale by energy compensation

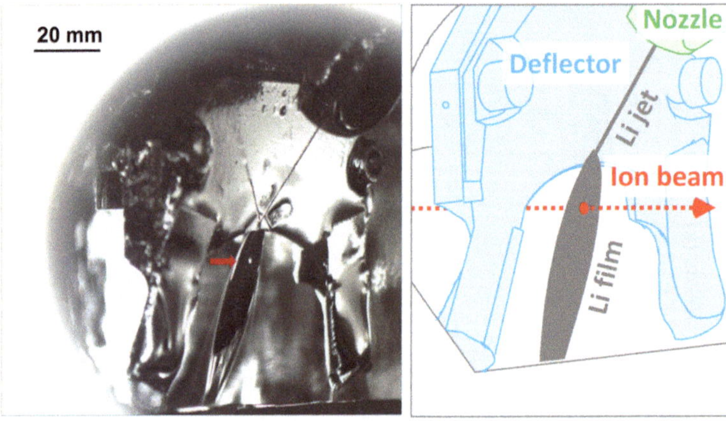

Fig. 8.43 Image of a uranium beam on the liquid-lithium film (left) established inside the charge stripper primary chamber and the corresponding schematics. (Courtesy T. Kanemura, MSU). The corresponding beam power on the production target is 30 kW. The film typically converts an uranium beam of average charge state from 35^+ to 75^+. The stripping efficiency of the 5 accelerated charge states (73^+, 74^+, 75^+,76^+,77^+) of the uranium beam is approximately 83%

and multi-second timescale by shifting of the impact point were implemented to counteract fluctuations in film thickness [140]. The stripping system has been used in operations on various ion beams, including uranium. An off-line test with a proton beam produced by the LEDA source demonstrated that power depositions similar to full-power FRIB uranium beams could be achieved without destroying the film.

8.8.2.3 Stripping with Gas

An alternative to survive high-power deposition during stripping is the use of gas media. The main drawback is associated with the lower density of the gas medium in comparison with that of solid or liquid.

RIKEN used helium gas in conjunction with differential pumping for the first stage of charge stripping, as shown in Fig. 8.44. Plasma windows were being tested to help establish a high gas density [142].

8.8.3 Future Developments

Laser tripping has been studied for the charge-exchange process of H^- beams [143–145]. Research and development focused on methods to reduce the demands on laser power [146, 147].

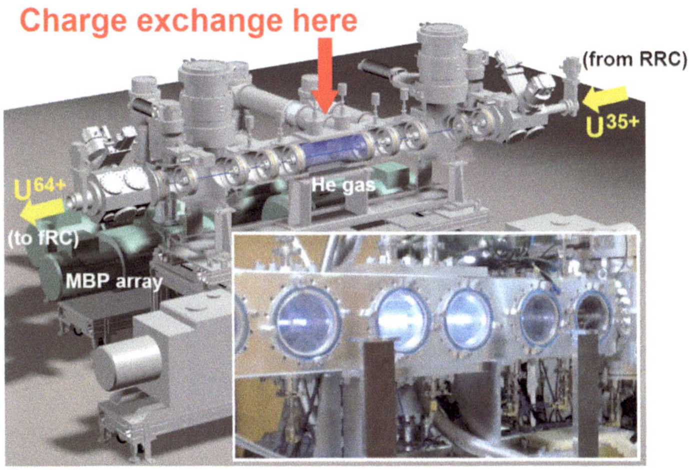

Fig. 8.44 Cross sectional view of the He gas stripper and picture of glowing 1-pμA uranium beams [141]. (Courtesy H. Imao, RIKEN)

Research and development on charge stripping with a liquid film focuses on several fronts. With the operating charge stripper using a liquid lithium film, the focus has been on gaining the experience required for maintenance to sustain extended operations for a user facility. In the meantime, feedback systems have been developed to successfully mitigate energy jitter caused by variations in film thickness [140]. Efforts are being made to increase the film thickness for increased stripping efficiency. Developments on materials other than lithium are also progressing. Films made of ethanol and mercury were formed stably in a vacuum [148].

RIKEN is pursuing a charge stripper ring (CSR) design [149] to increase the intensity of heavy-ion beams, particularly uranium ions, at RIBF. Unlike conventional single-pass strippers, the CSR allows multiple ions to circulate, undergoing repeated charge exchanges with an internal target for much higher efficiency, thus enabling CW operation and significant upgrades to beam intensity and power.

8.9 Targetry

Targetry, in general, refers to systems that purposefully intercept the accelerated beam. This section specifically discusses target and supporting systems for the generation of secondary beams.

The targetry configuration is designed to meet the requirements for secondary beam production. Targets for neutron production are designed to absorb most of the beam power of the primary protons; they also serve as beam dumps (Fig. 8.45) [150, 151]. The targets for the production of rare isotopes are designed to absorb

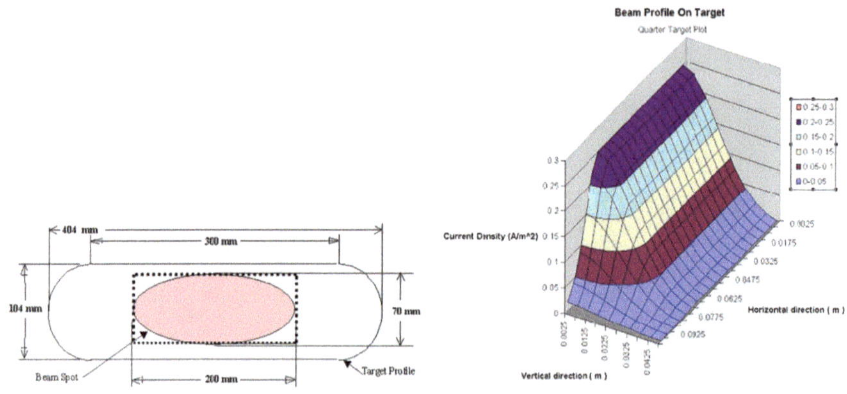

Fig. 8.45 (Left) Cross-section of the design proton beam spot on the SNS mercury target for neutron production. (Right) Design profile and the peak time-average current density of a 2 MW proton beam on the target face (quarter target plot)

~25% of the primary beam's power; separate beam dumps collect the unwanted beam [152]. High energy physics targets typically absorb <5% of the primary beam power; extended length beam dumps are needed, corresponding to high beam energy [153]. These also represent the three frontiers of targetry challenges [154].

The targeting system has been a performance-limiting bottleneck for many high-power hadron facilities. The high beam power enhances the corrosion and radiation damage that degrades the mechanical and thermal properties of the surrounding material. For pulsed beams, the targetry system must endure cycles of thermal and pressure shocks. While traditional targets use stationary solid material, recent high-power facilities use free-surface liquid metal (e.g. liquid lithium and mercury jets), confined liquid metal (e.g. lead bismuth eutectic and mercury), and rotating solid targets.

8.9.1 *Targetry for Neutron Production*

Neutron source targets are usually designed to intercept the entire incident ion beam for neutron production. To alleviate the thermal stress on the target material, the targetry optics can be designed to distribute the primary beam evenly over a sizable transverse dimension, as illustrated as an example in Fig. 8.45. The target material is selected to maximize the neutron yield.

Figure 8.46 shows evolutionary examples of the targets of secondary beam facilities [155]. Stationary targets are typically used for CW and long-pulse beams such as at SINQ, PSI [156] and at LASREF, LANL [157], and for pulsed beams with beam power below 0.5 MW [158] such as at LANSCE [159, 160], ISIS TS1 [161, 162] and TS2 [163], and CSNS [164].

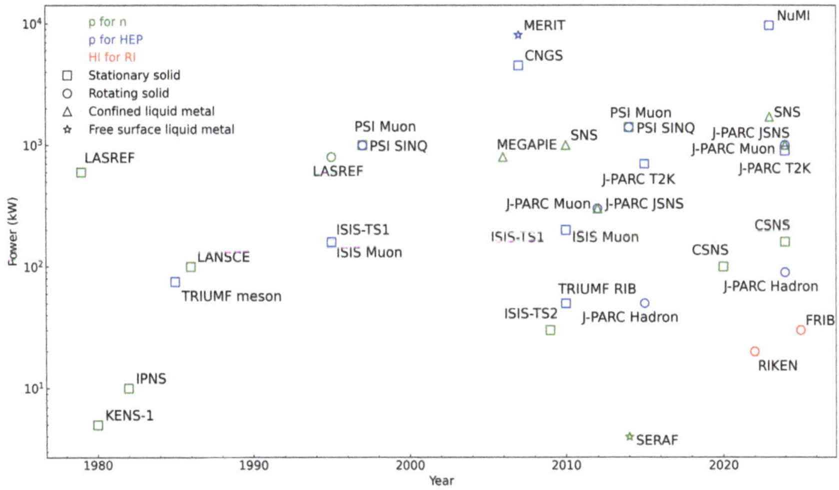

Fig. 8.46 Examples of secondary-beam production targets developed in the beam power frontier during the recent years. In most cases, the indicated date corresponded to the first year when the near-record performance was achieved at the facility

Non-stationary targets are typically used for pulsed neutron production at the MW level. As indicated in Table 1.1, both SNS [8] and J-PARC MLF [165] used confined liquid mercury for neutron production. Figures 8.47 and 8.48 show the SNS targetry system. Table 8.7 shows the beam interface requirements at the target. The problems of target surface pitting caused by the pulsed shock wave were mitigated in large part by treating the vessel surface along with mercury flow and bubble controls (Sect. 10.1.3.7). The pioneering MEGAPIE experiment [166] and the MYRRHA ADS target used liquid Pb-Bi eutectic (LBE) mixture fluid. For lower energy neutron production, both SARAF [167] and IFMIF [168] used free-surface liquid lithium. The MERIT high-power target experiment used a free-surface mercury jet [169, 170]. LANSCE used a rotating solid target. Lately, both the ESS [171] and the SNS second target station [172] have adopted the rotating target design for simplicity.

8.9.2 *Targetry for Nuclear and High Energy Physics Experiments*

The targetry system for nuclear and high-energy experiments includes production targets and beam dumps. Due to specific optical requirements on the secondary beams, the primary beams often need to meet stringent conditions on the target. For example, primary beams for rare isotope fragment separation and neutrino experiments should be focused down to the beam radius ~1 mm. The power density becomes significant.

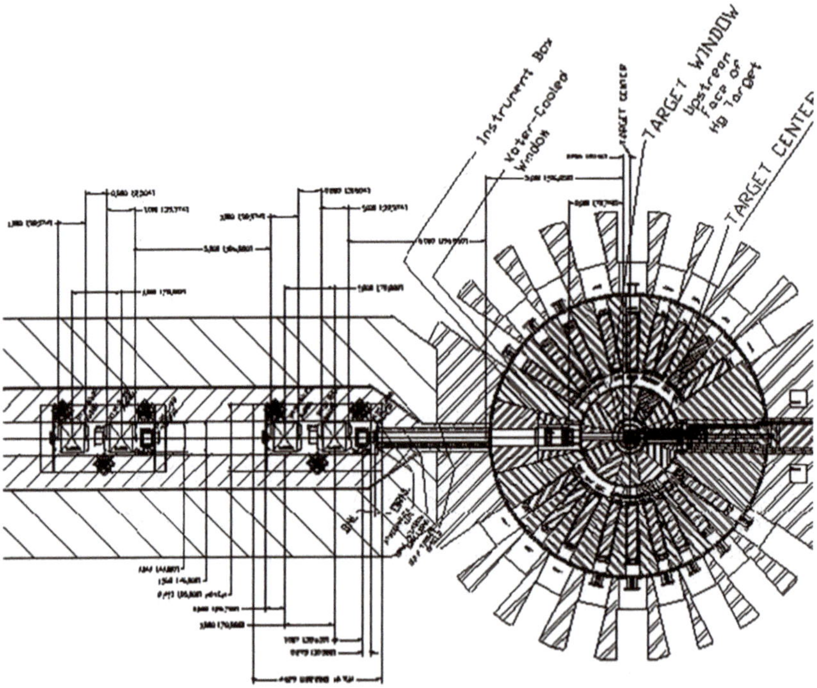

Fig. 8.47 Top-down view of the SNS ring-to-target transport (RTBT) beamline interfacing with the mercury target station. (Courtesy BNL and ORNL, USA)

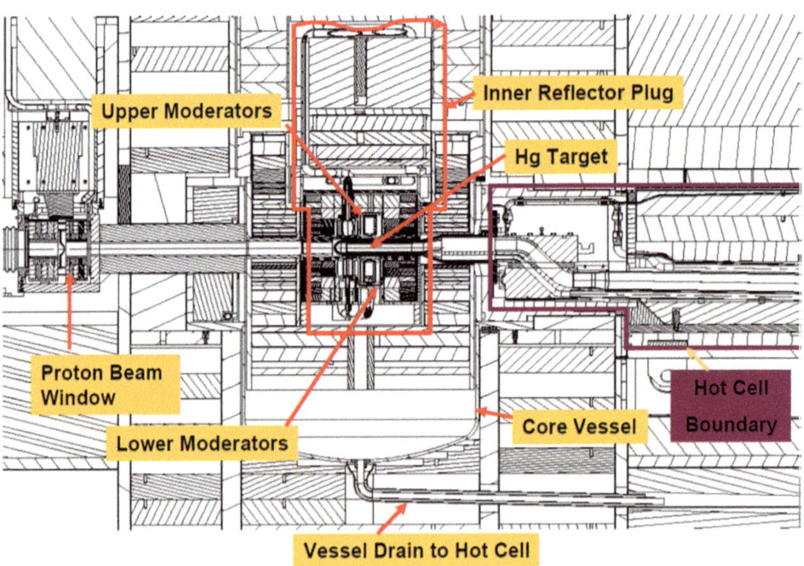

Fig. 8.48 SNS target station schematics including the spallation neutron target, the moderators and reflectors. (Courtesy ORNL, USA)

Table 8.7 Beam-target interface requirements for a 2 MW proton beam striking the SNS mercury target at 1 GeV energy with 60 Hz repetition rate. Possible failure scenarios of the targetry system include both mercury vessel leak and beam window rapture[a]

	Normal condition	Off-normal condition	Duration of fault
Target cross-section width	404 mm		
height	104 mm		
Beam spot full width	200 mm		
height	70 mm		
Beam power within nominal beam spot[b]	>90%	>50% outside of spot	≤2 pulses
Time-average beam current density over spot	≤0.143 A/m^2		
Peak time-average beam current density[c,d]	≤0.25 A/m^2	≤0.28 A/m^2	10 s
Peak single pulse density[e]	≤2.6×10^{16} p/m^2	≤3.2×10^{16} p/m^2	≤2 pulses
Tolerance on beam centroid			
horizontal	±2 mm		
vertical	±2 mm		
Unscheduled beam off[f,g]			
>5 s	≤50 per day		
>5 min	≤10 per day		

[a]A rupture-type failure of the water-cooled Inconel window will release water into the RTBT high vacuum for a single-wall failure, or helium for a double-wall or seal failure
[b]The power outside the spot in an off-normal condition represents beam error for two pulses. Target system components are to be designed to survive this failure
[c]The peak time-average current density is the most concentrated beam that the target must survive without causing a failure within the planned target lifetime
[d]The off-normal peak time-average current density allows a 10% increase over 10 seconds and assumes a failure that concentrates the beam for multiple pulses prior to detection. The target is designed to handle this concentrated heat load for the specified time within the beam spot
[e]The off-normal peak single pulse density allows two consecutive pulses with 25% higher than normal densities
[f]Missing beam that allows the target structure to cool down from normal operating temperature (>5 seconds) will produce fatigue cycles that need to be included in the design for expected fatigue life
[g]Missing beam that allows the entire mercury loop to cool down from normal operating temperature (>5 minutes) will produce fatigue cycles that need to be included in the design for expected fatigue life

8.9.2.1 Production Target

LANSCE [159], PSI [173] and J-PARC [174] used rotating solid targets with their primary proton beams for the production of mesons at the MW level. TRIUMF used a static target with the proton beam of the cyclotron for meson production at approximately 50 kW [175]. The Beam Dump Facility (BDF) is a proposed

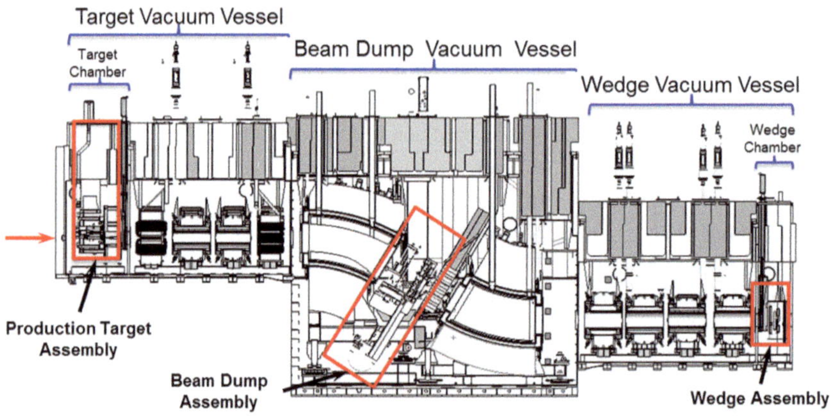

Fig. 8.49 FRIB rare isotope targetry system consisting of the production target, the beam dump, the wedge for fragment separation. (Courtesy MSU, USA). The red boxes indicate the remotely handled assembles of the production target, beam dump, and wedge systems

general-purpose facility at CERN designed for both fixed-target and beam dump experiments, utilizing the proton beam from the SPS with the primary goal of the Search for Hidden Particles (SHiP) experiment underground in the SPS North Area [176].

TRIUMF used a static target with the ISAC linac proton beam for rare isotope production by ISOL methods (Sect. 8.4.4) at approximately 75 kW beam power [177]. Compared with using the ISOL method with proton beams, the use of the IF method (Sect. 8.4.4) with heavy-ion beams for rare isotope production at lower energy per nucleon requires a much more challenging targetry system at a given beam power (Eqs. 8.3 and 8.4). RIKEN [178] and FRIB [179] both used rotating targets with their heavy-ion primary beams for the production of rare isotopes at 20 kW beam power level.

Figure 8.49 shows the FRIB target system for the production and separation of rare isotopes. For in-flight rare isotope production, FRIB was designed to focus 400 kW of a heavy-ion beam on an area of 1 mm rms diameter. By intercepting $\sim$25% of the total designed 400 kW of the primary beam for the production of rare isotopes, the power density deposited on the thin target would be $\sim$60 MW/cm^3. A radiation-cooled multi-slice graphite target with a diameter of 30 cm was designed to rotate at 5000 rpm, dissipating thermal stress.

Figure 8.50 shows the present single-slice graphite target, which consists of a graphite disk, a copper heat exchanger, a ceramic hub, and an invar spindle with bearings. A high-emissivity coating was applied to the interior of the heat exchanger to enhance the radiative cooling efficiency of the disk. High-temperature bearings and lubricants were also used.

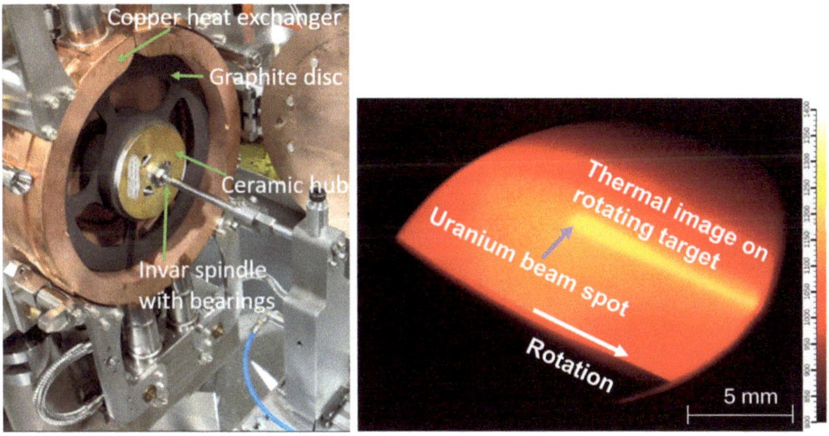

Fig. 8.50 (Left) Single-slice graphite target inside the water-cooled copper enclosure [179]. (Right) Thermal image of a 30 kW, 177 MeV/u uranium beam on the graphite disc target of 1.2 mm thickness rotating at 500 revolutions per minute [180, 181]. (Courtesy MSU, USA)

8.9.2.2 Beam Dump

Beam dumps for nuclear and high-energy physics experiments are demanding, as they must absorb a large fraction of beam power. Materials used for such beam dumps include blocks of copper, aluminum, carbon, beryllium, and tungsten, or thin shell vessels containing liquid to absorb beam power. When block dumps are used, the primary beam incident angle is often chosen to be small in order to spread the heating over a larger region of the block. For example, a beam dump for high-energy beams often uses water-cooled blocks with a conical hole several meters long. The internal beam dump for CERN's SPS is described in [182]. When liquid dumps are used, the container shell needs to be thin enough so that the Bragg peak of the primary beam falls within the liquid volume.

Figure 8.51 shows two FRIB beam dump designs: (1) a block dump for the current intermediate beam power, and (2) a rotating water filled dump for the ultimate high beam power ($\sim$75% of 400 kW). The static block dump was used with the incident beam at a 6° angle to reduce the deposited power density by up to approximately $\sim$20 kW at a primary beam energy of $\sim$200 MeV/u. An aluminum layer prevented the primary beam from reaching the cooling water channels of the dump to minimize water activation. The heat transfer was enhanced by using several mini-channels for more effective water cooling.

The ultimate FRIB beam dump was designed to be a rotating, thin-walled, water-filled drum, with circulating water serving both as a coolant for the wall and to stop the primary beam within the drum. The drum had a diameter of 700 mm and a height of 80 mm to distribute the heat and radiation damage. The shell was designed to be made of Ti-6Al-4V titanium alloy with a wall thickness as thin as 0.5 mm, so most of the power of the $\sim$200 MeV/u, $\sim$400 kW beam would be deposited in the water

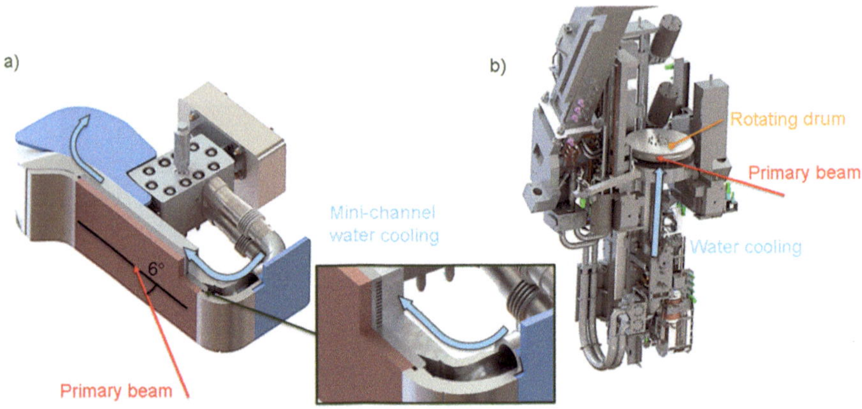

Fig. 8.51 FRIB beam dumps for different stages of beam power: (**a**) intermediate mini-channel beam dump [183, 184] and (**b**) rotating beam dump [185]. (Courtesy MSU, USA)

(Sect. 8.2). Fabricating this large-size shell would be challenging with conventional manufacturing processes due to its thin wall and the specific curved shape required to support the high water pressure inside the drum. The increase in primary beam energy from ~200 to ~400 MeV/u would significantly extend the stopping range, allowing for a thicker wall for more reliable operation.

8.9.3 Radiation Resistance, Remote Handling, and NCU

Ancillary systems are equally important in high-power targetry to facilitate operations and maintenance. In particular, radiation resistance, collimation, and shielding (Sect. 8.10), remote handling (Sect. 8.9.3), and non-conventional utilities (Sect. 8.13.2.4) require integrated design and implementation.

Radiation-induced damage to the electronics, embrittlement of the device material, and activation of the surroundings require careful consideration in the design. For example, magnets were wound with mineral-insulated cables as an integral part of the shielding in front of the SNS target. Quick disconnect vacuum flanges and remote water fittings allow for easy access. Computer simulations with programs such as MCNPX and MARS are often used to determine the transport of primary and secondary particles through the surrounding material and to estimate the level of radio-activation and the necessary protections. Mechanical analysis using computer simulation programs such as ANSYS [186] and prototype performance tests is essential, including resistance to thermal, radiation, and mechanical stresses, vacuum outgassing, vacuum leakage, and long-term fatigue.

Remote handling is mandatory for high-power targetry system maintenance. Remote handling design features often combine hands-on and fully remote oper-

Fig. 8.52 Two sets of remote handling manipulators viewed from outside of the FRIB target hall. (Courtesy M. Larmann, MSU)

ations to maximize efficiency and minimize operational costs. For example, FRIB operation required that the target hall be accessible to trained personnel for hands-on work 4 hours after the beam is turned off to prepare for subsequent fully remote operations to handle highly activated components. Remote handling (Fig. 8.52) has been performed routinely for the change of the target disk and the wedge (usually during every week-long experiment) during user operations and the annual maintenance of the beam dump. Tooling and procedures were also developed for specific repair handling, such as changing the target bearing cartridge and applying the high-emissivity coating to the target heat exchanger. Duplicate target and wedge assembly modules were developed to allow quick reconfiguration (<24 hours) during user experiments (Fig. 8.49).

Non-conventional utility (NCU) water provides cooling for the targetry devices and handles transferred radioactivity accordingly. As shown in Fig. 8.53, the FRIB NCU system includes two closed cooling loops, centrifugal pumps, heat exchangers, gas-liquid separators, drain tanks, and instrumentation; auxiliary systems include a hydrogen recombiner, a clean-up system, and a low-level liquid waste system. The hydrogen recombiner was designed to catalytically recombine H_2 produced by water radiolysis with O_2 and to capture gaseous radionuclides generated during operations. The clean-up system must remove particulate and ionic impurities from the cooling water, with remote handling capability and easier maintenance.

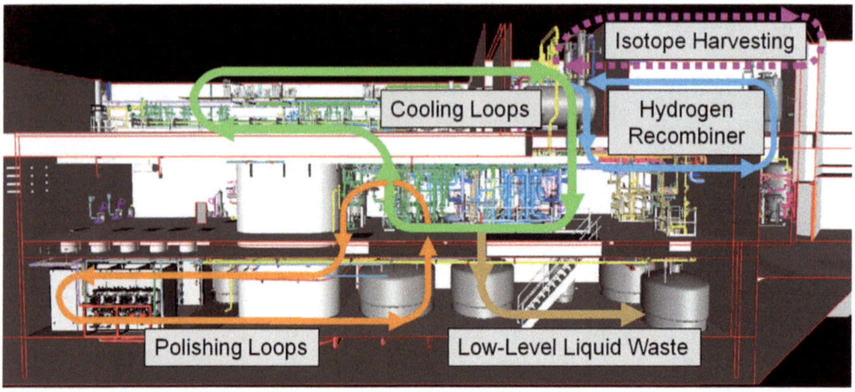

Fig. 8.53 FRIB NCU system with cooling loops (target loop and beam dump loop), polishing loops, hydrogen recombiner and low-level liquid waste. Isotope harvesting capability was recently added. (Courtesy F. Casagrande, MSU)

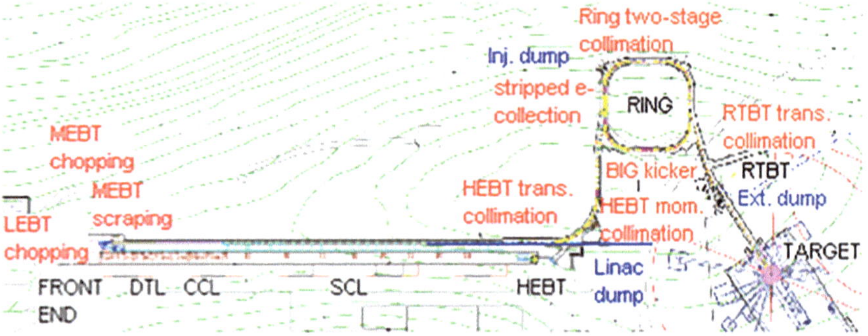

Fig. 8.54 Beam chopping and cleaning from the front end to the neutron target of the SNS accelerator complex

8.10 Other Beam-Interception Systems

In addition to those discussed in Sects. 8.8 and 8.9, other beam interception devices commonly used in high-power accelerators include choppers, beam-in-gap kickers, charge selectors, and collimators. They play an important role in high-power accelerators, reducing uncontrolled beam loss for hands-on maintenance. Figure 8.54 shows the beam cleaning system consisting of beam chopping, scraping, collecting, and cleaning devices from the front end to the neutron target of the SNS accelerator complex. Table 8.8 lists the estimated controlled beam loss mitigated by the beam cleaning system.

Table 8.8 Estimated controlled beam loss at the SNS with a 2 MW beam in the linac and the accumulator ring

Mechanism	Location	Estimated fraction
Front end		
Gap chopping	LEBT chopper	0.277
Gap chopping	MEBT chopper	0.043
HEBT		
H^0 from linac	Linac beam dump	10^{-5}
Linac transverse tail	HEBT transverse collimators	10^{-3}
Linac longitudinal tail	HEBT momentum collimator	10^{-3}
Ring		
Beam-in-gap	Beam-in-gap kicker	10^{-4}
Excited H^0 at foil	Ring collimator	1.3×10^{-5}
Partial ionization at foil	Injection beam dump	10^{-2}
Foil miss	Injection beam dump	10^{-2}
Energy straggling at foil	Ring collimator	3×10^{-6}
Ring beam halo	Ring collimator	1.9×10^{-3}
RTBT		
Kicker mis-fire	RTBT collimator	10^{-5}

8.10.1 Chopping and Beam-in-Gap Cleaning

Beam chopping refers to the periodic interruption of a beam, resulting in a desired non-continuous longitudinal bunch structure. Chopping reduces the average beam current and the beam "duty factor" without affecting the peak current.

Chopping and beam-in-gap cleaning both refer to manipulating the longitudinal dimension of the beam phase space. While RF devices such as RFQ and RF cavities create and maintain a bunched structure in the beam, choppers ensure that the beam gap is clear. Although chopping is preferably performed at low energy (typically with the relativistic factor of the beam $\beta < 0.1$) to minimize radio-activation, enforcement is often needed at multiple stages of beam energy.

Figure 2.26 shows the time structure of the SNS proton beam. The macropulse structure of 60 Hz was formed by the ion source, which was fed by a pulsed radio frequency power supply with a duty cycle of 6%. The 1.059 MHz minipulse structure was formed by excitation through the chopping of the LEBT chopper at a proton energy of 0.065 MeV [187] and the MEBT chopper [188] at a proton energy of 2.5 MeV. The micropulse structure of 402.5 MHz was formed by the RFQ bunching while the beam was accelerated simultaneously. To ensure a clean single-turn fast extraction from the accumulator ring, the minipulse beam gap was designed to be cleaned at the proton energy of 1 GeV by the beam-in-gap kicker for 300 ns at a frequency of 1.059 MHz.

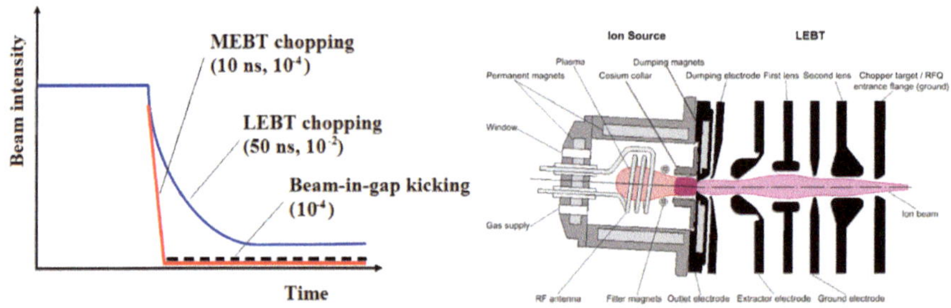

Fig. 8.55 (Left) Illustration of LEBT pre-chopping, MEBT chopping, and beam-in-gap cleaning of the SNS proton beam. (Right) SNS LEBT chopper downstream of the H^- ion source [187]. (Courtesy J. Staples, LBNL)

8.10.1.1 Chopping at Low and Medium Energy Beam Transports

Figure 8.55 (left) illustrates the three-step process by design of the mini-pulse structure at SNS, consisting of LEBT pre-chopping, MEBT chopping, and ring beam-in-gap cleaning. The goal was to create and maintain a beam gap of 300 ns with a distinction rate of 10^{-4} at a beam energy of 1 GeV. The LEBT pre-chopping was performed at a rise time of about 50 ns with a 10^{-2} distinction rate. At an energy of 0.065 MeV, the pre-chopping removes about 97% of the undesired beam. The remaining residual beam was cleaned by exciting the MEBT chopper at a rise time of 10 ns with a 10^{-4} distinction rate. Uncontrolled beam loss caused by the 4 to 5 pulses of partially deflected bunches at an energy of 2.5 MeV during the 10 ns rise time was tolerable. To ensure that after linac acceleration, ring injection, and accumulation, the gap remained free from residual beam at a 10^{-4} distinction rate, the beam-in-gap kicker was planned to be excited during beam accumulation so that uncontrolled beam loss would be minimized upon beam extraction when the extraction kickers were ramped up during the beam gap. During SNS operations, it was determined that beam-in-gap cleaning was not necessary, as the beam gap was sufficiently clean upon chopping in the linac. Furthermore, since LEBT chopping was improved by half the rise/fall time to 25 ns, and MEBT chopping experienced reliability problems, only LEBT chopping was routinely used in operations at the beam-on-target power level 1 MW.

The SNS LEBT chopper incorporated four solid-state switches ±3 kV that drive an einzel lens that was split into quadrants with a 4-phase chopping waveform. The suppressed beam was targeted at a four-segment Faraday cup, which provides online intensity and steering diagnostics. In the absence of space charge neutralization compensation, the electrostatic LEBT of the SNS was designed to be 12 cm long. More elaborate beam chopping was planned to be performed in the medium energy transport (MEBT), where more beamline space was available.

The SNS MEBT chopper was a traveling wave structure with a meander-type plate structure, as shown in Fig. 8.56 (left). The deflection plates, each 35 cm long, were separated by 1.8 cm. A maximum of ±2.4 kV applied to the plates deflected the

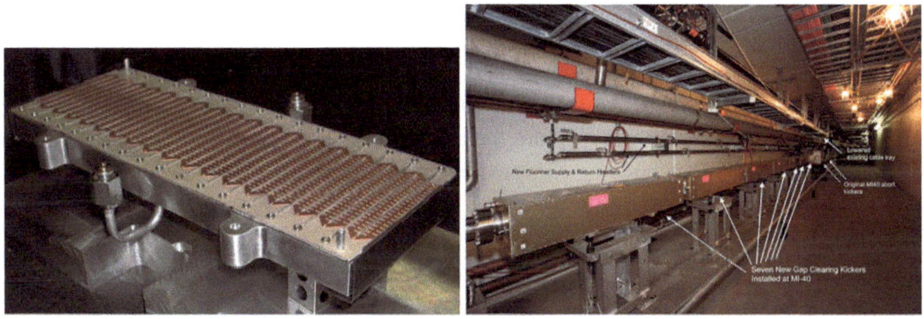

Fig. 8.56 (Left) SNS MEBT chopper [189]. (Courtesy S. Kurennoy, LANL). (Right) Beam-in-gap kicker magnets in the FNAL Main Injector tunnel [190]. (Courtesy I. Kourbanis, FNAL)

beam upward. The rise and fall time was $\sim$10 ns. A quadrupole triplet magnified the displacement and directed the beam onto the chopper target. The target was designed to handle an instantaneous power density of 200 kW/cm^2 and an average power of 400 W.

The FRIB driver linac has been operating in CW mode. The chopper was used during the CW mode to create a "notch" in the beam to engage diagnostic devices and, during commissioning, to create a pulsed beam to reduce the average beam power at the beam dump and avoid damage to the accelerator when the protection systems were not activated. The chopper typically functioned at 100 Hz to create 50 μs notches so that the current baseline could be periodically recovered with AC-coupled beam current transformers (ACCT). The chopper also functioned at 100 Hz to produce pulsed beams with a duty factor of 1% to 5% and a typical rise time of about 50 ns [191].

8.10.1.2 Beam-in-Gap Cleaning in Rings

Beam-in-gap cleaning usually refers to reinforcement cleaning at higher energy in the beam gap created by low-energy beam chopping. The beam-in-gap kicker can be used in both a non-accelerating accumulator ring and an energy-ramping synchrotron to mitigate uncontrolled beam loss due to residual beam in the beam gap.

The FNAL Main Injector (MI) synchrotron used beam-in-gap cleaning kicker magnets to remove residual particles in the injection gap during the high-power multi-batch slip stacking process [190]. As shown in Fig. 8.56 (right), the MI gap clearing kicker system inherited the features of fast kicker magnets for single-turn injection. The kicker magnet, placed outside of the vacuum chamber, made of a ferrite core, was part of a traveling-wave segmented LC network. The ceramic vacuum chamber alleviated the eddy current heating. A thyratron switch was used with the coaxial cable pulse forming line (PFL). The rise time was about 57 ns, with a maximum operating voltage of 55 kV in the thyratron [192].

8.10.2 *Single-Pass Collimation*

Collimation usually refers to manipulating the transverse dimension of the beam phase space [193]. Collimators are beam-limiting devices that narrow a beam of particles in the transverse dimension by reducing its spatial cross section. Collimators placed at a distance apart, allowing a phase advance in transverse motion, offer collimation of the beam in the transverse phase space of both spatial and angular amplitudes. Collimators placed in high-dispersion regions provide momentum cleaning of the longitudinal phase space.

The main sources of beam loss must be identified for collimation to be effective. Such sources include front-end optical aberrations; mismatches across the linac due to changes in the accelerating structure, frequency, and focusing strength; space-charge resonances; physical and momentum aperture limitations; and ring-specific items, including injection loss, resonances due to space-charge and magnetic errors, and collective instabilities.

Collimators range from simple aperture-limiting devices for low energy beams to sophisticated multistage implementations consisting of primary scrapers and secondary collectors for high energy (Sect. 8.10.3). Since collimators usually impose the limiting aperture on the beam, their coupling impedance to the beam, especially in the transverse directions, needs to be studied. Steps in vacuum chambers need to be tapered, and sharp bellows need to be avoided to reduce field emission and beam-coupling complications. The inner surface exposed to the beam often needs to be treated to reduce the secondary emission yield of the electron cloud.

The integrity and maintainability of the system are important design considerations. Due to exposure to high beam power, collimation components need to be resistant to heat, mechanical shock, and prompt radiation. Remote handling is often preferred.

8.10.2.1 Beam Attenuation, Degradation, Scraping, Contaminants Cleaning

Beam attenuators are used, usually immediately downstream of the ion source when the beam energy is low, to reduce peak beam intensity and power for beam commissioning and tuning, to allow the use of interceptive diagnostics, and to commission experimental systems that require the use of the primary beam. The attenuator can be made of screens with escape holes calibrated for the attenuation factor. If the attenuator is used as a safety device, the system must be under credited controls (Sect. 8.1.1).

Beam degraders may be used to reduce the final beam energy when it cannot be conveniently adjusted (e.g., from a cyclotron). Due to radio-activation concerns, they should be avoided for high-power operations.

At beam energies below the generation of neutrons, the device aperture is often used to limit the beam envelope so that the beam emittance at high energy requires minimal collimation. Low energy scraping has also been practiced in many existing machines: reactor grade graphite sleeves were fitted to the ISIS DTL to scrape the beam at 665 keV energy [124], and tungsten sleeves were used in the BNL DTL linac to scrape approximately 5% of the beam at 750 keV energy. The FRIB RFQ scraped beam halo at an energy of 0.5 MeV/u. Many proton linacs built in recent years have used adjustable scrapers in the MEBT downstream of the RFQ to reduce beam loss at higher energies. Adjustable vertical scrapers were installed in the SNS MEBT to clean the front end beam halo at 2.5 MeV.

"Contaminants" are various kinds of unwanted beams that coexist with the chosen accelerated beam. Contaminants with the same charge-to-mass ratio as the chosen beam can be removed through charge stripping because they are likely to be stripped to a different charge-to-mass ratio than the chosen beam and collected accordingly. Contaminants such as the neutral beam from the ion source that drifts downstream can be cleaned by bending the path of the chosen beam by design.

8.10.2.2 Charge Selection

Upon charge stripping, beams of unwanted charge states, including H^0 and H^-, downstream of H^- stripping and "wrong" charge states for heavy-ion stripping, must be removed from the beamline for proper collection, minimizing radio-activation. Figure 2.42 shows the injection of the SNS ring, where the H^- beam was stripped of electrons at 1 GeV kinetic energy [194]. Specially designed magnets, along with a second foil, guided the escape beams of the H^0 and H^- beams to a dedicated injection beam dump designed to collect nominally 2–4% and up to a maximum of 10% of the 1.4 MW design beam power.

The FRIB driver linac was designed to simultaneously accelerate up to 5 charge states after heavy ion beams were stripped at energies of 16–20 MeV/u. The other unwanted charge states of the beam must be intercepted by the charge selector. Due to the short range of the low-energy beam, the power density deposited into the selector surface has been high. For initial operations, the low-power charge selector consisted of water-cooled, adjustable jaws made of copper. For full-power operation, the intercepted beam power would be $\sim$42 kW. As shown in Fig. 8.57, the charge selector was designed with two high-speed rotating graphite disks to dissipate heat. The shielded assembly must allow remote handling for maintenance [195].

8.10.3 Two-Stage Collimation

At energies much higher than the neutron-generating Coulomb barrier, single-stage scraping is usually ineffective. In the Proton Storage Ring (PSR) at Los Alamos National Laboratory, which accumulated 3.5×10^{13} protons per pulse at 20 Hz repetition rate at 800 MeV energy, single-stage beam scraping was attempted but abandoned due to an excessive amount of out-scattering. Pre-injection beam-halo

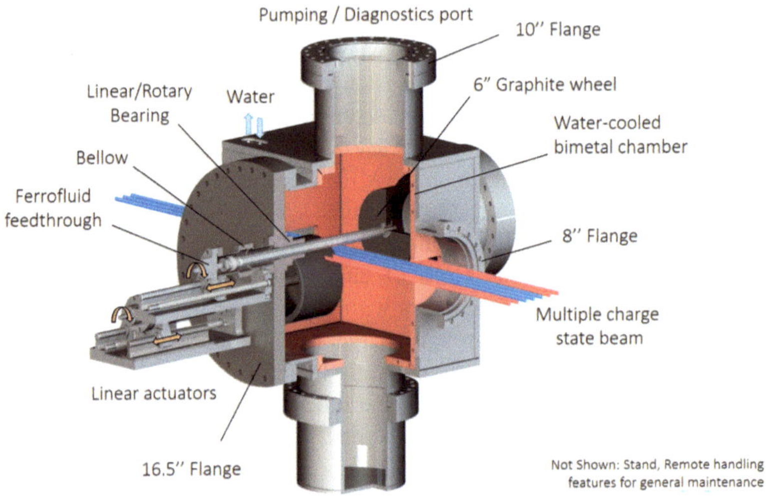

Fig. 8.57 FRIB rotating-wheel charge selector conceptual design (1″ =25.4 mm) [195]. Charge states are incident from the right with horizontal separation; the drum positions can be set to allowed transmission of 5 charge states (blue), while the rest are stopped (red) [195]. (Courtesy A.S. Plastun, MSU)

collimation was also attempted but abandoned since beam loss was dominated by foil scattering of the stored beam [196].

Two-stage collimation using primary scrapers and secondary collectors is used in high-energy beam collimation to improve efficiency [197]. Primary scrapers are usually thin, adjustable blades, the material and thickness of which are optimized for scattering, heating, and other engineering properties. The primary scraper of the SNS ring consisted of four tantalum blades, each 5 mm thick, as shown in Fig. 8.58 [200]. They were spaced in 45° angles, adjustable to the varying needs of the collimation aperture. The scraper assembly was shielded to contain radioactivity.

Although channeling [201] crystals have been considered possible candidates for primary scrapers, their channeling acceptance and resistance to heat and radiation must be tested for high-power beams.

Beam collectors, some of which function independently and some as secondary units downstream of the scrapers, are designed to stop and absorb the primary beam halo or secondary particles striking them. Collectors designed for high-power machines are typically long, complex elements. Unlike collimators meant for high-energy accelerators, which often consist of adjustable solid-metal jaws [202] due to stringent engineering requirements for stress and heat tolerance, thermal contraction, and shielding, these units are usually not adjustable in cross section [198] for operational reliability. Their lengths are designed to stop the primary beam-halo. Anticipating a relatively high beam power (typically 10 kW) and radiation level, they are cooled by closed-loop deionized water. Figure 8.59 shows the collector and shielding design for the SNS accumulator ring. The collector

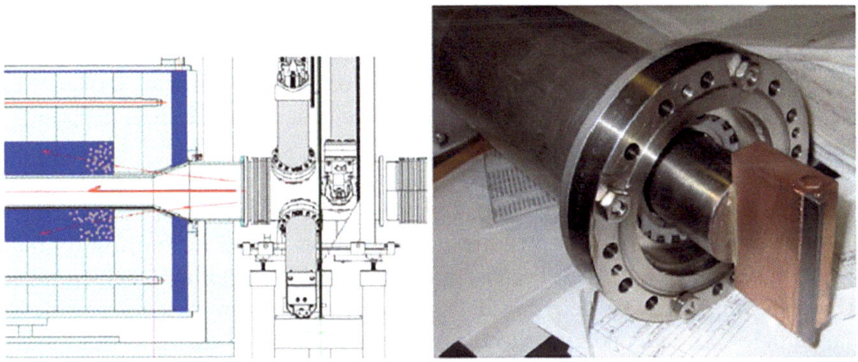

Fig. 8.58 (Left) Schematics of SNS ring primary scraper and shielding [198, 199]. (Courtesy H. Ludewig and N. Simos, BNL). (Right) The tantalum blade is mounted on a copper plate that is water cooled

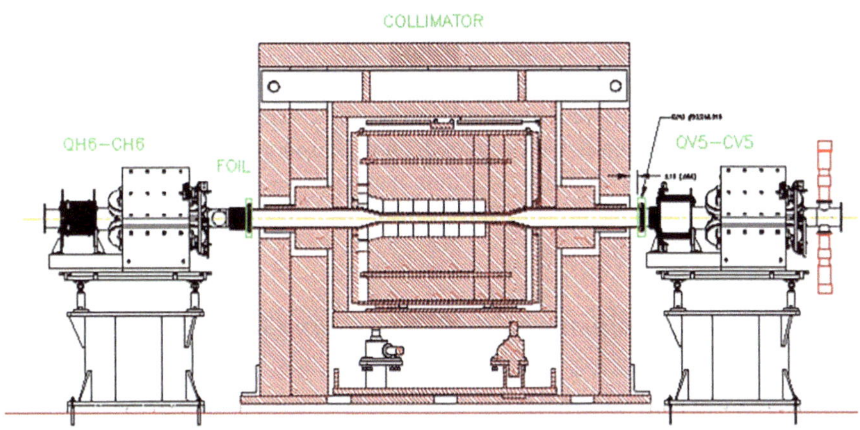

Fig. 8.59 Schematics of SNS collector and shielding. The vacuum chamber is made of double-layered stainless steel filled with helium gas between the layers to detect leaks [198, 199]. (Courtesy H. Ludewig, BNL). Due to stringent engineering requirements for stress and heat tolerance, thermal contraction, and shielding, for operational reliability these units are not adjustable in collimation cross section

consisted of layers of stainless steel blocks, a stainless steel particle bed submerged in cooling water, borated aluminum on the outside wall of the collimator vessel, and steel shielding to contain radiation. The vacuum chamber was made of double-layered Inconel-718, with helium gas in the 1 mm gap to detect leaks and copper winding for heat transfer between the two walls.

The experience with beam collimation and halo cleaning is growing among established facilities. At the ISIS neutron facility at RAL, the proton beam was accelerated from 70 to 800 MeV energy at a repetition rate of 50 Hz to reach a beam power of 160 kW. Collimation was provided in the transverse and longitudinal planes, with limited two-stage systems where possible [203]. The copper and graphite jaws used were adjustable and water-cooled. Controlling the dominant trapping loss (8%) below 100 MeV was crucial for machine operation. About 90% of the 2 kW loss was localized within 180° in the betatron phase from the primary jaw.

With J-PARC RCS, the transverse primary scraper is made of tungsten plates with a thickness of 1 mm in four directions [204]. The longitudinal primary scraper is made up of two horizontal plates of graphite sheets with a thickness of 0.1 mm. These materials were chosen because of their availability and high melting point. These thicknesses were determined based on the necessary scattering angle of the particles. The secondary collector plates were made of copper plates in four directions with a thickness of 200 mm to stop the beam injected at about 400 MeV. The material of the secondary collimator was chosen for its high density and high thermal conductivity. All collimator plates (transverse, longitudinal primaries, and all secondary collimators) could be moved independently. The entire system was air-cooled with cooling fins. The remote clamp system was used to facilitate the handling of the vacuum chamber. Radiation-resistant stepper motors, cables, and connectors were used to extend the radiation life of the system.

The SNS collimators were mounted on adjustable stands contained within steel shielding blocks. The units were cooled by a closed-loop water system to contain radio-activation. Quick disconnect vacuum flanges and water fixtures were used to facilitate maintenance handling in the high radio-activation region.

8.10.4 Crystal Collimation and Background Reduction

An innovative system using bent crystals is in operation at the CERN LHC during its runs as a heavy-ion collider, particularly for lead-ion beams. Bent crystals operating in channeling mode serve as primary collimators in the multistage hierarchy of the betatron collimation system. Compared to conventional primary collimators made of amorphous carbon-based composite materials, the crystal-based system improves collimation cleaning performance by more than a factor of five [205].

High-efficiency crystal channeling has also been demonstrated for high-energy proton beams [206]. However, bent crystals cannot be used during LHC operational modes with high-intensity proton beams. The present design must be updated to ensure electromagnetic compatibility with beams carrying up to 700 MJ, and new halo absorbers need to be developed to handle high-power halo losses. On the other hand, crystal collimation of proton beams has been successfully used operationally for background reduction in dedicated forward-physics runs at the LHC [207].

Bent crystals are also successfully deployed in the CERN-SPS to protect the slow-extraction septa from beam impact and to reduce overall beam losses during the extraction process. The crystals used in volume reflection mode and channeling deflect the beam that would impact the electrostatic septum blades in the extraction line, thus also improving the extraction efficiencies [208].

8.11 Beam Diagnostics

Beam diagnostics are essential for commissioning and operating a high-intensity synchrotron, not only to monitor and control the beam but also, equally importantly, as an integral part of a machine protection system (Sect. 8.2). It must accommodate a wide dynamic range of both beam intensity and timing structure, from low-intensity and duty cycle to full-power operations [209]. It must also cover both primary and secondary beams in different segments of the facility, including areas of high radiation damage.

8.11.1 Non-interceptive Diagnostics

Most non-interceptive beam measurements are based on the electro-magnetic interaction of fields induced by the beam with their environment. Beam current transformers, as well as beam position monitors, are based on this principle. The signals induced in the sensors must be amplified and shaped before they are converted to numerical values. These values are also treated numerically to extract meaningful machine parameter measurements. However, beam loss monitors are primarily based on the secondary emission from the lost particles.

Figure 8.60 shows the distribution of the main diagnostic devices in the FRIB driver linac for measurements of beam intensity and differential loss, beam offset, RF tuning (beam energy and phase), beam profile and tunes, transverse beam emittance, longitudinal phase space distribution, and beam losses. With the exception of beam profile and shape measurements, most measurements were performed with diagnostic devices that were non-interceptive to the core of the beam.

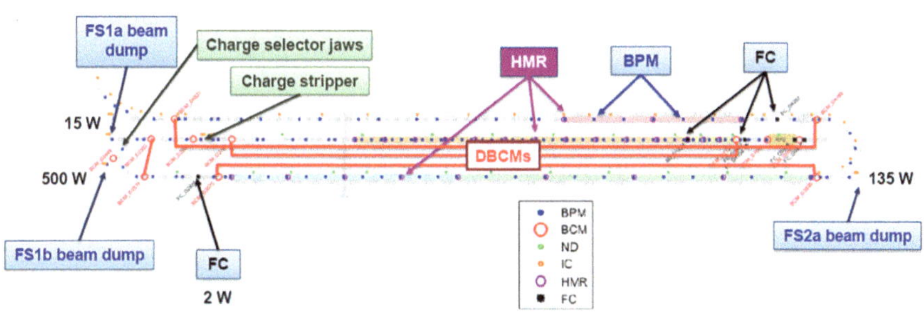

Fig. 8.60 FRIB driver linac beam instrumentation and diagnostics containing measurements of the beam intensity and differential loss using the beam current monitors (BCMs) and beam position monitors (BPMs), beam offset using BPMs, RF tuning (beam energy and phase) using BPMs and narrow-band receivers, beam profile and tune using the beam Loss monitors (BLMs), transverse beam emittance using profile monitors (PMs), longitudinal phase space distribution using BPM waveforms, and beam losses using the neutron detectors (NDs), ionization chambers (ICs), and differential BCMs [191]. (Courtesy S. Lidia, MSU)

8.11.2 Interceptive Diagnostics

Diagnostic instrumentation interceptive to the core of the beam is best conducted at low energy, below the Coulomb barrier, at a reduced duty cycle and average beam intensity, or with an interrupted measurement cycle. As alternatives to conventional profile measurements using wires at reduced intensity, ionization profile monitors [210] and laser profile monitors (for the H^- beam) [211] may be used at full beam intensity.

Figure 8.61 shows the FRIB front end beam instrumentation and diagnostics with the beam energy below 0.5 MeV/u. In addition to the devices described in Sect. 8.11.1, many interceptive devices were instrumented, including regular and fast Faraday cups, selecting slits, collimating apertures, attenuating screens, inserting viewer plates, silicon detectors, Allison emittance scanners, and pepper-pot emittance meters [191]. Beam characterization and machine protection for high-power heavy-ion beams require a full suite of diagnostics and instrumentation (Table 8.1).

Temporary intercepting diagnostic stations are often used during segmented beam commissioning to fully characterize machine and beam properties before downstream segments are installed. Figure 8.62 shows the FRIB D-station implemented during the commissioning of the front end and the segment of four

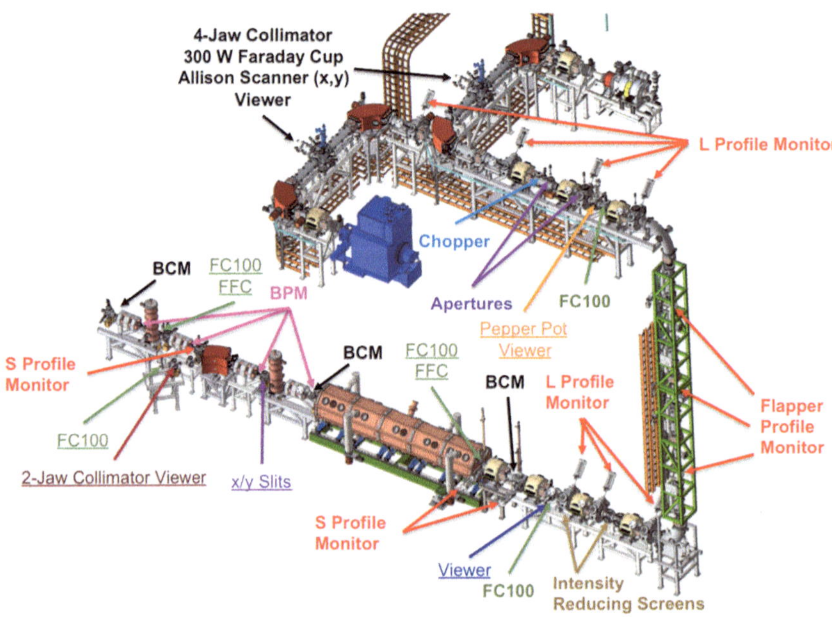

Fig. 8.61 FRIB front end beam instrumentation and diagnostics containing 12 types of 49 devices [191]. (Courtesy S. Lidia, MSU)

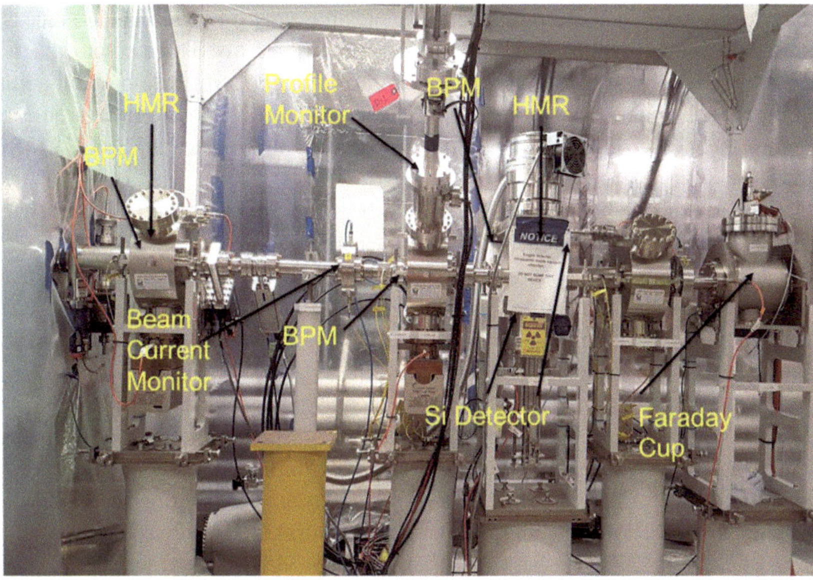

Fig. 8.62 The intercepting diagnostics station used during the commissioning of the FRIB front end and the segment of four $\beta_0 = 0.041$ cryomodules. (Courtesy MSU, USA)

$\beta_0 = 0.041$ cryomodules (Table 9.3). Subsequent segmental commissioning used diagnostic instruments associated with the corresponding beam dump.

8.11.3 Beam Diagnostics in Rings

Multiturn injection is a complex dynamic process during which the beam intensity can increase by three orders of magnitude, and the transverse beam size can grow by several tens of times. Diagnostic instrumentation [212] must possess a wide range of sensitivity and turn-by-turn capability to monitor the intensity, position, transverse and longitudinal profiles (e.g., ionization profile monitor, wire profile monitor, luminescence profile monitor, wall current monitor) and beam loss. Dual-plane beam position monitors are needed in all critical areas to monitor the orbit and local decoupling.

A credible measurement of the turn-by-turn evolution of the beam profile is difficult to obtain in a high-intensity ring. One candidate is the IPM, which measures the distribution of electrons freed by ionizing collisions of the beam with the residual gas in the beam line [210, 213]. The performance of the IPM is anticipated to be complicated by the presence of an electron cloud in the vacuum pipe. The IPM designed for the SNS ring uses a magnetic field to collect ionized electrons instead of ions, to avoid signal broadening due to the space-charge force. Sweeping electrodes are used, and the multichannel plates for electron collection are recessed from the vacuum chamber wall to avoid superfluous signals. Three electromagnets are used in each transverse direction to minimize optical perturbation and maintain operational robustness. Another alternative for turn-by-turn profile measurement is the luminescence profile monitor, which was successfully tested at the CERN SPS [214].

An understanding of the behavior of the beam in the presence of space charge effects and coupling impedance involves measuring the beam halo. With recently designed synchrotrons, measurements can be made by detecting the extent of beam loss scattered by the adjustable scrapers. For accumulator rings, the residual beam in the beam gap needs to be measured on a sub-microsecond time scale, preferably at a level of 10^{-5} of the peak beam density.

With electron cloud effects playing an increasing role in high-intensity rings, the measurement of electron density and energy spectrum becomes important [215]. From experience gained at PSR, the SNS implemented electron detectors at locations where significant beam loss and electron concentration were expected (e.g., near the injection foil, scrapers, and collimators) [194] Electron cloud effects were a concern at J-PARC, particularly during MR slow extraction [216]. Fourier transform analysis of high-harmonic betatron-sideband signals from a wideband beam position monitor was another technique that may be applicable for diagnosing electron-proton interactions. All time and frequency domain diagnostics would become considerably more complicated in rapid-cycling synchrotrons due to fast changes in beam velocity.

8.12 Survey and Alignment

Alignment ensures that the beam traverses along the design trajectory, e.g., through the magnetic centers of the magnets and the electrical centers of the resonators for linear and circular accelerators, minimizing undesired orbital deviation, magnetic feed-down effects, loss of aperture, and uncontrolled beam loss caused by misalignment. Complications arise in translating from the "flat" accelerator layout designed in Cartesian space to a physical, geocentric space where a gravity-based survey is conducted. Complications also arise from thermal contraction when the accelerator devices operate at low superconductivity temperatures, requiring a prescribed warm-cold correlation.

8.12.1 Survey Reference Frame

The geoid is a closed equipotential surface that is everywhere perpendicular to the direction of the gravity vector (the plumb line). The geoid surface is irregular due to local mass anomalies, rendering it unsuitable as a reference surface for calculations.

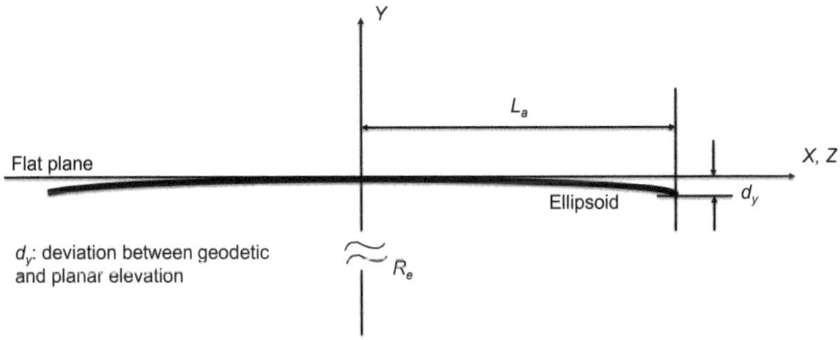

Fig. 8.63 Illustration of the deviation of the flat surface from the ellipsoid surface

However, it is the surface on which all survey measurements are made, as almost all survey instruments are set up with respect to gravity.

The geoid can be approximated by a theoretical smooth ellipsoidal surface or a "level" frame that follows the curvature of the earth. Such a "level" frame deviates from a "flat" Cartesian frame where an accelerator typically resides. As the dimension of the accelerator $2L_a$ extends, the vertical deviation d_y from the ellipsoidal surface becomes [217]

$$d_y \sim \frac{L_a^2}{2R_e} \tag{8.9}$$

as shown in Fig. 8.63, where R_e is the earth's radius that is latitude-dependent. For a typical high-power hadron facility, a Cartesian coordinate system will suffice as the survey reference frame, as long as the vertical coordinate is corrected for the curvature of the earth. Such a "flat" survey reference frame is often oriented so that the centers of most accelerator components are placed in the same horizontal plane, with the origin far away so that all coordinates are positive. For the FRIB facility, with a modest dimensional span of about 200 × 200 m (L_a=100 m) at a latitude that corresponds to $R_e \approx 6376$ km, the required vertical correction was up to ±0.78 mm.

8.12.2 Monuments

Survey reference monuments are designed to serve as a permanent reference for the facility, immune to ground settlement and disturbances. These monuments, often made of materials like invar, which has superior dimensional stability over temperature variations and sits directly on solid bedrock, can be position-calibrated by global navigation satellite systems. These reference markers should remain unperturbed throughout subsequent civil construction and facility establishment.

Network monuments are typically established along the walls and floors of the accelerator tunnel and surrounding buildings. They are periodically surveyed against reference monuments to determine possible ground movement and settlement trends. Based on these network monuments, the accelerator devices are placed according to their positions in the survey reference frame, transformed from their corresponding positions generated by the design lattice file.

The FRIB facility was established at a site adjacent to the previously existing NSCL facility. Before the new civil construction began, three reference monuments from the survey were "driven to refusal" into bedrock. The prior NSCL monuments were transferred to the FRIB survey reference frame, which connects the legacy facility with the new construction.

8.12.3 Fiducialization

The fiducialization process relates the internal component frame to the external alignment targets (fiducials) that can be seen by survey instruments. The internal component frame contains position and orientation information measured magnetically (i.e.the magnetic center and angle determined by magnetic field measurement), electrically (e.g. electrical centers of beam position monitors), and mechanically (e.g., with a gauge, a laser tracker, or a coordinate measuring machine, CMM).

For SC devices, a warm-cold correlation often needs to be established since measurements are typically conducted at room temperature, while the device operates at a much lower temperature of cryogen. This correlation may be based on theoretical predictions derived from mechanical models or from calibrated bench measurements at the operating temperature.

8.12.4 Alignment and Monitoring

The goal of alignment is to position the magnetic and electrical centers of important accelerator components at their design locations to achieve the best fit. As the measured magnetic and electrical centers may deviate from the presumed mechanical centers, corrections can be made during alignment positioning.

The alignment process may consist of absolute positioning and relative smoothing. Absolute positioning usually consists of a "blue line" survey laying out anchor bolt positions, rough positioning of components allowing subsequent adjustments, fine positioning setting positions relative to network monuments, and quality control resurvey.

Upon completion of the initial alignment, the as-built coordinates are registered as a reference. Subsequently, re-survey should be conducted both on the monuments and on fiducial positions to monitor the trends of possible building settlement and component movements.

For example, during the period when the FRIB driver linac was constructed and aligned, the tunnel floor and walls were also moving [218]. Four rounds of FRIB

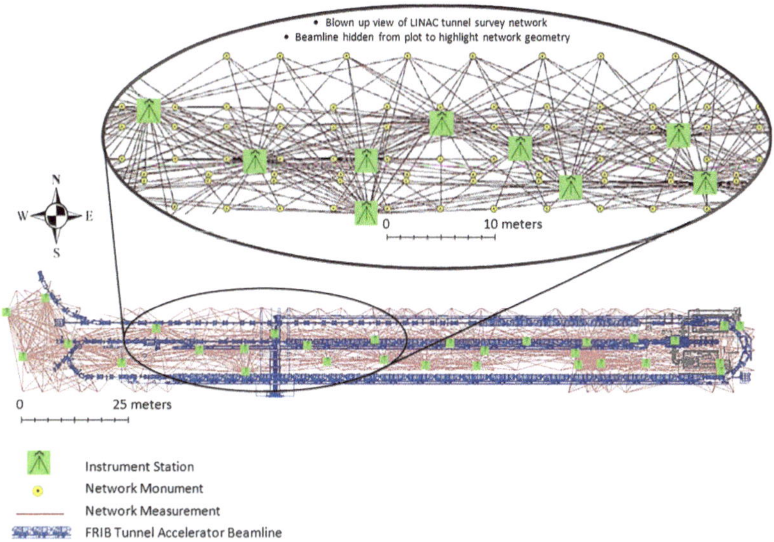

Fig. 8.64 FRIB tunnel driver linac survey network campaign, July 2016 epoch [218]. (Courtesy P. Manwiller, MSU)

network survey measurement campaigns were conducted during the acceleration installation process to monitor the deformation of the tunnel over time. Vertical deformations (i.e. "sinking") of up to 6 mm were measured in some areas of the existing facility near the construction site of the FRIB tunnel excavations. Figure 8.64 shows the measurements taken during the third such network measurement campaign in July 2016. Due to factors such as concrete curing, seasonal effects, and settlement, the coordinates of the network monuments' had to be updated after each network observation campaign. The changes in the monuments' coordinates between campaign epochs reflect the building's deformation during the interim periods. The tunnel's overall length changed during the beamline installation process.

Another example is given in Sect. 10.1.4.4 on the differential settlement (of up to approximately 20 cm) between the target floor and the RTBT tunnel floor, which was caused mainly by the recompression settlement of the ground beneath the tunnels. A comprehensive monitoring program was implemented to track the deformation of the tunnels [219].

8.13 Utilities

The essential utilities for accelerator facilities include electricity, water, cryogen for low-temperature operations, and air.

8.13.1 Electricity

High-power hadron accelerators consume power. The cost of electricity consumption often becomes a constraint that limits the operations of the accelerator facility [220]. Power consumption demands careful consideration starting from the conceptual design stage, affecting, e.g., the type of accelerators, the duty cycle, whether operating at superconducting temperature, and subsystem energy efficiency.

The PSI cyclotron accelerated continuous-wave proton beams to 0.59 GeV to strike the neutron target at 1.3 MW beam power. The total power consumption of 10.4 MW was roughly divided into 4.5 MW for the RF system, 2.6 MW for the magnet system, and 3.3 MW for the auxiliary systems and instruments. The SNS linac accelerated the proton beam to 1 GeV at 1.7 MW and accumulates with the ring for pulsed neutron generation. The accelerator power consumption of 20.5 MW was roughly divided into 16.3 MW for the linac and 4.2 MW for the ring. The multipurpose J-PARC facility used a linac to accelerate the proton beam to 400 MeV for injection into the RCS. The RCS accelerated the beam to 3 GeV, striking the neutron target with 1 MW design power for the MLF users. The beam could also be extracted from the MR operating at 30 GeV energy for both neutrino ($\sim$0.75 MW) and hadron ($\sim$0.1 MW) experiments. The total power consumption of 58.8 MW was divided as follows: 8.6 MW for the linac, 24 MW for the RCS, 21 MW for the MR, 2.8 MW for the MLF, and 2.4 MW for high-energy experiments.

The FRIB SRF linac was designed to accelerate all types of heavy ions to energies greater than 200 MeV/u, striking the target with 0.4 MW beam power. The design electrical power consumption was 18 MW for nominal operation and 21 MW after the planned upgrade to accelerate the ions above 400 MeV/u. Currently, the accelerator complex has been consuming 7 MW of electricity during normal operations and 4 MW during maintenance shutdown. The cryogenic refrigeration plant operated all year round, keeping the superconducting devices cold at all times.

Redundancy and backup are important aspects that should be considered in the planning stage. Figure 8.7 shows the electrical power schematics of the FRIB complex. Both the commercial supplier and the MSU power plant supplied electrical power to the MSU switch house. The two 25 MW circuits that originated in the MSU switch house were installed in an underground, concrete encased duct bank and were terminated in the FRIB switchgear room, supplying power (with a voltage of 13,200 V at an AC frequency of 60 Hz) to the FRIB accelerator complex. Both circuits were normally energized and delivered power in a split load bus arrangement, with two incoming circuit breakers normally closed to their respective serving circuits and an intervening normally open tie-breaker. If one circuit fails, the tie-breaker will automatically close and connect the load to the remaining functioning circuit. Switching between the two circuits is automatic in the building. Each circuit had the capacity to accommodate the entire load of the facility. In the event that outages occurred for both the commercial supplier and the MSU power plant, the following three emergency and/or backup power systems were used, as described in Sect. 8.3.2:

- Uninterruptible Power Supply (UPS) located on site for safety and control systems;
- Backup generators located on site for emergency lighting, the confinement systems, and the required utilities;
- The gas turbine generator located at the MSU campus power plant. provides cryogenic inventory protection supply to avoid the warm-up of the superconducting cavities and magnets.

8.13.2 Water

As only a small fraction of the electrical power supplied to the accelerator is converted to beam power, the rest of the consumed power must be dissipated. Water is used to remove thermal power and cool the accelerator devices and components.

The water that directly interacts with accelerator components or devices is usually contained in a closed primary loop. The closed loop ensures that water quality is under control and that any possible radioactivity is contained. Thermal power is transferred from the primary loop directly to the outside environment or through heat exchange with a secondary loop.

8.13.2.1 Chilled Water

Chilled water is a commodity that is often used to cool the air and equipment in a building. The chiller serves as the heart of the system, with the chilled water pipes as its arteries.

Chilled water in accelerator facilities is treated (not de-ionized) and is supplied in a closed loop as a cooling medium, similar to air cooled by a conventional air conditioning system. As part of a chilled water system, the condenser water absorbs heat from the refrigerant in the condenser barrel of the water chiller and is then sent via return lines to a cooling tower, which is a heat exchange device used to transfer waste heat to the atmosphere. The degree to which the cooling tower decreases the temperature depends on the outside temperature, relative humidity, and atmospheric pressure. The equipment required includes chillers, cooling towers, pumps, and electrical control equipment.

FRIB used two separate chilled water systems: one for a non-activated environment (CHW) and the other for an activated environment (CHA). The CHW system with a water temperature of $\sim$7 °C was used mainly for the HVAC of the service building (heating, ventilation, and air conditioning), air conditioning in the equipment room, heat rejection for the CHA system, and thermal exchange with water skids for the ion sources and RFQ cooling circuits. It rejected heat to the evaporator side of the chiller, which rejects heat to the cooling tower. The CHA system with a water temperature $\sim$12 °C was used mainly for the accelerator tunnel fan coil units and for thermal exchange with normal-conducting buncher cavities in the front end. It rejected heat to the CHW system.

8.13.2.2 Process Water

Process water is clean, treated water supplied in a closed loop; yet it is neither chilled nor de-ionized. Process water usually rejects heat from the cooling tower through thermal exchange.

FRIB used several process water loops with a temperature of ~30 °C. These loops provided cooling to the RF amplifier and power supply racks, the central cryogenic refrigeration plant, and the SRF building loads, including its RF racks, degassing furnaces, and dedicated cryogenic refrigeration plant.

8.13.2.3 Low Conductivity Water

The accelerator components made of copper are routinely cooled by deionized (DI) low-conductivity water (LCW) [221]. The release of copper oxides from contact surfaces is common. Releases comprise reddish-brown residue (Cu_2O) and, depending on the concentration of dissolved oxygen, a black residue (CuO). The LCW (or DI) water is supplied by a closed loop and maintained with low oxygen content, a specified pH value range, and a specific resistivity limit for corrosion and conductivity control. For water circuits that interact with different materials, such as copper, aluminum, and tungsten, the cooling treatment may also vary.

FRIB used two separate low-conductivity water systems: one for the non-activated environment (LCW) and the other for the activated environment (LCA), both operating at a temperature of ~30 °C. The LCW system with specific resistivity ~6 Ω·cm was primarily used for the front-end systems, including the RFQ amplifier system, magnets, diagnostics, and instrumentation. The LCA system with specific resistivity ~3 Ω·cm was primarily used for nearly all normal-conducting magnets in the accelerator tunnel, turbo vacuum pumps, low-power collimators, diagnostics, and instrumentation. Both LCW and LCA rejected heat to the cooling tower.

8.13.2.4 Non-Conventional-Utility Water

Cooling of components with extraordinarily high radioactivity, such as high-power targetry systems, often requires dedicated non-conventional utility water circuits. Radioactive waste, including spent resin, radioactive water, and tritium propagation, must be properly managed.

The FRIB NCU system included two closed loops of process water: one for the less-radioactive target and magnet coils, and the other for the beam dump and collimators. Filter cartridges containing resins to remove radioactive particulates from water must be replaced periodically. The beam dump loop also contained auxiliary systems such as the hydrogen recombiner, a cleaning system, and a low-level liquid waste system (Sect. 8.9.3). These primary water loops of the NCU reside in a restricted high-radiation area under strict access control (Sect. 8.1). These loops rejected heat to the secondary loop. The radioactivity survey has been mandatory before transporting any water or material outside of the area.

8.13.3 Cryogen

Liquid helium and nitrogen are the most common cryogens used in accelerator facilities with superconducting devices. The consumption of cryogens in legacy cryogenic systems due to leaks and the shortage of recycling and purification can be costly (Chap. 4).

Given the increasing cost of helium, every precaution is needed to safeguard the inventory. Redundant and backup electrical power supplies are mandatory (Sects. 8.3.2 and 8.13.1). Preferably, sufficient storage is in place to save the cryogen, even in its gaseous state.

Nitrogen is usually used in the shield circuit along the cryo-distribution lines. Compared to helium, nitrogen requires extra precautions due to the associated oxygen deficiency hazard (Sect. 8.1.3). The FRIB accelerator complex has not permitted the use of liquid nitrogen in the underground tunnel. Instead, helium gas with a temperature of 35 $\sim$55 K has been used.

8.13.4 Heating, Ventilation, Air-Conditioning

The temperature and humidity of the tunnel air must be controlled to minimize corrosion and ensure the stability of accelerator performance. Tunnel air should circulate in closed loops under a negative pressure relative to the atmosphere to prevent a possible uncontrolled release of radioactivity (Sect. 8.1.9). Radioactivity should be monitored in multiple places, including exhaust stacks before controlled release, in high efficiency particulate air (HEPA) filters, and near the HVAC shaft that penetrates the tunnel shielding.

The HEPA filtering system of central HVAC is generally specified to meet cleanliness requirements. When a high level of cleanliness is required, e.g., during maintenance periods when accessing SRF components, a localized clean room can be set up.

8.13.5 Other Commodities

Compressed air supplied through plumbed distribution lines is frequently used for accelerator instrumentation.

The instrumentation system must be designed to be fail-safe in case of electrical power outages.

Other types of gas frequently used in the accelerator tunnel include argon gas, which serves as a fire prevention measure, dry nitrogen for rust prevention during storage, and helium gas for leak checking.

8.13.6 Grounding

Grounding is essential not only to provide electrical safety requirements but also to mitigate possible electromagnetic interference (EMI) between electrical equipment and electronic systems [222]. An effective grounding system provides functions including a low-impedance reference potential for cable and equipment shields, local return paths for high-frequency noise currents, the ability to conduct high-frequency noise currents without producing significant electromagnetic radiation, and the capacity to absorb and dissipate significant levels of RF electrical energy.

FRIB used a low-impedance, facility-wide ground mesh embedded in floors and some facility walls, interconnected to the steel of the building and the perimeter grounds (Fig. 8.65). Access to this multi-point, quasi-equipotential-surface ground system mesh was provided through a combination of low-impedance perimeter ground busses and grounding plates located throughout the facility.

SNS enforced a ground-breaking separation between its linac, accumulator ring, and target station to avoid possible interference between the areas. This ground break was implemented at a significant cost, as the break also applied to rebars embedded in tunnel floors and walls.

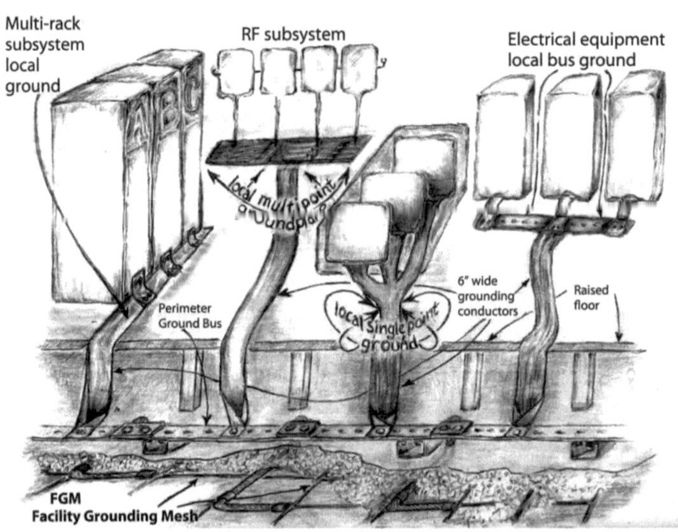

Fig. 8.65 FRIB distributed hybrid multi-point grounding system [222]. (Courtesy M. Thuot)

8.14 Radiation Impact Mitigation

As described in Sect. 1.4, the general design philosophy of high-power accelerators is based on minimizing uncontrolled beam loss. Mitigation of the effects of beam-induced radiation plays a major role in accelerator design.

8.14.1 *Sources of Radiation*

There are two main sources of radiation associated with high-power accelerators: those related to the beam in the accelerator and those from X-ray (bremsstrahlung) generating devices, such as ion sources and RF cavities. Radiations of concern for radiological protection include: (1) X-rays from electron or photon interactions generated in devices such as ion sources and RF cavities; (2) prompt radiation of neutrons produced by the beam intercepted by surrounding material; and (3) residual radiation of β- and γ-rays produced during the decay of radioactive materials.

8.14.2 *Prompt Radiation Impact Mitigation*

Mitigation of prompt-radiation impact needs to consider two situations: routine operations and accidental scenarios. During routine operations, controlled and uncontrolled beam losses, as described in Sect. 3.1, usually occur at well-defined locations with a repeatable pattern. For accidental or fault scenarios, the failure mode analysis determines the vulnerable conditions under which full beam loss occurs.

Shielding plays an important role in impact mitigation for both personnel and equipment protection. High-power accelerators are typically built underground, using earth for economical shielding. Other commonly used shielding materials include concrete, iron, lead, and marble. For routine operations, shielding must be designed to meet the requirements of the equivalent ambient dose rate according to the classification of the surrounding area outside the shielding. For accidental scenarios, interlocking radiation monitors must promptly inhibit the beam before the integrated accidental dose exceeds the acceptable level that can be received by personnel who may be present in the surrounding area. Shielding, access control, and radiation safety control systems comprise the main aspects of the personnel protection system described in Sect. 8.1.

As an integral part of the personnel protection system, shielding configuration control is important. The properties of the shielding material (such as the density and thickness of the concrete) must be certified for acceptance. Changes in shielding configuration (such as adding penetration for additional pipe or conduit routing) must go through a rigorous change control process, with compensatory measure analysis and documentation.

Radiation damage to materials must be taken into account in high-power accelerator design. Common practice includes removing electronics from the accelerator tunnel and minimizing the use of optical devices and other radiation-sensitive materials. Local shielding is used whenever possible to reduce the radiation dose absorbed by sensitive materials. The useful lifetime of the device must be assessed in accordance with the expected damage to organic materials (epoxy, polyimide, polyurethane, etc.), semiconductors, ceramics, vitreous materials (glass, scintillators), and metals. At locations of extremely high radiation, expensive radiation-hardened materials must be used.

8.14.3 Residual Radiation Impact Mitigation

Residual radiation is caused by induced radioactivity produced mainly by the interaction of primary charged particles and secondary neutrons with the environmental material. Neutrons are not affected by the Coulomb barrier of nuclei and can react at any energy to produce radioactive nuclides. Charged particles with energies above the Coulomb barrier also produce radioactive nuclides. Radioactive nuclides decay exponentially with different half-lives by emitting photons, nucleons, or light nuclei.

Because heavier elements in the material create more radioactive nuclides, lighter materials are preferred, in particular, in areas that may intercept the beam. In addition, reactions to trace elements in the material also need to be considered. From the accelerator design stage, the handling of radioactive waste must be strategically planned, including local shielding, short-term storage, removal, and disposal.

In addition to solid materials, induced radioactivity in liquids and air should also be considered. With high-power accelerators, liquids are used for both cooling (water for general cooling and liquid helium for superconducting materials, as discussed in Sect. 8.13) and beam-interception (liquid lithium for charge stripping and target, liquid mercury for the target). The level of radioactivity in liquids and air must be closely monitored and managed for public and environmental protection.

References

1. T. Kojima, T. Narita, K. Nishiyama, H. Sakaki, H. Takahashi, K. Tsutsumi, Development status of personnel protection system for IFMIF/EVEDA accelerator prototype, in *Proceedings of ICALEPCS'11* (JACoW Publishing, Geneva), pp. 1126–1129. https://jacow.org/icalepcs2011/papers/WEPMU028.pdf
2. U.S. Nuclear Regulatory Commission, 10 CFR, §20.1003: Definitions (2021). https://www.nrc.gov/reading-rm/doc-collections/cfr/part020/part020-1003.html
3. U.S. Nuclear Regulatory Commission, 10 CFR, §20.1201: Occupational dose limits for adults (1991). https://www.nrc.gov/reading-rm/doc-collections/cfr/part020/part020-1201.html

4. Michigan Occupational Safety and Health Administration, MIOSHA: MIOSHA Standards (2025). https://www.michigan.gov/leo/bureaus-agencies/miosha/standards
5. National Fire Protection Association, NFPA 70® National Electrical Code® (NEC®) (2024). https://www.nfpa.org/for-professionals/codes-and-standards/list-of-codes-and-standards/free-access
6. The American Society of Mechanical Engineers (ASME), BPVC: Boiler and pressure vessel code (2023). https://www.asme.org/codes-standards/bpvc-standards
7. The American Society of Mechanical Engineers (ASME), B31.3: Process piping (2024). https://www.asme.org/codes-standards/find-codes-standards/b31-3-process-piping
8. J. Haines, T. McManamy, T. Gabriel, R. Battle, K. Chipley, J. Crabtree, L. Jacobs, D. Lousteau, M. Rennich, B. Riemer, Spallation neutron source target station design, development, and commissioning. Nucl. Instrum. Methods Phys. Res. Sect. A **764**, 94–115 (2014). https://doi.org/10.1016/j.nima.2014.03.068. https://www.sciencedirect.com/science/article/pii/S0168900214003829
9. Environmental Protection Agency, CAA: Overview of the Clean Air Act and Air Pollution (2025). https://www.epa.gov/clean-air-act-overview
10. Environmental Protection Agency, NAAQS: Reviewing National Ambient Air Quality Standards: Scientific and Technical Information (2025). https://www.epa.gov/naaqs
11. C.Z. Antoine, A. Aspart, M. Berthelot, Y. Gasser, J.P. Poupeau, F. Valin, Morphological and chemical studies of Nb samples after various surface treatment, in *Proceedings of SRF'99* (JACoW Publishing, Geneva, 1999), pp. 295–303. https://jacow.org/SRF99/papers/TUP035.pdf
12. P. Kneisel, Surface preparation of niobium, in *1st Workshop on RF Superconductivity* (1980), pp. 27–40
13. K. Saito, F. Furuta, Sulfur generation mechanism during electropolishing with niobium cavities, in *Proceedings of SRF'09* (JACoW Publishing, Geneva, 2009), pp. 838–838. https://jacow.org/SRF2009/papers/THPPO090.pdf
14. E.M. Marszal, E.W. Scharpf et al., *Safety Integrity Level Selection: Systematic Methods Including Layer of Protection Analysis* (ISA, the Instrumentation, Systems, and Automation Society, North Carolina, 2002)
15. U.S. Nuclear Regulatory Commission, Regulatory Guide, 4.20: Constraints on releases of airborne radioactive materials to the environment for licensees other than power reactors (2021). https://www.nrc.gov/docs/ML1101/ML110120299.pdf
16. H. Bethe, J. Ashkin, *Experimental Nuclear Physics*, ed. by E. Segre (Wiley, New York, 1953), p. 253
17. J.F. Ziegler, M.D. Ziegler, J.P. Biersack, SRIM–The stopping and range of ions in matter (2010). Nucl. Instrum. Methods Phys. Res. Sect. B **268**(11–12), 1818–1823 (2010)
18. Z. Liu, J. Crisp, T. Russo, R. Webber, Y. Zhang, A new beam loss detector for low-energy proton and heavy-ion accelerators. Nucl. Instrum. Methods Phys. Res. Sect. A **767**, 262–266 (2014)
19. M. Ikegami, Personnel and machine protection for FRIB commissioning, operations, and power ramp up. *Presented at IPAC'25*, Taipei, Jun 2025, paper TUXD1, to be published
20. D. Chabot, M. Ikegami, M.G. Konrad, D.G. Maxwell, The FRIB run permit system, in *Proceedings of ICALEPCS'17* (JACoW Publishing, Geneva), pp. 646–648. https://doi.org/10.18429/JACoW-ICALEPCS2017-TUPHA098. https://jacow.org/icalepcs2017/papers/TUPHA098.pdf
21. M. Ikegami, D. Dudley, M.G. Konrad, Z. Li, G. Shen, V. Vuppala, Operation mode and machine state control for FRIB driver linac operation, in *Proceedings of LINAC'16, Linear Accelerator Conference*, no. 28 (JACoW Publishing, Geneva, 2017), pp. 956–958. https://doi.org/10.18429/JACoW-LINAC2016-THPLR045. https://jacow.org/linac2016/papers/THPLR045.pdf
22. R. Apsimon, G. Burt, A.C. Dexter, R.B. Appleby, P. Baudrenghien, K.N. Sjobak, Modeling the low level RF response on the beam during crab cavity quench, in *Proceedings of IPAC'17*

(JACoW Publishing, Geneva), pp. 1098–1101. https://doi.org/10.18429/JACoW-IPAC2017-MOPVA102. https://jacow.org/ipac2017/papers/MOPVA102.pdf
23. J. Branlard, V. Ayvazyan, O. Hensler, H. Schlarb, C. Schmidt, W. Cichalewski, Superconducting cavity quench detection and prevention for the European XFEL, in *Proceedings of ICALEPCS'13* (JACoW Publishing, Geneva), pp. 1239–1241. https://jacow.org/ICALEPCS2013/papers/THPPC072.pdf
24. Z.Y. Yao, R.E. Laxdal, W.R. Rawnsley, V. Zvyagintsev, Cryomodule protection for ARIEL e-Linac, in *Proceedings of SRF'15* (JACoW Publishing, Geneva), pp. 861–864. https://jacow.org/SRF2015/papers/TUPB103.pdf
25. T.T. Arkan et al., LCLS-II cryomodules production experience and lessons learned at Fermilab, in *Proceedings of SRF'19* (JACoW Publishing, Geneva, 2019), pp. 709–714. https://doi.org/10.18429/JACoW-SRF2019-TUP101. https://jacow.org/srf2019/papers/TUP101.pdf
26. Z.Y. Yao, R.E. Laxdal, Performance of contaminated superconducting linac after vacuum excursion, in *Proceedings of SRF'23, International Conference on RF Superconductivity*, no. 21 (JACoW Publishing, Geneva, 2023), pp. 332–336. https://doi.org/10.18429/JACoW-SRF2023-MOPMB092
27. R. Scrivens, Proton and ion sources for high intensity accelerators, in *Proceedings of EPAC'04, European Particle Accelerator Conference*, no. 9 (JACoW Publishing, Geneva, 2004), pp. 103–107. http://accelconf.web.cern.ch/e04/papers/TUYLH01.pdf
28. O.B. Morgan, G.G. Kelley, R.C. Davis, Technology of intense dc ion beams. Rev. Sci. Instrum. **38**(4), 467–480 (1967). https://doi.org/10.1063/1.1720740
29. I.N. Draganic, D. Kleinjan, G. Rouleau, H^+ and H^- ion beam injectors at LANSCE: beam production status and planned injector upgrades, in *Proceedings of IPAC'19* (JACoW Publishing, Geneva), pp. 2087–2089. https://doi.org/10.18429/JACoW-IPAC2019-TUPTS071. http://accelconf.web.cern.ch/ipac2019/papers/TUPTS071.pdf
30. R.R. Stevens Jr, E.P. Chamberlin, R.W. Hamm, J.R. McConnell, R.L. York, Injector operations at LAMPF, in *Proceedings of LINAC'79, Linear Accelerator Conference* (JACoW Publishing, Geneva), pp. 465–468. https://jacow.org/l79/papers/S8-13.pdf
31. E. Boltezar et al., The new CERN 50-MeV linac, in *Proceedings of LINAC'79, Linear Accelerator Conference* (JACoW Publishing, Geneva), pp. 66–77. https://jacow.org/l79/papers/S2-2.pdf
32. W. Schweizer, U. Ratzinger, B. Klump, K. Volk, A high intensity 200 mA proton source for the FRANZ-Project (Frankfurt-Neutron-Source at the Stern-Gerlach-Center). Rev. Sci. Instrum. **85**(2), 02A743 (2014). https://doi.org/10.1063/1.4842335
33. H.V. Smith, J.D. Schneider, R.L. Sheffield, Low-Energy Demonstration Accelerator (LEDA) test results and plans, in *Proceedings of PAC'01* (JACoW Publishing, Geneva), pp. 3296–3298. https://jacow.org/p01/papers/RPPH031.pdf
34. K. Masuda et al., Commissioning of IFMIF prototype accelerator towards CW operation, in *Proceedings of LINAC'22, International Linear Accelerator Conference*, no. 31 (JACoW Publishing, Geneva, 2022), pp. 319–323. https://doi.org/10.18429/JACoW-LINAC2022-TU2AA04
35. Y. He, Z. Wang, Z. Qin, H. Xu, L. Yang, Z. Wang, J. Zhang, X. Zhang, L. Gu, W. Zhan, Development of accelerator driven advanced nuclear energy (ADANES) and nuclear fuel recycle, in *Proceedings of IPAC* (2019), pp. 4389–4393
36. K.N. Leung, Research and development of H- ion sources. Nucl. Instrum. Methods Phys. Res. Sect. B **40**, 1028–1032 (1989)
37. K.N. Leung, 7.1.4 H^- ion source, in *Handbook of Accelerator Physics and Engineering*, ed. by A.W. Chao et al. (World Scientific, Singapore, 2022), pp. 675–678
38. K.W. Ehlers, B. Gavin, E.L. Hubbard, High-intensity negative ion sources. Nucl. Instrum. Methods **22**, 87–92 (1963)
39. V. Dudnikov, Hydrogen negative ion source with penning geometry, m, in *Proceedings of IV Allunion Conference for Charged Particles Accelerators*, p. 323 (1975)
40. P.W. Allison, A direct extraction H^- ion source, in *Proceedings of PAC'77* (JACoW Publishing, Geneva), pp. 1594–1597

41. H.V. Smith, P. Allison, J. Sherman, Pulsed H beams from Penning SPS sources equipped with circular emitters, in *Proceedings of LINAC'88, Linear Accelerator Conference* (JACoW Publishing, Geneva), pp. 164–166. https://jacow.org/l88/papers/MO3-47.pdf
42. P.J.S. Barratt, A.P. Letchford, M. Perkins, C.W. Planner, R. Sidlow, Operational experience of Penning H-Ion Sources at ISIS, in *Proceedings of EPAC'96, European Particle Accelerator Conference*, no. 5 (JACoW Publishing, Geneva, 1996), pp. 1525–1527
43. K. Leung, K. Ehlers, Self-extraction negative ion source. Rev. Sci. Instrum. **53**(6), 803–809 (1982)
44. R.R. Stevens, R. Kandarian, J.R. McConnell, R.L. York, Status of the new high intensity H^- injector at LAMPF, in *Proceedings of LINAC'84, Linear Accelerator Conference* (JACoW Publishing, Geneva), pp. 226–228. https://jacow.org/l84/papers/WEQ0004.pdf
45. Y. Mori, S. Fukumoto, A. Ueno, K. Ikegami, A. Takagi, Intense negative ion sources. Part. Accel. **32**, 9–14 (1990)
46. K. Leung, K. Ehlers, M. Bacal, Extraction of volume-produced h^- ions from a multicusp source. Rev. Sci. Instrum. **54**(1), 56–61 (1983)
47. R.L. York, R.R. Stevens Jr, K.N. Leung, K.W. Ehlers, Extraction of h- beams from a magnetically filtered multicusp source. Rev. Sci. Instrum. **55**(5), 681–686 (1984). https://doi.org/10.1063/1.1137821
48. R. Welton, M. Stockli, S. Murray, R. Keller, Recent advances in the performance and understanding of the SNS ion source, in *AIP Conference, Proceedings*, vol. 763 (American Institute of Physics, 2005), pp. 296–314
49. D. Yuan, R. Baartman, K. Kendall, M. McDonald, D. Mosscrop, P. Schmor, Extraction of H-ions from a DC cusp source, in *AIP Conference, Proceedings*, vol. 158 (American Institute of Physics, 1987), pp. 346–355
50. K. Jayamanna, F. Ames, I. Bylinskii, M. Lovera, B. Minato, A 60 mA DC H^- multi cusp ion source developed at TRIUMF. Nucl. Instrum. Methods Phys. Res. Sect. A **895**, 150–157 (2018)
51. R. Keller, The front-end systems for the spallation neutron source, in *AIP Conference. Proceedings*, vol. 576 (American Institute of Physics, 2001), pp. 362–365
52. A. Ueno, K. Ikegami, Y. Namekawa, H. Oguri, K. Ohkoshi, Status of the J-PARC negative hydrogen ion source, in *Proceedings of LINAC'10, Linear Accelerator Conference*, no. 25 (JACoW Publishing, Geneva), pp. 1016–1018. https://jacow.org/LINAC2010/papers/THP118.pdf
53. D-Pace, ion sources, beam diagnostic devices, and beamline systems. https://www.d-pace.com
54. J. Alessi, A. Pikin, 7.1.6 Multi-charged heavy ion source, in *Handbook of Accelerator Physics and Engineering*, ed. by A.W. Chao et al. (World Scientific, Singapore, 2022), pp. 680–683
55. R. Geller, Electron cyclotron resonance multiply charged ion sources. IEEE Trans. Nucl. Sci. **23**(2), 904–912 (1976). https://doi.org/10.1109/TNS.1976.4328376
56. C. Lyneis, D. Leitner, M. Leitner, C. Taylor, S. Abbott, The third generation superconducting 28 ghz electron cyclotron resonance ion source venus (invited)a). Rev. Sci. Instrum. **81**(2), 02A201 (2010). https://doi.org/10.1063/1.3271135
57. L. Sun, W. Lu, H. Zhao, J. Guo, W. Wu, B. Wu, X. Ou, E. Mei, S. Zheng, X. Zhang et al., Progress on the development of key technologies for the fourth generation ecr ion source FECR. J. Phys. Conf. Ser. **2244**, 012021 (2022). IOP Publishing
58. V. Kutner, Intense high charge state ion sources. Rev. Sci. Instrum. **65**(4), 1039–1044 (1994)
59. T. Miyata, T. Miyoshi, T. Sakuma, M. Yamamoto, A. Kitagawa, M. Muramatsu, Y. Sato, Improvement of the yield of highly charged ions by a gas-pulsing technique and the current status of the NIRS penning source. Rev. Sci. Instrum. **75**(5), 1863–1865 (2004)
60. E. Donets, Electron beam ion sources and associated physics at JINR. Nucl. Instrum. Methods Phys. Res. Sect. B **9**(4), 522–525 (1985). https://doi.org/10.1016/0168-583X(85)90359-3. https://www.sciencedirect.com/science/article/pii/0168583X85903593
61. J.G. Alessi, D. Barton, E. Beebe, S. Bellavia, O. Gould, A. Kponou, R. Lambiase, R. Lockey, A. McNerney, M. Mapes, Y. Marneris, M. Okamura, D. Phillips, A.I. Pikin, D. Raparia, J.

Ritter, L. Snydstrup, C. Theisen, M. Wilinski, The Brookhaven National Laboratory electron beam ion source for RHICa). Rev. Sci. Instrum. **81**(2), 02A509 (2010). https://doi.org/10.1063/1.3292937
62. A. Balabaev, S. Kondrashev, K. Konukov, A. Lozhkin, B. Sharkov, A. Shumshurov, A. Charushin, K. Makarov, Y. Satov, Y. Smakovskii et al., Laser ion source based on a 100 J/1 Hz CO 2-laser system. Rev. Sci. Instrum. **75**(5), 1572–1574 (2004)
63. B.Y. Sharkov et al., Experiments with Fe-ion beam generation acceleration and accumulation in ITEP-TWAC facility, in *Proceedings of EPAC'08, European Particle Accelerator Conference*, no. 11 (JACoW Publishing, Geneva, 2008), pp. 352–354. https://jacow.org/e08/papers/MOPC122.pdf
64. M. Okamura, T. Takeuchi, R.A. Jameson, S. Kondrashev, H. Kashiwagi, K. Sakakibara, T. Kanesue, J. Tamura, T. Hattori, Direct plasma injection scheme in accelerators (invited). Rev. Sci. Instrum. **79**(2), 02B314 (2008). https://doi.org/10.1063/1.2821590
65. I.G. Brown, J.E. Galvin, R.A. MacGill, High current ion source. Appl. Phys. Lett. **47**(4), 358–360 (1985). https://doi.org/10.1063/1.96163
66. A. Adonin, R. Hollinger, Progress on MEVVA source VARIS at GSI. Rev. Sci. Instrum. **89**, 052304 (2018)
67. R. Geller, Electron cyclotron resonance sources: historical review and future prospects. Rev. Sci. Instrum. **69**(3), 1302–1310 (1998)
68. C. Lyneis, D. Leitner, M. Leitner, C. Taylor, S. Abbott, The third generation superconducting 28 GHz electron cyclotron resonance ion source VENUS. Rev. Sci. Instrum. **81**(2) (2010)
69. W. Huang, D. Xie, L. Sun, Development of a new high temperature oven for the production of intense metal ion beams with ECR ion sources. Rev. Sci. Instrum. **90**, 123301 (2019)
70. F. Wenander, Charge breeding of radioactive ions with EBIS and EBIT. J. Instrum. **5**(10), C10004 (2010)
71. Argonne National Laboratory, ATLAS, Argonne Tandem Linac Accelerator System (2025). https://www.anl.gov/atlas
72. M. Thoennessen, B. Sherrill, From isotopes to the stars. Nature **473**(7345), 25–26 (2011)
73. Y. Blumenfeld, T. Nilsson, P. Van Duppen, Facilities and methods for radioactive ion beam production. Phys. Scr. **2013**(T152), 014023 (2013)
74. D.J. Morrissey, B.M. Sherrill, Radioactive nuclear beam facilities based on projectile fragmentation. Philos. Trans. A Math. Phys. Eng. Sci. **356**(1744), 1985–2006 (1998)
75. O. Kofoed-Hansen, K. Nielsen, Short-lived krypton isotopes and their daughter substances. Phys. Rev. **82**(1), 96 (1951)
76. R. Klapisch, R. Bernas, The detection of the short-lived lithium isotopes 8li and 9li by preliminary experiments with an on-line mass spectrometer. Nucl. Instrum. Methods **38**, 291–295 (1965)
77. J.R. Alonso, A. Chatterjee, C.A. Tobias, High purity radioactive beams at the Bevalac. IEEE Trans. Nucl. Sci. **26**(3), 3003–3005 (1979). https://doi.org/10.1109/TNS.1979.4329919
78. Y. Viyogi, T. Symons, P. Doll, D. Greiner, H. Heckman, D. Hendrie, P. Lindstrom, J. Mahoney, D. Scott, K. Van Bibber et al., Fragmentations of Ar 40 at 213 MeV/Nucleon. Phys. Rev. Lett. **42**(1), 33 (1979)
79. J.P. Dufour, A. Fleury, R. Bimbot, Secondary reactions as a tool to produce exotic nuclei. Phys. Rev. C **23**, 801–806 (1981). https://doi.org/10.1103/PhysRevC.23.801
80. B. Sherrill, D. Morrissey, J. Nolen Jr, J. Winger, The A1200 projectile fragment separator. Nucl. Instrum. Methods Phys. Res. Sect. B **56**, 1106–1110 (1991)
81. M. Lindroos, Review of ISOL-type radioactive beam facilities, in *Proceedings of EPAC'04, European Particle Accelerator Conference*, no. 9 (JACoW Publishing, Geneva, 2004), pp. 45–49. http://accelconf.web.cern.ch/e04/papers/TUXCH01.pdf
82. H. Ravn, Status and future development of ion sources for on-line mass separators. Nucl. Instrum. Methods Phys. Res. Sect. B **70**(1–4), 107–117 (1992)
83. S. Gales, The GANIL facility. Scholarpedia **5**(7), 9670 (2010)

84. F. Ames, P. Bricault, H. Heggen, P. Kunz, J. Lassen, A. Mjøs, S. Raeder, A. Teigelhöfer, Ion source developments for the production of radioactive isotope beams at TRIUMF. Rev. Sci. Instrum. **85**, 02B912 (2014)
85. J. Wei, Rare isotope beams and high-power accelerators, in *Proceedings of NAPAC'19* (JACoW Publishing, Geneva), pp. 993–999. https://doi.org/10.18429/JACoW-NAPAC2019-FRXBB1. https://jacow.org/napac2019/papers/FRXBB1.pdf
86. O. Kamigaito, Overview of the world-wide RIB facilities, in *Proceedings of IPAC'13* (JACoW Publishing, Geneva), pp. 4000–4004. https://jacow.org/IPAC2013/papers/FRXCB201.pdf
87. M. Hausmann, A. Aaron, A. Amthor, M. Avilov, L. Bandura, R. Bennett, G. Bollen, T. Borden, T. Burgess, S. Chouhan et al., Design of the advanced rare isotope separator ARIS at FRIB. Nucl. Instrum. Methods Phys. Res. Sect. B **317**, 349–353 (2013)
88. J. Grillenberger, J.M. Humbel, A.C. Mezger, M. Seidel, W. Tron, Status and further development of the PSI high intensity proton facility, in *Proceedings of Cyclotrons'13* (JACoW Publishing, Geneva, 2013), pp. 37–39. https://jacow.org/CYCLOTRONS2013/papers/MOPPT004.pdf
89. M. Seidel, P. Schmelzbach, A. Schmelzbach, Upgrade of the PSI cyclotron facility to 1.8 MW, in *Proceedings of Cyclotrons'07* (Paul Scherrer Institute, 2007), pp. 157–162
90. J. Grillenberger, Is a megawatt accelerator sustainable? in *Sustainable Accelerators Workshop* (2022)
91. A.V. Aleksandrov et al., Commissioning of the new SNS RFQ and 2.5 MeV beam test facility, in *Proceedings of IPAC'17* (JACoW Publishing, Geneva, 2017), pp. 2438–2440. https://doi.org/10.18429/JACoW-IPAC2017-TUPVA145. https://jacow.org/ipac2017/papers/TUPVA145.pdf
92. Y. Yamazaki, J-PARC Status, in *Proceedings of PAC'09, Particle Accelerator Conference*, no. 23 (JACoW Publishing, Geneva, 2009), pp. 18–22. https://jacow.org/PAC2009/papers/MO2BCI03.pdf
93. S.O. Schriber, High-beta linac structures, in *Proceedings of LINAC'79, Linear Accelerator Conference* (JACoW Publishing, Geneva, 1979), pp. 164–175. https://jacow.org/l79/papers/S3-7.pdf
94. J. Bernardin, J. Billen, S. Brown, N. Bultman, W. Fox, T. Hardek, R. Hardekopf, S. Hopkins, S. Nath, D. Rej, P. Tallerico, L. Young, Spallation Neutron Source (SNS). Technical Report, SNS 101020000-TR0001—R00, SNS Linac R&D Program (2002)
95. J.H. Billen, S.S. Kurennoy, L.M. Young, K.R. Crandall, Room-temperature linac structures for the Spallation Neutron Source, in *Proceedings of PAC'01* (JACoW Publishing, Geneva), pp. 1104–1106. https://jacow.org/p01/papers/MPPH139.pdf
96. K. Hasegawa, Commissioning of energy upgraded linac of J-PARC, in *Proceedings of LINAC'14, Linear Accelerator Conference*, no. 27 (JACoW Publishing, Geneva, 2014), pp. 417–422. https://jacow.org/LINAC2014/papers/TUIOB03.pdf
97. S.K. Esin, L.V. Kravchuk, A.I. Kvasha, P.N. Ostroumov, V.L. Serov, MMF linac upgrade possibilities for the pulsed neutron source, in *Proceedings of PAC'95* (JACoW Publishing, Geneva, 1995), pp. 1173–1174
98. D.E. Nagle: Status and performance of LAMPF, in *Proceedings of LINAC'72, Linear Accelerator Conference* (JACoW Publishing, Geneva, 1972), pp. 4–11. https://jacow.org/l72/papers/A02.pdf
99. L. De Forest, The audion: a new receiver for wireless telegraphy. Proc. Am. Inst. Elect. Eng. **25**(10), 719–747 (1906)
100. W. Schottky, *Physik der Glüheelektroden*, vol. 2. (Akademische verlagsgesellschaft mbh, 1928)
101. R.H. Varian, S.F. Varian, A high frequency oscillator and amplifier. J. Appl. Phys. **10**(5), 321–327 (1939)
102. A. Yano, M. Sakamoto, K. Tetsuka, H. Urakata, S. Miyake, H. Yonezawa, S. Fukuda, M. Kawamura, S. Anami, E. Chishiro, H. Yoshikawa, Development of the klystrons for the high intensity proton accelerator facility, in *Proceedings of APAC'01* (JACoW Publishing, Geneva, 2001)

103. M. McCarthy, D. Anderson, R. Fuja, P. Gurd, T. Hardek, Y. Kang, J. Bradley, K. Young, D. Rees, W. Roybal, Spallation neutron source high power RF installation and commissioning progress, in *Proceedings of the 2005 Particle Accelerator Conference (Proc. PAC'05)* (IEEE, 2005), pp. 3520–3522
104. M.S. Champion et al., Progress on the Proton Power Upgrade project at the Spallation Neutron Source, in *Proceedings of LINAC'22, International Linear Accelerator Conference*, no. 31 (JACoW Publishing, Geneva, 2022), pp. 390–393. https://doi.org/10.18429/JACoW-LINAC2022-TUPOJO19
105. D.E. Anderson, V. Peplov, D.J. Solley, M. Wezensky, Recent developments in the improvement campaign for the high voltage converter modulator at the Spallation Neutron Source, in *2014 IEEE International Power Modulator and High Voltage Conference (IPMHVC)* (IEEE, 2014), pp. 672–675
106. D.G. Morris et al., RF System for FRIB accelerator, in *Proceedings of IPAC'18* (JACoW Publishing, Geneva, 2018), pp. 1765–1770. https://doi.org/10.18429/JACoW-IPAC2018-WEXGBF3. http://accelconf.web.cern.ch/ipac2018/papers/WEXGBF3.pdf
107. A. Zaltsman et al., Progress in the development of high Level RF for the SNS ring, in *Proceedings of PAC'03* (JACoW Publishing, Geneva), pp. 1195–1197. https://jacow.org/p03/papers/TPAB007.pdf
108. I. Gardner, Ferrite dominated cavities (1992)
109. C. Ohmori et al., 20-year collaboration on synchrotron RF between CERN and J-PARC, in *Proceedings of IPAC'22, International Particle Accelerator Conference*, no. 13 (JACoW Publishing, Geneva, 2022),pp. 3130–3133. https://doi.org/10.18429/JACoW-IPAC2022-FROXSP1. https://jacow.org/IPAC2022/papers/FROXSP1.pdf
110. C. Ohmori, E. Ezura, K. Hara, K. Hasegawa, Y. Makida, R. Muto, M. Nomura, T. Ogitsu, T. Shimada, A. Schnase et al., Development of a high gradient RF system using a nanocrystalline soft magnetic alloy. Phys. Rev. Spec. Top. Accel. Beams **16**(11), 112002 (2013)
111. J.P. Burnet, Requirements for power converters. arXiv preprint, arXiv:1607.01597 (2016)
112. F. Bordry, Power converters for particle accelerators, in *2005 European Conference on Power Electronics and Applications* (2005), p. 30. https://doi.org/10.1109/EPE.2005.219762
113. K. Holland et al., Design and operational experience of FRIB magnet and electrostatic power supplies, in *Proceedings of HIAT2025, HIAT*, no. 16 (JACoW Publishing, Geneva, 2025), pp. 177–180. https://doi.org/10.18429/JACoW-HIAT2025-TUP23. https://proceedings.jacow.org/hiat2025/doi/jacow-hiat2025-tup23/
114. X. Qi, Z. Hao, W. Zhang, The Status of the CSNS/RCS Power Supply System, in *Proceedings of IPAC'18* (JACoW Publishing, Geneva), pp. 2850–2852. https://doi.org/10.18429/JACoW-IPAC2018-WEPML070. http://accelconf.web.cern.ch/ipac2018/papers/WEPML070.pdf
115. L. Li, W. Kang, C. Deng, S. Li, Y. Liu, J. Zhou, Y. Wu, X. Wu, B. Yin, Magnetic field measurement and analysis of the CSNS/RCS quadrupole magnets. IEEE Trans. Appl. Supercond. **28**(3), 1–4 (2018)
116. W. Huang, H. Sun, H. Zhao, Y. Liu, Y. Li, Z. Qiao, X. Li, C. Zhang, A prototype RF power source for CSNS/RCS. Sci. China-Phys. Mech. Astron. **54**, 218–221 (2011)
117. W. Sheng, F. Shou-Xian, F. Shi-Nian, L. Wei-Bin, O. Hua-Fu, Q. Qing, T. Jing-Yu, W. Jie, Introduction to the overall physics design of CSNS accelerators. Chin. Phys. C **33**(S2), 1–3 (2009). https://doi.org/10.1088/1674-1137/33/s2/001
118. IPNS upgrade, A feasibility study (1995). https://doi.org/10.2172/226408. https://www.osti.gov/biblio/226408
119. W. Zhang et al., SNS extraction fast kicker pulsed power system, in *Proceedings of EPAC'04, European Particle Accelerator Conference*, no. 9 (JACoW Publishing, Geneva, 2004), pp. 1810–1812. http://accelconf.web.cern.ch/e04/papers/WEPKF087.pdf
120. J.L. Mi et al., SNS extraction kicker system and first article BPFN test, in *Proceedings of EPAC'02, European Particle Accelerator Conference*, no. 8 (JACoW Publishing, Geneva, 2002), pp. 2652–2654. https://jacow.org/e02/papers/TUPLE068.pdf
121. J.L. Mi et al., SNS extraction kicker power supply manufacture status, in Proceedings of EPAC'04, European Particle Accelerator Conference, no. 9 (JACoW Publishing, Geneva, 2004), pp. 1571–1573. https://jacow.org/e04/papers/TUPLT188.pdf

122. K. Sondericker, Measurement of VAT fast gate valve closing time. VAT Inc. Private communication (2012)
123. O. Gröbner, R. Calder, Beam induced gas desorption in the CERN Intersecting Storage Rings. IEEE Trans. Nucl. Sci. **20**(3), 760–764 (1973)
124. B. Boardman, Spallation neutron source: description of accelerator and target. Technical report, Science Research Council (1982)
125. H. Hseuh, The AGS Booster systems. Vacuum **41**, 1993–1996 (1990)
126. M. Kinsho, D. Nishizawa, Y. Saito, H. Suzuki, H. Yokomizo, Vacuum system design for the 3 GeV-proton synchrotron of JAERI-KEK joint project. J. Vac. Sci. Technol. A **20**(3), 829–832 (2002). https://doi.org/10.1116/1.1468656
127. C. Luo, J. Meng, J. Yang, J. Jiao, G. Shen, C. Li, N. Wei, G. Zhu, W. Xie, Z. Chai et al., Study on the performance of titanium alloy-lined thin-walled vacuum chamber. Vacuum **221**, 112808 (2024)
128. R. Ellefson, A. Miiller, Recommended practice for calibrating vacuum gauges of the thermal conductivity type. J. Vacuum Sci. Technol. A: Vacuum Surf. Films **18**(5), 2568–2577 (2000)
129. D. Li, K. Jousten, Comparison of the stability of hot and cold cathode ionization gauges. J. Vacuum Sci. Technol. A: Vacuum Surf. Films **21**(4), 937–946 (2003)
130. K. Welch, *Capture Pumping Technology* (Elsevier, Amsterdam, 2001)
131. H. Hseuh, C. Lanni, Evaluation of Zr–V–Fe getter pump for UHV system. J. Vacuum Sci. Technol. A: Vacuum Surf. Films **A1**(2), 1283–1287 (1983)
132. H. Hseuh, H. Worwetz, Performance of BNL-TSTA compound cryopump. J. Vacuum Sci. Technol. A: Vacuum Surf. Films **18**(2), 1131–1135 (1981)
133. R. Davis, C. De La Parra, H. Hseuh, P. Mickaliger, D. Pate, Measurement of helium leak rates with temperature and pressure in RHIC. Vacuum **60**(1–2), 131–135 (2001)
134. M. Gulley, P. Keating, H. Bryant, E. MacKerrow, W. Miller, D. Rislove, S. Cohen, J. Donahue, D. Fitzgerald, S. Frankle et al., Measurement of H-, H 0, and H+ yields produced by foil stripping of 800-MeV H- ions. Phys. Rev. A **53**(5), 3201 (1996)
135. J. Wei, Synchrotrons and accumulators for high-intensity proton beams. Rev. Mod. Phys. **75**, 1383–1432 (2003). https://doi.org/10.1103/RevModPhys.75.1383
136. Y.N. Rao, R.A. Baartman, I.V. Bylinskii, V.A. Verzilov, J.M. Schippers, TRIUMF extraction foil developments and contamination reduction, in *Proceedings of Cyclotrons'13* (JACoW Publishing, Geneva), pp. 269–271. https://jacow.org/CYCLOTRONS2013/papers/TU3PB04.pdf
137. J.A. Nolen, F. Marti, Charge strippers of heavy ions for high intensity accelerators. Rev. Accel. Sci. Technol. **6**, 221–236 (2013)
138. Y. Momozaki, C.B. Reed, J.A. Nolen, J.R. Specht, D.B. Chojnowski, J.S. Song, F. Marti, P. Guetschow, J. Sherman, Proton beam-on-liquid lithium stripper film experiment. J. Radioanal. Nucl. Chem. **305**, 843–849 (2015)
139. T. Kanemura, M. LaVere, R. Madendorp, F. Marti, T. Maruta, Y. Momozaki, P.N. Ostroumov, A.S. Plastun, J. Wei, Q. Zhao, Experimental demonstration of the thin-film liquid-metal jet as a charge stripper. Phys. Rev. Lett. **128**, 212301 (2022). https://doi.org/10.1103/PhysRevLett.128.212301
140. T. Kanemura et al., Liquid lithium charge stripping technology: Achievement and lessons learned. *Presented at IPAC'25*, Taipei, Jun 2025, paper MOYD2, to be published
141. H. Imao et al., R&D of helium gas stripper for intense uranium beams, in *Proceedings of Cyclotrons'13* (JACoW Publishing, Geneva), pp. 265–268. https://jacow.org/CYCLOTRONS2013/papers/TU3PB03.pdf
142. H. Imao, H. Okuno, H. Kuboki, S. Yokouchi, N. Fukunishi, O. Kamigaito, H. Hasebe, T. Watanabe, Y. Watanabe, M. Kase et al., Charge stripping of U 238 ion beam by helium gas stripper. Phys. Rev. Spec. Top. Accel. Beams **15**(12), 123501 (2012)
143. I. Yamane, H^- charge-exchange injection without hazardous stripping foils. Phys. Rev. Spec. Top. Accel. Beams **1**(5), 053501 (1998)
144. V. Danilov, A. Aleksandrov, S. Assadi, J. Barhen, W. Blokland, Y. Braiman, D. Brown, C. Deibele, W. Grice, S. Henderson et al., Proof-of-principle demonstration of high efficiency

laser-assisted H-beam conversion to protons. Phys. Rev. Spec. Top. Accel. Beams **10**(5), 053501 (2007)
145. S. Cousineau, A. Rakhman, M. Kay, A. Aleksandrov, V. Danilov, T. Gorlov, Y. Liu, M. Plum, A. Shishlo, D. Johnson, First demonstration of laser-assisted charge exchange for microsecond duration H-beams. Phys. Rev. Lett. **118**(7), 074801 (2017)
146. S. Cousineau, A. Rakhman, M. Kay, A. Aleksandrov, V. Danilov, T. Gorlov, Y. Liu, C. Long, A. Menshov, M. Plum, A. Shishlo, A. Webster, D. Johnson, High efficiency laser-assisted h^- charge exchange for microsecond duration beams. Phys. Rev. Accel. Beams **20**, 120402 (2017). https://doi.org/10.1103/PhysRevAccelBeams.20.120402
147. A.V. Aleksandrov, S.M. Cousineau, T.V. Gorlov, Y. Liu, A. Rakhman, A.P. Shishlo, A crab-crossing scheme for laser-ion beam applications, in *Proceedings of NAPAC'19* (JACoW Publishing, Geneva), pp. 639–641. https://doi.org/10.18429/JACoW-NAPAC2019-WEYBB5. https://jacow.org/napac2019/papers/WEYBB5.pdf
148. K. Yamamoto, N. Ogiwara, M. Kuramochi, Characterization of liquid targets in vacuum condition. e-J. Surf. Sci. Nanotechnol. **21**(4), 359–364 (2023)
149. H. Imao, Charge stripper ring for RIKEN RI beam factory. J. Instrum. **15**(12), P12036 (2020)
150. G. Bauer, Overview on spallation target design concepts and related materials issues. J. Nucl. Mater. **398**(1–3), 19–27 (2010)
151. D. Winder, The path to high beam power targets, in *Proceedings of IPAC'24, IPAC'24—15th International Particle Accelerator Conference*, no. 15 (JACoW Publishing, Geneva, 2024). https://proceedings.jacow.org/ipac2024/pdf/WEZN2_talk.pdf
152. T. Kanemura, others, High power targetry devices at FRIB: challenges, status and plan, in *Proceedings of HIAT2025, HIAT*, no. 16 (JACoW Publishing, Geneva, 2025). https://doi.org/10.18429/JACoW-HIAT2025-MOZ03. https://indico.jacow.org/event/82/contributions/6682
153. M. Calviani, Challenges for beam intercepting devices at CERN's high-intensity and high-energy facilities, in *Proceedings of the 4th J-PARC Symposium (J-PARC Symposium 2024)* (2024). https://conference-indico.kek.jp/event/287/contributions/5442/attachments/3946/5393/MCa_JPARC_Symposium_2024_fin_v2.pdf
154. P. Hurh, The Radiation Damage in Accelerator Target Environments (RaDIATE) Collaboration R and D Program—Status and Future Activities, in *Proceedings of IPAC'17* (JACoW Publishing, Geneva), pp. 2550–2553. https://doi.org/10.18429/JACoW-IPAC2017-WEOCB3. https://jacow.org/ipac2017/papers/WEOCB3.pdf
155. M. Futakawa, Targetry overview—various target concepts and expectable next generation targets in power frontier applications, in *Proceedings of the 2nd International Symposium on Science at J-PARC—Unlocking the Mysteries of Life, Matter and the Universe* (2015), p. 001002
156. G. Bauer, Y. Dai, W. Wagner, Sinq layout, operation, applications and R&D to high power. Journal de Physique IV (Proceedings) **12**, 3–26 (2002). EDP sciences
157. W. Sommer, R. Werbeck, S. Maloy, M. Borden, R. Brown, Materials selection and qualification processes at a high-power spallation neutron source. Mater. Charact. **43**(2–3), 97–123 (1999)
158. M. Holding, T. Broome, The status of the ISIS targets, in *Proceedings of the Eleventh Meeting of the International Collaboration on Advanced Neutron Sources (ICANS-XI), KEK–90-25*, vol. 1 (1991), pp. 478–487
159. G. Russell, J. Gilmore, H. Robinson, G. Legate, A. Bridge, R. Sanchez, R. Brewton, R. Woods, H. Hughes, Lansce target system performance, in *Proceedings of International Collaboration on Advanced Neutron Sources (ICANS)*, Los Alamos, NM, 483–496 (1988)
160. L. Rybarcyk, Recent operational experience at the LANSCE facility, in *Proceedings of HB'10* (JACoW Publishing, Geneva), pp. 386–390. https://jacow.org/HB2010/papers/TUO2C03.pdf
161. D. Wilcox, L. Jones, Design and optimisation of the ISIS TS1 project target. J. Phys. Conf. Ser. **1021**, 012058 (2018). IOP Publishing
162. D. Coates, ISIS target station one upgrade project—an overview of the development work being undertaken to improve the Target, Reflector and Moderator (TRaM) support systems. J. Phys. Conf. Ser. **1021**, 012059 (2018). IOP Publishing

163. D.M. Jenkins, The ISIS second target station target, reflector and moderator design and operation, in *Proceedings of the 19th Meeting on Collaboration of Advanced Neutron Sources (ICANS XIX)* (2010)
164. F. Wang, T. Liang, W. Yin, Q. Yu, L. He, J. Tao, T. Zhu, X. Jia, S. Zhang, Physical design of target station and neutron instruments for China Spallation Neutron Source. Sci. China-Phys. Mech. Astron. **56**(12), 2410–2424 (2013)
165. M. Futakawa, K. Haga, T. Wakui, H. Kogawa, T. Naoe, Development of the Hg target in the J-PARC neutron source. Nucl. Instrum. Methods Phys. Res. Sect. A **600**(1), 18–21 (2009)
166. G. Bauer, M. Salvatores, G. Heusener, MEGAPIE, a 1 MW pilot experiment for a liquid metal spallation target. J. Nucl. Mater. **296**(1), 17–33 (2001). https://doi.org/10.1016/S0022-3115(01)00561-X. https://www.sciencedirect.com/science/article/pii/S002231150100561X. 4th Int. Workshop on Spallation Materials Technology
167. S. Halfon, A. Arenshtam, D. Kijel, M. Paul, L. Weissman, D. Berkovits, Y. Eisen, I. Eliyahu, G. Feinberg, A. Kreisel, I. Mardor, G. Shimel, A. Shor, I. Silverman, M. Tessler, First demonstration of high-power proton irradiation and neutron production with a liquid-lithium target, in *Conference of the Nuclear Societies in Israel* (2014). https://inis.iaea.org/records/nww02-6x165
168. H. Nakamura, P. Agostini, K. Ara, S. Cevolani, T. Chida, M. Ciotti, S. Fukada, K. Furuya, P. Garin, A. Gessii, D. Guisti, V. Heinzel, H. Horiike, M. Ida, S. Jitsukawa, T. Kanemura, H. Kondo, Y. Kukita, R. Lösser, H. Matsui, G. Micciche, M. Miyashita, T. Muroga, B. Riccardi, S. Simakov, R. Stieglitz, M. Sugimoto, A. Suzuki, S. Tanaka, T. Terai, J. Yagi, E. Yoshida, E. Wakai, Latest design of liquid lithium target in IFMIF. Fusion Eng. Des. **83**(7), 1007–1014 (2008). https://doi.org/10.1016/j.fusengdes.2008.06.014. https://www.sciencedirect.com/science/article/pii/S0920379608001622. Proceedings of the Eight International Symposium of Fusion Nuclear Technology
169. H.G. Kirk et al., The MERIT high-power target experiment at the CERN PS, in *Proceedings of EPAC'08, European Particle Accelerator Conference*, no. 11 (JACoW Publishing, Geneva, 2008), pp. 2886–2888. https://jacow.org/e08/papers/WEPP169.pdf
170. I. Efthymiopoulos et al., The MERIT (nTOF-11) high tntensity liquid mercury target experiment at the CERN PS, in *Proceedings of EPAC'08, European Particle Accelerator Conference*, no. 11 (JACoW Publishing, Geneva, 2008), pp. 262–264. https://jacow.org/e08/papers/MOPC087.pdf
171. R. Linander, U. Odén, Status update on the ESS target systems development. ICANS XXI paper, TM-P04 (2016)
172. T. McManamy, M.J. Rennich, R.K. Crawford, P.J. Geoghegan, J.G. Janney, Rotating target development for SNS second target station. Oak Ridge National Laboratory, ICANS-XIX (2010)
173. D. Kiselev, P. Baumann, P. Duperrex, S. Jollet, P.R. Kettle, A. Knecht, D. Laube, C. Nyfeler, A. Papa, D. Reggiani et al., Progress and challenges of the PSI meson targets and relevant systems, in *Proceedings of the 3rd J-PARC Symposium (J-PARC2019)* (2021), p. 011102
174. S. Makimura, Y. Kobayashi, Y. Miyake, N. Kawamura, P. Strasser, A. Koda, K. Shimomura, H. Fujimori, K. Nishiyama, M. Kato et al., Development of a muon rotating target for J-PARC/MUSE. Phys. Proc. **32**, 795–801 (2012)
175. E.W. Blackmore, A.S. Dowling, R. Ruegg, M.C. Stenning, Operating experience with meson production targets at TRIUMF, in *Proceedings of PAC'05, Particle Accelerator Conference*, no. 21 (JACoW Publishing, Geneva), pp. 1919–1921. https://jacow.org/p05/papers/RPPE026.pdf
176. E. Lopez Sola, M. Calviani, P. Avigni, M. Battistin, J. Busom Descarrega, J. Canhoto Espadanal, M.A. Fraser, S. Gilardoni, B. Goddard, D. Grenier, R. Jacobsson, K. Kershaw, M. Lamont, A. Perillo-Marcone, M. Pandey, B. Riffaud, S. Sgobba, V. Vlachoudis, L. Zuccalli, Design of a high power production target for the beam dump facility at CERN. Phys. Rev. Accel. Beams **22**, 113001 (2019). https://doi.org/10.1103/PhysRevAccelBeams.22.113001

177. P.G. Bricault, F. Ames, M. Dombsky, V. Hanemaayer, P. Kunz, J. Lassen, Progress towards New RI and Higher RIB Intensities at TRIUMF, in *Proceedings of Cyclotrons'10* (JACoW Publishing, Geneva), pp. 365–370. https://jacow.org/Cyclotrons2010/papers/THM1CIO04.pdf
178. A. Yoshida, K. Morita, K. Morimoto, D. Kaji, T. Kubo, Y. Takahashi, A. Ozawa, I. Tanihata, High-power rotating wheel targets at RIKEN. Nucl. Instrum. Methods Phys. Res. Sect. A **521**(1), 65–71 (2004)
179. F. Pellemoine, M. Avilov, W. Mittig, S. Fernandes, M. Schein, Development of a production target for FRIB: thermo-mechanical studies. J. Radioanal. Nucl. Chem. **299**, 933–939 (2014)
180. S.R. Esparza, I. Nesterenko, M. Hausmann, M. Patil, S. Lidia, Mechanical design of the thermal imaging system for the FRIB target, in *Proceedings of IPAC'24, International Particle Accelerator Conference*, no. 15 (JACoW Publishing, Geneva, 2024), pp. 2287–2290. https://doi.org/10.18429/JACoW-IPAC2024-WEPG36
181. M. Patil, J. Song, N. Bultman, M. Reaume, M. Larmann, R. Quispe, Thermal analysis of rotating single slice graphite target system for FRIB, in *Proceedings of IPAC'24* (2024), pp. 3827–3829
182. A.R. Francia, A.P. Marcone, S. Pianese, K. Andersen, G.A. Izquierdo, J.A. Briz, D.C. Perez, E. Carlier, T. Coiffet, L.S. Esposito, J.L. Grenard, D. Grenier, J. Humbert, K. Kershaw, J. Lendaro, A.O. Rolo, K. Scibor, D. Senajova, S. Sgobba, C. Sharp, D. Steyaert, F.M. Velotti, H. Vincke, V. Vlachoudis, M. Calviani, Design and early operation of a new-generation internal beam dump for CERN's super proton synchrotron. Phys. Rev. Accel. Beams **27**, 043001 (2024). https://doi.org/10.1103/PhysRevAccelBeams.27.043001
183. R. Quispe-Abad et al., Thermal-fluid analysis and operation of a low power water-cooled tilted beam dump at Facility for Rare Isotope Beams, in *Proceedings of IPAC'24, International Particle Accelerator Conference*, no. 15 (JACoW Publishing, Geneva, 2024), pp. 3823–3826. https://doi.org/10.18429/JACoW-IPAC2024-THPS40
184. J. Song, N. Bultman, T. Kanemura, G. Lee, S. Miller, M. Patil, R. Quispe-Abad, M. Reaume, J. Wei, Design and experimental thermal validation of the mini-channel beam dump for FRIB, in *16th International Conference on Heavy Ion Accelerator Technology (HIAT'25)* (2025)
185. M. Avilov, A. Aaron, A. Amroussia, W. Bergez, C. Boehlert, T. Burgess, A. Carroll, C. Colin, F. Durantel, P. Ferrante et al., Thermal, mechanical and fluid flow aspects of the high power beam dump for FRIB. Nucl. Instrum. Methods Phys. Res. Sect. B: Beam Interact. Mater. Atoms **376**, 24–27 (2016)
186. Ansys, Ansys Mechanical. https://www.ansys.com/products/structures/ansys-mechanical
187. J. Staples, J. Ayers, D. Cheng, J. Greer, M. Hoff, A. Ratti, The SNS four-phase LEBT chopper, in *Proceedings of PAC'99*, vol. 3 (IEEE, 1999), pp. 1961–1963. https://jacow.org/p99/papers/WEA32.pdf
188. R.A. Hardekopf, A.V. Aleksandrov, D.E. Anderson, S.S. Kurennoy, J. Power, Design, construction, and initial operation of the SNS MEBT chopper system, in *Proceedings of EPAC'04, European Particle Accelerator Conference*, no. 9 (JACoW Publishing, Geneva, 2004), pp. 150–152. http://accelconf.web.cern.ch/e04/papers/WEOBCH02.pdf
189. S.S. Kurennoy, J.F. Power, Meander-line current structure development for SNS fast chopper, in *Proceedings of LINAC'00, Linear Accelerator Conference*, no. 20 (JACoW Publishing, Geneva), pp. 932–934. https://jacow.org/l00/papers/THD12.pdf
190. I. Kourbanis, P. Adamson, J. Biggs, B. Brown, D. Capista, C. Jensen, G. Krafczyk, D. Morris, D. Scott, K. Seiya et al., A gap clearing kicker for main injector. arXiv preprint arXiv:1208.3414 (2012)
191. S.M. Lidia, Instrumentation design and challenges at FRIB, in *Proceedings of HB'14* (JACoW Publishing, Geneva), pp. 267–272. https://jacow.org/HB2014/papers/WEO2AB01.pdf
192. C. Jensen, R. Reilly, I. Terechkine, Gap clearing kicker magnet for main injector. Technical report, Fermi National Accelerator Lab. (FNAL), Batavia (2009)
193. J. Wei, Beam cleaning in high power proton accelerators, in *AIP Conference, Proceedings*, vol. 693 (American Institute of Physics, 2003), pp. 38–46

194. J. Wei, D.T. Abell, J. Beebe-Wang, M. Blaskiewicz, P.R. Cameron, N. Catalan-Lasheras, G. Danby, A.V. Fedotov, C. Gardner, J. Jackson, Y.Y. Lee, H. Ludewig, N. Malitsky, W. Meng, Y. Papaphilippou, D. Raparia, N. Tsoupas, W.T. Weng, R.L. Witkover, S.Y. Zhang, Low-loss design for the high-intensity accumulator ring of the Spallation Neutron Source. Phys. Rev. ST Accel. Beams **3**, 080101 (2000). https://doi.org/10.1103/PhysRevSTAB.3.080101
195. A. Plastun, N. Bultman, S. Carter, T. Ginter, T. Kanemura, M. Larmann, M. LaVere, F. Marti, S. Miller, I. Nesterenko et al., Advanced charge selector for stripped heavy ion beams. Target **8**, 11 (2024)
196. R.J. Macek, PSR experience with beam losses, instabilities and space charge effects, in *AIP Conference, Proceedings*, vol. 448 (American Institute of Physics, 1998), pp. 116–127
197. T. Trenkler, J.B. Jeanneret, The principles of two stage betatron and momentum collimation in circular accelerators. Part. Accel. **50**(CERN-SL-95-03-AP), 287–311 (1995)
198. H. Ludewig et al., Collimator systems for the SNS ring, in *Proceedings of PAC'99* (JACoW Publishing, Geneva), pp. 548–550. https://jacow.org/p99/papers/THCR4.pdf
199. N. Simos, H. Ludewig, D. Raparia, N. Catalan-Lasheras, J. Brodowski, G. Murdoch, SNS collimating system design—performance and integration, in *AIP Conference, Proceedings*, vol. 693 (American Institute of Physics, 2003), pp. 162–166
200. N. Catalan-Lasheras, Y. Lee, H. Ludewig, N. Simos, J. Wei, Optimization of the collimation system for the Spallation Neutron Source accumulator ring. Phys. Rev. Spec. Top. Accel Beams **4**(1), 010101 (2001)
201. V.M. Biryukov, Y.A. Chesnokov, V.I. Kotov, *Crystal Channeling and Its Application at High-Energy Accelerators* (Springer, Berlin, 2013)
202. D. Kalchev, Two-stage betatron and momentum collimation studies with applications to LHC, in *7th ICFA Miniworkshop on High Intensity High Brightness Hadron Beams: Beam Halo and Scraping* (2000), pp. 101–106
203. C.M. Warsop, Studies of beam loss control on the ISIS synchrotron, in *Proceedings of 9th European Particle Accelerator Conference (EPAC)*, Lucerne (2004), p. 1464
204. K. Yamamoto, S. Kato, H. Harada, M. Yoshimoto, Y. Yamazaki, N. Hayashi, P.K. Saha, H. Hotchi, M. Kinsho, Performance of the beam collimation system in the J-PARC RCS. Prog. Nucl. Sci. Tech. **4**, 243–246 (2014)
205. S. Redaelli et al., Crystal collimation of heavy-ion beams at the Large Hadron Collider. Phys. Rev. Accel. Beams **28**(5), 051001 (2025). https://doi.org/10.1103/PhysRevAccelBeams.28.051001
206. W. Scandale et al., Observation of channeling for 6500 GeV/c protons in the crystal assisted collimation setup for LHC. Phys. Lett. B **758**, 129–133 (2016). https://doi.org/10.1016/j.physletb.2016.05.004
207. D. Mirarchi et al., Reducing beam-related background on forward physics detectors using crystal collimation at the large hadron collider. Phys. Rev. Applied **14**(6), 064066 (2020). https://doi.org/10.1103/PhysRevApplied.14.064066
208. F. Velotti et al., Demonstration of non-local crystal shadowing at the CERN SPS, in *Proceedings of IPAC'23, International Particle Accelerator Conference*, no. 14 (JACoW Publishing, Geneva, 2023), pp. 271–274. https://doi.org/10.18429/JACoW-IPAC2023-MOPA098
209. A.V. Aleksandrov, The SNS beam diagnostics experience and lessons learned, in *Proceedings of BIW'10* (JACoW Publishing, Geneva), pp. 30–33 (2010). https://jacow.org/BIW2010/papers/3121MOIMNB02.pdf
210. R. Connolly, P. Cameron, T. Shea, R. Witkover, Ionization Profile Monitor for SNS. Technical report, Brookhaven National Lab.(BNL), Upton (1999)
211. W. Blokland et al., SNS laser profile monitor progress, in *Proceedings of EPAC'04, European Particle Accelerator Conference*, no. 9 (JACoW Publishing, Geneva, 2004), pp. 2852–2854. http://accelconf.web.cern.ch/e04/papers/THPLT167.pdf
212. R.L. Witkover, P.R. Cameron, T.J. Shea, R.C. Connolly, M. Kesselman, Beam instrumentation for the Spallation Neutron Source ring, in *Proceedings of PAC'99* (JACoW Publishing, Geneva), pp. 2250–2252. https://jacow.org/p99/papers/WEA169.pdf

213. W.H. DeLuca, Beam detection using residual gas ionization. IEEE Trans. Nucl. Sci. **16**(3), 813–822 (1969)
214. G. Burtin et al., The luminescence profile monitor of the CERN SPS, in *Proceedings of EPAC'00, European Particle Accelerator Conference*, no. 7 (JACoW Publishing, Geneva, 2000), pp. 256–258. https://jacow.org/e00/papers/TUOAF103.pdf
215. R. Rosenberg, K. Harkay, A rudimentary electron energy analyzer for accelerator diagnostics. Nucl. Instrum. Methods Phys. Res. Sect. A **453**(3), 507–513 (2000)
216. T. Toyama, Electron cloud effects in the J-PARC rings and related topics (2005)
217. H. Friedsam, M. Penicka, S. Zhao, Survey and alignment report on the primary control network for the APS. Technical report, Argonne National Lab.(ANL), Argonne (1993)
218. P.E. Manwiller, Three-dimensional network adjustment of laser tracker measurements for large-scale metrology applications. J. Surv. Eng. **147**(1), 05020009 (2021)
219. J.J. Error, D.R. Bruce, J.J. Fazekas, S.A. Helus, J.R. Maines, Deformation monitoring of the spallation neutron source (SNS) tunnels, in *Proceedings of PAC'05, Particle Accelerator Conference*, no. 21 (JACoW Publishing, Geneva), pp. 2509–2511. https://jacow.org/p05/papers/WPAE037.pdf
220. J. Grillenberger, S.H. Kim, M. Yoshii, M. Seidel, V. Yakovlev, Energy efficiency and saving potential analysis of the high intensity proton accelerator HIPA at PSI, in *Proceedings of IPAC'17* (JACoW Publishing, Geneva), pp. 2399–2402. https://doi.org/10.18429/JACoW-IPAC2017-TUPVA129. https://jacow.org/ipac2017/papers/TUPVA129.pdf
221. R. Dortwegt, Low-conductivity water systems for accelerators, in *Proceedings of PAC'03* (JACoW Publishing, Geneva), pp. 630–634. https://jacow.org/p03/papers/ROPB008.pdf
222. M. Thuot, Resolving EMI issues to optimize accelerator beam diagnostic performance, in *AIP Conference, Proceedings*, vol. 732 (American Institute of Physics, 2004), pp. 47–67

Chapter 9
Phases of an Accelerator Facility

Abstract This chapter describes the main phases of the life cycle of establishing a high-power hadron accelerator facility: conceptualization, design, construction, commissioning, and operation. Using the US Department of Energy accelerator project establishment process as an example, we focus on the key requirements, milestones, and deliverables of such a phase. In addition to understanding accelerator physics and comprehending the technology, these managerial aspects are essential to deliver a successful accelerator project.

A major accelerator facility usually takes more than 10 years to prepare and construct and then has an operational life span of more than 10 years. With proper maintenance and upgrades, some facilities have been in active service for 50 years or more. The accelerator facility typically undergoes periods of conceptualization, project construction, and facility operations. The construction project also undergoes periods of design, construction, and commissioning. This chapter describes the phases of these accelerator facilities and projects.

For example, major accelerator construction projects sponsored by the US Department of Energy generally have the following high-level milestones [1]: approve mission needs (Critical Decision 0, or CD-0), approve alternative selection and cost range (CD-1), approve performance baseline (CD-2), approve construction (CD-3), and approve operations (CD-4). With these milestones, the construction project can be defined in four phases: initiation (pre-CD-0), definition (from around CD-0 to around CD-1), execution (from after CD-1 to CD-4), and closure (on and after CD-4).

9.1 Preparation

This section discusses tasks conducted in preparation for a new accelerator facility. Actual accelerator project examples are used to illustrate the main points.

J. Wei et al., *High Power Hadron Accelerators*, Particle Acceleration and Detection,
https://doi.org/10.1007/978-3-032-17662-2_9

9.1.1 Mission Need

The mission need is to justify two aspects: (1) the facility to be established is likely to produce capacity that supports the strategic plan of the funding agency and meets the demands of the represented scientific community, and (2) there is a mission requirement that cannot be met through non-material means. The approval of the mission need is the authorization to develop alternative concepts and functional requirements.

9.1.2 Research and Development

Establishing a frontier accelerator almost always requires R&D in some key beam dynamics and technology aspects. Cutting-edge design and technology lead to performance breakthroughs and affordable costs. On the other hand, the risks associated with uncertainties in R&D progress and outcomes require mitigation plans.

The SNS project (Sect. 10.1) adopted mature technology as much as possible for its driver accelerators [2]. The NC linac design was mainly based on the LANSCE linac. The design of the SRF linac and associated cryogenics was primarily based on the JLab practices. Major accelerator R&D focuses on low-loss accelerator design and programs to mitigate uncontrolled beam loss: phase-space beam collimation, electron cloud mitigation, and ring extraction kicker coupling impedance reduction. The R&D on the liquid mercury target was carried out in collaboration with J-PARC to mitigate target vessel cavitation caused by high-power beam pulses.

The J-PARC project (Sect. 10.2) pursued a wide range of R&D programs that benefited the accelerator community [3]. The RCS was designed with an imaginary transition energy to avoid a high-intensity beam crossing the transition energy. Hardware R&D programs included high-gradient RF cavities using water-cooled magnetic alloys, laminated magnets with water-cooled coils, and stranded coils to avoid high eddy currents, and vacuum chambers with thin shielding that had a high impedance to eddy currents but a low impedance to the beam.

FRIB R&D (Sect. 10.3.3) mainly focused on four topics: low-β cryomodules containing accelerating resonators and focusing solenoids, along with modular cryogenic distribution; charge stripping with circulating liquid lithium and with helium gas confined by plasma windows; multi-layer, multi-time-scale machine protection for high intensity, low energy heavy ions with high stopping power and low detection sensitivity; and high power secondary beam production targets made of multi-slice rotating graphite [4].

9.1.3 Alternative Selection

Alternative selection includes site selection and pre-conceptual design scenario selection. Site selection often involves competition between candidate host facilities. The selection of the design scenario usually involves a comparison between different accelerator configurations and key technologies, which are almost always related to the performance envelope and technology R&D prioritization.

Before the SNS project was established, several national laboratories in the United States participated in the competition, advocating their own design scenarios. ORNL was selected to host the project in partnership with four other laboratories (Sect. 10.1.1): LBNL for the front end, LANL for the linac, BNL for the ring, ORNL for the target, and ANL for the instruments. Later, alternative selections were made that retained the accumulator ring design over the subsequently proposed rapid-cycling synchrotron design (Sect. 10.1.3.4) and selected the superconducting RF linac over the previous normal-conducting linac design (Sect. 10.1.3.2). With the change in linac technology, JLab was added as a new partner laboratory. The selected accelerator configuration best serves the single-purpose, high-power neutron production.

J-PARC selected a rapid-cycled synchrotron over an accumulator ring to take advantage of the energy gain that was needed for subsequent acceleration by the main ring, with beams injected from a normal-conducting linac (Sect. 10.2.3). The selected accelerator configuration best served a multipurpose facility that required beams for neutron production at a lower energy and for experiments at a higher energy.

Rare isotope accelerator facilities can be driven by a cyclotron (RIKEN), a synchrotron (FAIR), or a linac (SPIRAL2). FRIB project selected a CW SRF linac as its driver accelerator to achieve high beam power for secondary beam productivity [5] (Sect. 10.3).

9.1.4 Conceptual Design

A conceptual design report (CDR) documents the mission need, the conceptual design after alternative selections, the accelerator parameter range, the preliminary acquisition approach, and the project cost range based on the R&D and component prototyping progress. The conceptual design is expected to be site-specific, with environmental impact and geotechnical studies conducted. The report usually covers accelerator systems, experimental systems, and conventional facility design, as well as environmental, safety, health, and quality (ESH&Q) aspects.

The conceptual design of accelerator systems includes the accelerator lattice layout and beam dynamics analysis, as well as the physical, mechanical, and electrical considerations of the accelerator. The design is expected to cover all major subsystems, including global controls, personnel protection, and machine protection systems, along with performance parameters, expectations, and specifications.

For a major accelerator construction project, a conceptual design report (CDR) is sometimes preceded by a pre-conceptual design report (PDR) that documents design concepts before R&D is concluded, and followed by a technical design report (TDR) that documents the technical configurations during construction. The outline of the J-PARC accelerator TDR is shown as an example in Appendix II: J-PARC Accelerator Technical Design Report Outline.

9.1.5 Cost Control and Value Engineering

Cost control is vital to the successful establishment and completion of any large-scale accelerator project. Cost control is performed both "top-down" by management enforcement based on funding profiles and "bottom-up" by experienced system owners based on itemized estimates.

Before selecting the host and site of the project, competition among competing proposals is often the most effective method to control the overall costs of the project. Before the project baseline (Sect. 9.1.6), cost scrubbing and value engineering are necessary to confine the total project cost within the allocated funding budget. An effective cost scrubbing practice requires qualified technical experts to accept critical input from experienced peers working under pressure imposed by a conscientious project leader.

Before the establishment of the FRIB project (Sect. 10.3), fierce competition between two institutes in the RIA project led to the optimization of resource allocation and conceptual design, resulting in a reduction in project costs of almost 50%. FRIB is highly successful. Although the RIA project itself was not realized, its concepts and research contributed to the successful establishment of the FRIB project.

Similarly, before the establishment of the SNS project (Sect. 10.1), several national laboratories in the US competed with their own design concepts and efforts. Later, partnerships were formed among these laboratories to establish the SNS project. Section 10.1.4 describes the value engineering practice carried out with the SNS project.

9.1.6 Performance Baseline

The baseline refers to the scope, schedule, and cost of a project. The performance baseline of an accelerator construction project defines the scope to be delivered at a specified cost and schedule during the project period, from R&D to commissioning for operations.

Table 9.1 FRIB key performance parameters (KPP) at CD-4

System	Parameter	Performance criteria
Accelerator system	Accelerate heavy-ion beam	Measure FRIB driver linac ^{36}Ar beam with energy larger than 200 MeV/nucleon and a beam current larger than 20 pnA
Experimental systems	Produce a fast rare isotope beam of ^{84}Se	Detect and identify ^{84}Se isotopes in FRIB fragment separator focal plane
	Stop a fast rare isotope beam in gas and reaccelerate a rare isotope beam	Measure reaccelerated rare isotope beam energy larger than 3 MeV/nucleon
Conventional facilities	Linac tunnel	Beneficial occupancy of subterranean tunnel structure of approximately 500 ft path length (minimum) to house FRIB driver linear accelerator
	Cryogenic helium liquefier plant – building and equipment	Beneficial occupancy of the cryogenic helium liquefier plant building and installation of the helium liquefier plant complete
	Target area	Beneficial occupancy of target area and one beam line installed and ready for commissioning

The scope of the project includes the work that will be performed to achieve key performance parameters (KPP), marking the completion of the construction project and the start of operations. It also includes the deliverables (usually hardware) to meet the ultimate performance parameters (UPP). The work is usually organized in terms of the work breakdown structure (WBS), which is a hierarchical decomposition oriented toward deliverables of the work to be performed by the project team, organized in an organizational breakdown structure (OBS).

Table 9.1 lists the FRIB KPP that partially satisfies the CD-4 requirements of the construction project.

The project schedule can be arranged using the critical path method, constructing a model of the project that includes the following: (1) a list of all activities required to complete the project, as categorized within the work breakdown structure; (2) the time duration that each activity will take to complete; (3) the dependencies between the activities; and (4) logical endpoints such as milestones or deliverable items. This process determines the critical path as the sequence of activities in the project that adds up to the longest overall duration (Fig. 9.1).

The project cost is directly associated with each WBS item, with spending profiles determined by the scheduled activities. Figure 9.2 and Table 10.9 show the breakdown of the FRIB total project cost (TPC). The basis of estimate (BOE) is established based on the preliminary acquisition strategy, which includes direct costs, indirect costs (overhead), contingency, and escalation. The cost is divided into two general categories: labor and non-labor (Fig. 9.2). Labor costs refer to in-house efforts for project management, R&D, design, integration and coordination, in-house fabrication, installation, integrated tests, and beam commissioning. Labor

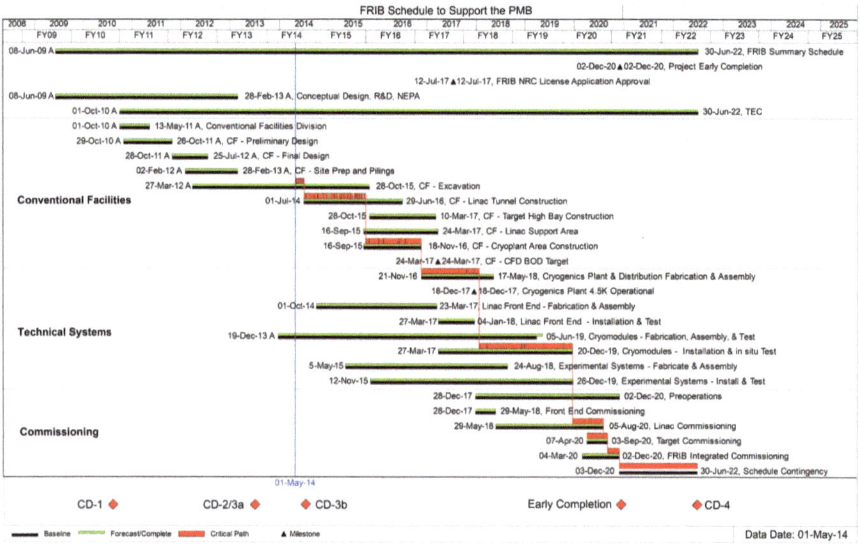

Fig. 9.1 FRIB project baseline schedule defined in May 2014 (Table 10.1). The critical path of the project is indicated in red color. The project officially achieved CD-4 in April 2022, two months ahead of the baseline schedule

hours are estimated based on previous experience, taking into account expert opinion, trend analysis, learning curves, and mass production savings. Non-labor costs include industrial procurements of technical components, work-for-others contracts, and civil construction contracts. Non-labor costs are estimated based on supplier quotes, engineering judgment, historical data, and cost databases/cost books. The costs of construction spares are also included in the procurements to be executed during the construction phase of the project.

Figure 9.3 shows the funding profile for the SNS construction project [6]. The prior year funding covered early R&D and conceptual design efforts before the project was baselined in December 1997 (US fiscal year 1998), as discussed in Sect. 10.1. The hosting laboratory, ORNL, distributed the funds from the US DOE to other partner laboratories, which covered both labor and materials for scope delivery. An agreement was in place in which reduced overhead rates were applied to both labor and material charges.

A major US DOE accelerator project is required to use the earned value management system (EVMS) (Table A.4) to measure and report the schedule and cost performance of the project compared to the approved baseline, ensuring that the project is on schedule and within budget.

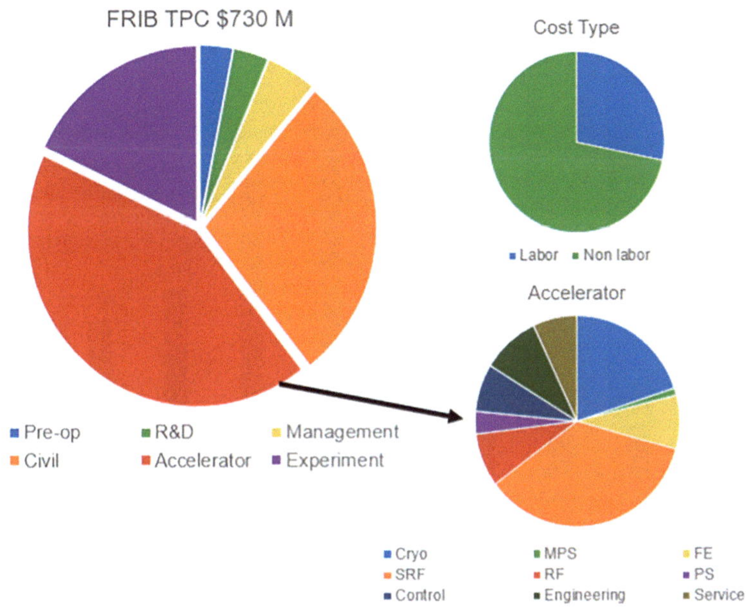

Fig. 9.2 FRIB total project cost (TPC) breakdown, accelerator cost breakdown, and cost type breakdown upon completion of the construction project in May 2022

9.1.7 Long-Lead Procurements

With the project baseline established and the preliminary design in place, long-lead procurements have been identified. Long-lead procurements can be initiated prior to the official start of technical systems (CD-3) to meet the project baseline schedule.

FRIB obtained approval for long-lead procurement (CD-3a) at the same time that the project baseline was approved (Fig. 10.7). Long-lead procurements were launched 2 years before the start of technical construction (CD-3b). They included the high-power ECR ion source, the continuous-wave radio frequency quadrupole (RFQ), the niobium material for the SRF cavities, the pre-production SRF cavities, the 4.5 K refrigeration "cold box," and the cryogenic distribution prototyping.

For some recent accelerator projects for which reaching CD-2 is challenging, CD-3a is obtained prior to CD-2, allowing long-lead procurements on parts of the scope that are adequately mature, particularly when funds are available to execute such procurements.

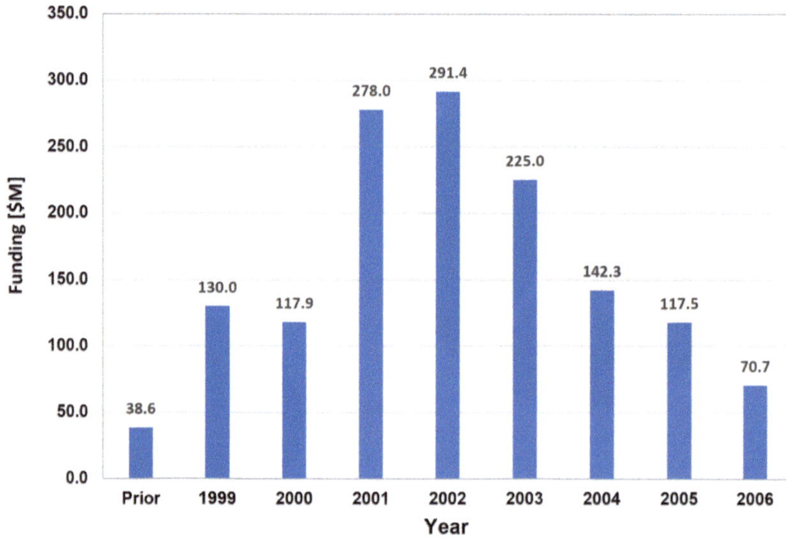

Fig. 9.3 SNS funding profile by US fiscal year upon completion of the construction project in 2006

9.1.8 Workforce Development and Compatibility

Recruiting and developing a qualified workforce with experienced subject matter experts (SMEs) is essential for the successful execution of a frontier accelerator project. Seasoned SMEs are crucial not only to the R&D progress, which must align with the project plan, but also for the delivery of the construction baseline. Strategic recruitment is often necessary to strengthen technical capabilities in targeted areas that meet project needs.

For example, pursuing a power-frontier accumulator ring (AR)-based facility likely requires expertise and a workforce in cryogenics and SRF for the establishment of a full-energy linac. SNS established a partnership based on the available expertise and relied on the workforce of partner laboratories to deliver the project on time (Sect. 10.1).

During the establishment of FRIB, approximately 70% of SMEs in various disciplines were recruited worldwide, including cryogenics, SRF, liquid metal, SC magnets, accelerator physics, rare isotope physics, controls, diagnostics, mechanical engineering, electrical engineering, ion sources, and high-power targetry. Acquiring and retaining world experts in key areas is essential to the success of FRIB in delivering the project with a minimum funding budget (Sect. 10.3.7).

The workforce needed for a construction project includes scientists, engineers, and technicians. The practice of workforce development differs across countries in the world. For example, in the United States, labor expenses are part of the total project cost. The ramp-up and ramp-down of the workforce need to match the funding profile. SNS addressed this challenge by utilizing the workforce of partner laboratories during the peak construction period (Sect. 10.1). In Japan, industrial partners and contractors are heavily involved in providing a workforce for operations and technical support.

9.1.9 In-House Production Infrastructure Readiness

Similar to the long-lead procurements, the establishment of the infrastructure necessary to carry out in-house construction work should be initiated early enough to meet the baseline schedule.

FRIB started field preparation and utility preparation (e.g., electrical substations) around the same time as the project baseline. Before starting technical construction, the 2500 m^2 SRF building (Fig. 9.4) was established on site to house the areas for material inspection, cavity coordinate measurements and alignment checks, vacuum furnace degassing, parts cleaning, chemical etching, high pressure water rinsing, SRF coupler conditioning, cavity Dewar test, cold mass assembly, and cryomodule testing. This facility, together with the cryomodule assembly area and the machine shop, supports the production throughput of testing five cavities per week and one cryomodule per month.

9.2 Construction

Upon approval from the funding agency, construction of the accelerator begins according to the established baseline. The execution of the project follows the established Project Execution Plan (PEP).

9.2.1 Project Execution Plan

The PEP is developed as the governing document that establishes the means to execute, monitor, and control the accelerator construction project. The PEP serves as the main communication vehicle to ensure awareness and knowledge of the project objectives and the general approach to project achievement. An outline of the FRIB PEP is shown as an example in Appendix I: FRIB Project Execution Plan Outline.

Fig. 9.4 The SRF building established at MSU for in-house processing and certification of SRF sub-components and cryomodules. The facility was established in 3 year, just in time for the FRIB cryomodule mass production. (Courtesy MSU, USA)

9.2.2 Configuration Management

Configuration management is essential for the execution of a baselined construction project to ensure the delivery of the specified scope on cost and on schedule.

9.2.2.1 Parameter List, Lattice File, CCDB

Given the complexity of a major accelerator project, it is important to maintain version-controlled parameter lists to document the machine design configuration that meets user expectations and project constraints, and to guide the establishment of system requirements (as documented in the system requirements documents, SRD) and interfaces (as documented by interface control documents, ICD). Correspondingly, version-controlled lattice files document the configuration of the optical devices along the beam trajectory.

FRIB uses its central configuration database (CCDB) to register the acceptance, installation, and readiness status of the devices for beam commissioning. Other configuration management means include design engineering drawings stored in Adept and control codes stored in AssetCentre (Table A.4 of Chap. A).

The outline of the FRIB parameter list is shown as an example in Appendix IV: FRIB Parameter List Outline. The lattice file and detailed system requirements are documented in the appendix of the parameter list. The SNS major parameter list is shown as another example in Appendix III: SNS Major Parameter List.

9.2.2.2 Project Change Request

The project change request (PCR) records changes in the scope, schedule, and cost of a WBS element. Depending on the level of impact on the project, major PCR often require approval from the funding agency to ensure that the scope, cost, and schedule of the project are not compromised.

9.2.3 Contingency, Risk Register, and Reverse Contingency

The project cost contingency is the amount of funds within the project TPC that is reserved to cover cost uncertainties and risk exposures. Accelerator projects in the United States typically carry a contingency of about 30% at the time of the project baseline (CD-2). FRIB maintained a contingency of >30% of the work remaining throughout the construction project period.

The project schedule contingency is a reserve in the schedule to cover potential delays or the time needed to resolve unexpected issues. As the workforce draws a salary (the "standing army" cost), an extension in the schedule almost always corresponds to an increase in the cost.

The risk register tracks potential risks, enabling proactive mitigation and ensuring project success. The risk items may be either retired or realized. The realized risk draws funds from the contingency.

Scope creep refers to the uncontrolled expansion of the scope of a project, including the addition of new features and "improvements" that were not agreed upon as part of the baseline. It often leads to potential delays and budget overruns; therefore, it must be avoided.

The project scope contingency refers to the work that can be performed if there are remaining funds upon project completion. The scope reverse contingency refers to a strategy in which, instead of allocating additional funds for potential project overruns, the project team identifies specific deliverables or features that can be deferred or reduced in scope if necessary to ensure budget and schedule adherence. The reverse contingency can only be justified if the baseline goals are not compromised.

9.2.4 *Engineering Design*

Following the conceptual design, the engineering design is performed prior to commercial procurement or in-house fabrication. Engineering design can be divided into conceptual (~30%), preliminary (~60%), and final (~90%) stages. At each stage, major design reviews are conducted.

In the conceptual design stage, the requirements, specifications, and interfaces of the systems are determined, and the safety hazards are identified and mitigated. Consequently, a series of documents is produced at this stage: the system requirements document (SRD), the interface control document (ICD), the hazard analysis (HA), and the safety assessment document (SAD).

The preliminary design stage corresponds to the project baseline. In the intermediate design stage, the acquisition strategy is defined. Commercial procurement can be build-to-specification, build-to-model, build-to-print, or commercially off-the-shelf (COTS). In-house fabrication can be based on individually procured sub-components. At this stage, device assembly drawings or system piping and instrumentation diagrams (P&ID) are usually completed, along with detailed engineering specifications and requirements.

In the final design stage, models and/or drawings are produced ready for procurement; the statement of work and the acceptance criteria list are developed; the procurement strategy is defined; and installation and commissioning plans are developed. For build-to-specification or build-to-model procurement orders, the final design review is often conducted together with the awarded commercial supplier. The maturity of the main subsystem design can be assessed according to the following six questions:

- Are the requirements and interfaces defined?
- Is the appropriate prototyping completed?
- Do prototypes meet specifications?
- Is the detailed design complete?
- Has the design been reviewed?
- Does a procurement/fabrication plan exist?

9.2.5 *Procurements*

Commercial procurements are carried out by business representatives in close coordination with the project's technical staff. Except for COTS products, accelerator equipment procurements usually consist of the following steps: (1) development of an acquisition strategy; (2) development of the statement of work, specification, and acceptance criteria listings; (3) issuance of a request for proposal (RFP) or request for information; (4) proposal selection and contracting; (5) contract execution; and (6) equipment acceptance.

The acquisition strategy includes build-to-print, build-to-specification, and those in-between, such as build-to-model:

- Build-to-specification is when a supplier is responsible for both the design and manufacturing of equipment that conforms to the requirements and specifications regarding its operation. The supplier has the liberty to utilize their design expertise and manufacturing skills to build to the given specifications. This approach is most appropriate when the supplier is experienced in delivering equipment with similar specifications and interface conditions. The project team must work closely with the supplier to optimize the specifications, ensure interface conformance, and perform system-level integration.
- Build-to-print is when a supplier produces work instructions, assembly drawings, and calls out detailed manufacturing practices used in building the equipment in accordance with the engineering drawings provided by the project. The project is in control of the intellectual property rights while having the flexibility to select appropriate suppliers. This approach is most appropriate when the project team has completed R&D and is ready to transfer the technology to the supplier for mass production.
- Build-to-model is compromising the two approaches mentioned above. By providing engineering models or manufacturing road maps instead of detailed engineering drawings, the project team saves engineering labor while allowing for engineering design iteration and optimization with the supplier.

For example, FRIB procured its 4.5 K cryogenic plant using the build-to-specification approach, benefiting from the similarity to the cryogenic plants at SNS and JLab, which were built by the same supplier. The SRF cavities are procured using the build-to-print approach, with multiple suppliers mass producing about 350 cavities. The NC magnets are procured using the build-to-model approach, with experienced magnet suppliers iterating the design model for engineering optimization.

Active engagement and timely interaction with a supplier are key to procurement success. In major procurements, daily communications, weekly teleconferencing, and frequent visits are common practices. Extra efforts are needed with a supplier in a foreign country, sometimes requiring local third-party engineering coordination and quality monitoring.

9.2.6 In-House Fabrication

While industrial providers are preferred for the mass production of accelerator equipment and components due to manufacturing efficiency, quality consistency, and resource optimization, in-house fabrication generally focuses on technology development, prototyping, critical processing and assembly, and one-of-a-kind device fabrication. In-house production quality fabrication requires rigorous work instruction and procedure establishment, work control and work plan development, and ESH&Q conformance assurances.

With the FRIB project, approximately 65% of the US$438.6 M scope of accelerator and experimental system construction (Table 10.9) was for materials and work-for-others contracts, and approximately 35% was for internal labor. At peak construction, a total of about 220 full-time equivalent staff managed the technical aspects of the procurements and performed in-house work in the areas of accelerator physics and integration, controls (hardware, high-level, personnel protection, machine protection, and global timing), mechanical engineering, electrical engineering (RF, power supply, and electronics), cryogenics, SRF and SC magnets, front end, high power targetry (charge stripping and selection, target, beam dump, remote handling, and non-conventional utility), beam instrumentation and measurement, and rare isotope separation. Major in-house FRIB fabrication included the processing and certification of SRF cavities, cryomodule assembly, SC magnet assembly, cryo-distribution fabrication for experimental systems, and high-power targetry device assembly.

9.2.7 *Digital Hygiene and Standardization*

The objective of digital hygiene is to reduce the vulnerability of network security and increase efficiency through standardization. Digital Hygiene applies throughout the life cycle of the device or software.

The FRIB digital hygiene program [7] applies to all FRIB control networks, including the accelerator system controls network, the experimental systems controls network, the cryogenics network, the building automation and control network (BACnet), the K500 SEE network, and the development network. The program determines the suitability of devices and software for installation in a FRIB control system network through risk-based analysis.

Standardization applies to both in-house developed and commercially procured devices and products. FRIB has developed the general purpose digital board (FGPDB) to control various aspects of the facility, including the LLRF controller [8], beam diagnostics [9], and machine protection [10]. This FPGA based board provides a standardized platform for controlling different parts of the accelerator and experimental setup. FRIB also purchases all programmable logic controllers (PLCs) from the same trusted brand.

9.2.8 *Receiving, Acceptance, Test, Storage, Installation*

Receiving, acceptance, testing, and storage (RATS) are steps in preparation for the installation of the equipment. Upon delivery, receiving, and timely acceptance of purchased equipment and components often require planned space and resource coordination. Storage planning for both purchased and internally manufactured equipment facilitates orderly installation.

The installation of equipment can start when the civil construction reaches the date of "ready for equipment," which can be well ahead of the "beneficial occupancy date" (BOD) when the full civil capacity is in place, including heating, ventilation, and air conditioning (HVAC). Installation usually starts with utility (electrical substations and distribution, water treatment facilities), mechanical (survey monuments, equipment stands), electrical (racks, cable trays, cabling, and control networks), and cryogenic (cryogenic refrigeration and distribution) systems, followed by staged technical equipment installation.

9.2.9 Project Reviews

Reviews by both the funding agency and peer subject matter experts are essential to ensure the proper execution of the project, achieving the design goals and project deliverables. During review preparation, project teams usually organize their thoughts, reflect on lessons learned, and adjust plans and work priorities to reduce technical risks and get back on track toward the baseline. Reviews by seasoned subject matter experts external to the project are particularly useful in addressing challenges and technical issues.

An example is given in Appendix VI: List of FRIB External Reviews and Workshops on Accelerator Systems from CD-1 (September 2010) to CD-4 (April 2022). This list includes final design reviews of major accelerator systems, accelerator readiness reviews, semiannual technical advisory reviews, and semiannual reviews by the funding agency.

9.3 Commissioning

The objective of accelerator commissioning is to perform an integrated test of accelerator systems upon completion of the construction and to accelerate charged particle beams to reach the key performance parameters (KPP), demonstrating that the constructed accelerator is capable of meeting design capabilities.

9.3.1 Safety and Safety Envelopes

Safety is a primary concern in accelerator commissioning and operation. The accelerator safety envelope (ASE) defines the boundary conditions and limitations for the safe and environmentally sound operations of an accelerator facility. For example, the ASE of the FRIB facility mainly includes radiological dose, radiological effluents, and groundwater activation limits under the regulation of the state of Michigan and the US Nuclear Regulatory Commission.

As described in Sect. 8.1, a hazard analysis (HA) must be performed and documented to provide a comprehensive assessment of the hazards of the facility and the accident scenarios that can compromise the ASE. The analysis needs to exhaustively include all foreseeable hazard scenarios associated with all operating scenarios over the entire life-cycle of the accelerator facility. Based on the HA, a safety assessment document (SAD) is prepared containing the mitigation measures for each identified hazard.

The operational safety envelope (OSE) and the conduct of operations (CO) are the usual means that accelerator physicists and operators use to mitigate identified hazards and prevent violations of the ASE (Table 9.2). The OSE is usually enforced with credited design features (such as shielding), credited engineered controls, and credited administrative controls (Sect. 8.1.1) with enforced beam parameter limits.

In addition to the ASE, HA, SAD, OSE, and CO, other important programs and procedures that are usually in place before the start of commissioning include the unreviewed safety issue procedure (USI) and the configuration management plan.

9.3.2 Roles, Responsibilities, Accountabilities and Authorities

R2A2 (roles, responsibilities, accountabilities, and authorities) is a framework that ensures team members understand the chain of command, their responsibilities, the accountability to related team members and for the results, and the decision-making authorities. A well-defined R2A2 helps avoid confusion and often leads to more efficient and effective teamwork. As an example, Table 9.2 shows the responsibilities and accountabilities related to safety during the commissioning and operation of the FRIB accelerator facility.

In addition to the line management structure defined by the OBS (Sect. 9.1.6), matrixed roles are often appointed to further clarify responsibilities. For example, the FRIB project appointed system builders and system owners for each system and the major devices, who are responsible for the construction and future operation of the system or device. Area managers were appointed to each major area, such as the cryoplant building, the SRF building, the front end building, the linac tunnel, the target hall, the fragment separator area, and the various experimental vaults. They were responsible for identifying hazards and enforcing personnel protection plans. Segment managers were appointed to each major beamline segment, such as the front end, the linac segments, the folding segments, the target beam delivery segment, the fragment separator segment, the secondary beam segments, and the reaccelerator segments, responsible for the integrated design, construction coordination, commissioning, and operational performance of the segment.

Work coordination becomes increasingly important as the project progresses from fabrication to the stages of installation, integrated testing, and commissioning. Daily work control meetings are often necessary to coordinate the safe execution of

Table 9.2 Safety responsibility and accountability for FRIB accelerator commissioning and operations

Responsible entity	Means and methods	Accountable to
University (MSU)	Accelerator Safety Envelope (ASE)	State of Michigan; NRC
Laboratory (FRIB)	Hazard Analysis (HA) Safety Assessment Document (SAD)	State of Michigan; NRC; MSU
Accelerator Systems Division (ASD)	Operational Safety Envelope (OSE) Conduct of operations	FRIB
ASD	Credited Engineered Controls: validation forms in DCC	
ASD	Credited Administrative Controls: record in Hour-log	
ASD	Shielding: validation forms in DCC	
ASD	Beam parameter limits: record in Hour-log	
AP[a]/Operations[b]	Energy	ASD
AP[a]/Operations[b]	Power (current & duty)	ASD
Front end[c]	Beam species	ASD
Fragment separation[d]	Secondary beam rigidity	ASD
ASD	Dispersible waste: survey, monitor, record	
Operations[b]	Staffing requirements	ASD

[a] Accelerator Physics Department
[b] Operations Department
[c] Front End Department
[d] Fragment Separation Department

planned work. Again, a well-defined R2A2 is essential: the line manager qualifies the workers and is responsible for their training; the system owner certifies the workers to conduct work on their system; and the area manager is responsible for authorizing the work to be carried out safely in their area.

9.3.3 Device Readiness and Integrated Tests

During the construction stage, individual devices and sub-systems are tested to validate their performance readiness before being installed in the accelerator vault/tunnel. The main purpose of an integrated test after installation is to ensure readiness for beam commissioning.

Device readiness reviews (DRR) are often conducted to assess hardware readiness. Such reviews focus on the mitigation of identified hazards to the environment, safety, and health, as well as on device operation readiness. Integrated tests are to be performed only after the hazards have been fully mitigated and the operating procedures have been fully developed.

Integrated tests involve devices installed mechanically and electrically, safety measures and hazard mitigation, device hardware control, configuration control, and operator training. The successful completion of the integrated tests marks the readiness of the system for beam commissioning.

9.3.4 Accelerator Readiness and Beam Commissioning

Beam commissioning is the final step in an accelerator construction project before the facility goes into operation.

Machine modes and beam modes are usually defined to characterize the accelerator configuration (Sect. 8.2.3). Machine modes describe accelerator operation scenarios that specify areas where beams are present. Beam modes describe beam characteristics, including particle species and charge states, beam structure, duty cycle, and intensity. Beam commissioning can be staged in phases according to machine and beam modes.

The approval of beam commissioning is usually contingent upon a successful accelerator readiness review (ARR). In the ARR, hardware, people, and system/documentation readiness are evaluated. Hardware readiness is best recorded through successful integrated tests. People readiness includes defining the roles, responsibilities, authority, and accountability of the "chain of command" of the operational personnel; staff training; developing the conduct of operations, including the qualification, certification, and authorization of the operators, as well as operator training. System/documentation readiness covers aspects such as quality assurance/quality control, radiation safety program and access control, credit controls enforcement, work control and work planning, training and recording, configuration control, document control, systems review and recommendation tracking, and accelerator and operational safety envelope conformance.

Recent accelerator projects often adopt the phased beam commissioning strategy to reduce overall project risk and optimize workforce efficiency [11, 12]. Phased commissioning usually requires strategically planned infrastructure support and rigorous safety precautions when project construction and beam commissioning proceed in parallel.

With the FRIB project, strategically designed segmentation in the cryo-distribution system facilitated phased commissioning with parallel installation and maximized machine availability and maintainability. The helium refrigeration system fed cryogens through four separate distribution lines, one for each of the three linac segments (LS1 to LS3) and the fourth for the experimental systems (ES) area. U-tube connections allow for flexible and robust configuration changes. Each CM has been connected via U-tubes, allowing for independent cool-down and warm-up. The segmented design provided the flexibility needed for phased

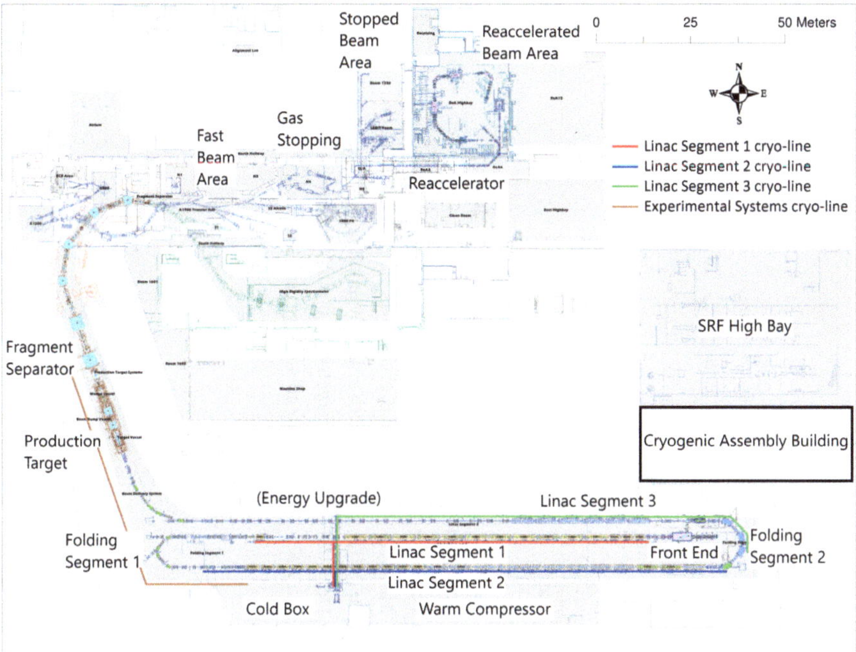

Fig. 9.5 Schematic layout of the FRIB accelerator complex for phased commissioning. The primary beam is generated from the front end and accelerated through linac segment 1 (LS1), folding segment 1 (FS1), LS2, FS2, and LS3, then strikes the production target for rare isotope production of fast, stopped, and reaccelerated beams [13]

commissioning. For example, in 2020, LS1 (Fig. 9.5, red) had remained cold since April 2019 after the LS1 beam commissioning; LS2 (Fig. 9.5, blue) was cooled in preparation for beam commissioning; LS3 (Fig. 9.5, green) was in fabrication and installation; and the cryogenic and magnet systems of the ES area (Fig. 9.5, brown) were still being designed (Table 9.3).

Phased commissioning demands even more stringent configuration control. In each phase of FRIB commissioning (Table 9.3), the ASE, OSE, and SAD were carefully reviewed and updated as the scope of the commissioning expanded, followed by the corresponding DRR and ARR.

9.4 Operations

The goal of accelerator operations is to operate the facility safely in accordance with the design performance parameters, ensuring high availability and user satisfaction.

Table 9.3 Beam commissioning stages for the FRIB accelerator complex

Run	Area with beam	Energy (MeV/u)	Species	Date	Main goals
1	Front end	0.5	Ar, Kr	Jul 2017	Front end and civil integration
2	+3 cryomodules	2	Ar, Kr	May 2018	Cryogenic integration
3	FE + LS1 + FS1	20	Ar, Kr, Xe, U	Feb 2019	QWR and charge stripping validation
4	+ FS1 + LS2	200	Ar, Kr, Xe	Mar 2020	2 K cryogenics and HWR validation
5	+ FS2 + LS3 + BDS	200	Ar, Kr, Xe	Apr 2021	Driver linac validation
6	+ target, beam dump		RI	Dec 2021	Targetry, RI production demonstration
7	+ ARIS	200	Ar, Kr	Jan 2022	Readiness for user operations

9.4.1 Conformance to Regulations

Accelerator operations must comply with federal (or central government) and state (or local) regulations. For example, current DOE accelerator facilities must comply with the Department of Energy Safety of Accelerators Order US DOE O 420.2D [14]. The order defines accelerators and establishes accelerator-specific safety requirements and approval authorities, which, when supplemented by other applicable safety and health requirements, promote safe operations to ensure the protection of workers, the public, and the environment.

9.4.2 Operation Hours

Accelerator operation hours usually consist of operating program hours and scheduled off hours. During scheduled off-hours, repairs and updates are made to improve machine availability and to enhance its performance.

The operating program hours usually consist of hours for user experiments, tuning and restoring the accelerator, and conducting machine and beam studies. During the machine studies period, further understanding is gained in beam dynamics and accelerator characteristics. For example, intrabeam stripping was a new mechanism confirmed during SNS operations. Understanding such beam dynamics mechanisms is essential both for raising operational performance and for guiding the design of future high intensity accelerators.

Table 9.4 shows the FRIB operation hours in Fiscal Year 2024, which was the second full year after the start of operations in May 2022. During this year, FRIB scheduled 4242 h of beam time for the scientific program, 1839 h of beam time for the industrial program, and 2703 h of scheduled off-hours. The scheduled off-hour includes the annual major maintenance shutdown and the holiday break.

The facility availability, which is defined as the ratio of the achieved to the scheduled program time, is a measure of machine reliability. Table 9.4 indicates that the general availability of the FRIB facility is 95.7%. Downtime analysis is a powerful tool to monitor trends in system performance and identify areas

Table 9.4 FRIB operation hours in Fiscal Year 2024. The total scheduled operation time is 6081 h. The overall facility availability is 95.7%. The science time efficiency is 62.1%

Scientific program					Industrial program			Scheduled	Total
User Research Hour	Tuning Restore Hour	Beam Study Hour	UOF[a] Hour	Scheduled Scientific Hour	User Hour	UOF[b] Hour	Scheduled Industrial Hour	Off	Hour
2488	557	961	236	4242	1816	23	1839	2703	8784

[a] Unscheduled off for the scientific program;
[b] Unscheduled off for the industrial program.

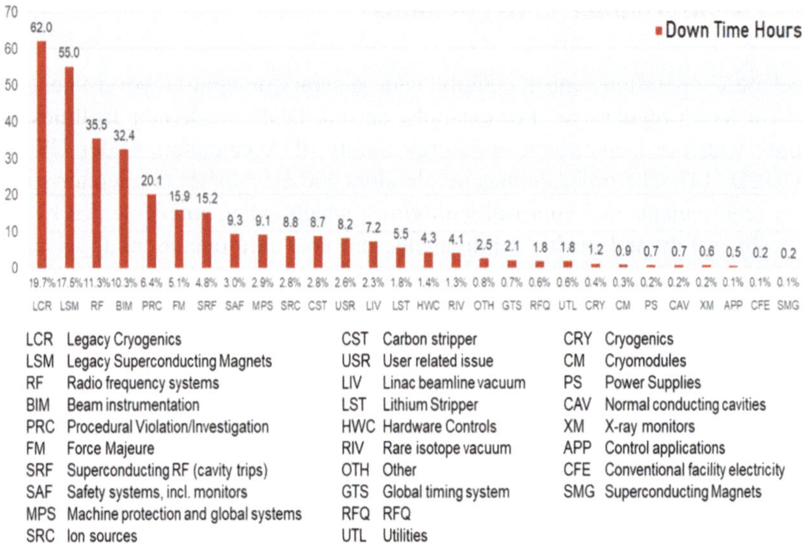

Fig. 9.6 FRIB facility downtime hours by system category during the first year of scientific user operations [15]

requiring improvement (Fig. 9.6). Routine maintenance and planned accelerator improvements ensure high availability for user operations.

9.4.3 Availability, Spares, and Accelerator Improvement Projects

The availability of the facility described in Sect. 9.4.2 can be expressed as the product of the availabilities of the systems (or devices),

$$\alpha = \prod_i \alpha_i, \qquad \alpha_i = \frac{(\mathrm{MTBF})_i}{(\mathrm{MTBF})_i + (\mathrm{MDT})_i} \tag{9.1}$$

where i denotes an individual system (or device) on which the operation of the facility depends.

The availability of a system (or device) is described in terms of the mean time between failures (MTBF) and the mean downtime (MDT) of the system (or device). The MTBF is the predicted time interval between inherent failures during normal system operation. MTBF can be calculated as the arithmetic mean time between failures of a system. The MDT is the average amount of time that a system is not functioning. It is a key metric used to evaluate the efficiency of repair and maintenance practices. Measures to increase MTBF include improving build quality and implementing preventive maintenance. Measures to reduce MDT include the establishment of spare and hot spare inventories, the implementation of compensatory measures, and the ensuring of readiness of support staff and the service call-in list.

Establishing a proper spare inventory, implementing instant swappable hot spares for critical devices, and executing accelerator improvement projects are important measures to reduce facility downtime and improve availability. For example, FRIB implemented "hot sparcs" for critical arcas such as the ion source and charge stripper systems. Figure 8.11 shows three high voltage platforms, each of which houses an ion source. Figure 8.42 shows the rotating wheel carbon charge stripper as a compromised hot spare for the high-performance liquid lithium charge stripper.

The spare inventory is usually established on the basis of the MTBF and MDT of the system (or device) and the corresponding failure mode and effects analysis (FMEA). For legacy systems, such data is usually well established. For newly developed systems like FRIB's liquid lithium charge stripper, the experience is usually lacking. Well-planned backup measures and hot spares are particularly important.

Figure 9.6 shows some of the leading categories of facility downtime during the first year of FRIB user operations: legacy RF, legacy cryogenics (LCR) and legacy SC magnets (LSM) systems. Major accelerator improvements include renovations of aging legacy systems associated with the experimental beamlines and system automation to improve operational efficiency and machine availability. Other major accelerator improvements, most of which involve technological developments, include the phased deployment of high-power beam-intercepting devices, such as the charge strippers, the charge selection systems, the production target, and the beam dump, along with support systems, including non-conventional utilities (NCU) and remote handling facilities.

9.4.4 Ramp Up to Full Performance Specifications

After the start of user operations (for example, after achieving the KPP and CD-4 for an accelerator project in the United States), a significant amount of effort and resources are still needed for most accelerator facilities to reach the full design specifications.

For example, since the start of facility operations in 2006, SNS took 8 years to ramp up the beam power, reaching the design level of 1.4 MW. During the subsequent years, efforts were made to ensure more reliable targetry system operations, as shown in Fig. 9.7. The availability also increased to about 90% in the 5 years since the start of facility operations, replacing broken accelerator components, understanding performance degradation, and controlling the errant beam [16, 17].

Figure 1.3 indicated the planned increase in beam power by FRIB in the coming years. Experience must be gained to safely handle the increased radiological

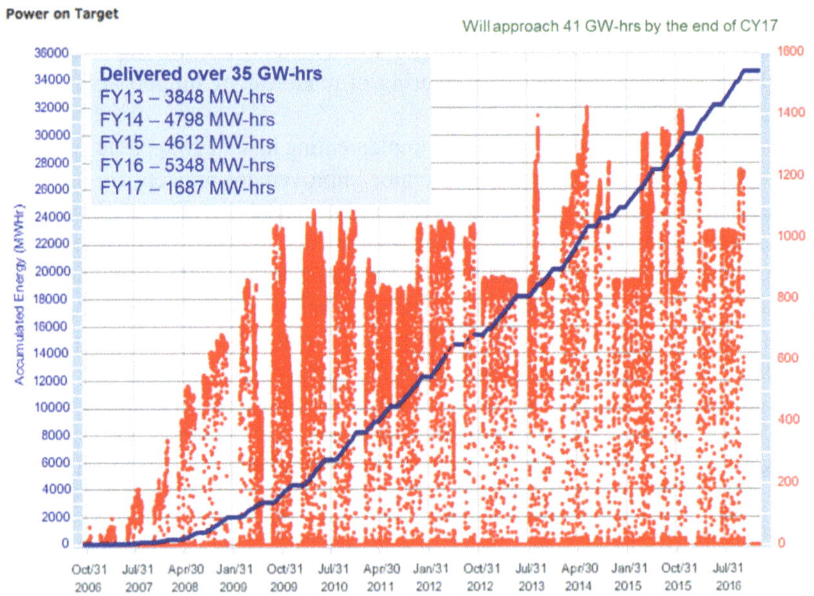

Fig. 9.7 SNS beam power on target and accumulated energy since the start of facility operations in 2006. (Courtesy J. Galambos [17])

impacts associated with high beam power; extensive machine studies and advanced beam tuning procedures are needed to minimize uncontrolled beam losses for the desired operating conditions [15]. Technological developments and accelerator improvements play an important role in achieving the ultimate design performance.

9.4.5 Upgrades

Accelerator upgrades should be planned in the very early stages when the facility layout is determined. The upgrade path determines the lifespan of the facility.

The baseline of the SNS project was to use 1 GeV high-intensity proton beams from the accumulator ring to produce pulsed neutrons. The upgrade path was defined at the initial design stage and included raising the driver linac H^- beam energy to 1.3 GeV so that higher intensity beams would be injected into the ring to double the output power [18]. After SNS started operating in 2006, the proton power upgrade (PPU) project was subsequently approved and essentially completed in 2025, with the aim of increasing the power of the proton beam to 2.8 MW (Sect. 10.1.5) [19, 20]. The project for the second neutron target station is also in progress [21–23].

The J-PARC upgrade covered not only more powerful neutron and muon production and more powerful high-energy experiments but also other programs, such as accelerator-driven subcritical systems [24] (Sect. 10.2.9).

FRIB's science-driven upgrade plans include doubling the linac output energy by filling the vacant slots in the linac tunnel with higher-β cryomodules [25], simultaneous heavy ion and light ion acceleration, and an ISOL option for rare isotope production (Sect. 10.3.8).

Appendix I: FRIB Project Execution Plan Outline

Appendix II: J-PARC Accelerator Technical Design Report Outline

Appendix III: SNS Major Parameter List

WBS	Parameter	Base value	Unit	Comment
1.0	Proton beam power on target	1.4	MW	
1.0	Proton beam kinetic energy on target	1.0	GeV	
1.0	Average beam current on target	1.4	mA	
1.0	Pulse repctition rate	60	IIz	
1.0	Protons per pulse on target	1.5×10^{14}	protons	
1.0	Charge per pulse on target	24	μC	
1.0	Energy per pulse on target	24	kJ	
1.0	Proton pulse length on target	695	ns	
1.0	Ion type (Front end, Linac, HEBT)	H minus		
1.0	Average linac macropulse H^- current	26	mA	from RFQ to injection foil
1.0	Linac beam macropulse duty factor	6	%	
1.0	Front end length	7.5	m	
1.0	Linac length	331	m	including 71 m for 9 empty cryomodule slots
1.0	HEBT length	170	m	
1.0	Accumulator ring circumference	248	m	
1.0	RTBT length	150	m	
1.0	Ion type (Ring, RTBT, Target)	proton		
1.0	Ring filling time	1.0	ms	
1.0	Ring revolution frequency	1.058	MHz	
1.0	Number of injected turns	1060		
1.0	Ring filling fraction	68	%	
1.0	Ring extraction beam gap	250	ns	
1.0	Maximum uncontrolled beam loss	1	W/m	
1.0	Target material	Hg		
1.0	Number of ambient/cold moderators	1/3		
1.0	Number of neutron beam shutters	18		
1.0	Initial number of instruments	5		

Appendix IV: FRIB Parameter List Outline

Contents	Page

Appendix V: FRIB Parameter List Appendix

Document Name	DCC Link
1. FRIB Lattice File	T10501-BL-000068
2. ASD Power Supply Requirements	T30102-SP-000290
3. Vacuum Requirements (ASD)	T31201-SP-000031
4. Diagnostics Requirements (ASD)	T31201-SP-000021
5. RF Requirements (ASD)	T31307-SP-000413
6. FRIB SRF Cavity Parameters Table	T31202-SP-000067
7. Beam Dump Requirements (ASD)	T31201-SP-000039
8. Project Requirements Document	T10100-BL-000025
9. Fragment Separator Diagnostics Requirements	T40302-SP-000077
10. Vacuum Requirements (ESD)	T40307-SP-000072
11. Summary of FRIB Cryogenic Heat Loads	T30200-TD-000244
12. Alignment Requirements (ESD)	T40300-SP-000074
13. PPS Requirements	T10500-CM-000115
14. MPS Requirements	T31201-SP-000040
15. Timing Requirements	T31201-SP-000038
16. High Level Controls Requirements	T31301-SP-000078
17. Optical Elements Requirements (ASD)	T30102-SP-000289
18. FRIB Fragment Separator Magnet Requirements	T40301-SP-000064
19. Global Coordinate System	T30000-MDE-0002-0000
20. FRIB Driver Linac Element Placement Requirements (ASD)	T31201-SP-000032
21. Charge Stripping Requirements	T30705-SP-000080
22. FRIB Key Beam List	U10600-BL-000036
23. FRIB site plan	Portal Standard Graphics
24. FRIB Layout	Portal Standard Graphics
25. ESD Power Supply Requirements	T40303-SP-000288
26. Cryogenic Interface Requirements for Superconducting Magnets in the FRIB Preseparator	T40503-SP-000558

Appendix VI: List of FRIB External Reviews and Workshops on Accelerator Systems from CD-1 (September 2010) to CD-4 (April 2022)

Date	Event name
2010-07	DOE OPA Review for CD-1
2010-10	Accelerator Alignment External Review
2011-01	Cryomodule Heat Source External Review
2011-02	Machine Protection, Diagnostics Timing
2011-02	Accelerator Systems Advisory Review
2011-03	DOE OPA Review
2011-04	Accelerator Lattice External Review
2011-05	Diagnostics External Review
2011-05	Electrical Grounding Workshop
2011-07	Controls External Review
2011-08	Cryogenic External Review
2011-08	SRF External Review
2011-08	Cryomodule (β_0 =0.53) External Review
2011-08	Front End External Review
2011-09	DOE OPA Review
2011-10	Magnets External Review
2011-10	Charge Stripper External Review
2011-11	Installation, Pre-ops, Commissioning
2011-11	Accelerator Systems Advisory Review
2011-12	RF & Power Supply External Review
2012-01	Diagnostics Preliminary Design Peer Review
2012-01	Personnel Protection System Workshop
2012-01	Vacuum, Beam Dump, Collimator Review
2012-02	Lithium Stripper Safety Review
2012-04	DOE OPA Review for CD-2
2012-06	Accelerator Systems Advisory Review
2012-07	SRF Subsystem Review
2012-09	Utility Interface External Review
2012-09	Final Engineering Design Approach Review
2012-10	DOE OPA Review
2012-11	HP Source Magnet/Coldmass Workshop
2012-11	FRIB Vacuum Workshop
2012-12	Accelerator Systems Advisory Review
2012-12	MSU Independent Review
2013-02	ReA coupler, SRF fabrication, TDCM, ETCM
2013-03	Magnet Shielding Workshop
2013-06	Accelerator Systems Advisory Review
2013-06	DOE OPA Review for CD-3a
2013-07	RF and Power Supplies Intermediate

2013-07	Personnel Protection System Intermediate
2013-08	ARTEMIS Ion Source/HV Platform Inter. Design
2013-09	ReA6 Cryomodule Project Status Review
2013-10	QWR Prototype Cryomodule Final Design
2013-10	Lithium Charge Stripper Intermediate Design
2013-10	MPS and GTS Intermediate Design Review
2013-10	Alignment Intermediate Design Review
2013-11	cryo-distribution Intermediate Design Review
2013-11	HWR Cryomodule Intermediate Design
2013-11	ASD Room-temperature Dipole Magnet Interm.
2013-11	ASD Superconducting Magnet Final Design
2013-11	ASD RT Optical Element Intermediate Design
2013-11	Diagnostics Intermediate Design Review
2013-11	MEBT Bunch Preproduction Review
2013-11	Low-level Controls Intermediate Design
2013-12	Accelerator Systems Advisory Review
2014-01	FRIB In-depth Technical Interface Series
2014-03	Cryomodule Maintenance Workshop
2014-03	Machine Protection Systems Workshop
2014-03	ESH Advisory Committee Review
2014-03	Prelim. Start-up & Commissioning Plan Review
2014-04	ARTEMIS Ion Source/HV Platform Final Design
2014-04	Accelerator Systems Advisory Review
2014-05	ASD RT Magnet Final Design Review
2014-06	β_0 =0.53 HWR Cryomodule Final Design Review
2014-06	Cryomodule SC solenoid Final Design review
2014-06	DOE OPA Review for CD-3b
2014-08	RF and Power Supply Final Design Review
2014-08	Lithium Stripper Final Design Review
2014-09	ECR Cold Mass Design Review
2014-09	Alignment Network Final Design Review
2014-09	Vacuum In-progress Design Review
2014-09	Cryoplant and cryo-distribution Final Design Review
2014-11	ICFA Workshop on High Power Hadron beams
2014-12	Accelerator Systems Advisory Review
2014-12	MSU Independent Review
2015-02	ASD Diagnostics Final Design Review
2015-02	ESHAC Review
2015-03	DOE OPA Review
2015-05	EVMS Review
2015-05	Vacuum Final Design Review
2015-05	Personnel Protection System In-progress Review
2015-05	Front End Controls Final Design Review
2015-06	Accelerator Systems Advisory Review
2015-07	Front End Diagnostics Intermediate Design Review

2015-08	Controls Network In-Process Review
2015-08	Time Critical Front End Controls In-Process Review
2015-09	Phase I Personnel Protection System 90
2015-09	MPS, GTS and Central Control 90
2015-09	Hardware Controls Final Design Review
2015-10	ASD Beam Dump Final Design Review
2015-11	DOE OPA Review
2015-12	Accelerator Systems Advisory Review
2016-01	DOE Review on FRIB Operations Cost
2016-01	β_0 =0.041 Cryomodule Final Design Review
2016-01	LEBT Beam Plug Final Design Review
2016-01	High Power ECR Cryostat Review
2016-03	ESHAC Review
2016-03	β_0 =0.085 Matching Cryomodule Intermediate Design Review
2016-04	EVMS Review
2016-05	ASD Diagnostics Electronics Final Design Review
2016-05	Accelerator Systems Advisory Review
2016-06	β_0 =0.085 Matching Cryomodule Final Design Review
2016-06	DOE OPA Review
2016-08	FRIB 2 K Cold Box Final Design Review
2016-08	β_0 =0.29 Cryomodule Preliminary Design Review
2016-09	ARTEMIS Device Readiness Review
2016-09	ECR Cryostat Final Design Review
2016-11	Accelerator Systems Advisory Review
2016-12	DOE OPA Review
2017-01	β_0 =0.53 Matching Cryomodule Intermediate Design Review
2017-03	β_0 =0.53 Matching Cryomodule Final Design Review
2017-04	IH Buncher Intermediate Design Review
2017-05	Accelerator Systems Advisory Review
2017-06	DOE OPA Review
2017-08	Accelerator Readiness Review (Front End) ARR01
2017-08	Folding Segment 2 Dipole Intermediate Design Review
2017-08	IH Buncher Final Design Review
2017-10	Folding Segment 2 Dipole Final Design Review
2017-11	High Performance ECR Ion Source Intermediate Design Review
2017-11	Accelerator Systems Advisory Review
2017-12	DOE OPA Review
2018-01	Lithium Charge Stripper Device Readiness Independent Review
2018-02	β_0 =0.65 Cavity Bead Pull Test Stand Intermediate Design Review
2018-02	Lithium Charge Stripper Offline Test Final Inspection
2018-02	Bead Pull Test Stand Intermediate Review
2018-04	Accelerator Systems Advisory Review
2018-05	High performance Ion Source Final Design Review
2018-05	Accelerator Readiness Review (Cryomodule 1-3) ARR02
2018-05	DOE OPA Review

2018-09	Superconducting Dipole 3 Intermediate Design Review
2018-10	Accelerator Systems Advisory Review
2018-11	DOE OPA Review
2018-12	Superconducting Dipole 3 Final Design Review
2019-01	Accelerator Readiness Review (Cryomodule 1–15) ARR03
2019-02	Cold Iron Quadrupole Triplet 3 (CIQT3) Intermediate Design Review
2019-04	Technical Systems Advisory Review
2019-05	CIQT Final Design Review
2019-05	DOE OPA Review
2019-06	ES Fragment Pre-separator cryo-distribution Final Design Review
2019-09	A1900 Transfer Hall Reconfiguration Plan Intermediate Review
2019-10	SCD-1 Final Design Review
2019-10	Technical Systems Advisory Review
2019-11	DOE OPA Review
2020-03	Accelerator Readiness Review (Cryomodule 1–39) ARR04
2020-04	Target Facility Radiation Safety System Final Design Review
2020-04	Technical Systems Advisory Review
2020-05	DOE OPA Review
2020-06	A1900 Beamline Vacuum Final Design Review
2020-09	Experimental Systems Fast Protection System Final Design Review
2020-09	Target Facility ODH System Post-90% Design Review
2020-09	Fragment Separator Diagnostics, DAQ System Final Design Review
2020-09	Technical Systems Advisory Review
2020-11	DOE OPA Review
2021-04	Accelerator Readiness Review (ReA6) ReA6 ARR
2021-04	Accelerator Readiness Review (FRIB driver linac) ARR05
2021-04	Technical Systems Advisory Review
2021-05	DOE OPA Review
2021-06	Accelerator Traineeship Advisory Panel Review
2021-09	FRIB Science and Technology (S&T) Review
2021-10	Technical Systems Advisory Review
2021-10	Accelerator Readiness Review (FRIB target hall) ARR06
2022-01	Accelerator Readiness Review (VPS, transfer hall) ARR07
2022-02	DOE OPA Review for CD-4
2022-04	Technical Systems Advisory Review
2022-05	DOE Operations Review

References

1. U.S. Department of Energy, Order O 413.3B, Program and project management for the acquisition of capital assets (2023). https://www.directives.doe.gov/directives-documents/400-series/0413.3-BOrder-B-chg7-ltdchg

2. N.R. Holtkamp, EPS-AG 2008 Gersh Budker Prize presentation: the successful construction and commissioning of the Spallation Neutron Source, in *Proceedings of the EPAC'08, no. 11 in European Particle Accelerator Conference* (JACoW Publishing, Geneva, 2008), pp. 2960–2962. https://jacow.org/e08/papers/THPPGM03.pdf
3. Y. Yamazaki, J-PARC status, in *Proceedings of the PAC'09, no. 23 in Particle Accelerator Conference* (JACoW Publishing, Geneva), pp. 18–22. https://jacow.org/PAC2009/papers/MO2BCI03.pdf
4. J. Wei, et al., Advances of the FRIB project. Int. J. Mod. Phys. E **28**(03), 1930003 (2019). https://doi.org/10.1142/S0218301319300030
5. H. Grunder, et al., ISOL task force report to NSAC (1999)
6. T.E. Mason, et al., Spallation Neutron Source project completion report. Prepared for the US Department of Energy Office of Science (June 2006)
7. S. Beher, Facility for rare isotope beams digital hygiene program. Internal report FRIB-S30105-AD-000634-R001 (2023)
8. S. Zhao, et al., The LLRF control design and validation at FRIB, in *Proceedings of the NAPAC'19* (JACoW Publishing, Geneva), pp. 667–669. https://doi.org/10.18429/JACoW-NAPAC2019-WEPLM03. https://jacow.org/napac2019/papers/WEPLM03.pdf
9. S.M. Lidia, et al., Overview of beam diagnostic systems for FRIB, in *Proceedings of the IBIC'15* (JACoW Publishing, Geneva), pp. 226–230. https://doi.org/10.18429/JACoW-IBIC2015-MOPB071. https://jacow.org/IBIC2015/papers/MOPB071.pdf
10. Z. Li, L.R. Dalesio, M. Ikegami, S.M. Lidia, L. Wang, S. Zhao, FRIB fast machine protection system: engineering for distributed fault monitoring system and light speed response, in *Proceedings of the LINAC'16, no. 28 in Linear Accelerator Conference* (JACoW Publishing, Geneva, 2017), pp. 959–961. https://doi.org/10.18429/JACoW-LINAC2016-THPLR046. https://jacow.org/linac2016/papers/THPLR046.pdf
11. J. Wei, et al., RHIC sextant test – physics and performance, in *Proceedings of the PAC'97* (JACoW Publishing, Geneva), pp. 69–71
12. M. Ikegami, et al., Commissioning plan for the FRIB driver linac, in *Proceedings of the LINAC'14, no. 27 in Linear Accelerator Conference* (JACoW Publishing, Geneva, 2014), pp. 152–154. https://jacow.org/LINAC2014/papers/MOPP041.pdf
13. J. Wei, et al., The FRIB SC-Linac – installation and phased commissioning, in *Proceedings of the SRF'19* (JACoW Publishing, Geneva), pp. 12–20. https://doi.org/10.18429/JACoW-SRF2019-MOFAA3. https://jacow.org/srf2019/papers/MOFAA3.pdf
14. U.S. Department of Energy, Order O 420.2D, Safety of accelerators (2022). https://www.directives.doe.gov/directives-documents/400-series/0420.2-BOrder-d
15. J. Wei, C. Alleman, H. Ao, B. Arend, D. Barofsky, S. Beher, G. Bollen, N. Bultman, F. Casagrande, W. Chang, et al., Technological developments and accelerator improvements for the FRIB beam power ramp-up. J. Instrum. **19**(05), T05011 (2024)
16. J. Galambos, SNS performance and the next generation of high power accelerators, in *Proceedings of the NAPAC'13* (JACoW Publishing, Geneva), pp. 1443–1447. https://jacow.org/PAC2013/papers/FRYAA1.pdf
17. J. Galambos, Operations experience of SNS at 1.4 MW and upgrade plans for doubling the beam power, in *Proceedings of the IPAC'19* (JACoW Publishing, Geneva), pp. 4380–4384. https://doi.org/10.18429/JACoW-IPAC2019-FRXPLM1. https://jacow.org/ipac2019/papers/FRXPLM1.pdf
18. J. Wei, The Spallation Neutron Source project – physical challenges, in *Proceedings of the EPAC'02, no. 8 in European Particle Accelerator Conference* (JACoW Publishing, Geneva, 2007), pp. 1064–1066. https://jacow.org/e02/papers/THPLE015.pdf
19. M. Champion, M. Connell, N. Evans, J.D. Galambos, M. Howell, G. Johns, S.H. Kim, J. Moss, G.S. Stephens, K.S. White, Completion of the proton power upgrade project at the Spallation Neutron Source (2024)
20. S.H. Kim, M. Doleans, J. Galambos, M. Howell, J. Mammosser, Superconducting linac upgrade plan for the Second Target Station project at SNS, in *Proceedings of the SRF'15* (2015), pp. 268–271

21. T.A. Gabriel, J.M. Barnes, L.A. Charlton, J. DiStefano, K. Farrell, J. Haines, J.O. Johnson, L.K. Mansur, S.J. Pawel, M. Siman-Tov, et al., Target systems overview for the Spallation Neutron Source. Nucl. Technol. **132**(1), 49–60 (2000)
22. J. Haines, T. McManamy, T. Gabriel, R. Battle, K. Chipley, J. Crabtree, L. Jacobs, D. Lousteau, M. Rennich, B. Riemer, Spallation neutron source target station design, development, and commissioning. Nucl. Instrum. Methods Phys. Res., Sect. A **764**, 94–115 (2014). https://doi.org/10.1016/j.nima.2014.03.068. https://www.sciencedirect.com/science/article/pii/S0168900214003829
23. J.D. Galambos, D.E. Anderson, D. Bechtol, K.L. Bethea, N. Brown, W. Carden, S.M. Chae, A. Clark, D.M. Counce, K. Craft, et al., Technical design report, second target station. Technical report, Oak Ridge National Lab (ORNL), Oak Ridge, TN (United States) (2015)
24. M. Kinsho, Status and future upgrade of J-PARC accelerators, in *Proceedings of the IPAC'16, no. 7 in International Particle Accelerator Conference* (JACoW Publishing, Geneva, 2016), pp. 999–1003. https://doi.org/10.18429/JACoW-IPAC2016-TUXA01. https://jacow.org/ipac2016/papers/TUXA01.pdf
25. FRIB400 whitepaper (2019). https://frib.msu.edu/_files/pdfs/frib400_final.pdf

Chapter 10
Examples of High Power Accelerator Facility Establishment

Abstract Following the process of establishment of accelerator facilities discussed in the previous chapter, this chapter reviews the history of three recently completed high-power hadron accelerator projects. SNS, J-PARC, and FRIB. We discuss design philosophy, design choices and alternative selection, key R&D outcomes, and important experiences. The emphasis is on the technical challenges and lessons learned.

In this chapter, examples are provided of the establishment of recent high-power accelerator facilities in the world: the Spallation Neutron Source (SNS), the Japan Particle Accelerator Research Complex (J-PARC), and the Facility for Rare Isotope Beams (FRIB).

10.1 Spallation Neutron Source

This section discusses both the SNS construction project (Sects. 10.1.1, 10.1.2, 10.1.3, and 10.1.4) that was completed in 2006 and the Proton Power Upgrade project (PPU, Sect. 10.1.5) that will be completed in 2025.

10.1.1 SNS Project Overview

As described in Sect. 1.6.2.1, the goal of the Spallation Neutron Source (SNS) project was to design, construct, and commission into operation an accelerator-based pulsed neutron research facility for studies of the structure and dynamics of materials that are substantially better than those of any other facility in the world (Figs. 1.19 and 10.1). The key requirements were to complete a facility capable of delivering primary beam power above 1 MW, which was much higher than that of existing pulsed neutron facilities. The US DOE Office of Basic Energy Sciences

J. Wei et al., *High Power Hadron Accelerators*, Particle Acceleration and Detection,
https://doi.org/10.1007/978-3-032-17662-2_10

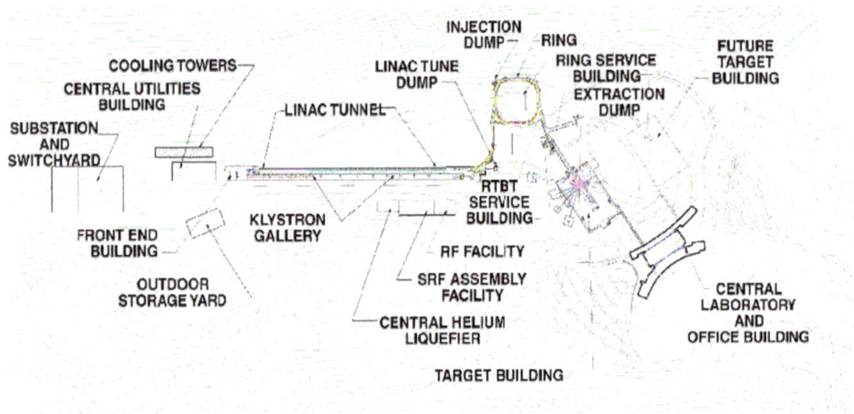

Fig. 10.1 Schematic layout of the SNS accelerator complex. The complex consists an ion source, an RFQ, a DTL, a CCL, and an SRF linac accelerating the H$^-$ beam up to 1.3 GeV, a proton accumulator ring and its transport lines, and the target station for neutron production

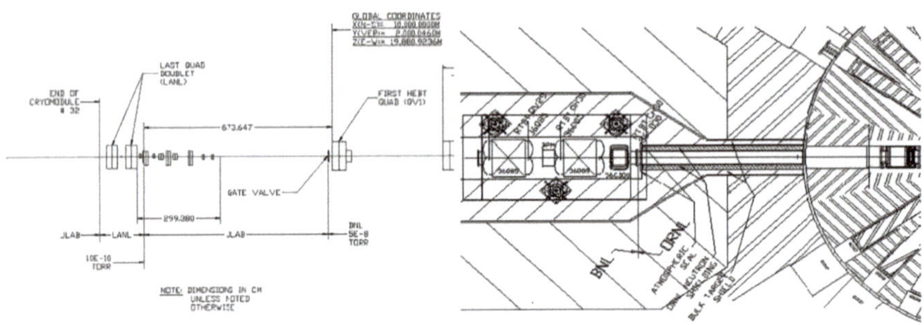

Fig. 10.2 Examples of engineering interface between partner laboratories during the construction of the SNS project (left) between LANL, JLab, and BNL at the linac-to-HEBT interface and (right) between BNL and ORNL at the RTBT-to-target interface

funded the construction of SNS at a total project cost of $1,411.7 M. Since the completion of the project in 2006, SNS has provided the most intense proton beams for pulsed neutron production in the world [1].

The construction of the SNS was a partnership among six national laboratories of the US Department of Energy (DOE) (Fig. 10.2). Lawrence Berkeley (LBNL) designed and built the front-end systems. Los Alamos (LANL) was responsible for the physics design and normal-conducting portions of the linear accelerator and all of the linac radio-frequency systems. Jefferson Laboratory (JLab) was responsible for the superconducting linac and associated cryogenic systems. Brookhaven (BNL) was responsible for the accumulator ring and beam transport. Argonne (ANL) together with Oak Ridge (ORNL) led the design of the initial suite of scattering instruments. ORNL was responsible for target systems, conventional facility construction, system integration, and overall project management, including risk and contingency management. ORNL also operates the SNS as a DOE Office of Science user facility. The collaboration was the largest of its kind in the US accelerator community and was used to bring together the best expertise from each partner laboratory, all sharing interests in the successful completion of this state-of-the-art high-power hadron accelerator project.

In 1977, the US National Academy of Sciences (NAS) published "Neutron Research on Condensed Matter," which recommended a high-flux pulsed spallation source [2]. In 1984, the NAS reviewed the needs for major facilities, and in the report Major Facilities for Materials Research and Related Disciplines (Seitz-Eastman Committee) recommended (1) the construction of a new high-flux, steady-state neutron source and (2) the development of a plan leading to a high-intensity pulsed neutron facility. Recommendations of the Seitz-Eastman Committee were reaffirmed in 1993 by the Basic Energy Sciences Advisory Committee (BESAC) Panel on Neutron Sources for America's Future, known as the Kohn Committee. Following the cancellation of the proposed Advanced Neutron Source reactor project, BESAC convened another panel in 1996 to reevaluate the need for neutron facilities in the United States. This panel strongly recommended that a 1 MW pulsed spallation source, which could be upgraded, be given the highest construction priority. In response, SNS was conceived, and the secretary of DOE approved Critical Decision (CD)-1, "Approval of Mission Need," and CD-2, "Approval of the Level 0 Project Baseline" for SNS in August 1996 and December 1997, respectively. After four years of R&D, the construction of the project started upon the approval of CD-3 in November 1999. The design, construction, and commissioning were concluded with the approval of CD-4 in May 2006. The first three instruments began their commissioning and were made available to the scientific community in August 2007. The proton beam on the neutron production target reached 1 MW in 2009 and 1.4 MW in 2013, as shown in Fig. 9.7. In 2017, a total of 20 instruments were completed, and the SNS hosts about 1,400 researchers per year. Table 10.1 summarizes key dates for the SNS project [3].

10.1.2 Accelerator Design Philosophy

The SNS project was designed to reach an average beam power above 1.4 MW for pulsed neutron production [4]. In the approved specification, the accelerator system operates at a repetition rate of 60 Hz and an average current of 1.6 mA. It consists of an H^- RF volume source of 48 mA peak current at 6% duty; a low-energy beam transport (LEBT) [5] housing a first-stage beam chopper with <50 ns rise/fall time; a 402.5 MHz, 4-vane radio-frequency quadrupole (RFQ) [6–8]; a medium-energy beam transport (MEBT) [9] housing a second-stage chopper with <10 ns rise/fall time, an adjustable beam-halo scraper, diagnostic devices, and matching quadrupoles; a 402.5 MHz, 6-tank drift-tube linac (DTL) [10] with permanent-magnet quadrupoles; an 805 MHz, 4-module coupled-cavity linac (CCL) [11]; an 805 MHz, superconducting RF (SRF) linac of medium-β and high-β cavities

Table 10.1 SNS major project dates

Event	Date
Conceptual design and R&D started	Oct 1995
Mission need approved (CD-1)	Aug 1996
Level 0 project baseline approved (CD-2)	Dec 1997
Line item project started	Oct 1998
Environmental impact statement record of decision	Jun 1999
Start construction approved (CD-3)	Nov 1999
Ground breaking	Dec 1999
Superconducting RF linac change approved	May 2000
First beam accelerated in front-end system	Dec 2002
1 mA beam accelerated through DTL tank 1	Nov 2003
Begin instrumentation installation	Mar 2004
Beam through warm linac (DTL and CCL)	Aug 2004
First beam on target integrated performance test accomplished	Apr 2006
Project work completed (CD-4)	May 2006

accelerating the beam to the full energy [12, 13]; a high-energy beam transport (HEBT) for diagnostics, transverse and longitudinal collimation, matching, energy correction, and painting; and an accumulator ring compressing the 1 GeV, 1 ms pulse to 650 ns for delivery onto the target through a ring-target beam transport (RTBT) (Fig. 8.54).

As shown in Appendix III of Chap. 9, the energy acceptance of the ring is about 50 MeV, mainly due to the conditions for a tolerable loss of the stripping of H^- and H^0. The back-up scenario corresponds to the case in which the surface field of the SRF cavity is lower than expected (37.5 MV/m). An extra tunnel space (71 m) is reserved to extend the linac length to obtain higher output energy for future upgrades to beam power (Sect. 10.1.5).

The primary concern is that radio-activation caused by excessive uncontrolled beam loss can limit machine availability and maintainability (Sect. 1.4). Based on operational experience, hands-on maintenance demands that the average uncontrolled beam loss does not exceed 1 W/m. Uncontrolled losses are usually attributed to (1) mismatch upon change of linac structure, lattice, and frequency; (2) space-charge effects, including envelope and parametric resonances, as well as non-equipartition in the linac, and resonance crossing and instability enhancement in the ring; (3) limited physical and momentum acceptance; (4) premature H^- and H^0 stripping and ring injection foil scattering; (5) magnetic errors, fringe fields, and misalignments; (6) instabilities (resistive impedances due to extraction kicker, and electron cloud); and (7) accidental loss due to system malfunction (ion source and linac, ring extraction kickers) (Fig. 10.3). Intrabeam stripping (Sect. 3.7.4) of the H^- beam in the linac is a mechanism that was discovered only after the SNS started to operate; it was mitigated by increasing the beam size.

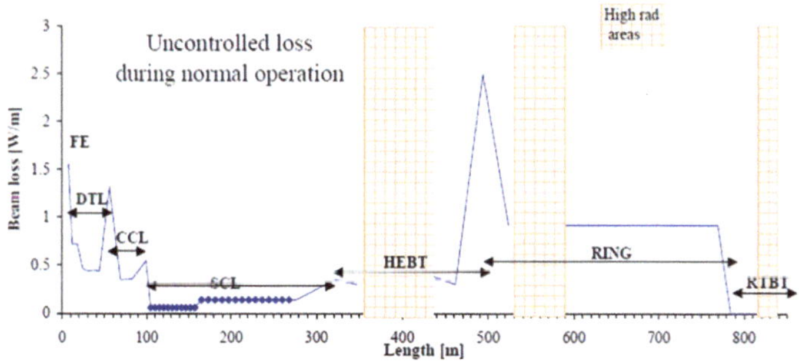

Fig. 10.3 Expected uncontrolled beam loss across the SNS accelerator complex

SNS addresses the above issues by adopting a low-loss design philosophy [4]. Above all, foreseen losses are localized to shielded areas using (1) adjustable scrapers in the MEBT, (2) transverse and momentum collimators in the HEBT prior to ring injection; (3) two-stage transverse collimation and momentum cleaning with a beam-in-gap (BIG) kicker in the ring; (4) collimator protection in the RTBT; and (5) beam-gap cleaning with LEBT and MEBT choppers and the ring BIG kicker.

The emphasis is also placed on the flexibility and reliability of the machine. The SRF linac allows operation with one failed cavity or klystron; the ring accepts $\sim$5% variation in linac output energy; a wide ring tuning range avoids resonances; a robust injection allows independent horizontal, vertical, and longitudinal painting; adjustable collimation systems accommodate variable beam sizes; and design reserve and redundancy ensure high availability (e.g., spare cryomodule for quick replacement, power supplies compatible with 1.3 GeV energy, multi-foil exchange, spare kicker power supply (PFN), and aperture clearance for one-kicker failure).

Finally, the facility is designed with the potential to reach a beam energy of up to 1.3 GeV and a beam power greater than 2 MW, capable of supplying a second neutron target station. The higher energy can be reached by upgrading the superconducting RF cavity gradient and klystron power supplies, and by filling the presently unoccupied linac tunnel spaces with up to nine additional cryomodules. The ring is capable of accommodating the increase in energy and power without extensive hardware changes; space is reserved for two additional extraction kickers and for the replacement of two injection chicane dipoles to satisfy the H^- stripping conditions.

10.1.3 Accelerator Design Choices

During the period around the start of construction (CD-3), several major technical debates and design optimizations were conducted that defined the project path forward. These included changing the normal-conducting linac to a superconducting

Table 10.2 Comparison of estimated SNS project cost after baseline in July 1999 and before completion in June 2005

WBS	1999–07 baseline [$M]	2005–06 EAC [$M]	Increase
1.02 Project support	50.5	72.1	42.8%
1.03 Front end systems	18.4	20.8	13.0%
1.04 Linac systems	222.0	314.9	41.8%
1.05 Ring & transfer system	125.3	143.9	14.8%
1.06 Target systems	79.9	114.8	43.7%
1.07 Instrument systems	97.3	63.8	−34.4%
1.08 Conventional facilities	256.6	391.1	52.4%
1.09 Integrated control systems	54.4	59.8	9.9%
Estimate at complete	904.4	1181.4	30.6%
Contingency	255.2	11.3	
Contingency percentage	28.2%	24.6%	
Total Estimated Cost (TEC)	1159.5	1192.7	2.9%
R&D		100.0	
Pre-Ops		119.0	
Other Project Cost	200.5	219.0	9.2%
Total Project Cost (TPC)	1360.0	1411.7	3.8%

linac structure from 186 MeV up to the full energy, changing from the all-FODO lattice to the FODO-doublet hybrid lattice, and the debate on possibly changing from the accumulator ring to a rapid-cycling synchrotron.

10.1.3.1 Integrated Controls

The SNS project developed its work breakdown structure (WBS) to facilitate multi-laboratory partnerships. As shown in Table 10.2, partner laboratories were placed at WBS level 2 (LBNL for WBS.1.03 front-end systems, LANL and JLab for WBS.1.04 linac systems, BNL for WBS.1.05 ring & transfer system, ORNL for WBS.1.06 target systems, and ANL for WBS.1.07 instrument systems) with a defined scope to deliver at baseline cost and schedule. The project management had the vision of forming an integrated control scope with baseline cost and schedule (WBS.1.09 integrated control systems), organized centrally and performed by individual teams distributed across various laboratories [14]. In a complex multi-laboratory framework, such arrangements facilitated the smooth delivery of controls on specifications, on schedule, and on budget.

The Experimental Physics and Industrial Control System (EPICS) [15] was selected as the supervisory control and data acquisition (SCADA) (Sect. 7.2.1). This toolkit provided a tool set for the configuration and integration of all signals for all devices. When SNS was under construction, it was not yet a de facto standard. At the time, four of the five partner laboratories involved in the SNS project, with the exception of BNL, had used EPICS. Additionally, a naming standard was developed as a guideline to establish a name space for all process variables (PVs) [14]. SNS also integrated their facility control into EPICS, which allowed all signals of the facility to be integrated into the control system.

XAL was selected as the physics application development environment [16]. All physics applications were developed in this environment, which included a physics model and applications for beam commissioning and operation [17].

One of the primary challenges for the control system was the development of hardware and software from multiple laboratories. The hardware challenges were to provide the instrumentation for the LLRF, a timing system [18], and a machine protection system [19].

LBNL, LANL, and BNL all developed LLRF controller boards. LBNL had to deliver the front-end LLRF as early as possible, while LANL was developing a new VXI-based LLRF system. Concerns had been raised at each technical review; yet, they were dismissed by the development team. In the last months before the LLRF was required for the linac, the LBNL design was modified and deployed for the linac [20]. As each team was responsible for delivering their part of the accelerator complex, the three teams could not coalesce on one solution. A more salient lesson is that technical reviews with competent and cooperative reviewers and developers need to be more introspective so that the projects could be executed in an effective and efficient manner.

With a beam rate of 60 Hz and a pulse length of 1 ms. The timing system developed by BNL was selected. It consisted of two distributed communication lines, one to distribute events and one to distribute real-time data. This system provides triggers for all beam-related instruments. For all other instruments, VME was chosen as the backplane for all IOs.

Beam characteristic instruments were developed outside of these standards. They were integrated into Labview. Beam control did not require coherent pulse-to-pulse data. The facility has operated successfully without this capability.

In summary, as lessons learned, the LLRF had issues that could have been mitigated with an appropriate focus on the concerns of the reviewers. The beam diagnostics met requirements and did not impact the commissioning or operation of the SNS; however, it could have been built with correlated data. Instead, the stated desire to be independent led to the choice of Labview. In general, physicists have always made the most of the tools that were provided, and engineering solutions benefit from early standards, cooperative developments, and critical reviews.

10.1.3.2 From Normal-Conducting to Superconducting RF Linac

Until 1999, the baseline SNS design consisted of a room-temperature linac accelerating the H^- beam to a full energy of 1 GeV, injecting it into an accumulator ring to realize the pulsed beam structure for neutron production. In spring 1999, about six months before the start of construction, the SNS management sponsored two month-long major alternative design studies: one by the linac team on an SRF linac [21, 22] and the other by the ring team on a rapid-cycle synchrotron design

[4]. The linac design studies favored the SRF linac over the room-temperature linac. The ring team favored an accumulator ring design with increased acceptance, increased circumference, and improved optical structure and matching to further reduce uncontrolled beam loss. The decision to maintain the baseline accumulator ring design made the investment in a new SRF linac more rewarding, as the linac was not shortened for the reduced injection energy of a rapid-cycling synchrotron.

In 2000, the main section of the SNS linac was officially changed from an NC structure to an SRF structure after many months of study and extensive reviews. The new baseline configuration of the linac consisted of a DTL that accelerates an incoming 2.5 MeV beam from an RFQ to 70 MeV, a CCL that accelerates the 70 MeV beam to 186 MeV [23], and a superconducting section that accelerates the 186 MeV beam to 1 GeV [13]. The superconducting portion consists of two parts: a medium-β section and a high-β section. Despite the fact that SRF linacs had been considered an option for various accelerator projects in Europe, Japan, and the US, the SNS linac was the first hadron (H^-) SRF linac constructed to deliver more than 1 MW of average beam power.

An initial study led by the Los Alamos team advocated the choice of an SRF design [21], based on advantages that include reductions in AC power, even though the RF duty is only 7%, in the length of the linac due to higher accelerating gradients, and in the number of klystrons and associated RF power systems. Furthermore, there is an increased safety margin against radio-activation from beam losses that would impede hands-on maintenance and limit availability. This latter advantage is the result of the higher gradients that increase the longitudinal focusing, larger bore radius that becomes economically possible, and the improved vacuum in the cryogenic environment of the superconducting accelerator.

The initial study proposed replacing the normal-conducting linac design of the pulsed linac, which has a duty factor of 6%, with SRF cavities at an energy of approximately 200 MeV. This section would comprise several subsections, each of which uses a periodic array of normal-conducting quadrupole magnets and superconducting cavities with identical shapes. The design takes advantage of the large velocity acceptance of the superconducting cavities to cover the velocity range from about 200 MeV to 1 GeV with only two elliptical cavity shapes. Compared with the baseline 805 MHz normal-conducting linac design, the superconducting design saves about 50% of the 2.5 MW klystrons, reduces the overall length by approximately 20%, and decreases the AC power by about 10 MW. The design parameter choices were intended to provide the maximum benefits from SRF technology at the lowest cost, consistent with minimal schedule impact for the project, which already had a baseline normal-conducting linac design. The design uses 2.5 MW, 805 MHz klystrons that have already been chosen for the project, implying that each klystron will drive more than one cavity.

In early summer 1999, a working group led by the SNS management was formed, comprising participants from ANL, Cornell University, JLab, LANL, and ORNL from the USA, as well as DESY, INFN Milano and JAERI from abroad. The working group met in August 1999 at Argonne and in September 1999 at Newport News. The design report was completed in November 1999 [24], and the design decision was approved by a peer review committee in December 1999. The formal baseline change took place in January 2000, and the SRF linac project began on February 1, 2000, with JLab newly joining the SNS partnership responsible for the design and construction of the linac SRF structures.

With the final design, the SRF linac operating at 805 MHz frequency accelerates the H^- beam from 186 MeV to the top energy. Compared with the original NC CCL linac, the SRF linac provides a high accelerating gradient (11–16 MV/m) capable of reaching a higher beam energy, encounters less beam loss and halo scraping due to its larger bore radius, is immune to one cavity/klystron failure, operates at a better vacuum, and has higher reliability and availability.

The selection of two types of SRF cavities ($\beta_0 = 0.61$ and 0.81, respectively) allows for economic savings and future energy upgrades. However, a relatively large phase slip requires a detailed error-sensitivity analysis. The choice of cavity geometric value β_0 is based on a smooth transition from the warm-section linac, maximized final output energy, and a comfortable transition from medium-β to high-β section with tolerance for one cavity failure. Constant-gradient continuous focusing is chosen to maximize the accelerating field strength.

Based on JLab's operational experience, SNS linac cryomodules are individually connected to the cryo-distribution system with U-tubes, allowing for easy maintenance. Quadrupole doublet magnets for transverse focusing operate at room temperature, outside of the cryomodules. Optimization of the number of cavities per cryomodule and the number of cells per cavity is based on minimizing the number of warm-cold transitions and associated heat loads, minimizing the number of cavities and associated components such as RF input power couplers, and considering the ease of cavity fabrication. Two types of cryomodules are designed: medium-β with a 0.61 cryomodule, each containing 3 cavities, and high-β with a 0.81 cryomodule, each containing 4 cavities. Six-cell cavities were chosen for ease of handling during fabrication and for the required longitudinal acceptance. Taking into account the tight construction schedule, a moderate peak surface field of 27.5 MV/m was chosen for the medium-β cavity. Using electro-polishing, a higher peak field of 35 MV/m is assumed for the high-β cavity.

To reduce uncertainties in the RF controls of an ion beam ($\beta < 1$) under Lorentz detuning, microphonics, beam transients, and injection energy offset, a decision was made to drive each cavity with its own klystron using independent amplitude and phase control. New 550 kW klystrons were individually procured, which supply power to each SRF cavity. The high-power RF coupler is based on the KEK B-factory coupler design and scales from KEK's 508 MHz to SNS's 805 MHz frequency.

The SNS linac was not designed with a low-β section to replace the normal-conducting DTL and CCL, due to the lack of time for sufficient R&D caused by the large physical dimensions of the cavities and the lack of mechanical strength to handle the Lorentz de-tuning force.

SNS operation has confirmed the design choices of the SRF linac; the two types of cavities reached similar gradients in operation despite the design assumptions, as shown in Fig. 10.4. The SRF linac has demonstrated superior flexibility and

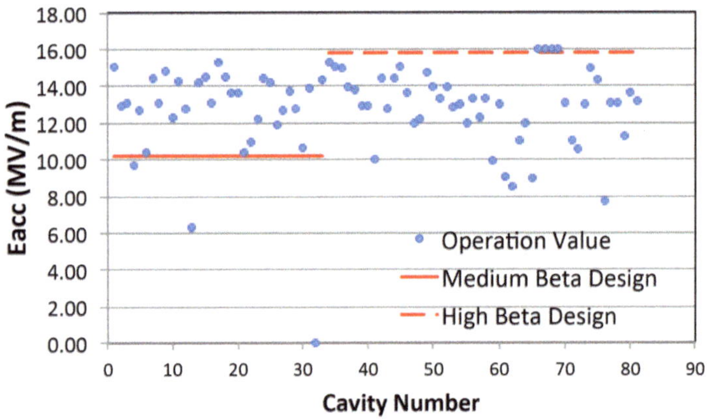

Fig. 10.4 Measured accelerating gradients of the 81 SNS $\beta_0 = 0.61$ (medium) and $\beta_0 = 0.81$ (high) cavities in 23 cryomodules in comparison with the design goals [25]. (Courtesy J. Galambos, ORNL)

availability with automatic tuning, as the use of one klystron per cavity enables operation with one or two cavities turned off by re-phasing the neighboring cavities. This capability is due to the fact that the superconducting linac has a large velocity acceptance and reserve capability [25].

In selecting a superconducting linac over a normal conducting linac, the management of SNS had a vision beyond a comparison of technical merits, in particular because a duty factor of 7% did not necessarily warrant the choice of SC technology. However, by choosing "building the first (full-scale hadron) cold linac" over "building the last (full-scale hadron) warm linac" in the world, the management endeavored to attract worldwide talent to ORNL for an exciting new accelerator project with cutting-edge technology.

10.1.3.3 Cryogenics

As previously explained in Sect. 10.1.3.2, JLab was chosen for the design of the SNS SRF linac and the cryogenic system in 1999 based on the operational experience gained from CEBAF [26]. The aggressive SNS schedule left very little time for the evaluation of the improvement options, except to use the CEBAF designs as much as possible with the lessons learned in the establishment and operation of the CEBAF accelerator and to reduce the risk associated with a new design. Based on JLab's operational experience, SNS linac cryomodules are individually connected to the cryo-distribution system with U-tubes to allow easy commissioning and maintenance. Similarly to the CEBAF design, only the RF cavities are in the cryomodule; the quadrupole doublet magnets for transverse focusing operate at room temperature and are located outside of the cryomodule.

Fresh memories of the long and difficult lessons from the JLab experience of making the large 2 K cold compression system fully operational (without subatmospheric compression at room temperature) played a significant role in setting the direction of the SNS cryogenic system. In addition to distributing the 4 K to 2 K heat exchanger from a single large heat exchanger to multiple smaller ones located within each cryomodule to reduce the effect of heat leakage and improve the operational stability of the system, the designs of the cryogenic refrigeration and distribution systems were largely based on the JLab cryogenic system to minimize risk to the project. The adoption of the JLab cryogenic system and cryomodule design greatly reduced the construction and commissioning schedule of the project. However, it did not add any significant improvements to the cryogenic system, other than the 4 K to 2 K heat exchanger, nor did it improve the efficiency of the system or reduce operational utility costs compared to the JLab cryogenic system.

10.1.3.4 Debate on Rapid-Cycling Synchrotron Alternative

During the same period when the initial SRF linac study was conducted, a study was carried out that compared the baseline structure of the full-energy linac plus an accumulator ring with a rapid-cycling synchrotron (RCS) design: a 60 Hz, 400 MeV linac feeds two vertically stacked RCSs that accelerate the proton beam to 2 GeV energy [27]. The biggest challenge to the RCS design is the stringent beam loss criterion (1 W/m): although it has been relaxed by a factor of 5, only 0.4% uncontrolled loss is allowed for a beam power of 2 MW assuming a collimation efficiency of 90%. However, among the existing rings, the lowest loss of approximately 0.3% had been achieved at LANL's PSR, an 800 MeV accumulator, compared to typical losses of a few to tens of percent in RCSs (e.g., ISIS, FNAL and AGS Boosters). This situation was only changed later after the J-PARC RCS was successfully built and commissioned, with minimal beam loss throughout the acceleration cycle.

As discussed in Sect. 10.2, the later built J-PARC invested greatly in new technologies that enabled rapid-cycling, allowing the use of the sinusoidal RF waveform for acceleration in the synchrotron. Lacking R&D time for an already baselined project, the SNS RCS study relied on existing ring technology. Without significant R&D investment, the SNS RCS study conservatively relied on a programmable RF waveform for acceleration. The SNS RCSs operating at 30 Hz require a high RF voltage (about 400 kV per ring at 1.4–1.9 MHz) for fast acceleration, a large magnet aperture to accommodate the space charge at a lower energy, ceramic vacuum pipes with detailed RF screening, and high-performance power supplies. Minimization of magnetic errors due to eddy currents, ramping, saturation, and power-supply tracking is not trivial. The study concluded that the required RCS design is technically more demanding and less cost-effective.

As shown in Fig. 10.5 (left), the RCS design comprises a 400 MeV linac that is injected into two vertically stacked synchrotrons, each accelerating beam pulses of 10^{14} protons at a repetition rate of 30 Hz from 400 MeV to 2 GeV, producing

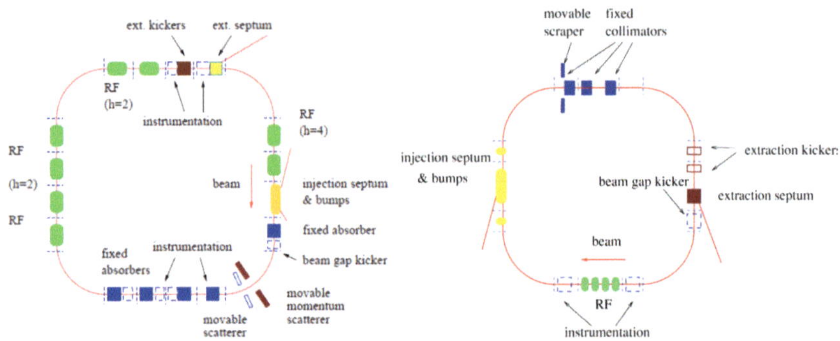

Fig. 10.5 (Left) Schematic layout of one of the two rings of the vertically stacked rapid-cycling synchrotrons of the alternative SNS RCS design [27]. (Right) Schematic layout of the SNS accumulator ring. The ring circumference is 248 m. The four straight sections are designed for injection, collimation, the RF system, and extraction [28]

the same total beam power of 2 MW. Each synchrotron has a four-fold symmetry comprising four FODO arcs and four dispersion-free straights. The four straight sections house injection, collimation, RF, and extraction systems. RF cavities of harmonics 2 and 4 occupy 50% of the available straights. The beam injection is entirely contained in one straight section, followed immediately by absorbers to protect against injection malfunction and beam halo. The beam cleaning and collimation section starts with a beam gap kicker located 90° in phase upstream of the primary scraper, followed by absorbers covering a 2π betatron phase. Beam loss is localized to shielded locations (collimation and injection regions) using multi-stage collimators. Momentum cleaning is achieved using a multi-turn beam gap kicker. With a design collimation efficiency of 95%, the total allowed fractional beam loss on the collimators is 1%. For engineering reliability, collimators are designed to withstand an average 10^{-2} beam power and to endure two full beam pulses for commissioning and emergency handling. Various measures (a foil interchange mechanism, quick disconnect flanges, crane, etc.) are adopted to enable the rapid removal and replacement of devices in high-radiation areas. To prevent beam loss under off-normal conditions, redundancy is built into critical areas. The extraction system is designed so that the beam is contained within the acceptance, even when a kicker is misfired. The RTBT is designed with a suitable phase advance so that a kicker misfire does not cause a change in the beam location on the target.

The synchrotron lattice has a hybrid structure with FODO arcs and triplet straights. It combines the simplicity and ease of correction of the FODO with the flexibility of the triplet of long drifts. Each arc consists of four 8 m long FODO cells. Five of the arc quadrupoles at sites of large dispersion are sandwiched between a chromatic sextupole and an orbit-correction dipole. The other quadrupoles are sandwiched by corrector packages containing both linear elements for orbit correction and decoupling, and nonlinear elements for resonance correction.

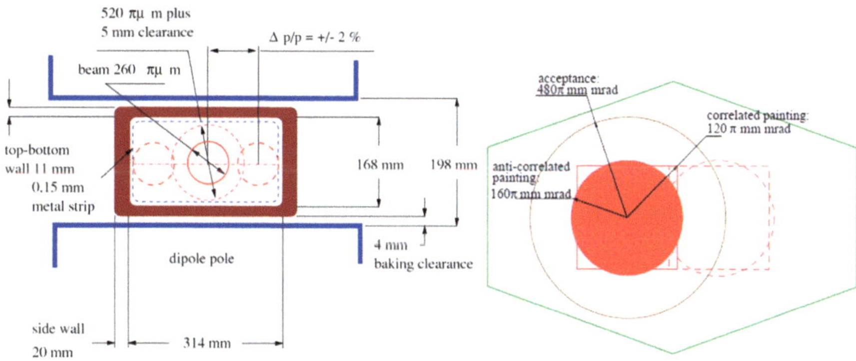

Fig. 10.6 (Left) Schematic drawing of the ceramic vacuum chamber cross section in the dipole region of the proposed RCS version of SNS [27]. The horizontal acceptance accommodates a momentum deviation of $\Delta p/p = \pm 2\%$. Beam is collimated between 300 and 400 πmm·mr in dispersion-free regions. (Right) Schematic illustration of the transverse dimensions in the SNS accumulator ring [28]. The green hexagonal boundary represents the cross section of the dipole vacuum chamber. The brown circular line corresponds to the minimum aperture in the ring at 480 πmm·mr. The beam is represented by the red circle for anticorrelated painting and the red square for correlated painting. The displacement of the beam centroid in the dipole (indicated by the dashed outlines) corresponds to 1% momentum deviation

The beam is injected dispersion-free for independent transverse and longitudinal manipulation. Longitudinally, the RF cavities in the HEBT are used to correct the centroid and broaden the momentum for instability control. Transversely, painting is performed using either correlated or anti-correlated orbit bumps. The injection chicane and bump magnets (collapsed after injection) are contained in a drift section independent of lattice tuning. The maximum magnetic field of 5.3 kG limits the stripping loss to below 10^{-7} per meter. The pole surface of the two dipoles near the stripping foil is specially shaped for electron collection.

To minimize eddy currents and maintain good field quality, programmable power supplies are used to moderate the ramp rate below 30 T/s. The linear ramp-up takes 24 ms, and the ramp-down takes 8 ms. The synchronous RF phase is below 20°, and the bucket area is increased monotonically to minimize beam loss. Once accelerated, the two beam bunches, each 150 ns long and 550 ns apart, are kicked vertically in a single turn by 8 fast ferrite kickers and then horizontally extracted by a Lambertson septum.

The vacuum chamber provides an acceptance of more than three times the full beam emittance (Fig. 10.6 (left)), and twice the full momentum spread. The transverse clearance allows for efficient two-stage collimation, and the longitudinal clearance allows for the collection of momentum halo using the BIG kicker. Laminated dipole, quadrupole, sextupole, and multipole magnets are designed with a peak field below saturation ($\sim$1 T). The dipoles and quadrupole poles are shaped to minimize integrated errors (sextupole and do-decapole, respectively) to the 10^{-4} level. Errors in the fringe field, whose fractional contribution is proportional to the ratio between beam emittance and magnet length (about 10^{-3}), can be compensated for by the correctors. Separate trim quadrupoles are designed to compensate for the tracking error between the main dipole and the quadrupole field. Arc sextupoles powered in four families are used for chromaticity adjustment and off-momentum matching. Correctors up to an octupole are used for orbit correction, decoupling, and resonance compensation. Ceramic vacuum chambers about 1 cm thick, with internal silver strips 0.15 mm thick, terminated at one end with resistor-capacitor networks, provide high impedance to the eddy current but low impedance to the image current. The steps are tapered, and the ports and bellows are shielded to reduce impedance. The inner surface of the chamber is coated with TiN to reduce secondary emission and potential electron-cloud effects. The space for a fast damper system is reserved as a future remedy.

10.1.3.5 Accumulator Optimization with Hybrid Lattice and Larger Acceptance

The baseline SNS accumulator ring lattice after several rounds of design optimization is shown in Fig. 2.30 [28]. Notably, early in the optimization, all-FODO lattices were changed to a hybrid lattice with FODO cells in the arcs and doublets over the straight sections. Compared with FODO structures, the doublet structure offers a long, uninterrupted drift space that facilitates efficient injection and beam collimation. The four-fold symmetric ring lattice contains four dispersion-free straight sections, housing injection, collimation, RF, and extraction. Each achromatic arc consists of 4 FODO cells with a 90° horizontal phase advance. After optimization, the ring lattice combines the simplicity and ease of correction of the FODO structure with the flexibility of the doublet structure for dedicated injection and collimation manipulation. Injection in a dispersion-free region allows for independently adjustable painting in the transverse (with orbit bumps in the ring) and longitudinal (with an energy-spreading phase-modulated RF cavity in the HEBT) directions for robust operation [29]. The 12.5 m uninterrupted straight section with flexible phase advance further improves the collimation efficiency [30]. Compared with the original all-FODO lattice, the matching between the arcs and the straights increases the arc acceptance by 50% with the same magnet aperture.

Another area of optimization that has significant cost implications is the transverse and momentum acceptances. An adequate acceptance-to-emittance ratio is essential to allow for collimating the beam and minimizing uncontrolled beam loss. Collimation studies indicate that efficient cleaning of a beam extending out to 240 πmm·mr requires a ring acceptance of 480 πmm·mr. This acceptance will allow particles scattered by the collimation system to return to the collimators without being lost elsewhere in the ring. Along the arcs, where the dispersion is non-zero, the same acceptance of 480 πmm·mr is maintained for particles with momentum deviation $\Delta p/p_0 = \pm 1\%$. Figure 10.6 (right) illustrates schematically the transverse dimensions of the beam, collimators, and beam pipe around the ring.

Finally, to allow longitudinal collimation with the BIG kicker (see Sect. 8.10.1.2), the longitudinal acceptance of the arc is set to be $\Delta p/p_0 = \pm 2\%$ for a beam extending out to 160 πmm·mr. Then off-momentum particles can reach the gap between bunches without hitting the vacuum chamber in high-dispersion areas. This longitudinal acceptance will allow one to make chromaticity measurements and upgrade the RF to cure instabilities.

A design change with implications for radiation protection was the transition from an alpha to an omega configuration for HEBT, ring, and RTBT, avoiding the crossover of the two transport lines [31]. Along with this change, crane coverage is provided for the ring and transfer lines, eliminating the need for personnel to be near activated equipment for repair and removal. The tunnel width was increased by 0.6 m allowing easier access to failed equipment. In addition, a personnel entrance was added to the ring between the HEBT and RTBT, so that personnel will not have to enter this area directly through the HEBT or RTBT.

10.1.3.6 Electron Cloud Preventive Mitigation

The SNS accumulator ring team made every effort to prevent performance-limiting mechanisms that haunted prior machines, such as PSR. One example is the preventative measures to mitigate the electron-cloud effects, as discussed in Sect. 3.5.2.

To minimize the SEY from the chamber surfaces, thus avoiding electron multipacting, the entire AR surfaces facing the beam were coated with $\sim 0.1\,\mu$m thick titanium nitride (TiN). The SEY, which peaked at $E_e \sim 300$ eV and ranged from 1.5 to 2.5 for most chamber materials, was lowered to ~ 1 on TiN coated surfaces [32, 33]. DC magnetron sputtering was developed to coat the curved 4 m half-cell chambers using a cylindrical Ti cathode with a 25 mm diameter and AlNiCo magnets inserted into a 38 mm diameter titanium tube, as shown in Fig. 10.7. The magnets were stacked with 13 mm spacers, resulting in a looping magnetic field of several hundred gauss. The 13 mm diameter hole in the center of the magnets allows for water cooling of the cathode during the high-power plasma discharge. A gas distribution line for Ar and N_2 was attached along the length to control the nitrogen pressure distribution [34].

Fig. 10.7 (**a**) A 4 m long curved half-cell chamber of the SNS accumulator ring; (**b**) the Ti cathode with magnets and gas line; and (**c**) the discharge plasm with bright rings at the location of spacers between the magnets. (Courtesy H. Hseuh, BNL)

Fig. 10.8 (**a**) The SNS ring injection kicker ceramic chambers that was first coated with 0.7 μm Cu, and then with 0.1 μm TiN, using anode screen which smoothed out the electrical field. (**b**) One of the 14 extraction kicker modules assembled with 1 cm by 5 cm opening masks for TiN coating, to cover >90% ferrite surfaces. (Courtesy H. Hseuh, BNL)

As shown in Fig. 10.8, two layers of coating were applied to the ceramic injection kicker chambers: a 0.7 μm thick copper layer to carry the image current, followed by a 0.1 μm thick TiN layer for a low SEY, resulting in an end-to-end resistance of 0.04 Ω on each chamber, allowing the passage of the image current [35]. For the 14 extraction ferrite kickers, masks with 1 cm by 5 cm openings and 1 mm gaps were installed to coat the inner surface with $> 90\%$ TiN coverage [35]. The resistance across the coated strips was $> 100\,\Omega$. No difference was measured in the field distortion between coated and uncoated kickers.

10.1.3.7 Targetry System

Mercury, rather than solid heavy metal cooled with water, was selected as the target material for SNS primarily because of its potential for increased power handling capability and greatly reduced waste stream. Another significant advantage is that no active cooling for the mercury is required after shutdown because the decay heat is distributed over a large mass. Mercury was also selected as the reference liquid target material because it (1) is a liquid at room temperature, (2) has good heat transport properties, and (3) has a high atomic number and mass density, resulting in a high neutron yield and source brightness [36].

SNS, in collaboration with J-PARC and other institutes, invested heavily in understanding the potential limitations of the target system [37]. Initial studies indicated that cavitation damage [38, 39] can limit the life of the targets, and the damage rate can scale with the fourth power of the beam power [40, 41]. During the evolution of the SNS design, the injection of small bubbles of gas into mercury was identified as a potential method to improve the useful lifetime and performance of the target [42], but the injection of gas was not included in the initial design of the SNS targets.

Table 10.3 SNS accelerator system phased commissioning

Segment	Date
Front end	October 2002
DTL tank 1	August 2003
DTL tanks 1–3	March 2004
DTL tanks 4–6 and CCL modules 1–3	August 2004
CCL module 4 and SCL	July 2005
HEBT and ring (to extraction dump)	December 2005
RTBT and target	April 2006

10.1.4 Technical Challenges and Lessons Learned

Partnership among six DOE national laboratories, with clear responsibilities and centralized budget authority, has been successful for the SNS project. In general, partner labs were responsible for subsystem R&D, design, procurement, testing, and delivery. ORNL was responsible for installation, subsystem integrated testing, and commissioning. The work breakdown structure (WBS, Chap. 9) level 2 elements consist of accelerator systems, experimental systems, and conventional facility systems. The accelerator systems consist, as WBS level 3 elements, of partner laboratories (the front end, the warm part of the linac, the SRF linac and cryogenics, and the ring and transport), together with the integrated controls system and ORNL integration.

Early, planned, and phased beam commissioning from 2002 led to the establishment of the facility on time in 2006 (Table 10.3). The front end systems of the ion source, LEBT, and RFQ were beam commissioned at LBNL in 2002 before they were delivered to ORNL. Seven Accelerator Readiness Reviews (ARRs) were conducted at ORNL prior to the beam commissioning of the corresponding sections of the accelerator complex.

The R&D funded as part of the SNS project resolved many of the early identified technical challenges. For example, beam collimation, H^- injection, electron cloud, and collective instability mitigation were among the early R&D activities for the ring system. These potential performance-limiting mechanisms were thoroughly studied, leading to design improvements and implementations. The decision to use a superconducting linac at the SNS was made relatively late in the construction project. Despite the limited time available for R&D, all of the design prospects came to fruition, including a high accelerating gradient, low beam halo, uncontrolled beam loss, and operational flexibility and robustness.

The energy upgrade was planned from the beginning, allowing a 30% increase in beam energy and a doubling of beam power on the neutron target. The geometry of the linac tunnel was laid out to accommodate space for nine additional cryomodules to provide an additional 300 MeV of beam energy. The cryogenic distribution system was installed to support the entire 1.3 GeV linac. The circumference of the accumulator ring was extended to house future longer, lower-field injection magnets to accommodate the lower Lorentz stripping threshold of a higher-energy H^- beam.

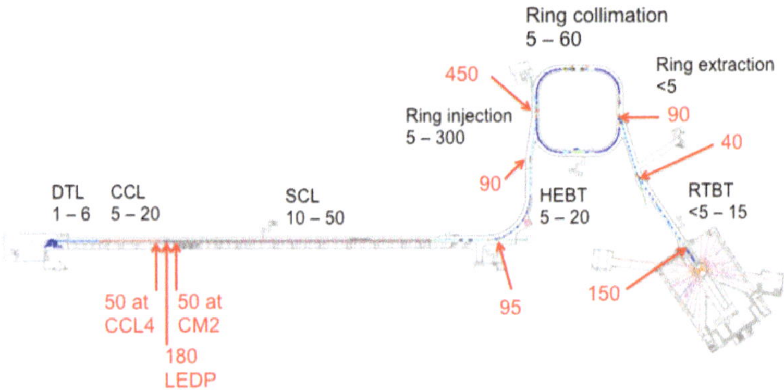

Fig. 10.9 Typical activation levels after 1 MW operations followed by ∼48 hours of low-power studies (Courtesy M. Plum). All numbers are mrem/h (10 μSv/h) at 30 cm from the beam line. The numbers in red indicate localized hot spots [43]

Value engineering is a common practice for reducing project costs to fit a preset budget. However, if poorly executed, this can irreversibly compromise accelerator performance. In 1999, two major value engineering exercises were conducted: one to reduce the aperture of the normal-conducting linac structure, and the other to possibly reduce the circumference and the transverse beam chamber aperture of the accumulator ring. The ring value engineering concluded that neither the circumference nor the aperture should be reduced, given the challenges of minimizing uncontrolled beam loss. On the other hand, linac value engineering led to a reduction in the CCL aperture from 4 cm to 3 cm to save costs. Later, when the SRF linac was adopted, simulations predicted beam loss near the end of the CCL as the focusing strength was reduced to match the SRF optics. During operations, high beam loss occurs in the region between the CCL and the SCL around the CCL beam stop, which damages the gate valves near the end of the CCL section [43], as shown by the relatively high level of radio-activation in the region of the low-energy differential pumping (LEDP) section of Fig. 10.9. Due to such a beam loss, the "LEDP upgrade" has been on the machine improvement agenda since the start of operations.

In general, the SNS construction project was successfully completed on time and within budget. Table 10.2 shows the estimate at completion (EAC) performed in July 1999 after the project baseline and in June 2005 before the project completion. In July 1999, the project contingency of \$255.2 M was 28.2% of the EAC. In June 2005, 11 months before the project achieved CD-4, the estimate to complete (ETC) was about \$45.9 M. The remaining project contingency was \$11.3 M. The comparison indicated the following:

- The cost performance of the front end systems (LBNL scope), ring & transfer systems (BNL scope), and the integrated control systems (ORNL-led collaboration) was healthy.

- Centralized controls work proceeded mostly smoothly as planned, as discussed in Sect. 10.1.3.1.
- The cost of linac systems (LANL and JLab scope) was significantly higher than the baseline value by $> 40\%$. Contributing factors included both the switchover from the NC to the SC linac and the challenges encountered with the NC section of the linac (Sect. 10.1.4.2).
- The cost of target systems (ORNL scope) was significantly higher than the baseline by $> 40\%$ due to the technical complexities associated with the breakthrough technology of a high-power liquid mercury system.
- The excess in conventional facilities (CF), including office buildings, is visible at $>50\%$ of the baseline. The cost of CF is approximately 33% of the EAC.
- The effort of project support was significantly underestimated at the baseline. Building a strong home team is essential to integrate overall design efforts, integrate partner laboratories efforts, mitigate issues and challenges encountered throughout the project, and comprehend installation, commissioning, and transition to operations.
- Because the total project cost (TPC) was capped, the scope of the instrument systems was significantly reduced. This reduction slowed the initial scientific output of the facility.

Lessons learned at SNS are fundamentally important for any accelerator project that relies on a multi-laboratory partnership. In the following sections, the challenges and lessons learned in the subsystems of the facility are discussed in detail.

10.1.4.1 Front End and NC Linac

The development of the front-end system was launched well in advance at LBNL, prior to the start of the construction of the project. The front end system, including both the ion source and the RFQ, was commissioned with the H^- beam at LBNL before it was delivered to ORNL in 2002. The early establishment of the front end was the basis for the successful phased commissioning of the SNS project.

Tight optical focusing used for chopping and antichopping in a long MEBT is a source of beam-halo generation. Studies show that even without the antichopper, partially deflected particles are still mostly contained by the envelope of the nominal unchopped beam. The MEBT quadrupoles are thus made independently adjustable so that alternative optics can be realized, avoiding tight focusing at the antichopper or MEBT chopper.

A key challenge in linac performance is minimizing beam emittance growth and centroid jitter in both transverse and longitudinal directions upon ring injection, thereby reducing foil traversal, scattering, and radio-activation. The warm DTL operating at 402.5 MHz is expected to be less sensitive to vibrational noise than most existing linacs operating at 200 MHz. Tight RF control ($<0.5\%$ amplitude and $0.5°$ phase error) warrants tolerable energy variation before the beam enters the energy correction and spreading cavities in the HEBT.

Permanent magnet quadrupoles are used in the DTL due to the tight geometry (402.5 MHz starting at 2.5 MeV); however, electromagnetic quadrupoles could be used at DTL tank three and beyond. During the construction stage, improper shielding of the magnetic field from these permanent magnets complicated electron-beam welding, resulting in weld distortion and leaks in the drift tubes. Subsequent repairs exceeded the Curie temperature of some magnets, compromising the magnetic field quality. A task force was formed to repair the damaged devices and restore the project schedule.

The high voltage converter modulator (HVCM) system was a major contributor to accelerator downtime until 2009 [44]. Fire trucks had been on site from time to time when HVCM faults occurred. Thorough investigation and analysis of the failure modes helped to identify root causes and solutions for subsystem enhancements.

The lessons learned from the LLRF development and re-work were summarized in Sect. 10.1.3.1 [20].

10.1.4.2 SRF Linac

The insightful decision to adopt modern SRF technology in the establishment of the SNS linac has been validated by the robust operations of the facility since 2006 and the relatively quick increase in beam power to 1 MW. Subsequent development of advanced maintenance technologies, such as plasma processing, helps maintain the state-of-the-art performance of the SRF systems.

As the decision was made late during the project construction cycle, after the project baseline, there was no time for project-specific R&D. Inevitable complications during mass production consumed budgetary and schedule contingencies. As some of the cryomodules were not tested after assembly until after they were installed in the SNS tunnel, the commissioning at ORNL was relatively challenging. Until reprocessing was extensively applied, the SNS linac often operated below the design energy of 1 GeV.

Using only two types of cavity β for more than 800 MeV of acceleration compromises the equipartition law. During the design stage, it was speculated that space-charge coupling can cause transverse and longitudinal emittance exchange when the emittance ratio meets resonance conditions. In addition, depending on the level of initial mismatch, a space-charge parametric halo may develop in the linac. Efforts were made to reserve an economically affordable large aperture and to ensure tuneability in the MEBT, CCL, and SRF linac.

The effect of higher-order modes (HOM) on the cavities is another issue. The overlapping of the beam and HOM spectrum is possible due to the pulsed time structure of the beam and the fact that the beam frequency shifts with variable ring energy and repetition rate (e.g., for some two-target operation scenarios). Fortunately, transverse and longitudinal (beam break-up) instabilities are minor issues for an ion beam in the presence of a cavity-to-cavity frequency spread. HOM dampers are used solely for power dissipation.

The large bore aperture of the SRF linac allowed for the mitigation of the newly discovered beam loss mechanism of intrabeam stripping (Sect. 3.7.4) [45]. The large aperture allowed SNS to reduce the loss by expanding the beam (limited by the beam halo). SNS would have been far more radioactive if it had been a traditionally normal conducting copper linac, as originally designed.

10.1.4.3 Ring and Transport

The low-loss design philosophy of the SNS accumulator ring was successfully achieved, leading to the successful operation of a ring that held a world record in both intensity (the number of charged particles per pulse) and power (both average and peak) for protons. Implementations to suppress mechanisms that limit the performance of previous machines have been successful. Per-forming as originally designed, the ring further supports facility upgrades with a 30% increase in injection energy and a final average beam power above 2 MW.

The decision in 1999 against switching the design from an accumulator ring to rapid-cycling synchrotrons avoided delays in the project schedule and potential escalation of the project budget. As witnessed by the RCS development at J-PARC, an extensive amount of R&D has been needed to pave the way toward establishing the rapid-cycling technology that meets the requirements of a megawatt class facility. The decision against the reduction in the transverse aperture and beam admittance again avoided possible short-sighted savings that may compromise the path towards future intensity and power upgrades.

Accelerator R&D started well before the project baseline and was important for the smooth establishment of the facility. The "PSR instability" that limited the performance of the Proton Storage Ring at LANL was a subject of early R&D by the SNS ring team. Measures to mitigate electron cloud effects have been proven to be successful in avoiding anticipated instabilities. The optimization of the design and the reduction of the beam coupling impedance associated with the extraction kickers and pulse-forming network were highly effective. In comparison, studies in the J-PARC RCS showed possible complications caused by such devices. Conditions for the onset of collective instabilities and instability damping using chromaticity sextupoles were satisfactorily bench-marked during later beam studies.

Provisions were implemented in the design of the ring and transport to ensure a straightforward path to future upgrades [46–48]. The ring optical lattice allows space for upgraded magnets operating at lower fields to avoid premature Lorentz stripping. The extensive, multi-stage beam collimation is likely to be essential for higher beam power operations. The energy corrector and spreader removed from the project baseline during the value engineering process may also be re-installed in future campaigns to ramp up beam intensity and power.

Solid steel, as opposed to laminated steel, was selected for most ring and transport magnet cores for cost savings [46]. Individually, a good field quality ($<10^{-4}$ relative error at full acceptance) is achieved. However, excessive (up to 0.25%) magnet-to-magnet variation is found in the dipole transfer function and its current dependence. These dipoles are shimmed to achieve a variation below 10^{-4} for 1 GeV operation and are sorted according to 1.3 GeV measurement data to minimize the orbit corrector strength, ensuring satisfactory performance for both SNS baseline and upgrade operations.

10.1.4.4 Conventional Facilities

A lesson learned with SNS conventional facility construction was on the differential ground settlement between the target floor and the RTBT tunnel floor caused by (1) Over-excavation: In one area of the tunnel, the ground was over-excavated by several meters before construction. This created a section with significantly weaker underlying soil, which experienced the most severe settlement of approximately 20 cm. (2) Backfill weight: After completing the accelerator tunnels, builders placed five meters of native soil on top for radiation shielding. This substantial weight recompressed the soil underneath the tunnels, which had expanded after the original excavation. (3) Geological conditions: The construction site presented geological challenges, including the possibility of sinkhole formation in the area.

The ground settlement caused not only vertical displacement but also deformations and cracks in the tunnel floor and wall. The SNS survey and alignment team implemented a comprehensive monitoring program to track the deformation of the tunnels [49]. Monitoring results eventually showed that the rate of settlement decreased significantly over time. The persistent deformation of the survey network required repeated re-survey campaigns to ensure the alignment of the accelerator components. A continuous waterproof membrane was placed over the entire tunnel system before the final backfill. This prevented water leakage, even as cracks developed in the concrete lining.

10.1.5 SNS Proton Power Upgrade Project

The Proton Power Upgrade (PPU) project followed the provisions conceived during the design of the SNS project [4]. The PPU project doubles the beam power capability of the SNS facility from 1.4 MW to 2.8 MW to facilitate new discovery capabilities, including the accommodation of a second neutron target station [50]. Seven cryomodules containing 28 SRF cavities built by JLab have been added to the reserved space at the end of the existing linac to achieve an energy capability of 1.3 GeV. Three high-voltage converter modulators (HVCM) were added to drive the 28 new klystrons to power the new cavities. Using the space reserved for the straight section of the accumulator ring, two injection chicane magnets and two injection dump magnets have been upgraded with lower peak magnetic fields but higher integrated strengths. The capabilities of the mercury target station were increased using gas injection and jet flow technologies. A tunnel extension downstream of the

Table 10.4 SNS PPU major project dates

Event	Date
Mission need approved (CD-0)	Nov 2004
Alternative selection & cost range approved (CD-1)	Apr 2018
Long lead procurement approved (CD-3a)	Oct 2018
Long lead procurement approved (CD-3b)	Sep 2019
Performance baseline approved (CD-2)	Oct 2020
Start of construction approved (CD-3)	Oct 2020
SNS operation outage I for installation	Aug–Nov 2022
SNS operation outage II for installation	Mar–Jun 2023
SNS operation outage III for installation	Aug 2023–Jun 2024
Project key-performance-parameters (KPP) achieved	2024
Planned early completion of project (CD-4)	2025

ring was added to enable the future tie-in of the second target station beam transport line without interrupting the operations of the first target station.

The mission need (CD-0) for the PPU project was approved in November 2004, prior to the completion of the SNS project in May 2006. As listed in Table 10.4, the PPU project achieved CD-1 in 2018 and CD-2 in 2020, respectively [51]. Two phases of long-lead procurements, CD-3a in 2018 and CD-3b in 2019, included superconducting RF cavities, fundamental RF power couplers, cryomodules, RF systems, and the construction of conventional facilities in the existing klystron gallery. The installation of the technical systems was carried out during three planned periods of SNS operation outages. In July 2023, SNS reached a world record of 1.7 MW average proton beam power after the first four PPU cryomodules were installed, allowing the beam to be accelerated to 1.05 GeV energy. In 2024, all key performance parameters of the project were achieved in preparation for the project's early completion (CD-4) in 2025.

10.2 Japan Proton Accelerator Research Complex

In contrast to the Spallation Neutron Source (SNS) project [52] dedicated to pulsed neutron production, the Japan Proton Accelerator Research Complex (J-PARC) [53] is a multi-purpose facility, as shown in Figs. 1.7 and 10.10. Not only neutrons, but also other secondary particles such as muons, kaons, and neutrinos are produced by the MW-class proton accelerators and are fully utilized for studying materials, life sciences, and high-energy physics experiments.

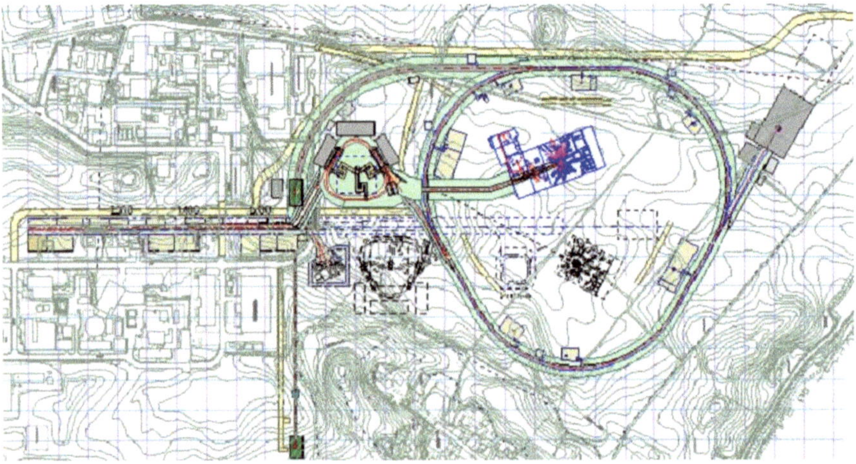

Fig. 10.10 Schematic layout of the J-PARC accelerator complex. The complex consists a 400 MeV H^- NC linac, an RCS ring accelerating the proton beam to 400 MeV for neutron and muon production, and a synchrotron to further accelerate the proton beam to 30 GeV for high-energy experiments

10.2.1 J-PARC Facility Overview

The J-PARC project was conceived to meet the growing demand in Japan (as in many other countries) for high-power proton accelerators in a wide variety of scientific and engineering fields, including industrial applications. Pulsed spallation neutron sources with a proton beam power of 1 MW or more and a pulse length on the order of 1 μs are powerful tools for studying materials and life sciences, and muons are becoming popular for the same purpose in a complementary way. Also, the neutron and muon themselves are still very interesting fundamental particles to study in physics beyond the standard model. For example, the electrical dipole moment of the former and $(g_\mu - 2)$ of the latter can be measured precisely using high-intensity secondary beams. Kaons are used for fundamental particle experiments by measuring their rare decays, and they can be used to produce hypernuclei, which will open a new horizon in nuclear physics by introducing a new degree of freedom: strangeness. In addition, long-baseline neutrino experiments are promising in the field of fundamental particle physics by measuring lepton flavor mixing. CP violation in the lepton sector would then fall within the measurement range. Finally, a basic study of an accelerator-driven system for nuclear waste transmutation (ADS) is possible by using a high-power proton linac.

The J-PARC project started by joining the Japan Hadron Facility (JHF) project [54–57] and the Neutron Science Project (NSP) [58], promoted respectively by the High Energy Accelerator Research Organization (KEK) and the former Japan Atomic Energy Research Institute (JAERI). In 2005, JAERI joined with the Japan Nuclear Cycle Development Institute (JNC) to form the current Japan Atomic

Table 10.5 J-PARC facility major dates

Event	Date
Funded for J-PARC construction	Apr 2000
Linac accelerating beam to 181 MeV for commissioning	Jan 2007
RCS accelerating beam to 3 GeV for commissioning	Oct 2007
Start of neutron experiment	May 2008
MR accelerating beam to 30 GeV for commissioning	Dec 2008
MR beam slow-extracted to the hadron experimental facility	Jan 2009
MR beam fast-extracted to the neutrino production target	Apr 2009
Funded for linac beam energy upgrade to 400 MeV	Apr 2009
Funded for earthquake recovery	Apr 2011
Linac beam accelerated to 400 MeV	Dec 2013
Funded for MR magnet power supply upgrade	Apr 2019
Neutrino user operation with 750 kW (FX) beam power	Apr 2023
Neutron user operation with 1 MW equivalent RCS beam power	Apr 2024

Energy Agency (JAEA). Here, the JHF project was developed from the Japan Hadron Project (JHP) [59] planned by the Institute of Nuclear Study (INS) of the University of Tokyo, together with the High Energy Physics Laboratory (formerly KEK). For this purpose, the present KEK was formed by joining the former KEK and INS.

As indicated in Table 10.5, funding was provided to KEK and JAEA in the Japanese fiscal year 2000 to allow the J-PARC construction project, which consists of three accelerators: linac, rapid-cycling synchrotron (RCS), and main ring (MR), to officially start in 2001. Despite many years of preparation and R&D, development of innovative technology on several fronts continued to challenge the project schedule. The scope of the project was adjusted by lowering the linac output energy from 400 MeV to 181 MeV and the MR output energy from 50 GeV to 30 GeV. With the adjusted scope and schedule, accelerator construction progressed towards the beam commissioning of all three accelerators, and the three facilities started on schedule. The J-PARC project was successfully completed in 2009 with the start of user operations at the Materials and Life Science Experimental Facility (MLF), the neutrino production target, and the Hadron Experimental Facility.

10.2.2 Accelerator Scheme

As discussed in Sect. 1.6.1.1 and illustrated in Fig. 10.11, the J-PARC complex consists of three accelerators: linac, RCS, and MR. The linac accelerates the H^- beam to 181−400 MeV. This is injected into the RCS by charge stripping, and the resulting proton beam is accelerated to 3 GeV with a repetition rate of 25 Hz (the beam power is 0.6−1 MW). Most of the beam is extracted to the MLF,

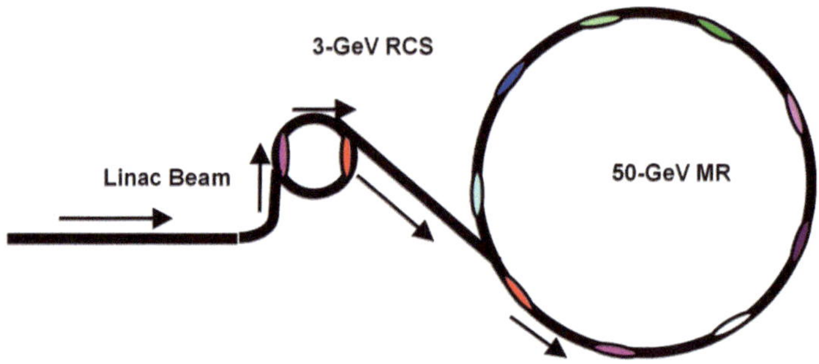

Fig. 10.11 Injection schemes from the linac to the 3 GeV RCS and from the 3 GeV RCS to the 30–50 GeV MR. The linac medium bunches indicated by the red and pink color in Fig. 10.12 (left) are injected every other times to form the two bunches in the RCS ring. Every three seconds four successive pairs of two bunches are transferred and injected to the MR

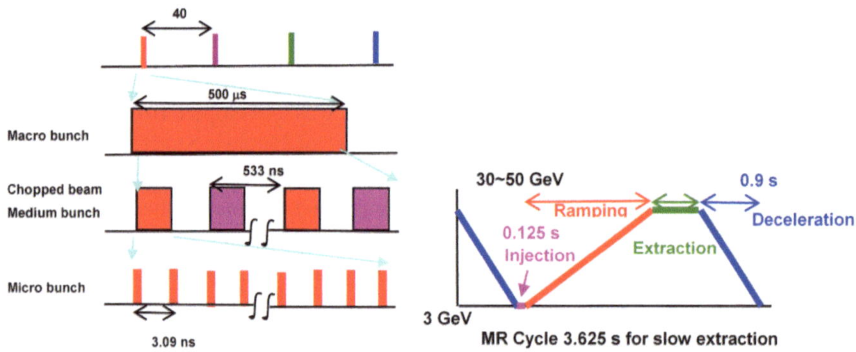

Fig. 10.12 (Left) The time structure of the J-PARC linac beam. The uppermost figure shows that the macro bunches are accelerated every 40 ms, that is, with a repetition rate of 25 Hz (this is upgradable to 50 Hz for future ADS experiment). The next figure shows the bunch length of the macro bunch. The third figure shows the medium bunches formed by the chopper in the MEBT. The lowest figure shows the micro bunches produced by the 324 MHz RFQ. (Right) The MR cycle for slow extraction. It takes 0.125 s for injection, 1.9 s for ramping, 0.7 s for slow extraction, and finally 0.9 s for deceleration

where the muon production target and the neutron production target are located in series. However, every three seconds, four successive pairs of two beam bunches are extracted from the RCS into the MR, which accelerates them to 30–50 GeV. The beams are then either fast-extracted to the neutrino production target with a power of 0.75 MW or slow-extracted (Fig. 10.12, right) to the Hadron Experimental Facility with a power of 0.1 MW. The neutrinos thus produced are sent to the Super Kamiokande detector, which is located 295 km west of J-PARC. At the Hadron Experimental Facility, kaon rare decay experiments, hypernuclei experiments, etc., are conducted. The 400 MeV linac beams are planned to be used for basic research on ADS after the linac repetition rate is upgraded from 25 Hz to 50 Hz.

In general, the number of secondary beams is proportional to the beam power if the beam energy exceeds the threshold for their production. To provide many types of users with the appropriate high-power beams, the 3 GeV, 1 MW RCS is placed between the 400 MeV injector linac and 50 GeV MR, and it is used to generate pulsed neutrons and muons in the MLF, as shown in Fig. 1.7. It may be said that this accelerator scheme is a version of the KEK PS, upgraded by approximately an order of magnitude in both beam energy and beam current (the KEK PS comprised a 40 MeV linac, a 500 MeV RCS, and a 12 GeV MR). The distinctive features of the J-PARC accelerator, as detailed in the following sections, mostly arise from the multi-purpose concept.

10.2.3 RCS Versus AR for Pulsed Neutron Source

As mentioned in the previous section, "multi-purpose" is key to the J-PARC specification. This requires the generation of proton beams with energies of both several tens of GeV for high-energy physics experiments and a few GeV for pulsed neutron and muon production. J-PARC needs an RCS as an efficient injector to the MR, and the RCS can also be used to generate pulsed neutrons and muons. This is the first reason why J-PARC chose an RCS scheme instead of an accumulation ring (AR) scheme with a full-energy linac, such as SNS and LANSCE (see Sect. 1.3.2). The second reason is that an RCS-based neutron source may have some advantages compared to an AR-based one, including savings on the linac cost and avoiding large-scale SRF and cryogenic deployments. However, the cost of an RCS is usually much greater than that of a comparable AR.

The spallation mechanism, in which a proton beam bombards a heavy metal target, is the most efficient way to generate high-intensity neutron beams. The neutron production rate is approximately proportional to the proton beam energy if the proton energy is between 500 MeV and 3 GeV; thus, the number of neutrons is approximately proportional to the proton beam power in this energy range. As the time-of-flight method is widely used to measure neutron energy, a further requirement for the proton beam is a pulse length of around 1 μs, together with a low repetition rate to cover the low-energy neutron spectra. Typical parameters are listed in Table 10.6. Since no ion source can produce the required number of protons per 1 μs, a linac macro bunch of approximately 1 ms is accumulated in a ring with a circumference of around 300 m and then fast-extracted to the neutron production target. Here, the issue is whether the protons in the ring accelerate or not. The naive answer is that, provided the difficult processes of injection into and extraction from the ring can be managed successfully, one should accelerate the beam since, by doing so, one can increase the beam power or generate the same beam power with lower beam current.

In other words, the RCS scheme has an advantage over the AR scheme regarding a lower beam current for the same beam power. For example, the beam current of a 1 MW, 3 GeV RCS is 333 μA, which is one third as high as that of a 1 MW, 1 GeV

Table 10.6 Parameters in comparison of J-PARC rapid-cycling synchrotron and SNS accumulator ring

Parameter	J-PARC RCS	SNS AR
Beam stored energy per pulse [kJ]	40	24 (40)
Number of protons per pulse [10^{13}]	8.3	15 (25)[a]
Beam pulse length [μs]	<1	<1
Beam energy [GeV]	3	1
Beam power [MW]	1	1.4 (1)
Beam current [mA]	0.333	1.4 (1)
Repetition [Hz]	25	60 (25)
Injection energy [GeV]	0.4	1
$\beta^2\gamma^3$	1.475	6.750
Beam emittance at painting [πmm·mr]	216	91[b] (142)
Laslette tune shift	−0.16	−0.15 (−0.16)
Linac peak current [mA]	50	38 (75)
Linac beam pulse length [μs]	500	1000
Beam-on rate after chopping [%]	56	68 (56)

[a] The values in the parentheses are scaled from the SNS ones by assuming the same repetition rate and beam power as those of J-PARC for comparison between the RCS and AR schemes

[b] This value is estimated from the tune shift, using the same bunching factor and form factor as those for J-PARC

AR (1 mA). Here, note that a 3 GeV AR is practically impossible for the neutron source with a pulse length of 1 μs. When an H^- beam is injected into a ring, the outermost valence electrons of the H^- ions will undergo Lorentz stripping if the H^- ions pass through a strong magnetic field. To avoid Lorentz stripping, a weak magnetic field must be used, which requires a long straight section to inject the high-energy H^- beams. Since the circumference of the ring is limited by the pulse length of the extracted beam (required for the neutron beam), the maximum possible injection beam energy is limited to around 1.3 GeV.

The low beam current for the RCS implies that the RCS scheme is relatively immune to beam instabilities. In addition, the output beam energy of the linac is lower, which reduces the cost of linac construction. The injection energy of the RCS is determined by the following trade-off between the space-charge effect and the construction cost. The reduction in injection energy, and hence the linac cost, increases the space-charge effect, which must be compensated for by enlarging the RCS transverse aperture and, consequently, increasing the RCS cost. The trade-off can be exemplified with the space charge figure of merit of the Laslette tune shift (Sect. 3.3.2.2) which is proportional to $\frac{N}{\epsilon\beta^2\gamma^3}$, where N and ϵ are the number of particles and the beam emittance, respectively. This parameter is similar for both J-PARC and SNS, as seen from the Laslette tune shift in Table 10.6. Therefore, the beam emittance, accelerator admittance, and magnet aperture can be smaller for a higher injection energy. In general, the RCS demands a larger ring admittance compared to the AR, as shown in Table 10.6.

As most uncontrolled beam loss in an RCS occurs at injection energy (Sect. 3.4.3.2), a lower injection energy compared to an AR implies that more beam loss can be tolerated in an RCS. Combined with the reduced beam intensity in an RCS, the tolerable fractional beam loss is even greater. Table 10.6 shows that the RCS allows the beam loss rate to be 7.5 times higher than in the AR case. For example, the 4 kW beam collimator system of the J-PARC RCS can accept a beam loss of 3% during injection, while the same system can only accept a beam loss of 0.4% in the AR case.

It is vital to the RCS scheme that beam loss is minimized beyond injection, since beam loss at higher energy gives rise to greater radioactivity. However, the normal FODO lattice implies that the beam must pass through the transition energy during the course of acceleration. To avoid the transition and associated beam loss (see Sect. 3.8), the beam optics of the RCS and MR have been designed based on the high-γ_T lattice and the imaginary-γ_T one, respectively, by using the missing bend technique [60]. Although this kind of optics has been attempted for the electron/positron synchrotron [61] or has been used to design a proton synchrotron [62], J-PARC is the first operational proton synchrotron to employ a high or imaginary transition energy to eliminate beam loss during acceleration.

In addition to the transition-free lattice design, the RCS should have the following features. First, rapid acceleration requires a high field gradient RF accelerating system. Second, rapid acceleration requires rapid ramping of the magnetic field, which will induce eddy currents on the metallic surface. To suppress eddy current effects, ceramic vacuum chambers with RF screening should be used, magnetic cores should be laminated with slit core ends, and magnetic coils should be stranded. Third, the lower injection energy, in turn, implies a higher space-charge effect. To mitigate the effect, large aperture magnets are required, giving rise to large fringing fields. In addition, it is difficult to inject the beam uniformly into the large aperture together with the charge exchange through a foil, and extraction is also difficult to manage. These challenging developments and the mass production of these components are reported in Sects. 10.2.4 and 10.2.5.

10.2.4 Magnetic Alloy Loaded Cavity

Rapid acceleration requires a high accelerating field gradient, as mentioned above. Since the J-PARC 3 GeV RCS (25 Hz, 0.18–3 GeV) accelerates the beam twice as fast as the ISIS RCS (50 Hz, 70–800 MeV), which was the world's fastest previously, it needs a maximum field of 23 kV/m, which is twice as high as that for conventional ring-RF systems. Here, the high-γ_T lattice, which requires longer arc sections than those of a conventional FODO lattice, shortens the straight sections available for RF acceleration. It is again noted that the circumference of the RCS for a pulsed neutron source is quite limited by its short pulse length. This high field gradient at the accelerating gap can be electromagnetically induced only by high RF magnetic flux density in the magnetic core. Ferrite, which has been conventionally used in proton synchrotrons, drastically decreases its $\mu Q f_{RF}$ value in this high RF magnetic flux density, which is typically necessary beyond a field gradient of 10 kV/m (the $\mu Q f_{RF}$ value is proportional to the shunt impedance; see Sect. 8.5.3). In other words, ferrite cannot practically be used in this range of magnetic flux density. However, a higher field gradient is potentially feasible using the magnetic alloy (MA) core, since the $\mu Q f_{RF}$ value of the MA core has an almost flat response to the magnetic flux density, which generates an electric field gradient of the order of 100 kV/m [63, 64].

The MA-loaded cavity has another advantage over the conventional ferrite-loaded ones. Since medium-energy protons significantly vary their velocities during the course of acceleration, the accelerating frequency must follow the change in the revolution time. This is one of the reasons why the ferrite-loaded cavities have commonly been used for proton synchrotrons. The permeability of the ferrite can be adjusted by applying an appropriate DC-bias magnetic field to the ferrite, such that the inductance and, thus, the resonant frequency can be changed. However, this tuning mechanism for the ferrite-loaded cavity complicates its low-level RF control system (LLRF). On the other hand, no tuning system is necessary for the MA-loaded one since the Q value can be extremely low (a value of less than one is possible). The choice of the MA thus significantly simplifies the LLRF. Another advantage of the low-Q cavity is that the second harmonics of the RF power can also be fed into the accelerating cavity [65]. In other words, two types of cavities (one for the acceleration and the other for the second harmonic) are not required. However, its high R/Q (low stored energy) with low value Q requires wideband beam loading compensation through feed-forward control. As a result, even the high power system should be wide-band.

To minimize the bandwidth necessary for compensation, the value of Q is optimized by adjusting the gap between the MA cores, which are cut radially, as shown in Fig. 10.13 (right) [66]. The Q values thus optimized are 2.9 (with 1.5 mm gap) for 400 MeV injection or 2.0 (with 1 mm gap) for 181 MeV injection for the RCS, and 10.0 (with 10 mm gap) for the MR, respectively. A J-PARC ring RF system, driven by one power amplifier, comprises three accelerating gaps. The electric accelerating field at each gap is induced by magnetic flux in six MA cores, as shown in Fig. 10.13 (left). The RCS acceleration is performed by ten RF systems, with one spare system added, while the MR acceleration is carried out by six RF systems. The MA core is formed by winding a MA tape with a width of 35 mm and a thickness of 18 μm. The MA tape is coated with 2 μm thick silica on one side to electrically insulate one layer from another.

When high RF power was fed to the cavity, the cross-sectional surface of the cut cores was severely damaged. Thus, both the Q value and the resonant frequency of the RCS cavity with uncut cores were proposed to be adjusted by adding inductance in parallel with the cavity [67]. However, even the uncut cores were damaged by the transverse (radial) electric field on the MA cores near the acceleration gap. Subsequently, it was found that the electric field and the induced current melted the shorts between the MA layers. The shorts occurred because the silica insulator

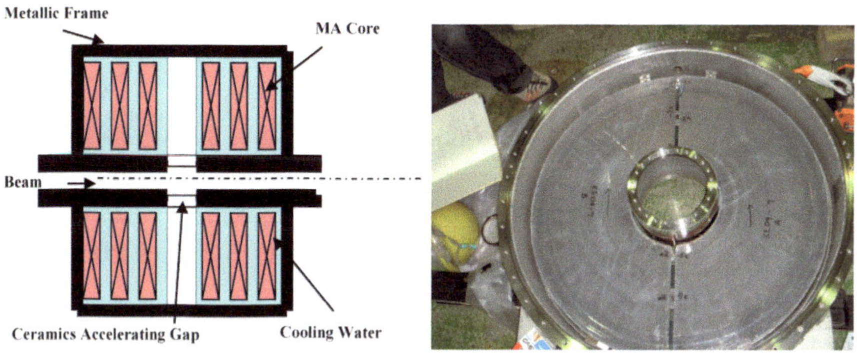

Fig. 10.13 (Left) Schematic view of the J-PARC synchrotron RF accelerating cavity. (Right) Cut MA cores of the RF cavity. (Courtesy J-PARC, Japan)

was scratched during the process of winding the MA tapes to form the cores. This problem was finally solved by improving the manufacturing process in the factory [67, 68].

The MR RF system is such that the Q value and the resonant frequency of the MR system are higher than those of the RCS system. Thus, no value of added inductance could simultaneously meet the high Q value and the high resonant frequency. Therefore, the development of the MA core cutting technique for the MR system had to continue. Cut cores have now been developed that can handle the designed power load, incorporating the above improvement in MA core insulation and diamond polishing of the cut surface [66].

Both the uncut cores for the RCS and the cut cores for the MR performed well in operation, typically for over 2,000 hours, contributing to the successful beam commissioning of both the RCS and the MR. However, it was found after 3,000 hours of high-power RF operation that one of the eleven RF cavity systems in the RCS revealed a degradation in its impedance. During 3 years of operation since 2007, impedance reductions in RF cavities resulting from the buckling of MA cores were detected three times. The MA core buckling was found to be caused by thermal stress, which was improved by impregnation with low viscosity epoxy resin [69]. Subsequently, one-by-one replacements were made, and the RF cavities were loaded with the improved MA cores to operate properly.

10.2.5 *Technical Issues in RCS Other Than RF*

10.2.5.1 Ceramics Vacuum Chambers

All the vacuum chambers of the RCS that are exposed to fast varying magnetic fields have been made of alumina ceramic. In addition, the chambers must be electromagnetically shielded to prevent the radiation of high-power microwaves

Fig. 10.14 (Left) The ceramic vacuum chamber for the RCS bending magnet. (Right) The ceramic vacuum chamber with a tennis-racket shape aperture

arising from the circulating charged particles. For this purpose, copper strips were electroformed in the ceramic chambers [70, 71]. The titanium flanges were then silver-brazed to the chambers, and the brazing surfaces of the chambers were metallized by nickel striking plating.

The vacuum chambers for the quadrupole magnets (QM's) were relatively easy to manufacture, having a circular cross section. However, the cross section of the racetrack shape, as shown in Fig. 10.14 (left), was chosen for the bending magnets (BM) in order to maintain the large physical aperture at a reasonable cost. For the injection section, vacuum chambers with an aperture of tennis-racket shape [72], as shown in Fig. 10.14 (right), are needed to allow the injected beam from the linac to merge into the circulating beam. Much more time than expected was required for the development and mass production of these ceramic vacuum chambers, particularly for those with special shapes. Some chambers were delivered to the J-PARC site just in time for the scheduled beam commissioning.

10.2.5.2 Large Aperture Magnets

In order to mitigate the space-charge defocusing effect, the QMs should have a large aperture, be short, and be close to the neighboring ones for frequent focusing. As a result, the fringing field is quite strong and can significantly interfere with neighboring QMs. Together with the saturation effects at the core ends, multipole components higher than quadrupole might substantially reduce the dynamic aperture. If this had happened, the compensating octupole magnets would have been installed to recover the dynamic aperture. Fortunately, the dynamic aperture proved to be sufficiently large in the beam simulation, which took into account the measured data for the actual magnet layouts.

10.2.5.3 Magnetic Field Tracking

From the beginning, the electromagnetic compatibility issue was expected to be very difficult to solve for the RCS magnet system. For this reason, a year was scheduled for powering and controlling the tests of the magnet system in situ. The QM's are driven by seven series-resonant circuit networks, while the BM's are driven by a single parallel-resonant one. Insulated Gate Bipolar Transistor (IGBT)-based power supplies are used because the fast switching of the IGBT is required for the precise tracking of all the magnetic fields with their respective saturation patterns. Here, IGBT stands for Insulated Gate Bipolar Transistor. Fast switching implies high-frequency components that can be easily coupled with other networks through incomplete grounding. In fact, some chassis or grounds revealed several hundred volts at certain frequency components. Although almost all the electromagnetic compatibility issues have been solved by improving the grounding and/or by detuning the resonant frequencies of the power-absorbing parasitic resonant circuits, the issue for the shift-bump system still remains unsolved. In situ efforts were also made to improve the signal-to-noise ratios of almost all beam diagnostic systems by filtering the noise, in addition to enhancing electromagnetic compatibility.

10.2.5.4 Injection and Extraction

Injection and extraction devices with large apertures are difficult to develop and manufacture. In particular, the decay of the magnetic fields in the injection bump magnets should be faster than 100 μs for a beam power of 1 MW [73]. This is required to reduce the number of hits of the circulating beams on the charge-exchange foil. The capacitors installed in the ceramic vacuum chamber, through which the mirror current of the beam passes, were damaged by this fast switching noise.

The extraction kicker system is designed in such a way that the full aperture beam can be extracted to the beam transport to prevent even the halo beam from hitting the beam pipe. The beam transport to the muon and neutron production targets also has the same aperture as the ring [74]. This is an important design policy to make full use of the advantages of the RCS by allowing beam loss only at the low-energy stage [75].

10.2.6 Unique Features of J-PARC Linac

A high-quality injector linac is vital for a high-intensity accelerator complex, since low-emittance beams with small halos transversely and longitudinally must be stably delivered to the RCS to minimize beam loss at injection. In particular, the stability of the beam energy is very important. For this purpose, the linac scheme was chosen as follows (Fig. 10.15 (left)).

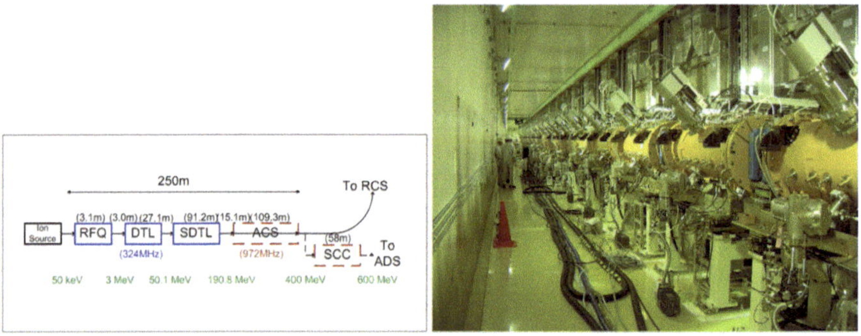

Fig. 10.15 (Left) Scheme of the J-PARC Linac. (Right) The SDTL's installed in the J-PARC linac. (Courtesy J-PARC, Japan)

The H^- ions from the volume production type ion source [76] are guided through the 50 keV LEBT to the 3 MeV, 324 MHz radio frequency quadrupole (RFQ) [77]. The beam is then transported by the 3 MeV MEBT [78], where a 324 MHz RF chopper [79] and rebunchers are installed, to the 50 MeV, 324 MHz drift tube linac (DTL) [80]. Finally, the 324 MHz separated DTL (SDTL) [81] and the 972 MHz annular-ring coupled structure (ACS) [82, 83] linac accelerates the beam to 191 MeV and 400 MeV, respectively. The ratio of the accelerating frequency of the DTL and SDTL to that of the ACS is an odd integer to maintain the possibility of simultaneously accelerating both protons and H^- ions for future use.

The RFQ energy of 3 MeV and the accelerating frequency of 324 MHz were chosen by optimizing the following factors. Here, the RFQ energy is the same as that of the MEBT and the first cell of the DTL, while the RFQ frequency is the same as that of the DTL (theoretically speaking, one can choose a multiple of 324 MHz). In general, it is preferable to increase the RFQ energy as high as possible to reduce both the emittance growth and halo formation, since the RFQ is ideal for this with its frequent and strong focusing in both the longitudinal and transverse planes. It is also preferable to start the DTL at as high an energy as possible for the following reasons. The low-energy DTLs has the shortest period among all DTLs, implying the densest mechanical structure. Nevertheless, the highest magnetic field should be applied to cope with the strongest space charge force, requiring more space. Thus, if the DTL can start at a higher energy, the DTL becomes easier to design and manufacture, both mechanically and electromagnetically.

However, the high-energy, high-frequency RFQ proved difficult to tune because the operational quadrupole mode becomes mixed with the dipole mode. In general, an RFQ operates in its fundamental quadrupole mode (TE21n) with $n = 0$, where n is the number of nodes of longitudinal magnetic fields or transverse electric fields. The operational frequency of 324 MHz is a few MHz above the fundamental dipole mode (TE11n) with $n = 0$. If the RFQ is elongated in order to accelerate the beam beyond 2 MeV for example, the frequency of the dipole mode with $n = 1$ is lowered, coming close to the operational frequency. Then, the operational quadrupole mode is mixed with this dipole mode, even though there is a small amount of deviation

Fig. 10.16 (Left) The J-PARC RFQ with PISL's. (Right) Electromagnetic coil with intrinsic water cooling channels produced by means of the electroforming technique and the wire-cutting. (Courtesy J-PARC, Japan)

from quadrupole symmetry. The π-mode stabilizing loop (PISL) [77] shown in Fig. 10.16 (left) was invented to raise the frequency of the dipole mode far above the operational mode. In this way, the fields in the high-energy, high-frequency RFQ can be stabilized. The first PISL stabilized RFQ had been manufactured at a frequency of 432 MHz for JHP, accelerating the H^- beam to 3 MeV, which was the highest in the world at that time. Now, PISL's have been installed in both the SNS RFQ and J-PARC RFQ.

The beam, which cannot be accepted longitudinally by the RCS RF system, must be chopped before injection. A beam chopper with very fast rise and fall times (ideally much shorter than the bunch spacing) is installed in the MEBT. This is one of the most difficult devices to realize. Higher energy beams are more difficult to chop, but if the beam energy is too low, fast chopping will be challenging, since the velocity of the beam is too slow. The MEBT energy of 3 MeV is thus the result of the trade-off between these two conflicting factors.

For the beam chopping system, it is quite common to use a traveling wave deflector chopper based on a meander-line design (for example, see [84]). Such a system was under development at J-PARC, but the fast power supply seemed difficult to realize once the chopper design had to be fixed. For this reason, the newly designed RF chopper was attempted [79, 85] (Fig. 10.17). The performance of the beam choppers of these two types is just sufficient for the present purposes (the rise and fall times are on the order of 10 ns, which is a few times the spacing of the bunches). The beams kicked during the rise or fall times are partly lost somewhere after the chopper, giving rise to some radioactivity.

The choice of the accelerating frequency is again the result of a trade-off. High frequency is preferable for the following two factors. First, if the frequency is

Fig. 10.17 The J-PARC RF chopper (Courtesy J-PARC, Japan). The beam comes from left to right. The transverse electric field (TE110-mode) generated between the two electrodes (upper to lower in the right figure) kicks the beam. The two adjacent RF cavities are used (the left figure)

typically higher than around 350 MHz, klystrons can be used, which are more stable and powerful RF power sources. Second, the high-frequency linac is more advantageous with respect to space-charge forces. This is a little controversial and sometimes sounds paradoxical since the bunches formed by the high-frequency RFQ are smaller both longitudinally and transversely (thus, the transverse charge density is higher). However, a shorter focusing period implies stronger focusing against the space-charge repulsive force.

On the other hand, at higher frequencies, the size of the drift tubes in the DTL decreases, making it difficult to use quadrupole electromagnets in the drift tubes at the start of the linac (this is why permanent magnets are used at SNS). As J-PARC wanted to retain flexibility to match the beams with variable space charge forces between the MEBT and the DTL, and optimize the focusing parameters within the DTL, effort has been devoted to the development of the most compact water-cooled quadrupole electromagnets (Fig. 10.16 (right)), making full use of the electroforming technique and wire cutting. The outcome is that 324 MHz is the highest possible frequency for the DTL, starting from 3 MeV. Thus, a klystron vendor was contracted to develop the new 324 MHz, 3 MW klystron with J-PARC pulse length and repetition in collaboration with the J-PARC team. Both developments were successful.

The 324 MHz, 3 MW klystrons developed for the J-PARC DTL and RFQ have been in stable and reliable operation with a repetition rate of 50 Hz and a pulse length of 600 μs. For this reason, a frequency of 324 MHz and an MEBT energy of 3 MeV have been chosen for many other projects (FNAL Project X [86], RAL [87], GSI FAIR [88], and CSNS [89]). Here, the superconducting cavity technology is considered for the future GeV-class proton linac in some cases. Since the frequency of 324 or 325 MHz is one-quarter of the L band chosen for the superconducting International Linear Collider (ILC), one could choose a 325 MHz front-end linac and an L-band high-energy superconducting linac to gain the full benefit from the development of the worldwide ILC collaboration.

The use of an SDTL (Fig. 10.15, right) is also a new initiative. At a medium energy of several tens of MeV, frequent focusing is no longer necessary. Thus, the quadruple magnets can be taken outside the DTL tank. Here, note that drift tubes with built in magnets are much more expensive than those without them. In addition, the shunt impedance of the SDTL can be optimized by eliminating the drift tube size/shape condition to accommodate the quadrupole magnets. In addition, the transverse focusing transition is located between the DTL and the SDTL, while the longitudinal focusing transition is between the 324 MHz SDTL and the 972 MHz high-energy structure (the transverse focusing of the SDTL region can be the same as or similar to that of the high-energy structure region). Emittance growth, and thus beam loss, usually arises from mismatches at these kinds of transitions. Simultaneous matching of both the transverse and longitudinal planes (as is the conventional case without the SDTL in between the DTL and the high-energy structure) is difficult to achieve; however, this is much easier if the matching is separated. This is another advantage of the SDTL concept.

10.2.7 Developments for the 400 MeV Linac

The possible candidates for the high-energy structure were extensively discussed in [90]. On this basis, the angular-ring coupled structure (ACS) (Fig. 10.18) was

Fig. 10.18 The lowest-energy J-PARC ACS cavity under power test [91]. (Courtesy H. Ao)

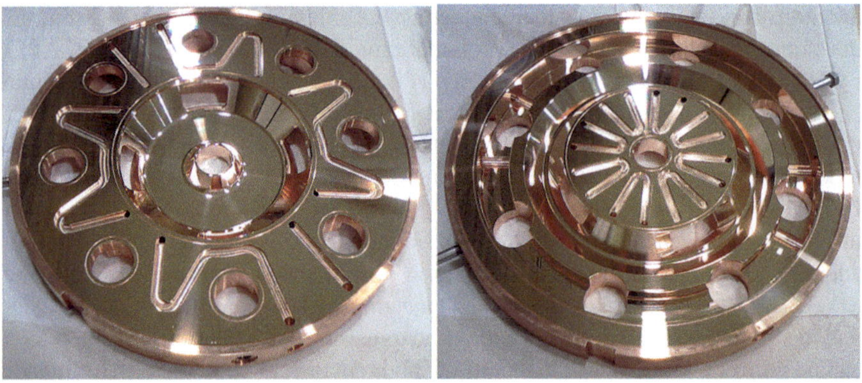

Fig. 10.19 The J-PARC ACS half cells viewed from the accelerating cell side (left) and the coupling cell side (right), respectively. The four slots couple the former with the latter. (Courtesy J-PARC, Japan)

chosen to accelerate the beam from 190 MeV to 400 MeV. The ACS may be considered an axially symmetric version of the side-coupled structure (SCS), geometrically speaking. The axial symmetry around the beam axis keeps the electric field therein from deflecting the beam transversely. It is noted that the use of the axially symmetric structure of the ACS may reduce halo formation in the high-energy region compared with the asymmetric structure of the SCS, where the transverse field in the SCS is approximately 1% as high as the longitudinal one.

Note that the axial symmetry of the ACS makes it possible to machine the structure with a lathe, which is ideal for high-precision machining. Although the coupling slots require three-dimensional machining, the proportion of this machining is less than a few percent of the total machining time. The silver brazing of the ACS is simple and reliable, as all half cells, as shown in Fig. 10.19, are vertically stacked one by one to form the entire structure in a brazing furnace.

Although the geometrical structure of the ACS can be obtained by rotating the SCS around the beam axis, its resonant frequency spectra are entirely different from those of the SCS. Since the coupling cavity of the ACS is coaxial, it has many parasitic modes just above the coupling mode. This is the reason why its development was not successful long after the structure was proposed [92]. In order to eliminate the mixing of these parasitic modes with the coupling mode, the symmetry was improved by increasing the number of coupling slots to four instead of two.

A 1296 MHz ACS was developed and tested successfully for JHP. For the J-PARC linac, the ACS frequency was reduced to 972 MHz, resulting in a much larger size of the structure. Attempts have been made to reduce the size of the 972 MHz ACS to that of the 1296 MHz one, and the present version was also developed with an improved coupling of 6%.

Power tests of the J-PARC ACS structure were performed for two bunchers and the lowest energy structure, with an accelerating field 10% higher than designed (4.1 MV/m) and a full duty factor of 3%. Here, the bunchers were installed in the transition between the 324 MHz SDTL and the 972 MHz ACS. In order to perform the high power test, a 972 MHz klystron was developed. After the ACS was further modified to be suitable for mass production [83], the manufacture of the 400 MeV linac began.

In addition to the ACL, efforts to restore the linac energy from 181 to 400 MeV also included further RFQ developments, with the aim of injecting more beam current into the ring. The first RFQ was designed for the JHF with an RCS beam power of 0.6 MW, which required a linac peak current of 30 mA. The J-PARC dedicated RFQ and its spare unit were developed to accelerate a peak current of 50 mA, capable of achieving the RCS beam power of 1 MW.

10.2.8 Beam Commissioning and Initial Operations

The early results of the RCS beam commissioning are detailed in [93]. A pulse of 2.6×10^{13} protons (which corresponds to a beam power of 310 kW when operated at 25 Hz) was sent from the RCS to the neutron production target. Here, a peak current of 15 mA with a macro pulse length of 500 μs and a medium pulse length of 600 ns (chopped in the MEBT) was delivered from the 181 MeV linac to the RCS (Fig. 10.12). Beam loss of 1% is concentrated during the beam injection period. This fact indicates that the transition-free optics design is effective in suppressing beam loss during acceleration. It is noted that the Laslett space-charge tune shift of -0.16 at injection is the same as that of the 1 MW RCS beam case with 400 MeV beam injection into the RCS.

The beam was painted during injection at an emittance of 150 πmm·mr both horizontally and vertically [94]. In addition, the second harmonic of the RF field, which is 80% as high as the fundamental one, was applied to the beam together with the phase sweep of $-80°$ to 0°. A momentum offset of -0.1% was also used. It is again emphasized that the low-Q MA-loaded cavities can be excited by the second harmonic of RF power, together with the fundamental one, without requiring a second harmonic cavity. These RF gymnastics are detailed in [95], while the comparisons with the beam simulation results are presented in [96].

A slight beam loss in the high-dispersion section was still observed during acceleration. This is perhaps because the sextupole magnets are excited only during injection for chromaticity compensation. This slight beam loss is expected to be eliminated by compensating for the chromaticity during acceleration. Beam powers of 100 kW and 200 kW were also demonstrated from the RCS for one hour and 7 s, respectively. In this way, beam commissioning made very efficient progress until the end of September 2008, enabled by all the accelerator components operating reliably and stably. In particular, the linac operation was very reliable and stable as an injector. The availability of linac beams amounted to 90% up to that point, and the linac beam commissioning had progressed very efficiently [97]. When linac beam commissioning was started, phase and amplitude scans were necessary for all the DTL and SDTL cavities. During this tuning, the beam was guided down to a beam dump approximately 300 m downstream. No steering was needed, demonstrating good alignment throughout all the quadrupole magnets.

During acceleration by the DTL, the rms emittances for a peak current of 30 mA increased significantly (nearly doubled), while no halo formation was observed beyond the Gaussian charge distribution. However, the rms emittances grew negligibly throughout the SDTL acceleration, whereas some halo formation was observed. This was perhaps because the accurate matching had not yet been completed.

After the end of September 2008, the discharge in the RFQ became the most serious issue regarding J-PARC beam operation. For example, the unscheduled downtime in May and June 2008 was only 5.4% of the total linac operation of 508 hours, while that in December 2008 and January 2009 amounted to 27.0% of the operation of 863 hours, of which 24.8% was due to discharge in the RFQ. Long-term conditioning of the RFQ was necessary by closing the gate valve between the RFQ and the ion source. The conditioning was performed using a short pulse and a high voltage, or sometimes the power was set to the level where multipacting appeared to increase the vacuum pressure. The pumping speed has recently been upgraded by installing the cryopump and other components. The back-up RFQ was also under design in parallel.

The results of MR beam commissioning are reported in [98]. Here, it is emphasized that the MR beam commissioning was also carried out efficiently, in part because the injector operation was reliable and stable. The imaginary gamma lattice design for the MR was also successful in eliminating beam loss during acceleration.

10.2.9 Performance Ramp Up and Upgrades

The J-PARC performance ramp-up has been diligently pursued. As indicated in Table 10.5, several major investments restored and enabled the capability of the facility. In 2009, J-PARC was funded to upgrade the linac energy from 181 MeV to the originally designed 400 MeV, which allows for the injection of a higher beam current and beam power into the RCS. The increased injection energy of the RCS reduced the performance-limiting effects of space charge that scale with $1/\beta^2\gamma^3$ (Sect. 3.3.2.2). In 2011, a heroic effort was made to recover from the Great East Japan earthquake that occurred on 11 March 2011 [99], with operations resuming during the same year.

The J-PARC team has demonstrated the capability of the linac and RCS to reach the ultimate performance parameters, with impressive beam loss control. Challenges with the liquid mercury target system delayed progress towards delivering the design beam power from the J-PARC linac and RCS to the MLF. Major accelerator improvements include the replacement of the ion source and RFQ to allow the linac to produce a beam current above 50 mA, as well as an upgrade to the RCS RF system [100]. In 2024, the RCS successfully delivered 1 MW equivalent beam power to the

MLF as originally designed. Issues such as insufficient cooling water capacity still limit year-round high-power operations.

The planned approach to increase the beam power of the 30 GeV slow-cycling synchrotron is to reduce the MR cycle time rather than increase the MR top energy to the originally designed value of 50 GeV [101] This has been endorsed by the Accelerator Systems Advisory Committee (A-TAC). In 2019, J-PARC was funded to upgrade the MR magnet power supply to shorten the MR acceleration cycle. In 2023, the MR successfully delivered 750 kW of fast-extracted beam for the operation of the neutrino user program, as originally designed. Currently, in 2025, the MR delivers a 900 kW beam to neutrino users and 90 kW of a slow-extracted beam to the Hadron Experimental Facility. A further upgrade is planned for the MR fast-extracted beam to reach 1.3 MW beam power for neutrino experiments, mainly by reinforcing the RF system.

10.3 Facility for Rare Isotope Beams

As introduced in Sect. 1.6.1.2, the purpose of rare isotope beam facilities is to (1) understand the origin of elements and model extreme astrophysics environments (nuclear astrophysics); (2) develop a comprehensive model of atomic nuclei (nuclear structure); (3) use atomic nuclei to test fundamental symmetries and search for new particles (fundamental physics); and (4) search for new applications of isotopes and solutions to societal problems, including medicine, energy, material sciences, and national security (applications). Adopting the in-flight fragmentation separation (IF) method, FRIB uses high-energy heavy projectiles (e.g., uranium beams) to bombard a thin target of a relatively light element (e.g., graphite).

10.3.1 FRIB Facility Overview

The Facility for Rare Isotope Beams (FRIB) was built as a scaled-down successor to the more ambitious and ultimately canceled Rare Isotope Accelerator (RIA) project. Following the growing recognition of the need for advanced rare-isotope beam facilities in the US, the Nuclear Science Advisory Committee of the US DOE/NSF convened a task force that unanimously recommended the construction of RIA, a unique facility designed to use a powerful superconducting linac to produce intense beams of stable ions for various applications [102]. The proposed RIA facility was planned to utilize both IF and ISOL methods. In early 2006, the RIA project was canceled due to budget constraints.

The 2007 NSAC Rare Isotope Beam (RIB) Task Force was created to define the science agenda and technical specifications for a future US Facility for Rare Isotope Beams (FRIB) [103]. Its work informed the 2007 Nuclear Science Long-Range Plan, which gave FRIB the highest priority for new construction. In 2008,

Table 10.7 FRIB major project dates

Event	Date
DOE-SC and MSU sign coopeartive agreement	Jun 2009
Alternative selection and cost range approved (CD-1), DOE issues NEPA FONSI	Sep 2010
Performance baseline (CD-2) approved	Aug 2013
Start of civil construction and long-lead procurement (CD-3a) approved	Aug 2013
Ground breaking and start civil construction	Mar 2014
Start of technical construction approved (CD-3b)	Aug 2014
SRF building construction completed for cryomodule mass production	Sep 2016
Ar beam accelerated through RFQ	Sep 2017
4 K liquid helium produced by the FRIB cryoplant in the installed cryomodules	Nov 2017
Ar, Kr beam accelerated through first three cryomodules with energy >2 MeV/u	Jul 2018
Liquid lithium circulated in the charge stripper unit by the electromagnetic pump	Aug 2018
2 K liquid helium produced by the FRIB cryoplant in the installed cryomodules	Dec 2018
Ar, Kr beam accelerated through linac segment 1 with energy >20 MeV/u	Feb 2019
Ar, Kr, Xe beam accelerated with energy >200 MeV/u before COVID-19 stand down	Mar 2020
FRIB project construction completion (CD-4)	Apr 2022
Start of user operation at 1 kW beam power	May 2022

after fierce competition, Michigan State University was selected by the DOE to host FRIB.

The primary goal of the FRIB project is to design, construct, and commission into operation a rare isotope research facility based on a high-power heavy-ion driver linac that is substantially better than any other heavy-ion facility in the world. The key requirements were to complete a facility capable of delivering the primary heavy-ion beam power above 400 kW, which is more than an order of magnitude higher than that of existing facilities. The US DOE Office of Nuclear Physics and Michigan State University funded the construction of FRIB at a total project cost of US$730 M. Upon completion of the project by 2022, FRIB is expected to provide the most demanded rare isotope beams in the world for scientific research and applications.

Table 10.7 lists the major project dates since 2008. The "mission need" (CD-0) was established in 2008. Civil preparation started in 2012 as soon as the environmental assessment was conducted and the project was baselined (Fig. 10.20). In 2014, DOE-SC approved the start of technical construction (CD-3b) (Fig. 10.21).

The design and construction of the FRIB accelerator have been facilitated under individual work-for-others agreements with many DOE-SC national laboratories, including ANL, BNL, FNAL, JLab, LANL, LBNL, ORNL, and SLAC, and in collaboration with institutes worldwide, including BINP, KEK, IHEP, IMP, INFN, INR, RIKEN, TRIUMF, and Tsinghua University. However, the FRIB project team at MSU is solely responsible for managing and organizing the work for all accelerator systems, including in-house design, construction, commissioning,

Fig. 10.20 Aerial view of the facility for rare isotope beams in 2014 when the ground was broken for the construction of the driver linac tunnel. (Courtesy MSU, USA)

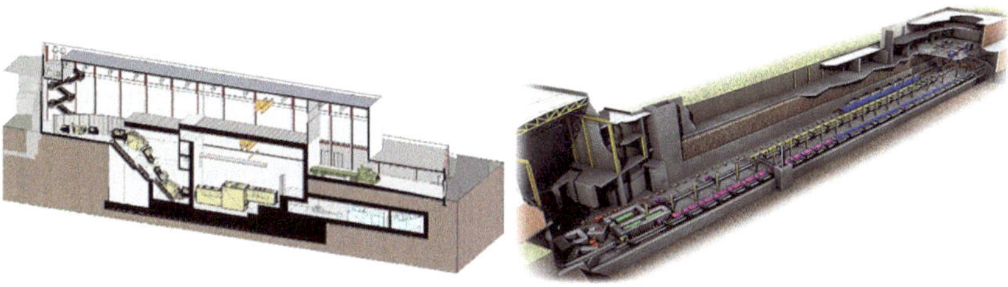

Fig. 10.21 Schematics of (right) the driver linac and (left) the production target and vertical pre-separator of the FRIB construction project. (Courtesy MSU, USA)

procurements with industrial providers, and work for-others agreements with other laboratories.

The FRIB project was completed in April 2022, on scope, at cost, and ahead of the baseline schedule established about 10 years prior. The ramp up to the ultimate design beam power of 400 kW is progressing; 1 kW was delivered for initial user runs, and 5 kW was delivered in February 2023, as shown in Fig. 1.6. In December 2023, a record high beam power of 10.4 kW uranium was delivered to the target, producing rare isotopes that were separated and analyzed in the Advanced Rare Isotope Separator (ARIS). Primary beams delivered for rare isotope production include ^{16,18}O, ^{22}Ne, ^{28}Si, ^{36,40}Ar, ^{48}Ca, ^{64,70}Zn, ^{82}Se, ^{86}Kr, ^{124}Xe, ^{198}Pt and ^{238}U.

Starting in late 2024, the beam power on the target reached 20 kW for primary beams, including uranium. The facility operates with an availability of over 92%, annually delivering more than 6,000 user hours, of which 4,000 hours are for scientific users and 2,000 hours are for industrial users.

10.3.2 Accelerator Design Philosophy

In creating this new one-of-a-kind facility, FRIB builds on existing facilities at the National Superconducting Cyclotron Laboratory (NSCL), a National Science Foundation (NSF) user facility at MSU. Starting in 2014, the reaccelerator (ReA3), consisting of a radio-frequency quadrupole (RFQ) and a superconducting radio-frequency (SRF) linac, was constructed and commissioned to accelerate beams of rare isotopes. The scope of the FRIB project thus only includes a high-power driver accelerator, a high-power target, and fragment separators, which better match the project budget allocation. The drawback of such an arrangement is that FRIB must be built within the constraints of the existing infrastructure. To physically fit the long driver linac in the center of the campus without prohibitive civil costs, the accelerator was folded twice into a paper clip shape, as shown in Fig. 9.5. Such a design resulted in additional expenses for incorporating two second-order achromats as the folding segments and added the complexity of interferences between the lower- and higher-energy segments of the linac. The design took advantage of the first folding segment to clean up the beam loss after charge stripping.

Full-energy, continuous-wave (CW) linac technology, instead of cyclotron-based or synchrotron-based technology, is chosen to deliver the primary beam that meets the FRIB requirements for rare-isotope productivity and separation accuracy. Up to 400 kW of beams are focused to a diameter of 1 mm (90%), an energy spread of 1% (95% peak-to-peak), and a bunch length of <3 ns on the target. In-flight fragmentation separation produces rare isotope beams with the required resolution.

SRF technology is the energy-efficient choice for the CW linac. The FRIB driver linac uses SRF technology to accelerate charged beams of all stable isotopes, from hydrogen to uranium, at a low energy of 500 keV/u with practically sized cavity bores by housing both cavities and solenoids in a cryomodule. A two-gap scheme is chosen throughout the entire linac, providing both efficient acceleration and focusing. Developments in digital low-level RF control and solid-state RF amplifier technologies have made individual cavity powering and control reliable and cost-effective. Four types of accelerating cryomodules are used, all operating at 2 K: $\beta_0 = 0.041$ and 0.085 QWRs at 80.5 MHz, and $\beta_0 = 0.29$ and 0.53 HWRs at 322 MHz. The cryomodules also contain superconducting (SC) solenoid packages that operate at 4.5 K for transverse focusing and steering.

High availability, maintainability, reliability, tunability, and upgradability are especially required for the FRIB accelerator to operate as a national scientific user facility.

- Availability: The accelerator is designed with high beam-on-target availability, accommodating normal, alternative, and fault scenarios. In the normal scenario, a liquid lithium stripper is used to raise the average charge state of ^{238}U beam to 78+ for efficient acceleration. Fault scenarios include situations where SC cavities under-perform by up to 20% of the design gradients. Furthermore, key components and subsystems are implemented with spares and design redundancies.

- Maintainability: The average uncontrolled beam loss is limited to below the 1 W/m level for all ion species, from proton to uranium, to facilitate hands-on maintenance. For a proton beam at high energy, this level corresponds to an average activation of about 1 mSv/h measured at a distance of 30 cm from the beam chamber surface, 4 hours after operations shutdown. For heavy ions like uranium at low energies, activation and radiation shielding are of less concern; the 1 W/m limit addresses concerns regarding damage to SC cavity surfaces and cryogenic heat load. To facilitate the maintenance of individual cryomodules, warm interconnect sections are used between cryomodules, and U-tubes with bayonet connections are employed for cryogenic distribution. Furthermore, identical, interchangeable cryomodules are used for each β_0 type of cavity.
- Reliability: A Machine Protection System (MPS) minimizes component damage and operational interruptions caused by both acute (fast) and chronic (slow) beam losses. Upon acute beam loss, the MPS response time is 35 μs (including diagnostics, signal processing, and residual beam dumping). The MPS response to slow beam loss is complicated by the low sensitivity of conventional ion chamber loss detection to low-energy heavy ions and the beam loss signal background from adjacent linac segments, where beam energies are significantly different. Beam-halo-monitor rings in the warm interconnect sections and possible thermal sensors in cold regions are planned for more sensitive loss detection.
- Tuneability: The accelerator is designed to be easily tuneable during both beam commissioning and operation. In linac segment 1, where the beam transverse-phase advance is large, cold beam-position monitors (BPM) are implemented in the cryomodules. Efforts are made to establish beam-tuning strategies based on virtual accelerators and on-line models under normal and fault conditions.
- Upgradeability: Space is reserved in linac segment 3 to house another 11 cryomodules to readily increase the energy of ^{238}U beam ~400 MeV/u. The linac tunnel allows for future expansion so that a dedicated light-ion injector can be added to support rare isotope production using the isotope separation on-line (ISOL) method. Using an RF deflector cavity and a Lambertson septum magnet, $^{3}He^{+}$ beam supplying protons to the ISOL target can share cycles with the ^{238}U beam feeding the fragmentation target; thus, simultaneous users are supported. Further-more, space is reserved to house instrumentation, including non-destructive diagnostics and sub-harmonic bunchers that are compatible with future user experimental demands.

Table 10.8 lists the primary FRIB parameters. The FRIB driver accelerator is designed to accelerate all stable ions to energies >200 MeV/u with a beam power on the target of up to 400 kW. The driver linac consists of a 47 m long front end

Table 10.8 FRIB primary parameters for the baseline design

Parameter	Value
Primary beam ion species	H to ^{238}U
Beam kinetic energy on target [MeV/u]	>200
Maximum beam power on target [kW]	400
Macropulse duty factor [%]	100
Beam current on target (^{238}U) *e*mA	0.7
Beam radius on target (90%) [mm]	0.5
Driver linac beam path length [m]	517
Average uncontrolled beam loss [W/m]	<1

containing ECR ion sources and an RFQ, followed by a 472 m long SRF linac with QWR of $\beta_0 = 0.041$ and 0.085, and HWR of $\beta_0 = 0.29$ and 0.53 in a folded layout to facilitate charge stripping and beam collimation, and to accommodate the limited real estate footprint in the center of the MSU campus.

10.3.3 R&D and Long-Lead Procurements

R&D in preparation for the establishment of a rare isotope facility began at MSU around 2000, focusing on SRF technology. R&D programs supported by the FRIB project since 2008 have focused on SRF, liquid lithium charge stripping, and high-power targetry for heavy ions. Later identified focuses include integrated cryogenics and machine protection for high-power, low-energy heavy-ion beams.

10.3.3.1 Superconducting RF R&D

It took MSU more than 15 years to build the SRF infrastructure to support the mass production of the SRF system in house. As detailed in Sect. 10.3.5, initial development encountered numerous difficulties and challenges. In 2010, the weekly SRF teleconference involving world experts was launched and has continued until today. Since then, FRIB SRF development has benefited greatly from the expertise of the low-β SRF community. FRIB has been collaborating with INFN/LNL on resonator development and with ANL on RF coupler and tuner development. It was assisted by JLAB on cryomodule design, KEK on superconducting solenoids, and FNAL and JLab on cavity treatments.

The low-β cryomodules built for the MSU 3 MeV/u re-accelerator (ReA3) use a traditional top-down design, with the cavities and solenoids hanging from a strong back. To facilitate efficient assembly, simplify alignment, and allow U-tube cryogenic connections for maintainability, FRIB developed an innovative "bottom-up" cryomodule design and spent several years confirming the design through the validation of the alignment mechanism using mock-up cavities and the validation of the prototype cryomodule. The cold mass is mounted on supporting posts on linear

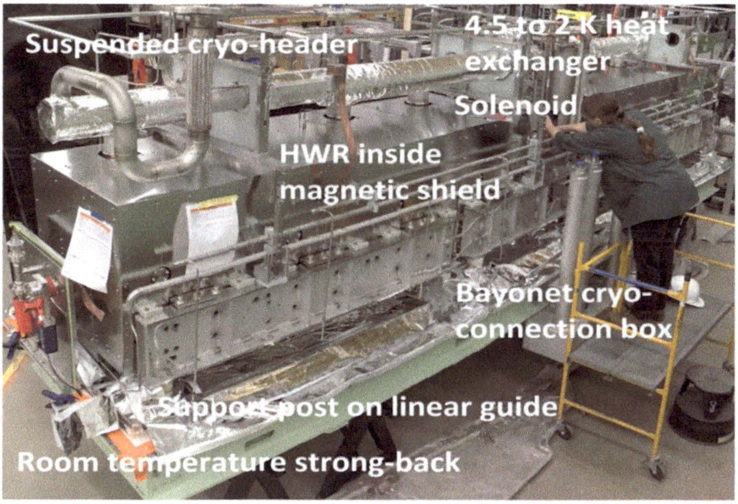

Fig. 10.22 Partially assembled "bottom-up" FRIB $\beta_0 = 0.53$ SRF cryomodule, with 8 SRF cavities operating at 2 K and a solenoid operating at 4.5 K temperature, respectively. (Courtesy MSU, USA)

roller bearings to streamline the assembly and alignment of the cryomodule. The cryogenic headers are suspended from the top for microphonics suppression. Local magnetic shielding allows solenoids to operate at $\sim$8 T without compromising the performance of the resonator. This design was successfully tested and validated in 2015 (Fig. 10.22).

10.3.3.2 Integrated Cryogenics System R&D

The initial plan for a "turnkey" approach to a cryogenic helium refrigeration system from the industry exposed the project to serious risks in both budget and scope. The approach failed since a large-scale, non-standard system is not a manufacturer's standard product; and many of the required subsystems are not within the expertise of the primary contractor. As detailed in Sect. 10.3.4, this led FRIB to utilize the design and operational experience of JLab and SNS and to acquire expertise, forming an internal team. In 2013, plans for an integrated design of cryo-plant, cryo-distribution, and cryomodules were developed, which proved to be the key to efficient SRF operations (Fig. 10.23).

10.3.3.3 Liquid Lithium Charge Stripping R&D

Early development of liquid lithium charge stripping has been carried out at ANL since 2005. In 2010, a liquid lithium film was established in the prototype lithium

Fig. 10.23 FRIB cryoplant operating at both 4.5 K and 2 K. Segmented distribution lines are connected to the refrigeration system with U-tubes. (Courtesy MSU, USA)

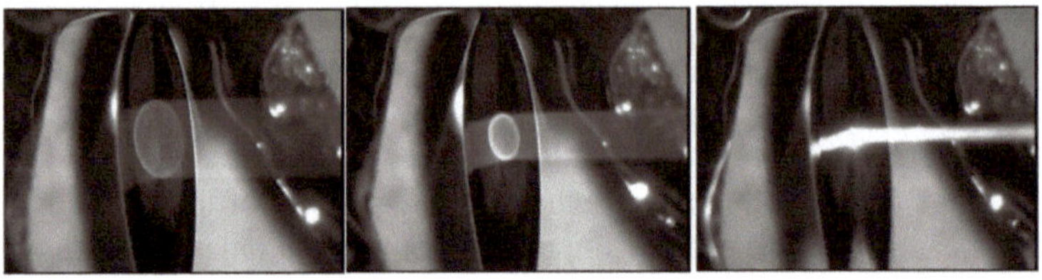

Fig. 10.24 Liquid lithium film flowing at high speed (~50 m/s) intercepting a proton beam of about 60 kV at ANL [105]. The test produced power deposition densities similar to those of the full power uranium beams from FRIB

stripper, with its thickness measured by an electron beam [104]. Subsequently, the powerful LEDA proton source from LANL, capable of producing 100 mA current CW beam, was restored at MSU. In 2013, tests with the proton beam produced by the LEDA source demonstrated that power depositions similar to those of the full power uranium beams from FRIB are achievable without destroying the film, as shown in Fig. 10.24 [105].

Construction of the production lithium stripper system started at MSU around 2013. Significant efforts were made to mitigate fire safety hazards related to liquid lithium when it is exposed to air or water. In 2018, liquid lithium was circulated with an electromagnetic pump. In 2019, the liquid lithium device operated unattended for an extended period of more than 10 days without human intervention. Subsequently, the liquid lithium charge stripper was successfully used to strip heavy-ion beams during FRIB commissioning and operation [106].

10.3.3.4 Multi-layered Machine Protection R&D

Machine protection is challenging for the intense low energy heavy-ion beams of FRIB due to low detection sensitivity and high power concentration in a short range.

Machine protection is challenging for FRIB's intense low-energy heavy ion beams due to the low detection sensitivity and high power concentration over a short range. Furthermore, due to the folded geometry of the driver linac, beam loss at high energy interferes with loss detection for low-energy beams [11]. FRIB adopts multi-layer, multi-time scale approaches (Sect. 8.2 and Table 8.1). The fast protection system (FPS) is required to mitigate beam loss within 35 μs (10 μs for detection, 10 μs for controls processing, 5 μs for inhibit action, and 10 μs for beam clearing) to prevent damage to beamline components. Primary detection methods include low-level RF (LLRF) monitoring, differential beam current monitoring, halo monitor rings for high-sensitivity loss detection, and current-monitoring modules for critical magnet power supply inhibition. The fast beam inhibit is achieved by independently disabling both the upper and lower 90° electrostatic bends in the LEBT. An additional mitigation at a slower time scale is to disable the ECR ion source high voltage.

10.3.3.5 High Power Targetry R&D

Early R&D on the high-power targetry system includes a prototype rotating target and a rotating beam dump. The prototype target, which comprises a rotating carbon disk with multiple slices and heat dissipation through radiative cooling (Fig. 10.25), was designed, built, and successfully tested with an intense electron beam that simulates the thermal and mechanical conditions for target operation [107]. The development of a thin titanium shell water beam dump has encountered problems with bearings and water seals.

10.3.3.6 Long-Lead Procurements

With the flexibility offered by multiple funding sources, long-term procurements for the FRIB project began even before CD-3a authorization. These included the cryogenic refrigeration 4.5 K "cold box," the cryogenic distribution, the niobium material for SRF cavities, the pre-production of SRF cavities, the radio frequency quadrupole (RFQ), and the superconducting electron cyclotron resonance (ECR) ion source.

10.3.4 Integrated Cryogenic Systems and Their Planning

For superconducting particle accelerators, the time required for the procurement and commissioning of the cryogenic system plays an important role in the duration of

Fig. 10.25 FRIB rotating multi-slice graphite target of 30 cm diameter rotating at the design speed of 5000 rpm. The power density on the target surface would be 60 MW/cm^3 if the target were stationary. (Courtesy MSU, USA)

the construction phase of the overall project, as it is essential for the commissioning of other major systems. However, the proper design of this critical system to meet the requirements of the accelerator is technically challenging. It is complex and energy-intensive, and it requires addressing modes of operation, off-design modes, commissioning and maintenance plans, and system safety concerns. Given all of these factors, the cryogenic system directly affects the short- and long-term availability and the operating costs of the entire project.

Primary load operating requirements, such as cryomodule cavity operating temperature and pressure, power coupler liquefaction, and magnet refrigeration, are the most influential parameters for the design of the cryogenic refrigeration system. These requirements establish the cost and complexity of the cryogenic system. So, to achieve an efficient, reliable, and safe system at minimal cost, the primary loads (i.e., cryomodule cryogenic process), distribution, and refrigeration systems must be designed and optimized collectively, rather than in isolation [108]. This includes incremental commissioning plans for all cryogenic subsystems and the other parts of the accelerator system.

It is not uncommon to use inaccurate and ill-defined estimates for the cryogenic system specifications and project schedule. This gives the initial illusion of a short project construction phase schedule. In addition, the effect of these poorly understood estimates on the project implementation results in a system that incurs higher operational expenses for the life of the accelerator and reduces the availability of the facility. For project management, the "turnkey" model is attractive, as it transfers most of the risk in system design and development to a third party (i.e., the supplier). This approach is often 'sold' under the misnomer that these large systems are mature or well-established technologies, readily available from industry. However, these large-scale cryogenic systems are usually significantly different from the manufacturer's standard product line. For an experienced supplier, this risk is reflected in an expensive project cost proposal. The initial choice of a "turnkey" approach was abandoned by FRIB toward the end of 2011, after two years of effort, due to the prohibitive cost proposal from the industry. After this weakness of the "turn-key" approach was recognized, an engineering design team was established at the beginning of 2012, comprising people with design and operational experience in similar systems, such as at CEBAF and SNS [109].

The layout of the FRIB accelerator was driven by the requirement to integrate the existing cyclotron at NSCL and its experimental areas while keeping it on the MSU campus within the cyclotron/FRIB site. This posed considerable challenges to the layouts of civil construction and accelerator systems, including the cryogenic system. The interrelationship in the design of modern superconducting accelerator subsystems involving the cryomodule, cryo-distribution, and cryogenic plant to achieve an overall efficient system is critical [110]. In addition, the advantages of using a cryogenic coupling, similar to those used at JLab and the SNS, to connect subsystems (e.g., cryomodule and cryo-distribution) are also well established [108]. The cryogenic operational requirements for the cool-down, warm-up, commissioning, regular operation, quench recovery, and maintenance of each device, such as the niobium resonators and the SC magnets (solenoids) housed within the same cryomodule, must be established well in advance of the cryogenic system planning. For the FRIB project, the cryomodule and cryo-distribution design concepts were finalized by the end of 2012. A standardized cryogenic design approach was adopted for the six types of cryomodules and their interfaces to the cryo-distribution system [110]. This minimized the design effort and the production cost by using common components and the same (or very similar) design details. During the same period, the decision was made to separate the cryo-distribution to the linac into three segments, designated LS1, LS2 and LS3. There is also a fourth cryo-distribution branch to the Experimental Systems (ES) area SC magnets. This plan assisted in the independent incremental installation and commissioning, which were critical to meeting the aggressive FRIB schedule.

As mentioned earlier, one of the main challenges in planning the cryogenic system is understanding the load requirements of the superconducting devices that must be supported. This includes requirements for normal steady loads, normal transient loads, and atypical conditions such as cool-down, warm-up, commissioning, and maintenance. The next challenge is the process and mechanical design of the cryogenic system to support these loads, which should be cost-effective and efficient. It can be expected to be executed on schedule and within budget based on experience. To achieve all the project goals, the FRIB cryogenic design team made the following major decisions in conjunction with the FRIB management in February 2012 [111]:

- Decision 1: The laboratory would assume responsibility for the planning, process design, integration, and commissioning of the cryogenic subsystems.
- Decision 2: The laboratory would assume responsibility for the design of most of the cryogenic subsystems. In many cases, this was a print-to-order

design obtained from the industry. Other subsystems would be procured using a comprehensive equipment specification based on the process design. This approach provided the initial information needed by the FRIB conventional facilities (CF) for the building and the cryogenic system utilities without the delay of waiting for a preliminary design from suppliers. This information guided the building layout, foundations, trenches, door openings, cranes, conventional utilities, conduits, and equipment installation strategy.
- Decision 3: The cryomodule designs would be revised based on a detailed approach at the part level, incorporating JLab/SNS-style cryogenic couplings and individual 4.5 K to 2 K heat exchangers in each cryomodule [108]. In addition, the linac cryo-distribution system would be divided into three linac segments to support a stepwise commissioning and, if required, to support operation under different operating conditions.
- Decision 4: An overall detailed cryogenic system design plan developed at the component level was based on the revised cryomodule cryogenic design requirements (from Decision 3) and the revised cryo-distribution [110]. In addition, a plan for long lead items and subsystem procurements would be established to meet the overall project schedule and regain some of the lost time, taking into account the availability and funding profile provided by the project team at the end of 2012.

These key decisions represented a fundamental change in direction from the initial plan of a "turnkey" procurement and a single linac distribution system with two main cryogenic valve boxes and all welded connections to the linac cryomodules [109]. In addition, the decision to keep all designs and documentation in FRIB and to handle all procurements through the FRIB procurement department was critical to the execution of the project.

Designing a flexible cryogenic system is very important for operating the system at maximum efficiency and, thereby, achieving low operating costs during lengthy commissioning periods and long-term operations at maximum capacity; or at part loads due to maintenance needs. This is important since actual loads are typically somewhat different from estimates, which may include various margins in the planning, unplanned loads, and unplanned modes of operation to support the commissioning of a partially populated linac. The FRIB cryo-distribution is designed to support the 4.5 K helium supply to 4.5 K and 2 K loads at any desired pressure from 1.2 bar to 4 bar. This is unlike the JLab and SNS cryo-distribution systems, which operate at supercritical supply and are believed to be more stable, leading to fewer vibrations, especially during the RF cavity operations. At FRIB, during LS1 commissioning, there were difficulties in locking-in the niobium resonators. Using the liquid helium Dewar supply at 1.6 bar, the vibrations preventing the locking of the resonators were eliminated. This is also a more efficient 2 K process [108]. The efficiency and flexibility of the floating pressure process of the Ganni cycle [112] were demonstrated with the cryogenic refrigeration system that can operate at the cold box supply pressure anywhere between 6.0 bar and 20.0 bar, with a nearly constant exergetic efficiency of the system ranging from 30% to

100% of the maximum refrigeration capacity under various combinations of loads [113, 114]. This includes the incremental commissioning stages. The FRIB 4.5 K and 2.0 K refrigeration systems were successfully commissioned in 2017 and 2018, respectively [111, 115].

The ES cryo-distribution system, along with a few of the 14 superconducting magnets in the experimental system, was commissioned in March 2021. The three FRIB linac segments (LS1, LS2, and LS3), which include 46 cryomodules, four SC bending magnets, and the ES distribution system, were maintained at 4.5 K with less than 1.40 MW of compressor power and less than 100 liters per hour (lph) of liquid nitrogen (used for the refrigerator precooler), operating the refrigerator at less than 7.0 bar. The refrigeration system is also designed and tested to support the three linacs at 4.5 K without liquid nitrogen precooling to avoid thermal cycling and to prevent loss of helium inventory in case there are problems with liquid nitrogen delivery. Typically, the commissioning periods for these large accelerators are quite long. For example, the delivery of the initial cryogenics needed in the accelerator to reach user operation typically requires four to five years; for FRIB, it was 2017 to 2022. Since the Ganni-cycle floating pressure process allows for near constant efficiency and supports these commissioning loads, the utility use is around 30% of the design, resulting in significant utility savings and reduced wear on the cryo-system equipment.

The FRIB cryosystem demonstrated the feasibility and value of careful planning, which can help reduce the commissioning time of various accelerator SC components. It also demonstrated the feasibility of designing flexible and efficient cryogenic refrigeration and distribution systems that can be implemented if carefully planned from the beginning. In summary, despite the challenging design, schedule, and cost constraints of the FRIB accelerator, the beam was delivered in January 2022, largely due to the careful planning and execution of the cryo-plant and cryo-distribution system, which efficiently supported the phased commissioning.

10.3.5 Superconducting RF System Establishment

Superconducting RF development at MSU started in 2001, within the framework of the RIA project [116] for the construction of a latest generation radioactive beam facility in the United States. Suitable SRF linac layouts with multicharge beam transport capability have been developed and proposed first by ANL [117, 118] and then by MSU [119], both supported by extensive cavity and cryomodule R&D programs. The construction of the new facility, renamed FRIB after its evolution from the initial design, was finally assigned by the DOE to the MSU proposal in 2009. Several prototypes of superconducting cavities (multi-cell elliptical, QWRs, and HWRs [120]) were built and tested at MSU in collaboration with other laboratories that have well-established SRF experience (INFN Legnaro, JLab, INFN Milano, and later TRIUMF and ANL). The first $\beta_0 = 0.041$ QWR cryomodule, a kind of proof of principle for the FRIB ones, was installed in the ReA postaccelerator for radioactive beams at NSCL and put into operation in 2010 [121]. This success prefigured a similar one for the following $\beta_0 = 0.085$ QWR cryomodule, designed with the same principles, with the cavity design obtained by scaling the diameter of the $\beta_0 = 0.041$ one. Contrary to expectations, production cavities showed puzzling behavior and a lack of performance, missing an important FRIB milestone and casting a shadow over the entire project. This prompted a greater effort in the FRIB cavity and cryomodule R&D in 2011, which included a thorough review of all cavities and cryomodule designs and ultimately led to the current FRIB SRF configuration.

10.3.5.1 The Strange Case of the ReA QWRs

Subtle phenomena can sometimes cause great trouble and long delays. The FRIB QWR experience and its positive conclusion may serve as an instructive example. In the initial roadmap, scaling, after its successful testing, of the $\beta_0 = 0.041$ QWR to a $\beta_0 = 0.085$ appeared to be the safest way to proceed in a timely manner toward the finalization of the QWR linac design. This was obtained by increasing the low-β cavity outer diameter to fit the new β_0, while preserving its mechanical design philosophy during the overall re-optimization of the design parameters (see Fig. 10.26).

The first $\beta_0 = 0.085$ prototype without a helium vessel ("naked") was tested immersed in liquid helium, with a thick niobium bottom plate in place of the final thin tuning plate to hold the pressure. The SRF test confirmed the good RF results of the low-β ReA3 cavities, suggesting that the mechanical design could also be

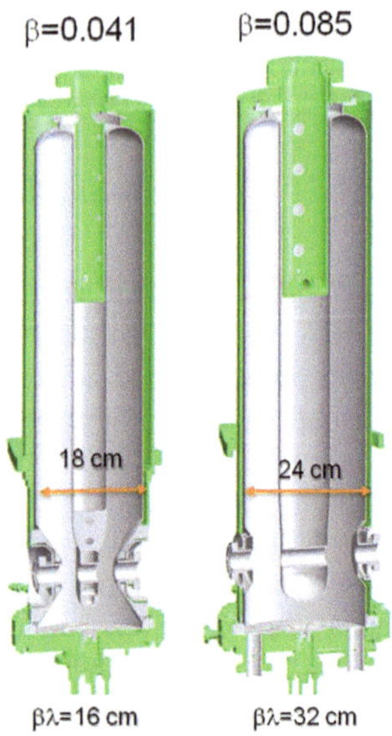

Fig. 10.26 The ReA QWRs initial design. The higher β_0 design was obtained by scaling the diameter of the lower β_0 one. (Courtesy MSU, USA)

considered validated to a large extent by previous tests of the nearly identical lower beta cavities. In this promising scenario, the decision was made to skip further R&D and go straight to pre-production, resulting in considerable savings in time and cost. A group of 10 additional cavities, fully equipped with helium vessels and tuning plates, was produced to be installed in the first $\beta_0 = 0.085$ cryomodule of the ReA linac, which was also planned to be the FRIB cryomodule demonstrator.

Unexpectedly, all individual tests at 4.2 K yielded very disappointing, non-reproducible results, with low Q_0, a strong Q-slope, and early cavity quenching at low gradient. After the installation of the helium vessel, even the previously well-performing prototype has started to fail. Several possible causes have been suspected: Q-disease, unwanted changes in the surface treatment and cavity preparation, construction and welding defects, overheating of the tuning plate from the RF coupler and normal conducting transition, high residual magnetic field from new components, or poor RF contact at the tuning plate.

Q-disease was previously observed in the 4 K naked cavity testing when a slow cooling speed was applied. However, in dressed resonators, even fast cooling could not solve the problem, ruling out Q-disease as the main cause of low Q. Construction and welding defects were excluded because they were not compatible with the favorable initial results of the naked prototype. Surface treatments based on Chemical Polishing (CP) and High Pressure Rinsing (HPR) were thoroughly checked, and all chemicals and procedures were analyzed step-by-step. They were finally ruled out because they were producing good results with different resonators, such as the FRIB half-wave ones. The problem was then related to the overall cavity design.

The possibility of tuning plate overheating by the RF power coupler was first confirmed by computer simulations and subsequently in tests with the final thin Nb tuning plate. Even after cooling the stainless steel backing plate directly with liquid helium (see Fig. 10.27), the results were unsatisfactory. However, the different behavior of the $\beta_0 = 0.085$ cavity compared to the $\beta_0 = 0.041$ cavity could only be partially explained by its larger plate diameter and higher RF power requirements, since overheating and partial transition to the normal conducting state should have occurred only at sufficiently high RF power, while low Q was observed even at low power.

The residual magnetic field was ruled out by measurements, and because the only material added to the dressed prototype, compared to the naked one, was the Titanium vessel, which had no possibility of magnetization.

The quality of the RF contact between the tuning plate and the outer conductor bottom flange was checked by mounting the tuning plate assembly with pressure films on the contact surface and analyzing them after dismounting the plate. This test showed that, unlike the lower β_0 cavity, the pressure on the RF contact was rather unreliable. The contact was improved by trying different types of rings (spring type, indium, blunt edge, etc.); however, this resulted in unpredictable and unsatisfactory outcomes when tested under SRF conditions. After this first testing campaign, which greatly improved the understanding of design problems, the SRF results were still disappointing, and the real culprit remained unknown. To identify the

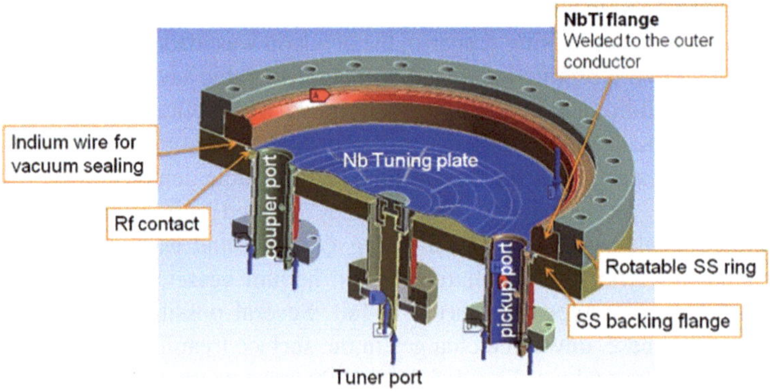

Fig. 10.27 Details of the initial tuning plate subsystem design with RF ports [122]. The design worked well in low-β_0 QWRs for ReA but was severely flawed in the similar high-β_0 ones with larger diameter

issue, all available old measurement data were re-analyzed, and new measurements were performed in a more systematic way to verify every design assumption and calculation. This campaign required a small team to perform approximately three months of continuous testing, with one to two cryogenic tests per week, along with all related cavity surface treatments and preparations. In parallel, mechanical modifications of prototypes were also conducted, including electron beam welding.

The conclusions of this demanding but fruitful campaign were as follows:

- The poor results of the first lot of $\beta_0 = 0.085$ QWR were caused by a subtle mechanical effect related to differential thermal contraction, several hundred μm, between the cavity and the stainless steel backing plate. This created a deformation at SRF temperature that released the pressure on the RF contact between the tuning plate and the NbTi bottom flange, allowing some RF current to flow through the stainless steel itself and causing a degradation of the SRF performance at low field. In the $\beta_0 = 0.041$ cavities, differential thermal contraction also existed but did not reach critical values due to the smaller dimensions of the components involved. To solve this point, the stainless steel material of the backing plate had to be replaced with a material with thermal contraction characteristics close to those of Nb.
- The thermal conductivity of NbTi, the material of which the original bottom flange was made, although sufficient in the $\beta_0 = 0.041$ cavities, was too low to guaranty reliable cooling by contact in the larger Nb tuning plate of the $\beta_0 = 0.085$ cavities. To solve this point, NbTi had to be replaced by high residual resistivity ratio (RRR) Nb.
- The RF coupler mounted on the tuning plate delivered a thermal load that was difficult to remove through the thin Nb sheet, especially in the larger cavity. The solution was to move the RF ports from the tuning plate to the well-cooled side of the cavity.

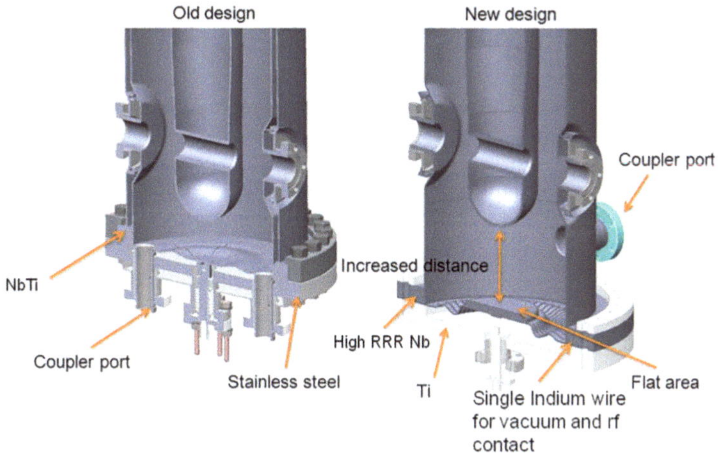

Fig. 10.28 Comparison between the original and the refurbished ReA QWR design [122]

- Due to the increased diameter of the new cavity, the distance between the inner conductor tip and the tuning plate caused an increase in the magnetic field and RF losses in the RF joint compared to the lower β_0 cavity. To gain a sufficient safety margin, this distance must be increased while preserving adequate tuning capability.
- A new design with an additional indium wire between the bottom flange and the tuning plate could successfully provide vacuum sealing, thermal linkage, and RF compatibility at the same time. This configuration would prevent the flow of the RF current over the stainless steel backing flange, and the presence of a good vacuum would also guaranty the presence of good RF contact.
- Since all problems were localized in the tuning plate complex, the ten existing low-performance cavities built for ReA3 could have been brought to full performance by modifying only their end parts. This action would have been relatively inexpensive compared to the construction of new cavities and would have been equally effective.

These findings provided a clear roadmap for the renovation of the 10 ReA resonators and a new design for the FRIB ones.

The ReA3 resonator design was modified as follows (Fig. 10.28):

- To increase the distance from the tuning plate to the inner conductor tip, the outer conductor was elongated to match both the low magnetic field at the RF joint (0.5 mT at the operating gradient) and sufficient tuning sensitivity (>2 kHz/mm);
- The NbTi/SS rotatable bottom flange was replaced with a fixed, high RRR Nb one;
- The old tuning plate with RF ports was replaced by a new one, which lacks RF ports and features longer slots and larger deformation capabilities (±10 mm) to reach a ±20 KHz tuning range;

- The SS end flange with RF and tuner ports was replaced by a Ti one with only the tuner port to remove differential thermal contraction problems (Nb and Ti have similar coefficients);
- The RF ports have been moved to the side of the resonator, to a position that was optimized for minimum RF losses in the coupler.

These modifications solved all previous problems, and ten under-performing QWRs could be fully recovered and installed on the ReA linac, bringing a bad adventure to a happy end.

In conclusion, the apparently innocuous scaling of a working design induced modifications that brought the cavity to a completely different operational regime. The apparently safe shortcut in R&D validation procedures prevented the timely discovery of this subtle effect, causing further delays instead of advancing the schedule. Since then, R&D and quality control procedures in the FRIB project have been further reinforced to reduce risk, becoming particularly strict and rigorous. This probably contributed to the early delivery of the FRIB linac one year ahead of the initial project schedule.

10.3.5.2 FRIB SRF Design Finalization

The design of the FRIB QWRs was finalized by taking advantage of the experience gained from the refurbished ReA resonators and seeking cost reductions, which are critical in a large machine like FRIB. To take into account the high beam current of FRIB, the original 30 mm aperture of the ReA cavities was increased to 36 mm for the FRIB QWRs. The diameter of the $\beta_0 = 0.085$ ReA cavity was increased from 240 mm to 270 mm (while keeping the same flange to flange distance along the beam line) to increase R_{sh} and reduce B_p/E_a. This allowed the nominal operation gradient in the $\beta_0 = 0.085$ section to be increased by 10% and the number of cryomodules to be reduced by one. Significant cost reduction was also achieved by replacing the well-performing but rather expensive high-RRR Nb bottom flange with a new low cost one of the same shape, primarily made of Ti, with only a small 3 mm thick Nb insert. This insert is cooled directly by liquid helium and is located in the area exposed to RF, hosting the indium seal. This results in excellent cooling of the indium and, through it, of the tuning plate.

In addition, electromagnetic, cryogenic, and mechanical designs for all FRIB cavities were re-optimized, improving their performance and allowing for a larger safety margin in achieving their specifications. The total number of cavities in the linac could be slightly reduced from the original design. The initial, rather complex HWR tuner was replaced by an ANL-type pneumatic one [123], with a newly designed interface to the cavity that allows easy installation in the limited space available between the resonators after cold mass assembly. Both fundamental power couplers for QWRs and HWRs were modified and upgraded. A 4 kW, double window adjustable FPC design developed at ANL was adopted for FRIB QWRs. In the single-window SNS type FPC used for FRIB HWRs, the antenna impedance profile was modified to reduce multipacting conditioning time. Resonator construction and tuning procedures, surface processing, and thermal treatments were also updated. A rigorous validation procedure, including integrated resonator and cryomodule testing under operational-like conditions, was set and enforced in the production chain to prevent any possible installation of flawed components in the linac. This rather intensive campaign required considerable effort but succeeded in meeting all FRIB requirements in due time.

The cryomodule design converged to a unique scheme for all types of cavities after realizing that the assembly time of the first-generation top-down cryomodule prototypes exceeded, by a large margin, the time scheduled for FRIB. A new bottom-up design [124] allows for an easier, faster, and more reproducible assembly procedure, providing better access to cavities and auxiliary components during critical mounting operations (Fig. 10.22). In the FRIB cryomodule, the rather long cold mass alignment rails, split into either two or three independent sections to limit thermal displacements, sit on short G10 posts that stand directly on the outer vessel baseplate, which hosts all RF, cryogenic, and beam vacuum interfaces. A clever system of linear bearings, placed between posts and rails, provides automatic alignment of the cold mass to the beam axis during cooldown, with excellent precision and reproducibility. The outer vessel top cover has only the functions of a vacuum vessel and a stiff support for the helium reservoir to prevent low-frequency mechanical vibrations. To validate this new design and the required long o-ring with three-way connections, a special cryomodule mock-up ("engineering test cryomodule", ETCM) was built and tested, achieving the planned kinematic support and post-cooldown alignment accuracy.

The 46 required cryomodules, containing 324 QWR and HWR cavities, were commissioned at a rate of about 1 cryomodule per day during an 8-hour shift. The high-power RF operation of the QWR SRF cavities was validated at 4.5 K; a more demanding condition than the baseline design of 2 K with respect to microphonics. All cryomodules have been installed in the linac and have met the design accelerating gradient, with no major field emission or multipacting issues [125].

In the meantime, an upgrade program was initiated, which had been planned since the beginning of the project, with the aim of increasing the final beam energy to 400 MeV/u for uranium ions. Eleven additional cryomodules, each containing five 5-cell elliptical cavities at 644 MHz [126], will be located in an 80 m section of the linac tunnel reserved for this purpose. It should be noted that the vast FRIB SRF program could be successfully pursued because of the construction at MSU of the new SRF highbay, one of the largest and best-equipped infrastructures worldwide for SRF resonators and cryomodule treatment, assembly, and characterization.

Fig. 10.29 An iron-dominated quadrupole magnet for an FRIB experiment beamline with a room temperature bore of 170 mm. (Courtesy MSU, USA)

10.3.6 Superconducting Magnet Rectification and Development

Superconducting magnets for the FRIB ARIS and beamline are iron-dominated. The design strategy for magnets beyond the target hall builds on and scales the experience gained from NSCL beamline magnet designs [127, 128]. Advantages of the NSCL beamline magnet design include a compact configuration with higher-order correctors nested inside quadrupoles and effective thermal management, resulting in low heat loads. For the target hall, high-temperature superconducting (HTS) magnets were initially proposed to address the high radiation dose, with early development and prototyping efforts described in Sect. 6.1.4. However, the HTS design was ultimately not selected due to technical risks, costs, and limitations in the availability of conductor lengths. Instead, NbTi coils with cyanate ester epoxy were adopted. Some challenges and lessons learned worth noting are the following:

- Iron-dominated quadrupoles provide a cost-effective solution up to certain sizes. The low operational-to-critical current ratio (20%) eliminates the need for high-quality layer winding and vacuum impregnation. Figure 10.29 illustrates an example of an assembled iron-dominated quadrupole. The coils are wound using 0.5 mm diameter NbTi wire with saddle-shaped ends. Small voids often occur within the coils; however, these are generally acceptable due to the ample current margin provided by the conductor. However, a notable challenge is the thermal mismatch between 1006 steel and copper. The thermal contraction of 1006 steel is approximately 60% that of copper (coil), making it critical to account for this difference during the design phase to prevent the coils from loosening. This issue becomes increasingly problematic as the diameter of the coil expands by a factor of two or more (from 200 mm to 300 mm up to 600 mm bore sizes). Combined

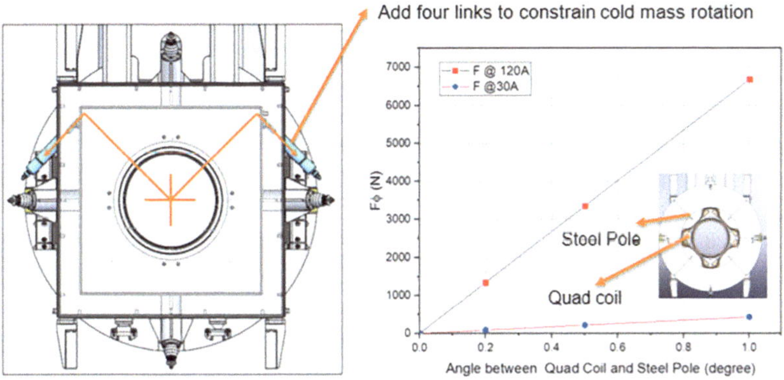

Fig. 10.30 Warm iron quadrupole in the FRIB Target Hall. (Left) Anti-rotation links added and (Right) tangential force as a function of current and misalignment angle

with the random-wind technique, the mismatch leads to a high number of training quenches, sometimes preventing magnets from achieving their design fields.

- High-order corrector nesting. Nesting correctors inside quadrupoles allows a compact lattice design. However, this advantage comes with increased complexity regarding electromagnetic (EM) forces and unintended dipole fields resulting from interactions between sextupole and quadrupole iron components. Due to the iron yoke design, correctors must be placed inside quadrupoles, pushing the quadrupole coils outward, away from the bore. The resultant EM forces are substantial and must be supported separately, posing significant design challenges, particularly as bore sizes scale above 400 mm. Additional mechanical supports are necessary to limit the deformation of both the higher-order correctors and the supporting bore tubes.
- To minimize the generation of radioactive waste materials, a significant design alteration was made for the target hall magnets. Specifically, the pole and yoke steel were placed outside the cryostat—unlike the integrated approach depicted in Fig. 10.29. This separation enables the reuse of the steel if a magnet needs to be replaced. However, this change means that all coil-steel interaction forces must now be managed entirely by the cryostat support links. Consequently, the magnet's center is no longer fixed due to the strain on these support links under EM loads. Figure 10.30 illustrates an example encountered for the warm iron quadrupoles 4, 5, and 7 in the target hall. The initial design had four links at each end of the magnet to support the cold mass. In ideal conditions, there is no tangential force to rotate the cold mass, and the link configuration has no provision against any torque. However, as shown in the figure, any angle misalignment will cause a force to rotate the cold mass, and the force increases as the cold mass rotates. This created two issues: (1) the magnet center changes as the magnet ramps up, and (2) a large tangential force on the vertical and horizontal links prevents the magnets from reaching full field. The issue was resolved by adding two anti-rotation links per side. This was a perfect example

of designing a magnet without conducting envelope analysis, as described in Sect. 6.2.

- The selected magnet technology can inherently constrain future upgrade paths. For a large facility, the cryogenic system is tailored to meet the needs of the accelerator. Magnet technology choices significantly impact cryogenic system designs. For example, vapor-cooled current leads take 4 K helium vapor, returning it as a liquefaction load, whereas HTS and conduction-cooled leads introduce considerable heat loads at elevated temperatures (50–80 K). While iron-dominated quadrupoles have the benefit of operating at lower currents, this can pose limitations for power supplies and DC cabling, especially if a facility later chooses to upgrade to coil-dominated magnets. Therefore, it is advisable to integrate potential future upgrade scenarios into initial design considerations.
- Condensation issues significantly affect the long-term operation of magnets; yet, they are frequently overlooked during the design phase. Typically, current leads are designed in over-cooling regions to allow for safety margins, resulting in condensation when magnets are not energized. Stationary water often leads to the corrosion of feedthroughs, occasionally causing ground faults. Forced-air circulation or increasing the conductor's surface area for better air exchange are commonly adopted solutions. Addressing these condensation issues during the design phase can prevent considerable operational challenges later. Additionally, vapor-cooled leads are susceptible to contamination by air or water vapor. NSCL magnets previously utilized plastic feedthroughs for electrical isolation, which tended not to be helium leak tight, drawing air into helium return lines and contaminating cryogenic plants. Figure 10.31 shows an example of a lead can, illustrating water-induced corrosion. Nowadays, ceramic feedthroughs have largely been used to address the sealing aspect but remain vulnerable to corrosion caused by water condensation.

Based on these experiences, the iron-dominated quadrupole design has clear merits and is an effective solution for quadrupoles with a bore diameter of around 100 mm. However, for quadrupoles with room-temperature bores of 300 mm or larger, alternative designs should be evaluated. For this reason, FRIB is developing coil-dominated superconducting quadrupole magnets intended as future replacements for existing fragment separator magnets and High Rigidity Spectrometer (HRS) magnets.

Coil-dominated quadrupoles can utilize either a pure $\cos 2\theta$ current distribution, as discussed in Chap. 6, or a tilted $\cos\theta$ distribution known as Canted-Cosine-Theta (CCT) [129, 130]. A discrete $\cos\theta$ current distribution can be achieved by appropriately placing racetrack coil sections or by precisely winding the wire according to a pattern that adheres to the $\cos\theta$ distribution, such as that proposed by Walstrom [131]. The latter approach provides superior field quality, with wires placed in accurately machined grooves. In this approach, field quality is guaranteed primarily by machining tolerances rather than coil winding or assembly tolerances,

Fig. 10.31 NSCL experiment beamline magnet vapor cooled lead can with plastic fittings. (Courtesy MSU, USA)

as shown in Fig. 10.32 [132]. The bobbin is constructed from aluminum 6061 and is e-coated for electrical insulation. Additionally, the bobbin serves as a thermal conduction path to allow the use of a conduction cooling scheme and aids in quench protection by accelerating normal zone development and absorbing part of the dissipated energy through induced eddy currents. This design also allows higher-order correctors to be positioned outside the quadrupole coils, enabling coil placement to be closer to the bore. This configuration achieves the desired quadrupole gradient with reduced coil fields and eliminates unwanted dipole components induced by interactions between the quadrupole iron poles and sextupole coils.

A prototype quadrupole has been built in-house at FRIB and has been successfully tested up to full current without training quenches [133, 134]. On the current scale, the costs of iron-dominated and coil-dominated magnets are approximately comparable. As magnet sizes increase, coil-dominated designs become economically advantageous because the cost of iron scales quadratically with magnet size. The same design concept was applied to the GANIL Super Separator Spectrometer triplet [135]. Field measurements showed unexpected higher-order errors caused by misalignment of the external yoke. This issue was mitigated by removing the yoke [136].

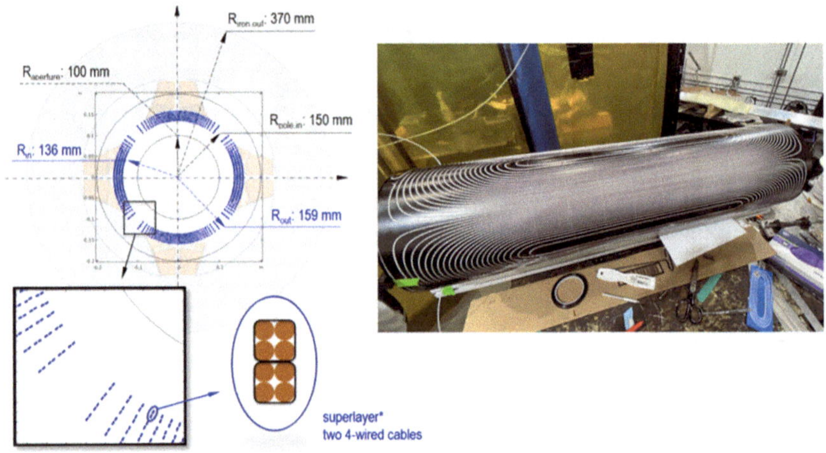

Fig. 10.32 Coil-dominated quadruple design equivalent to a A1900 triplet. (Left) The comparison of cross-section with the iron-dominated design in the back and (Right) the prototype coil built in-house at FRIB. (Courtesy MSU, USA)

10.3.7 Technical Challenge Summary and Lessons Learned

Most of the technical challenges associated with the FRIB driver linac were identified in the early design stage [137]. These included physics challenges such as beam loss control and multiple charge-state acceleration, as well as technological challenges such as large-scale SRF systems, large-scale cryogenics, loss detection and machine protection, high-power charge-stripping, and high-power targetry systems. Complications are mostly associated with beam-intercepting systems, in particular, the beam dump and associated diagnostic systems.

The lessons from the FRIB technical design, construction, commissioning, and early operations include [138]:

- Recruit worldwide and retain key subject matter experts (own the best people);
- Balance and phase the mandated R&D to allow the project to progress on schedule and ramp up as planned (balanced R&D)
- Develop and mature key technologies in time to support the project schedule (own the technology);
- Align interests for infrastructure investment to support key construction steps and future research (align interests, invest in infrastructure);
- Closely collaborate with US national labs and worldwide partners for knowledge transfer and project support; rigorously manage collaboration (collaborate without losing control);
- Strategically facilitate phased commissioning to stagger the demand on the workforce, validate the design principles, provide feedback on improvements, and meet the schedule (phase the scope for optimization);

- Conduct rigorous external reviews, inviting the best experts to critique the work (review rigorously);
- Engage with industrial providers through exchange visits, weekly meetings, and extended stays (intimately engaging with vendors);
- The original "turn-key" approach to procuring the large-scale cryogenic helium system from industry exposed the project to serious risks in budget and scope (avoid "turn-key" in large-scale cryogenics);
- Early shortcuts taken in QWR sub-component validation proved costly (avoid shortcuts);
- Shared vacuum vessels in the target area complicate maintenance (consider maintenance);
- The lack of diagnostics and correctors in the 3-D geometric layout complicates fragment separation (ensure adequate diagnostics and adjustments);
- Conduct systematic R&D for novel technology, e.g., bottom-up cryomodule design (systematic R&D);
- Thorough testing is needed for all major technical equipment, e.g., SRF sub-components, cryomodules, and superconducting magnets (test thoroughly);
- Pro-actively facilitate critical system validation, e.g., for liquid Li stripper (facilitate critical validation).

Some of the lessons learned are associated with technical challenges related to the FRIB targetry systems that were not fully identified in the early design stage.

In general, the FRIB construction project was successfully completed on time and within budget. Table 10.9 shows the estimate at completion (EAC) performed in April 2014 after the project baseline and in February 2022 before the project completion. In April 2014, the project contingency of \$172.8 M was 31%. In February 2022, the \$730 M project was 99.5% complete. The comparison indicated the following:

- The cost performance of the management and support (TPC.T.1), conventional facilities (TPC.T.2), and accelerator system (TPC.T.3) was healthy.
- The cost of the experimental system (TPC.T.4) was significantly higher than the baseline value by 131.5%. The complexity of the targetry system was underestimated.
- The pre-operations and transition-to-operations were smooth due to well-planned phased commissioning (Table 9.3).

The cost experience (Table 10.9) and the lessons learned from the FRIB project are rather different from those of the SNS project (Table 10.2).

10.3.8 Commissioning, Operations, Power Ramp Up, and Upgrades

FRIB beam commissioning was conducted in seven stages that started in 2017 and ended in 2022 (Table 9.3), as discussed in Sect. 9.3.4. Upon delivering the CD-4

Table 10.9 Comparison of estimated FRIB project cost after baseline in April 2014 and before completion in February 2022

WBS	2014–04 baseline [$M]	2022–02 EAC [$M]	Increase
TPC.T.1 Management & support	38.6	36.0	−6.8%
TPC.T.2 Conventional facilities	165.8	208.1	25.5%
TPC.T.3 Accelerator system	237.9	312.0	31.1%
TPC.T.4 Experimental system	54.7	126.6	131.5%
Estimate at complete	497.0	682.0	37.4%
Contingency	172.8		
Contingency percentage	31.0%		
Total Estimated Cost (TEC)	669.8	682.0	1.8%
TPC.R Conceptual design, R&D, NEPA	24.7	24.7	
TPC.P Pre-operations	35.5	22.7	
Other Project Cost	60.2	47.4	−21.3%
Total Project Cost (TPC)	730.0	730.0	0%

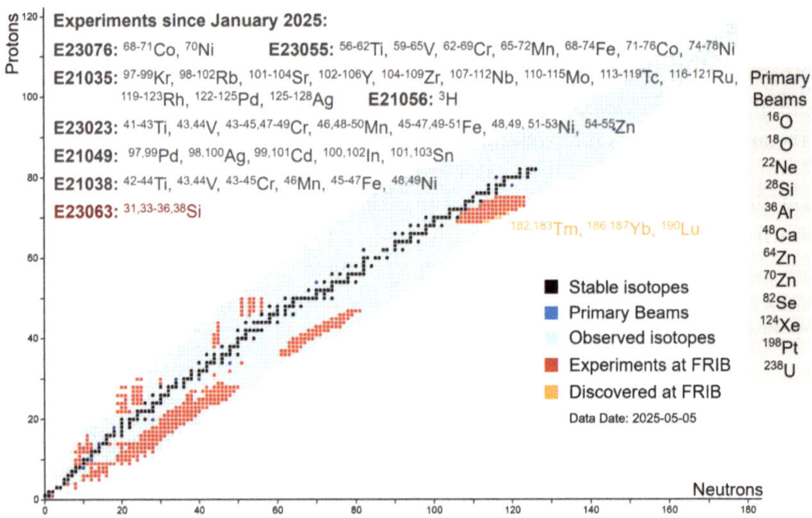

Fig. 10.33 Rare isotopes delivered to FRIB scientific users during the first three years of operations until April 2025

milestone, the user program started in May 2022. FRIB has served both the nuclear physics community and industrial users, with typical operating hours indicated in Table 9.4. Figure 10.33 shows the scientific experiments conducted.

During the first three years of operations, the primary beam power increased by a factor of 20, from the initial 1 kW to 20 kW for heavy ions ranging from oxygen to uranium. Since the start of operations, four more ARRs have been conducted by increasing the beam power to 3 kW, 5 kW, 10 kW, and 20 kW, respectively.

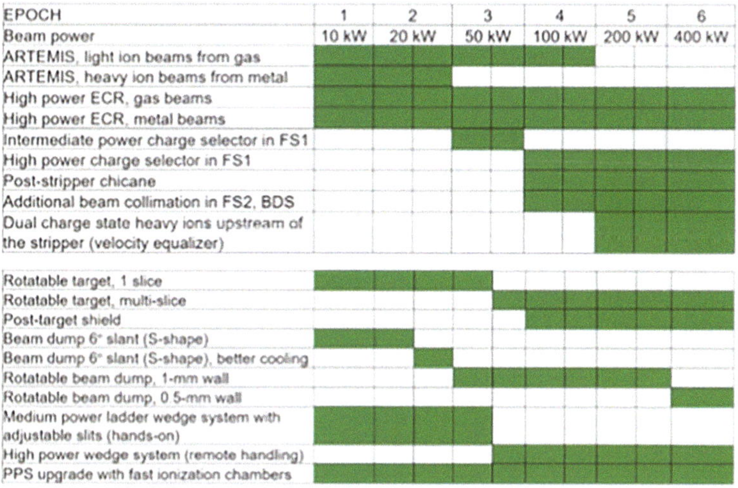

Fig. 10.34 Phased deployment of system improvements for the beam power ramp up

Figure 10.34 shows the technological path to ultimate performance at the 400 kW beam power, which is another factor of 20 from the current operating level.

Meanwhile, the High Rigidity Spectrometer project (HRS) has passed its own CD-2 and CD-3 DOE milestones [139]. The Isotope Harvesting project is also in progress, with the aim of using isotopes generated as by-products during the primary research of FRIB, which could have applications in medicine, plant science, and other fields [140].

The proposed FRIB energy upgrade (FRIB400) would increase the energy on target from 200 MeV/u to 400 MeV/u for uranium ion beams [141]. The upgrade would further extend the scientific reach and discovery potential of FRIB, in alignment with the overarching intellectual challenges articulated in the 2015 Nuclear Science Advisory Committee (NSAC) Long Range Plan and the National Research Council (NRC) Decadal Study. FRIB400 uses 644 MHz, $\beta_0 = 0.65$ 5-cell elliptical cavities [142]. As part of the development effort, two unjacketed 5-cell prototype cavities have been built to achieve $Q_0 = 3.5 \times 10^{10}$ at $E_{\mathrm{acc}} = 17.5$ MV/m with nitrogen doping, which is the first demonstration of this type of cavity [143]. The next steps for FRIB400 R&D include (1) achieving a high Q_0 in a prototype cryomodule, (2) developing subsystems such as the FPC and tuner, and (3) integrated testing of a cavity with the tuner and FPC. A total of 55 cavities in 11 cryomodules are needed for FRIB400; 80 m of the space in the linac tunnel has been allocated for FRIB400 cryomodules.

Problems

10.1 (Rapid-Cycling Synchrotron Design) Two identical, vertically stacked rapid-cycling synchrotrons are housed in the same tunnel. The circumference is 300 m. The proton beams are injected at 400 MeV, and extracted at 2 GeV. The repetition rate is 30 Hz for each ring. The pulse in each ring contains 10^{14} particles. The RF system operates at the harmonic $h = 2$, and the pulse contains two bunches.

1. What is the total output beam power? What is the total average beam current of the facility?
2. What is the tolerable fractional uncontrolled beam loss in each ring?
3. What is the range of RF frequency swing?
4. The beam gap reserved for the extraction kicker rise is a minimum 200 ns. Assume that the density distribution of the bunch is parabolic. What is the maximum bunching factor? What is the average and peak current in the ring?

References

1. N.R. Holtkamp, EPS-AG 2008 Gersh Budker Prize presentation: the successful construction and commissioning of the Spallation Neutron Source, in *Proceedings of the EPAC'08, No. 11 in European Particle Accelerator Conference* (JACoW Publishing, Geneva, 2008), pp. 2960–2962. https://jacow.org/e08/papers/THPPGM03.pdf
2. N.R. Council, et al., Neutron research on condensed matter: a study of the facilities and scientific opportunities in the United States (1977)
3. T.E. Mason, et al., Spallation Neutron Source project completion report. Prepared for the US Department of Energy Office of Science (June 2006)
4. J. Wei, et al., Design optimization and the path towards a 2 MW Spallation Neutron Source, in *Proceedings of the PAC'01* (JACoW Publishing, Geneva), pp. 319–321. https://jacow.org/p01/papers/WOAA006.pdf
5. R. Keller, The front-end systems for the spallation neutron source, in *AIP Conference Proceedings*, vol. 576 (American Institute of Physics, 2001), pp. 362–365
6. A. Ratti, R. DiGennaro, R. Gough, M. Hoff, R. Keller, K. Kennedy, R. MacGill, J. Staples, S. Virostek, R. Yourd, The fabrication and initial testing of the SNS RFQ, in *Proceedings of the PAC'01*, vol. 5 (IEEE, 2001), pp. 3930–3932
7. S. Virostek, J. Staples, Analysis of thermally induced frequency shift for the spallation neutron source RFQ (2000). arXiv preprint physics/0008169
8. D. Li, J.W. Staples, S.P. Virostek, Detailed modeling of the SNS RFQ structure with CST microwave studio, in *Proceedings of the LINAC'06, No. 23 in Linear Accelerator Conference* (JACoW Publishing, Geneva), pp. 580–582. https://jacow.org/l06/papers/THP008.pdf
9. J. Staples, D. Oshatz, T. Saleh, Design of the SNS MEBT, in *Proceedings of the LINAC'00, No. 20 in Linear Accelerator Conference* (JACoW Publishing, Geneva), pp. 250–252. https://jacow.org/l00/papers/MOD18.pdf
10. T. Ilg, et al., Mechanical design of the drift-tube linac (DTL) for the Spallation Neutron Source, in *Proceedings of the PAC'03* (JACoW Publishing, Geneva), pp. 2841–2843. https://jacow.org/p03/papers/RPAB051.pdf
11. N. Bultman, J. Billen, Z. Chen, M. Collier, D. Richards, L.M. Young, Fabrication and tuning of the SNS CCL hot model, in *Proceedings of the PAC'03* (JACoW Publishing, Geneva), pp. 2844–2846. https://jacow.org/p03/papers/RPAB052.pdf
12. J. Stovall, S. Nath, J. Billen, L. Young, M. Lynch, D. Rees, J. Galambos, D. Jeon, D. Raparia, J. Wei, et al., Superconducting linac for the SNS. Technical report, Thomas Jefferson National Accelerator Facility (TJNAF), Newport News, VA (2000)
13. C.H. Rode, The SNS superconducting Linac system, in *Proceedings of the PAC'01* (JACoW Publishing, Geneva), pp. 619–623. https://jacow.org/p01/papers/ROPB008.pdf

14. D.P. Gurd, Management of a large, distributed control system development project, in *Proceedings of the ICALEPCS'01* (JACoW Publishing, Geneva). https://jacow.org/ica01/papers/TUDI002.pdf
15. L.R. Dalesio, J.O. Hill, M. Kraimer, S. Lewis, D. Murray, S. Hunt, W. Watson, M. Clausen, J. Dalesio, The experimental physics and industrial control system architecture: past, present, and future. Nucl. Instrum. Methods Phys. Res. Sect. A **352**(1), 179–184 (1994). https://doi.org/10.1016/0168-9002(94)91493-1, https://www.sciencedirect.com/science/article/pii/0168900294914931
16. C.M. Chu, J. Galambos, J. Wei, C.K. Allen, P. McGehee, SNS application programming plan, in *Proceedings of the ICALEPCS'01* (JACoW Publishing, Geneva). https://jacow.org/ica01/papers/THAP060.pdf
17. J. Galambos, et al., XAL application programming structure, in *Proceedings of the PAC'05, No. 21 in Particle Accelerator Conference*, pp. 79–83 (JACoW Publishing, Geneva). https://jacow.org/p05/papers/ROPA001.pdf
18. R. Nelson, B. Oerter, T. Shea, C. Sibley, SNS timing system, in *Proceedings of the ICALEPCS'01* (JACoW Publishing, Geneva). https://jacow.org/ica01/papers/FRAT001.pdf
19. C. Sibley III, D.J. Armstrong, A. Jones, T.A. Justice, D.H. Thompson, The SNS machine protection system: early commissioning results and future plans, in *Proceedings of the PAC'05, No. 21 in Particle Accelerator Conference* (JACoW Publishing, Geneva), pp. 1727–1729. https://jacow.org/p05/papers/RPPE021.pdf
20. L. Doolittle, A. Ratti, M. Monroy, M. Champion, H. Ma, Operational performance of the SNS LLRF interim system, in *Proceedings of the PAC'03*, vol. 3 (IEEE, 2003), pp. 1464–1466
21. T.P. Wangler, et al., Design of a proton superconducting linac for a neutron spallation source, in *Proceedings of the SRF'99* (JACoW Publishing, Geneva), pp. 336–344. https://jacow.org/SRF99/papers/WEA001.pdf
22. Y. Cho, Evolution of the SNS superconducting proton linac, in *Proceedings of the SRF'01* (JACoW Publishing, Geneva). https://jacow.org/srf01/papers/TL023.pdf
23. J. Stovall, et al., Expected beam performance of the SNS linac, in *Proceedings of the PAC'01* (JACoW Publishing, Geneva), pp. 446–450. https://jacow.org/p01/papers/WOPB007.pdf
24. Preliminary design report: superconducting radio frequency linac for the Spallation Neutron Source. SNS-SRF-99-101 (1999)
25. J. Galambos, Operational experience with high power beams at the SNS superconducting linac, in *Proceedings of the LINAC'08, No. 24 in Linear Accelerator Conference* (JACoW Publishing, Geneva), pp. 710–713. https://jacow.org/LINAC08/papers/WE202.pdf
26. V. Ganni, D.M. Arenius, B.S. Bevins, W.C. Chronis, J.D. Creel, J.D. Wilson, Jr., Design, fabrication, commissioning, and testing of a 250 g/s, 2-K helium cold compressor system. AIP Conf. Proc. **613**(1), 288–304 (2002). https://doi.org/10.1063/1.1472033
27. J. Wei, et al., Low-loss rapid-cycling-synchrotron design for the Spallation Neutron Source, in *Proceedings of the EPAC'00, No. 7 in European Particle Accelerator Conference* (JACoW Publishing, Geneva, 2000), pp. 981–983. https://jacow.org/e00/papers/THP2B13.pdf
28. J. Wei, D.T. Abell, J. Beebe-Wang, M. Blaskiewicz, P.R. Cameron, N. Catalan-Lasheras, G. Danby, A.V. Fedotov, C. Gardner, J. Jackson, Y.Y. Lee, H. Ludewig, N. Malitsky, W. Meng, Y. Papaphilippou, D. Raparia, N. Tsoupas, W.T. Weng, R.L. Witkover, S.Y. Zhang, Low-loss design for the high-intensity accumulator ring of the Spallation Neutron Source. Phys. Rev. ST Accel. Beams **3**, 080101 (2000). https://link.aps.org/doi/10.1103/PhysRevSTAB.3.080101
29. J. Wei, et al., Injection choice for Spallation Neutron Source ring, in *Proceedings of the PAC'01* (JACoW Publishing, Geneva), pp. 2560–2562. https://jacow.org/p01/papers/WPPH050.pdf
30. N. Catalan-Lasheras, Y.Y. Lee, H. Ludewig, N. Simos, J. Wei, Optimization of the collimation system for the Spallation Neutron Source accumulator ring. Phys. Rev. ST Accel. Beams **4**, 010101 (2001). https://link.aps.org/doi/10.1103/PhysRevSTAB.4.010101
31. M. Haire, ALARA review of the Spallation Neutron Source accumulator ring and transfer lines. Technical report, Oak Ridge National Lab. (ORNL), Oak Ridge, TN (United States) (2003)

32. V. Baglin, et al., The secondary electron yield of technical materials and its variaton with surface treatments, in *Proceedings of the EPAC'00, No. 7 in European Particle Accelerator Conference* (JACoW Publishing, Geneva, 2000), pp. 217–221. https://jacow.org/e00/papers/THXF102.pdf
33. P. He, et al., Secondary electron emission measurements for TiN coating on stainless steel of SNS accumulator ring vacuum chamber, in *Proceedings of the EPAC'04, No. 9 in European Particle Accelerator Conference* (JACoW Publishing, Geneva, 2004), pp. 1804–1806. http://accelconf.web.cern.ch/e04/papers/WEPKF085.pdf
34. R.J. Todd, P. He, H.C. Hseuh, D. Weiss, Summary on titanium nitride coating of SNS ring vacuum chambers, in *Proceedings of the PAC'05, No. 21 in Particle Accelerator Conference* (JACoW Publishing, Geneva), pp. 3088–3090. https://jacow.org/p05/papers/RPPE049.pdf
35. H.C. Hseuh, et al., Physical and electromagnetic properties of customized coatings for SNS injection ceramic chambers and extraction ferrite kickers, in *Proceedings of the PAC'05, No. 21 in Particle Accelerator Conference* (JACoW Publishing, Geneva), pp. 3028–3030. https://jacow.org/p05/papers/RPPE048.pdf
36. J. Haines, T. McManamy, T. Gabriel, R. Battle, K. Chipley, J. Crabtree, L. Jacobs, D. Lousteau, M. Rennich, B. Riemer, Spallation neutron source target station design, development, and commissioning. Nucl. Instrum. Methods Phys. Res., Sect. A **764**, 94–115 (2014). https://doi.org/10.1016/j.nima.2014.03.068, https://www.sciencedirect.com/science/article/pii/S0168900214003829
37. D.E. Winder, Evolution of the high-power spallation neutron mercury target at the SNS, in *Proceedings of the IPAC'21* (JACoW Publishing, Geneva), pp. 3735–3739. https://doi.org/10.18429/JACoW-IPAC2021-THXC03, https://jacow.org/ipac2021/papers/THXC03.pdf
38. T.A. Gabriel, J.R. Haines, T.J. McManamy, Overviewof the target systems for the Spallation Neutron Source (1998). https://www.osti.gov/biblio/1715
39. M. Kass, J. Whealton, N. Clapp, J. DiStefano, J. DeVan, J. Haines, M. Akerman, T. Gabriel, Nonlinear cavitation erosion of stainless steel in mercury versus applied power. Tribol. Lett. **5**, 231–234 (1998)
40. M. Futakawa, H. Kogawa, R. Hino, H. Date, H. Takeishi, Erosion damage on solid boundaries in contact with liquid metals by impulsive pressure injection. Int. J. Impact Eng. **28**(2), 123–135 (2003). https://doi.org/10.1016/S0734-743X(02)00054-4, https://www.sciencedirect.com/science/article/pii/S0734743X02000544
41. J.R. Haines, B. Riemer, D.K. Felde, J.D. Hunn, S.J. Pawel, C.C. Tsai, Summary of cavitation erosion investigations for the SNS mercury target. J. Nucl. Mater. **343**(1–3), 58–69 (2005)
42. K. Okita, S. Takagi, Y. Matsumoto, Propagation of pressure waves, caused by a thermal shock, in liquid metals containing gas bubbles. J. Fluid Sci. Technol. **3**(1), 116–128 (2008)
43. M.A. Plum: Beam loss mitigation in the Oak Ridge Spallation Neutron Source, in *Proceedings of the HB'12* (JACoW Publishing, Geneva), pp. 329–333. https://jacow.org/HB2012/papers/TUO3C04.pdf
44. V.V. Peplov, D.E. Anderson, R.I. Cutler, M. Wezensky, J.D. Hicks, R.B. Saethre, SNS linac modulator operational history and performance, in *Proceedings of the PAC'11, No. 24 in Particle Accelerator Conference* (JACoW Publishing, Geneva, 2011), pp. 1340–1342. https://jacow.org/PAC2011/papers/TUP275.pdf
45. A. Shishlo, J. Galambos, A. Aleksandrov, V. Lebedev, M. Plum, First observation of intrabeam stripping of negative hydrogen in a superconducting linear accelerator. Phys. Rev. Lett. **108**, 114801 (2012). https://doi.org/10.1103/PhysRevLett.108.114801
46. J. Tuozzolo, et al., Design and manufacturing of SNS accumulator ring and transport system DC magnets, in *Proceedings of the PAC'03* (JACoW Publishing, Geneva), pp. 2156–2158. https://jacow.org/p03/papers/WPAB052.pdf
47. J.Y. Tang, H.C. Hseuh, L.A. Smart, Design and development of the SNS ring vacuum instrumentation and control systems, in *Proceedings of the PAC'01* (JACoW Publishing, Geneva), pp. 779–781. https://jacow.org/p01/papers/MPPH006.pdf
48. D. Raparia, Y. Lee, W. Weng, J. Wei, Beam transfer lines for the spallation neutron source, in *AIP Conference Proceedings*, vol. 642 (American Institute of Physics, 2002), pp. 130–132

49. J.J. Error, D.R. Bruce, J.J. Fazekas, S.A. Helus, J.R. Maines, Deformation monitoring of the Spallation Neutron Source (SNS) tunnels, in *Proceedings of the PAC'05, No. 21 in Particle Accelerator Conference* (JACoW Publishing, Geneva), pp. 2509–2511. https://jacow.org/p05/papers/WPAE037.pdf
50. J.D. Galambos, D. Anderson, C. Barbier, K.B. Bekar, D. Bunch, J. Bullman, Jr., M. Buchanan, G. Capps, M. Champion, A.E. Coleman, et al., Final design report. Proton Power Upgrade Project. Technical report, Oak Ridge National Laboratory (ORNL), Oak Ridge, TN (United States) (2020)
51. M. Champion, M. Connell, N. Evans, J.D. Galambos, M. Howell, G. Johns, S.H. Kim, J. Moss, G.S. Stephens, K.S. White, Completion of the proton power upgrade project at the Spallation Neutron Source (2024)
52. S. Henderson, SNS progress, challenges and upgrade options, in *Proceedigns of the EPAC'08, No. 11 in European Particle Accelerator Conference* (JACoW Publishing, Geneva, 2008), pp. 2892–2896. https://jacow.org/e08/papers/THXG01.pdf
53. Y. Yamazaki, M. Mizumoto, Accelerator complex for the joint project of KEK/JHF and JAERI/NSP, in *Proceedings of the PAC'99* (JACoW Publishing, Geneva), pp. 513–517. https://jacow.org/p99/papers/THDL1.pdf
54. Proposal for Japan hadron facility (1997)
55. JHF accelerator design study report (1998)
56. Y. Mori, The Japanese hadron facility, in *Proceedings of the PAC'97* (JACoW Publishing, Geneva), pp. 920–924
57. Y. Yamazaki, The Japan hadron facility accelerator, in *Proceedings of the APAC'98* (JACoW Publishing, Geneva), pp. 314–318
58. M. Mizumoto, et al., A proton accelerator for neutron science project at JAERI, in *Proceedings of the APAC'98* (JACoW Publishing, Geneva), pp. 309–313
59. Y. Yamazaki, M. Kihara, Development of the high-intensity proton linac for the Japanese Hadron Project, in *Proceedings of the LINAC'90, Linear Accelerator Conference* (JACoW Publishing, Geneva), pp. 543–547. https://jacow.org/l90/papers/TH101.pdf
60. S. Machida, Y. Mori, Y. Ishi, S. Shibuya, Design and beam dynamics issues of the JHF synchrotrons, in *Proceedings of the APAC'98* (JACoW Publishing, Geneva), pp. 381–383
61. A. Ando, S. Hashimoto, Y. Shoji, Negative alpha-p operation at NewSUBARU. AIP Conf. Proc. **705**(1), 13–16 (2004). https://doi.org/10.1063/1.1757722
62. M.K. Craddock, R. Baartman, J. Beveridge, E.W. Blackmore, J.I.M. Botman, R. Burge, W. Cameron, J. Doornbos, T.A. Hodges, D.E. Lobl, G.H. Mackenzie, C.J. Oram, A.J. Otter, D. Raparia, P.A. Reeve, J.R. Richardson, G. Stinson, I.M. Thorson, R.E. Worsham, W.K. Lacey, C.W. Planner, F. Pedersen, H.O. Schoenauer, T. Suzuki, C. Yamaguchi, The TRIUMF KAON factory. IEEE Trans. Nucl. Sci. **32**(5), 1707–1709 (1985). https://doi.org/10.1109/TNS.1985.4333699
63. C. Ohmori, et al., High field gradient cavity for JAERI-KEK joint project, in *Proceedings of the EPAC'02, No. 8 in European Particle Accelerator Conference* (JACoW Publishing, Geneva, 2007), pp. 257–259. https://jacow.org/e02/papers/WEALA001.pdf
64. C. Ohmori, E. Ezura, K. Hara, K. Hasegawa, Y. Makida, R. Muto, M. Nomura, T. Ogitsu, T. Shimada, A. Schnase, et al., Development of a high gradient rf system using a nanocrystalline soft magnetic alloy. Phys. Rev. Spec. Top. Accel. Beams **16**(11), 112002 (2013)
65. A. Schnase, et al., Higher harmonic voltages in J-PARC RCS operation, in *Proceedings of the PAC'09, No. 23 in Particle Accelerator Conference* (JACoW Publishing, Geneva), pp. 1988–1990. https://jacow.org/PAC2009/papers/WE5PFP003.pdf
66. M. Yoshii, et al., Proton beam acceleration with MA loaded RF systems in J-PARC RCS and MR synchrotron, in *Proceedings of the PAC'09, No. 23 in Particle Accelerator Conference* (JACoW Publishing, Geneva), pp. 770–772. https://jacow.org/PAC2009/papers/TU4RAC04.pdf
67. A. Schnase, et al., MA cavities for J-PARC with controlled Q-value by external inductor, in *Proceedings of the PAC'07, No. 22 in Particle Accelerator Conference* (JACoW Publishing, Geneva, 2007), pp. 2131–2133. https://jacow.org/p07/papers/WEPMN040.pdf

68. M. Nomura, et al., Impedance measurements of MA loaded RF cavities in J-PARC synchrotrons, in *Proceedings of the PAC'09, No. 23 in Particle Accelerator Conference* (JACoW Publishing, Geneva), pp. 1985–1987. https://jacow.org/PAC2009/papers/WE5PFP002.pdf
69. M. Nomura, M. Yamamoto, A. Schnase, T. Shimada, H. Suzuki, F. Tamura, K. Hara, K. Hasegawa, C. Ohmori, M. Toda, M. Yoshii, The origin of magnetic alloy core buckling in J-PARC 3GeV RCS. Nucl. Instrum. Methods Phys. Res. Sect. A **623**(3), 903–909 (2010). https://doi.org/10.1016/j.nima.2010.08.111, https://www.sciencedirect.com/science/article/pii/S0168900210018991
70. M. Kinsho, D. Nishizawa, Y. Saito, H. Suzuki, H. Yokomizo, Vacuum system design for the 3 GeV-proton synchrotron of JAERI-KEK joint project. J. Vac. Sci. Technol. A **20**(3), 829–832 (2002). https://doi.org/10.1116/1.1468656
71. M. Kinsho, Y. Saito, Z. Kabeya, N. Ogiwara, Titanium flanged alumina ceramics vacuum duct with low impedance. Vacuum **81**(6), 808–811 (2007). https://doi.org/10.1016/j.vacuum.2005.11.056. https://www.sciencedirect.com/science/article/pii/S0042207X0500401X. Proceedings of the European Vacuum Conference (EVC-9)
72. Y. Saito, M. Kinsho, Z. Kabeya, Production process of alumina-ceramic vacuum chambers for J-PARC. J. Phys. Conf. Ser. **100**(9), 092020 (2008). https://doi.org/10.1088/1742-6596/100/9/092020
73. T. Takayanagi, et al., Performance of the bump system for the painting injection at J-PARC, in *Proceedings of the PAC'09, No. 23 in Particle Accelerator Conference* (JACoW Publishing, Geneva), pp. 1507–1509. https://jacow.org/PAC2009/papers/TU6PFP091.pdf
74. S.I. Meigo, et al., Beam commissioning of Spallation Neutron and muon source in J-PARC, in *Proceedings of the PAC'09, No. 23 in Particle Accelerator Conference* (JACoW Publishing, Geneva), pp. 1439–1441. https://jacow.org/PAC2009/papers/TU6PFP066.pdf
75. J. Kamiya, T. Takayanagi, M. Watanabe, Performance of extraction kicker magnet in a rapid cycling synchrotron. Phys. Rev. ST Accel. Beams **12**, 072401 (2009). https://doi.org/10.1103/PhysRevSTAB.12.072401
76. H. Oguri, A. Ueno, K. Ikegami, Y. Namekawa, K. Ohkoshi, Status of the Japan proton accelerator research complex H. Ion source. Phys. Rev. ST Accel. Beams **12**, 010401 (2009). https://doi.org/10.1103/PhysRevSTAB.12.010401
77. A. Ueno, Y. Yamazaki, New field stabilization method of a four-vane type RFQ. Nucl. Instrum. Methods Phys. Res. Sect. A **300**(1), 15–24 (1991). https://doi.org/10.1016/0168-9002(91)90701-Q, https://www.sciencedirect.com/science/article/pii/016890029190701Q
78. M. Ikegami, et al., Beam commissioning of the J-PARC linac medium energy beam transport at KEK, in *Proceedings of the PAC'03* (JACoW Publishing, Geneva), pp. 1509–1511. https://jacow.org/p03/papers/TPAG035.pdf
79. T. Kato, S. Fu, M. Ikegami, S. Wang, Beam study with RF choppers in the MEBT of the J-PARC proton linac, in *Proceedings of the PAC'03* (JACoW Publishing, Geneva), pp. 1455–1457. https://jacow.org/p03/papers/TPAG014.pdf
80. F. Naito, et al., Development of the 50-MeV DTL for the JAERI/KEK joint project, in *Proceedings of the LINAC'00, No. 20 in Linear Accelerator Conference* (JACoW Publishing, Geneva), pp. 563–565. https://jacow.org/l00/papers/TUD08.pdf
81. T. Ito, et al., RF characteristics of the SDTL for the J-PARC, in *Proceedings of the LINAC'06, No. 23 in Linear Accelerator Conference* (JACoW Publishing, Geneva), pp. 740–742. https://jacow.org/l06/papers/THP068.pdf
82. T. Kageyama, K. Yoshino, Y. Yamazaki, A new annular-coupled structure suppressing higher order modes' mixing with the pi/2 coupling mode. Part. Accel. **32**(KEK-89-94), 33–38 (1989)
83. H. Ao, et al., Status of mass production of the ACS cavity for the J-PARC linac energy upgrade, in *Proceedings of the IPAC'10* (JACoW Publishing, Geneva), pp. 618–620. http://accelconf.web.cern.ch/IPAC10/papers/MOPEC066.pdf
84. R.A. Hardekopf, A.V. Aleksandrov, D.E. Anderson, S.S. Kurennoy, J. Power, Design,construction, and initial operation of the SNS MEBT chopper system, in *Proceedings of the EPAC'04, No. 9 in European Particle Accelerator Conference* (JACoW Publishing, Geneva, 2004), pp. 150–152. http://accelconf.web.cern.ch/e04/papers/WEOBCH02.pdf

85. S. Fu, T. Kato, F. Naito, K. Yoshino, 324-MHz RF deflector design and test, in *Proceedings of the LINAC'98, Linear Accelerator Conference* (JACoW Publishing, Geneva), pp. 585–587. https://jacow.org/l98/papers/TU4079.pdf
86. S.D. Holmes, Project X at Fermilab: prospects and plans, in *Proceedings of the PAC'09, No. 23 in Particle Accelerator Conference* (JACoW Publishing, Geneva), pp. 4241–4245. https://jacow.org/PAC2009/papers/FR1GRI02.pdf
87. A.P. Letchford, et al., Status of the RAL front end test stand, in *Proceedings of the EPAC'08, No. 11 in European Particle Accelerator Conference* (JACoW Publishing, Geneva, 2008), pp. 3437–3439. https://jacow.org/e08/papers/THPP029.pdf
88. R.M. Brodhage, et al., Prototype construction of a coupled CH-DTL proton linac for FAIR, in *Proceedings of the PAC'09, No. 23 in Particle Accelerator Conference* (JACoW Publishing, Geneva), pp. 4908–4910. https://jacow.org/PAC2009/papers/FR5REP060.pdf
89. S. Fu, et al., Status of the China Spallation Neutron Source project, in *Proceedings of the PAC'09, No. 23 in Particle Accelerator Conference* (JACoW Publishing, Geneva), pp. 3053–3057. https://jacow.org/PAC2009/papers/TH1GRI02.pdf
90. Y. Yamazaki, Recent technological developments in accelerating structures, in *Proceedings of the LINAC'92, Linear Accelerator Conference* (JACoW Publishing, Geneva), pp. 580–584. https://jacow.org/l92/papers/TH1-04.pdf
91. H. Ao, et al., First annular-ring coupled structure cavity for the Japan Proton Accelerator Research Complex linac. Phys. Rev. ST Accel. Beams **15**, 051005 (2012). https://doi.org/10.1103/PhysRevSTAB.15.051005
92. V.G. Andreev, V.M. Belugin, V.G. Kulman, E.A. Mirochnik, B.M. Pirozhenko, Study of high-energy proton linac structures, in *Proceedings of the LINAC'72, Linear Accelerator Conference* (JACoW Publishing, Geneva), pp. 114–118. https://jacow.org/l72/papers/E04.pdf
93. H. Hotchi, et al., Beam commissioning of the 3-GeV rapid cycling synchrotron of the Japan Proton Accelerator Research Complex. Phys. Rev. ST Accel. Beams **12**, 040402 (2009). https://doi.org/10.1103/PhysRevSTAB.12.040402
94. P.K. Saha, Y. Shobuda, H. Hotchi, N. Hayashi, T. Takayanagi, H. Harada, Y. Irie, Direct observation of the phase space footprint of a painting injection in the Rapid Cycling Synchrotron at the Japan Proton Accelerator Research Complex. Phys. Rev. ST Accel. Beams **12**, 040403 (2009). https://doi.org/10.1103/PhysRevSTAB.12.040403
95. F. Tamura, et al., Longitudinal painting studies in the J-PARC RCS, in *Proceedings of the PAC'09, No. 23 in Particle Accelerator Conference* (JACoW Publishing, Geneva), pp. 1445–1447. https://jacow.org/PAC2009/papers/TU6PFP068.pdf
96. H. Hotchi, et al., High-intensity demonstrations in the J-PARC 3-GeV RCS, in *Proceedings of the PAC'09, No. 23 in Particle Accelerator Conference* (JACoW Publishing, Geneva), pp. 1504–1506. https://jacow.org/PAC2009/papers/TU6PFP090.pdf
97. M. Ikegami, Progress in the beam commissioning of J-PARC linac and its upgrade path, in *Proceedings of the LINAC'08, No. 24 in Linear Accelerator Conference* (JACoW Publishing, Geneva), pp. 16–20. https://jacow.org/LINAC08/papers/MO201.pdf
98. H. Kobayashi, Beam commissioning of the J-PARC main ring, in *Proceedings of the PAC'09, No. 23 in Particle Accelerator Conference* (JACoW Publishing, Geneva), pp. 1823–1827. https://jacow.org/PAC2009/papers/WE1GRI02.pdf
99. K. Hasegawa, M. Kinsho, H. Oguri, T. Koseki, Status of J-PARC accelerator facilities after the great east Japan earthquake, in *Proceedings of the IPAC'11, No. 2 in International Particle Accelerator Conference* (JACoW Publishing, Geneva, 2011), pp. 2727–2729. https://jacow.org/IPAC2011/papers/WEPS095.pdf
100. K. Yamamoto, K. Moriya, H. Okita, I. Yamada, M. Chimura, P.K. Saha, Y. Shobuda, F. Tamura, M. Yamamoto, T. Morishita, et al., J-PARC linac and RCS operational status and upgrade plan to 2 MW. J. Neutron Res. **26**(2–3), 59–67 (2024)
101. T. Koseki, et al., Upgrade plan of J-PARC MR-toward 1.3 MW beam power, in *Proceedings of the IPA'18* (2018), pp. 966–969
102. H. Grunder, et al., ISOL task force report to NSAC (1999)
103. J. Symons, et al., Report to NSAC of the rare-isotope beam task force (2007)

104. J.A. Nolen, F. Marti, Charge strippers of heavy ions for high intensity accelerators. Rev. Accel Sci. Technol. **6**, 221–236 (2013)
105. Y. Momozaki, C.B. Reed, J.A. Nolen, J.R. Specht, D.B. Chojnowski, J.S. Song, F. Marti, P. Guetschow, J. Sherman, Proton beam-on-liquid lithium stripper film experiment. J. Radioanal. Nucl. Chem. **305**, 843–849 (2015)
106. T. Kanemura, M. LaVere, R. Madendorp, F. Marti, T. Maruta, Y. Momozaki, P.N. Ostroumov, A.S. Plastun, J. Wei, Q. Zhao, Experimental demonstration of the thin-film liquidmetal jet as a charge stripper. Phys. Rev. Lett. **128**, 212301 (2022). https://doi.org/10.1103/PhysRevLett.128.212301
107. M. Avilov, S. Fadeev, S. Fernandes, M. Golkovsky, W. Mittig, F. Pellemoine, M. Schein, A 50-kW prototype of the high-power production target for the FRIB. J. Radioanal. Nucl. Chem. **305**, 817–823 (2015)
108. V. Ganni, P. Knudsen, Helium refrigeration considerations for cryomodule design. AIP Conf. Proc. **1573**(1), 1814–1821 (2014). https://doi.org/10.1063/1.4860928
109. V. Ganni, P. Knudsen, D. Arenius, F. Casagrande, Application of JLab 12GeV helium refrigeration system for the FRIB accelerator at MSU. AIP Conf. Proc. **1573**(1), 323–328 (2014). https://doi.org/10.1063/1.4860718
110. V. Ganni, P. Knudsen, F. Casagrande, S. Jones, Cryogenic design of FRIB cryomodule and distribution system and the present status. IOP Conf. Ser.: Mater. Sci. Eng. **755**(1), 012091 (2020). https://doi.org/10.1088/1757-899X/755/1/012091
111. F. Casagrande, V. Ganni, P. Knudsen, M. Wright, S. Jones, C. Nguyen, A. Fila, N. Hassan, G. Vargas, N. Joseph, S. Stanley, FRIB cryogenic system status. IOP Conf. Ser.: Mater. Sci. Eng. **755**(1), 012089 (2020). https://doi.org/10.1088/1757-899X/755/1/012089
112. V. Ganni, P. Knudsen, Optimal design and operation of helium refrigeration systems using the Ganni cycle. AIP Conf. Proc. **1218**(1), 1057–1071 (2010). https://doi.org/10.1063/1.3422267
113. P. Knudsen, V. Ganni, N. Hasan, M. Wright, F. Casagrande, G. Vargas, N. Joseph, FRIB helium refrigeration system commissioning and performance test results. IOP Conf. Ser.: Mater. Sci. Eng. **755**, 012090 (2020). http://doi.org/10.1088/1757-899X/755/1/012090
114. P. Knudsen, V. Ganni, N. Hasan, M. Wright, F. Casagrande, G. Vargas, N. Joseph: FRIB helium refrigeration system commissioning and performance test results. IOP Conf. Ser.: Mater. Sci. Eng. **755**(1), 012090 (2020). https://doi.org/10.1088/1757-899X/755/1/012090
115. P. Knudsen, V. Ganni, F. Casagrande, A. Fila, N. Hasan, M. Wright, G. Vargas, N. Joseph, Design, fabrication, commissioning and testing of FRIB 2 K cold compressor system. IOP Conf. Ser.: Mater. Sci. Eng. **755**(1), 012092 (2020). http://doi.org/10.1088/1757-899X/755/1/012092
116. C.W. Leemann, The Rare-isotope Accelerator (RIA) facility project, in *Proceedings of the LINAC'00, No. 20 in Linear Accelerator Conference* (JACoW Publishing, Geneva), pp. 331–335. https://jacow.org/l00/papers/TU103.pdf
117. P.N. Ostroumov, K.W. Shepard, Correction of beam-steering effects in low-velocity superconducting quarter-wave cavities. Phys. Rev. ST Accel. Beams **4**, 110101 (2001). https://doi.org/10.1103/PhysRevSTAB.4.110101
118. K.W. Shepard, The U.S. RIA project SRF linac, in *Proceedings of the SRF'01* (JACoW Publishing, Geneva). https://jacow.org/srf01/papers/TA003.pdf
119. J. Kim, D. Gorelov, F. Marti, H. Podlech, X. Wu, R.C. York, Design study of a superconducting linac for RIA, in *Proceedings of the PAC'01* (JACoW Publishing, Geneva), pp. 1741–1743. https://jacow.org/p01/papers/TPPH035.pdf
120. T.L. Grimm, et al., Superconducting RF activities for the rare isotope accelerator at Michigan State University, in *Proceedings of the SRF'03* (JACoWPublishing, Geneva), pp. 32–36. https://jacow.org/SRF2003/papers/MOP03.pdf
121. W. Hartung, et al., Production cavities and cryomodules for a heavy ion re-accelerator at Michigan State University, in *Proceedings of the SRF'09* (JACoW Publishing, Geneva), pp. 201–204. https://jacow.org/SRF2009/papers/TUPPO011.pdf

122. A. Facco, et al., Faced issues in ReA3 quarter-wave resonators and their successful resolution, in *Proceedings of the SRF'13* (JACoW Publishing, Geneva), pp. 873–877. https://jacow.org/SRF2013/papers/THIOD02.pdf
123. S. Stark, et al., FRIB HWR tuner development, in *Proceedings of the LINAC'16, No. 28 in Linear Accelerator Conference* (JACoW Publishing, Geneva, 2017), pp. 535–537. https://doi.org/10.18429/JACoW-LINAC2016-TUPLR029, https://jacow.org/linac2016/papers/TUPLR029.pdf
124. S.J. Miller, et al., Low-beta cryomodule design optimized for large-scale linac installations, in *Proceedings of the SRF'13* (JACoW Publishing, Geneva), pp. 825–829. https://jacow.org/SRF2013/papers/THIOA04.pdf
125. J. Wei, et al., FRIB construction, installation, SRF and beam commissioning: experience and lessons learned, in *Proceedings of the of LINAC'20*, Liverpool (2020)
126. M. Xu, et al., Design of $\beta = 0.65$, 5 cells, 644 MHz elliptical cavity for FRIB upgrade, in *Proceedings of the IPAC'18* (JACoW Publishing, Geneva), pp. 3712–3715. https://doi.org/10.18429/JACoW-IPAC2018-THPAL035, http://accelconf.web.cern.ch/ipac2018/papers/THPAL035.pdf
127. A.F. Zeller, J.C. DeKamp, M.J. Dubois, J.A. Nolen, R.T. Swanson, Design and construction of a low current, cryogenically efficient beam line quadrupole, in *Proceedings of the Cyclotrons'84* (JACoW Publishing, Geneva, 1984), pp. 79–81. https://jacow.org/c84/papers/B16.pdf
128. A.F. Zeller, et al., Magnetic elements for the a1900 fragment separator at the NSCL, in *Advances in Cryogenic Engineering, Advances in Cryogenic Engineering*, ed. by P. Kittel, vol. 43 (Springer, Boston, 1998). https://doi.org/10.1007/978-1-4757-9047-4_28
129. R.B. Meinke, M.J. Ball, C.L. Goodzeit, Superconducting double-helix accelerator magnets, in *Proceedings of the PAC'03* (JACoW Publishing, Geneva), pp. 1996–1998. https://jacow.org/p03/papers/WPAE025.pdf
130. W. Wu, et al., Multipole magnets for the HIAF fragment separator using the Canted-Cosine-Theta (CCT) geometry, in *Proceedings of the HIAT'18* (JACoW Publishing, Geneva), pp. 170–174. https://doi.org/10.18429/JACoW-HIAT2018-THOBA01, https://jacow.org/hiat2018/papers/THOBA01.pdf
131. P.L. Walstrom, Design of end turns in current-dominated dipole and quadrupole magnets for fields with low higher-harmonic content, in *Proceedings of the EPAC'02, No. 8 in European Particle Accelerator Conference* (JACoW Publishing, Geneva, 2007), pp. 2448–2450. https://jacow.org/e02/papers/MOPLE008.pdf
132. D. Greene, et al., Design of coil-dominated quadrupole triplet for high rigidity isotope beams. IEEE Trans. Appl. Supercond. **33**(5), 1–5 (2023)
133. Y. Choi, et al., The high transmission beamline superconducting magnets for high rigidity spectrometer project at FRIB: design status. IEEE Trans. Appl. Supercond. **35**(5), 1–5 (2025)
134. D. Greene, et al., Status of coil-dominated discrete-cosine-theta quadrupole prototype for high rigidity isotope beams, in *Proceedings of the IPAC'24* (2024). https://doi.org/10.18429/JACoW-IPAC2024-WEPS67
135. J.A. Nolen, et al., Design and status of the super separator spectrometer for the GANIL-SPIRAL2 project, in *Proceedings of the HIAT'12* (JACoW Publishing, Geneva), pp. 23–23. https://jacow.org/HIAT2012/papers/MOB03.pdf
136. A. Esper, et al., Superconducting multipole triplet field measurements, in *Proceedings of the IPAC'23, No. 14 in International Particle Accelerator Conference* (JACoW Publishing, Geneva, 2023), pp. 4051–4054. https://doi.org/10.18429/JACoW-IPAC2023-THPA043
137. J. Wei, et al., Accelerator challenges of hadron linacs and the facility for rare isotope beams-extending high beam power from protons to heavy ions, in *Proceedings of the HB'14* (JACoW Publishing, Geneva), pp. 12–17. https://jacow.org/HB2014/papers/MOZLR07.pdf
138. J. Wei, et al.: FRIB transition to user operations, power ramp up, and upgrade perspectives, in *Proceedings of the SRF'23, No. 21 in International Conference on RF Superconductivity* (JACoW Publishing, Geneva, 2023), pp. 1–8. https://doi.org/10.18429/JACoW-SRF2023-MOIAA01

139. T. Glasmacher, A. Gade, G. Bollen, J. Wei, The facility for rare isotope beams: providing new opportunities for science. Nucl. Phys. News **34**(3), 5–12 (2024)
140. E.P. Abel, M. Avilov, V. Ayres, E. Birnbaum, G. Bollen, G. Bonito, T. Bredeweg, H. Clause, A. Couture, J. DeVore, et al., Isotope harvesting at FRIB: additional opportunities for scientific discovery. J. Phys. G: Nucl. Part. Phys. **46**(10), 100501 (2019)
141. FRIB400 whitepaper (2019). https://frib.msu.edu/_files/pdfs/frib400_final.pdf
142. P. Ostroumov, C. Contreras, A. Plastun, J. Rathke, T. Schultheiss, A. Taylor, J. Wei, M. Xu, T. Xu, Q. Zhao, et al., Elliptical superconducting RF cavities for FRIB energy upgrade. Nucl. Instrum. Methods Phys. Res. Sect. A **888**, 53–63 (2018)
143. S.H. Kim, et al., SRF R&D for FRIB linac energy upgrade with high-performance medium-beta elliptical cavity CW cryomodules. *Presented at SRF'23*, Grand Rapids, MI, June 2023, paper FRIBA01, unpublished (2023)

Appendix A
List of Common Software for Hadron Accelerators

This section lists examples of common software used in the design, study, and management of high-power hadron accelerators.

Table A.1 lists examples of such codes used in design optimization and beam dynamics analysis for the past high power hadron accelerators. The list includes the envelope simulation for linac modeling (TRACE-3D [1] and FLAME [2]), macro-particle-based tracking code (TRACK [3], BEAMPATH [4], DYNAMION [5], PARMELA [6], TraceWin [7], GPT [8] and IMPACT [9]), the lattice design packages (SYNCH [10], MAD [11, 12] and TEAPOT [13]), simulation including space charge (ACCSIM [14], SIMPSONS [15], and ORBIT [16, 17]), dedicated tools for the longitudinal phase space (ESME [18], TIBETAN [19], BLonD [20]), special tools for electron cloud effects (CLOUDLAND [21], POSINST [22], PyECLOUD [23]), and other stochastic effects simulation including intrabeam scattering, cooling, and crystallization (BBFP [24], SOLID [25]). Several codes are available for the design and simulation of Radio Frequency Quadrupole (RFQ) accelerators: PARMTEQ [26, 27], DesRFQ [28], TOUTATIS [29], DYNAMION [5], TRACK [3]. The codes TRANSPORT [30], COSY Infinity [31], MARYLIE [32] are useful for the design, optimization and simulation of beam transport systems.

Table A.2 lists examples of software codes for accelerator engineering and analysis, including mechanical engineering (Ansys [44], AutoCAD [45], and SolidWorks [46]), electromagnetic analysis (CST Studio [47], MAFIA [48], OPERA [49], Superfish [50], and TBCI [51]), stopping in material (SRIM [52]), and civil engineering interface (BIM [53]).

J. Wei et al., *High Power Hadron Accelerators*, Particle Acceleration and Detection,
https://doi.org/10.1007/978-3-032-17662-2

Table A.1 Examples of software codes used in hadron accelerator design and beam dynamics studies

Name	Main purpose	Reference
ACCSIM	A program to simulate the accumulation of intense proton beams	[14]
AVAS	A code for 3-D multiparticle beam dynamics simulation in a linac with improved S-PICNIC algorithm for space charge effects and combined t-code/z-code particle motion modes. Its GPU version AVASx provides ultra high speed dynamics simulation with three levels of acceleration capability	[33]
BBFP	Bunched-beam Fokker Planck solver for longitudinal bunch evolution under intrabeam scattering and stochastic cooling	[24]
BEAMPATH	A code library BEAMPATH for 2-D and 3-D space charge dominated beam dynamics study in linear particle accelerators and beam transport lines is developed	[4]
BLonD	A simulation suite for longitudinal particle tracking in synchrotron	[20]
CLOUDLAND	Electron cloud simulation with 3-D electron space-charge, beam-electron interaction, and various magnetic and electric fields	[21]
COSY Infinity	A differential-algebra-based code that allows the computation of high order maps including parameter dependencies for systems consisting of electric, magnetic, and circular elements, including their fringe fields	[31]
DesRFQ	Interactive optimization of the RFQ parameters by particle tracking with space charge	[28]
DYNAMION	3-D equation of particle motion, multi-charge ion beams, PIC model for the space charge, includes RFQ, DTL...	[5]
ESME	To calculate the evolution of a distribution of particles in energy and azimuth as it is acted upon by the radio frequency system of a synchrotron or storage ring	[18]
FLAME	Multi-charge ion beam envelope optimization. FLAME is used for FRIB real-time modeling	[2]
GPT	Multiparticle simulation with space charge in 3-D fields	[8]
IMPACT	An object-oriented parallel particle-in-cell code for beam dynamics simulation in linear accelerators	[9]
MAD8	Methodical Accelerator Design, version 8	[11]
MAD-NG	A standalone multi-platform tool for linear and non-linear optics design and optimisation	[34]
MAD-X	Methodical Accelerator Design, version X	[12]
MARYLIE	A program for charged particle beam transport based on Lie algebraic methods	[32]
ORBIT	A particle-in-cell tracking code in 6-D phase space that transports bunches of interacting particles through a series of nodes representing elements, dynamic effects, or diagnostics	[16]
PARMELA	Phase and radial motion in electron linear accelerators, expanded to track particles of any mass and charge	[6]
PARMTEQ	RFQ design and simulation code, Los Alamos Scientific Laboratory	[26, 27]
POSINST	Simulation of the build-up and dissipation of the electron cloud in a representative field-free section of the vacuum chamber	[22]

(continued)

Table A.1 (continued)

Name	Main purpose	Reference
PyECLOUD	A 2D macro-particle code for the simulation of electron cloud effects in particle accelerators	[23]
PyORBIT	A particle-in-cell simulation tool for cyclical and linear particle accelerators	[17]
SAD	A computer program complex for accelerator design developed in KEK since 1986	[35]
SCTR	A family of codes to simulate the complex dynamics of charged particle beams in accelerators, with a special focus on space-charge effects and beam-beam interactions	[36]
SIMPSONS	6-D phase space tracking with space charge	[15, 37]
SOLID	A molecular dynamics program for the study of beam cooling and crystallization in circular rings	[25, 38]
SYNCH	A program for design and analysis of synchrotrons and beamlines	[10]
TEAPOT	A thin element accelerator program for optics and tracking	[13]
TEBATAN	Tracking of ions bunches in their environment using time-domain analysis in particular during transition crossing	[19]
TOUTATIS	RFQ simulation code, CEA Saclay	[29]
TRACE 3-D	A beam-dynamics program that calculates the envelopes of a bunched beam through a user-defined transport system	[1]
TraceWin	Well known in the community of high-intensity linear accelerators, the transport code TraceWin is able to simulate a beam from the source to the target using either simple linear model or multiparticle simulations including 2D or 3D space-charge	[7]
TRACK	Design, simulation and commissioning of hadron linacs. The code supports multi-component ion beams, electron beams, end-to-end simulation in 3D external and 3D space charge fields using CIC method	[3]
TRANSOPTR	A beam transport design code that incorporates autoriatic optimization of a beam transport system under general constraints	[39]
TRANSPORT	A computer program for designing charged particle beam transport systems	[30]
UAL	Unified Accelerator Libraries	[40]
WARP	An open-source particle-in-cell Python package designed for simulation of plasmas and high current particle beams in a range of applications	[41]
Xsuite	An integrated beam physics simulation framework	[42, 43]

Table A.2 Examples of software codes used in electromagnetic field and engineering analysis

Name	Main purpose	Reference
Ansys	Finite Element Analysis (FEA) software for structural engineering	[44]
AutoCAD	AutoDesk CAD	[45]
BIM	Building Information Modeling, an intelligent, 3-D model-based tool that provides users with a digital representation of a facility's physical and functional characteristics	[53]
CAD	Computer-aided design, a software technology that allows designers and engineers to create 2-D and 3-D models of products	
COMSOL®	A multi-physics platform enabling coupled analysis and accepting CAD models	[54]
CST Studio	CST Microwave Studio®: 3-D EM simulation software tool designed for analyzing and designing high-frequency devices and systems	[47]
MAFIA	A three-dimensional electromagnetic CAD system for magnets, RF structures, and transient wake-field calculations	[48]
OPERA	Finite Element Analysis (FEA) software suite (TOSCA, ELEKTRA, SCALA, CARMEN, SOPRANO and TEMPO) that allows users to perform electromagnetic (EM) and electromechanical system simulations in 2-D and 3-D	[49]
SRIM	The stopping and range of ions in matter	[52]
SolidWorks	3-D CAD software used for designing and developing products, from 3-D modeling to creating 2-D drawings and simulations	[46]
POISSON/SUPERFISH	A suite of computer programs used to calculate static electric and magnetic fields, as well as EM fields in resonant structures, like RF cavities	[50, 55]
TBCI	Three-dimensional beam cavity interaction	[51]

Table A.3 lists examples of software codes including radiation transport (FLUKA [56], GEANT4 [57], MARS [58], MCNP [59], MCNPX [60], PHITS [61], and STRUCT [62]) and rare isotope production (LISE++ [63]).

Table A.4 lists examples of software commonly used for accelerator project management, quality assessment and monitoring, and configuration documentation.

Table A.3 Examples of software codes used in radiation transport and rare isotope production analysis

Name	Main purpose	Reference
FLUKA	A multipurpose MonteCarlo code which can transport a variety of particles over a wide energy range in complex geometries	[56]
GEANT4	A toolkit for simulating particles through matter	[57]
LISE++	A software package to simulate rare isotope production with fragment-separators	[63]
MARS	A particle and heavy ion transport code system	[58]
MCNP	The Monte Carlo N-Particle transport computer code	[59]
MCNPX	Monte Carlo N-Particle eXtended, a general-purpose Monte Carlo radiation transport code with 3-D geometry and continuous-energy transport of 34 particles and light ions	[60]
PHITS	A particle and heavy Ion transport code system	[61]
STRUCT	Particle tracking and interaction with material of collimators in synchrotrons and beam lines	[62]

Table A.4 Examples of software codes used in accelerator project management, quality assessment and monitoring, and configuration documentation

Name	Main purpose	Reference
Adept	A platform from Synergis Software that can be used to create a central repository for documents across an enterprise	[64]
AssetCentre	A centralized tool for securing, managing, versioning, tracking and reporting automation-related asset information	[65]
DCC	Document control center, a process that manages documents from their creation to their distribution and storage	
EVMS	US DOE G 413.3-10A, Earned Value Management System	
MagStat	Evaluation of magnetic field quality and trends based on magnet measurement data	[66]
Primavera P6	Project codes used to organize and report on projects	[67]
SurvStat	Evaluation of alignment quality and trends based on magnet survey data	[68]

References

1. K.R. Crandall, Trace 3-D documentation. Technical report, Los Alamos National Laboratory (1987)
2. Z.Q. He, et al., Development status of FRIB on-line model based beam commissioning application, in *Proceedings of the LINAC'16, No. 28 in Linear Accelerator Conference* (JACoW Publishing, Geneva, 2017), pp. 100–103. https://doi.org/10.18429/JACoW-LINAC2016-MOPRC015, https://jacow.org/linac2016/papers/MOPRC015.pdf
3. P. Ostroumov, V. Aseev, B. Mustapha, TRACK–a code for beam dynamics simulation in accelerators and transport lines with 3D electric and magnetic fields. ANL, March **27** (2006)
4. Y.K. Batygin, Particle-in-cell code BEAMPATH for beam dynamics simulations in linear accelerators and beamlines. Nucl. Instrum. Methods Phys. Res. Sect. A **539**(3), 455–489 (2005). https://doi.org/10.1016/j.nima.2004.10.029, https://www.sciencedirect.com/science/article/pii/S0168900204023435
5. S. Yaramyshev, W. Barth, S. Lauber, M. Miski-Oglu, A. Rubin, U. Scheeler, H. Vormann, M. Vossberg, DYNAMION—A powerful beam dynamics software package for the development of ion linear accelerators and decelerators. Appl. Sci. **13**(14) (2023). https://doi.org/10.3390/app13148422. https://www.mdpi.com/2076-3417/13/14/8422
6. L. Young, J. Billen, The particle tracking code PARMELA, in *Proceedings of the PAC'03*, vol. 5 (2003), pp. 3521–3523
7. D. Uriot, N. Pichoff, Status of TraceWin code, in *Proceedings of the IPAC'15* (JACoW Publishing, Geneva), pp. 92–94. https://doi.org/10.18429/JACoW-IPAC2015-MOPWA008. https://jacow.org/IPAC2015/papers/MOPWA008.pdf
8. General Particle Tracer (GPT). https://www.pulsar.nl/gpt/
9. J. Qiang, R.D. Ryne, S. Habib, V. Decyk, An object-oriented parallel particle-in-cell code for beam dynamics simulation in linear accelerators, in *Proceedings of the 1999 ACM/IEEE Conference on Supercomputing (Proc. SC '99)* (1999), pp. 55–59
10. A. Garren, A. Kenney, E. Courant, A. Russell, M. Syphers, SYNCH: a program for design and analysis of synchrotrons and beamlines–users guide. Technical report, Brookhaven National Laboratory (BNL), Upton, NY (1993)
11. H. Grote, F.C. Iselin, The MAD program (Methodical accelerator design), version 8.1 – user's reference manual. Technical report, CERN, Geneva (1990). https://cds.cern.ch/record/90480
12. H. Grote, F. Schmidt, MAD-X-an upgrade from MAD8, in *Proceedings of the PAC'03*, vol. 5 (IEEE, 2003), pp. 3497–3499
13. L. Schachinger, R. Talman, TEAPOT: a thin element accelerator program for optics and tracking. Technical report, Lawrence Berkeley Laboratory, CA (USA). SSC Central Design Group (1985)
14. F.W. Jones, H.O. Schönauer, G. MacKenzie, ACCSIM: a program to simulate the accumulation of intense proton beams. Part. Accel. **31**(TRI-PP-89-73), 199–204 (1989)
15. S. Machida, The Simpsons program 6-D phase space tracking with acceleration, in *AIP Conference Proceedings*, vol. 297 (American Institute of Physics, 1993), pp. 459–466
16. J. Galambos, J. Holmes, D. Olsen, A. Luccio, Beebe-Wang, J., ORBIT user's manual, Oak Ridge National Laboratory. Technical report, SNS/ORNL/AP Technical Note (1999)
17. A. Shishlo, S. Cousineau, J. Holmes, T. Gorlov, The particle accelerator simulation code PyORBIT. Proc. Comput. Sci. **51**, 1272–1281 (2015)
18. J.A. MacLachlan, J.F. Ostiguy, User's guide to ESME 2002. Fermilab (Nov 2002)
19. J. Wei, Longitudinal dynamics of the nonadiabatic regime on alternating gradient synchrotrons. Ph.D. thesis, SUNY, Stony Brook (1990)
20. H. Timko, J. Esteban Müller, A. Lasheen, D. Quartullo, Benchmarking the beam longitudinal dynamics code BLonD p. WEPOY045 (2016). https://doi.org/10.18429/JACoW-IPAC2016-WEPOY045, https://cds.cern.ch/record/2207431
21. L. Wang, D. Raparia, J. Wei, S. Zhang, Mechanism of electron cloud clearing in the accumulator ring of the spallation neutron source. Phys. Rev. Spec. Top. Accel. Beams **7**(3), 034401 (2004)

22. M. Pivi, M. Furman, Electron cloud development in the Proton Storage Ring and in the Spallation Neutron Source. Phys. Rev. Spec. Top. Accel Beams **6**(3), 034201 (2003)
23. G. Iadarola, E. Belli, K. Li, L. Mether, A. Romano, G. Rumolo, Evolution of Python tools for the simulation of electron cloud effects, p. THPAB043 (2017). https://doi.org/10.18429/JACoW-IPAC2017-THPAB043, https://cds.cern.ch/record/2289165
24. J. Wei, A.G. Ruggiero, Beam life-time with intrabeam scattering and stochastic cooling, in *Proceedings of the PAC'91* (JACoW Publishing, Geneva), pp. 1869–1872
25. J. Wei, X.P. Li, A.M. Sessler, Low-energy states of circulating stored ion beams: Crystalline beams. Phys. Rev. Lett. **73**(23), 3089 (1994)
26. K.R. Crandall, PARMTEQ with image charges. Technical report AT-1-92-213, Los Alamos National Laboratory (1991)
27. F. Krawczyk, Los Alamos National Laboratory. http://laacg.lanl.gov, laacg@lanl.gov
28. A.A. Kolomiets, et al., The RFQ design software DesRFQ. Technical note, ITEP/ANL (2005)
29. R. Ferdinand, N. Pichoff, R. Duperrier, J.M. Lagniel, TOUTATIS, the CEA/Saclay RFQ code, in *Proceedings of the LINAC'00, Linear Accelerator Conference* (JACoW Publishing, Geneva), pp. 839–841. https://jacow.org/l00/papers/THB19.pdf
30. K.L. Brown, F. Rothacker, D.C. Carey, C. Iselin, TRANSPORT-A computer program for designing charged particle beam transport systems. Technical report, European Organization for Nuclear Research (1980)
31. M. Berz, K. Makino, COSY INFINITY version 8.1. User's guide and reference manual (2002)
32. A. Dragt, D. Douglas, F. Neri, E. Forest, L. Healy, P. Schutt, J. van Zeijts, MARYLIE 3.0 user's manual (1999)
33. C. Jin, Z. Wang, X. Qi, Y. He, Z. Li, K. Sun, X. Chen, C. Feng, X. Zhao, Y. Hu, et al., Advanced virtual accelerator software: a linear accelerator simulation code. Phys. Rev. Accel. Beams **28**(4), 044602 (2025)
34. L. Deniau, MAD-NG, a standalone multiplatform tool for linear and non-linear optics design and optimisation (2025). https://arxiv.org/abs/2412.16006
35. K. Oide, et al., SAD. https://acc-physics.kek.jp/SAD/
36. K. Ohmi, et al., Sctr. https://journals.aps.org/prab/abstract/10.1103/PhysRevAccelBeams.27.101001. Described in Phys. Rev. Accel. Beams **27**, 101001 (2024), and in KEK slide deck: https://conferenceindico.kek.jp/event/75/contributions/1473/attachments/979/1037/BeambeamcodeOhmi.pdf
37. S. Machida, et al., SIMPSON. https://conference-indico.kek.jp/event/75/contributions/1477/attachments/991/1050/machida20190919_v5.pdf
38. J. Wei, H. Okamoto, A.M. Sessler, Necessary conditions for attaining a crystalline beam. Phys. Rev. Lett. **80**(12), 2606 (1998)
39. M. De Jong, E. Heighway, A first order space charge option for transoptr. IEEE Trans. Nucl. Sci. **30**(4), 2666–2668 (2007)
40. N. Malitsky, R. Talman, User guide for UAL (C++ interface). Technical report, Relativistic Heavy Ion Collider (RHIC), Brookhaven National Laboratory (BNL), Upton, NY (2008). https://doi.org/10.2172/1061913, https://www.osti.gov/biblio/1061913
41. A. Friedman, D.P. Grote, et al., WARP. https://warp.lbl.gov/home
42. Xsuite. https://xsuite.readthedocs.io/en/latest/index.html#
43. G. Iadarola, et al., Xsuite: an integrated beam physics simulation framework, in *Proceeings of the HB'23, No. 68 in ICFA Advanced Beam Dynamics Workshop on High-Intensity and High-Brightness Hadron Beams* (JACoW Publishing, Geneva, 2023), pp. 73–80. https://doi.org/10.18429/JACoW-HB2023-TUA2I1
44. Ansys, Ansys Mechanical. https://www.ansys.com/products/structures/ansys-mechanical
45. Autodesk, CAD software. https://www.autodesk.com/solutions/cad-software
46. SolidWorks, Dassault systèmes software. https://www.solidworks.com/
47. SIMULIA, CST Microwave Studio ®. https://www.3ds.com/products/simulia/cst-studio-suite
48. R. Klatt, F. Krawczyk, W. Novender, C. Palm, T. Weiland, B. Steffen, T. Barts, M. Browman, R. Cooper, C. Mottershead, et al., MAFIA-A three-dimensional electromagnetic CAD system for magnets, RF structures, and transient wake-field calculations, in *1986 Linear Accelerator Conference Proceedings, SLAC*, vol. 303 (1986)

49. V. Fields, *Opera-3d Reference Manual*, vol. 3010, No. 10 (Vector Fields Limited, England, 2004)
50. K. Halbach, Superfish-A computer program for evaluation of RF cavities with cylindrical symmetry (1976)
51. T. Weiland, TBCI and URMEL-new computer codes for wake field and cavity mode calculations. IEEE Trans. Nucl. Sci. **30**(4), 2489–2491 (1983)
52. J.F. Ziegler, M.D. Ziegler, J.P. Biersack, SRIM–The stopping and range of ions in matter. Nucl. Instrum. Methods Phys. Res. Sect. B **268**(11–12), 1818–1823 (2010)
53. Navisworks, Building information modeling. https://www.autodesk.com/products/navisworks/
54. COMSOL Multiphysics® v. 6.3: COMSOL AB, Stockholm, Sweden. https://www.comsol.com
55. M. Menzel, H.K. Stokes, Users guide for the poisson/superfish group of codes. Technical report, Los Alamos National Laboratory (LANL), Los Alamos, NM (1987)
56. G. Battistoni, T. Boehlen, F. Cerutti, P.W. Chin, L.S. Esposito, A. Fasso, A. Ferrari, A. Lechner, A. Empl, A. Mairani, et al., Overview of the FLUKA code. Ann. Nucl. Energy **82**, 10–18 (2015)
57. G. Collaboration, S. Agostinelli, et al., GEANT4–a simulation toolkit. Nucl. Instrum. Meth. A **506**(3), 250–303 (2003)
58. N.V. Mokhov, C.C. James, The MARS code system user's guide version 15 (2016). Technical report, Fermi National Accelerator Laboratory (FNAL), Batavia, IL (2017)
59. J.A. Kulesza, T.R. Adams, J.C. Armstrong, S.R. Bolding, F.B. Brown, J.S. Bull, T.P. Burke, A.R. Clark, R.A.A. Forster III, J.F. Giron, et al., MCNP® code version 6.3. 0 theory & user manual. Technical report, Los Alamos National Laboratory (LANL), Los Alamos, NM (2022)
60. L.S. Waters, G.W. McKinney, J.W. Durkee, M.L. Fensin, J.S. Hendricks, M.R. James, R.C. Johns, D.B. Pelowitz, The MCNPX Monte Carlo radiation transport code, in *AIP Conference Proceedings*, vol. 896 (American Institute of Physics, 2007), pp. 81–90
61. T. Sato, Y. Iwamoto, S. Hashimoto, T. Ogawa, T. Furuta, S.i. Abe, T. Kai, P.E. Tsai, N. Matsuda, H. Iwase, et al., Features of particle and heavy ion transport code system (PHITS) version 3.02. J. Nucl. Sci. Technol. **55**(6), 684–690 (2018)
62. A. Drozhdin, N. Mokhov, STRUCT program user's reference manual (1994)
63. O. Tarasov, D. Bazin, M. Hausmann, M. Kuchera, P. Ostroumov, M. Portillo, B. Sherrill, K. Tarasova, T. Zhang, LISE cute++, the latest generation of the LISE++ package, to simulate rare isotope production with fragment-separators. Nucl. Instrum. Methods Phys. Res. Sect. B **541**, 4–7 (2023)
64. Adept, Synergis Technologies, LLC. https://www.synergissoftware.com/
65. AssetCentre, Factorytalk®assetcentre software. https://www.rockwellautomation.com/
66. J. Wei, et al., Field quality evaluation of the superconducting magnets of the Relativistic Heavy Ion Collider, in *Proceedings of the PAC'95* (JACoW Publishing, Geneva), pp. 461–463
67. Primavera: P6 enterprise project portfolio management. https://www.oracle.com/
68. J. Wei, G. Ganetis, R. Gupta, M. Harrison, M. Hemmer, A. Jain, F. Karl, S. Peggs, S. Tepikian, P. Thompson, D. Trbojevic, P. Wanderer, Misalignment evaluation of superconducting magnets in the Relativistic Heavy Ion Collider, in *Proceedings of the EPAC'96, No. 5 in European Particle Accelerator Conference* (JACoW Publishing, Geneva, 1996), pp. 2222–2224

Solutions

Problems of Chap. 2

2.1 The error of these relations can be estimated by expanding both sides of the relation

$$\gamma^{-2} \equiv 1 - \beta^2 \tag{A.1}$$

with $\gamma = \gamma_0 + \Delta\gamma$ and $\beta = \beta_0 + \Delta\beta$, while assuming $\Delta\beta/\beta_0 \ll 1$ and $\Delta\gamma/\gamma_0 \ll 1$. One obtains

$$\beta_0^2\gamma_0^2\left(\frac{\Delta\beta}{\beta_0}\right) = \frac{\Delta\gamma}{\gamma_0}\left\{1 - \left(\frac{3}{2} + \frac{1}{2\beta_0^2\gamma_0^2}\right)\left(\frac{\Delta\gamma}{\gamma_0}\right) + O\left[\left(\frac{\Delta\gamma}{\gamma_0}\right)^2\right]\right\}. \tag{A.2}$$

The relative error of such estimates are thus of the order of

$$\left(\frac{3}{2} + \frac{1}{2\beta_0^2\gamma_0^2}\right)\left(\frac{\Delta\gamma}{\gamma_0}\right) + O\left[\left(\frac{\Delta\gamma}{\gamma_0}\right)^2\right]. \tag{A.3}$$

2.2 ([1]; also Courtesy of F. Marti)

In an isochronous cyclotron, the cyclotron frequency (Eq. 2.1)

$$\omega_c = \frac{qeB_z}{Am_0\gamma} \tag{A.4}$$

is a constant. At the extraction radius R_{ext}

$$R_{\text{ext}} = \frac{\beta_{\text{max}}c}{\omega_c} = \frac{Am_0c}{qeB_{z,\text{ext}}}\beta_{\text{max}}\gamma_{\text{max}} \propto \beta_{\text{max}}\gamma_{\text{max}}\ , \tag{A.5}$$

for given magnetic field $B_{z,\text{ext}}$ at the cyclotron extraction radius.

J. Wei et al., *High Power Hadron Accelerators*, Particle Acceleration and Detection,
https://doi.org/10.1007/978-3-032-17662-2

To keep the particle isochronous, the velocity of the particle is linearly proportional to the radius r,

$$\beta(r) = \frac{\omega_c r}{c} . \tag{A.6}$$

The orbit separation is given by

$$\Delta R = \frac{c}{\omega_c} \Delta\beta = \frac{1}{A m_0 c \omega_c} \frac{\Delta W}{\beta\gamma^3} , \tag{A.7}$$

where ΔW is the energy gain per turn. If a minimum orbit gap $\Delta R_{\rm min}$ is to be kept between turns at the extraction radius,

$$\frac{1}{A m_0 c \omega_c} \frac{\Delta W}{\beta\gamma^3} > \Delta R_{\rm min} , \tag{A.8}$$

then

$$\Delta W_{\rm ext} > q e c B_{z,\rm ext} \Delta R_{\rm min} \beta_{\rm max} \gamma_{\rm max}^2 \propto \beta_{\rm max} \gamma_{\rm max}^2 . \tag{A.9}$$

Thus, the minimum energy gain per turn is proportional to $\beta_{\rm max}\gamma_{\rm max}^2$. Compared with a linac where the energy gain is proportional to γ, the scaling of the energy gain with γ^2 (Eq. A.9) indicates the limitation of a cyclotron in accelerating beam to high energy.

2.3 For the transverse particle motion in a ring, the transverse matrix from a location 0 of zero dispersion $D(0) = D'(0) = 0$ to location s is given as

$$\begin{bmatrix} x \\ x' \\ \Delta p/p \end{bmatrix}_s = \begin{bmatrix} C & S & D \\ C' & S' & D' \\ 0 & 0 & 1 \end{bmatrix} \begin{bmatrix} x \\ x' \\ \Delta p/p \end{bmatrix}_0 \tag{A.10}$$

where

$$C(s) = \frac{\beta(s)}{\beta(0)} \left[\cos\Delta\mu + \alpha(s)\sin\Delta\mu\right], \quad S(s) = \sqrt{\beta(s)\beta(0)}\sin\Delta\mu, \tag{A.11}$$

where β and α are the Courant-Snyder parameters and $\Delta\mu$ is the phase advance. The dispersion $D(s)$ that satisfies the relation

$$D''(s) + K(s)D(s) = \frac{1}{\rho(s)} \tag{A.12}$$

has solution given by

$$D(s) = S(s)\int_0^s \frac{C(s')}{\rho}(s')ds - C(s)\int_0^s \frac{S(s')}{\rho(s')}ds. \tag{A.13}$$

1. Equation A.13 can be used to solve the problem. With the arc consisting of 4 FODO cells, each of length L_C and phase advance $\mu_C = \pi/2$, we start $s = 0$ from the focusing quadrupole magnet QF just upstream of the first dipole magnet of the arc. Let μ_D denote the phase advance to the first dipole magnet, and $\theta_D = \int_{D-}^{D+} \frac{ds}{\rho}$ the bending angle at each dipole magnet. Thus, the peak dispersion at the next focusing quadrupole magnets $s = L_C$ and $s = 2L_C$ are

$$\begin{aligned} D(L_C) &= \theta_D\beta_D\left[\cos\mu_D + \cos\left(\mu_D + \frac{\pi}{4}\right)\right] \\ D(2L_C) &= \theta_D\sqrt{\beta_{QF}\beta_D}\left[\sin\mu_D + \sin\left(\mu_D + \frac{\pi}{4}\right) + \sin\left(\mu_D + \frac{\pi}{2}\right)\right. \\ &\quad \left. + \sin\left(\mu_D + \tfrac{3\pi}{4}\right)\right]. \end{aligned} \tag{A.14}$$

Obviously, $\theta_D = \pi/16$. Let $\mu_D = \pi/8$. Using the β-function expression of the matched FODO cell Eq. 2.59, one obtains with $\mu_C = \pi/2$

$$D_{\max,FODO} = D(2L_C) \approx 0.670\, L_C \tag{A.15}$$

2. To start the arc with a defocusing quadrupole magnet QD, one may repeat the steps above shifting the starting point $s = 0$ to the focusing quadrupole QF upstream of this defocusing QD. The arc now contains two dispersion peaks at the locations of $s = 2L_C$ and $s = 3L_C$,

$$D_{\max,\text{DOFO}} = D(2L_C) = D(3L_C) \approx 0.479\, L_C. \tag{A.16}$$

3. Comparing with the peak dispersion of a matched FODO cell Eq. 2.59,

$$D_{\max,\text{matched}} \approx 0.265\, L_c. \tag{A.17}$$

4. Comparing Eqs. A.15, A.16, and A.17, one observe that the missing-dipole (half-field) scheme results in the lowest peak dispersion and thus demands least horizontal aperture due to dispersion. However, the total arc length is extended due to the segments of half field. The FODO scheme with a single dispersion peak at the center of the arc demands largest horizontal aperture due to the highest dispersion. The lattice of the SNS accumulator ring (Sect. 10.1) adopted the DOFO scheme with intermediate dispersion yet compact arc.

2.4 Denote the kicks applied at location of the three-bump process at 1, 2, and 3 as θ_1, θ_2 and θ_3, respectively. The first kick to the closed orbit occurs at location 1 with

$$x_1 = 0, \quad x_1' = \theta_1. \tag{A.18}$$

For a closed bump, the closed orbit must return to the unperturbed state upon the third kick θ_3. Thus, before the third kick is applied,

$$x_3 = 0, \quad x_3' = -\theta_3 \tag{A.19}$$

The transfer matrix from 1 to 2 is given by

$$\begin{cases} x_2 = \sqrt{\beta_1\beta_2}\sin\Delta\mu_{21}\,\theta_1 \\ x_2' = \sqrt{\dfrac{\beta_1}{\beta_2}}\,[\cos\Delta\mu_{21} - \alpha_2\sin\Delta\mu_{21}]\,\theta_1. \end{cases} \tag{A.20}$$

The back-transfer matrix from 3 to 2 is given by

$$\begin{cases} x_{2,b} = \sqrt{\beta_3\beta_2}\sin\Delta\mu_{23}\,(-\theta_3) \\ x_{2,b}' = \sqrt{\dfrac{\beta_3}{\beta_2}}\,[\cos\Delta\mu_{23} - (-\alpha_2)\sin\Delta\mu_{23}]\,(-\theta_3). \end{cases} \tag{A.21}$$

Relating Eqs. A.20 and A.21 with the kick θ_2 at location 2,

$$x_2 = x_{2,b} \quad x_2' + \theta_2 = -x_{2,b}', \tag{A.22}$$

one obtains

$$\frac{\theta_1\sqrt{\beta_1}}{\sin\Delta\mu_{32}} = \frac{\theta_2\sqrt{\beta_2}}{\sin\Delta\mu_{13}} = \frac{\theta_3\sqrt{\beta_3}}{\sin\Delta\mu_{21}}. \tag{A.23}$$

2.5 Answer: $X_C(t) = \sqrt{\dfrac{t}{t_{\max}}}R, \quad 0 \le t \le t_{\max}$.

2.6 Use the normalized transverse phase space of the ring in terms of the normalized variables X, X' and phase μ:

$$X \equiv \frac{x}{\sqrt{\beta_x}}, \quad X' \equiv \frac{dX}{d\mu} = \frac{\alpha_x x + \beta_x x'}{\sqrt{\beta_x}}, \quad \mu(s) = \int^s \frac{ds'}{\beta(s')}\,. \tag{A.24}$$

where the ring C-S parameters are β_x and α_x.

Upon multi-turn injection of a beam with emittance ϵ_i and C-S parameters β_i and α_i, injecting with the center of the input beam (x_i, x_i') relative to the instantaneous injection orbit bump. The ring beam emittance is ϵ_x, and ring C-S parameters are β_x and α_x at injection.

1. Multi-turn injection must occur at the moment when x reached the maximum value for a given transverse ellipse of the ring at injection. This corresponds to

$$X' = 0\,, \tag{A.25}$$

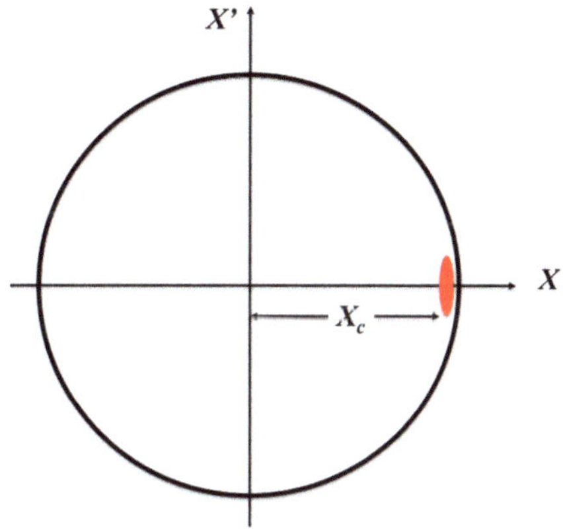

Fig. A.1 Normalized phase space illustration of multi-turn injection in a synchrotron ring

and with the injecting beam also upright (Fig. A.1). This implies, together with Eq. A.24,

$$\frac{\alpha_i}{\beta_i} = \frac{\alpha_x}{\beta_x} = -\frac{x_i'}{x_i} . \tag{A.26}$$

2. The injecting beam follows the elliptical contours

$$\gamma_i x^2 + 2\alpha_i x x' + \beta_i x'^2 \leq \frac{\epsilon_i}{\pi} \tag{A.27}$$

Using Eqs. A.24 and A.26, Eq. A.27 becomes

$$\frac{\beta_i}{\beta_x} X'^2 + \frac{\beta_x}{\beta_i} X^2 \leq \frac{\epsilon_i}{\pi} . \tag{A.28}$$

Therefore, the injecting beam ellipse can be parametrized as

$$X = \sqrt{\frac{\beta_i \epsilon_i}{\pi \beta_x}} \cos\theta, \quad X' = \sqrt{\frac{\beta_x \epsilon_i}{\pi \beta_i}} \sin\theta, \quad 0 < \theta \leq 2\pi . \tag{A.29}$$

Let X_c be the injection closed orbit center relative to ring beam origin in the normalized phase space of the ring. Using Eq. A.29, the ring emittance circle corresponding to the injecting beam ellipse can be parameterized as

$$\frac{\epsilon(\theta)}{\pi} = \left(X_c + \sqrt{\frac{\beta_i}{\pi \beta_x}\epsilon_i} \cos\theta \right)^2 + \frac{\beta_x}{\pi \beta_i}\epsilon_i \sin^2\theta . \tag{A.30}$$

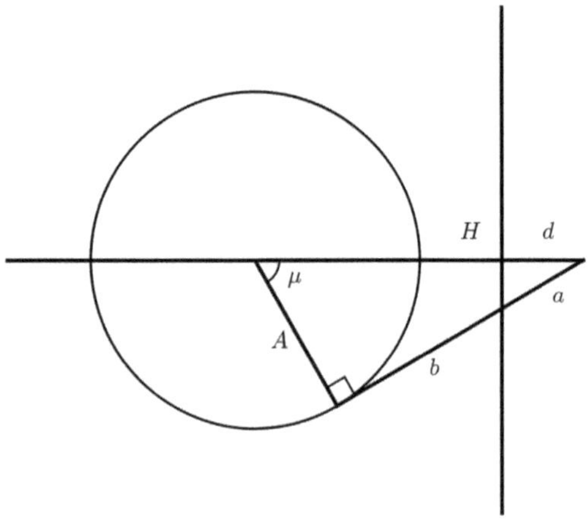

Fig. A.2 Illustration of the two-stage collimation for efficiency optimization in rings

The optimal condition is reached when

$$\left.\frac{d^2\epsilon(\theta)}{d\theta^2}\right|_{\theta=0} = 0\,. \tag{A.31}$$

With the assumption that

$$\epsilon_i \ll \epsilon_x\,, \tag{A.32}$$

Equation A.31 is met when the condition

$$\frac{\beta_i}{\beta_x} \approx \left(\frac{\epsilon_i}{\epsilon_x}\right)^{\frac{1}{3}} \tag{A.33}$$

is satisfied, where

$$\frac{\epsilon_x}{\pi} \approx X_c^2\,. \tag{A.34}$$

This conclusion is given in [2] (Courtesy of C. Prior and G. Rees).

2.7 Figure 2.50 illustrates the two-stage collimation process in the transverse phase space of a beam in a ring. Figure A.2 shows the details in the normalized phase space (X, X'). The goal is to minimize the length of section b which represents the escaped particles. Obviously,

$$b = A\tan\mu - a, \quad a = \frac{d}{\cos\left(\frac{\pi}{2}-\mu\right)}, \quad d = \frac{A}{\cos\mu} - (A+H). \tag{A.35}$$

Thus,

$$b = -\frac{\cos\mu}{\sin\mu}A + \frac{A+H}{\sin\mu}. \tag{A.36}$$

Let $\left.\frac{db}{d\mu}\right|_{\mu=\mu_1} = 0$, one obtains

$$\cos\mu_1 = \frac{A}{A+H}. \tag{A.37}$$

With similarly reasoning, one obtains the optimal phase advance of the second collector

$$\cos\mu_2 = \pi - \mu_1. \tag{A.38}$$

In terms of the real space parameters, Eq. A.37 becomes

$$\cos\mu_1 = \frac{a_S}{a_C}\sqrt{\frac{\beta_{x,S}}{\beta_{x,C}}} \tag{A.39}$$

where a_S and a_C denote the aperture of the scraper and the collector, respectively, and $\beta_{x,S}$ and $\beta_{x,C}$ denote the Courant-Snyder parameters at the corresponding locations.

2.8 The essence of the proof is an evaluation of the contribution of the longitudinal field B_z from the vicinity of one magnet end since, along a path parallel to the magnet axis such as path BC in Fig. A.3, there is no contribution to the longitudinal field integral either from well inside or well outside the magnet.

According to the principle of linear superposition, the fringe field at the end of the magnet can, in accordance with Eq. 2.77, be written as

$$\mathbf{B_F} = \sum_{n=1}^{\infty}(\mathbf{B}_{\mathbf{F}n} + \mathbf{A}_{\mathbf{F}n}) \tag{A.40}$$

where $\mathbf{B}_{\mathbf{F}n}$ and $\mathbf{A}_{\mathbf{F}n}$ are the fringe field that correspond to b_n and a_n, respectively.

As shown in Fig. A.3, for any multipole b_n or a_n, there exists a closed curve C_1 extending from inside the body of the magnet to infinity without enclosing any electric current. The cross angle θ between the plane containing curve C_1 and the horizontal plane is chosen to be

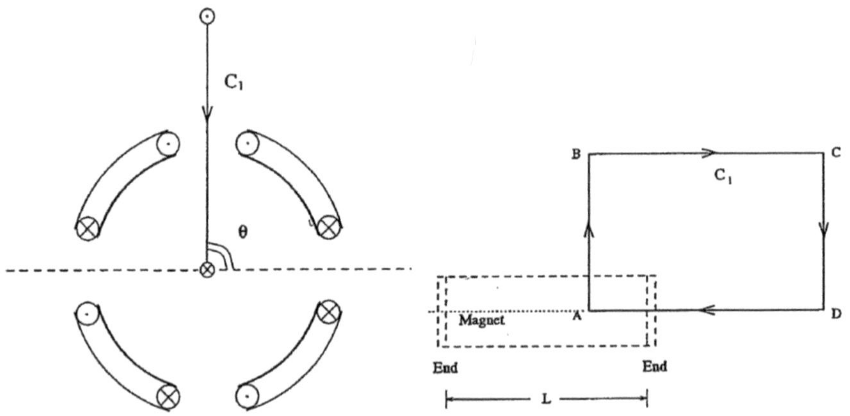

Fig. A.3 (Left) Front and (Right) side view of the closed curve C_1 of integration extending from inside the body of the magnet to infinity without enclosing any electric current

$$\theta = \begin{cases} \dfrac{2k-1}{2n}\pi, & \text{for } a_n \\ \\ \dfrac{k}{n}\pi, & \text{for } b_n, \end{cases} \tag{A.41}$$

where $1 \leq k \leq n$ is an integer, so that

$$\int_A^B \mathbf{B} \cdot d\mathbf{l} = 0. \tag{A.42}$$

Because the magnet field at infinity is equal to zero, one obtains from Ampere's law,

$$\int_D^A \mathrm{B}_{\mathrm{F}n,z}(x = y = 0; z)dz = \int_D^A \mathrm{A}_{\mathrm{F}n,z}(x = y = 0; z)dz = \int_D^A \mathbf{B} \cdot d\mathbf{l} = 0. \tag{A.43}$$

Hence,

$$\int_D^A B_z(x = y = 0; z)dz = 0. \tag{A.44}$$

Suppose that the particle moves across the end of the magnet at a transverse displacement $x_\perp$, as shown by the trajectory BC in Fig. A.4. Using Eq. A.44 and the closed curve of integration C_2, we obtain

$$\int_B^C B_z(x_\perp)dz = -\int_A^B \mathbf{B}_\perp(\mathbf{x}'_\perp) \cdot d\mathbf{x}'_\perp. \tag{A.45}$$

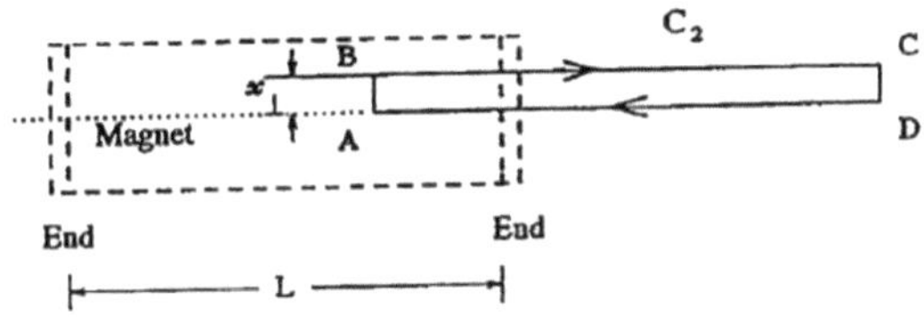

Fig. A.4 Side view of the closed curve C_2 for integration extending from inside the body of the magnet to infinity without enclosing any electric current

Hence, the transverse deflection $\Delta p_{\parallel}$ produced by the longitudinal component of the fringe field is

$$|\Delta p_{\parallel}| = e|\int_{B}^{C} \mathbf{v}_{\perp} \times \mathbf{B}_{\mathbf{z}}(x_{\perp})dz| = e|v_{\perp}\int_{A}^{B} \mathbf{B}_{\perp}(\mathbf{x}'_{\perp}) \cdot d\mathbf{x}'_{\perp}| \approx e|v_{\perp}x_{\perp}B_{\perp}(x_{\perp})|. \tag{A.46}$$

The total deflection Δp_0 produced by the main magnet body is

$$|\Delta p_0| = e|\int_0^L \mathbf{v}_{\mathbf{z}} \times \mathbf{B}_{\perp} dz|. \tag{A.47}$$

The factor $v_{\perp}x_{\perp}$ can be averaged as follows. By the standard "pseudo-harmonic" description of betatron motion, letting $S = \sin\psi$, $C = \cos\psi$,

$$x = \sqrt{\epsilon_{\perp}\beta_{\perp}}C, \quad x' = \sqrt{\frac{\epsilon_{\perp}}{\beta_{\perp}}}\left(S - C\frac{\beta'_{\perp}}{2}\right). \tag{A.48}$$

Using the results $\langle C^2S^2\rangle = 1/8$, $\langle C^4\rangle = 3/8$, and $\langle C^2S\rangle = 0$, one obtains

$$\sqrt{\langle x^2x'^2\rangle} = \epsilon_{\perp}\sqrt{\frac{1 + 0.75\beta'^2_{\perp}}{8}} \tag{A.49}$$

and the formula stated in the theorem of Eq. 2.78 [3].

Problems of Chap. 3

3.1 Hint: Use the definition of the mean decay length λ_s,

$$N = N_0 \exp\left(-\frac{\ln 2}{\lambda_s}s\right), \tag{A.50}$$

where N_0 and N are the initial and remaining number of particles, respectively, and s is the path length in the laboratory frame.
Answer:

1. 0.30 T; hence, the allowed maximum magnetic field to transport a 1 GeV H^- beam is low.
2. 3.7 m.

3. 15.1 T/m.
4. 0.33 T; hence, the order-of-magnitude difference in beam loss criteria only corresponds to 10% difference in allowed maximum magnetic field.
5. 79 m; hence, design of a high energy (up to 8 GeV) H^- linac is practically difficult.

Problems of Chap. 6

6.1 Most non-solenoidal magnets for particle accelerators use long coils with current in the magnet body flowing parallel to the magnet axis along the center trajectory of particle motion. The magnetic field can be approximated as 2-D in the direction transverse to the axis (see Eq. 3.40 and Problem 2.8) [4] as

$$B_y + iB_x = B_0 \sum_{n=1}^{\infty} (b_n + ia_a)(x+iy)^{n-1} = B_0 \sum_{n=1}^{\infty} (b_n + ia_a) r^{n-1} \exp[i(n-1)\theta], \tag{A.51}$$

where n is the multipole order, and $r\exp(i\theta) \equiv x + iy$, Express the magnetic field in terms of the polar components yield

$$\begin{aligned} B_r(r,\theta) &= \sum_{n=1}^{\infty} r^{n-1}(b_n \sin n\theta + a_n \cos n\theta) \\ B_\theta(r,\theta) &= \sum_{n=1}^{\infty} r^{n-1}(b_n \cos n\theta - a_n \sin n\theta). \end{aligned} \tag{A.52}$$

For a magnet of $2m$ pole, where $m = 1$ is a dipole magnet, $m = 2$ is a quadrupole magnet, etc. the corresponding magnet symmetry is defined as a total flip of the polar field components upon rotation of angle π/m,

$$B_r(r,\theta) = -B_r\left(r, \theta + \frac{\pi}{m}\right), \quad B_\theta(r,\theta) = -B_\theta\left(r, \theta + \frac{\pi}{m}\right). \tag{A.53}$$

Thus, the multipoles n that are allowed by the symmetry are

$$n = m(2k+1) \quad k = 0, 1, 2, \tag{A.54}$$

For a quadrupole magnet, $m = 2$, the allowed multipoles are $n = 2, 6, 10, \ldots$ (quadrupole, 12-th pole, 20-th pole, etc.)

6.2 Hint: Use Ampere's law

$$NI = \oint \frac{\mathbf{B}}{\mu} \cdot d\mathbf{l}. \tag{A.55}$$

Due to the high relative permeability of the iron, the only length of integration of relevance is across the magnet gap with vacuum permeability of $\mu_0 = 4\pi \times 10^{-7}$ [T m A^{-1}].

Answer: 15 turns.

6.3 Hint: Use Eq. 6.12

$$\sigma_m = \frac{2[\text{T}] \times 1000[\text{A}] \times 2[\text{m}]}{2 \times 10^{-6}[\text{m}^2]} = 2\,[\text{Gpa}]. \tag{A.56}$$

Answer: The Hoop stress is 2 Gpa and much higher than the yield stress of 60 Mpa. The conductor will be damaged. To reduce the stress with a safety margin of 1.5 (2/3 of the yield stress), the cross section of the conductor should be 100 mm^2.

Problems of Chap. 10

10.1 Hints: (1) The rule of thumb for the condition of hands-on maintenance is to allow an average uncontrolled beam loss no more than 1 W/m along the accelerator tunnel; (2) Uncontrolled beam loss in rapid cycle synchrotrons occurs mostly at the beam injection stage of the acceleration cycle.

The average beam power $\langle P \rangle$ from the ring is given by

$$\langle P \rangle = E_k \langle I \rangle = eE_k f_R N_p \tag{A.57}$$

where E_k is the kinetic energy of the particle extracted from the ring, $\langle I \rangle = ef_R N_p$ is the average beam current in the ring, f_R is the beam pulse repetition rate in the ring, and N_p is the number of particles per pulse.

For a parabolic longitudinal bunch current distribution $I(\phi)$ with the maximum phase extend of $\hat{\phi}$,

$$I(\phi) = \left[1 - \left(\frac{\phi}{\hat{\phi}}\right)^2\right] \hat{I}, \quad |\phi| < \hat{\phi} \le \pi, \tag{A.58}$$

where the full RF bucket corresponds to phase span of $\pm\pi$, the peak beam current $\hat{I}$ in the ring and the bunching factor B_R are given by

$$\hat{I} = \frac{3\pi}{2\hat{\phi}} \langle I \rangle, \quad B_R \equiv \frac{\langle I \rangle}{\hat{I}} = \frac{2\hat{\phi}}{3\pi} = \frac{2}{3} \frac{C_R - L_{\text{gap}}}{C_R}, \tag{A.59}$$

respectively, where C_R is the ring circumference, and L_{gap} is the total length of beam gap reserved for extraction kicker excitation.

1. The total output beam power is about 2 MW. The total average beam current of the facility is about 1 mA.
2. The tolerable fractional uncontrolled beam loss in each ring is about 7.5×10^{-4}.
3. The range of RF frequency swing is from 1.426 MHz to 1.89 MHz.

4. The maximum bunching factor is 0.41. The average beam current in the ring is 15.0 A. The peak beam current in the ring is 36.6 A.

Comment: This was the alternative rapid-cyclying-synchrotron design proposal to the SNS baseline accumulator ring design during the 1999 spring study (Sect. 10.1).

References

1. R.A. Baartman, in *Proceedings of the Cyclotrons'13* (JACoW Publishing, Geneva), pp. 305–309. https://jacow.org/CYCLOTRONS2013/papers/WE2PB01.pdf
2. C. Prior, G. Rees, Nucl. Instrum. Methods Phys. Res. Sect. A **415**(1–2), 357 (1998)
3. J. Wei, R. Talman, Part. Accel. **55**, 93 (1996)
4. A. Jain, P. Wanderer, in *Handbook of Accelerator Physics and Engineering*, ed. by A.W. Chao, et al. (World Scientific, Singapore, 2022), pp. 633–639

Index

J. Wei et al., *High Power Hadron Accelerators*, Particle Acceleration and Detection,
https://doi.org/10.1007/978-3-032-17662-2

J

K

T

MIX
Papier aus verantwortungsvollen Quellen
Paper from responsible sources
FSC® C105338

If you have any concerns about our products,
you can contact us on
ProductSafety@springernature.com

In case Publisher is established outside the EU,
the EU authorized representative is:
Springer Nature Customer Service Center GmbH
Europaplatz 3, 69115 Heidelberg, Germany

Printed by Libri Plureos GmbH
in Hamburg, Germany